Origin and Evolution of the Ontong Java Plateau

Special Publication reviewing procedures

The Society makes every effort to ensure that the scientific and production quality of its books matches that of its journals. Since 1997, all book proposals have been refereed by specialist reviewers as well as by the Society's Books Editorial Committee. If the referees identify weaknesses in the proposal, these must be addressed before the proposal is accepted.

Once the book is accepted, the Society has a team of Book Editors (listed above) who ensure that the volume editors follow strict guidelines on refereeing and quality control. We insist that individual papers can only be accepted after satisfactory review by two independent referees. The questions on the review forms are similar to those for *Journal of the Geological Society*. The referees' forms and comments must be available to the Society's Book Editors on request.

Although many of the books result from meetings, the editors are expected to commission papers that were not presented at the meeting to ensure that the book provides a balanced coverage of the subject. Being accepted for presentation at the meeting does not guarantee inclusion in the book.

Geological Society Special Publications are included in the ISI Index of Scientific Book Contents, but they do not have an impact factor, the latter being applicable only to journals.

More information about submitting a proposal and producing a Special Publication can be found on the Society's web site: www.geolsoc.org.uk.

It is recommended that reference to all or part of this book should be made in one of the following ways:

FITTON, J. G., MAHONEY, J. J., WALLACE, P. J. & SAUNDERS, A. D. (eds) 2004. *Origin and Evolution of the Ontong Java Plateau.* Geological Society, London, Special Publications, **229**.

KROENKE, L. W., WESSEL, P. & STERLING, A. 2004. Motion of the Ontong Java Plateau in the hotspot frame of reference: 122 Ma–present. *In*: FITTON, J. G., MAHONEY, J. J., WALLACE, P. J. & SAUNDERS, A. D. (eds) 2004. *Origin and Evolution of the Ontong Java Plateau.* Geological Society, London, Special Publications, **229**, 9–20.

GEOLOGICAL SOCIETY SPECIAL PUBLICATION NO. 229

Origin and Evolution of the Ontong Java Plateau

EDITED BY

J. GODFREY FITTON
University of Edinburgh, UK

JOHN J. MAHONEY
University of Hawaii, USA

PAUL J. WALLACE
University of Oregon, USA

and

ANDREW D. SAUNDERS
University of Leicester, UK

2004
Published by
The Geological Society
London

THE GEOLOGICAL SOCIETY

The Geological Society of London (GSL) was founded in 1807. It is the oldest national geological society in the world and the largest in Europe. It was incorporated under Royal Charter in 1825 and is Registered Charity 210161.

The Society is the UK national learned and professional society for geology with a worldwide Fellowship (FGS) of 9000. The Society has the power to confer Chartered status on suitably qualified Fellows, and about 2000 of the Fellowship carry the title (CGeol). Chartered Geologists may also obtain the equivalent European title, European Geologist (EurGeol). One fifth of the Society's fellowship resides outside the UK. To find out more about the Society, log on to www.geolsoc.org.uk.

The Geological Society Publishing House (Bath, UK) produces the Society's international journals and books, and acts as European distributor for selected publications of the American Association of Petroleum Geologists (AAPG), the American Geological Institute (AGI), the Indonesian Petroleum Association (IPA), the Geological Society of America (GSA), the Society for Sedimentary Geology (SEPM) and the Geologists' Association (GA). Joint marketing agreements ensure that GSL Fellows may purchase these societies' publications at a discount. The Society's online bookshop (accessible from www.geolsoc.org.uk) offers secure book purchasing with your credit or debit card.

To find out about joining the Society and benefiting from substantial discounts on publications of GSL and other societies worldwide, consult *www.geolsoc.org.uk*, or contact the Fellowship Department at: The Geological Society, Burlington House, Piccadilly, London W1J 0BG: Tel. +44 (0)20 7434 9944; Fax +44 (0)20 7439 8975; E-mail: enquiries@geolsoc.org.uk.

For information about the Society's meetings, consult *Events* on www.geolsoc.org.uk. To find out more about the Society's Corporate Affiliates Scheme, write to enquiries@geolsoc.org.uk.

Published by The Geological Society from:
The Geological Society Publishing House
Unit 7, Brassmill Enterprise Centre
Brassmill Lane
Bath BA1 3JN,
UK

(*Orders*: Tel. +44 (0)1225 445046
Fax +44 (0)1225 442836)
Online bookshop: http://bookshop.geolsoc.org.uk

British Library Cataloguing in Publication Data
A catalogue record for this book is available from the British Library.

ISBN 186239-157-2

Typeset by Type Study, Scarborough, UK
Printed by MPG Books Ltd, Bodmin, UK

Distributors

USA
AAPG Bookstore
PO Box 979
Tulsa
OK 74101-0979
USA
Orders: Tel. +1 918 584-2555
Fax +1 918 560-2652
E-mail bookstore@aapg.org

India
Affiliated East–West Press PVT Ltd
G-1/16 Ansari Road, Daryaganj,
New Delhi 110 002
India
Orders: Tel. +91 11 2327-9113
Fax +91 11 2326-0538
E-mail affiliat@nda.vsnl.net.in

Japan
Kanda Book Trading Company
Cityhouse Tama 204
Tsurumaki 1-3-10
Tama-shi
Tokyo 206-0034
Japan
Orders: Tel. +81 (0)423 57-7650
Fax +81 (0)423 57-7651
E-mail geokanda@ma.kcom.ne.jp

Contents

FITTON, J. G., MAHONEY, J. J., WALLACE, P. J. & SAUNDERS, A. D. Origin and evolution of the Ontong Java Plateau: introduction 1

Geological evolution and palaeomagnetism

KROENKE, L. W., WESSEL, P. & STERLING, A. Motion of the Ontong Java Plateau in the hot-spot frame of reference: 122 Ma–present 9

ANTRETTER, M., RIISAGER, P., HALL, S., ZHAO, X. & STEINBERGER, B. Modelled palaeolatitudes for the Louisville hot spot and the Ontong Java Plateau 21

RIISAGER, P., HALL, S., ANTRETTER, M. & ZHAO, X. Early Cretaceous Pacific palaeomagnetic pole from Ontong Java Plateau basement rocks 31

ZHAO, X., ANTRETTER, M., RIISAGER, P. & HALL, S. Rock magnetic results from Ocean Drilling Program Leg 192: implications for Ontong Java Plateau emplacement and tectonics of the Pacific 45

PETTERSON, M. G. The geology of north and central Malaita, Solomon Islands: the thickest and most accessible part of the world's largest (Ontong Java) ocean plateau 63

Biostratigraphy

SIKORA, P. J. & BERGEN, J. A. Lower Cretaceous planktonic foraminiferal and nannofossil biostratigraphy of Ontong Java Plateau sites from DSDP Leg 30 and ODP Leg 192 83

BERGEN, J. A. Calcareous nannofossils from ODP Leg 192, Ontong Java Plateau 113

Petrology and geochemistry

TEJADA, M. L. G., MAHONEY, J. J., CASTILLO, P. R., INGLE, S. P., SHETH, H. C. & WEIS, D. Pin-pricking the elephant: evidence on the origin of the Ontong Java Plateau from Pb–Sr–Hf–Nd isotopic characteristics of ODP Leg 192 basalts 133

FITTON, J. G. & GODARD, M. Origin and evolution of magmas on the Ontong Java Plateau 151

HERZBERG, C. Partial melting below the Ontong Java Plateau 179

SANO, T. & YAMASHITA, S. Experimental petrology of basement lavas from Ocean Drilling Program Leg 192: implications for differentiation processes in Ontong Java Plateau magmas 185

CHAZEY, W. J., III & NEAL, C. R. Large igneous province magma petrogenesis from source to surface: platinum-group element evidence from Ontong Java Plateau basalts recovered during ODP Legs 130 and 192 219

ROBERGE, J., WHITE, R. V. & WALLACE, P. J. Volatiles in submarine basaltic glasses from the Ontong Java Plateau (ODP Leg 192): implications for magmatic processes and source region compositions 239

BANERJEE, N. R., HONNOREZ, J. & MUEHLENBACHS, K. Low-temperature alteration of submarine basalts from the Ontong Java Plateau 259

Volcaniclastic rocks

THORDARSON, T. Accretionary-lapilli-bearing pyroclastic rocks at ODP Leg 192 Site 1184: a record of subaerial phreatomagmatic eruptions on the Ontong Java Plateau 275

WHITE, R. V., CASTILLO, P. R., NEAL, C. R., FITTON, J. G. & GODARD, M. Phreatomagmatic eruptions on the Ontong Java Plateau: chemical and isotopic relationship to Ontong Java Plateau basalts 307

CHAMBERS, L. M., PRINGLE, M. S. & FITTON, J. G. Phreatomagmatic eruptions on the Ontong Java Plateau: an Aptian $^{40}Ar/^{39}Ar$ age for volcaniclastic rocks at ODP Site 1184 325

SHAFER, J. T., NEAL, C. R. & CASTILLO, P. R. Compositional variability in lavas from the Ontong Java Plateau: results from basalt clasts within the volcaniclastic succession at Ocean Drilling Program Site 1184 333

CASTILLO, P. R. Geochemistry of Cretaceous volcaniclastic sediments in the Nauru and East Mariana basins provides insights into the mantle sources of giant oceanic plateaus 353

Index 369

Origin and evolution of the Ontong Java Plateau: introduction

J. GODFREY FITTON[1], JOHN J. MAHONEY[2], PAUL J. WALLACE[3] & ANDREW D. SAUNDERS[4]

[1]*School of GeoSciences, University of Edinburgh, Grant Institute, West Mains Road, Edinburgh EH9 3JW, UK (e-mail: Godfrey. Fitton@ed.ac.uk)*

[2]*School of Ocean and Earth Science and Technology, University of Hawaii, Honolulu, HI 96822, USA*

[3]*Department of Geological Sciences, 1272 University of Oregon, Eugene, OR 97403–1272, USA*

[4]*Department of Geology, University of Leicester, Leicester, LE1 7RH, UK*

This volume summarizes the results of recent research on the Ontong Java Plateau (OJP) in the western Pacific Ocean (Fig. 1). The plateau is the most voluminous of the world's large igneous provinces (LIPs) and represents by far the largest known magmatic event on Earth. LIPs are formed through eruptions of basaltic magma on a scale not seen on Earth at the present time (e.g. Coffin & Eldholm 1994; Mahoney & Coffin 1997). Continental flood basalt provinces are the most obvious manifestation of LIP magmatism, but they have oceanic counterparts in volcanic rifted margins and giant submarine ocean plateaus. LIPs have also been identified on the Moon, Mars and Venus, and may represent the dominant form of volcanism in the solar system (Head & Coffin 1997). The high magma production rates (i.e. large eruption volume and high eruption frequency) involved in LIP magmatism cannot be accounted for by normal plate tectonic processes. Anomalously hot mantle often appears to be required, and this requirement has been a key consideration in the formulation of the currently favoured plume-head hypothesis in which LIPs are formed through rapid decompression and melting in the head of a newly ascended mantle plume (e.g. Richards *et al.* 1989; Campbell & Griffiths 1990). Eruption of enormous volumes of basaltic magma over short time intervals, especially in the subaerial environment, may have had significant effects on climate and the biosphere, and LIP formation has been proposed as one of the causes of mass extinctions (e.g. Wignall 2001).

Several issues need to be addressed in order to understand LIP formation. These include: the timing and duration of magmatism; the size, timing and duration of individual eruptions; the eruption environment of the magmas (subaqueous or subaerial); the magnitude of crustal uplift accompanying their emplacement; and the composition and temperature of their mantle sources. The study of continental LIPs can address these to a large extent, and considerable progress has been made in these areas. Petrological and geochemical studies on the sources of continental flood basalt, however, are always compromised by the possibility of contamination of the magma by the continental crust and lithospheric mantle through which it passes. Basalt from plateaus that formed entirely in an oceanic environment, being free of such contamination, offers a clear view of LIP mantle sources, but is difficult and expensive to sample. Nevertheless, the basaltic basement of several ocean plateaus has been sampled in the course of Deep Sea Drilling Project (DSDP) and Ocean Drilling Program (ODP) legs.

The Ontong Java Plateau

The OJP covers an area of about 2.0×10^6 km^2 (comparable in size with western Europe), and OJP-related volcanism extends over a considerably larger area into the adjacent Nauru, East Mariana, and possibly the Pigafetta and Lyra, basins (Fig. 1). With a maximum thickness of crust beneath the plateau of 30–35 km (e.g. Gladczenko *et al.* 1997; Richardson *et al.* 2000), the volume of igneous rock forming the plateau and filling the adjacent basins could be as high as 6×10^7 km^3 (e.g. Coffin & Eldholm 1994).

Seismic tomography experiments show a rheologically strong, but seismically slow, upper mantle root extending to about 300 km depth beneath the OJP (e.g. Richardson *et al.* 2000; Klosko *et al.* 2001). Gomer & Okal (2003) have measured the shear-wave attenuation in this root and found it to be low, implying that the slow seismic velocities must be due to a compositional, rather than thermal, anomaly in the

From: FITTON, J. G., MAHONEY, J. J., WALLACE, P. J. & SAUNDERS, A. D. (eds) 2004. *Origin and Evolution of the Ontong Java Plateau*. Geological Society, London, Special Publications, **229**, 1–8. 0305-8719/$15.00

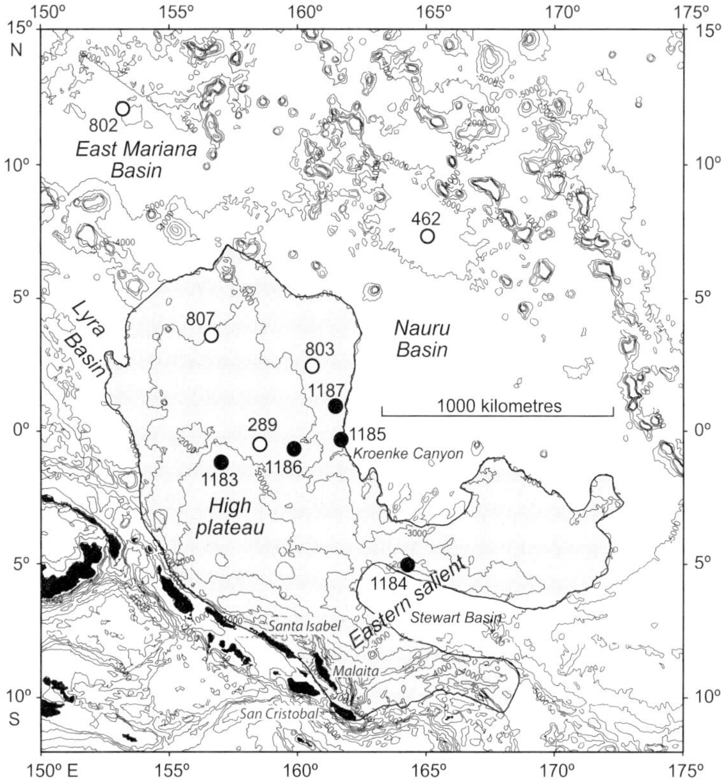

Fig. 1. Predicted bathymetry (after Smith & Sandwell 1997) of the Ontong Java Plateau and surrounding areas showing the location of DSDP and ODP basement drill sites. Leg 192 drill sites are marked by black circles; open circles represent pre-Leg 192 drill sites. The edge of the plateau is defined by the –4000 m-contour, except in the SE part where it has been uplifted through collision with the Solomon arc.

mantle. The nature and origin of this compositional anomaly has not yet been established.

The OJP seems to have been formed rapidly at around 120 Ma (e.g. Mahoney *et al.* 1993; Tejada *et al.* 1996, 2002; Chambers *et al.* 2002; Parkinson *et al.* 2002), and the peak magma production rate may have exceeded that of the entire global mid-ocean ridge system at the time (e.g. Tarduno *et al.* 1991; Mahoney *et al.* 1993; Coffin & Eldholm 1994). Degassing from massive eruptions during the formation of the OJP could have increased the CO_2 concentration in the atmosphere and oceans (Larson & Erba 1999), and led to, or at least contributed significantly to, a world-wide oceanic anoxic event accompanied by a 90% reduction in nannofossil palaeoflux (Erba & Tremolada 2004).

Collision of the OJP with the old Solomon arc has resulted in uplift of the OJP's southern margin to create on-land exposures of basaltic basement in the Solomon Islands (Fig. 1), notably in Malaita, Santa Isabel and San Cristobal (e.g. Petterson *et al.* 1999). In addition to these exposures, the basaltic basement on the OJP and surrounding Nauru and East Mariana basins has been sampled at 10 DSDP and ODP drill sites. However, the most recent drilling leg (ODP Leg 192 in September–November 2000) was the first designed specifically to address the origin and evolution of the OJP (Mahoney *et al.* 2001). Earlier research on the OJP has been reviewed by Neal *et al.* (1997). The principal aim of the present volume is to present the results of research that has followed from ODP Leg 192, and most of the papers in it were written or co-authored by participants in this leg. The volume complements the recent thematic set of papers on the origin and evolution of the Kerguelen Plateau, the world's second largest oceanic LIP, published in *Journal of Petrology* (Wallace et al. 2002).

Geological evolution and palaeomagnetism

Several authors (e.g. Mahoney & Spencer, 1991; Richards *et al.* 1991; Tarduno *et al.* 1991) have favoured the starting plume head of the Louisville hot spot (now at *c.* 52°S) as the source of the OJP. In the first paper of the volume, **Kroenke *et al.*** use a new model of Pacific absolute plate motion, based on the fixed hot-spot frame of reference, to track the palaeogeographic positions of the OJP from its present location on the Equator back to 43°S at the time of its formation (*c.* 120 Ma). This inferred original position is 9° north of the present location of the Louisville hot spot, and suggests that this hot spot was not responsible for the formation of the OJP or, alternatively, that the hot spot has drifted significantly relative to the Earth's spin axis (as the Hawaiian hot spot appears to have done; e.g. Tarduno *et al.* 2003). **Kroenke *et al.*** also note the presence of linear gravity highs in the western OJP, which they speculate may indicate formation of the OJP close to a recently abandoned spreading centre. **Antretter *et al.*** point out that the palaeomagnetic palaeolatitude of the OJP (*c.* 25°S) determined by **Riisager *et al.*** (and Riisager *et al.* 2003) further increases the discrepancy with the location of the Louisville hot spot. **Zhao *et al.***'s investigation of the rock-magnetic properties of basalt from the OJP shows that original and stable magnetic directions are preserved, allowing robust estimates of palaeolatitude. The discrepancy between the palaeolatitudes calculated from the palaeomagnetic data and from the fixed-hot-spot reference frame is interpreted by **Riisager *et al.*** as evidence for movement between hot spots. **Antretter *et al.*** show that the Louisville hot spot may have moved southwards over the past 120 Ma, and that taking account of both hot-spot motion and true polar wander reduces the discrepancy and makes the formation of the OJP by the Louisville hot spot barely possible, if still unlikely.

The thickest exposures of the OJP basement rocks in the Solomon Islands are found on the remote island of Malaita (Fig. 1). **Petterson** presents the results of geological surveys that reveal a monotonous succession of Early Cretaceous tholeiitic pillow basalt, sheet flows and sills (the Malaita Volcanic Group) 3–4 km thick. Rare and very thin interbeds composed of laminated pelagic chert or limestone suggest high eruption frequency and emplacement into deep water. The Malaita Volcanic Group is conformably overlain by a 1–2 km-thick Cretaceous–Pliocene pelagic sedimentary cover sequence, punctuated by alkaline basalt volcanism during the Eocene and by intrusion of alnöite during the Oligocene.

Age and biostratigraphy

The age and duration of OJP magmatism has not yet been established with any certainty. OJP basalts are difficult to date by the widely used $^{40}Ar/^{39}Ar$ method because of their very low potassium contents. Published $^{40}Ar/^{39}Ar$ data (Mahoney *et al.* 1993; Tejada *et al.* 1996, 2002) suggest a major episode of OJP volcanism at *c.* 122 Ma and a minor episode at *c.* 90 Ma. $^{40}Ar/^{39}Ar$ analysis (Chambers *et al.* 2002; L. M. Chambers unpublished data) of samples from ODP Leg 192 Sites 1185, 1186 and 1187 (Fig. 1) gives ages ranging from 105 to 122 Ma. Chambers *et al.* (2002) suggest that their younger apparent ages (and, by implication, the data on which the 90 Ma episode is based) are the result of argon recoil and therefore represent minimum ages. Biostratigraphic dating based on foraminifera and nannofossils (**Sikora & Bergen**; **Bergen**) contained in sediment intercalated with lava flows at ODP Sites 1183, 1185, 1186 and 1187 suggests that magmatism on the high plateau extended from latest early Aptian on the plateau crest to late Aptian on the eastern edge. This corresponds to age ranges of 122–112 Ma (Harland *et al.* 1990) or 118–112 Ma (Gradstein *et al.* 1995). However, Re–Os isotopic data on basalt samples from these same four drill sites define a single isochron with an age of 121.5±1.7 Ma (Parkinson *et al.* 2002).

The oldest sediment overlying basement on the crest of the OJP occurs within the upper part of the *Leupoldina cabri* planktonic foraminiferal zone and corresponds with a prominent $\delta^{13}C$ maximum (**Sikora & Bergen**). This result shows that eruption of basaltic lava flows continued through much of Oceanic Anoxic Event 1a, of which the formation of the plateau is a postulated cause (e.g. Larson & Erba 1999). Nannofossil studies (**Bergen**) reveal six unconformities in the Lower Aptian–Miocene pelagic cover sequence recovered during Leg 192.

Petrology and geochemistry

The Malaita Volcanic Group (**Petterson**) has been divided by Tejada *et al.* (2002) into two chemically and isotopically distinct stratigraphic units: the Kwaimbaita Formation (>2.7 km thick) and the overlying Singgalo Formation (*c.* 750 m maximum exposed thickness). Basalt of the Kwaimbaita Formation was found to be compositionally similar to the basalt forming units C–G at ODP Site 807, on the northern flanks of the OJP (Fig. 1), whereas the Singgalo Formation is similar to the overlying unit A at Site 807. Thus, Kwaimbaita-type and Singgalo-type basalt flows with the same stratigraphic relationship are found at two sites 1500 km apart on the plateau (Tejada *et al.* 2002). A third basalt type, with higher MgO and lower concentrations of incompatible elements than any previously reported from the OJP, was recognized during ODP Leg 192 at Sites 1185 and 1187 on the eastern edge of the plateau (Mahoney *et al.* 2001). We propose the term *Kroenke-type basalt* because it was discovered on the flanks of the submarine Kroenke Canyon at Site 1185 (Fig. 1).

Tejada *et al.* use radiogenic-isotope (Sr, Nd, Pb, Hf) ratios to show that Kwaimbaita-type basalt is found at all but one of the OJP drill sites and therefore represents the dominant OJP magma type. Singgalo-type basalt, on the other hand, appears to be volumetrically minor. Significantly, Kroenke-type basalt is isotopically identical to Kwaimbaita-type basalt (**Tejada *et al.***) and may therefore represent the parental magma for the bulk of the OJP. Age-corrected radiogenic-isotope ratios in Kroenke- and Kwaimbaita-type basalts show a remarkably small range. **Tejada *et al.*** model the initial Sr-, Nd-, Pb- and Hf-isotope ratios in these two basalt types as representing originally primitive mantle that experienced a minor fractionation event (e.g. the extraction of a small amount of partial melt) at about 3 Ga or earlier.

The remarkable homogeneity of OJP basalts is also seen in their major- and trace-element composition (**Fitton & Godard**). **Fitton & Godard** use geochemical data to model the mantle source composition and hence to estimate the degree of partial melting involved in the formation of the OJP. Incompatible-element abundances in the primary OJP magma can be modelled by around 30% melting of a peridotitic primitive mantle source from which about 1% by mass of average continental crust had previously been extracted. The postulated depletion is consistent with the isotopic modelling of **Tejada *et al.*** To produce a 30% melt requires decompression of very hot (potential temperature >1500°C) mantle beneath thin lithosphere. Thin lithosphere is consistent with the suggestion by **Kroenke *et al.*** that the OJP may have formed close to a recently abandoned spreading centre. Alternatively, lithospheric thinning could have resulted from thermal erosion caused by the upwelling of hot plume material.

An independent estimate of the degree of melting is provided by **Herzberg**, who uses a forward- and inverse-modelling approach based on peridotite phase equilibria. He obtains values of 27 and 30% for fractional and equilibrium melting, respectively. Further support for large-degree melting is provided by the platinum-group element (PGE) concentrations determined by **Chazey & Neal**. The PGEs are highly compatible in mantle phases and sulphides, so their abundance is sensitive to degree of melting and sulphur saturation. Concentrations of PGEs in the OJP basalts are rather high, and consistent with around 30% melting of a peridotite source from which sulphide phases had been exhausted during the melting process. Some basalt samples have PGE abundances that are too high to be accounted for by a standard model peridotite source, and an additional source of PGEs appears to be needed. **Chazey & Neal** speculate that a small amount of material from the Earth's core may have been involved in the generation of OJP magmas.

Derivation of the dominant, evolved, Kwaimbaita magma type through fractional crystallization of the primitive Kroenke-type magma is consistent with the isotopic (**Tejada *et al.***) and geochemical (**Fitton & Godard**) evidence, and with melting experiments carried out by **Sano & Yamashita**. **Sano & Yamashita**'s results show that the variations in phenocryst assemblage and whole-rock basalt major-element compositions can be modelled adequately by fractional crystallization in shallow (<6 km) magma reservoirs.

Glass from the rims of basaltic pillows was recovered from most drill sites on the OJP, and this glass preserves a record of the volatile content of the magmas at the time of eruption. **Roberge *et al.*** show that water contents in the glasses are uniformly low and imply water contents in the mantle source that are comparable with those in the source of mid-ocean ridge basalt. This is an important observation because it shows that the large degrees of melting estimated for the OJP magmas cannot have been caused by the presence of water but require high temperatures. The sulphur contents of OJP glasses confirm **Chazey & Neal**'s inference of sulphur undersaturation in the magmas. The water depth of lava emplacement controls the CO_2 content of the glasses, and data obtained by **Roberge *et al.*** imply depths ranging from about 1000 m on the crest of the OJP to about 2500 m on its eastern edge. The amount of CO_2 released during formation of the OJP is difficult to determine without reliable information on primary magmatic CO_2 contents and precise knowledge of the duration of volcanism, but **Roberge *et al.*** estimate a maximum value that is around 10 times the flux from the global mid-ocean ridge system. Erba & Tremolada (2004) estimate that the 90% reduction in nannofosil palaeofluxes that they link to emplacement of the OJP requires a three- to sixfold increase in volcanogenic CO_2.

The submarine emplacement of most of the OJP resulted in low-temperature alteration of the basalts through contact with sea water. The alteration ranges from slight to complete, and unaltered olivine and glass were found in some of the basaltic lava flows sampled in the drill cores. A detailed study of the alteration processes is reported by **Banerjee *et al.***, who show that alteration started soon after emplacement and is indistinguishable from that affecting mid-ocean ridge basalt. There is no evidence for high-temperature alteration in any of the basalt recovered from the OJP. The initial and most pervasive stage of alteration resulted in the replacement of olivine and interstitial glass with celadonite and smectite. Later interaction between basalt and cold, oxidizing sea water caused local replacement of primary phases and mesostasis by smectite and iron oxyhydroxides. Glass shards in tuffs at Site 1184 show clear textural evidence of microbial alteration (Banerjee & Muehlenbachs 2003).

Volcaniclastic rocks

One of the most exciting discoveries of ODP Leg 192 was a thick succession of basaltic volcaniclastic rocks at Site 1184 on the eastern salient of the OJP (Fig. 1). Drilling at this site penetrated 337.7 m of tuff and lapilli tuff, before the site had to be abandoned through lack of time. A detailed volcanological study by **Thordarson** concludes that the volcaniclastic succession was the result of large phreatomagmatic eruptions in a subaerial setting. This setting contrasts strikingly with that of the lava flows sampled on the main plateau and in the Solomons, which were all erupted under deep water (**Roberge *et al.***; **Petterson**). **Thordarson** divides the succession into six subunits or members, each representing a single massive eruptive event. Fossilized or carbonized wood fragments were found near the bottom of four of the eruptive members (Mahoney *et al.* 2001). The volcaniclastic succession at Site 1184 provides the only evidence so far for significant amounts of subaerial volcanism on the OJP.

Three of the six eruptive members at Site 1184 contain blocky glass clasts with unaltered cores, and these cores allow the reliable determination of the composition of the erupted magma. **White *et al.*** used microbeam techniques to determine the major- and trace-element compositions of samples of the glass. The glasses are very similar in composition to the Kwaimbaita- and Kroenke-type basalts sampled on the high plateau. Each member has a distinct glass composition and there is no intermixing of glass compositions between them, confirming **Thordarson**'s conclusion that each is the result of one eruptive phase, and that the volcaniclastic sequence has not been reworked. **White *et al.***'s major- and trace-element data for the glass clasts suggest that the voluminous subaerially erupted volcaniclastic rocks at Site 1184 belong to the same magmatic event as that responsible for the construction of the main plateau. Thus, the OJP would have been responsible for volatile fluxes into the atmosphere in addition to chemical fluxes into the oceans. Both factors may have influenced the contemporaneous oceanic anoxic event (**Sikora & Bergen**; Erba & Tremolada 2004).

The geochemical evidence (**White *et al.***; **Fitton & Godard**) linking the phreatomagmatic eruptions recorded at Site 1184 to the formation of the main plateau is supported by the Early Cretaceous age implied by the steep (–54°) magnetic inclination preserved in the volcaniclastic rocks (**Riisager *et al.***). However, this evidence appears to be contradicted by the presence of rare Eocene nannofossils at several levels within the succession (**Bergen**). In an attempt to resolve this paradox, **Chambers *et al.*** applied the $^{40}Ar/^{39}Ar$ dating method to feldspathic material separated from two basaltic clasts, and to individual plagioclase crystals separated from the matrix of the volcaniclastic rocks. The clasts gave *minimum* age estimates of *c.* 74 Ma, and the plagioclase crystals a mean value of 123.5±1.8 (1σ) Ma.

Shafer *et al.* analysed a suite of 14 basaltic clasts extracted from four of the volcaniclastic units and, despite their extensive alteration, showed that the clasts were derived from a source similar to that of the Kroenke- and Kwaimbaita-type basalts on the main plateau. Significantly, the composition of the clasts (**Shafer *et al.***) varies with the bulk composition of their host volcaniclastic units (**Fitton & Godard**), showing that they must be cognate. **Chambers *et al.*** conclude that both the clasts and the plagioclase crystals that they used in their $^{40}Ar/^{39}Ar$ dating belong to the same magmatic episode as the host volcaniclastic rocks and are not xenoliths or xenocrysts from older basement. Thus, the combined $^{40}Ar/^{39}Ar$, geochemical and palaeomagnetic evidence favours an early Cretaceous age for the volcaniclastic succession. **Thordarson** and **Chambers *et al.*** suggest that the Eocene nannofossils were introduced later, possibly along fractures.

Volcaniclastic rocks (recovered during previous ODP legs) are also found in thick, reworked and redeposited successions overlying Early Cretaceous basalt in the Nauru and East Mariana basins to the east and north of the OJP, respectively. Much of the volcaniclastic material consists of hyaloclastite and this, together with the presence of wood and shallow-water carbonate fragments, suggests that both the East Mariana Basin and the Nauru Basin volcaniclastic rocks were derived from once-emergent volcanic sources or from relatively shallow water. In the final paper in this volume, **Castillo** presents chemical and isotopic data for these volcaniclastic rocks and compares them with data for the OJP. The Nauru Basin volcaniclastic rocks have incompatible trace-element and Nd-isotope compositions typical of the Kwaimbaita-type tholeiitic lavas of the OJP, suggesting that these deposits were shed from the plateau itself. On the other hand, the East Mariana Basin volcaniclastic rocks have high concentrations of incompatible trace-elements, and Nd- and Pb-isotope ratios typical of alkalic ocean island basalts, and are unrelated to the OJP.

A mantle plume origin for the Ontong Java Plateau?

One of the principal objectives of ODP Leg 192 was to test the plume-head hypothesis for the formation of giant ocean plateaus, and many of the results presented in this volume are consistent with such an origin. The discovery of high-MgO Kroenke-type basalt allows us to calculate the composition of the primary magma and hence deduce the nature of the mantle source and the degree of melting. Isotopic (**Tejada *et al.***) and chemical (**Fitton & Godard**; **Chazey & Neal**) data are consistent with a mildly depleted peridotite mantle source, and phase-equilibria (**Herzberg**) and trace-element (**Fitton & Godard**; **Chazey & Neal**) modelling independently constrain the degree of melting of this peridotite source to around 30%. Melting to this extent can only be achieved by decompression of hot (potential temperature >1500°C) peridotite beneath thin lithosphere. To achieve an *average* of 30% melting requires that the mantle is actively and rapidly fed into the melt zone, and a start-up mantle plume provides the most obvious mechanism. A plume-head impinging on thin lithosphere theoretically should have caused uplift of a sizeable part of the plateau above sea level, as in Iceland, and, indeed, at least part of the eastern salient was emergent (**Thordarson**). However, the abundance of essentially non-vesicular submarine lava and the absence of any basalt showing signs of subaerial weathering show that all the other sampled portions of the OJP were emplaced below sea level (e.g. Neal *et al.* 1997; Mahoney *et al.* 2001). Volatile concentrations in quenched pillow-rim glasses suggest eruption depths ranging from 1100 m at Site 1183 to 2570 m at Site 1187 (**Roberge *et al.***).

We have not yet been able to resolve the paradox of apparent high mantle potential temperature coupled with predominantly submarine emplacement. Widespread melting of the mantle following the impact of an asteroid provides a possible means of avoiding uplift (e.g. Ingle & Coffin 2004), but the resulting magma would be generated entirely within the upper mantle and should normally be expected to have the chemical and isotopic characteristics of Pacific mid-ocean ridge basalt. OJP basalt is isotopically (**Tejada *et al.***) and chemically (**Fitton & Godard**) distinct from Pacific mid-ocean ridge basalt. Furthermore, no mass extinction occurred at the time of OJP formation, even though the required asteroid would have had a diameter significantly greater than that thought to have been responsible for the extinctions at the Cretaceous–Tertiary boundary (Ingle & Coffin 2004). A more detailed case against an impact origin for the OJP is set out by **Tejada *et al.*** An eclogitic source does not provide an alternative to the plume hypothesis because the high-Mg parental magma would require almost total melting, and consequently a very high potential temperature would still be needed to provide the latent heat of fusion. We can also rule out a hydrous mantle source because the magmas have very low H_2O contents (**Roberge *et al.***).

The papers collected together into this volume show the progress that has been made in our understanding of the origin and evolution of the Ontong Java Plateau following its successful drilling during ODP Leg 192. We now have a much clearer view of the range and distribution of basalt types on the plateau, and we have identified a potential parental magma composition represented by Kroenke-type basalt. The age and duration of magmatism is still uncertain because we have still only scratched the surface of the 30–35 km-thick OJP crust. However, it now seems plausible that almost the entire plateau formed in a single, widespread magmatic event at *c.* 120 Ma. The identification of a thick succession of volcaniclastic rocks at Site 1184 shows that at least part of the plateau was erupted in a subaerial environment. We conclude that the start-up plume hypothesis appears to fit more of the observations than do any of the alternative hypotheses, but the lack of uplift of the magnitude predicted by the plume hypothesis and the lack of an obvious hot-spot track remain to be explained.

We thank the ODP and Transocean/Sedco-Forex staff on board the *JOIDES Resolution* for their considerable contribution to the success of ODP Leg 192. ODP is sponsored by the US National Science Foundation (NSF) and participating countries under management of Joint Oceanographic Institutions (JOI), Inc. We also thank the many colleagues who reviewed papers in this volume. They are acknowledged individually in the respective papers. Finally, we are indebted to S. Oberst of the Geological Society Publishing House for the care and patience with which she handled the production of this volume.

References

BANERJEE, N.R. & MUEHLENBACHS, K. 2003. Tuff life: Bioalteration in volcaniclastic rocks from the Ontong Java Plateau. *Geochemistry, Geophysics, Geosystems*, **4**, 2002GC000470.

CAMPBELL, I.H. & GRIFFITHS, R.W. 1990. Implications of mantle plume structure for the evolution of flood basalts. *Earth and Planetary Science Letters*, **99**, 79–93.

CHAMBERS, L.M., PRINGLE, M.S. & FITTON, J.G. 2002. Age and duration of magmatism on the Ontong Java Plateau: ^{40}Ar–^{39}Ar results from ODP Leg 192. Abstract V71B-1271. *Eos, Transactions of the American Geophysical Union*, **83**, F47.

COFFIN, M.F. & ELDHOLM, O. 1994. Large igneous provinces: crustal structure, dimensions, and external consequences. *Reviews of Geophysics*, **32**, 1–36.

ERBA, E. & TREMOLADA, F. 2004. Nannofossil carbonate fluxes during the Early Cretaceous: phytoplankton response to nutrification episodes, atmospheric CO_2 and anoxia. *Paleoceanography*, **19**, 1–18.

GLADCZENKO, T.P., COFFIN, M.F. & ELDHOLM, O. 1997. Crustal structure of the Ontong Java Plateau: modeling of new gravity and existing seismic data. *Journal of Geophysical Research*, **102**, 22 711–22 729.

GOMER, B.M. & OKAL, E.A. 2003. Multiple-ScS probing of the Ontong Java Plateau. *Physics of the Earth and Planetary Interiors*, **138**, 317–331.

GRADSTEIN, F.M., AGTERBERG, F.P., OGG, J.G., HARDENBOL, J., VAN VEEN, P., THIERRY, J. & HUANG, Z. 1995. A Triassic, Jurassic and Cretaceous time scale. *In*: BERGGREN, W.A., KENT, D.V., AUBRY, M.P. & HARDENBOL, J. (eds) *Geochronology, Time Scales and Global Stratigraphic Correlation*. SEPM, Special Publication, **54**, 95–126.

HARLAND, W.B., ARMSTRONG, R.L., COX, A.V., CRAIG, L.E., SMITH, A.G. & SMITH, D.G. 1990. *A Geologic Time Scale 1989*. Cambridge University Press, Cambridge.

HEAD, J.W. & COFFIN, M.F. 1997. Large igneous provinces: a planetary perspective. *In*: MAHONEY, J.J. & COFFIN, M.F. (eds) *Large Igneous Provinces: Continental, Oceanic, and Planetary Flood Volcanism*. American Geophysical Union, Geophysical Monograph, **100**, 411–438.

INGLE, S. & COFFIN, M.F. 2004. Impact origin for the greater Ontong Java Plateau? *Earth and Planetary Science Letters*, **218**, 123–124.

KLOSKO, E.R., RUSSO, R.M., OKAL, E.A. & RICHARDSON, W.P. 2001. Evidence for a rheologically strong chemical mantle root beneath the Ontong–Java Plateau. *Earth and Planetary Science Letters*, **186**, 347–361.

LARSON, R.L. & ERBA, E. 1999. Onset of the Mid-Cretaceous greenhouse in the Barremian–Aptian: igneous events and the biological, sedimentary, and geochemical consequences. *Paleoceanography*, **14**, 663–678.

MAHONEY, J.J. & COFFIN, M.F. (eds). 1997. *Large Igneous Provinces: Continental, Oceanic, and Planetary Flood Volcanism*. American Geophysical Union, Geophysical Monograph, **100**.

MAHONEY, J.J. & SPENCER, K.J. 1991. Isotopic evidence for the origin of the Manihiki and Ontong–Java oceanic plateaus. *Earth and Planetary Science Letters*, **104**, 196–210.

MAHONEY, J.J., FITTON, J.G., WALLACE, P.J. *et al.* 2001. *Proceedings of the Ocean Drilling Program, Initial Reports*, 192 (online). Available from World Wide Web: http://www-odp.tamu.edu/publications/192_IR/192ir.htm

MAHONEY, J.J., STOREY, M., DUNCAN, R.A., SPENCER, K.J. & PRINGLE, M. 1993. Geochemistry and geochronology of the Ontong Java Plateau. *In*: PRINGLE, M., SAGER, W., SLITER, W. & STEIN, S. (eds) *The Mesozoic Pacific. Geology, Tectonics, and Volcanism*. American Geophysical Union, Geophysical Monograph, **77**, 233–261.

NEAL, C.R., MAHONEY, J.J., KROENKE, L.W., DUNCAN, R.A. & PETTERSON, M.G. 1997. The Ontong Java Plateau. *In*: MAHONEY, J.J. & COFFIN, M.F. (eds) *Large Igneous Provinces: Continental, Oceanic, and Planetary Flood Volcanism*. American Geophysical Union, Geophysical Monograph, **100**, 183–216.

PARKINSON, I.J., SCHAEFER, B.F. & ARCULUS, R.J. 2002. A lower mantle origin for the world's biggest LIP? A high precision Os isotope isochron from Ontong Java Plateau basalts drilled on ODP Leg 192. *Geochimica et Cosmochimica Acta*, **66**, Supplement, A580.

PETTERSON, M.G., BABBS, T., NEAL, C.R., MAHONEY, J.J., SAUNDERS, A.D., DUNCAN, R.A., TOLIA, D., MAGU, R., QOPOTO, C., MAHOA, H. & NATOGGA, D. 1999. Geological–tectonic framework of Solomon Islands, SW Pacific: crustal accretion and growth within an intra-oceanic setting. *Tectonophysics*, **301**, 35–60.

RICHARDS, M.A., DUNCAN, R.A. & COURTILLOT, V. 1989. Flood basalts and hot-spot tracks: plume heads and tails. *Science*, **246**, 103–107.

RICHARDS, M.A., JONES, D.L., DUNCAN, R.A. & DEPAOLO, D.J. 1991. A mantle plume initiation model for the Wrangellia and other oceanic flood basalt provinces. *Science*, **254**, 263–266.

RICHARDSON, W.P., OKAL, E.A. & VAN DER LEE, S. 2000. Rayleigh-wave tomography of the Ontong Java Plateau. *Physics of the Earth and Planetary Interiors*, **118,** 29–61.

RIISAGER, P., HALL, S., ANTRETTER, M. & ZHAO, X. 2003. Paleomagnetic paleolatitude of Early Cretaceous Ontong Java Pleateau basalts: implications for Pacific apparent and true polar wander. *Earth and Planetary Science Letters*, **208**, 235–252.

SMITH, W.H.F. & SANDWELL, D.T. 1997. Global seafloor topography from satellite altimetry and ship depth soundings. *Science*, **277,** 1956–1962.

TARDUNO, J.A., DUNCAN, R.A., SCHOLL, D.W., COTTRELL, R.D., STEINBERGER, B., THORDARSON, T., KERR, B.C., NEAL, C.R., FREY, F.A., TORII, M. & CARVALLO, C. 2003. The Emperor Seamounts: southward motion of the Hawaiian hotspot plume in Earth's mantle. *Science*, **301**, 1064–1069.

TARDUNO, J.A., SLITER, W.V., KROENKE, L.W., LECKIE, M., MAHONEY, J.J., MUSGRAVE, R.J., STOREY, M. & WINTERER, E.L. 1991. Rapid formation of the Ontong Java Plateau by Aptian mantle plume volcanism. *Science*, **254,** 399–403.

TEJADA, M.L.G., MAHONEY, J.J., DUNCAN, R.A. & HAWKINS, M.P. 1996. Age and geochemistry of basement and alkalic rocks of Malaita and Santa Isabel, Solomon Islands, southern margin of Ontong Java Plateau. *Journal of Petrology*, **37,** 361–394.

TEJADA, M.L.G., MAHONEY, J.J., NEAL, C.R., DUNCAN, R.A. & PETTERSON, M.G. 2002. Basement geochemistry and geochronology of Central Malaita, Solomon Islands, with implications for the origin and evolution of the Ontong Java Plateau. *Journal of Petrology*, **43,** 449–484.

WALLACE, P.J., FREY, F.A., WEIS, D. & COFFIN, M.F. (eds). 2002. *Origin and Evolution of the Kerguelen Plateau, Broken Ridge and Kerguelen Archipelago. Journal of Petrology*, **43**, 1105–1413.

WIGNALL, P.B. 2001. Large igneous provinces and mass extinctions. *Earth Science Reviews*, **53**, 1–33.

Motion of the Ontong Java Plateau in the hot-spot frame of reference: 122 Ma–present

L. W. KROENKE, P. WESSEL & A. STERLING

School of Ocean and Earth Science and Technology, University of Hawaii, Honolulu, HI 96822, USA (e-mail: kroenke@soest.hawaii.edu)

Abstract: A new model of Pacific absolute plate motion between 140 and 0 Ma, generated in the fixed hot-spot frame of reference, has been used to track palaeogeographic positions of the Ontong Java Plateau (OJP) from the time (*c.* 122 Ma) and location (*c.* 43°S) of its formation to its present location north of the Solomon Islands. The resulting OJP seafloor flow-line suggests that changes in Pacific plate motion, passage over hot spots and Pacific Rim tectonism all have influenced the continuing structural development and deformation of the plateau. Satellite-derived gravity, bathymetry and Rayleigh-wave tomography potentially reveal the structural fabric of the OJP and adjoining Nauru Basin, including the orientation of probable fracture zones, location of possible relict spreading centres and the presence of a thick lithospheric root, as well as possible later hot-spot-related modification of the fabric. The most recent phase of OJP deformation, which began about 6 Ma, accelerated at 2.6 Ma and continues today, has resulted in the uplift of the islands of Malaita and Santa Isabel, and the formation of the Malaita Anticlinorium, with slip along the old fracture zones possibly triggering submarine canyon formation on the NE side of the OJP. This collision-related deformation also is probably responsible for the ongoing uplift and tilting of the islands of Nauru and Banaba NE of the OJP high plateau.

The OJP (Fig. 1) is the largest Pacific oceanic plateau, the size of Greenland, which formed in the Early Cretaceous. The tectonic setting, composition and age of emplacement of the OJP, as summarized by Neal *et al.* (1997) and the Ocean Drilling Program (ODP) Leg 192 Shipboard Scientific Party (2001), suggest that the OJP basement lavas were formed from a hot-spot-type mantle source and emplaced on relatively thin (and thus probably relatively young) lithosphere. Plate reconstructions back to 125 Ma (Neal *et al.* 1997), in a fixed hot-spot frame of reference (Morgan 1972), placed the location of the OJP hot spot at 42°S, 159°W, well north of the Louisville hotspot (*c.* 52°S, *c.* 139°W) and somewhat off-axis from a possible spreading ridge joined via a triple junction to the Pacific–Phoenix spreading centre. These reconstructions also assumed that a post-plateau-emplacement spreading event occurred in the Stewart and Ellice basins.

In contrast to a prolonged Icelandic type of emplacement along a slow but active spreading ridge, formation of the OJP appears to have occurred relatively rapidly following cessation of Pacific–Phoenix spreading shortly after Chron M1. Basement ages over most of the plateau, obtained by ^{40}Ar–^{39}Ar and Re–Os dating, cluster around 122 Ma (Mahoney *et al.* 1993; Tejada *et al.* 1996, 2002; Parkinson *et al.* 2001). A few ages from the margins of the OJP and in surrounding basins, although recently questioned (Chambers *et al.* 2002), are suggestive of relatively minor post-emplacement eruptions around 110 and 90 Ma (Castillo *et al.* 1994; Neal *et al.* 1997). Basement rocks across the OJP are low-K tholeiitic basalt with, surprisingly for its immense size, only a small range of major-element, trace-element and Nd–Pb–Sr isotopic composition (e.g. Fitton & Godard 2004; Tejada *et al.* 2004).

A recent study of Pacific absolute plate motion (APM) by Wessel *et al.* (2003), using the hot-spotting technique of Wessel & Kroenke (1997) and polygon finite rotation method of Harada & Hamano (2000), allows numerous Pacific seamount chains to be fitted simultaneously without invoking hot-spot drift. The Pacific trails fitted by this new APM model are shown in Figure 2. The hot-spot locations for these trails are given in Table 1. The finite rotations derived for the APM model by Wessel *et al.* (2003) are presented in Table 2. Wessel *et al.* conclude that the bend in the Hawaiian–Emperor seamount chain (HEB) is also expressed in related coeval bends in numerous other chains across the Pacific (Fig. 3), and thus represents a significant change in Pacific

From: FITTON, J. G., MAHONEY, J. J., WALLACE, P. J. & SAUNDERS, A. D. (eds) 2004. *Origin and Evolution of the Ontong Java Plateau*. Geological Society, London, Special Publications, **229**, 9–20. 0305-8719/$15.00

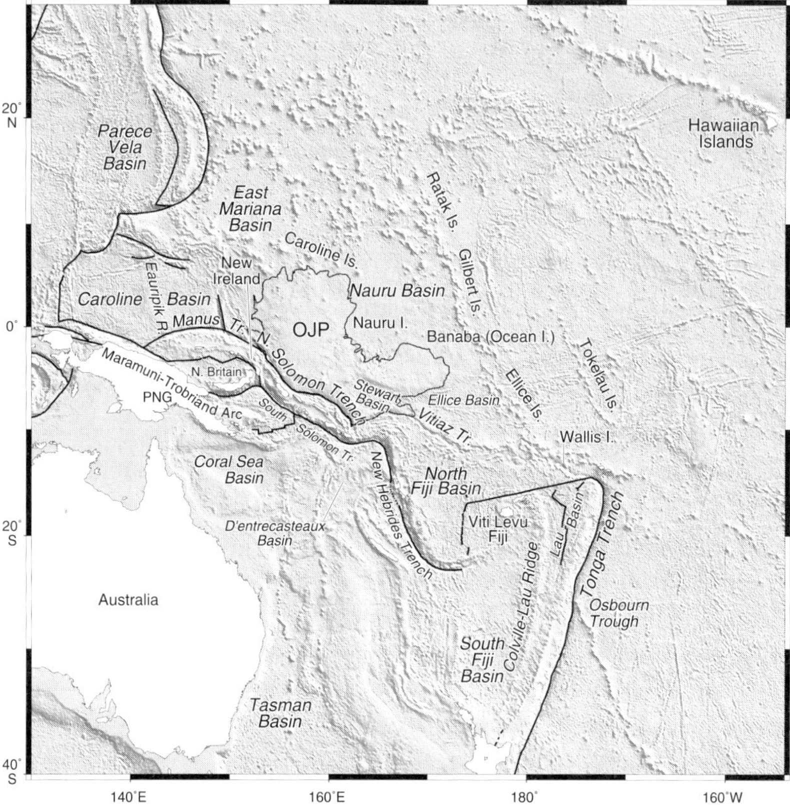

Fig. 1. Location of the Ontong Java Plateau and physiographical features discussed in the text. (A more detailed view of the region surrounding the plateau is shown in Figure 6.)

APM. This Pacific-wide expression of the HEB strongly supports the concept of a fixed Pacific hot-spot reference frame. Therefore, this APM is believed to accurately reflect the motion of the Pacific plate over the fixed hot spots.

The purpose of this paper is to use the new Pacific APM model to track palaeogeographic positions of the OJP from its time and place of origin to its present location north of the Solomon Islands, and to document regional tectonic events that may be associated with concomitant changes in the APM of the plateau since its emplacement. It should be noted at this juncture that recent changes in the Pacific APM are based on the averaging of nine seamount chains (Wessel *et al.* 2003) and the most recent changes are associated with the largest uncertainties (Harada & Hamano 2000; Wessel *et al.* 2004). Thus, the timing and geometry of the most recent changes in the APM model remain somewhat questionable. In the following sections, please refer to Figures 1 and 6 for geographical locations of the features discussed in the text.

OJP trail

A pronounced change in Pacific APM, from southwestward to northwestward, occurred at *c*. 124 Ma (Sager *et al.* 1999), shortly after Chron M1, when the Pacific–Phoenix spreading centre shut down (Nakanishi & Winterer 1998). Extrapolation of magnetic anomalies to the west of the M1 reversal (at the western end of the Nova–Canton Trough in Fig. 4), assuming no ridge jumps, would project the failed spreading centre south of the OJP's eastern salient (Fig. 4). However, seamount, volcanic ridge and submarine canyon alignments across the eastern OJP and southern Nauru Basin, visible in both the gravity and bathymetric maps (Figs 5 and 6), as well as basement offsets across these features (Gladczenko *et al.* 1997), are suggestive of a fracture zone pattern (black lines in Figs 5 and 6)

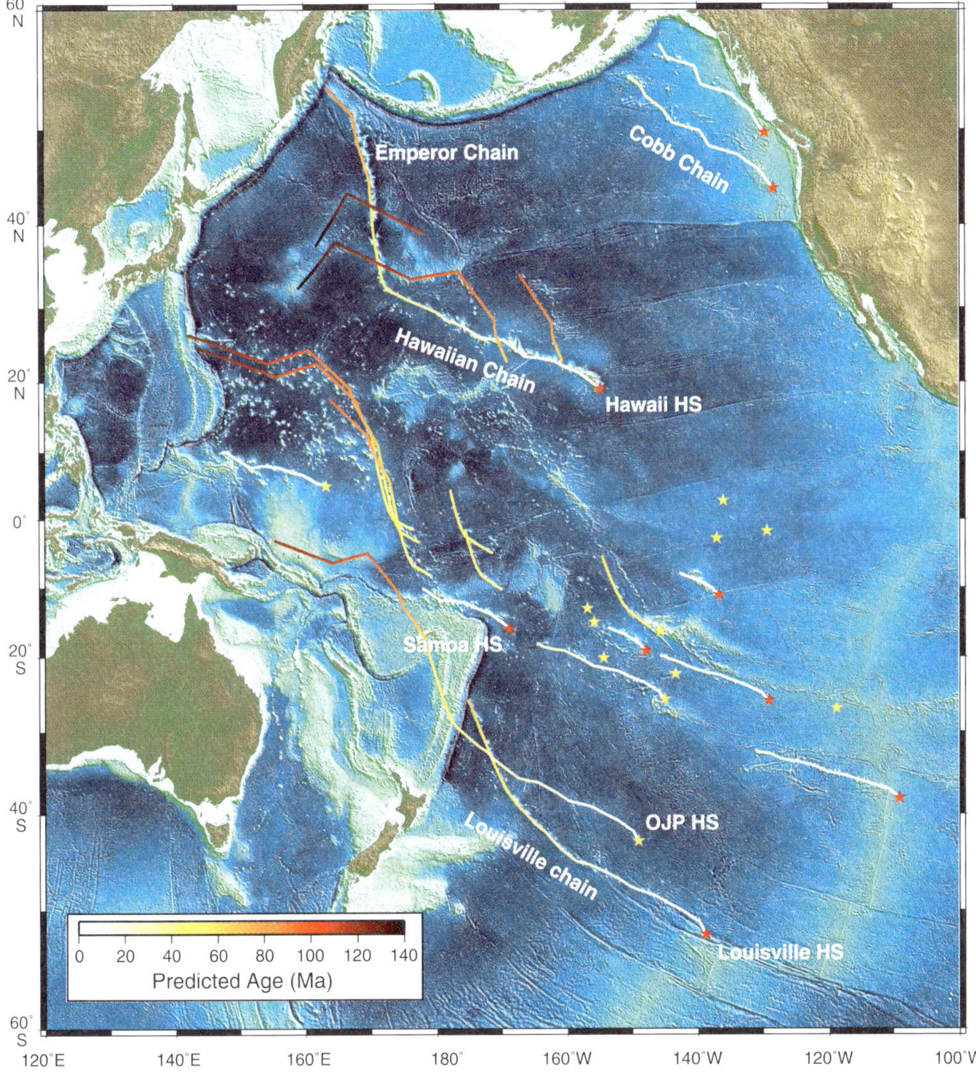

Fig. 2. Pacific hot-spot trails based on the Pacific APM model using the hot-spotting technique (Wessel & Kroenke 1997) and polygon finite rotation method (Harada & Hamano 2000) from Wessel *et al.* (2003). Stars show active or recently active (red) and inactive (yellow) hot-spot locations. The predicted OJP hot-spot trail, based on the new Pacific APM model and the assumption that the eruptive site was originally centred over the thickest part of the OJP root, is shown together with the other Pacific hot-spot trails. The APM model locates the OJP hot-spot at *c*. 43°S, 147°W.

that would require the presence of an intervening triple junction between this pattern and the nearby Phoenix lineations (Fig. 4). Moreover, we speculate that the presence of linear gravity highs orthogonal to this pattern on the western side of the high plateau may reveal the location of pre-OJP spreading centres (red lines in Figs 5 and 6) that ceased spreading shortly before the *c*. 122 Ma OJP emplacement. These spreading centres, although inactive by 122 Ma, probably would still have been bordered by very thin, young lithosphere, and thus might have been an important reason for the large amounts of partial melting estimated for the OJP basalts (e.g. Fitton & Godard 2004). Vertical tomographic cross-sections (Richardson *et al.* 2000) reveal an approximately 300 km-deep mantle 'root' beneath the OJP. The thickest part of the

Table 1. *Pacific hot-spot locations*

Hot spot	Latitude	Longitude
Hawaii	19°02′N	154°48′W
Louisville	52°15′S	138°45′W
Cobb	43°58′N	128°15′W
Caroline	05°00′N	163°18′E
Marquesas	11°12′S	136°42′W
Pitcairn	26°00′S	129°00′W
Foundation	38°12′S	109°00′W
Society	49°40′N	129°38′W
Samoa	52°51′N	135°09′W
Kodiak	22°16′S	143°16′W
Tokelau S	16°21′S	145°36′W
Tokelau N	29°30′S	139°42′W
Austral	25°45′S	144°57′W
Wake N	21°12′S	159°48′W
Wake S	22°27′S	151°18′W
Tuvalu	19°08′S	147°50′W
Tuamotu	16°00′S	168°45′W
Musicians	13°06′S	156°48′W
Shatsky N	15°04′S	155°50′W
Shatsky S	20°06′S	154°21′W

Table 2. *Finite rotations parameters*

Latitude	Longitude	Time (Ma)	Angle (°)
45°15′N	38°20′W	0.6	1.63
53°30′N	61°41′W	2.2	3.88
59°30′N	63°47′W	5.5	6.12
60°44′N	66°11′W	7.7	7.88
64°57′N	65°58′W	10.6	10.12
67°35′N	58°55′W	11.9	11.92
66°21′N	64°37′W	13.0	13.61
65°19′N	73°14′W	17.4	16.23
67°31′N	67°19′W	19.4	18.03
68°23′N	68°49′W	21.2	19.70
67°14′N	69°55′W	26.0	22.18
67°03′N	65°36′W	28.5	24.57
66°08′N	65°48′W	30.3	25.83
65°58′N	65°13′W	33.8	27.60
65°27′N	62°48′W	38.8	30.25
64°36′N	61°04′W	45.0	33.22
63°06′N	63°15′W	47.9	34.36
60°34′N	66°28′W	51.4	35.76
56°52′N	69°43′W	55.8	37.89
53°32′N	72°41′W	58.6	39.86
50°42′N	74°45′W	66.2	41.85
48°29′N	78°21′W	76.0	46.00
46°50′N	79°37′W	85.0	48.01
46°00′N	78°35′W	96.0	55.88
50°51′N	77°58′W	110.0	59.66
53°06′N	81°36′W	125.0	71.16
58°53′N	96°56′W	140.0	75.38

root occurs at *c.* 3°S, 157°E (green cross in Fig. 6), very close to a postulated relict spreading centre segment on the high plateau. For modelling purposes, we assume this site was centred over the plume head at the time of emplacement of the OJP; this allows the calculation of a hypothetical OJP hot-spot trail based on the new Pacific APM model (Fig. 2). This predicted trail locates the OJP hot spot at around 43°S, 147°W, approximately 1000 km east of the position (42°S, 159°W) proposed by Neal *et al.* (1997) but still well north of the Louisville hot spot (*c.* 52°S, *c.* 139°W), as shown in Figure 2.

OJP flow line

The predicted trail shown in Figure 2 delineates the expected path of volcanism produced by the OJP hot spot if the hot spot had remained active to the present. This trail, however, does not reflect the actual motion of the OJP through time; i.e. this path does not give the displacement history of the plateau relative to the hot spot. Instead, the actual trajectories of the plateau need to be retraced using a seafloor flow line (Lancelot & Larson 1975; Wessel & Kroenke 1997). The OJP flow line based on the new Pacific APM model is shown in Figure 7. The times of changes in flow-line trajectories (bends), visible in both Figure 2 and Figure 7, appear to correlate well with tectonic events in the vicinity of the OJP and across the Pacific Basin. Notably, during some of the APM changes, as described below, the OJP appears to have been deformed or otherwise modified by the change, while at other times the OJP's immense size and deep mantle root may have contributed to initiating the APM change.

Sequence of events

Based on the OJP flow line and hypothetical hot-spot trail, both of which are shown in Figure 7, a sequence of regional tectonic events can be documented. As noted above, for example, a pronounced change in Pacific APM, from a SW to a NW trajectory, occurred at approximately 124 Ma (Kroenke & Wessel 1997; Sager *et al.* 1999), about the time, or shortly before, the OJP was emplaced. This was followed by a less pronounced change in the Pacific APM, from a NW to a WSW trajectory (Kroenke & Wessel 1997), which occurred at about 110 Ma, about the time when flood basalts may have been emplaced in the East Mariana and northern Nauru basins (Castillo *et al.* 1994). Another pronounced change in Pacific APM, from a WSW to a NNW trajectory, occurred at *c.* 96 Ma, when slow spreading began between Australia and Antarctica (Cande & Mutter 1982; Gaina *et al.* 1998).

Fig. 3. Arrows point to coeval bends in Pacific seamount trails at 48–49 Ma (HEB time), after Wessel *et al.* (2003). Note the abrupt decrease in volcanic flux after the bend in the Hawaiian chain and the simultaneous terminations in the two Tokelau and the three Ratak–Gilbert–Ellice collinear trails. A possible hot-spot trail is also shown on the southern side of the OJP; the hot spot may presently be located near Wallis Island. Volcanic ridges (VR) on the northern side of the eastern salient that extend into the Nauru Basin appear to be very close to the bend of a projected Samoan trail.

This change may have been followed by another phase of OJP basalt emplacement (*c.* 90 Ma) along the eastern flank of the plateau, which perhaps formed part of the eastern salient and parts of the future islands of Santa Isabel and San Cristobal (Tejada *et al.* 1996). It should be

Fig. 4. Map of the west-central Pacific showing locations of magnetic anomaly lineations and fracture zones (after Nakanishi *et al.* 1992). Westward extrapolation of the Phoenix M1 reversal at the western end of the Nova–Canton Trough, assuming no ridge jumps, projects the extinct spreading centre south of the OJP's eastern salient.

noted, however, that Chambers *et al.* (2002) recently have questioned the younger ^{40}Ar–^{39}Ar ages for both the OJP and surrounding basin lavas.

A series of smaller changes in Pacific plate motion appear to have occurred between approximately 85 and 65 Ma. At *c.* 85 Ma, a modest change to a more northward trajectory occurred; a major change also occurred in Australia plate motion, as spreading began in the Tasman Basin (Royer & Rollet 1997; Gaina *et al.* 1998), which may also have been concomitant with initiation of rifting and spreading in the Stewart and Ellice basins (Neal *et al.* 1997) and along the Osbourn Trough east of the Tonga Trench (Billen & Stock 2000). At about 76 Ma, another modest change occurred in Pacific APM, to a more northward trajectory, as spreading ended in the Stewart and Ellice basins and possibly along the Osbourn Trough (Billen &

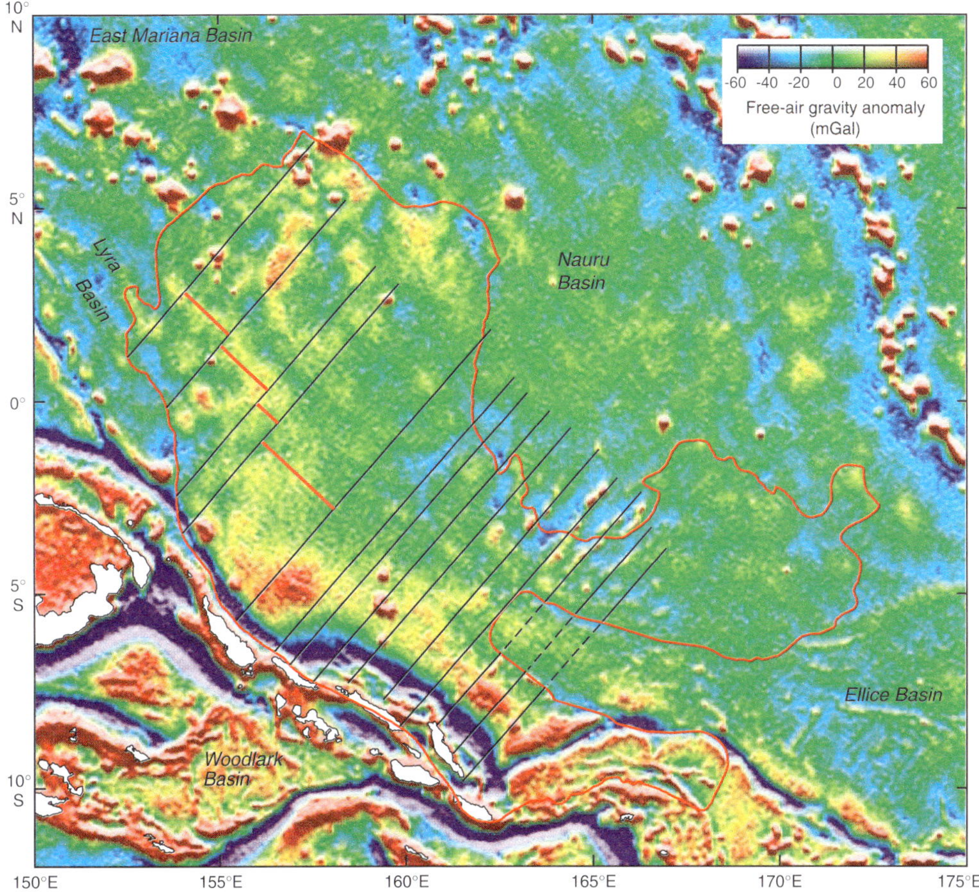

Fig. 5. Satellite-derived free-air gravity field (after Smith & Sandwell 1997) of the OJP. The plateau is outlined by a thin red line. Gravity lows on the eastern side of the plateau are suggestive of a fracture zone pattern (black lines), whereas the orthogonal linear gravity highs on the western side of the plateau (red lines) may indicate the location of former spreading centres.

Stock 2000). At about 65 Ma, still another modest change occurred in Pacific APM concomitant with a major change in Australia plate motion and the beginning of spreading in the Coral Sea Basin (Veevers & Li 1991).

Another pronounced change in Pacific plate motion, from a northward to a WNW trajectory, occurred at 48–49 Ma (HEB time), based on new ^{40}Ar–^{39}Ar plateau ages (Sharp & Clague 2002). Following this APM change, intra-plate subduction began along the Manus–North Solomon–Vitiaz (Melanesian) Trench (Kroenke 1984), which was aligned parallel to the post-HEB Pacific APM. Subduction was heralded by metamorphic events in the Florida Group and Eastern Belt Islands of the Solomon Islands that occurred between approximately 44 and 35 Ma (Neef & McDougall 1976), after which the earliest Lemau Intrusives were emplaced between 38 and 32 Ma in New Ireland (Stewart & Sandy 1988) near the western end of the Melanesian Arc. A seamount trail on the southern side of the OJP also can be fitted by the new APM model, which displays a bend appropriate for HEB time (Fig. 3). The predicted hot-spot location for this trail lies near Wallis Island, which was active as recently as *c.* 0.8 Ma (Duncan 1985). If the Samoan trail is also projected back to HEB time (Fig. 3), an interesting coincidence becomes apparent; i.e. the short volcanic ridge and seamount trails on the northern side of the OJP's eastern salient (VR in Fig. 3) lie near the bend in the Samoan trail, suggesting that opening or cracking of old fracture zones

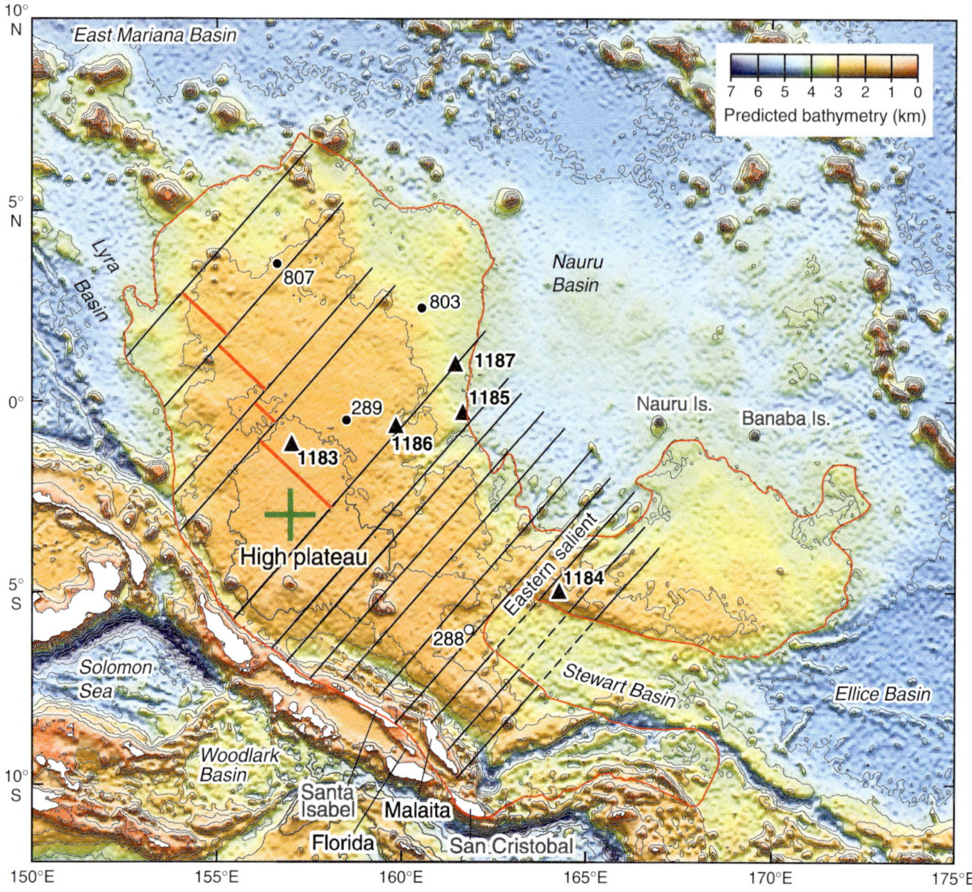

Fig. 6. Predicted bathymetry (after Smith & Sandwell 1997) showing the seamount, volcanic ridge and submarine canyon alignments across the OJP and southern Nauru Basin that are visible in both the gravity field (Fig. 5) and bathymetry. These patterns are suggestive of a fracture zone pattern (black lines). Also shown are the orthogonal linear gravity highs on the western side of the high plateau (red lines from Fig. 5) and the location of the thickest part of the OJP (green cross) based on vertical tomographic cross-sections of the OJP's mantle root (Richardson *et al.* 2000). Islands on the southern side of the high plateau may be part of the Wallis trail shown in Fig. 3. The triangles indicate Leg 192 drill sites; dots represent pre-Leg 192 drill sites that reached basement (except for Site 288).

may have occurred there as the Pacific plate changed direction. Such cracking may have allowed Samoan plume-derived magmas to form the volcanic ridges (see Fig. 7).

At about 34 Ma, Pacific APM was slowed as OJP lithosphere encountered the Melanesian Trench, and alnöites were emplaced on the southern margin of the OJP as the plateau lithosphere near the subduction zone was flexed and stretched (Coleman & Kroenke 1981). At the same time spreading began in the Caroline Basin (Hegarty & Weissel 1988) west of the OJP and in the D'entrecasteaux and South Fiji basins. At approximately 27 Ma, spreading ended in the Caroline Basin as the buoyant, young basin lithosphere encountered the Manus Trench, shutting down Melanesian subduction, forcing the OJP 'soft' docking (i.e. with little deformation) against the North Solomon block of the Melanesian Forearc (Petterson *et al.* 1997) and causing the Pacific plate to be displaced northward by the Australia plate. The Eauripik Ridge was formed, and the Parece Vela Ridge was pushed to the north, initiating northward ridge propagation in the Parece Vela Basin between 27 and 23 Ma (Okino *et al.* 1998, 1999); at the same time, noticeable southward offsets were produced in the Hawaiian and Cobb

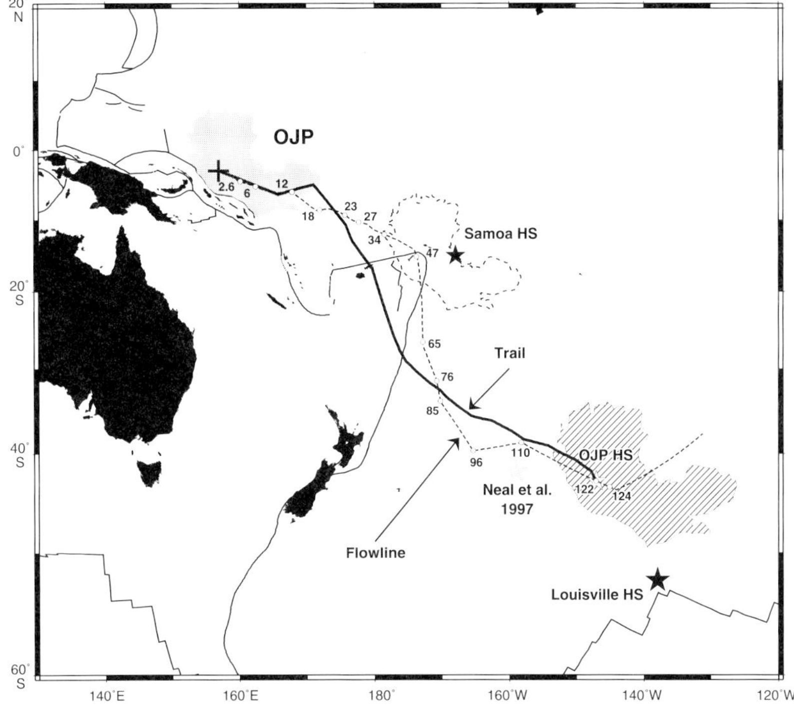

Fig. 7. The OJP flow line (dashed line), based on the new Pacific APM model, showing times of Pacific APM changes and locations of the OJP at 122 (shading with diagonal lines) and 47 Ma (unshaded). The solid dark line shows the hypothetical OJP hot-spot trail. The largest uncertainties in APM changes appear to be associated with the most recent changes (Wessel *et al.* 2003). The white star marks the location of the postulated OJP hot spot. The grey star marks Neal *et al.*'s (1997) postulated OJP hot-spot location. Dark stars mark the locations of the Samoa and Louisville hot spots.

seamount chains, as well as a minor offset in the Louisville chain. Spreading ended in the D'entrecasteaux and South Fiji basins, and subduction began along the Tonga Trench, followed by Tonga (Lau–Colville) Arc volcanism at about 22 Ma (Kroenke 1984).

At *c.* 23 Ma, the Pacific plate resumed a WNW motion and a minor change occurred in Australia plate motion. Maramuni–Trobriand subduction began, followed by Maramuni Arc volcanism at about 20 Ma (Hill & Raza 1999). At about 18 Ma, arc volcanism was briefly rejuvenated along the Melanesian Arc. The northward motion of the Australia plate coupled with the drag of the western end of the Melanesian backarc (Sepik Arc) along the developing Maramuni Forearc may be responsible for the 'sinistral wrenching' along the Papua New Guinea (PNG) margin that occurred from 23 to 12 Ma (Hill & Raza 1999). The increasing drag exerted by the accreting Sepik terrain, in turn, may be responsible for forcing the Pacific plate northward again between 18 and 12 Ma, and also for the extension described by Hill & Raza as occurring along the Maramuni Arc from 17 to 14 Ma. The ensuing northward clockwise rotation of the Pacific plate produced noticeable southward offsets in the Caroline and Hawaiian chains, and a minor offset in the Louisville chain, and pushed the Parece Vela Basin northward, initiating counterclockwise spreading-ridge rotation in the Parece Vela Basin (Okino *et al.* 1998, 1999).

At *c.* 12 Ma, South Solomon (New Britain–San Cristobal–New Hebrides) Trench subduction began, followed by South Solomon (Manus) Arc volcanism at about 8.8 Ma (Francis 1988) and, APM model uncertainties notwithstanding, the Pacific plate again resumed WNW motion. At about 6 Ma, the Australia plate slab that was being subducted beneath the South Solomon Arc apparently collided with the OJP root (Wessel & Kroenke 2000), causing Pacific plate motion to change slightly, producing a slight bend in the Hawaiian chain at Kauai. Concomitantly, the

Woodlark Basin began to open (Taylor *et al.* 1999) and collide with the South Solomon Forearc, producing the NE–SW compression that triggered the OJP 'hard' docking (Petterson *et al.* 1997), evidenced by the extensive deformation, thrust faulting and obduction of OJP crust in the Solomon Islands. At the same time, Lau Basin rifting west of the Tonga Arc was initiated (Karig 1970; Hawkins 1974), as Viti Levu, driven by the North Fiji Basin opening, collided with the northern end of the Tonga Arc (Yan & Kroenke 1993).

Finally, at about 2.6 Ma, the Australia plate–OJP collision intensified. The OJP was now lodged in the Australia slab window; the result was to 'pin' the Pacific plate at the OJP. Both the Solomon Sea and Australia plates were deflected, and the Pacific plate (including the OJP) rotated counterclockwise about a low-latitude rotation pole near the OJP (Wessel & Kroenke 2000). This rotation, in turn, produced a prominent bend in the Hawaiian chain at the western end of Molokai, as well as bends in other Pacific seamount trails. As Woodlark Basin spreading continued, the South Solomon Forearc was pushed NE causing the adjacent OJP crust to be folded and obducted onto the Solomon Islands to form the islands of Malaita and Santa Isabel, and the Malaita Anticlinorium (Petterson *et al.* 1997). This collisional crustal shortening appears to be continuing today and may be responsible for recent submarine canyon formation (Kroenke 1972; Berger *et al.* 1976) and other large-scale mass-wasting (Kroenke *et al.* 1971; Berger & Johnson 1976) on the NE side of the plateau. This mass-wasting is probably caused by recurring slip on the old fracture zones shown in Figure 6, as well as the active seafloor structural deformation NE of the plateau in the vicinity of Nauru Island (Kroenke & Walker 1986). This collision-related deformation also is probably responsible for the ongoing uplift of Nauru Island (Jacobson *et al.* 1997), and the uplift and tilting of Banaba Island (Owen 1923) to the NE of the OJP high plateau.

Concluding remarks

From the foregoing discussion, it appears that many regional tectonic events that have occurred during and since OJP emplacement can be correlated with the changes in the absolute motion of the Pacific plate. However, many problems remain. The suggested pre-OJP crustal fabric must be evaluated in much greater detail before it can be accepted as fact. Also, for example, was there more than one period of OJP formation; e.g. did most of the eastern salient form well after the high plateau? Did a post-emplacement spreading event occur in the Stewart Basin? If so, what was the age of basin formation? What are the eruptive ages of the seamounts on the southern side of the OJP? Do they, indeed, constitute a hot-spot trail? What is the composition and age of the volcanic ridges on the northern side of the eastern salient? Were they formed as a result of decompressional melting, possibly with involvement of Samoan plume material, during an APM change? Further research will be required, particularly much more sampling and dating of basement and seamounts over and around the OJP, before much progress can be made towards answering many of these questions.

We thank D. Müller and D. Scheirer for helpful critical reviews, and N. Hulbirt for help with the illustrations. We also thank our Leg 192 shipboard colleagues and the ODP and Transocean/Sedco-Forex staff of the *JOIDES Resolution* for a very interesting and successful cruise. Funding for modelling studies and manuscript preparation was through ODP and NSF grants F001300, F001347 and OCE99-06773. ODP is sponsored by the National Science Foundation and participating countries under management of Joint Oceanographic Institutions (JOI), Inc. SOEST Contribution 6227.

References

BERGER, W.H. & JOHNSON, T.C. 1976. Deep-sea carbonates: dissolution and mass wasting on the Ontong Java Plateau. *Science*, **192**, 785–787.

BERGER, W.H., JOHNSON, T.C. & HAMILTON, E.L. 1976. Sedimentation on Ontong Java Plateau; observations on a classic 'carbonate monitor'. *In*: ANDERSON, N.R. & MALAHOFF, A. (eds) *The Fate of Fossil Fuel CO_2 in the Oceans. Marine Science*, **6**, 543–567.

BILLEN, M.I. & STOCK, J. 2000. Morphology and origin of the Osbourn Trough. *Journal of Geophysical Research*, **105**, 13 481–13 489.

CANDE, S.C. & MUTTER, J.C. 1982. A revised identification of the oldest sea-floor spreading anomalies between Australia and Antarctica. *Earth and Planetary Science Letters*, **58**, 151–160.

CASTILLO, P.R., PRINGLE, M.S. & CARLSON, R.W. 1994. East Mariana Basin tholeiites; Cretaceous intraplate basalts or rift basalts related to the Ontong Java plume? *Earth and Planetary Science Letters*, **123**, 139–154.

CHAMBERS, L.M., PRINGLE, J.G. & FITTON, J.G. 2002. Age and duration of magmatism on the Ontong Java Plateau: ^{40}Ar–^{39}Ar results from ODP Leg 192. *Eos, Transactions of the American Geophysical Union*, **83**, 47, F1444.

COLEMAN, P.J. & KROENKE, L.W. 1981. Subduction without volcanism in the Solomon Islands Arc, *Geo-Marine Letters*, **1**, 129–134.

DUNCAN, R. 1985. Radiometric ages from volcanic rocks along the New Hebrides–Samoa Lineament.

In: BROCHER, T.M. (ed.) *Geological Investigations of the Northern Melanesian Borderland*. Circum-Pacific Council for Energy and Mineral Resources, Earth Science Series, **3**, 67–76.

FITTON, J.G. & GODARD, M. 2004. Origin and evolution of magmas on the Ontong Java Plateau. *In*: FITTON, J.G., MAHONEY, J.J., WALLACE, P.J. & SAUNDERS, A.D. (eds) *Origin and Evolution of the Ontong Java Plateau*. Geological Society, London, Special Publications, **229**, 151–178.

FRANCIS, G. 1988. Stratigraphy of Manus Island, western New Ireland Basin, Papua New Guinea. *In*: MARLOW, M.S., DADISMAN, S.V. & EXON, N.F. (eds) Geology and offshore resources of Pacific island arcs – New Ireland and Manus region, Papua New Guinea. *Circum-Pacific Council for Energy and Mineral Resources*, Earth Science Series, **9**, 31–40.

GAINA, C., MULLER, D.R., ROYER, J.-Y., STOCK, J., HARDEBECK, J. & SYMONDS, P. 1998. The tectonic history of the Tasman Sea: A puzzle with 13 pieces. *Journal of Geophysical Research*, **103**, 12 413–12 433.

GLADCZENKO, T.P., COFFIN, M.F. & ELDHOLM, O. 1997. Crustal structure of the Ontong Java Plateau: Modeling of new gravity and existing seismic data. *Journal of Geophysical Research*, **102**, 22 711–22 729.

HARADA, Y. & HAMANO, Y. 2000. Recent progress on the plate motion relative to hotspots, history and dynamics of global plate motions. *In*: RICHARDS M.A., GORDON, R.G. & VAN DER HILST, R.D. (eds) *The History and Dynamics of Global Plate Motions*. American Geophysical Union, Geophysical Monograph, **121**, 327–338.

HAWKINS, J.W. 1974. Geology of the Lau Basin, a marginal sea behind the Tonga Arc. *In*: BURK, C.A. & DRAKE, C.L. (eds) *The Geology of Continental Margins*. Springer, Berlin, 505–520.

HEGARTY, K.A. & WEISSEL, J.K. 1988. Complexities in the development of the Caroline plate region, western equatorial Pacific. *In*: NAIRN, A.E.M., STEHLI, F.G. & UYEDA, S. (eds) *The Ocean Basins and Margins*, **7B**, *The Pacific Ocean*. Plenum, New York, 277–301.

HILL, K.C. & RAZA, A. 1999. Arc–continent collision in Papua New Guinea: Constraints from fission track thermochronology. *Tectonics*, **18**, 950–966.

JACOBSON, G., HILL, P.J. & GHASSEMI, F. 1997. Geology and hydrogeology of Nauru Island. *In*: VACHER, H.L. & QUINN, T.M. (eds) *Geology and Hydrogeology of Carbonate Islands*. Developments in Sedimentology, **54**, 707–742.

KARIG, D.E. 1970. Ridges and basins of the Tonga–Kermadec island arc system. *Journal of Geophysical Research*, **75**, 239–254.

KROENKE, L.W. 1972. *Geology of the Ontong Java Plateau*. Hawaii Institute of Geophysics Technical Report, **HIG 72–5**.

KROENKE, L.W. 1984. *Cenozoic Tectonic Development of the Southwest Pacific*. UN ESCAP, CCOP/SOPAC Technical Bulletin, **6**.

KROENKE, L.W. & WALKER, D.A. 1986. Evidence for the formation of a new trench in the Western Pacific. *Eos, Transactions of the American Geophysical Union*, **67**, 12, 145–146.

KROENKE, L.W. & WESSEL, P. 1997. Pacific Plate motion between 125 and 90 Ma and the formation of the Ontong Java Plateau. *Eos, Transactions of the American Geophysical Union*, **78**, Fall Meeting Supplement, 726.

KROENKE, L.W., MOBERLY, R., WINTERER, E.L. & HEATH, G.R. 1971. Lithologic interpretation of continuous reflection profiling. *In*: WINTERER, E.L., RIEDEL, W.R. *et al.*, *Initial Reports of the Deep Sea Drilling Project*, **7**, 1161–1226.

LANCELOT, Y. & LARSON, R.L. 1975. Sedimentary and tectonic evolution of the northwestern Pacific. *Initial Reports of the Deep Sea Drilling Project*, **33**, 925–939.

MAHONEY, J.J., STOREY, M., DUNCAN, R.A., SPENCER, K.J. & PRINGLE, M.S. 1993. Geochemistry and age of the Ontong Java Plateau. *In*: PRINGLE, M.S., SAGER, W.W., SLITER, W.V. & STEIN, S. (eds) *The Mesozoic Pacific: Geology, Tectonics, and Volcanism*. American Geophysical Union, Geophysical Monograph, **77**, 233–262.

MORGAN, W.J. 1972. Plate motions and deep mantle convection. *Geological Society of America Memoir*, **132**, 7–22.

NAKANISHI, M. & WINTERER, E.L. 1998. Tectonic history of the Pacific–Farallon–Phoenix triple junction from Late Jurassic to Early Cretaceous: An abandoned Mesozoic spreading system in the Central Pacific Basin. *Journal of Geophysical Research*, **103**, 12 453–12 468.

NAKANISHI, M., TAMAKI, K. & KOBAYASHI, K. 1992. Magnetic anomaly lineations from late Jurassic to early Cretaceous in the west-central Pacific Ocean. *Geophysical Journal International*, **109**, 701–719.

NEAL, C.R., MAHONEY, J.J., KROENKE, L.W., DUNCAN, R.A. & PETTERSON, M.G. 1997. The Ontong Java Plateau. *In*: MAHONEY, J. & COFFIN, M. (eds) *Large Igneous Provinces: Continental, Oceanic, and Planetary Flood Basalt Volcanism*. American Geophysical Union, Geophysical Monograph, **100**, 183–216.

NEEF, G. & MCDOUGALL, I. 1976. Potassium–argon ages on rocks from Small Nggela Island, British Solomon Islands, S.W. Pacific. *Pacific Geology*, **11**, 81–85.

OKINO, K., KASUGA, S. & OHARA, Y. 1998. A new scenario of the Parece Vela Basin genesis. *Marine Geophysical Researches*, **20**, 21–40.

OKINO, K., OHARA, Y., KASUGA, S. & KATO, Y. 1999. The Philippine Sea: New survey results reveal the structure and the history of the marginal basins. *Geophysical Research Letters*, **26**, 2287–2290.

OWEN, L. 1923. Note on the phosphate deposit of Ocean Island; with remarks on the phosphates of the equatorial belt of the Pacific Ocean. *Quarterly Journal of the Geological Society of London*, **LXXIX**.

PARKINSON, I.J., SCHAEFER, B.F. & THE ODP LEG 192 SHIPBOARD SCIENTIFIC PARTY. 2001. A lower mantle origin for the worlds biggest LIP? A high precision Os isotope isochron from Ontong Java

Plateau basalts drilled on ODP Leg 192. *Eos, Transactions of the American Geophysical Union*, **82**, 47, F1398.

PETTERSON, M.G., NEAL, C.R., MAHONEY, J.J., KROENKE, L.W., SAUNDERS, A.D., BABBS, T.L., DUNCAN, R.A., TOLIA, D. & MCGRAIL, B. 1997. Structure and deformation of north and central Malaita, Solomon Islands: tectonic implications for the Ontong Java Plateau–Solomon Arc collision, and for the fate of oceanic plateaus. *Tectonophysics*, **283**, 1–33.

RICHARDSON, W.P., OKAL, E.A. & VAN DER LEE, S. 2000. Rayleigh-wave tomography of the Ontong Java Plateau. *Physics of the Earth and Planetary Interiors*, **118**, 29–51.

ROYER, J.Y. & ROLLET, N. 1997. Plate tectonic setting of the Tasmanian region. *In*: EXON, N.F. & CRAWFORD, A.J. (eds) *West Tasmanian Margin and Offshore Plateaus; Geology, Tectonic and Climactic History, and Resource Potential. Australian Journal of Earth Sciences*, **44**, 543–560.

SAGER, W.W., KIM, J., KLAUS, A., NAKANISHI, M. & KHANKISHIEVA, L.M. 1999. Bathymetry of Shatsky Rise, northwest Pacific Ocean: Implications for ocean plateau development at a triple junction. *Journal of Geophysical Research*, **104**, 7557–7576.

SHARP, W.D. & CLAGUE, D.A. 2002. An older slower Hawaii–Emperor Bend. *Eos, Transactions of the American Geophysical Union*, **83**, 47, F1282.

SHIPBOARD SCIENTIFIC PARTY. 2001. Leg 192 Summary. *In*: MAHONEY, J.J., FITTON, J.G., WALLACE, P.J. *et al. Proceedings of the Ocean Drilling Program, Initial Reports*, **192**, College Station, TX (Ocean Drilling Program), 1–75.

SMITH, W.H.F. & SANDWELL, D.T. 1997. Global seafloor topography from satellite altimetry and ship depth soundings. *Science*, **227**, 1956–1962.

STEWART, W.D. & SANDY, M.J. 1988. Geology of New Ireland and Djaul Islands, northeastern Papua New Guinea. *In*: MARLOW, M.S., DADISMAN, S.V. & EXON, N.F. (eds) *Geology and Offshore Resources of Pacific Island Arcs – New Ireland and Manus Region, Papua New Guinea*. Circum-Pacific Council for Energy and Mineral Resources, Earth Science Series, **9**, 13–30.

TAYLOR, B., GOODLIFFE, A.M. & MARTINEZ, F. 1999. How continents breakup: Insights from Papua New Guinea. *Journal of Geophysical Research*, **104**, 7497–7512.

TEJADA, M.L.G., MAHONEY, J.J., CASTILLO, P.R., INGLE, S.P., SHETH, H.C. & WEIS, D. 2004. Pin-Pricking the elephant: evidence on the origin of the Ontong Java Plateau from Pb-Sr-Hf-Nd isotopic characteristics of ODP Leg 192 basalts. *In*: FITTON, J.G., MAHONEY, J.J., WALLACE, P.J. & SAUNDERS, A.D. (eds) *Origin and Evolution of the Ontong Java Plateau*. Geological Society, London, Special Publications, **229**, 133–150.

TEJADA, M.L.G., MAHONEY, J.J., DUNCAN, R.A. & HAWKINS, M. 1996. Age and geochemistry of basement and alkalic rocks of Malaita and Santa Isabel, Solomon Islands, southern margin of the Ontong Java Plateau. *Journal of Petrology*, **37**, 361–394.

TEJADA, M.L.G., MAHONEY, J.J., NEAL, C.R., DUNCAN, R.A. & PETTERSON, M.G. 2002. Basement geochemistry and geochronology of central Malaita, Solomon Islands, with implications for the origin and evolution of the Ontong Java Plateau. *Journal of Petrology*, **43**, 449–484.

VEEVERS, J.J. & LI, Z.X. 1991. Review of seafloor spreading around Australia; II, Marine magnetic anomaly modeling. *Australian Journal of Earth Sciences*, **38**, 391–408.

WESSEL, P. & KROENKE, L. 1997. A geometric technique for relocating hot spots and refining absolute plate motions. *Nature*, **387**, 365–369.

WESSEL, P. AND KROENKE, L.W. 2000. The Ontong Java Plateau and Late Neogene changes in Pacific Plate motion. *Journal of Geophysical Research*, **105**, 28 255–28 278.

WESSEL, P., HARADA, Y., KROENKE, L.W. & STERLING, A. 2004. The Hawaiian–Emperor bend: clearly a record of Pacific plate motion change. *Eos, Transactions of the American Geophysical Union*, **84**, Fall Meeting Supplement, Abstract V32A–0990.

YAN, C.Y. & KROENKE, L.W. 1993. A plate tectonic reconstruction of the Southwest Pacific 0–100 Ma (animated). *In*: BERGER, W.H., KROENKE, L.W., MAYER, L. *et al. Proceedings of the Ocean Drilling Program, Scientific Results*, **130**, 697–709.

Modelled palaeolatitudes for the Louisville hot spot and the Ontong Java Plateau

MARIA ANTRETTER[1], PETER RIISAGER[2], STUART HALL[3], XIXI ZHAO[4] & BERNHARD STEINBERGER[5]

[1]*Department of Earth and Environmental Sciences, University of Munich, Theresienstrasse 41, D-80333 München, Germany (e-mail: maria@geophysik.uni-muenchen.de)*

[2]*Danish Lithosphere Centre, Øster Voldgade 10, DK-1350 Copenhagen K, Denmark*

[3]*Department of Geosciences, University of Houston, Houston, TX 77204-5007, USA*

[4]*Earth Sciences Department, University of California at Santa Cruz, Santa Cruz, CA 95064, USA*

[5]*IFREE, JAMSTEC, 2-15 Natsushima-cho, Yokosuka 237–0061, Japan*

Abstract: Formation of the Ontong Java Plateau (OJP), a large igneous province in the western Pacific, has been attributed to a rising plume head in the initial stage of the Louisville hot spot, approximately 120–125 Ma ago. However, the Neal *et al.* plate reconstruction suggests that the plateau formed approximately 9° north of the current location of this hot spot at 51°S . The magnetization of the plateau's basement records a palaeolatitude of approximately 25°S which further increases the discrepancy with the plume-head model. Modelling the motion of the Louisville hot spot for the last 120 Ma yields a possible southward motion of up to about 6°. True polar wander (TPW) models also shift the predicted palaeolatitudes of the plateau farther north. Taking into account both hot-spot motion and TPW, formation of the OJP by the Louisville hot spot remains a possibility.

The Ontong Java Plateau (OJP) in the western Pacific is the largest of the large igneous provinces (LIPs). Ar–Ar dating of basalts from the plateau indicates that it started to form approximately 120 Ma ago (Mahoney *et al.* 1993; Chambers *et al.* 2003). Most current models ascribe oceanic plateaus to the initial 'plume-head' stage of hot-spot development (e.g. Richards 1991). Richards (1991) and Tarduno *et al.* (1991) favoured the initial plume head of the Louisville hot spot (now at 51°S, 138°W) as the source of the OJP, but recent plate reconstructions suggest that the plateau was formed well to the NE of the current location of this hot spot (Neal *et al.* 1997; Kroenke *et al.* 2004). According to these reconstructions, the centre of the OJP at approximately 125 Ma was at 42°S, 159°W, and therefore 9° north and 21° west of the Louisville hot spot, approximately 1600 km distant (Fig. 1). This reconstruction is partly based on hot-spot tracks (including Louisville) and would therefore change if hot-spot motions were considered. The most recent palaeomagnetic results confirm that the OJP was far north of the present position of the Louisville hot spot at the time of the plateau's formation. Palaeomagnetic investigations on basalts from ODP Leg 192 to the OJP result in a palaeolatitude of $23.8\,^{-2.0}/_{+1.9}$°S for the centre of the OJP 120 Ma ago (Riisager *et al.* 2003). These results have been combined with the palaeomagnetic investigation of volcaniclastic sediments from ODP Leg 192, resulting in a palaeolatitude of approximately 25°S (Riisager *et al.* 2004).

Mantle plumes have frequently been thought to be fixed in the Earth's mantle. If true polar wander (TPW) does not occur and the geomagnetic axial dipole hypothesis holds, then basalts from stationary mantle plumes would always be produced at the same place in the hot-spot reference frame as well as in the magnetic reference frame. Apart from palaeosecular variations, fresh basalts would always memorize the same direction of magnetization when cooling below their blocking temperature, corresponding to the present hot-spot latitude. Previous palaeomagnetic investigations (Mayer & Tarduno 1993) and the palaeomagnetic results from ODP Leg 192 to the OJP (Riisager *et al.* 2003, 2004) are inconsistent with these assumptions. Explanations for a discrepancy between palaeolatitudes and present-day hot-spot latitude include:

From: FITTON, J. G., MAHONEY, J. J., WALLACE, P. J. & SAUNDERS, A. D. (eds) 2004. *Origin and Evolution of the Ontong Java Plateau*. Geological Society, London, Special Publications, **229**, 21–30. 0305-8719/$15.00

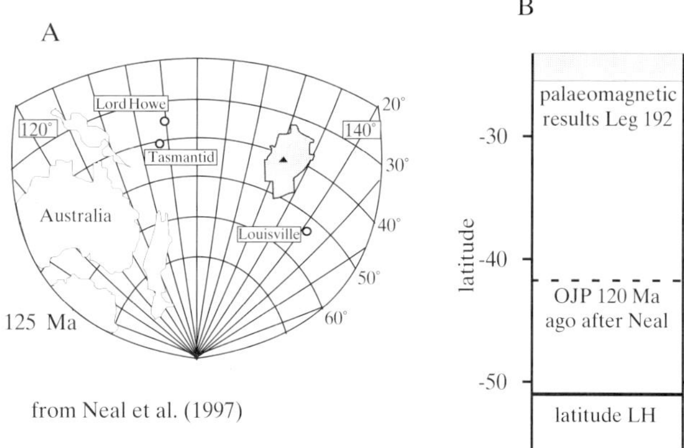

Fig. 1. (**A**) Approximate location of the Ontong Java Plateau (OJP) at 125 Ma after the plate reconstruction of Neal *et al.* (1997). The triangle represents the inferred location of the OJP plume centre beneath the crest of the high plateau. The reconstruction uses the Pacific plate Euler poles of Kroenke & Wessel (1997) (after Neal *et al.* 1997). (**B**) Sketch presenting the discrepancy between the present-day latitude of the Louisville hot spot, the latitude of the OJP 120 Ma ago after Neal *et al.* (1997) and the palaeolatitudes obtained by recent palaeomagnetic investigations on basalts from the OJP.

- mantle plumes can be advected when rising through the convecting mantle, resulting in hot-spot motion on the Earth's surface (e.g. Steinberger & O'Connell 1998);
- TPW, the relative motion between the mantle (with the hot spots) and the rotation axis of the Earth, may affect the latitude of a hot spot;
- non-dipole geomagnetic field components may contribute significantly to the Earth's magnetic field (e.g. Coupland & Van der Voo 1980; McElhinny *et al.* 1996) yielding palaeolatitudes that differ from those predicted by a dominantly geo-axial dipole (GAD) field.

Torsvik *et al.* (2001) and Van der Voo & Torsvik (2001) state that the octupole component is the most dominant non-dipole component. Introducing octupole contributions would rectify the latitude offset that is observed between the Louisville hot spot and the palaeomagnetic results from the OJP. Estimates of the magnitude of the octupole/dipole ratio indicate that paleolatitudes determined with the dipole formula could be about 7.5° too low at mid-latitudes (e.g. Van der Voo & Torsvik 2001).

In this chapter, we continue the discussion of the possible link between the OJP and the Louisville hot spot. We study the effect of a moving Louisville hot spot and the effect of TPW on palaeolatitudes, and discuss the consequence for the possible link between plateau and hot spot. However, there will be no further discussion of any contributions made by octupole fields.

Modelling of hot-spot motion

The motion of mantle plumes in a convecting mantle due to large-scale mantle flow can be estimated by geodynamic modelling (Steinberger & O'Connell 1998). The method that we use for the calculation has been previously explained in detail (Steinberger & O'Connell 1998, 2000; Steinberger 2000*a*; Antretter *et al.* 2002) and is briefly described here.

A large-scale mantle flow field is computed (Hager & O'Connell 1979, 1981) using models of internal density heterogeneities and surface plate motions. Density heterogeneities are inferred from seismic tomography or subduction history. For the scaling factors $(\delta\rho/\rho)/(\delta v_S/v_S)$ to convert seismic velocity (v) to present-day density (ρ) variations, we consider three cases: (1) scaling factor 0.2, where only mantle density anomalies below 220 km depth are included; (2) scaling factor 0.3, where only mantle density anomalies below 220 km depth are included; and (3) scaling factor 0.2, where all mantle density anomalies are included. A scaling factor between about 0.2 and 0.3 has been inferred from laboratory experiments, in combination with theoretical arguments (e.g. Karato 1993). Cases (1) and (2) disregard the uppermost 220 km in order to exclude seismic velocity variations that may be due to lithospheric roots and not therefore related to density variations that drive mantle flow. In most cases density variations are advected in the flow field backward in time for the past 68 Ma.

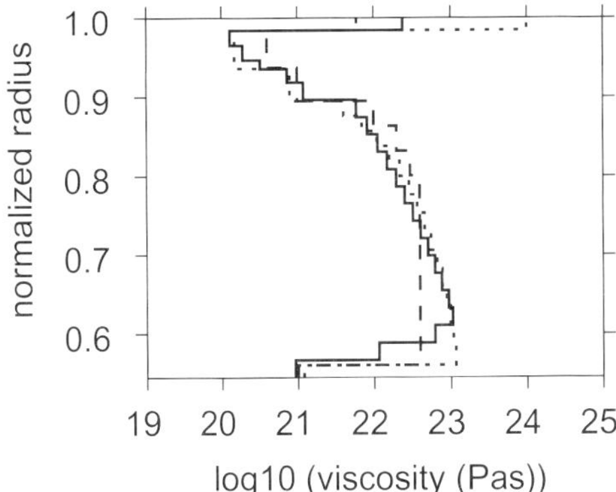

Fig. 2. Three radial viscosity models used in this paper. Dashed line, 'model A' (Steinberger & O'Connell 1998); dotted line, 'model B' (Steinberger & O'Connell 2000); continuous line, 'model C' (Steinberger & Calderwood 2001).

In most cases, relative plate motions are adopted from Mueller *et al.* (1993, 1997); for plates and times not included there or before 83 Ma, we follow Gordon & Jurdy (1986) (0–64 Ma) and Lithgow-Bertelloni *et al.* (1993) (64–120 Ma). Absolute motions of the African and Pacific plates are re-determined for times 0–74 Ma. We also re-computed plate boundaries in the Pacific Basin in 2 Ma intervals during the past 120 Ma based on the digital isochrons of Mueller *et al.* (1997), available online at ftp://ftp.agg.nrcan.gc.ca/products/agegrid/isochrons.dat.gz. Elsewhere, plate boundary locations are adopted from the compilation by Lithgow-Bertelloni *et al.* (1993).

The inferred past plate boundary locations in the Pacific Basin depend on the 'absolute' plate motions, which in turn depend on hot-spot motion. In principle, it requires an iteration to obtain mutually consistent hot-spot motions, plate motions and plate boundaries (Tarduno *et al.* 2003). Here, for simplicity, we use the same plate motions and boundaries that were computed for *one specific* model of hot-spot motion – the 'moving source' model of Tarduno *et al.* (2003) – that yields a good fit to Pacific hot-spot tracks. This procedure is appropriate as, for models of hot-spot motion that yield a good fit to both the Hawaiian and Louisville hot-spot track, the Louisville hot spot is located in an intra-plate location on the Pacific plate. Except for the very recent past, the Louisville hot spot is located far away from plate boundaries and, as long as this is the case, the computed Louisville hot-spot motion depends very little on the exact plate boundary conditions.

Furthermore, radial mantle viscosity structures (Steinberger & O'Connell 1998, 2000; Steinberger & Calderwood 2001) are assumed for the calculations. These are shown in Figure 2, and will be referred to as viscosity models A, B and C.

Most models were computed for an incompressible mantle without phase boundaries. Some models consider compressibility and phase boundaries, as in Steinberger (2000*a*). In particular, if phase boundaries are considered, they are assumed to be in thermal equilibrium. Density anomalies at the depth of the phase boundaries inferred from three-dimensional (3-D) tomography models are converted to temperature anomalies which, in turn, are converted to phase boundary deflections that are treated as sheet mass anomalies. Based on the work of Akaogi *et al.* (1989) and Akaogi & Ito (1999) we use conversion factors from density to sheet mass anomalies of 132 km at depth 400 km and –58 km at depth 670 km. Roughly speaking, this means that a density anomaly layer of thickness 132 km around 400 km depth is counted twice, whereas a density layer of 58 km thickness at depth 670 km is disregarded. To calculate the motion of hot spots, we assume that initially (for the Louisville hot spot, at 120 Ma) vertical plume conduits are subsequently distorted in large-scale mantle flow. Besides being distorted in large-scale flow we also allow for buoyant 'Stokes' rising of plume conduits through mantle flow. The resulting hot spot motion depends on both the large-scale flow field and the buoyant rising of the plume conduit. Buoyant rising velocity is computed from a modified Stokes formula: $u(z) = u_0 \times (r/r_0)^2 \times (\eta_0/\eta(z))$, where $u_0 = 51$ mm year^{-1}, $r_0 = 100$ km, $\eta_0 = 10^{21}$ Pa s, z is the depth, $\eta(z)$ is the ambient mantle viscosity and r is plume conduit radius. The value $u_0 = 51$ mm year^{-1} corresponds to rising speed obtained in laboratory experiments (Richards & Griffiths 1988) scaled to Earth dimensions, and for a reasonable density contrast of 30 kg m^{-2} between plume and ambient mantle.

We use three different assumptions about plume conduit radius.

(1) Independent of depth, $r = (B/B_0)^{1/4} \times r_1$, where B is anomalous plume mass flux, $B_0 = 10^3$ kg s^{-1}, and $r_1 = 41.0$ km for viscosity models B and C (i.e. most cases), and $r_1 = 62.5$ km for model A. This assumption follows Steinberger (2000*a*) and Steinberger & O'Connell (2000), where it is further discussed.

(2) Dependent on depth, $r = ((B/B_0) \times (\eta(z)/\eta_0))^{1/4} \times r_1$ with $r_1 = 65.9$ km for viscosity model B and 68.2 km for model C – such that conduit radius immediately

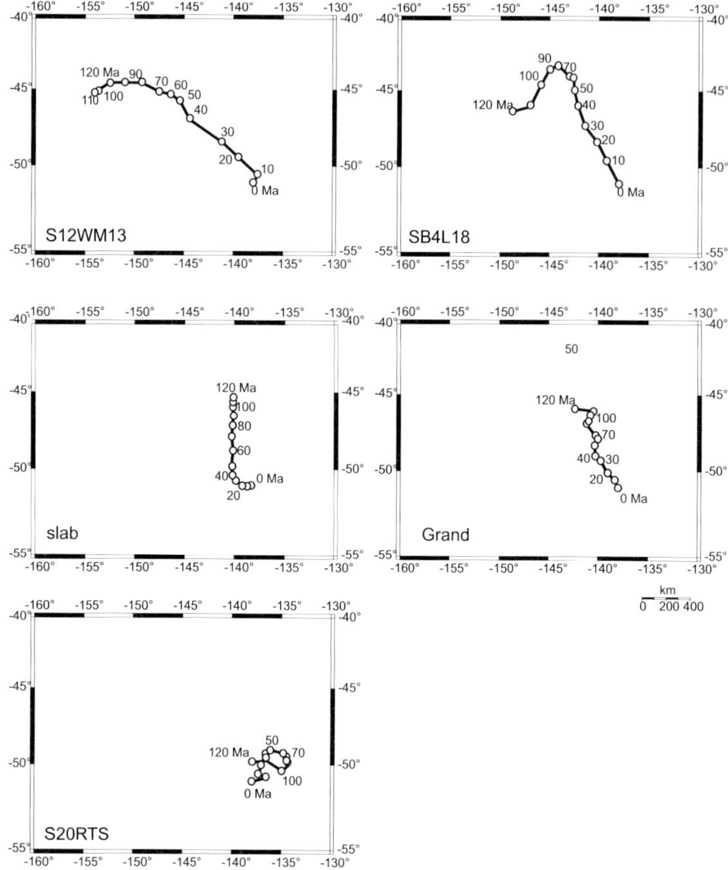

Fig. 3. Modelled drift of the Louisville hot spot assuming its initiation at 120 Ma. The dots give the position of the hot spot every 10 Ma, indicated also by numbers. The present-day position of the Louisville hot spot is assumed at 51°S at 0 Ma. The tomographic models S12WM13 (Su *et al.* 1994) and SB4L18 (Masters *et al.* 2000), and the slab model of Steinberger (2000*b*), have been used in combination with mantle viscosity model B to obtain a mantle flow field in which the rising mantle plume is advected; the tomographic models of Grand (2002) and S20RTS (Ritsema & Van Heijst 2000) have been used with viscosity model C.

below the lithosphere is again 41.0 km for $B = B_0$. This corresponds to assuming relative viscosity variations with depth inside the conduit are the same as in the ambient mantle.

(3) Dependent on depth, $r = ((B/B_0) \times (\eta(z)/\eta_0))^{1/4} \times r_1 + r_{th}$, with $r_1 = 21.6$ km and $r_{th} = 40$ km. Different from (2), the conduit is surrounded by a 'thermal halo' of thickness r_{th}. This assumption is intermediate between (1) and (2).

We use $B = 2.0 \times 10^3$ kg s^{-1}, the average between 0.9×10^3 kg s^{-1} (Sleep 1990) and 3.0×10^3 kg s^{-1} (Davies 1988). Computations were also carried out for these two values, which produced very similar results.

We assume a present hot-spot location at 50.9°S, 138.1°W, such that the predicted track approximately matches location (50.4°S, 139.2°W) and age 0.5 ± 0.2 Ma (Watts *et al.* 1988) of the youngest seamount of the Louisville chain. This location is also supported by geochemical evidence (Vlastelic *et al.* 1998), but differs from the location (53.8°S, 140.2°W) proposed by Wessel & Kroenke (1997).

Results: motion of the Louisville hot spot

Some representative results for hot-spot motion are shown in Figure 3. The tomographic models S12WM13 (Su *et al.* 1994) and SB4L18 (Masters *et al.* 2000), and the slab model of Steinberger (2000*b*), have been used in combination with

Table 1. *Parameters and results of numerical model runs, in addition to those where results are displayed in Figure 3*

#	TM	SF	MVM	PVM	td	adv	com	pb	HS120 (long.)	HS120 (lat.)
1	1	1	A	1	y	y	n	n	−149.4	−50.2
2	1	1	A	1	y	y	y	n	−150.1	−51.1
3	1	1	A	1	y	y	n	y	−149.1	−50.7
4	1	1	B	1	y	y	n	n	−147.3	−48.5
5	1	3	B	1	y	y	n	n	−153.4	−45.1
6	1	1	B	1	y	n	n	n	−140.2	−45.9
7	1	1	B	1	n	n	n	n	−144.0	−45.3
8	1	1	C	1	y	y	n	n	−151.1	−50.5
9	1	1	C	2	y	y	n	n	−138.4	−52.3
10	1	1	C	3	y	y	n	n	−146.5	−51.1
11	2	1	B	1	y	y	n	n	−148.1	−46.0
12	2	1	B	2	y	y	n	n	−142.6	−45.7
13	2	1	C	1	y	y	n	n	−151.4	−48.2
14	2	1	C	1	y	y	y	y	−151.3	−50.6
15	2	1	C	1	y	n	n	n	−146.8	−42.3
16	2	2	C	1	y	y	n	n	−154.3	−50.2
17	2	2	C	1	y	y	y	n	−155.8	−52.6
18	2	3	C	1	y	y	n	n	−158.4	−44.7
19	2	3	C	1	y	n	n	n	−146.8	−42.0

Abbreviations: #, model case number. TM, tomography model: 1, S12WM13 (Su *et al.* 1994); 2, SB4L18 (Masters *et al.* 2000). SF, scaling factors $(\delta\rho/\rho)/(\delta v_s/v_s)$ to convert seismic velocity to density variations: 1, scaling factor 0.2, only mantle density anomalies below 220 km depth are included; 2, scaling factor 0.3, only mantle density anomalies below 220 km depth are included; 3, scaling factor 0.2, all mantle density anomalies are included. MVM, mantle viscosity model (see text): A, Steinberger (2000); B, Steinberger & O'Connell (1998); C, Steinberger & Calderwood (2001). PVM, plume viscosity model: 1, viscosity inside plume conduit and plume conduit radius are constant; plume conduit rising speed is inversely proportional to viscosity of the mantle surrounding the conduit; 2, viscosity inside plume conduit and plume conduit radius increase with depth; plume conduit rising speed is inversely proportional to the square root of viscosity of the mantle surrounding the conduit; 3, viscosity inside plume conduit and plume conduit radius increase with depth; plume conduit is surrounded by a thermal halo (see text for more details). td, time-dependent plate motion boundary condition (y/n) (y, as explained in the text; n, constant present-day plate motions and boundaries). adv, advection of density heterogeneities for 68 Ma (y/n). com, compressible mantle (y/n). pb, phase boundaries (y/n). HS120, computed hot-spot location (°) at 120 Ma.

mantle viscosity model B to obtain a mantle flow field in which the rising mantle plume is advected. The tomographic models of Grand (2002; an updated model based on Grand *et al.* 1997) and S20RTS (Ritsema & Van Heijst 2000) have been used with viscosity model C. The first scaling factor relation (including only density anomalies below depth 220 km) is used in all cases except in combination with S12WM13 (Su *et al.* 1994), where the third relation (including all density anomalies, as in Steinberger & O'Connell 1998) is used. All cases shown are for an incompressible mantle without phase boundaries, time-dependent plate motion boundary conditions, backwards advection of density heterogeneities as explained and the first assumption about plume conduit radius.

Calculations yield a southward motion of 5° for the model based on SB4L18 (Masters *et al.* 2000) and 6° for the model based on S12WM13 (Su *et al.* 1994), and the slab model of Steinberger (2000*b*). The calculations with SB4L18 (Masters *et al.* 2000) and S12WM13 (Su *et al.* 1994) give approximately 10° and 15° eastward motion, respectively. The former gives a position of the hot spot 120 Ma ago at 45°S and 153°W, very close to the centre of the OJP at the same time, according to the plate reconstruction of Neal *et al.* (1997). However, the two positions cannot be directly compared, as the reconstruction of Neal *et al.* (1997) is based on the assumption of hot-spot fixity. The hot spot moves approximately 5° southward and 4° eastward using the model of Grand (2002), and total motion is only about 1° southward and 1° eastward when using the tomographic model of Ritsema & Van Heijst (2000). To test the effect of other modelling assumptions on our results, several input parameters have been varied for two representative tomographic input models (Su *et al.* 1994; Masters *et al.* 2000), and the results are shown in Table 1.

Discussion

Motion of plume conduits in our model is dominated by advection in the lower part of the mantle, and buoyant rising in the upper part of the mantle. Thus, computed surface hot-spot motion is frequently similar to flow at mid-mantle depth (i.e. upper part of the lower mantle) where the transition between 'low' and 'high' viscosity occurs (Steinberger & O'Connell 2000). All flow models considered show downward flow surrounding the Pacific, and a large-scale upwelling beneath the Pacific, which tends to be more focused for the models based on tomography and more diffuse for the one based on subduction history. Horizontal flow at mid-mantle depth has an outward flow component away from the large-scale upwelling, which is southward in the area of the Louisville hot spot. Another component is plate return flow towards the Pacific–Antarctic ridge, which is southeastward in this area for more recent times. Thus, from these two flow components we can qualitatively expect a southward component of Louisville hot-spot motion, but its speed should not exceed horizontal flow speeds at mid-mantle depths (*c*. 1 cm year^{-1}).

Details of mantle density structure in the south Pacific are not well known, hence there are considerable differences among models of flow and hot-spot motion (Fig. 2). For the present-day flow fields computed for models S12WM13 and SB4L18, the large upwelling beneath the Pacific is connected with a regional upwelling SE of New Zealand, SW of the Louisville hot spot. This yields, for those cases where advection of density heterogeneities is not considered, a mid-mantle flow in a SE direction, superimposed on plate return flow in the same direction that results in a SE motion of the hot spot. If density anomalies derived from these two tomography models are advected back in time, the upwelling SE of New Zealand tends to become stronger and a NE hot-spot motion is predicted for older times (Fig. 2A and C). Thus, in cases 4, 13 and 18 of Table 1, which include advection, the total southward hot-spot motion is less than in the corresponding cases 6, 15 and 19 without advection.

Comparison between cases 6 and 7 of Table 1 shows the effect of different plate motion boundary conditions. More recently, the hot spot has been closer to the ridge, hence more affected by (SE) plate return flow than in the more distant past. Hence, using constant present-day boundary conditions yields stronger hot-spot motion.

Comparison between cases 1–3, 13 and 14, as well as 16 and 17, shows that considering phase boundaries and compressibility leads to a predicted hot-spot location farther south at 120 Ma. Comparison between cases 4 and 5, as well as 13, 16 and 18, shows the effect of different scaling factors. Stronger density anomalies lead to stronger flow and hence more hot-spot motion. Comparison between cases 1, 4 and 8, as well as 11 and 13, shows the effect of radial ambient mantle viscosity structure. Comparison between cases 8–10, as well as 11 and 12, shows the effect of different buoyant plume rising speed. In the cases with higher rising speed, the computed total hot-spot motion tends to be less, but with a variable effect on the N–S component.

Altogether, in Figure 3 and Table 1, hot-spot motion varies between 9° southward (6° if we only count cases that consider advection of density heterogeneities) and 1° northward. The result used by Tarduno *et al.* (2003), which was obtained for a 'mean' tomography model, is also within that range.

Obviously, it is also possible to construct models with substantially larger Louisville hot-spot motion (e.g. Steinberger & O'Connell 1998); however, such models are not consistent with observations, such as hot-spot tracks globally, whereas our calculations suggest that all model parameters used here are broadly consistent with global observations.

In contrast to mid-mantle flow, flow at the base of the mantle is mostly towards large-scale and regional upwellings. Thus, in our model, plume conduits tend to get tilted, with their bases closer to these upwellings than the top. If such a tilted conduit rises to the surface it straightens up again, and in the process moves *towards*, rather than away from, the upwelling. This effect is, for example, responsible for the change in direction of hot-spot motion computed during the past 10 Ma for the model using S12WM13, shown in Figure 2. However, motion away from lower-mantle upwellings tends to occur far more frequently in our models for the Louisville hot spot. Frequently, a rather strong tilt is computed for the Louisville plume conduit for the past few 10 Ma. In the real mantle, a strongly tilted plume conduit might not remain intact but break up into several drops (Whitehead 1982). A strong tilt may be responsible for Louisville hot-spot volcanism becoming rather episodic during the more recent past, producing individual seamounts with larger spacing.

In contrast, Tarduno *et al.* (2003) discussed how straightening up of a tilted conduit may have caused rapid southward motion of the Hawaiian hot spot, i.e. motion *towards* the large-scale upwelling. Qualitatively, a number of reasons can be given why this may have happened for Hawaii, but not for Louisville:

- the Hawaiian plume is stronger, hence has higher buoyancy and a greater tendency to straighten up;
- the Hawaiian plume may be older and, hence, in a stage of straightening up again, whereas the Louisville hot spot may have not yet reached that stage;
- because the flow field related to the large-scale upwelling under the central Pacific is superposed on plate return flow, which is in a SE direction beneath both hot spots, the transition from flow towards the upwelling to flow away from the upwelling occurs at a shallower depth in the case of Hawaii. Hence, the Hawaiian hot spot is more likely to move towards the upwelling than the Louisville hot spot.

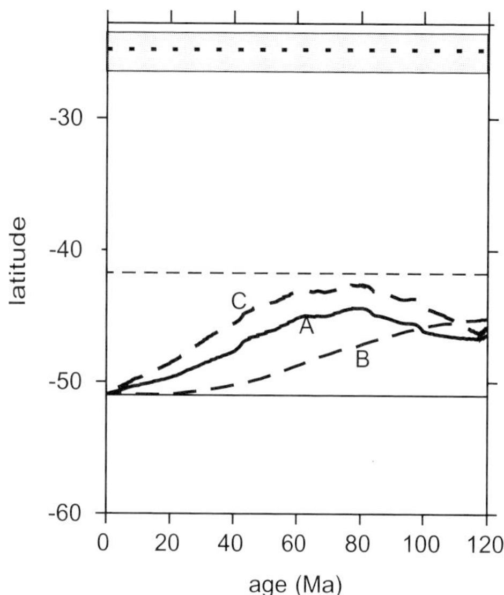

Fig. 4. Palaeolatitudes v. time for the Louisville hot spot for models that give southward hot-spot motion. A, Su *et al.* (1994); B, Steinberger (2000*b*); and C, Masters *et al.* (2000). The estimated present-day position of the Louisville hot spot, as well as the centre of the OJP suggested by Neal *et al.* (1997), are indicated by a solid and dashed horizontal line, respectively. The palaeomagnetically determined palaeolatitude and uncertainties are represented by the shaded area.

Effect of hot-spot motion on palaeolatitudes

Figure 4 shows the palaeolatitudes for the Louisville hot spot v. time for some of the calculations shown in Figure 3, which yield appreciable southward motion. For comparison, the horizontal dashed line indicates the centre of the plateau 120 Ma ago, as suggested by Neal *et al.* (1997). A southward motion of the hot spot could explain a large part of the discrepancy between the plate reconstruction of Neal *et al.* (1997) and the Louisville hot-spot latitude. However, its amount is not large enough to explain the low palaeolatitude of approximately 25°S obtained by Riisager *et al.* (2003, 2004) from basalts and volcaniclastic sediments from the OJP.

Effect of TPW on palaeolatitudes

True polar wander is defined as the relative motion between the Earth's mantle and the rotation axis of the Earth. The most recent TPW curves by Prévot *et al.* (2000) and Besse & Courtillot (2002) both use a fixed hot-spot reference frame. As discussed previously (Antretter *et al.* 2002), this should be modified if hot-spot motion is considered. However, we found that for models similar to the one shown here such a modification produces no more than a 2° difference. To keep the model simple we therefore did not implement such a modification.

The change in hot-spot latitude with time that results from TPW is shown in Figure 5. For both TPW curves the effect on the latitude is small for the past *c.* 100 Ma. Prior to that, TPW provides palaeolatitudes for the Louisville hot spot up to 11° farther north than the present-day hot-spot position (Fig. 5), thus further reducing the discrepancy between the present-day latitude of the Louisville hot spot and the recent palaeomagnetic results from the OJP.

Conclusion

Recent palaeomagnetic results from the OJP (Riisager *et al.* 2003, 2004) yield a palaeolatitude of approximately 25°S, whereas the present-day latitude of the Louisville hot spot is approximately 51°S. The effect of TPW on the palaeolatitudes can explain up to 11° of this difference. Our models for the motion of the Louisville hot spot yield up to 6° of southward motion (9°, if present-day density heterogeneities are used for all times), depending on the tomographic models and parameters used. Taking into account both TPW and hot-spot motion, and considering the largest southward shifts resulting from both

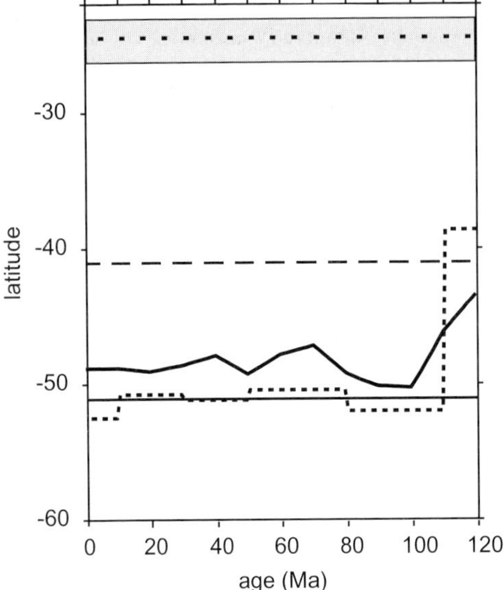

Fig. 5. The effect of TPW on the palaeolatitude of the Louisville hot spot for two different TPW paths (solid line, Besse & Courtillot 2002; dotted line, Prévot *et al.* 2000). The centre of the OJP suggested by Neal *et al.* (1997) and the palaeomagnetically determined palaeolatitude (with uncertainties) are shown as a dashed horizontal line and a shaded area, respectively.

effects, we can explain a latitudinal shift of approximately 17°–20°. Considering the effect of long-term octupole contributions may explain an additional shift of approximately 7.5°. Thus, a favourable combination of all three effects is just sufficient to explain the discrepancy of approximately 26° between the palaeomagnetic results and the present-day hot-spot position. Thus, it remains possible that the Louisville hot-spot was the source of the OJP.

Funding for this research was provided by the Deutsche Forschungsgemeinschaft DFG, ODP/Germany, Project Number So72/70–1 (M. Antretter, B. Steinberger), the Danish National Research Foundation (P. Riisager), US Science Support Program of JOI (S. Hall), and NSF grants EAR 443549–22178 and EAR 443747–22250 (X. Zhao). The authors would like to thank the crew and scientific party of ODP Leg 192 for their invaluable work and comments.

References

AKAOGI, M. & ITO, E. 1999. Calorimetric study on majorite–perovskite transition in the system $Mg_4Si_4O_{12}$–$Mg_3Al_2Si_3O_{12}$: transition boundaries with positive pressure–temperature slopes. *Physics of the Earth and Planetary Interiors*, **114**, 129–140.

AKAOGI, M., ITO, E. & NAVROTSKY, A. 1989. Olivine-modified spinel–spinel transitions in the system Mg_2SiO_4–Fe_2SiO_4: Calorimetric measurements, thermochemical calculations, and geophysical application. *Journal of Geophysical Research*, **94**, 15 671–15 685.

ANTRETTER, M., STEINBERGER, B., HEIDER, F. & SOFFEL, H. 2002. Paleolatitudes of the Kerguelen hotspot: new paleomagnetic results and dynamic modeling. *Earth and Planetary Science Letters*, **203**, 635–650.

BESSE, J. & COURTILLOT, V. 2002. Apparent and true polar wander and the geometry of the geomagnetic field in the last 200 Myr. *Journal of Geophysical Research*, **107**, 2300. doi: 10.1029/2000JB000050.

CHAMBERS, L., PRINGLE, M. & FITTON, G. 2003. Age and duration of magmatism on the Ontong Java Plateau. *Geophysical Research Abstracts*, **5**, 09657.

COUPLAND, D.H. & VAN DER VOO, R. 1980. Long-term non-dipole components in the geomangetic field during the last 130 m.y. *Journal of Geophysical Research*, **85**, 3529–3549.

DAVIES, G.F. 1988. Ocean bathymetry and mantle convection, 1, Large-scale flow and hotspots. *Journal of Geophysical Research*, **93**, 10 467–10 480.

GRAND, S.P., VAN DER HILST, R.D. & WIDIYANTORO, S. 1997. Global seismic tomography: A snapshot of convection in the Earth. *Geological Society of America*, **7**, 1–7.

GRAND, S.P. 2002. Available online at ftp: //bratsche.geo.utexas.edu/outgoing/steveg/.

GORDON, R.G. & JURDY, D. 1986. Cenozoic global plate motions. *Journal of Geophysical Research*, **91**, 12 389–12 406.

HAGER, B.H. & O'CONNELL, R.J. 1979. Kinematic models of large-scale mantle flow. *Journal of Geophysical Research*, **84**, 1031–1048.

HAGER, B.H. & O'CONNELL, R.J. 1981. A simple global model of plate dynamics and mantle convection. *Journal of Geophysical Research*, **86**, 4843–4867.

KARATO, S. 1993. Importance of anelasticity in the interpretation of seismic tomography. *Geophysical Research Letters*, **20**, 1623–1626.

KROENKE, L.W. & WESSEL, P. 1997. Pacific plate motion between 125 and 90 Ma and the formation of the Ontong Java Plateau (abstract). American Geophysical Union Chapman Conference on Global Plate Motions, Marshall, CA, **22**.

KROENKE, L.W., WESSEL, P. & STERLING, A. 2004. Motion of the Ontong Java Plateau in the hot-spot frame of reference: 122 Ma–present. *In*: FITTON, J.G., MAHONEY, J.J., WALLACE, P.J. & SAUNDERS, A.D. (eds) *Origin and Evolution of the Ontong Java Plateau*. Geological Society, London, Special Publications, **229**, 9–20.

LITHGOW-BERTELLONI, C., RICHARDS, M.A., RICARD, Y., O'CONNELL, R.J. & ENGEBRETSON, D.C. 1993.

Toroidal–poloidal partitioning of plate motions since 120 Ma. *Geophysical Research Letters*, **20**, 375–378.

MAHONEY, J.J., STOREY, M., DUNCAN, R.A., SPENCER, K. J. & PRINGLE, M. 1993. Geochemistry and geochronology of the Ontong Java Plateau. *In*: PRINGLE, M., SAGER, W., SLITER, W. & STEIN, S. (eds) *The Mesozoic Pacific: Geology, Tectonics, and Volcanism*. American Geophysical Union, Geophysical Monograph, **77**, 233–261.

MCELHINNY, M.W., MCFADDEN, P.L. & MERRILL, R.T. 1996. The time averaged paleomagnetic field 0–5 Ma. *Journal of Geophysical Research B: Solid Earth*, **101**, 25 007–25 028.

MUELLER, R.D., ROEST, W.R., ROYER, J.-Y., GAHAGAN, L.M. & SCLATER, J.G. 1997. Digital isochrons of the world's ocean floor. *Journal of Geophysical Research*, **102**, 3211–3214.

MUELLER, R.D., ROYER, J.-Y. & LAWVER, L.A. 1993. Revised plate motions relative to the hotspots from combined Atlantic and Indian Ocean hotspot tracks. *Geology*, **21**, 275–278.

MASTERS, G., LASKE, G., BOLTON, H. & DZIEWONSKI, A. 2000. The relative behaviour of shear velocity, bulk sound speed, and compressional velocity in the mantle: implications for chemical and thermal structure. *In*: KARATO, S. (ed.) *Seismology and Mineral Physics*. American Geophysical Union, Geophysical Monograph, **117**, 63–87.

MAYER, H. & TARDUNO, J.A. 1993. Paleomagnetic investigation of the igneous sequence, Site 807, Ontong Java Plateau, and a discussion of pacific true polar wander. *In*: BERGER, W.H., KROENKE, L.W., MAYER, L.A., *et al.* (eds) *Proceedings of ODP, Scientific Results*, **130**, 51–59, College Station, TX (Ocean Drilling Program).

NEAL, C.R., MAHONEY, J.J., KROENKE, L.W., DUNCAN, R.A. & PETTERSON, M.G. 1997. The Ontong Java Plateau. *In*: MAHONEY, J.J. & COFFIN, M.F. (eds) *Large Igneous Provinces: Continental, Oceanic and Planetary Flood Basaltic Volcanism*. American Geophysical Union, Geophysical Monograph, **100**, 183–216.

PRÉVOT, M., MATTERN, E., CAMPS, P. & DAIGNIÈRES, M. 2000. Evidence for a 20° tilting of the Earth's rotation axis 110 million years ago. *Earth and Planetary Science Letters*, **179**, 517–528.

RICHARDS, M.A. 1991. Hotspots and the case against a uniform viscosity composition mantle. *In*: SABADINI, R. & LAMBECK, K. (eds) *Glacial Isostasy, Sea Level and Mantle Rheology*. Kluwer, Dordrecht, 571–587.

RICHARDS, M.A. & GRIFFITHS, R.W. 1988. Deflection of plumes by mantle shear flow: Experimental results and a simple theory. *Geophysical Journal*, **94**, 367–376.

RICHARDS, M.A., JONES, D.L., DUNCAN, R.A. & DEPAOLO, D.J. 1991. A mantle plume initiation model for the Wrangellia flood basalt and other oceanic plateaus. *Science*, **254**, 263–267.

RIISAGER, P., HALL, S., ANTRETTER, M. & ZHAO, X. 2003. Paleomagnetic paleolatitude of Early Cretaceous Ontong Java Plateau Basalts: implications for pacific apparent and true polar wander. *Earth and Planetary Science Letters*, **208**, 235–252.

RIISAGER, P., HALL, S., ANTRETTER, M. & ZHAO, X. 2004. Early Cretaceous Pacific palaeomagnetic pole from Ontong Java Plateau basement rocks. *In*: FITTON, J.G., MAHONEY, J.J., WALLACE, P.J. & SAUNDERS, A.D. (eds) *Origin and Evolution of the Ontong Java Plateau*. Geological Society, London, Special Publications, **229**, 31–44.

RITSEMA, J. & VAN HEIJST, H.J. 2000. Seismic imaging of structural heterogeneity in Earth's mantle: Evidence for large-scale mantle flow. *Science Progress*, **83**, 243–259.

SLEEP, N. 1990. Hotspots and mantle plumes: Some phenomenology. *Journal of Geophysical Research*, **95**, 6715–6736.

STEINBERGER, B. 2000*a*. Plumes in a convecting mantle; models and observations for individual hotspots. *Journal of Geophysical Research*, **10**, 11 127–11 152.

STEINBERGER, B. 2000*b*. Slabs in the lower mantle – results of dynamic modelling compared with tomographic images and the geoid. *Physics of the Earth and Planetary Interiors*, **118**, 241–257.

STEINBERGER, B. & CALDERWOOD, A.R. 2001. Mineral physics constraints on viscous flow models of mantle flow. *Journal of Conference Abstracts*, Cambridge Publications, Abstracts for EUG XI, 423–424.

STEINBERGER, B. & O'CONNELL, R.D. 1998. Advection of plumes in mantle flow: implications for hotspot motion, mantle viscosity and plume distribution. *Geophysical Journal International*, **132**, 412–434.

STEINBERGER, B. & O'CONNELL, R.D. 2000. Effects of mantle flow on hotspot motion. *In*: RICHARDS, M.A., GORDEON, R.G. & VAN DER HILST, R.D. (eds) *The History and Dynamics of Global Plate Motions*. American Geophysical Union, Geophysical Monograph, **121**, 377–398.

SU, W.-J., WOODWARD, R.L. & DZIEWONSKI, A.M. 1994. Degree 12 model of shear velocity heterogeneity in the mantle. *Journal of Geophysical Research*, **99**, 6945–6980.

TARDUNO, J.A., DUNCAN, R.A., SCHOLL, D.W., COTTRELL, R.D., STEINBERGER, B., THORDARSON, T., KERR, B.C., NEAL, C.R., FREY, F.A., TORII, M. & CARVALLO, C. 2003. The Emperor Seamounts: Southward motion of the Hawaiian hotspot plume in Earth's mantle. *Science*, **301**, 1064–1069.

TARDUNO, J.A., SLITER, W.V., KROENKE, L.W., LECKIE, M., MAHONEY, J.J., MUSGRAVE, R.J., STOREY, M. & WINTERER, E.L. 1991. Rapid formation of the Ontong Java Plateau by Aptian mantle plume volcanism. *Science*, **254**, 399–403.

TORSVIK, T.H., MOSAR, J. & EIDE, E.A. 2001. Cretaceous–Tertiary geodynamics: A North Atlantic exercise. *Geophysical Journal International*, **146**, 850–867.

VAN DER VOO, R. & TORSVIK, T.H. 2001. Evidence for late Paleooic and Mesozoic non-dipole fields

provides an explanation for the Pangea reconstruction problems. *Earth and Planetary Science Letters*, **187**, 71–81.
VLASTELIC, I., DOSSO, L., GUILLOU, H., BOUGAULT, H., GELI, L., ETOUBLEAU, J. & JORON, J.L. 1998. Geochemistry of the Hollister Ridge: relation with the Louisville hotspot and the Pacific–Antarctic Ridge. *Earth and Planetary Science Letters*, **160**, 777–793.
WATTS, A.B., WEISSEL, J.K., DUNCAN, R.A. & LARSON, R.L. 1988. Origin of the Louisville Ridge and its relationship to the Eltanin Fracture Zone System. *Journal of Geophysical Research*, **93**, 3051–3077.
WESSEL, P. & KROENKE, L.W. 1997. A geometric technique for relocating hotspots and refining absolute plate motions. *Nature*, **387**, 365–369.
WHITEHEAD, J.A. 1982. Instabilities of fluid conduits in a flowing Earth – are plates lubricated by the asthenosphere? *Geophysical Journal of the Royal Astronomical Society*, **70**, 415–433.

Early Cretaceous Pacific palaeomagnetic pole from Ontong Java Plateau basement rocks

PETER RIISAGER[1,2], STUART HALL[3], MARIA ANTRETTER[4] & XIXI ZHAO[2]

[1]*Danish Lithosphere Centre, Øster Voldgade 10, DK-1350 Copenhagen K, Denmark*

[2]*Earth Sciences Department, University of California at Santa Cruz, Santa Cruz, CA 95064, USA*

[3]*Department of Geosciences, University of Houston, Houston, TX 77204-5007, USA*

[4]*Institut für Geophysik, University of München, Theresienstrasse 41, D-80333 München, Germany*

Abstract: We present new palaeomagnetic data from Ocean Drilling Program Site 1184 on the eastern salient of the Ontong Java Plateau (OJP) where 337.7 m of Early Cretaceous (*c.* 120 Ma) volcaniclastic rocks were drilled. Alternating field and thermal demagnetizations were equally effective in removing secondary components, allowing the characteristic remanent magnetization directions from a total of 173 samples (out of 183) to be defined. All samples have negative inclinations (normal polarity), and by treating each sample as an independent reading of the palaeomagnetic field a site-mean inclination of –53.9° (N = 173; α_{95} = 1.0°, k = 109) was obtained. The corresponding palaeo-colatitude is in excellent accordance with previously published time-averaged palaeo-colatitudes from contemporaneous basalts drilled at OJP and the Nauru Basin. Based on the intersection of the seven palaeo-colatitudes a new Early Cretaceous (*c.* 120 Ma) Pacific palaeomagnetic pole was obtained with co-ordinates 63.0°N, 10.1°E (95% confidence ellipse with a minor semi-axis of 2.9° with an azimuth of 32° and a major semi-axis of 47.7° with an azimuth of 122°). This pole is far more easterly than previously published Early Cretaceous Pacific palaeomagnetic poles. Based on published Pacific palaeogeographic reconstructions in the fixed hot-spot reference frame we were able to calculate different Pacific true polar wander (TPW) poles. All Pacific TPW poles are found to be statistically different from contemporaneous TPW poles obtained in the Indo-Atlantic realm, illustrating motion between the two groups of hot spots.

During the Ocean Drilling Program (ODP) Leg 192, to the Ontong Java Plateau, approximately 120 Ma old basement rocks (Chambers *et al.* 2002) were retrieved at five sites (Fig.1). Time-averaged site-mean palaeomagnetic inclination estimates have been obtained from the subaqueous basaltic lava flows and pillow units recovered at Sites 1183, 1185, 1186 and 1187 (Riisager *et al.* 2003*b*). In the present study we present new palaeomagnetic data from the 337.7 m of volcaniclastic rocks drilled at Site 1184. These new data, together with previously published ODP and Deep Sea Drilling Project (DSDP) data, allow us to derive a new Early Cretaceous palaeomagnetic pole for the Pacific. In the following we will compare this new approximately 120 Ma palaeomagnetic pole with the Pacific apparent polar wander path (APWP), and discuss its significance in relation to the true polar wander, hot-spot fixity and Pacific plate reconstructions.

Geological setting

The Ontong Java Plateau

The Ontong Java Plateau (OJP; Fig. 1) is the world's largest volcanic oceanic plateau, with a surface area of 2.0×10^6 km^2 and an estimated volume of 4×10^7–5×10^7 km^3 (e.g. Eldholm & Coffin 2000). Existing age constraints point to the formation of mostly all of the OJP in a single geologically brief period around 120 Ma (e.g. Tarduno *et al.* 1991; Parkinson *et al.* 2001; Chambers *et al.* 2002), making it the largest magmatic event on Earth during the last 200 Ma. The enormous total partial melt volume (Coffin & Eldholm 1994), as well as the composition of lavas (Tejada *et al.* 2002), is best explained by the plume-head or plume-impact model, in which widespread basaltic flood eruptions occur as the inflated head of a rising new mantle plume approaches the base of the lithosphere (Richards

From: FITTON, J. G., MAHONEY, J. J., WALLACE, P. J. & SAUNDERS, A. D. (eds) 2004. *Origin and Evolution of the Ontong Java Plateau*. Geological Society, London, Special Publications, **229**, 31–44. 0305-8719/$15.00

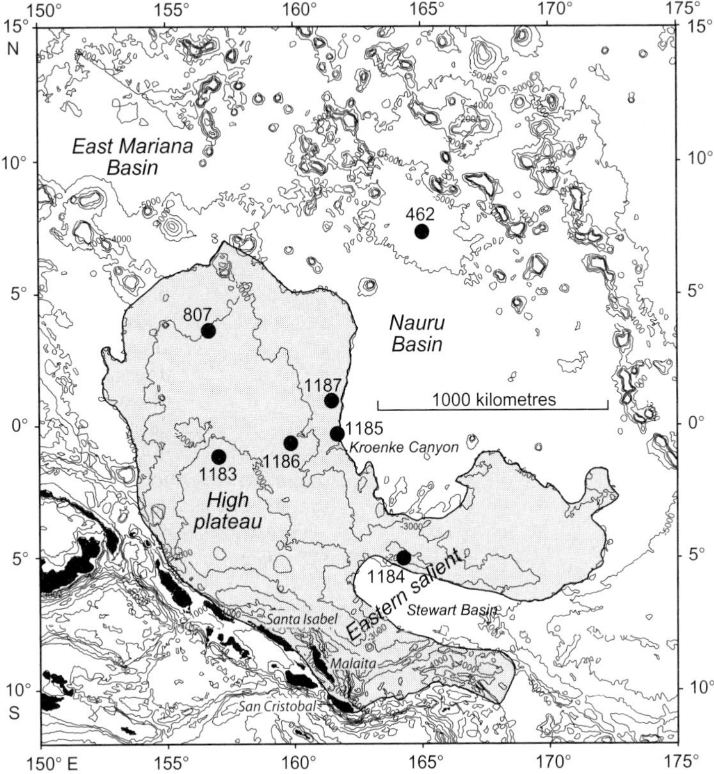

Fig. 1. Predicted bathymetry (after Sandwell & Smith 1997) of the OJP showing the locations of Site 1184 and other sites (Sites 462, 807, 1183, 1185, 1186 and 1187), for which a time-averaged mean inclination estimate has been obtained. The OJP is outlined. The bathymetric contour interval is 1000 m.

et al. 1989). The oceanic crust south of the OJP has been subducted and with it the possible plume-tail trace (hot-spot track). It is therefore not possible to directly link the OJP to any known hot-spot source (e.g. Tejada *et al.* 2002).

The maximum extent of Ontong Java-related volcanism may go well beyond the plateau proper, as the Early Cretaceous lava flows in the Nauru Basin appear to be related to the plateau (e.g. Neal *et al.* 1997), and similarly the Manihiki Plateau to the SE may have been constructed from the same broad source (e.g. Tarduno *et al.* 1991).

Site 1184

Site 1184 is located on the eastern salient of the OJP (Fig. 1) on a fault block with a small *c.* 4° dip (down-dip azimuth *c.* 18°) of the upper surface of the basement, as determined from two crossing seismic lines (Mahoney *et al.* 2001). A 337.7 m-thick sequence of volcaniclastic rocks was recovered at Site 1184 (82.6% average recovery), consisting of ash- to lapilli-size lithic clasts and vitric shards, accretionary lapilli, armoured lapilli, crystal fragments (plagioclase and clinopyroxene), and matrix and/or cement (Mahoney *et al.* 2001). Based on chemical composition and lithology the sequence has been divided into six eruption units, with each eruption unit containing a variable number of subunits or depositional sets, which most probably reflect multiple explosions during eruption (Thordarson 2004). The occurrence of ash-fall layers, wood and organic-rich intervals document significant temporal hiatuses in the drilled sequence (Mahoney *et al.* 2001). All rocks have been affected by low-temperature alteration processes in amounts similar to those found in normal seafloor basalts (Banerjee *et al.* 2004). Palaeomagnetic samples were obtained from pieces long enough to ensure that their up–down orientation was preserved during rotary coring. From the 337.7 m of volcaniclastics 183 azimuthally unoriented palaeomagnetic samples were obtained.

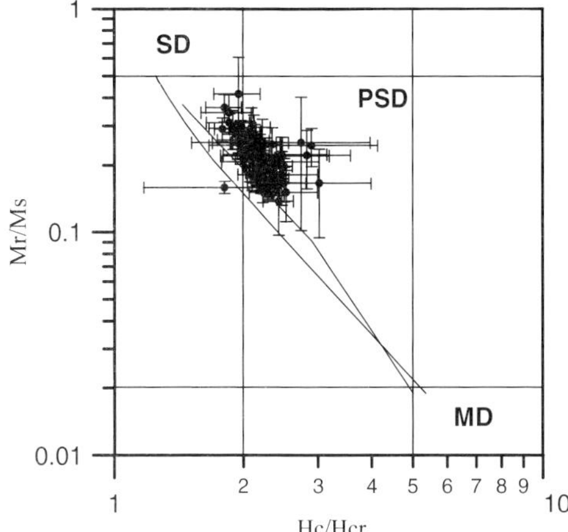

Fig. 2. Day plot of averaged hysteresis parameters of the samples studied. Five subsamples were measured for each sample; error bars are 1σ. Solid lines are the mixing curves for magnetite (Dunlop 2002). SD is single domain; PSD, pseudo-single domain; MD, multi-domain.

Rock and palaeomagnetic properties of Site 1184 volcaniclastics

The rock magnetic experiments include measurements of the temperature dependence of weak field susceptibility, $\kappa(T)$, in an argon atmosphere using a KLY-2 kappabridge, and hysteresis measurements performed on a Micromag alternating gradient force magnetometer. All $\kappa(T)$ curves are irreversible, showing destruction of a ferromagnetic phase at *c.* 200–300°C, most probably maghemite, in addition to an apparently reversible phase with a Curie temperature of *c.* 570°C, most probably magnetite. Hysteresis experiments were performed on five samples of approximately 20 mg from each palaeomagnetic specimen in order to obtain averaged hysteresis parameters for the otherwise inhomogeneous volcaniclastic rocks. The averaged hysteresis parameters with corresponding standard errors (Fig. 2) are slightly shifted to the right of the single domain–multi-domain mixing curves for magnetite (Dunlop 2002), suggesting a small contribution of superparamagnetic grain sizes. Further details concerning the rock magnetic properties are given in Zhao *et al.* (2004).

Detailed alternating field (AF) and thermal step-wise demagnetizations were performed at the palaeomagnetic laboratories at the University of California at Santa Cruz, the University of Houston and the University of Munich. Data from different laboratories were found to be in excellent agreement with each other. Based on the demagnetization data, the characteristic remanent magnetization (ChRM) directions were calculated using principal component analysis (Kirschvink 1980), accepting only ChRM components based on five or more demagnetization steps, pointing towards the origin and having a maximum angular deviation (MAD) of less than 10° (Fig. 3, Table 1). Thermal and AF demagnetizations yielded similar ChRM inclinations (Fig. 4), confirming that both methods are equally effective in removing secondary components. Out of the total of 183 palaeomagnetic specimens, 173 satisfied the above-mentioned acceptance criteria; further details are listed in Table 1.

The variation in ChRM inclinations within the individual eruption units (Fig. 5) suggests that the units were emplaced over a sufficiently long period to record palaeosecular variation. The deposition rates are, however, expected to have varied widely during the eruptions (Thordarson 2004) and it is difficult to more precisely discern the temporal relationship between consecutive samples based solely on the palaeomagnetic data. In the following each sample is treated as an independent reading of the palaeomagnetic field, and we obtained a site-mean inclination of –53.9° (N = 173; α_{95} = 1.0°, k = 109) using the statistical procedures of McFadden & Reid (1982). A similar site-mean inclination of –54.0° ($N = 5$, $\alpha_{95} = 1.6°$, $k = 3064$) is obtained if samples are grouped in eruption units. The bulk of the volcaniclastic succession is interpreted as subaerially deposited (Thordarson 2004), and the remanent magnetization is therefore most probably of thermal origin.

Early Cretaceous palaeomagnetic palaeo-colatitudes for the Pacific

The palaeomagnetic palaeo-colatititude of Site 1184 is shown in Figure 6 together with the palaeo-colatitudes for other OJP drill sites where sufficient material was recovered to average out secular variation (Sites 807, 1183, 1185, 1186 and 1187; Mayer & Tarduno 1993; Riisager *et al.* 2003*b*). The excellent agreement of all OJP palaeo-colatitudes is interpreted to indicate that the studied basalts have suffered little or no tectonic disturbance since their emplacement. Another site considered here is Site 462 in the Nauru Basin (Fig.1), where more than 600 m of Early Cretaceous basalts and sills were recovered during DSDP Legs 61 and 89. The available $^{40}Ar/^{39}Ar$ ages for Site 462 range

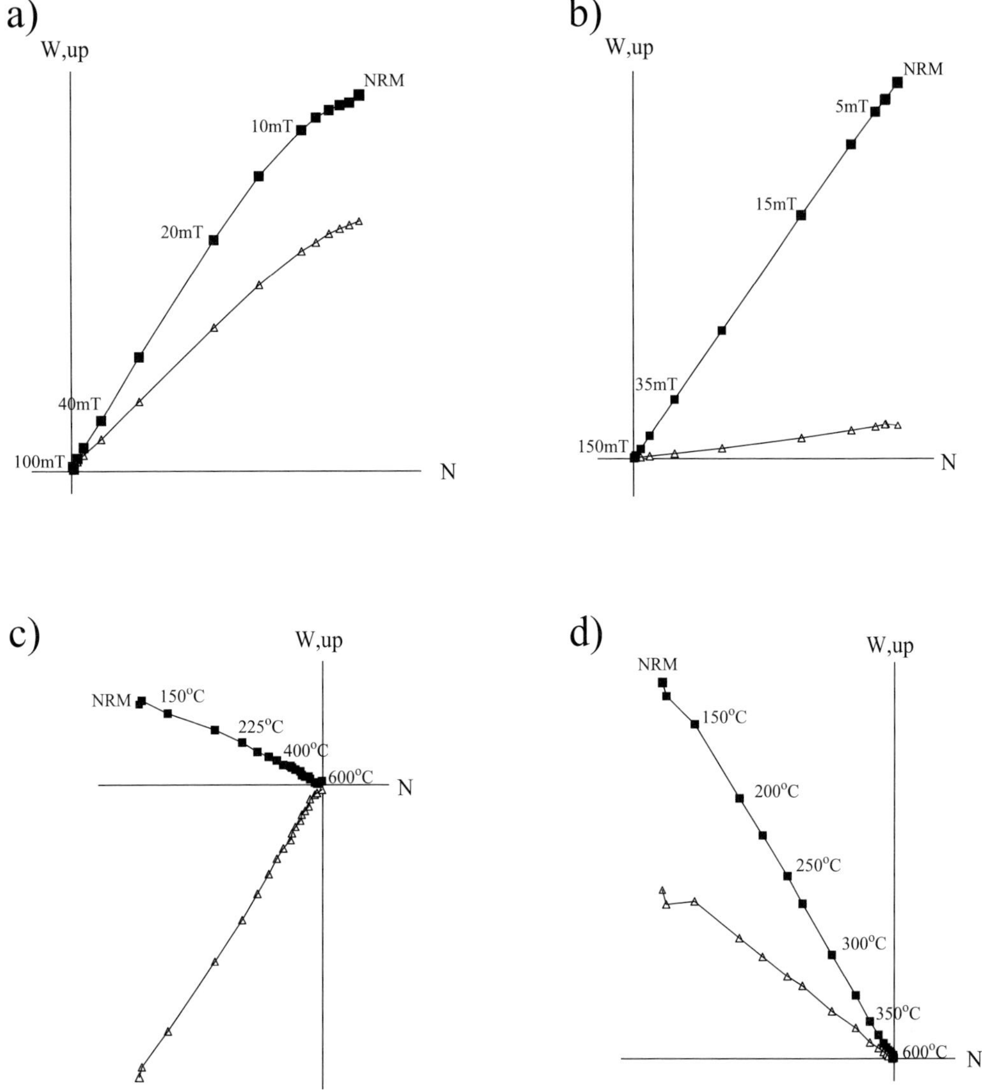

Fig. 3. Orthogonal vector plots of selected samples. Typical examples of AF demagnetization data: (**a**) 1184A-12R-01W, 114–116 cm; (**b**) 1184A-35R-03W, 109–111 cm; and thermal demagnetisation data; (**c**) 1184A-15R-01W, 98–100 cm; (**d**) 1184A-31R-05W, 11–13 cm.

from *c.* 110 Ma (Ozima *et al.* 1981; Castillo *et al.* 1994) to *c.* 130 Ma (Takigami *et al.* 1986); however, revision of the stratigraphic age strongly indicates that the lava flows are Aptian (Moberly *et al.* 1986), i.e. the same age as the OJP. Combining the Site 462 palaeomagnetic inclinations for igneous units 22, 23, 26, 29, 30 and 34 drilled during Leg 61 (Steiner 1981) and units 45–54 of Leg 89 (Ogg 1986), we obtain a new site-mean inclination of –33.2° ($N = 15$; $\alpha_{95} = 7.1°$, $k = 28$). The OJP and Nauru Basin site-mean inclinations and their statistical parameters are listed in Table 2. The seven OJP and Nauru Basin palaeomagnetic palaeo-colatitudes suggest that the central part of the OJP formed at approximately 25°S, significantly north of the present-day latitude, 51°S, of the Louisville hot spot (see also Antretter *et al.* 2004).

Table 1. *Inclinations of the characteristic remanent magnetization component*

Core	Section	Interval (cm)	Depth (mbsf)	Inclination (°)	Intensity (A/m)	MAD (°)	Demagnetization type*
9R	1W	79–81	201.89	–61.0	0.14	5.0	Th
10R	1W	36–38	203.26	–56.1	0.17	2.3	Th
11R	1W	75–77	207.15	–58.5	0.11	3.3	Th
11R	2W	70–72	208.60	–50.9	0.26	3.8	Th
11R	3W	21–23	209.27	–53.2	0.23	1.9	AF
12R	1W	114–116	212.04	–47.7	0.39	0.8	AF
12R	3W	84–86	214.61	–53.3	0.24	0.8	AF
12R	4W	89–91	216.13	–65.1	0.69	1.4	AF
12R	5W	73–75	217.29	–59.6	0.95	0.6	AF
13R	1W	80–82	221.20	–53.8	0.42	0.5	AF
13R	2W	87–89	222.71	–55.6	0.34	2.1	AF
13R	3W	97–99	224.04	–50.2	0.23	2.3	AF
13R	4W	125–127	225.83	–50.1	0.11	2.5	Th
13R	5W	9–11	226.14	–50.6	0.15	1.9	AF
14R	1W	130–132	231.30	–49.1	0.49	1.1	AF
14R	2W	133–135	232.76	–52.6	0.16	2.5	Th
14R	3W	96–98	233.80	–49.2	1.56	1.0	AF
14R	4W	46–48	234.63	–58.5	0.42	1.8	Th
14R	5W	117–119	236.70	–57.3	0.11	2.1	AF
14R	6W	20–22	237.14	–61.2	0.13	2.2	Th
15R	1W	98–100	240.58	–54.5	0.61	1.5	Th
15R	1W	133–135	240.93	–44.5	0.63	0.9	AF
15R	2W	41–43	241.49	–54.3	0.53	2.7	Th
16R	1W	58–60	244.98	unstable	0.35		Th
16R	1W	68–70	245.08	–55.1	0.93	2.5	AF
16R	2W	76–78	246.66	–56.6	0.25	1.7	AF
16R	2W	90–92	246.80	unstable	0.26		Th
16R	3W	27–29	247.65	–50.3	0.56	1.9	Th
16R	3W	65–67	248.03	unstable	0.31		Th
16R	4W	42–44	248.91	–58.0	0.30	6.5	Th
17R	1W	24–26	249.44	–62.3	0.30	2.2	AF
17R	2W	25–27	250.84	–63.8	0.29	1.1	Th
17R	3W	4–6	252.14	–55.2	0.40	2.2	Th
17R	4W	99–101	254.52	–62.6	0.87	3.4	AF
17R	5W	79–81	254.52	–60.2	0.25	6.3	Th
17R	6W	104–106	257.12	–54.8	0.34	3.7	AF
17R	7W	46–48	258.02	–58.9	0.27	1.4	Th
18R	2W	85–87	260.79	–44.6	0.56	6.0	Th
18R	2W	97–99	260.91	–58.1	0.23	3.2	AF
18R	3W	45–47	261.75	–55.2	0.41	2.1	AF
18R	4W	89–90	263.71	–65.7	0.24	1.7	Th
18R	5W	86–88	265.06	unstable	0.22		AF
18R	6W	23–25	265.93	–55.3	0.41	0.9	AF
19R	1W	67–69	269.17	–59.0	1.92	1.7	AF
19R	2W	87–89	270.18	–58.6	1.37	1.4	Th
19R	3W	61–63	271.21	–55.4	0.16	5.4	Th
19R	4W	50–52	272.16	–56.8	0.11	6.4	AF
19R	5W	14–16	273.21	–56.8	1.92	0.7	AF
19R	6W	50–52	274.83	–47.3	0.74	2.9	Th
19R	8W	17–19	277.52	–54.1	0.70	1.3	Th
20R	1W	66–68	278.76	–59.5	0.40	4.7	AF
20R	2W	81–83	280.41	unstable	0.15		Th
20R	3W	120–122	282.11	–61.4	0.20	6.6	Th
20R	4W	74–76	283.13	–56.4	1.55	2.7	AF
20R	5W	67–69	284.15	–62.3	0.23	2.9	AF
21R	1W	71–73	288.41	–56.9	0.10	2.7	AF
21R	2W	145–147	290.59	–50.6	0.07	8.0	Th
21R	3W	33–35	290.98	–37.5	0.21	2.8	AF
21R	4W	37–39	292.22	–35.5	0.11	2.8	AF
21R	5W	77–79	293.45	–65.9	0.12	7.6	AF
21R	6W	21–23	294.39	–61.8	0.13	4.5	AF
22R	1W	68–70	297.98	–47.8	2.27	2.2	AF

Table 1. *continued*

Core	Section	Interval (cm)	Depth (mbsf)	Inclination (°)	Intensity (A/m)	MAD (°)	Demagnetization type*
22R	2W	2–4	298.12	–40.4	0.16	1.1	Th
22R	3W	124–126	300.36	–56.0	0.22	1.4	Th
22R	4W	39–41	300.96	unstable	0.07		Th
22R	5W	116–118	303.07	–34.6	0.24	1.4	AF
24R	2W	111–113	318.46	–54.7	2.83	2.8	Th
24R	3W	73–75	319.42	–53.6	2.92	3.0	AF
24R	4W	107–109	320.84	–52.9	1.64	1.4	AF
24R	5W	117–119	322.41	–51.0	2.03	2.2	Th
24R	6W	108–110	323.72	–54.9	4.52	2.9	Th
24R	7W	108–110	324.98	–50.0	3.56	1.3	AF
24R	8W	116–118	326.51	–50.4	3.37	1.2	AF
25R	1W	11–13	326.31	–52.6	5.23	1.6	AF
25R	3W	11–13	328.89	–49.5	1.73	1.7	AF
25R	4W	13–15	329.64	–49.3	1.81	2.2	Th
25R	5W	21–23	330.90	–50.6	2.79	0.7	AF
25R	6W	21–23	332.16	–55.0	2.50	1.2	AF
25R	7W	16–18	333.61	–56.3	2.52	1.1	AF
26R	1W	27–29	336.17	–50.2	3.85	0.6	AF
26R	2W	27–29	337.64	–50.7	6.89	0.6	AF
26R	3W	98–100	339.55	–54.4	2.73	1.2	Th
26R	4W	25–27	340.13	–52.7	6.10	1.7	AF
26R	5W	67–69	342.04	–50.6	1.80	4.0	AF
26R	6W	33–35	343.20	–53.1	1.05	2.7	AF
26R	7W	23–25	344.54	–57.3	1.16	1.5	AF
27R	1W	21–23	345.81	–50.9	1.96	1.3	AF
27R	2W	22–24	347.33	–58.0	1.22	3.4	AF
27R	3W	36–38	348.98	–47.9	2.23	1.8	AF
27R	4W	95–97	351.03	–59.1	1.40	1.9	Th
27R	5W	16–18	351.74	–54.3	2.41	1.3	AF
28R	1W	34–36	355.64	–51.4	4.04	1.2	Th
28R	2W	113–115	357.93	–61.4	2.07	2.4	AF
28R	3W	31–33	358.40	–50.5	2.61	1.5	AF
28R	4W	17–19	359.73	–42.9	1.54	4.2	Th
28R	5W	16–18	361.22	unstable	1.77		Th
28R	6W	38–40	362.94	–61.5	1.98	3.0	AF
28R	7W	17–19	364.23	–38.3	1.58	1.5	AF
29R	1W	134–136	366.34	–54.0	2.26	2.5	AF
29R	2W	2–4	366.52	–57.6	1.90	2.0	AF
29R	3W	147–149	369.35	–52.8	1.43	1.5	Th
29R	4W	2–4	369.40	–56.6	2.10	1.9	AF
29R	5W	66–68	371.54	–55.3	4.31	1.2	AF
30R	1W	71–73	375.43	–52.3	1.56	2.1	AF
30R	2W	126–128	377.46	–48.2	7.25	1.4	Th
30R	3W	17–19	377.87	–52.6	10.13	0.8	Th
30R	4W	1–3	379.21	–47.8	3.58	1.0	AF
30R	5W	88–90	381.27	–59.6	0.08	2.6	Th
30R	6W	114–116	382.79	–46.3	2.02	1.8	AF
30R	7W	36–38	383.42	–53.6	0.75	1.7	Th
31R	1W	141–143	385.81	–54.6	2.53	2.5	AF
31R	4W	135–137	389.31	–51.3	0.77	2.2	Th
31R	5W	11–13	389.54	–52.4	0.72	0.7	Th
31R	6W	26–28	391.06	–49.5	2.00	1.3	AF
31R	7W	16–18	392.48	–55.0	1.42	1.1	Th
32R	1W	71–73	394.81	–47.4	1.90	2.7	AF
32R	2W	8–10	395.11	–52.9	3.34	0.6	Th
32R	3W	88–90	397.41	–47.7	1.73	1.6	AF
32R	4W	137–139	399.40	–54.3	4.85	1.0	AF
32R	5W	130–132	400.85	–54.5	2.09	1.5	Th
32R	6W	129–131	402.34	–55.2	2.37	0.7	AF
32R	7W	69–71	403.24	–53.3	2.62	0.7	AF
33R	1W	118–120	404.98	–58.6	1.68	3.6	AF
33R	2W	25–27	405.44	–55.5	1.20	2.3	Th

Table 1. *continued*

Core	Section	Interval (cm)	Depth (mbsf)	Inclination (°)	Intensity (A/m)	MAD (°)	Demagnetization type*
33R	3W	22–24	406.51	-57.8	1.73	1.0	AF
33R	4W	11–13	407.90	-52.4	2.78	1.4	AF
33R	5W	113–115	410.13	-58.1	1.89	1.7	Th
33R	6W	105–107	411.55	-54.5	2.33	1.1	AF
34R	2W	25–27	414.99	-54.5	2.46	1.4	Th
34R	3W	122–124	417.14	-54.3	2.55	0.5	AF
34R	4W	59–61	417.79	-55.6	3.48	0.9	AF
34R	5W	28–30	418.98	-55.7	2.40	2.2	AF
34R	6W	12–14	420.17	-46.7	1.92	0.6	AF
34R	7W	7–9	421.27	-54.5	1.00	1.8	Th
34R	8W	63–65	422.78	-47.5	2.45	5.0	AF
35R	1W	95–97	423.97	-53.5	4.67	2.0	Th
35R	2W	17–19	424.72	-51.7	2.92	1.1	AF
35R	3W	108–110	426.86	-53.8	4.67	0.3	AF
35R	4W	13–15	427.38	-56.3	2.82	0.8	AF
35R	5W	20–22	428.68	-53.1	1.09	1.5	Th
35R	6W	63–65	430.29	-46.1	0.35	6.5	AF
35R	7W	30–32	431.46	-47.3	1.26	7.9	AF
35R	8W	11–13	432.09	-52.8	1.71	1.2	Th
36R	1W	50–52	433.40	-60.0	0.12	3.7	AF
36R	2W	56–58	434.90	-52.1	0.59	2.6	Th
36R	3W	26–28	435.85	-48.3	0.12	1.7	AF
36R	4W	80–82	437.69	-57.5	0.29	2.5	Th
36R	5W	119–121	439.59	-48.5	0.12	5.2	AF
36R	6W	96–98	440.86	-54.7	0.11	1.9	AF
36R	7W	10–12	441.10	-37.0	0.05	2.2	AF
37R	1W	3–5	442.63	-57.2	0.02	5.1	Th
37R	2W	100–102	444.62	-48.0	0.67	3.1	Th
37R	3W	2–4	445.14	unstable	0.51		AF
38R	2W	84–86	454.66	-61.0	0.32	1.4	AF
38R	3W	4–6	455.34	-65.5	0.18	4.1	AF
39R	1W	148–150	463.38	-59.9	0.00	2.0	AF
39R	2W	103–105	464.43	-52.1	0.16	2.4	Th
39R	3W	65–67	465.55	unstable	0.15		AF
39R	4W	110–112	467.36	-56.2	0.27	1.6	Th
39R	5W	85–87	468.52	-60.5	0.40	2.6	AF
40R	1W	39–41	471.89	-53.0	0.53	1.1	AF
40R	2W	126–128	474.15	-52.4	0.28	4.3	Th
40R	3W	55–57	474.85	-60.0	0.37	1.9	AF
40R	4W	118–120	476.98	-46.8	0.20	2.1	AF
40R	5W	49–51	477.79	-59.1	0.28	1.3	Th
41R	3W	9–11	483.73	-51.3	0.13	1.9	AF
41R	4W	134–136	486.46	-50.5	0.63	0.5	AF
41R	5W	11–13	486.72	-55.4	0.57	1.4	Th
42R	3W	15–17	493.85	-55.8	0.33	1.2	AF
42R	5W	13–15	496.37	-56.6	0.14	3.1	Th
42R	7W	6–8	498.97	-57.3	0.21	1.9	AF
43R	1W	144–146	501.74	-54.3	0.23	2.4	AF
43R	3W	119–121	503.92	-55.9	0.16	2.6	Th
43R	4W	85–87	505.08	-59.0	0.43	1.5	AF
44R	1W	24–26	510.14	-57.5	0.39	2.0	AF
44R	2W	55–57	511.97	-55.6	0.16	1.3	Th
44R	3W	99–101	513.74	-58.0	0.36	1.2	AF
44R	6W	56–58	517.06	-49.8	3.99	1.4	AF
44R	8W	61–63	519.30	unstable	0.30		Th
45R	1W	110–112	520.60	-55.8	0.29	2.8	AF
45R	2W	83–85	521.48	-59.9	0.31	1.4	Th
45R	4W	65–67	524.10	-50.8	0.16	4.7	Th
45R	5W	12–14	524.94	-58.7	0.27	5.5	AF

*Demagnetization type: AF, alternating field; Th, thermal; mbsf, metres below seafloor.

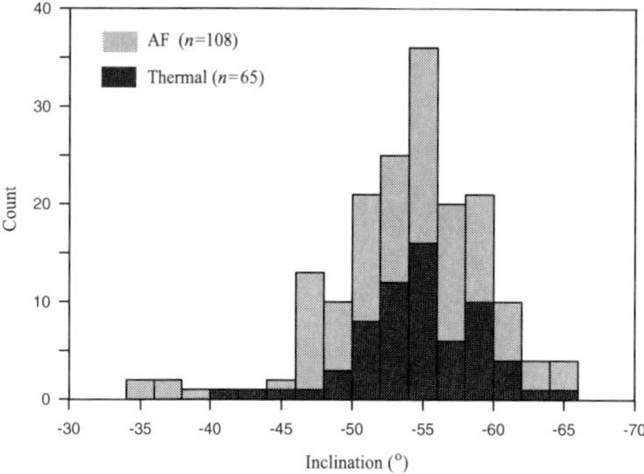

Fig. 4. Histogram of ChRM inclination values derived from AF demagnetization data (light grey) and thermal demagnetization data (black) from Hole 1184A volcaniclastic sediment samples.

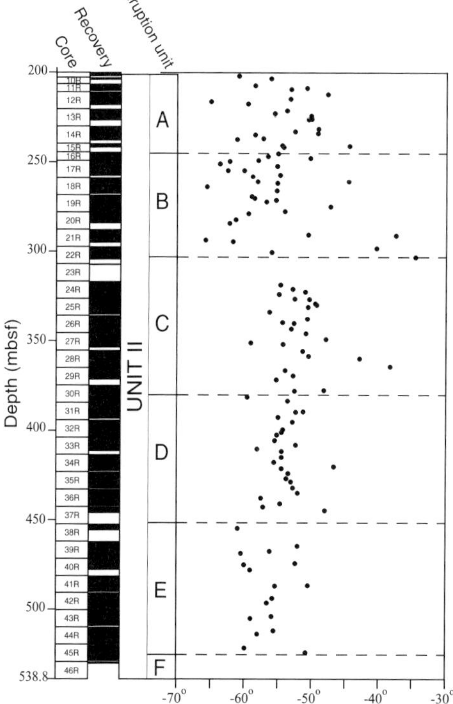

Fig. 5. Site 1184 downhole variation in ChRM inclination.

Early Cretaceous palaeomagnetic pole for the Pacific

Based on the procedure introduced by Cox & Gordon (1984), a new Early Cretaceous

Table 2. *Site mean inclinations and palaeo-colatitudes for OJP and Nauru Basin ODP and DSDP sites*

Site	Site-latitude (°N)	Site-longitude (°E)	*N*/*n*	Inc (°)	α_{95} (°)	*k*	ASD (°)	Age (Ma)	θ (°)	σ (°)	Reference
1184	−5.011	164.233	173/173	−53.9	1.0	109	9.4	121	124.4	*c.* 5	This study
1183	−1.177	157.015	5/39	−46.7	14.3	40	13.9	121	117.9	7.2	Riisager *et al.* (2003*b*)
1185	−0.358	161.668	15/93	−40.8	4.3	74	9.4	121	113.3	2.2	Riisager *et al.* (2003*b*)
1186	−0.680	159.844	7/39	−43.2	6.9	88	8.9	121	115.2	3.5	Riisager *et al.* (2003*b*)
1187	−0.943	161.451	14/99	−39.2	4.5	74	9.3	121	112.2	2.3	Riisager *et al.* (2003*b*)
462	7.242	165.032	15/68	−33.2	7.1	28	14.1	121	108.1	3.6	Steiner (1981); Ogg (1986)
807	3.600	152.600	14/75	−32.9	6.6	34.8	12.7	121	107.9	3.3	Mayer & Tarduno (1993)

N/*n* is number of inclination groups/samples; Inc is the mean inclination of the site; *k* and α_{95} are statistical parameters; ASD is angular standard deviation; θ and σ are palaeo-colatitude and standard deviation, respectively, based on which the palaeomagnetic pole is calculated. Note that the Site 1184 standard deviation is set to 5° to reflect the uncertainty related to post-emplacement tectonic rotation.

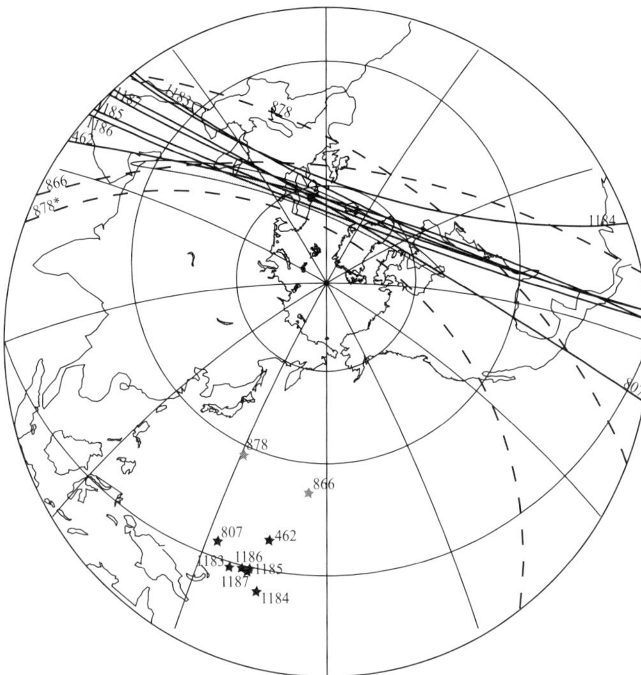

Fig. 6. The ODP drill sites studied (stars) and their corresponding palaeo-colatitudes (small circles). The excellent agreement for the seven different OJP and Nauru Basin drill sites (solid lines) is evident from the palaeo-colatitudes lying close to each other (and crossing each other). The palaeomagnetic pole, 63.0°N, 10.1°E, is based on the intersection of the palaeo-colatitudes. Also shown is the palaeo-colatitudes of ODP Sites 866 and 878 (hatched lines). For Site 878 changing the polarity assignment yields a palaeo-colatitude (878*) in better agreement with the OJP and Nauru Basin data.

palaeomagnetic pole was calculated based on the intersection of the seven OJP and Nauru Basin palaeo-colatitudes (Table 2). The pole, 63.0°N, 10.1°E, has a 95% confidence ellipse with a minor semi-axis of 2.9° with an azimuth of 32°, and a major semi-axis of 47.7° with an azimuth of 122° (Fig. 6). The pole fits all available palaeomagnetic palaeo-colatitudes quite well, with the largest misfit of 7.2° for Site1183 and a low χ^2 value of 2.8 (5 degrees of freedom). The major axis of the 95% confidence ellipse is large due to the near-parallel strike of the palaeo-colatitudes (Fig. 6).

The M-series magnetic lineations in basins east and north of the OJP (Nakanishi *et al.* 1992) show that since its formation the plateau has remained a coherent part of the Pacific plate. Ideally, the OJP and Nauru Basin palaeomagnetic data therefore could be combined with Early Cretaceous palaeo-colatitudes from other parts of the Pacific to obtain a better constrained pole. However, the only other time-averaged Early Cretaceous basaltic palaeomagnetic data, ODP Sites 866 (Tarduno & Sager 1995) and 878 (Nakanishi & Gee 1995), are statistically distinct from the OJP and Nauru Basin data (Fig. 6). The approximately 15° difference between our palaeomagnetic pole and the Site 878 palaeo-colatitude (Fig. 6) is difficult to explain in terms of non-rigidity of the Pacific plate as the difference is an order of magnitude larger than the estimated upper bound, *c.* 2 mm year^{-1}, on present-day intra-plate motion across stable plate interiors (Gordon 1998). The palaeo-colatitude of Site 878 relies on a polarity assignment of negative/positive inclinations to normal/reverse polarity, which is inconsistent with the positive inclinations recorded in early Aptian (i.e. emplaced during the Cretaceous normal superchron) limestone overlying the basalts (Premoli Silva *et al.* 1993). Changing the polarity assignment for Site 878 brings the palaeo-colatitude in much better agreement with the data presented here (Fig. 6). More palaeomagnetic data from Early Cretaceous Pacific basement rocks are necessary to elucidate whether the apparent difference between Early Cretaceous palaeo-colatitudes are

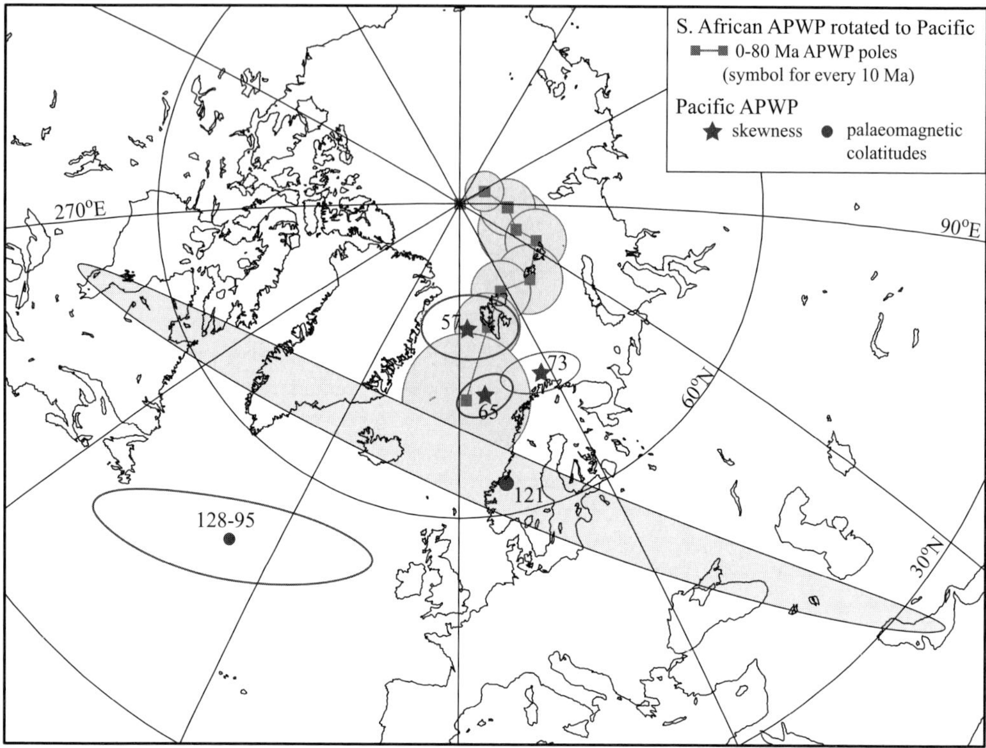

Fig. 7. The new Early Cretaceous Pacific palaeomagnetic pole and corresponding 95% confidence interval (grey area). Also shown are previously published palaeomagnetic poles of the Pacific plate based on magnetic lineation skewness (stars: Acton & Gordon 1991; Petronotis *et al.* 1994; Petronotis & Gordon 1999), and palaeomagnetic palaeo-colatitudes (circle: Tarduno & Sager 1995). Squares represent the 0, 10, 20, 30, 40, 50, 60, 70 and 80 Ma South African APWP poles (Besse & Courtillot 2002) rotated to the Pacific plate using the rotation parameters listed in Norton (2000). Our new *c.* 120 Ma palaeomagnetic pole is statistically distinct at the 95% confidence level from the 128–95 Ma palaeomagnetic palaeo-colatitude pole suggesting that the latter is less reliable than indicated by its 95% confidence ellipse (see text).

genuine. The following discussion will rely solely on the palaeomagnetic pole obtained from OJP and Nauru Basin sites.

Discussion

The Pacific apparent polar wander path

The continual accumulation of palaeomagnetic poles and improved relative plate motion circuits have led to well-defined APWPs for almost all major lithospheric plates (Besse & Courtillot 2002). The Pacific is the main exception because: (i) it is difficult to retrieve ocean-floor basalts suitable for direct palaeomagnetic studies; and (ii) the dominantly convergent plate boundaries make it problematic to transfer palaeomagnetic poles of other plates to the Pacific (Acton & Gordon 1994). The Pacific APWP is therefore traditionally based on indirect palaeomagnetic poles from either seamount magnetic anomaly modelling (Sager & Koppers 2000) and/or magnetic skewness data (Petronotis & Gordon 1999). The palaeomagnetic poles from seamount anomaly modelling may, however, be biased by viscous magnetizations (Gee *et al.* 1989), and recent analyses have cast serious doubt on their reliability (Tarduno & Cottrell 1997; Cottrell & Tarduno 2000; Riisager *et al.* 2003*b*), cautioning against the use of these data to draw geological and geodynamic inferences. We will not consider palaeomagnetic poles from seamount anomaly modelling in the following.

In Figure 7 our new pole is shown together with previously published Pacific palaeomagnetic poles consisting of three magnetic skewness poles (Acton & Gordon 1991; Petronotis *et*

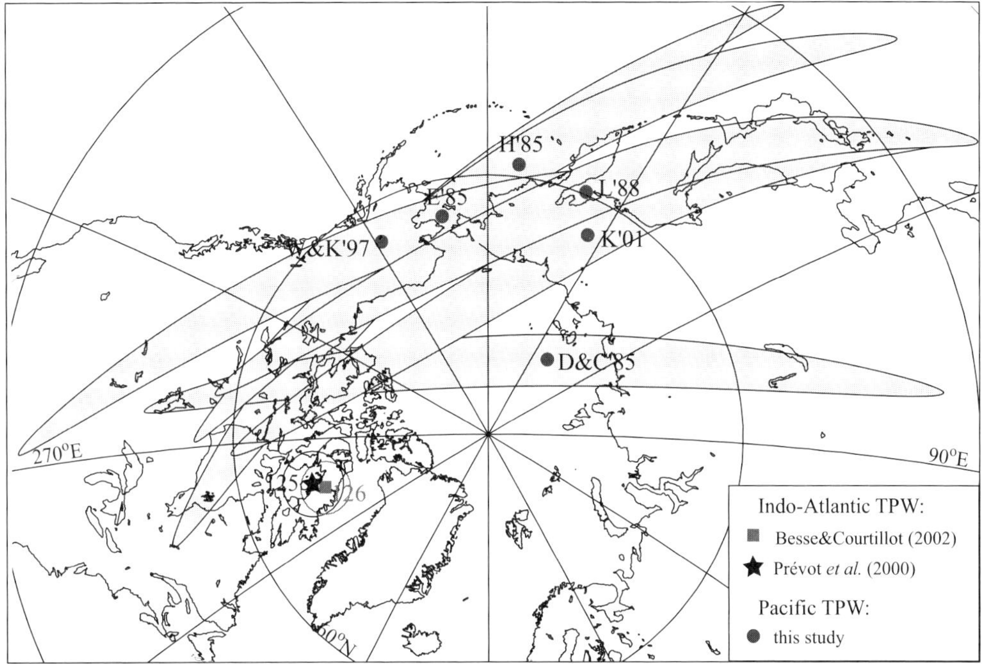

Fig. 8. The new palaeomagnetic pole rotated according to different Pacific hot-spot reconstructions: D&C'85, Duncan & Clague (1985); E'85, Engebretson *et al.* (1985); H'85, Henderson (1985); L'88, Lonsdale (1988); W&K'97, Wessel & Kroenke (1997); and K'01, Koppers *et al.* (2001). Also shown are contemporaneous Indo-Atlantic TPW estimates (Prévot *et al.* 2000; Besse & Courtillot 2002). All Pacific and Indo-Atlantic TPW estimates are statistically distinct on the 95% confidence level documenting relative motion between Pacific and Indo-Atlantic hot spots (see text).

al. 1994; Petronotis & Gordon 1999) and a single pole based on direct palaeomagnetic palaeo-colatitude data from seven short drill cores (Tarduno & Sager 1995). Also shown is the 0–80 Ma part of the South African APWP (Besse & Courtillot 2002) rotated to the Pacific using new improved plate circuits (Cande *et al.* 2000). We interpret the relatively good accordance between the 0–80 Ma rotated APWP and contemporaneous Pacific skewness poles to illustrate the reliability of the latter data set. On the other hand, there is poor agreement between our new pole and the 128–95 Ma palaeomagnetic palaeo-colatitude pole (Fig. 7). This difference may reflect non-averaged palaeosecular variation in the 128–95 Ma pole, where several palaeo-colatitudes are based on very short drill cores with only one to three independent palaeomagnetic field readings. Non-averaged palaeosecular variation is particularly problematic for palaeomagnetic pole determinations yielding too small confidence limits (Riisager *et al.* 2003*a*). Our new pole is relatively closer to the 73 Ma Pacific pole, as well as the rotated South African 80 Ma pole (Fig. 7), and there is therefore no need for the very fast Pacific plate motion drift rate of about 20 cm year^{-1} in the Early Cretaceous suggested by the 128–95 Ma pole (Cottrell & Tarduno 2003).

True polar wander or motion between hot spots?

Assuming the Pacific hot spots have remained fixed, it is possible to determine the past plate motion by backtracking the Pacific plate along its hot-spot tracks (Kroenke *et al.* 2004). Several such 'fixed hot-spot' reconstructions have previously been proposed based on different selections of more or less well-defined Pacific hot-spot tracks (Duncan & Clague 1985; Engebretson *et al.* 1985; Henderson 1985; Lonsdale 1988; Wessel & Kroenke 1997; Koppers *et al.* 2001). In Figure 8 we show our new palaeomagnetic pole rotated according to these different hot-spot reconstructions. None of the reconstructions brings the new palaeomagnetic pole in accordance with the geographic pole, demonstrating motion between the Pacific hot spots and the Earth's spin axis. Differences between palaeomagnetic and hot-spot reconstruction is traditionally interpreted in terms of true polar wander (TPW; i.e. rotation of the entire solid Earth). However, comparing our Pacific TPW poles with the TPW poles obtained from Indo-Atlantic hot spots and palaeomagnetic data (Prévot *et al.* 2000; Besse & Courtillot 2002), it is clear that the Pacific TPW is inconsistent with the Indo-Atlantic one (Fig. 8), supporting the suggestion of significant motion between Pacific and Indo-Atlantic hot spots (Tarduno & Gee 1995; Norton 2000).

Conclusion

New palaeomagnetic data from ODP Site 1184 are in good accordance with previously published time-averaged palaeo-colatitudes from basalts recovered at the OJP and the Nauru Basin. The combined analysis of all OJP and Nauru Basin data gives a palaeomagnetic pole that is far more easterly than previously published Early Cretaceous Pacific palaeomagnetic poles. Based on different Pacific fixed hot-spot reconstructions we calculate Pacific TPW poles, all of which are statistically different from TPW poles obtained in the Indo-Atlantic realm, illustrating motion between the two groups of hot spots.

This research used samples provided by the Ocean Drilling Program (ODP). ODP is sponsored by the US National Science Foundation (NSF) and participating countries under management of Joint Oceanographic Institution (JOI), Inc. Funding for this research was provided by the Danish National Research Foundation (P. Riisager), US Science Support Program of JOI (S. Hall), ODP/Germany Project Number So 72/70-1 (M. Antretter), and NSF grants EAR 443549-22178 and EAR 443747-22250 (X. Zhao). E. G. Laverty and D. Ogden performed parts of the laboratory experiments. We wish to thank R. Gordon and W. Sager for supplying and helping to implement the software for palaeomagnetic pole calculation. We are grateful to J. Tarduno, E. Herrero-Bervera, G. Fitton and J. Mahoney for their constructive comments on the original manuscript. This is contribution number 460 of Center for the Study of Imaging and Dynamics of the Earth, Institute of Geophysics and Planetary Physics at the University of California Santa Cruz.

References

ACTON, G.A. & GORDON, R.G. 1991. A 65 Ma paleomagnetic pole for the Pacific plate from the skewness of magnetic anomlies 27r–31. *Geophysical Journal International*, **106**, 407–420.

ACTON, G.A. & GORDON, R.G. 1994. Paleomagnetic tests of Pacific plate reconstructions and implications for motion between hotspots. *Science*, **263**, 1246–1254.

ANTRETTER, M., RIISAGER, P., HALL, S., ZHAO, X. & STEINBERGER, B. 2004. Modelled palaeolatitudes for the Louisville hot spot and the Ontong Java Plateau. *In*: FITTON, J.G., MAHONEY, J.J., WALLACE, P.J. & SAUNDERS, A.D. (eds) *Origin and Evolution of the Ontong Java Plateau*. Geological Society, London, Special Publications, **229**, 21–30.

BANERJEE, N.R., HONNOREZ, J. & MUEHLENBACHS, K. 2004. Low-temperature alteration of submarine basalts from the Ontong Java Plateau. *In*: FITTON, J.G., MAHONEY, J.J., WALLACE, P.J. & SAUNDERS, A.D. (eds) *Origin and Evolution of the Ontong Java Plateau*. Geological Society, London, Special Publications, **229**, 259–273.

BESSE, J. & COURTILLOT, V. 2002. Apparent and true polar wander and the geometry of the geomagnetic field in the last 200 Myr. *Journal of Geophysical Research*, **107**, 2300, 10.1029/2000JB000050.

CANDE, S.C., STOCK, J.M., MÜLLER, D. & ISHIHARA, T. 2000. Cenozoic motion between East and West Antarctica. *Nature*, **404**, 145–150.

CASTILLO, P.R., PRINGLE, M.S. & CARLSON, R.W. 1994. East Mariana basin tholeiites – Cretaceous intraplate basalts of rift basalts related to the Ontong Java Plume. *Earth and Planetary Science Letters*, **123**, 139–154.

CHAMBERS, L.M., PRINGLE, M.S. & FITTON, J.G. 2002. Age and duration of magmatism on the Ontong Java Plateau: $^{40}Ar/^{39}Ar$ results from ODP Leg 192. *Eos Transactions of the American Geophysical Union*, **83**, Fall Meeting Supplement, Abstract V71B-1271.

COFFIN, M.F. & ELDHOLM, O. 1994. Large igneous provinces: crustal structure, dimensions, and external consequences. *Reviews of Geophysics*, **32**, 1–36.

COTTRELL, R.D. & TARDUNO, J.A. 2000. Late Cretaceous true polar wander: Not so fast, *Science*, **288**, 2283a.

COTTRELL, R.D. & TARDUNO, J.A. 2003. A Late Cretaceous pole for the Pacific plate: implications for apparent and true polar wander and the drift of hotspots. *Tectonophysics*, **362**, 321–333.

COX, A. & GORDON, R.G. 1984. Paleolatitudes determined from paleomagnetic data from vertical cores. *Reviews of Geophysics and Space Physics*, **22**, 47–72.

DUNCAN, R.A. & CLAGUE, D.A. 1985. Pacific plate motion recorded by linear volcanic chains. *In*: NAIRN, A.E.A., STEHLI, F.L. & UYEDA, S. (eds) *The Ocean Basins and Margins, The Pacific Ocean 7A*. Plenum Press, New York, 89–121.

DUNLOP, D.J. 2002. Theory and application of the Day plot (Mrs/Ms versus Hcr/Hc) 1. Theoretical curves and tests using titanomagnetite data. *Journal of Geophysical Research*, **107**, 2056, 10.1029/2001JB000486.

ELDHOLM, O. & COFFIN, M. 2000. Large igneous provinces and plate tectonics. *In*: RICHARDS, M.A.

et al. (eds), *The History and Dynamics of Global Plate Motions*. American Geophysical Union, Geophysical Monograph, **121**, 309–326.

ENGEBRETSON, D.C., COX, A. & GORDON, R.G. 1985. *Relative Motions Between Oceanic and Continental Plates in the Pacific Basin*. Geological Society of America, Special Paper, **206**.

GEE, J., STAUDIGEL, H. & TAUXE, L. 1989. Contributions of induced magnetization to magnetization of seamounts. *Nature*, **342**, 170–173.

GORDON, R.G. 1998. The plate tectonic approximation: Plate nonrigidity, diffuse plate boundaries, and global plate reconstructions. *Annual Review of Earth and Planetary Sciences*, **26**, 615–642.

HENDERSON, L.J. 1985. *Motion of the Pacific plate relative to the hotspots since the Jurassic and model of oceanic plateaus of the Farallon Plate*. Ph.D. thesis, Northwestern University, USA.

KIRSCHVINK, J.L. 1980. The least squares line and plane and the analysis of paleomagnetic data. *Geophysical Journal of the Royal Astronomical Society*, **62**, 699–718.

KOPPERS, A.A.P., MORGAN, J.P., MORGAN, J.W. & STAUDIGEL, H. 2001. Testing the fixed hotspot hypothesis using Ar-40/Ar-39 age progressions along seamount trails. *Earth and Planetary Science Letters*, **185**, 237–252.

KROENKE, L.W., WESSEL, P. & STERLING, A.I. 2004. Motion of the Ontong Java Plateau in the hot-spot frame of reference: 120 Ma–present. *In*: FITTON, J.G., MAHONEY, J.J., WALLACE, P.J. & SAUNDERS, A.D. (eds) *Origin and Evolution of the Ontong Java Plateau*. Geological Society, London, Special Publications, **229**, 9–20.

LONSDALE, P. 1988. Geography and history of the Louisville hotspot chain in the southwest Pacific. *Journal of Geophysical Research*, **93**, 3078–3104.

MAHONEY, J.J., FITTON, J.G., WALLACE, P.J. *et al.* 2001. *Proceedings of the Ocean Drilling Program, Initial Reports*, **192**.

MAYER, H. & TARDUNO, J.A. 1993. Paleomagnetic investigation of the igneous sequence, Site 807, Ontong Java plateau, and a discussion of Pacific true polar wander. *Proceedings of the Ocean Drilling Program, Scientific Results*, **130**, 51–59.

MCFADDEN, P.L. & REID, A.B. 1982. Analysis of paleomagnetic inclination data. *Geophysical Journal of the Royal Astronomical Society*, **69**, 307–319.

MOBERLY, R. & SCHLANGER, S.O. *et al.* 1986. *Initial Reports of the Deep Sea Drilling Project*, **89**, 157–191.

NAKANISHI, M. & GEE, J.S. 1995. Paleomagnetic investigations of volcanic rocks: paleolatitudes of the northwestern Pacific guyots. *Proceedings of the Ocean Drilling Program, Scientific Results*, **144**, 585–604.

NAKANISHI, M., TAMAKI, K. & KOBAYASHI, K. 1992. Magnetic anomaly lineations from late Jurassic to early Cretaceous in the west-central Pacific Ocean. *Geophysical Journal International*, **109**, 701–719.

NEAL, C.R., MAHONEY, J.J., KROENKE, L.W., DUNCAN, R.A. & PETTERSON, M.G. 1997. The Ontong Java Plateau. *In*: MAHONEY, J.J. & COFFIN, M.F. (eds) *Large Igneous Provinces: Continental, Oceanic, and Planetary Flood Basalts*. American Geophysical Union, Geophysical Monograph, **100**, 183–216.

NORTON, I. 2000. Global hotspot reference frames and plate motion. *In*: RICHARDS, M.A., GORDON, R.G. & VAN DER HILST, R.D. (eds) *The History and Dynamics of Global Plate Motions*. American Geophysical Union, Geophysical Monograph, **121**, 339–357.

OGG, J.G. 1986. Paleolatitudes and magnetostratigraphy of Cretaceous and lower Tertiary sedimentary rocks, Deep Sea Drilling Project Site 585, Marian Basin, Western Central Pacific. *Initial Reports of the Deep Sea Drilling Project*, **89**, 629–645.

OZIMA, M., SAITO, K. & TAKIGAMI, Y. 1981. 40Ar/39Ar geochronological studies on rocks drilled at Holes 462 and 462A. *Initial Reports of the Deep Sea Drilling Project*, **61**, 701–703.

PARKINSON, I.J., SCHAEFER, B.F. & ODP LEG 192 SHIPBOARD SCIENTIFIC PARTY. 2001. A lower mantle origin for the world's biggest LIP? A high precision Os isotope isochron from Ontong Java Plateau basalts drilled on ODP Leg 192. *Eos, Transactions of the American Gophysical Union*, **82**, Fall Meeting Supplement, Abstract V51C-1030.

PETRONOTIS, K.E. & GORDON, R.G. 1999. A Maastrichtian palaeomagnetic pole for the Pacific plate from skewness analysis of marine magnetic anomaly 32. *Geophysical Journal International*, **139**, 227–247.

PETRONOTIS, K.E., GORDON, R.G. & ACTON, G.D. 1994. A 57 Ma Pacific plate paleomagnetic pole determined from a skewness analysis of crossings of marine magnetic anomaly 25r. *Geophysical Journal International*, **118**, 529–554.

PREMOLI SILVA, I., HAGGERTY, J.A., RACK, F. *et al.* 1993. *Proceedings of the Ocean Drilling Program, Initial Reports*, **144**.

PRÉVOT, M., MATTERN, E., CAMPS, P. & DAIGNIERES, M. 2000. Evidence for a 20° tilting of the Earth's rotation axis 110 million years ago. *Earth and Planetary Science Letters*, **179**, 517–528.

RICHARDS, M.A., DUNCAN, R.A. & COURTILLOT, V.E. 1989. Flood basalts and hot-spot tracks – Plume heads and tails. *Science*, **246**, 103–107.

RIISAGER, J., RIISAGER, P. & PEDERSEN, A.K. 2003*a*. Paleomagnetism of large igneous provinces: case-study from West Greenland, North Atlantic igneous province. *Earth and Planetary Science Letters*, **214**, 409–425.

RIISAGER, P., HALL, S., ANTRETTER, M. & ZHAO, X. 2003*b*. Paleomagnetic paleolatitude of Early Cretaceous Ontong Java Plateau basalts: Implications for Pacific apparent and true polar wander. *Earth and Planetary Science Letters*, **208**, 235–252.

SAGER, W.W. & KOPPERS, A.A.P. 2000. Late Cretaceous polar wander of the Pacific plate: Evidence of a rapid true polar wander event. *Science*, **287**, 455–459.

SANDWELL, D.T. & SMITH, W.H.F. 1997. Marine gravity anomaly from Geosat and ERS-1 satellite altimetry. *Journal of Geophysical Research*, **102**, 10 039–10 054.

STEINER, M.B. 1981. Paleomagnetism of the igneous complex, Site 462. *Initial Reports of the Deep Sea Drilling Project*, **61**, 717–729.

TAKIGAMI, Y., AMARI, S., OZIMA, M. & MOBERLY, R. 1986. 40Ar/39Ar geochronological studies of basalts from Holes 462A. *Initial Reports of the Deep Sea Drilling Project*, **89**, 519–522.

TARDUNO, J.A. & COTTRELL, R.D. 1997. Paleomagnetic evidence for motion of the Hawaiian hotspot during formation of the Emperor seamounts. *Earth and Planetary Science Letters*, **153**, 171–180.

TARDUNO, J.A. & GEE, J. 1995. Large-scale motion between Pacific and Atlantic hotspots. *Nature*, **378**, 477–480.

TARDUNO, J.A. & SAGER, W.W. 1995. Polar standstill of the Mid-Cretaceous Pacific plate and its geodynamic implications. *Science*, **269**, 956–959.

TARDUNO, J.A., SLITER, W.V., KROENKE, L., LECKIE, M., MAYER, H., MAHONEY, J.J., MUSGRAVE, R., STOREY, M. & WINTERER, E.L. 1991. Rapid formation of Ontong Java Plateau by Aptian mantle plume volcanism. *Science*, **254**, 399–403.

TEJADA, M.L.G., MAHONEY, J.J., NEAL, C.R., DUNCAN, R.A. & PETTERSON, M.G. 2002. Basement geochemistry and geochronology of central Malaita, Solomon islands, with implications for the origin and evolution of the Ontong Java Plateau. *Journal of Petrology*, **43**, 449–484.

THORDARSON, T. 2004. Accretionary-lapilli-bearing pyroclastic rocks at ODP Leg 192 Site 1184: a record of subaerial phreatomagmatic eruptions on the Ontong Java Plateau. *In*: FITTON, J.G., MAHONEY, J.J., WALLACE, P.J. & SAUNDERS, A.D. (eds) *Origin and Evolution of the Ontong Java Plateau*. Geological Society, London, Special Publications, **229**, 275–306.

WESSEL, P. & KROENKE, L. 1997. A geometric technique for relocating hotspots and refining absolute plate motions. *Nature*, **387**, 365–369.

ZHAO, X., ANTRETTER, M., RIISAGER, P. & HALL, S. 2004. Rock magnetic results from Ocean Drilling Program Leg 192: Implications for Ontong Java Plateau emplacement and tectonics of the Pacific. *In*: FITTON, J.G., MAHONEY, J.J., WALLACE, P.J. & SAUNDERS, A.D. (eds) *Origin and Evolution of the Ontong Java Plateau*. Geological Society, London, Special Publications, **229**, 45–61.

Rock magnetic results from Ocean Drilling Program Leg 192: implications for Ontong Java Plateau emplacement and tectonics of the Pacific

XIXI ZHAO[1], MARIA ANTRETTER[2], PETER RIISAGER[1,3] & STUART HALL[4]

[1]*Center for Study of Imaging and Dynamics of the Earth, Institute of Geophysics and Planetary Physics, University of California, Santa Cruz, CA 95064, USA (e-mail: xzhao@es.ucsc.edu)*

[2]*Institut für Geophysik, University of München, Theresienstrasse 41, D-80333 München, Germany*

[3]*Danish Lithosphere Centre, Øster Voldgade 10, DK-1350 Copenhagen K, Denmark*

[4]*Department of Geosciences, University of Houston, Houston, TX 77204–5007, USA*

Abstract: A rock magnetic study has been performed on rock samples recovered at Ocean Drilling Program Leg 192 sites on the Ontong Java Plateau in the western Pacific. Igneous rocks from the five Leg 192 sites displayed variable rock magnetic properties. The differences in the rock magnetic properties are a function of mineralogy and alteration. Titanomagnetite and titanomaghemite are present in the Ontong Java rocks. Samples with titanomagnetite exhibit Verwey transitions in the vicinity of 120K. Low-temperature curves for samples with multiple magnetic phases do not clearly show the Verwey transition. The inversion of titanomaghemite to a strongly magnetized magnetite is shown by the irreversible thermomagnetic-cooling curve. Despite the geographically widespread locations of the drill sites, variations in rock magnetic properties closely resemble each other, consistent with the fundamental results of the leg that the basement rocks were derived from homogeneous Kwaimbaita-type magma with a single age of approximately 120 Ma. The rock magnetic investigation provides constraints to evaluate the fidelity of the natural magnetic memory in the basalt rocks and corroborates the palaeomagnetic palaeolatitudes determinations for the Ontong Java Plateau. The generally good quality of rock magnetic data exhibited by Leg 192 rocks supports the inference that the characteristic directions of magnetization isolated from the Cretaceous Ontong Java Plateau sites were acquired near the onset of the Cretaceous Long Normal Superchron about 120 Ma. The portion of the Pacific plate containing the Leg 192 sites was in the southern hemisphere during the mid-Cretaceous volcanism.

The mid-Cretaceous (*c.* 124–90 Ma, please note: the word 'mid' here refers to mid, not Middle) was a period of great unrest in global magmatism, with enormous production of large igneous provinces (LIPs) and seamounts (e.g. Larson 1991; Anderson 1994). Two of the most voluminous LIPs in the world, the Ontong Java Plateau (OJP) in the western Pacific and Kerguelen Plateau–Broken Ridge in the southern Indian Ocean, erupted during this period. The volcanic carbon dioxide emissions released by the Cretaceous LIPs may have helped to cause the unusual global warmth in Late Cretaceous time (e.g. Tarduno *et al.* 1998) and triggered from two to seven oceanic anoxic events (OAEs) during the mid-Cretaceous (Erbacher & Thurow 1997; Leckie *et al.* 2002). Geological evidence indicates that during the mid-Cretaceous climate was considerably warmer, sea level was significantly higher and extinctions were more common (e.g. Coffin & Eldholm 1994; Tarduno *et al.* 1998; Eldholm & Coffin 2000). In addition, Earth's magnetic field was uncharacteristically steady, in that it did not switch from normal to reversed polarity for approximately 40 Ma (Aptian–Santonian, 121–83 Ma), and some researchers argue that true polar wander rates were higher (Prévot *et al.* 2000; Sager & Koppers 2000). These remarkable geological and geophysical signals in the Cretaceous have excited great interest in the geoscience community, leading many to suggest a connection between all of these phenomena and deep-mantle convection (e.g. Vogt 1975; Sheridan 1983; Courtillot & Besse 1987; Gubbins 1987; Larson 1991; Larson & Olson 1991; Heller *et al.* 1996).

The Ocean Drilling Program (ODP) and its predecessor the Deep Sea Drilling Project

From: FITTON, J. G., MAHONEY, J. J., WALLACE, P. J. & SAUNDERS, A. D. (eds) 2004. *Origin and Evolution of the Ontong Java Plateau*. Geological Society, London, Special Publications, **229**, 45–61. 0305-8719/$15.00

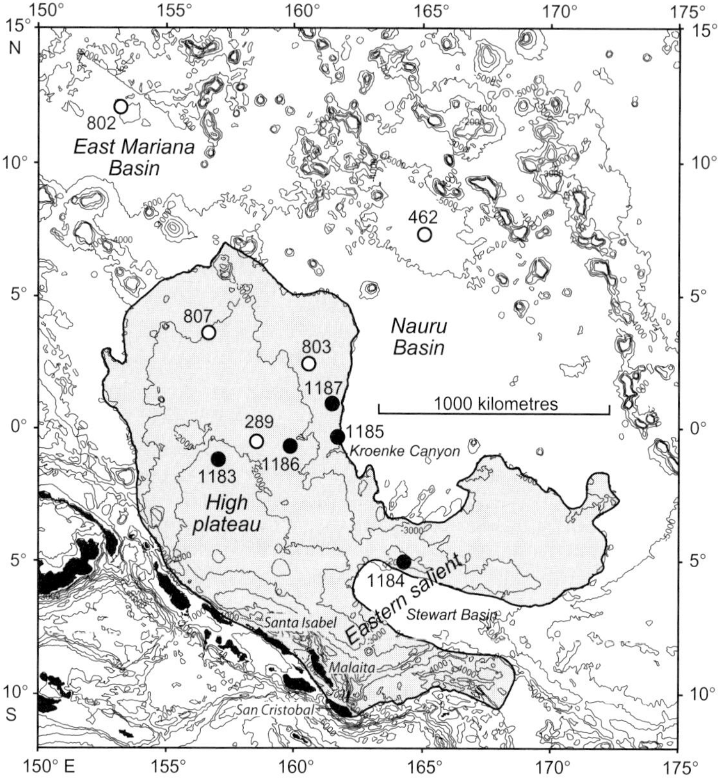

Fig. 1. Bathymetry map of the Ontong Java Plateau and surrounding areas showing the location of DSDP and ODP basement drill sites. Leg 192 drill sites are marked by black circles; open circles represent pre-Leg 192 drill sites. The edge of the plateau is defined by the –4000 m-contour, except in the SE part where it has been uplifted through collision with the Solomon arc (after Fitton & Godard 2004).

(DSDP) have made critical and irreplaceable contributions to understanding this list of Cretaceous phenomena and provided a quantum leap in our understanding of oceanic LIPs. In particular, ODP Leg 192 was especially designed to investigate the history of the OJP in the western Pacific, which is the largest oceanic LIP in the world (Fig. 1). Understanding the origin and evolution history of the OJP is of particular importance because it is the best manifestation of the mid-Cretaceous volcanism in the Pacific. Many researchers have favoured the model that the initial plume head of the Louisville hot spot (now at *c.* 51°S) was the source of the OJP (e.g. Mahoney & Spencer 1991; Richards *et al.* 1991; Tarduno *et al.* 1991; Phinney *et al.* 1999). Other scientists, opposing the idea of a deep-mantle plume source, have suggested that rapid growth and reorganization of the Pacific plate drove the mid-Cretaceous volcanism (e.g. Anderson 1994). Despite the huge size of the OJP (larger than Alaska), and its potential role in contributing to our understanding of mantle circulation and environmental change in the past, it is still a poorly understood lithospheric feature (Mahoney & Coffin 1997).

Prior to ODP Leg 192, published $^{40}Ar–^{39}Ar$ results suggested that the OJP formed largely in two discrete episodes at about 122 and 90 Ma, while samples from obducted sections on the Solomon Islands also indicate minor igneous activity at about 60, 44 and 34 Ma (Tejada *et al.* 2002 and references therein). Leg 192 recovered basement and sediment cores at five widely spaced sites in previously unsampled areas of the OJP (Fig. 1). Newly reported whole-rock $^{40}Ar–^{39}Ar$ analyses on the basaltic basement indicate that all Leg 192 basalts are approximately 120 Ma (Chambers *et al.* 2002), which is in excellent accordance with a Re–Os isochron age of 121.5 ± 1.7 Ma derived from the basement sites (Parkinson *et al.* 2001). Using these constraints, it now appears that an immense part of the OJP may have formed in a single event at

Table 1. *Simplified summary of Leg 192 drill sites that recovered volcanic rocks*

Site Hole	Location	Coordinates	Water depth (m)	Basement recovery (m)	Lithological description	Number of units	Age (Ma)*
1183A	Crest of the main OJP	1°10.6189′S 157°0.8988′E	1804.7	44.20	Submarine basaltic lava flows	8	120–122
1184A	Eastern Lobe	5°0.6653′S 164°13.9771′E	1661.5	278.90	Submarine volcaniclastic rocks	5	*c.* 120
1185A	Eastern edge of the main OJP	0°21.4560′S 161°40.0619′E	3898.9	11.17	Submarine pillow basalt	5	120–122
1185B	Eastern edge of the main OJP	0°21.4559′S 161°40.0511′E	3898.9	90.68	Submarine pillow basalt	12	120–122
1186A	Eastern slope of the main OJP	0°40.7873′S 159°50.6519′E	2728.7	39.36	Submarine pillow basalts	4	120–122
1187A	Eastern edge of the main OJP, 146 km north of Site 1185	0°56.5518′N 161°27.0784′E	3803.6	100.87	Submarine pillow lava flows	12	120–122

*New basement ages reported by Parkinson *et al.* (2001) and Chambers *et al.* (2002).

about 120 Ma, in contrast with the Kerguelen Plateau–Broken Ridge in the southern Indian Ocean, where the plateau appears to have formed in several intense episodes of volcanism over a long period of time (Coffin *et al.* 2000).

The original palaeolatitude of the OJP is still debatable, as no well-defined magnetic lineations have been found on the plateau. In the plate reconstruction of Neal *et al.* (1997), the OJP is placed at approximately 40°S in the early Aptian. During our shipboard and post-cruise studies for Leg 192, we have characterized the stable components of natural remanent magnetization of basalt flows at the Leg 192 drill sites, which *all* have negative inclinations, suggesting that the entire basalt sequence is of normal magnetic polarity (southern hemisphere). The normal magnetic polarity is consistent with eruption during the Cretaceous Normal Superchron. A mean palaeolatitude of 24°S for the OJP at the time of emplacement was also obtained (Riisager *et al.* 2003, 2004). This value is about 20° north of the latitude suggested by Neal *et al.*'s (1997) and Kroenke *et al.*'s (2004) reconstructions in an assumed fixed hot-spot reference frame (Engebretson *et al.* 1985; Lonsdale 1988; Wessel & Kroenke 1997; Koppers *et al.* 2001).

Rock magnetic properties can be used to correctly identify the magnetic mineralogy and particle sizes, to subdivide or to correlate different cooling units, to explore the relationship between magnetic properties and lithology and alteration, and to help assess the origin and stability of remanent magnetization used for establishing the palaeolatitudes of the OJP. In this paper, we present the results of such a rock magnetic investigation carried out on cores recovered from the Leg 192 sites, concentrating on the basement rocks. Additional work on samples of these sites is presented by Antretter *et al.* (2004) and Riisager *et al.* (2004). We will first briefly introduce some background information about the Leg 192 drill sites, and then describe the results of rock magnetic measurements. We then summarize the data and explore their implications for understanding the volcanic and tectonic history of the OJP.

Site setting and basement lithology

Igneous basement was penetrated at five drill sites during Leg 192. Four of the five sites (1183, 1185, 1186 and 1187) successfully reached basaltic basement that comprised a thick sequence of basaltic pillow and massive lavas. In contrast, a sequence of volcaniclastic rocks was recovered from Site 1184. The site locations and drilling results are documented in detail in the site chapters of the Leg 192 *Initial Reports* volume (see Mahoney *et al.* 2001). In the interest of brevity, we only briefly mention the basement lithology and palaeoenvironment of the drill sites here. A simplified summary of the drill sites and their ages is given in Table 1.

Site 1183 is located in the crest of the main OJP (Fig. 1). Drilling at Site 1183 recovered eight basaltic units, ranging in thickness from

0.36 to 25.70 m (Mahoney *et al.* 2001). All eight units contain pillow basalt with several thin limestone interbeds containing Aptian microfossils. Except near pillow margins, the basalt is essentially non-vesicular, implying a palaeodepth greater than 800 m (Moore & Schilling 1973). Shipboard geochemical analyses show that the basalt flows at Site 1183 are tholeiitic and very similar in composition to those forming the >2.7 km-thick Kwaimbaita Formation on Malaita, nearly 1000 km to the south (Fig. 1). Site 1183 was the only Leg 192 site at which much of the Miocene–Aptian sedimentary succession was cored.

At Site 1184 on the unnamed northern ridge of the eastern salient of the OJP, drilling penetrated 337.7 m of a volcaniclastic sequence, which contains five subunits of coarse lithic vitric tuff, lapilli tuff and lapillistone. Deep subaerial weathering at the top of the volcaniclastic section, coupled with a zeolite assemblage typically formed in non-marine environments, indicates that this part of the eastern salient was above sea level initially. Proximity to land also is suggested by wood fragments found in organic-rich ash layers in the volcaniclastic sequence (Mahoney *et al.* 2001). Compositionally, the volcaniclastic rocks are similar to Kwaimbaita-type basalt. Initial shipboard biostratigraphy suggested that the volcaniclastic rocks from Site 1184 are within Zone NP16 (or 41–43 Ma); however, a newly completed post-cruise study indicates these volcaniclastic rocks are very probably Cretaceous (Chambers *et al.* 2004), which makes the volcaniclastic sequence recovered at Site 1184 a major new research focus to test whether its magnetic profile (and palaeolatitude) resembles those of the basement sites.

Site 1185 was drilled on the eastern edge of the main OJP, at the northern side of the Kroenke Canyon, which is an enormous submarine canyon that extends from Ontong Java and Nukumanu atolls into the Nauru Basin (Fig. 1). Two holes were drilled at Site 1185 (Table 1). In Hole 1185A, 16.7 m of pillow basalt was cored. On the basis of apparent limestone interbeds, five eruptive units were defined. In Hole 1185B (20 m west of Hole 1185A) 216.6 m of basaltic basement were penetrated, and the section was divided into 12 units ranging in thickness from 1 to 65 m. Units 1, 3, 4 and 6–9 were identified as pillow basalt on the basis of glassy rims and grain-size variations, and Units 2 and 5 are more massive lava flows with pillowed tops and bases. Units 1–9 are separated by thick (as much as 70 cm) intervals of hyaloclastite breccia and Units 10–12 are massive flows. Seawater-derived fluids have interacted pervasively at low temperatures with the basaltic basement, and the basement section can be divided into two groups of flows with different alteration characteristics. One group consists of all the basement units of Hole 1185A and Units 1–9 of Hole 1185B. Alteration in these units occurred under highly oxidizing conditions and with high water/rock ratios. The second group includes Units 10–12 at Hole 1185B. The brecciated top of Unit 10 consists of pervasively altered angular basalt fragments. Such severe alteration is likely to be the result of exposure of very permeable basaltic seafloor to bottom sea water for an extended period (Mahoney *et al.* 2001; Banerjee *et al.* 2004).

Site 1186 is situated on the eastern slope of the main OJP, 206 km west of Site 1185 (Fig. 1). A total of 65.4 m of basement was penetrated in Hole 1186A, which consists of basalt lava flows with thin interbeds of sandstone, limestone and conglomerate. On the basis of limestone and hyaloclastite interbeds and downward changes in character from massive to pillowed, four units were defined, ranging from 10 to >26 m in thickness. Basement Unit 1 consists entirely of pillow lava, whereas Units 2–4 have massive interiors. The entire section of basaltic basement at Site 1186 has undergone low-temperature water–rock interactions. The overall alteration and chemical composition of the basalt are similar to that in the lower group of basalt flows at Site 1185, and especially to that at Site 1183. Shipboard chemical analysis also showed the basalt from Site 1186 to be of the Kwaimbaita type.

Site 1187, which also lies on the eastern edge of the main plateau, is 146 km north of Site 1185 (Fig. 1). Drilling recovered 135.8 m of basaltic basement, which includes 12 units (ranging in thickness from 0.7 to 41.3 m). Most of the sequence consists of pillow-lava flows that are compositionally indistinguishable from the basalt forming the upper groups of flows at Site 1185 (Mahoney *et al.* 2001; Fitton & Godard 2004). The only unequivocally massive portion is the fine-grained, 9 m-thick base of Unit 6. The level of basaltic alteration at Site 1187 is generally greater than at other Leg 192 sites, although fresh glass is present in many of the pillow rims. Overall, the secondary mineral assemblages and characteristics of basalt alteration at all Leg 192 sites strongly resemble those of typical ocean crust formed at spreading centres (Mahoney *et al.* 2001; Banerjee *et al.* 2004).

Laboratory methods

A total of 136 discrete palaeomagnetic samples were used for shore-based rock magnetic studies. The rock magnetic data presented in this paper are from measurements performed at the palaeomagnetism laboratories at the University of California at Santa Cruz (UCSC), and at the Institute for Rock Magnetism of the University of Minnesota. For rock magnetic characterization, samples were subjected to a wide range of magnetic measurements. These included: (1) Curie temperature determinations using both low and high applied fields (0.05 mT and 1 T, respectively); (2) hysteresis loop parameters measurement: saturation magnetization (J_s), saturation remanence (J_r), coercivity (H_c) and remanent coercivity (H_{cr}) determined from 50 to 300K; and (3) saturation isothermal remanent magnetization as a function of temperature (19–300K). A brief description of each experiment is given below.

Curie temperatures were determined by measurement of low-field magnetic susceptibility or induced moment v. temperature (using both the Kappabridge susceptometer at the University of California at Santa Cruz and the Princeton MicroMag vibrating sample magnetometer at the University of Minnesota). To avoid oxidation that could lead to chemical alteration, we conducted thermomagnetic analyses in an inert helium or argon atmosphere on samples chosen to be representative of the Leg 192 cores. We used a graphic method (Grommé *et al.* 1969) to determine the Curie temperature; the method uses the intersection of two tangents to the thermomagnetic curve that bounds the Curie temperature. This method is most straightforward to do by hand, even though it tends to underestimate Curie temperatures compared with the two other methods presented by Moskowitz (1981) and Tauxe (1998).

In this study, hysteresis loops and the associated parameters J_r/J_s, H_c and H_{cr} were obtained using alternating gradient magnetometers (AGFM; Princeton Measurements Corporation) capable of resolving magnetic moments as small as 5×10^{-8} emu (Flanders 1988). Saturation magnetization (J_s) is the largest magnetization a sample can have. The coercivity (H_c) is a measure of magnetic stability. The two ratios, J_r/J_s and H_{cr}/H_c, are commonly used as indicators of domain states and, indirectly, grain size. For magnetite, high values of J_r/J_s (>0.5) indicate small (<0.1 μm or so) single-domain (SD) grains, and low values (<0.1) are characteristic of large (>15–20 μm) multi-domain grains (MD). The intermediate regions are usually referred to as pseudo-single domain (PSD). H_{cr}/H_c is a much less reliable parameter, but conventionally SD grains have a value close to 1.1, and MD grains should have values >3–4 (Day *et al.* 1977; Dunlop 2002). For a selected group of samples we also examined the change of hysteresis loops during warming from 50 to 300K (at intervals of 25K).

Low-temperature measurements were made on 33 representative samples to help characterize the magnetic minerals and understand their rock magnetic properties. These measurements were designed to determine the Néel temperature and other critical temperatures of a magnetic substance, and were made from 10K to room temperature on 100–300 mg subsamples in a Quantum Design Magnetic Property Measurement System (MPMS) at the University of Minnesota. Samples were given a saturation isothermal remanent magnetization (SIRM) in a steady magnetic field of 2.5 T at room temperature (300K) and then cooled in a zero field to 19K, during which the remanence was measured at 5K intervals. The sample was then given a SIRM in a field of 2.5 T before warming it to 300K in zero field, while measuring the remanence value every 5K in sweep measurement fashion. Unlike high-temperature measurements, low-temperature measurement carries no risk of oxidation of a sample.

Results

Curie temperature determination of samples

Curie temperature determinations of samples from all sites are presented in Table 2. Strong field thermomagnetic curves were obtained to determine the magnetic phases in the samples. According to Curie temperatures (Table 2), three different groups of Leg 192 samples can be recognized. Group 1 (Fig. 2A) is characterized by a single ferromagnetic phase with Curie temperatures between 480 and 580°C, compatible with that of Ti-poor titanomagnetites. The cooling and heating curves are reasonably reversible. Most flows from Site 1185 belong to this group. Several rocks from Site 1184 also show this behaviour. Group 2 has lower Curie temperatures (260–280°C) that are typical of Ti-rich titanomagnetite (such as TM60) or low-temperature oxidized titanomaghemites. Group 2 curves (Fig. 2B) were mainly observed for pillow basalts from Site 1187. Other rock samples that can be included in this group are from the lower parts of Site 1185, which exhibit similar low Curie temperatures. Samples belonging to Group 3 have multiple magnetic phases (Fig. 2C). The irreversible thermomagnetic curve of Sample 192-1185A-24R-2, 79–81 cm, displays

Table 2. *Summary of Curie and Verwey transition temperatures of minicore samples from the Leg 192 sites*

Core, section, interval (cm)	Depth (mbsf)	T_c (°C) (Kappa bridge)	T_c (°C) (Micro VSM)	T_b (K)	T_v (K)	Rock type	Alteration
192–1183A-							
54R-5, 36–38	1132.49	570		30	c	Pillowed basalt	Moderate
55R-3, 131–133	1140.45	460				Pillowed basalt	Slight–moderate
57R-1, 131–133	1152.31	580				Pillowed basalt	Moderate
59R-1, 60–62	1161.2	575		52	120	Pillowed basalt	Slight–moderate
60R-1, 136–138	1167.86	485				Pillowed basalt	Slight–moderate
62R-1, 63–65	1181.63	560				Pillowed basalt	Slight
64R-1, 81–83	1192.81	570				Pillowed basalt	Slight–moderate
65R-1, 45–47	1195.85	580				Massive basalt	Slight
66R-3, 12–14	1203.06	570		50	105	Massive basalt	Slight
67R-1, 2–4	1204.92	580				Massive basalt	Slight–moderate
67R-3, 52–54	1208.36			32	108	Massive basalt	Slight–moderate
68R-1, 40–42	1210.10	480		45	120	Massive basalt	Slight–moderate
192-1184A-							
12R-5, 73–75	217.29	590				Lithic vitric tuff	
17R-4, 99–101	254.52	550				Lithic vitric tuff	
19R-3, 61–63	271.22			29	118	Lithic vitric tuff	
19R-8, 17–19	277.52	570				Lithic vitric tuff	
24R-5, 117–119	322.42			30	115	Vitric lithic tuff	
24R-8, 116–118	326.51	565				Lithic vitric tuff	
30R-5, 88–90	381.28		578	32	117	Vitric lithic tuff	
31R-7, 16–18	392.49			37	115	Vitric lithic tuff	
34R-6, 12–14	420.17	585				Lithic vitric tuff	
41R-4, 134–136	486.46	580				Lithic vitric tuff	
44R-6, 56–58	517.06	590				Lithic vitric tuff	
46R-1, 104–106	530.24	570				Lithic vitric tuff	
192-1185A-							
9R-3, 14–16	320.02	490				Pillowed basalt	Slight–moderate
9R-3, 126–128	321.25	550		30	c	Pillowed basalt	Moderate
10R-2, 7–9	324.05			60	b	Pillowed basalt	Moderate
192-1185B-							
4R-4, 111–113	323.64		485	45	c	Massive basalt	Moderate
4R-6, 76–78	327.09			40	c	Massive basalt	Moderate
5R-6, 24–26	335.37	250				Massive basalt	Slight
6R-2, 43–45	339.90	480				Massive basalt	Slight
7R-2, 102–104	350.62	580				Massive basalt	Moderate–complete
7R-3, 16–18	351.26	580				Massive basalt	Moderate–complete
8R-1, 96–98	358.66	560				Pillowed basalt	Moderate–high
9R-3, 126–128	371.42	560				Pillowed basalt	Moderate
10R-2, 7–9	378.43		570			Pillowed basalt	Moderate
10R-2, 27–29	378.63	550				Pillowed basalt	Moderate
14R-1, 85–87	406.55			40	b	Pillowed basalt	High
16R-1, 40–42	425.30	570	560	28	120	Pillowed basalt	Slight–high
16R-1, 64–66	425.54			25	115	Pillowed basalt	Slight–high
17R-2, 84–86	436.54	490				Massive basalt	Slight–moderate
19R-1, 37–39	449.47		280	60	110	Massive basalt	Slight
19R-4, 72–74	453.76	260				Massive basalt	Slight
21R-7, 36–38	471.58			70	c	Massive basalt	Slight
24R-2, 79–81	493.98	500	530	42	120	Massive basalt	Moderate
28R-1, 18–20	517.78	580				Massive basalt	Slight
192-1186A-							
31R-1, 16–18	970.16			42	115	Pillowed basalt	Slight
31R-3, 56–58	973.32	530				Pillowed basalt	Slight
32R-2, 7–9	977.74	540				Pillowed basalt	Slight–moderate
32R-3, 87–89	980.04		450	40	b	Pillowed basalt	Slight–moderate
33R-1, 17–19	981.17	525		40	c	Pillowed basalt	Slight
34R-3, 117–119	989.86	510				Massive basalt	Slight

Table 2. *continued*

Core, section, interval (cm)	Depth (mbsf)	T_c (°C) (Kappa bridge)	T_c (°C) (Micro VSM)	T_b (K)	T_v (K)	Rock type	Alteration
37R-1, 43–45	1015.23	430				Massive basalt	Slight
39R-4, 38–40	1028.73			60	110	Massive basalt	Slight
39R-4, 98–100	1029.33	420				Massive basalt	Slight
192-1187A-							
2R-2, 86–88	367.54	590				Pillowed basalt	Moderate–high
3R-2, 54–56	376.50	575			118	Submarine glass	Fresh
4R-5, 65–67	390.51	580				Pillowed basalt	Slight–high
5R-4, 121–123	399.32	590				Pillowed basalt	Moderate
6R-1, 9–11	403.49			52	c	Pillowed basalt	High
6R-6, 56–58	410.09	300				Pillowed basalt	Slight–moderate
7R-4, 32–34	417.24	585				Pillowed basalt	Moderate–high
8R-2, 64–66	424.63	520				Pillowed basalt	Slight–complete
8R-3, 59–61	425.92	320				Pillowed basalt	Moderate
9R-3, 1–3	435.10			30	c	Pillowed basalt	Slight–moderate
9R-3, 91–93	436.00	295		55	c	Pillowed basalt	Slight–moderate
10R-5, 40–42	447.81	305				Pillowed basalt	Slight–high
11R-3, 2–4	454.22	570				Pillowed basalt	Slight–moderate
11R-5, 31–33	456.91	510				Pillowed basalt	Slight–moderate
12R-3, 46–48	464.48	300				Massive basalt	Moderate
13R-2, 31–33	472.30	290		52	c	Massive basalt	Slight
13R-6, 98–100	478.15		460	25	120	Pillowed basalt	Slight–complete
14R-1, 35–37	480.65			35	122	Pillowed basalt	Moderate
14R-2, 36–38	482.16	580				Pillowed basalt	High–complete
14R-3, 28–30	483.46	570				Pillowed basalt	Moderate–high
15R-3, 88–90	493.00	480				Pillowed basalt	Slight–high
15R-4, 20–22	493.67	490				Pillowed basalt	Slight–moderate

Notes: The rock type and degrees of alteration are from Mahoney *et al.* (2001).
T_c, Curie temperature; T_b, unblocking temperature; T_v, Verwey transition; b, c, groups of low-temperature magnetometry, see text.

one magnetic phase with a Curie temperature of around 330°C on heating, most probably maghemite (Fig. 2C). The second high Curie temperature phase is observed around 530°C. The relatively large difference between heating and cooling of the sample suggests that a low-temperature oxidized titanomagnetite is the main magnetic mineral. Many volcaniclastic samples from Site 1184 and basalt flows from Site 1186 belong to this group. For brevity, we list only the high Curie temperatures in Table 2.

Hysteresis loop parameters

The samples analysed in this study indicate that submarine pillow basalt samples from Sites 1183, 1185, 1186 and 1187 show the dominance of the pseudo-single-domain size (M_r/M_s ratio (remanent magnetization/saturation magnetization) between 0.42 and 0.15, Table 3), probably indicating a mixture of multi-domain and a significant amount of single-domain grains (Fig. 3). Examples of room-temperature hysteresis loops for representative OJP basalt samples are shown in Figure 4.

For the volcaniclastic samples from Site 1184, hysteresis experiments were performed on 105 minicore samples; each was cut into five (*c.* 20 mg) subsamples in order to obtain averaged hysteresis parameters for the otherwise inhomogeneous volcaniclastic rocks (Table 3). The averaged hysteresis parameters with corresponding standard errors are slightly shifted to the right side of the mixing curves for multi-domain and single-domain grains of magnetite (Dunlop 2002), suggesting a small contribution of superparamagnetic grain sizes (Riisager *et al.* 2004).

We also examined the change of hysteresis behaviour as a function of temperature (50–300°K) for 12 representative samples to detect changes in domain state at low temperature. For all the samples, saturation magnetization increases from 50 to 300K, whereas coercivity and saturation remanence

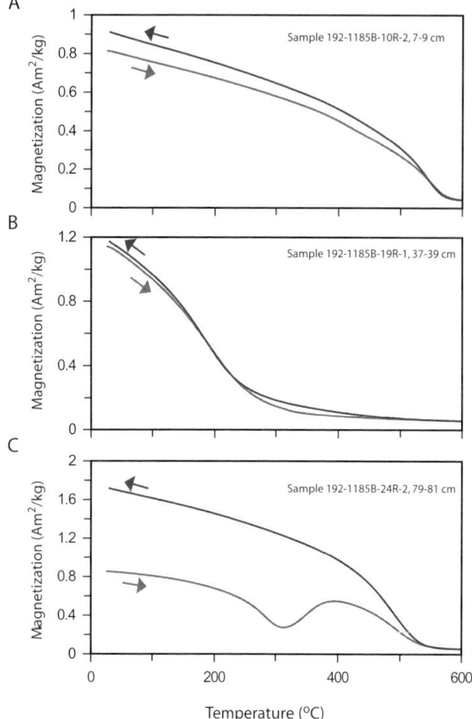

Fig. 2. Typical thermomagnetic curves for Leg 192 basalts. (**A**) Pillowed basalt Sample 192-1185B-10R-2, 7–9 cm; (**B**) massive basalt Sample 192-1185B-19R-1, 37–39 cm; and (**C**) massive basalt Sample 192-1185B-24R-2, 79–81 cm. The directions of arrows indicate heating and cooling curves.

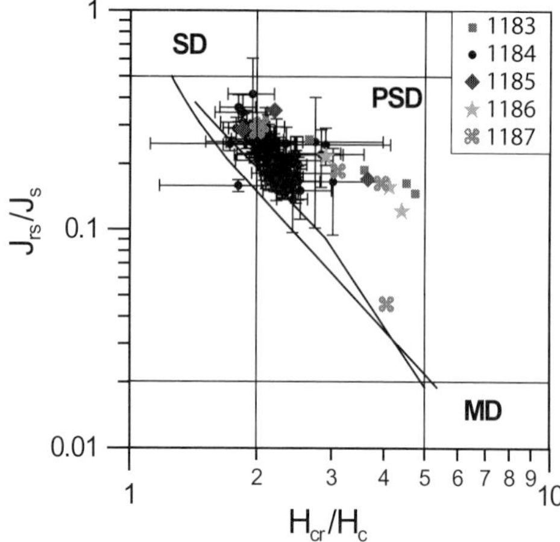

Fig. 3. Hysteresis ratios plotted on a Day *et al.* (1977)-type diagram for basement samples from Leg 192 drill sites. J_s is saturation magnetization, J_r is saturation remanent magnetization, H_c is coercivity and H_{cr} is remanent coercive force. The plot is usually divided into regions: single domain (SD) for $J_s/J_r > 0.5$ and $H_{cr}/H_c < 1.5$; multi-domain (MD) for $J_s/J_r < 0.05$ and $H_{cr}/H_c > 4$; and pseudo-single-domain in between (PSD). Error bars are 1σ for samples from Site 1184 (five subsamples were measured for each sample). Solid lines are the mixing curves for magnetite (Dunlop 2002).

systematically decrease at higher temperatures. Figure 5 shows selected hysteresis loops for two representative samples.

Low-temperature properties

As summarized in Table 2 and shown in Figure 6, the low-temperature curves of SIRM, both in zero field warming and cooling, display a variety of features. These include an unblocking temperature in the vicinity of 40–50K, most probably caused by superparamagnetic magnetite particles (Moskowitz *et al.* 1993), and a decrease in remanence in the 100–120K range, which is most probably caused by the Verwey transition (Verwey *et al.* 1947).

Figure 6A shows cooling and warming curves for Sample 192-1187A-14R-4, 35–36 cm, which indicates SD grain size. Remanence is lost at about 120–130K, both as the sample cools and warms through the Verwey transition. Only a few samples (mostly from fresh basalt or submarine glass) showed this behaviour.

Sample 192-1185B-10R-2, 7–9 cm (Fig. 6B), has a smaller grain size as indicated by hysteresis measurement (M_r/M_s = 0.35). A rapid decrease in remanence between 50 and 70K is observed (Fig. 6B). A similar phenomenon was observed in the study of oxidized synthetic magnetite by Özdemir *et al.* (1993) and was attributed to the presence of an ultrafine-grained superparamagnetic phase with very low unblocking temperature. No pronounced remanence transition is observed, although the sample displays a more rapid decrease of remanence between 50 and 120K. In comparison with the 'classic' Verwey transition, the remanence transition for this sample is blurred over a broad temperature interval and is shifted toward lower temperatures. These data suggest that the amount of SIRM loss at the Verwey transition decreases with decreasing particle size. Özdemir *et al.* (1993) also observed this behaviour and attributed it to surface oxidation. The majority of Leg 192 samples (including many volcaniclastic samples from Site 1184) show this type of low-temperature characteristic (Fig. 6B).

Figure 6C shows a third behaviour. No obvious Verwey transition is observed for pillow basalt Sample 192-1183A-54R-5, 36–38 cm

Table 3. *Summary of hysteresis properties of minicore samples from the Leg 192 sites*

Site	Core	Section	Interval (cm)	Depth (mbsf)	H_{cr}/H_c	M_r/M_s
1183A	54R	5	36–38	1132.49	2.65	0.25
1183A	59R	1	60–62	1161.2	4.52	0.16
1183A	66R	3	12–14	1203.06	4.74	0.15
1183A	68R	1	40–42	1210.1	3.58	0.19
1184A	9R	1	79–81	201.89	2.17	0.17
1184A	10R	1	36–38	203.26	2.06	0.21
1184A	11R	1	75–77	207.15	2.18	0.18
1184A	11R	2	70–72	208.6	2.29	0.17
1184A	11R	3	21–23	209.27	2.17	0.22
1184A	12R	1	114–116	212.04	2.07	0.22
1184A	12R	2	130–132	213.7	2.08	0.23
1184A	12R	3	84–86	214.61	2.05	0.23
1184A	12R	4	89–91	216.13	2.12	0.26
1184A	12R	5	73–75	217.29	2.19	0.18
1184A	13R	1	80–82	221.2	2.24	0.17
1184A	13R	2	87–89	222.71	2.31	0.19
1184A	13R	3	97–99	224.04	2.07	0.22
1184A	13R	4	125–127	225.83	2.19	0.19
1184A	14R	2	133–135	232.76	2.15	0.20
1184A	14R	3	96–98	233.8	1.81	0.16
1184A	14R	4	46–48	234.63	2.15	0.19
1184A	14R	5	117–119	236.7	2.21	0.20
1184A	14R	6	20–22	237.14	2.08	0.18
1184A	15R	1	98–100	240.58	2.23	0.18
1184A	15R	1	133–135	240.93	2.16	0.18
1184A	15R	2	41–43	241.49	2.16	0.19
1184A	16R	1	58–60	244.98	2.18	0.19
1184A	16R	2	76–78	246.66	2.43	0.14
1184A	16R	2	90–92	246.8	2.22	0.18
1184A	16R	3	65–67	248.03	2.14	0.19
1184A	16R	4	42–44	248.91	2.14	0.19
1184A	17R	1	24–26	249.44	2.17	0.20
1184A	17R	2	25–27	250.84	2.36	0.17
1184A	17R	3	4–6	252.14	2.15	0.20
1184A	17R	4	99–101	254.52	2.47	0.22
1184A	17R	5	79–81	254.52	2.36	0.15
1184A	17R	6	104–106	257.12	2.23	0.19
1184A	17R	7	46–48	258.02	2.28	0.19
1184A	18R	3	45–47	261.75	2.357	0.15
1184A	18R	4	89–91	263.71	2.28	0.17
1184A	18R	6	23–25	265.93	2.27	0.19
1184A	19R	3	61–63	271.21	2.37	0.18
1184A	19R	4	50–52	272.16	1.73	0.25
1184A	19R	8	17–19	277.52	2.16	0.20
1184A	20R	1	66–68	278.76	2.09	0.30
1184A	20R	2	81–83	280.41	2.19	0.17
1184A	21R	2	145–147	290.59	2.2	0.16
1184A	21R	5	77–79	293.45	2.29	0.18
1184A	21R	6	21–23	294.39	2.48	0.22
1184A	22R	4	39–41	300.96	2.53	0.15
1184A	22R	5	116–118	303.07	2.05	0.23
1184A	23R	1	69–71	307.59	1.95	0.42
1184A	24R	4	107–109	320.84	2.3	0.16
1184A	24R	5	117–119	322.41	2.31	0.18
1184A	24R	8	116–118	326.51	2.25	0.18
1184A	25R	3	11–13	328.89	2.1	0.21
1184A	25R	4	13–15	329.64	2.12	0.22
1184A	25R	7	16–18	333.61	2.5	0.20
1184A	25R	8	12–14	334.69	2.28	0.19
1184A	26R	3	98–100	339.55	2.16	0.24

Table 3. *continued*

Site	Core	Section	Interval (cm)	Depth (mbsf)	H_{cr}/H_c	M_r/M_s
1184A	26R	6	33–35	343.2	2.14	0.21
1184A	27R	4	95–97	351.03	1.89	0.29
1184A	27R	5	16–18	351.74	1.85	0.34
1184A	28R	3	31–33	358.4	2.4	0.21
1184A	28R	4	17–19	359.73	1.98	0.26
1184A	28R	7	17–19	364.23	2.08	0.26
1184A	29R	3	147–149	369.35	2.26	0.22
1184A	30R	2	126–128	377.46	1.79	0.29
1184A	30R	5	88–90	381.27	2.23	0.18
1184A	30R	7	36–38	383.42	1.95	0.27
1184A	31R	5	11–13	389.54	2.03	0.25
1184A	31R	7	16–18	392.48	1.95	0.28
1184A	32R	2	8–10	395.11	2.34	0.25
1184A	32R	5	130–132	400.85	2.34	0.19
1184A	32R	6	129–131	402.34	2.5	0.17
1184A	33R	2	25–27	405.44	2.24	0.21
1184A	33R	5	113–115	410.13	2.19	0.20
1184A	33R	6	105–107	411.55	2.22	0.22
1184A	34R	2	25–27	414.99	2.18	0.20
1184A	34R	6	12–14	420.17	2.13	0.25
1184A	34R	7	7–9	421.27	2.35	0.18
1184A	35R	3	108–110	426.86	2.16	0.22
1184A	35R	5	20–22	428.68	2.27	0.18
1184A	35R	8	11–13	432.09	1.81	0.36
1184A	36R	2	56–58	434.9	1.97	0.26
1184A	36R	4	80–82	437.69	1.97	0.29
1184A	36R	6	96–98	440.86	2.04	0.23
1184A	37R	1	3–5	442.63	2.22	0.21
1184A	39R	2	103–105	464.43	2.33	0.19
1184A	39R	4	110–112	467.36	2.262	0.22
1184A	40R	2	126–128	474.15	2.069	0.28
1184A	40R	5	49–51	477.79	1.854	0.31
1184A	41R	4	134–136	486.46	2.15	0.20
1184A	41R	5	11–13	486.72	2.47	0.18
1184A	42R	5	13–15	496.37	2.05	0.27
1184A	43R	1	144–146	501.74	2.3	0.18
1184A	43R	3	119–121	503.92	1.94	0.25
1184A	43R	4	85–87	505.08	2.18	0.22
1184A	44R	1	24–26	510.14	2.19	0.23
1184A	44R	2	55–57	511.97	3.02	0.17
1184A	44R	3	99–101	513.74	2.08	0.29
1184A	44R	6	56–58	517.06	2.9	0.24
1184A	44R	8	61–63	519.3	2.22	0.25
1184A	45R	1	110–112	520.6	2.83	0.22
1184A	45R	2	83–85	521.48	2.12	0.28
1184A	45R	4	65–67	524.1	2.34	0.18
1184A	45R	5	12–14	524.94	2.41	0.16
1184A	45R	6	58–60	526.7	2.19	0.26
1184A	46R	1	104–106	530.24	2.74	0.25
1185A	9R	3	66–68	320.54	1.84	0.28
1185B	4R	4	111–113	323.64	3.66	0.18
1185B	10R	2	7–9	378.43	2.2	0.35
1186A	31R	1	16–18	970.16	4.12	0.16
1186A	33R	1	17–19	981.17	2.9	0.21
1186A	39R	4	38–40	1028.73	4.4	0.12
1187A	3R	2	54–56	376.5	2.04	0.30
1187A	9R	3	91–93	436	3.94	0.16
1187A	13R	2	31–33	472.3	4.06	0.05
1187A	15R	3	88–90	493.01	3.1	0.18
1187A	15R	4	20–22	493.68	2.0	0.28

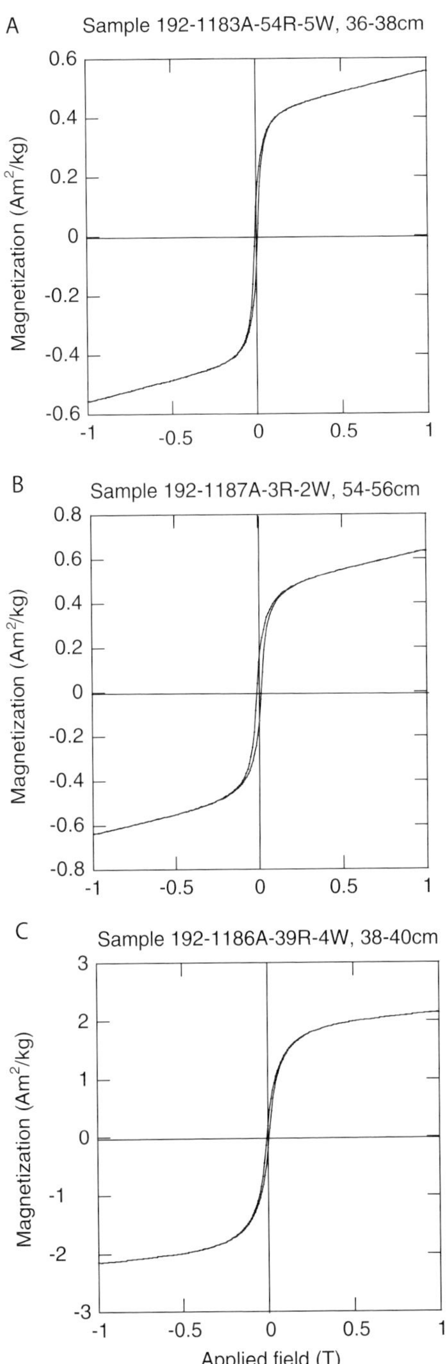

Fig. 4. Diagram of room-temperature hysteresis loops for representative basalt samples from Leg 192 drill sites. (**A**) Sample 192-1183A-54R-5, 36–38 cm; (**B**) Sample 192-1187A-3R-1, 54–56 cm; and (**C**) Sample 192-1186A-39R-4, 38–40 cm. The horizontal axis is applied field up to 1 T. The vertical axis is mass-specific magnetization (not corrected for slope).

during cooling to 20K. On warming from 20K, however, a distinctive bend in remanence occurs near 100K. With continued warming, the decay of remanence is almost linear all the way to room temperature (Fig. 6C). Interestingly, the low-temperature properties of the OJP basalt are very similar to those of basalt recovered from the Kerguelen Plateau–Broken Ridge in the southern Indian Ocean (Zhao *et al.* 2002).

Summary and discussion

The Leg 192 cores recovered from the five drill sites displayed variable rock magnetic signatures. Three general types of behaviour were found in the rock magnetic measurements.

One group has a single phase of Ti-poor titanomagnetite with Curie temperatures between 480 and 580°C (Table 2). The thermomagnetic curves of these samples exhibit very little difference between heating and cooling of the samples. Samples with titanomagnetite also exhibit a strong Verwey transition in the vicinity of 110K. These results are in good agreement with the hysteresis ratios, suggesting that the bulk magnetic grain size is in the PSD boundary (e.g. with lower H_{cr}/H_c values). Therefore, the basalts from this group are most probably good palaeomagnetic recorders and are likely to have preserved original and stable magnetic remanences.

The second group is observed exclusively in pillow lavas and is characterized by a Curie temperature of 260–280°C, which is typical of Ti-rich titanomagnetite (such as oxidized TM60) or low-temperature oxidized titanomaghemite (Dunlop & Özdemir 1997). Samples in this group apparently went through low-temperature oxidation, which is one of the two forms of alteration for titanomagnetite (Dunlop & Özdemir 1997). The low-temperature curves for this group do not show the Verwey transition. The rock-magnetically-inferred fine-grain size indicates a rapid cooling environment for the pillow lavas. Altogether, these rock-magnetic data seem to be sensitive indicators of low-temperature oxidation and support the contention that Ti-rich titanomagnetite is responsible for the magnetic signatures displayed in the pillow basalts.

The third group has more than one Curie temperature, which suggests the presence of multiple magnetic phases. Although the hysteresis ratios for rocks in this group still fall in the PSD region, the cluster is centred toward the MD region (with higher H_{cr}/H_c ratios). Low-temperature curves do not clearly show the Verwey transition (Fig. 6C). The thermomagnetic signature indicates the inversion of

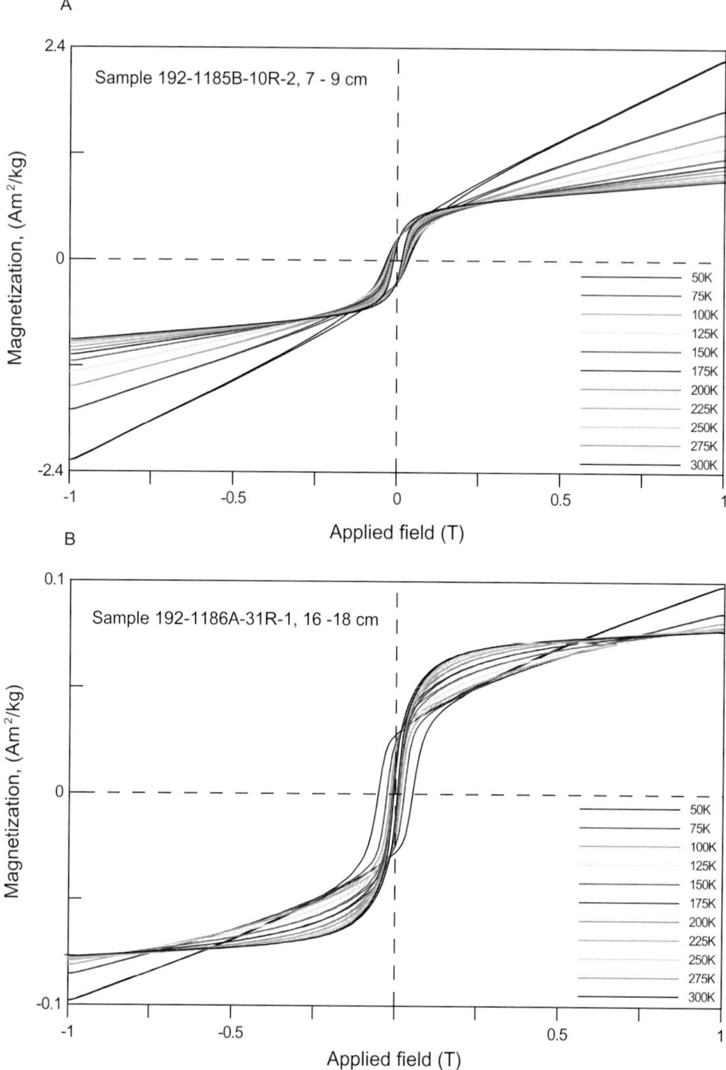

Fig. 5. Diagram of hysteresis behaviour for selected samples during warming from temperature 50 to 300K. (**A**) Pillowed basalt Sample 192-1185B –10R-2, 7–9 cm and (**B**) pillowed basalt Sample 192-1186A-31R-1, 16–18 cm. The horizontal axis is applied field up to 1 T. The vertical axis is magnetic moment.

titanomaghemite to a strongly magnetized magnetite, as shown by the irreversible cooling curve (Fig. 2C). Chemical remanent magnetization resulting from oxidation of titanomagnetite and inversion of titanomaghemite has been shown to parallel the original thermoremanence (Johnson & Merrill 1974; Hall 1977; Özdemir & Dunlop 1985; Dunlop & Özdemir 1997), which appears to be the case for Leg 192 samples. From data collected in our shore-based palaeomagnetic studies (Riisager *et al.* 2003, 2004), the same mean characteristic inclinations recorded by samples from this group (compared with those in the other two groups) suggest that these rocks retain stable remanent magnetization.

Leg 192 rock magnetic characteristics are consistent with shipboard observation that OJP basement sites show evidence of only low-temperature sea-water-mediated alteration in either the lava flows or overlying sediments (Mahoney *et al.* 2001; Banerjee *et al.* 2004). The lack of higher-temperature hydrothermal alteration is in turn consistent with the inference of significant distances between site locations and eruptive vent systems. Higher-temperature hydrothermal alteration systems would be

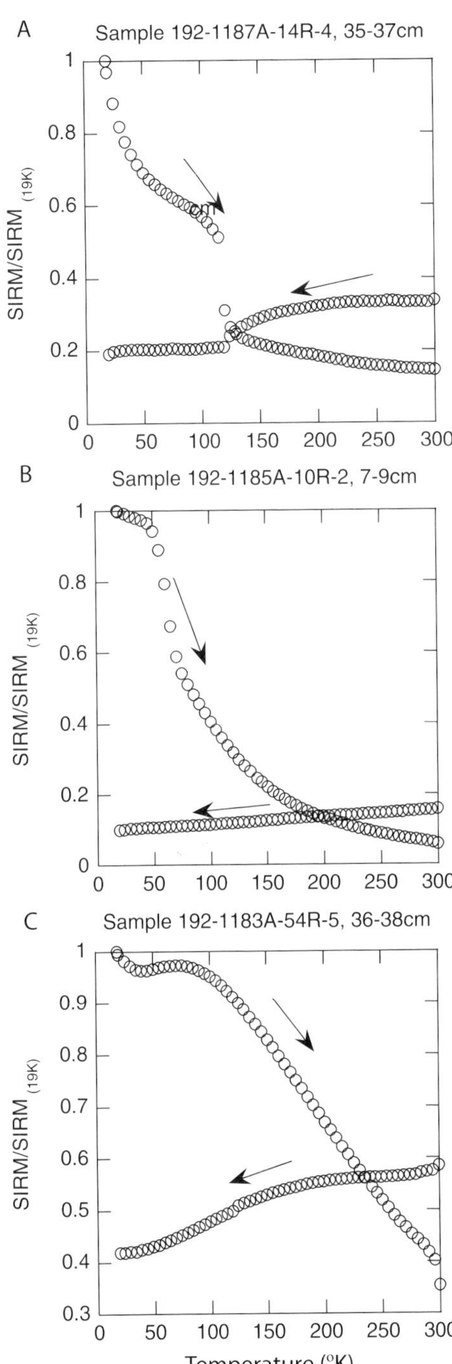

Fig. 6. Normalized low-temperature variation of saturation isothermal remanence (SIRM) for several representative samples during zero field cooling from 300 to 19K and zero field warming back to 300K. (**A**) Sample 192-1187A-14R-4, 35–37 cm; (**B**) Sample 192-1185A-10R-2, 7–9 cm; and (**C**) Sample 192-1183A-54R-5, 36–38 cm.

expected to be centred around major eruptive loci (Tejada *et al.* 1996, 2002). Knowledge of the primary magnetic mineralogy and subsequent mineralogical changes (and accompanying changes in magnetization) caused by secondary processes is critical for understanding the role of hydrothermal alteration, and knowledge of the nature and vigour of hydrothermal activity within the volcanic pile is key for understanding the oceanographic and climatic effects of plateau formation.

Rock magnetic properties are useful in evaluating the fidelity of the magnetic memory in the rocks. The generally good magnetic stability and other properties exhibited by Leg 192 rocks support the inference that the characteristic directions of magnetization isolated from the OJP sites were acquired during the Cretaceous Normal Superchron. The stable inclinations identified from these samples are therefore useful for plate tectonic studies. As we documented in our completed post-cruise palaeomagnetic study (Riisager *et al.* 2003, 2004), the mean palaeomagnetic inclination of the five Leg 192 sites and one Leg 130 site agree remarkably well, demonstrating not only that palaeosecular variation has been averaged out but also there is no significant tectonic tilting of the basalt units. The corresponding palaeolatitude for the OJP (24°S) is approximately 20° north of the latitude suggested by reconstructions (Neal *et al.* 1997; Kroenke *et al.* 2004) in an assumed fixed hotspot reference frame (Engebretson *et al.* 1985; Lonsdale 1988; Wessel & Kroenke 1997; Koppers *et al.* 2001).

Turning to the questions we mentioned in the Introduction, although the palaeomagnetically determined palaeolatitude data do not appear to support a *stationary* Louisville hot-spot origin for the OJP, these reliable basement palaeolatitude data are useful to constrain the history of Pacific plate motion. No marine magnetic anomaly skewness data are available for the Cretaceous Long Normal Superchron, which is a key piece in the mid-Cretaceous puzzle of the coupling of mantle and surface events (Larson 1991). However, palaeolatitudes of the OJP clearly show that the portion of the Pacific plate containing the Leg 192 sites was in the southern hemisphere at 120 Ma, which is essentially coincident with the onset of the Cretaceous Long Normal Superchron (Opdyke & Channel 1996). After 120 Ma, the Pacific plate continued to move northwards through the remainder of the Cretaceous and the Tertiary (Yan & Kroenke 1993). Eventually, the OJP collided with the old Solomon arc, and obducted pillowed and massive OJP basalts have been found in extensive river sections on several of the eastern

Solomon Islands (Fig. 1). Recent ^{40}Ar–^{39}Ar results for the majority of these basalts give ages averaging about 122 Ma; others yield ages around 90 Ma (e.g. Parkinson *et al.* 1996; Tejada *et al.* 1996, 2002). The *c.* 120 Ma peak of OJP volcanism is also manifested in Manihiki Plateau and in parts of the Kerguelen Plateau (e.g. Mahoney *et al.* 1993, 2001; Tejada *et al.* 1996; Frey *et al.*1999; Larson & Erba 1999; Duncan 2002). An important consequence of this volcanism was increased hydrothermal activity and widespread burial of marine organic matter during the early Aptian OAE1 (*c.* 120.5 Ma, Bralower *et al.* 1994; Erba 1994; Larson & Erba 1999). The *c.* 90 Ma volcanism peak is repeated in the Kerguelen and Caribbean plateaus (Sinton *et al.* 1998; Frey *et al.* 1999; Duncan 2002). The OAE2 at the Cenomanian–Turonian boundary (*c.* 93.5 Ma) was suggested to have related to this volcanism, including the formation of the Caribbean Plateau (Sinton & Duncan 1997; Kerr 1998; Snow & Duncan 2001). Interestingly, the *c.* 90 Ma phase of volcanic activity in the Pacific *roughly* corresponds with the end of the Cretaceous Long Normal Superchron (83 Ma, Opdyke & Channell 1996). It appears that an apparent episodicity of large plume activity may result from a mantle process that operates at a global scale and that large plumes are perhaps more frequent than had been thought.

Conclusions

This study has demonstrated that a combination of rock magnetic and palaeomagnetic studies on the same cores can provide important information that has considerable relevance to understanding the evolution of the OJP. Based on palaeomagnetic and rock magnetic results obtained during our post-cruise study of Leg 192 cores, we draw the following conclusions.

- The majority of pillow basalt samples from the Leg 192 sites underwent low-temperature oxidation. Nevertheless, these basalts are good palaeomagnetic recorders that preserve original and stable magnetic directions. The generally good magnetic stability exhibited by the titanomagnetite-bearing rocks suggests that the stable inclinations and corresponding palaeolatitudes are reliable for tectonic reconstruction of the Pacific plate.
- The rock magnetic and physical properties data from Leg 192 rocks support the inference that the characteristic directions of magnetization isolated from the Cretaceous OJP sites were acquired near the onset of the Cretaceous Long Normal Superchron about 120 Ma ago. The portion of the Pacific plate containing the Leg 192 sites was in the southern hemisphere during the mid-Cretaceous volcanism.

We thank J. G. Fitton, J. J. Mahoney, W. Williams and M. Steiner for insightful reviews and constructive suggestions on the original manuscript. This research used samples and data provided by the Ocean Drilling Program (ODP). The ODP is sponsored by the US National Science Foundation (NSF) and participating countries under management of Joint Oceanographic Institutions (JOI), Inc. Funding for this research was provided by the US Science Support Program of JOI and NSF grants EAR 443549-22178 and EAR 443747-22250 to X. Zhao, the Danish National Research Foundation to P. Riisager, the US Science Support Program of JOI to S. Hall and ODP/Germany Project Number So 72/70–1 to M. Antretter. Funding was also provided by the Center for the Study of Imaging and Dynamics of the Earth, Institute of Geophysics and Planetary Physics at the University of California Santa Cruz, contribution number 462. The authors acknowledge the invaluable assistance and skill provided by all members of the Leg 192 shipboard scientific party, the ODP marine technicians, and the crew of the *JOIDES Resolution*. We also wish to express our appreciation to the shore-based ODP staff for all of their pre- and post-cruise efforts.

References

ANDERSON, D.L. 1994. Superplumes or supercontinents? *Geology*, **22**, 39–42.

ANTRETTER, M., RIISAGER, P., HALL, S., ZHAO, X. & STEINBERGER, B. 2004. Modelled palaeolatitudes for the Louisville hot spot and the Ontong Java Plateau. *In*: FITTON, J.G., MAHONEY, J.J., WALLACE, P.J. & SAUNDERS, A.D. (eds) *Origin and Evolution of the Ontong Java Plateau*. Geological Society, London, Special Publications, **229**, 21–30.

BANERJEE, N.R., HONNOREZ, J. & MUEHLENBACHS, K. 2004. Low-temperature alteration of submarine basalts from the Ontong Java Plateau. *In*: FITTON, J.G., MAHONEY, J.J., WALLACE, P.J. & SAUNDERS, A.D. (eds) *Origin and Evolution of the Ontong Java Plateau*. Geological Society, London, Special Publications, **229**, 259–273.

BRALOWER, T.J., ARTHUR, M.A., LECKIE, R.M., SLITER, W.V., ALLARD, D. & SCHLANGER, S.O. 1994. Timing and paleoceanography of oceanic dysoxia/anoxia in the late Barremian to early Aptian. *Palaios*, **9**, 335–369.

CHAMBERS, L.M., PRINGLE, M.S. & FITTON, J.G. 2002. Age and duration of magmatism on the Ontong Java Plateau: $^{40}Ar/^{39}Ar$ results from ODP Leg 192. *Eos, Transactions of the American Geophysical Union*, **83**, Fall Meeting Supplement, Abstract V71B-1271.

CHAMBERS, L.M., PRINGLE, M.S. & FITTON, J.G. 2004. Phreatomagmatic eruptions on the Ontong Java Plateau: an Aptian $^{40}Ar/^{39}Ar$ age for volcaniclastic rocks at ODP Site 1184. *In*: FITTON, J.G., MAHONEY, J.J., WALLACE, P.J. & SAUNDERS, A.D.

(eds) *Origin and Evolution of the Ontong Java Plateau*. Geological Society, London, Special Publications, **229**, 325–331.

COFFIN, M. & ELDHOLM, O. 1994. Large igneous provinces: crustal structure, dimensions, and external consequences. *Review of Geophysics*, **32**, 1–36.

COFFIN, M.F., FREY, F.A., WALLACE, P.J. & THE LEG 183 SHIPBOARD SCIENTIFIC PARTY. 2000. Kerguelen Plateau–Broken Ridge, Sites 1135–1142. *Proceedings of the Ocean Drilling Program, Initial Reports*, **183** (CD-ROM).

COURTILLOT, V. & BESSE, J. 1987. Magnetic field reversals, polar wander, and core–mantle coupling. *Science*, **237**, 1140–1147.

DAY, R., FULLER, M. & SCHMIDT, V.A. 1977. Hysteresis properties of titanomagnetites: grain-size and compositional dependence. *Physics of the Earth and Planetary Interiors*, **13**, 260–266.

DUNCAN, R. A. 2002. A time frame for construction of the Kerguelen Plateau and Broken Ridge. *Journal of Petrology*, **43**, 1109–1119.

DUNLOP, D. & ÖZDEMIR, Ö. 1997. *Rock-magnetism, Fundamentals and Frontiers*. Cambridge University Press, Cambridge.

DUNLOP, D.J. 2002. Theory and application of the Day plot (M_{rs}/M_s versus H_{cr}/H_c) 1. Theoretical curves and tests using titanomagnetite data. *Journal of Geophysical Research*, **107**, 2056, 10.1029/2001JB000486.

ELDHOLM, O. & COFFIN, M. 2000. Large igneous provinces and plate tectonics. *In*: RICHARDS, M.A., GORDON, R.G. & VAN DER HILST (eds) *The History and Dynamics of Global Plate Motions*. American Geophysical Union, Geophysical Monograph, **121**, 309–326.

ENGEBRETSON, D.C., COX, A. & GORDON, R.G. 1985. *Relative Motions Between Oceanic and Continental Plates in the Pacific Basin*. Geological Society of America, Special Paper, **206**.

ERBA, E. 1994. Nannofossils and superplumes: The early Aptian 'nannoconid crisis'. *Paleoceanography*, **B**, 483–501.

ERBACHER, J. & THUROW, J. 1997. Influence of oceanic anoxic events on the evolution of mid-Cretaceous radiolaria in the North Atlantic and western Tethys. *Marine Micropaleontology*, **30**, 139–158.

FITTON, J.G. & GODARD, M. 2004. Origin and evolution of magmas on the Ontong Java Plateau. *In*: FITTON, J.G., MAHONEY, J.J., WALLACE, P.J. & SAUNDERS, A.D. (eds) *Origin and Evolution of the Ontong Java Plateau*. Geological Society, London, Special Publications, **229**, 151–178.

FLANDERS, P.J. 1988. An alternating-gradient magnetometer. *Journal of Applied Physics*, **63**, 3940–3945.

FREY, F., COFFIN, M. & WALLACE, P.J. 1999. Origin and evolution of a submarine large igneous province: The Kerguelen Plateau and Broken Ridge, southern Indian Ocean. *Eos Transactions of the American Geophysical Union*, **80**, F1103, Fall Meeting Supplement, Abstract F1103.

GROMMÉ, C.S., WRIGHT, T.L. & PECK, D.L. 1969. Magnetic properties and oxidation of iron-titanium oxide minerals in Alae and Makaopuhi lava lakes, Hawaii. *Journal of Geophysical Research*, **74**, 5277–5293.

GUBBINS, D. 1987. A mechanism for geomagnetic reversals. *Geophysical Journal of the Royal Astronomical Society*, **89**, 470.

HALL, J.M. 1977. Does TRM occur in oceanic layer 2 basalts? *Journal of Geomagnetism and Geoelectricity*, **29**, 411–419.

HELLER, P., ANDERSON, D., & ANGEVINE, C.L. 1996. Is the middle Cretaceous pulse of rapid sea-floor spreading real or necessary? *Geology*, **24**, 491–494.

JOHNSON, H.P. & MERRILL, R.T. 1974. Low temperature oxidation of a single-domain magnetite. *Journal of Geophysical Research*, **79**, 5533–5534.

KERR, A.C. 1998. Oceanic plateau formation: A cause of mass extinction and black shale deposition around the Cenomanian–Turonian boundary. *Journal of the Geological Society, London*, **155**, 619–626.

KOPPERS, A.A.P., MORGAN, J.W., MORGAN, J.P. & STAUDIGEL, H. 2001. Testing the fixed hotspot hypothesis using $^{40}Ar/^{39}Ar$ age progressions along seamount trails. *Earth and Planetary Science Letters*, **185**, 237–252.

KROENKE, L.W., WESSEL, P. & STERLING, A.I. 2004. Motion of the Ontong Java Plateau in the hot-spot frame of reference: 122 Ma–present. *In*: FITTON, J.G., MAHONEY, J.J., WALLACE, P.J. & SAUNDERS, A.D. (eds) *Origin and Evolution of the Ontong Java Plateau*. Geological Society, London, Special Publications, **229**, 9–20.

LARSON, R.L. 1991. Latest pulse of Earth: Evidence for a mid-Cretaceous superplume. *Geology*, **19**, 547–550.

LARSON, R.L. & ERBA, E. 1999. Onset of the mid-Cretaceous greenhouse in the Barremian–Aptian: Igneous events and the biological, sedimentary, and geochemical responses. *Paleoceanography*, **14**, 663–678.

LARSON, R.L. & OLSON, P. 1991. Mantle plumes control magnetic reversal frequency. *Earth and Planetary Science Letters*, **107**, 437–447.

LECKIE, R.M. BRALOWER, T.J. & CASHMAN, R. 2002. Oceanic anoxic events and plankton evolution: Biotic response to tectonic forcing during the mid-Cretaceous. *Paleoceanography*, **17**, 13-1/13–29, 10.1029/2001PA000623.

LONSDALE, P. 1988. Geography and history of the Lousville hotspot chain in the southwest Pacific. *Journal of Geophysical Research*, **93**, 3078–3104.

MAHONEY, J.J. & SPENCER, K.J. 1991. Isotopic evidence for the origin of the Manihiki and Ontong Java oceanic plateaus. *Earth and Planetary Science Letters*, **104**, 196–210.

MAHONEY, J.J. & COFFIN, M.F. (eds) 1997. *Large Igneous Provinces: Continental, Oceanic, and Planetary Flood Volcanism*. American Geophysical Union, Geophysical Monograph, **100**, 183–216.

MAHONEY, J.J. STOREY, M., DUNCAN, R.A., SPENCER, K.J. & PRINGLE, M.S. 1993. Geochemistry and age of the Ontong Java Plateau. *In*: PRINGLE, M.S., SAGER, W.W., SLITER, W.V. & STEIN, S. (eds) *The Mesozoic Pacific: Geology, Tectonics, and*

Volcanism. American Geophysical Union, Geophysical Monograph, **77**, 233–262.

MAHONEY, J.J., FITTON, J.G., WALLACE, P.J. *et al.* 2001. *Proceedings of the Ocean Drilling Program, Initial Reports*, **192** (CD-ROM). Available from Ocean Drilling Program, Texas A&M University, College Station TX 77845-9547, USA.

MOORE, J.G. & SCHILLING, J.-G. 1973. Vesicles, water and sulfur in Reykjanes Ridge basalts. *Contributions to Mineralogy and Petrology*, **41**, 105–118.

MOSKOWITZ, B.M. 1981. Methods for estimating Curie temperatures of titanomaghemites from experimental J_s–T data. *Earth and Planetary Science Letters*, **53**, 84–88.

MOSKOWITZ, B.M., FRANKEL, R.B. & BAZYLINSKI, D.A. 1993. Rock magnetic criteria for the detection of biogenic magnetite. *Earth and Planetary Science Letters*, **120**, 283–300.

NEAL, C.R., MAHONEY, J.J. KROENKE, L.W. DUNCAN, R.A. JAIN, J.C. & PETTERSON, M.G. 1997. The Ontong Java Plateau, *In*: MAHONEY, J.J. & COFFIN, M.F. (eds) *Large Igneous Provinces: Continental, Oceanic, and Planetary Flood Volcanism*. American Geophysical Union, Geophysical Monograph, **100**, 183–216.

OPDYKE, N.D. & CHANNELL, J.E.T. 1996. *Magnetic Stratigraphy*. International Geophysics Series, **64**. Academic Press, New York.

ÖZDEMIR, Ö. & DUNLOP, D. 1985. An experimental study of chemical remanent magnetization of synthetic monodomain titanomaghemites with initial thermoremanent magnetizations. *Journal of Geophysical Research*, **90**, 11 513–11 523.

ÖZDEMIR, Ö., DUNLOP, D.J. & MOSKOWITZ, B.M. 1993. The effect of oxidation on the Verwey transition in magnetite. *Geophysical Research Letters*, **20**, 1671–1674.

PARKINSON, I.J., ARCULUS, R.J. & DUNCAN, R.A. 1996. Geochemistry of Ontong Java Plateau basalt and gabbro sequences, Santa Isabel, Solomon Islands. *Eos, Transactions of the American Geophysical Union*, **77F**, 715.

PARKINSON, I.J., SCHAEFER, B.F. & ODP LEG 192 SHIPBOARD SCIENTIFIC PARTY. 2001. A lower mantle origin for the world's biggest LIP? A high precision Os isotope isochron from Ontong Java Plateau basalts drilled on ODP Leg 192. *Eos, Transactions of the American Geophysical Union*, **82**, Fall Meeting Supplement, F1398.

PHINNEY, E.J., MANN, P., COFFIN, M.F. & SHIPLEY, T.H. 1999. Sequence stratigraphy, structure, and tectonic history of the southwestern Ontong Java Plateau adjacent to the North Solomon Trench and Solomon Islands arc. *Journal of Geophysical Research*, **104**, 20 449–20 466.

PRÉVOT, M., MATTERN, E., CAMPS, P. & DAIGNIERES, M. 2000. Evidence for a 20° tilting of the Earth's rotation axis 110 million years ago. *Earth and Planetary Science Letters*, **179**, 517–528.

RICHARDS, M.A., JONES, D.L., DUNCAN, R.A. & DEPAOLO, D.J. 1991. A mantle plume initiation model for the Wrangellia flood basalt and other oceanic plateaus. *Science*, **254**, 263–267.

RIISAGER, P., HALL, S., ANTRETTER, M. & ZHAO, X. 2003. Palaeomagnetic palaeolatitude of Early Cretaceous Ontong Java Plateau basalts: Implications for Pacific apparent and true polar wander. *Earth and Planetary Science Letters*, **208**, 235–252.

RIISAGER, P., HALL, S., ANTRETTER, M. & ZHAO, X. 2004. Early Cretaceous Pacific palaeomagnetic pole from Ontong Java plateau basement rocks. *In*: FITTON, J.G., MAHONEY, J.J., WALLACE, P.J. & SAUNDERS, A.D. (eds) *Origin and Evolution of the Ontong Java Plateau*. Geological Society, London, Special Publications, **229**, 31–44.

SAGER, W.W. & KOPPERS, A.A.P. 2000. Late Cretaceous polar wander of the Pacific plate: Evidence of a rapid true polar wander event. *Science*, **287**, 455–459.

SHERIDAN, R.E. 1983. Phenomena of pulsation tectonics related to the breakup of the eastern North American continental margin. *Initial Reports of the Deep Sea Drilling Project*, **76**, 897–909.

SINTON, C.W. & DUNCAN, R.A. 1997. Potential links between ocean plateau volcanism and global ocean anoxia at the Cenomanian–Turonian boundary. *Economic Geology*, **92**, 836–842.

SINTON, C.W., DUNCAN, R.A., STOREY, M., LEWIS, J. & ESTRADA, J.J. 1998. An oceanic flood basalt province within the Caribbean plate. *Earth and Planetary Science Letters*, **155**, 221–235.

SNOW, L.J. & DUNCAN, R.A. 2001. Hydrothermal links between ocean plateau formation and global anoxia, *Eos, Transactions of the American Geophysical Union*, **82**, Fall Meeting Supplement, Abstract OS41A-0437.

TARDUNO, J.A., SLITER, W.V., KROENKE, L., LECKIE, M., MAYER, H., MAHONEY, J.J., MUSGRAVE, R., STOREY, M. & WINTERER, E.L. 1991. Rapid formation of Ontong Java Plateau by Aptian mantle plume volcanism. *Science*, **254**, 399–403.

TARDUNO, J.A., BRINKMAN, D.B., RENNE, P.R., COTTRELL, R.D., SCHER, H. & CASTILLO, P. 1998. Evidence for extreme climatic warmth from Late Cretaceous Arctic vertebrates. *Science*, **282**, 2241–2244.

TAUXE, L. 1998. *Paleomagnetism, Principles and Practice*. Kluwer, Dordrecht.

TEJADA, M.L.G., MAHONEY, J.J., DUNCAN, R.A. & HAWKINS, M.P. 1996. Age and geochemistry of basement and alkalic rocks of Malaita and Santa Isabel, Solomon Islands, southern margin of the Ontong Java Plateau. *Journal of Petrology*, **37**, 361–394.

TEJADA, M.L.G., MAHONEY, J.J., NEAL, C.R., DUNCAN, R.A. & PETTERSON, M.G. 2002. Basement geochemistry and geochronology of central Malaita, Solomon islands, with implications for the origin and evolution of the Ontong Java Plateau. *Journal of Petrology*, **43**, 449–484.

VERWEY, E.J., HAAYMAN, P.W. & ROMEIJN, F.C. 1947. Physical properties and cation arrangements of oxides with spinal structure. *Journal of Chemical Physics*, **15**, 181–187.

VOGT, P.R. 1975. Changes in geomagnetic reversal frequency at times of tectonic change: Evidence for coupling between core and upper mantle processes. *Earth and Planetary Science Letters*, **25**, 313–321.

WESSEL, P. & KROENKE, L. 1997. A geometric technique for relocating hotspots and refining absolute plate motions. *Nature*, **387**, 365–369.

YAN, C.Y. & KROENKE, L.W. 1993. A plate tectonic reconstruction of the southwest Pacific, 0–100 Ma. *In*: BERGER, W.H., KROENKE, L.W., MAYER, L.A. *et al.* (eds) *Proceedings of the Ocean Drilling Program, Scientific Results*, **130**, 697–709.

ZHAO, X., ANTRETTER, M., SOLHEID, P. & INOKUCHI, H. 2002. Identifying magnetic carriers from rock magnetic characterization of Leg 183 basement cores. *Proceedings of the Ocean Drilling Program, Scientific Results*, **183** (CD-ROM). Available from Ocean Drilling Program, Texas A&M University, College Station, TX 77845-9547, USA.

The geology of north and central Malaita, Solomon Islands: the thickest and most accessible part of the world's largest (Ontong Java) ocean plateau

MICHAEL G. PETTERSON

British Geological Survey, Keyworth, Nottingham NG12 5GG, UK (e-mail: mgp@bgs.ac.uk)

Abstract: This paper presents the most complete results yet published of geological surveys in Malaita, north of latitude 9°05'S between 1990 and 1995. The geology of Malaita reflects its position as an obducted part of the Alaska-size Ontong Java Plateau (OJP). The geology comprises a monolithological Cretaceous basalt basement sequence up to 3–4 km thick, termed the Malaita Volcanic Group (MVG), conformably overlain by a 1–2 km-thick Cretaceous–Pliocene pelagic sedimentary cover sequence. Cretaceous–Pliocene pelagic sedimentation was punctuated by alkaline basalt volcanism during the Eocene and by intrusion of alnöites during the Oligocene. Basement and cover sequences were both deformed by an intense, but short, middle Pliocene event. A number of localized, Upper Pliocene–Pleistocene, shallow-marine–subaerial, predominantly clastic formations overlie the middle Pliocene unconformity surface. The MVG comprises a monotonous sequence of pillowed and non-pillowed tholeiitic basalt lavas and sills with a predominant clinopyroxene–plagioclase–glass–opaques ± olivine mineralogy. The basaltic plateau morphology of the MVG is reflected in the presence of trap-like topographic features exposed in numerous river sections. Remarkably little sediment is present between basalt flows (most inter-lava contacts are basalt–basalt), indicating high to very high effusion rates. When present, inter-lava sediment is laminated pelagic chert or limestone, millimetres to centimetres thick, reflecting emplacement of the basalt in deep water (near or below the calcite compensation depth). Gabbro intrusions, dolerite dykes and an unusual spherulitic dolerite facies are locally present. The deep-water eruptive environment of the MVG probably was defined by the accumulation of voluminous eruptions from a multi-centred, submarine, possibly fissure-fed, volcanic source. The Malaitan cover sequence largely comprises a series of foraminifera-rich, pelagic calcilutites and calcisiltites with chert and, in the younger formations, arc-derived mudstone interbeds at various stratigraphic levels.

The bulk of this volume describes and interprets recently drilled core taken from submarine portions of the Ontong Java Plateau (OJP). This paper summarizes and discusses the main findings of the most recent on-land geological surveys of northern and central Malaita (1990–1995). An abridged version of these surveys was published by Petterson *et al.* (1997), but this is the first time the fuller results, particularly for the volcanic basement, have been presented within the more accessible literature. It is hoped that this work will form a useful basis for comparison with future drilling on the OJP and on-land surveys alike.

Regional geo-tectonics of the SW Pacific and Solomon Islands

The Solomon Islands is an archipelago situated between longitudes 156° and 170°E, and latitudes 5° and 12°S (Fig. 1a, b). The larger Solomon islands, which form a NW–SE-trending double chain, consist of Choiseul, the New Georgia Group, Santa Isabel, Guadalcanal, Malaita and Makira (San Cristobal). The eastern Santa Cruz Group (to the SE of the area shown in Fig. 1a) are, in geological terms, part of the Vanuatan arc system. The islands of New Britain, Bougainville, Solomon Islands and Vanuatu are termed the Greater Melanesian Arc, which marks the collisional zone between the Australian and Pacific plates.

Key tectonic elements of the greater Solomon region include the following.

- The Solomon arc and Solomon Islands are bounded to the NE by the Pacific plate, which is moving in a NW direction at about 10 cm $year^{-1}$, and the Australian plate to the south, which is moving NE at about 7 cm $year^{-1}$ (Yan & Kroenke 1993).
- The NW–SE-trending Solomon block is more immediately bounded by two trench systems:

From: FITTON, J. G., MAHONEY, J. J., WALLACE, P. J. & SAUNDERS, A. D. (eds) 2004. *Origin and Evolution of the Ontong Java Plateau*. Geological Society, London, Special Publications, **229**, 63–81. 0305-8719/$15.00

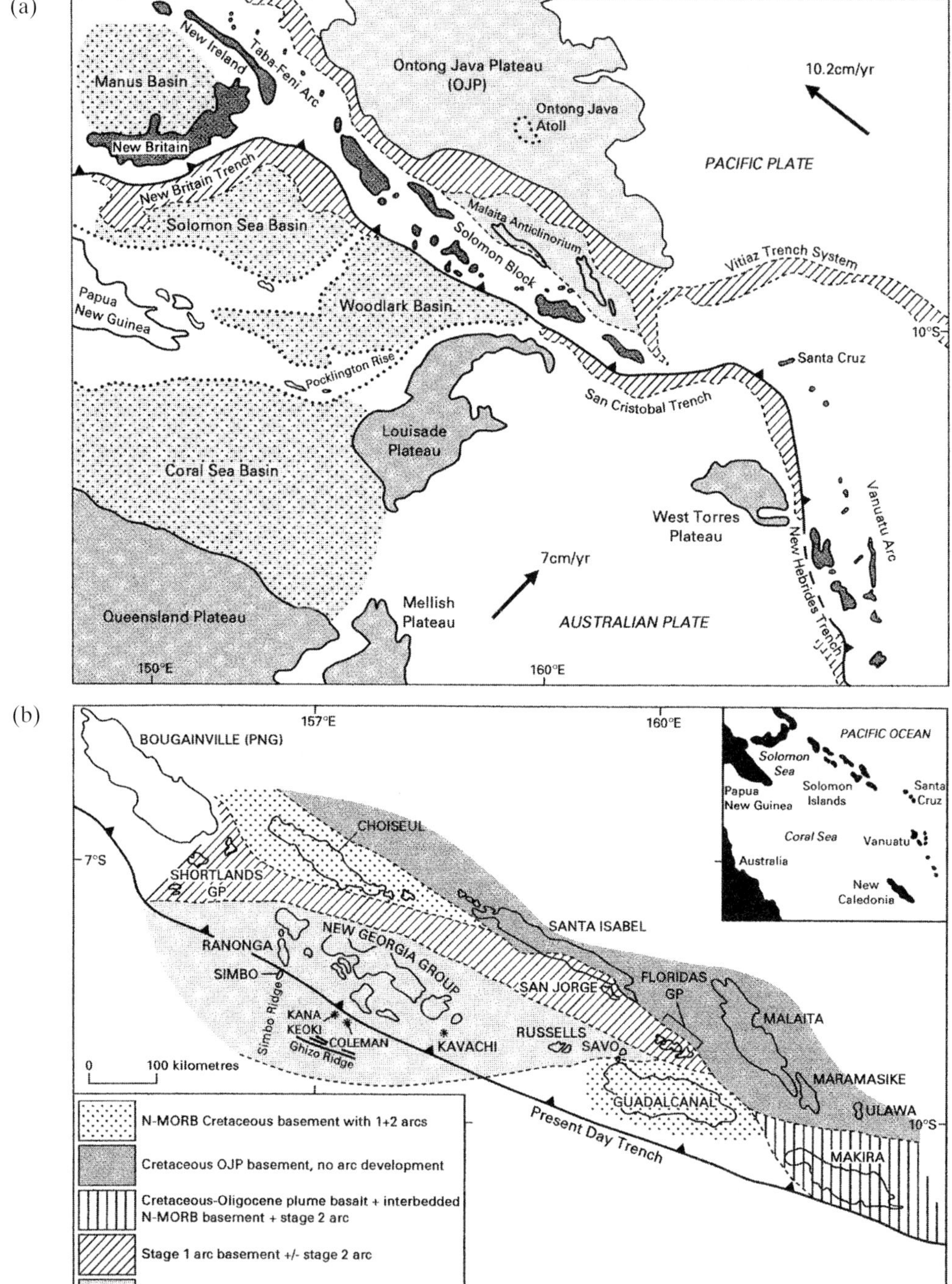

Fig. 1. (**a**) Generalized tectonic setting of the SW Pacific. Note the relative position of the OJP and the detached part of the OJP (Malaita Anticlinorium), which has sutured with the Solomon block. (**b**) Detailed geological setting of Solomon Islands, showing the different geological terrains (Petterson *et al.* 1999). Note the position of arc volcanoes Kana Keoki and Coleman south of the present-day active trench. Both figures from Petterson *et al.* (1999*a*), reproduced courtesy of Elsevier Publications.

the Vitiaz and South Solomon (New Britain–San Cristobal). The South Solomon trenches mark the contact between the Solomon block and the Australian plate, and are the topographical expression of the currently most active subduction zone system, which has been active since about 12 Ma (Kroenke 1984). The Vitiaz trench marks the contact between the Solomon block and the Pacific plate and/or the autochthonous part of the OJP. This trench system was mainly active in this region from the Eocene to about 12–8 Ma (Yan & Kroenke 1993), during which time the Pacific plate subducted southwards beneath a NE-facing Solomon arc. The collision between the OJP and the Solomon arc essentially terminated the active period of subduction at the Vitiaz trench, although the trench currently remains intermittently active. The Solomon block itself is broken up into a series of topographic highs and lows. The highest parts of the block are, by definition, the islands; transpression has produced a series of intra-block basins bounded by rhombohedral faults (e.g. Petterson *et al.* 1999*a*).

The OJP is the world's largest ocean plateau, with an areal extent equivalent to Alaska and a thickness of up to 35 km. First contact between the Solomon arc and the OJP probably began at about 20–25 Ma but collision was not forceful until about 4 Ma (e.g. Petterson *et al.* 1997; Kroenke *et al.* 2004). The OJP then began to partly subduct, partly detach, with a décollement surface situated some 5–10 km deep within the OJP (Petterson *et al.* 1999*a*). The detached allochthonous part of the OJP now forms a component of the Solomon terrain collage; it is currently detaching north of Choiseul, along most of Santa Isabel, Malaita, Ulawa and, probably, Makira (Furumoto *et al.* 1970; Mahoney *et al.* 1993, 2001; Gladczenko *et al.* 1997; Neal *et al.* 1997; Petterson *et al.* 1997, 1999*a*; Phinney *et al.* 1999; Tejada *et al.* 2002).

The Woodlark Basin (Fig. 1a) is a young (*c.* 5 Ma) triangular-shaped marginal basin that impinges directly on the South Solomon trench system between the New Britain and San Cristobal trench segments. This area coincides with a major trench shoaling: in places the physiographical definition of the 'trench' is weak and ocean depths are abnormally shallow for an oceanic trench. This characteristic is only one of a number of unusual tectonic phenomena associated with the Woodlark Basin (Auzende *et al.* 1994; Crook & Taylor 1994; Mann *et al.* 1998; Petterson *et al.* 1999*a*).

The fractured Solomon terrain collage

Petterson *et al.* (1999*a*) proposed that an amalgamated collage of terrains (Fig. 1b) has resulted from mid-ocean ridge volcanism, two stages of arc development, and collision and suturing with the OJP. The Solomon arc has formed since the Eocene as an intra-oceanic arc. There have been two main arc-forming stages: a stage 1 arc developed between the Eocene and Early Miocene as a NE-facing arc, whereas the stage 2 arc is the Late Miocene–present SW-facing arc. These two arc stages correspond to the switch in subduction polarity that occurred after the collision between the OJP and the Solomon arc, from SW-dipping to NE-dipping. Crustal units of varying provenance within the Solomon collage are separated by NW–SE and NE–SW linear structures that are essentially arc-parallel and arc-normal, respectively. These structures have developed as a result of a protracted period of highly oblique transpression.

Late Tertiary–present-day arc volcanism in the Solomon Islands

There are three subaerial volcanoes in this region: Simbo, Kavachi and Savo. The island of Simbo contains a fumarolically active vent that may be part of a still active volcano. The frequency and level of volcanic activity are unknown. Kavachi is a shallow submarine volcano that temporarily forms a small island when it erupts, which is every 4–10 years, most recently in 2002. After eruption the volcano deflates and the temporary volcanic island sinks beneath sea level. Kavachi's eruptive style (Fig. 2) varies from quiescent–effusive to highly explosive, with frequent phreatomagmatic eruptions (e.g. Johnson & Tuni 1987; Petterson *et al.* 1999*b*; McInnes 2000). Savo has erupted on a number of occasions in the historical past (Petterson *et al.* 2003), most recently between 1830 and 1840. It is a stratocone with a gabbro-basalt–basaltic andesite volcanic core and thick sequences of acid andesite–dacite, block-rich ash flows. The most recent volcanic activity on Savo has involved crater-centred pyroclastic eruptions that have deposited a concentrically symmetrical series of block and ash deposits around the island. The only other subaerial volcano within the country of the Solomon Islands is the Tinakula volcano in the Santa Cruz Group, which is more closely linked to the Vanuatan arc.

The submarine stratovolcanoes of Kana Keoki and Coleman are situated south of the

Fig. 2. The volcano of Kavachi displaying Surtseyan style explosive eruptions. Kavachi is the most spectacular example of present-day Solomon arc volcanism.

New Georgia Group, and south of the South Solomon trench. These volcanoes appear to be geologically young. They form part of an arc-related volcanic field (which also includes Simbo and the Ghizo ridges) that is built upon young Woodlark Basin oceanic crust (Taylor & Exon 1987; Crook & Taylor 1994). Although the current triple junction of the Woodlark spreading system with the trench is believed to be a trench–trench–transform junction, subduction has occurred along what is now the transform zone in the past. This arc-related volcanic field is a most unusual form of arc volcanism, because it occurs on the down-going, subducting plate (Fig. 1b). The older, more typical volcanic structures of the current stage of Solomon arc volcanism are Miocene–Pliocene–Pleistocene in age and extend from Choiseul and the New Georgia Group in the NW to western Guadalcanal in the SE. These volcanic rocks form a series of strato-volcanoes and volcanic shields that often coalesce to form composite volcanic fields.

Outline of the geology of Malaita

Malaita is one of the larger Solomon islands and until recently was the most populous, now rivalled by Guadalcanal. It is on Malaita where the most complete and the thickest sections of OJP sequences are to be found on land. The 1990–1995 surveys demonstrated that once the key structure was understood (a series of en echelon asymmetrical NW–SE periclinal structures that form blind tips to upper crustal décollement thrusts), it was possible to sample and observe the thickest and most extensive subaerial outcrops of the OJP. A 3–4 km-thick pile of OJP basalts is exposed, as well as a complete cover sequence of sedimentary and less-abundant alkaline volcanic rocks. The cover rocks are intruded by 34 Ma alnöites. Pipe-like intrusions that are probably similar to the alnöites have been imaged seismically offshore within the main body of the OJP (e.g. Nixon 1980). The sedimentary section and structural geology of the island also provide a record of the collision of the OJP with the Solomon arc. Thus, probably more than any other exposure of OJP rocks in the Solomons, Malaita provides key sections that aid interpretation of the widespread submarine drill holes and seismic lines across the Earth's largest ocean plateau.

In broad terms, the geology of Malaita is simple. A Cretaceous basement sequence of basalt 3–4 km thick and much smaller amounts of basic intrusive rock is conformably overlain by a Cretaceous–Quaternary sedimentary cover sequence some 1–3.5 km thick. Tectonic thickening has occurred and is particularly evident in the cover sequence. The sedimentary sequence is dominated by fine- to very-fine-grained pelagic calcareous sediment with minor mudstones and cherts, is punctuated by Eocene alkaline basaltic volcanic rocks and, as noted above, is intruded by Oligocene alnöites. Only the youngest Upper Pliocene–Recent sediments (calcisiltites, reef limestones, conglomerates and sandstones) show evidence of deposition within shallow-water–subaerial environments. Over 95% of the Malaitan stratigraphic column was formed by basaltic volcanism followed by slow sedimentation within a deep ocean environment at a considerable distance from continents or even small islands.

The Cretaceous–Early and middle Pliocene rocks have been extensively folded and thrusted during a short period of intense deformation in the middle Pliocene. This deformation event was characterized by NE–SW compression with synchronous, predominantly sinistral, strike-slip. A thin Upper Pliocene–Recent sedimentary sequence unconformably overlies the older deformed rocks.

Previous stratigraphic work

A number of studies have significantly advanced our knowledge of the stratigraphy of Malaita. Rickwood (1955, 1957) mapped parts of northern and NW Malaita, and produced the first geological map of northern and central Malaita. He also produced the first stratigraphic column of the island. Many of the stratigraphic terms he used are retained within the present study. Rickwood recognized the main twofold division of Malaitan stratigraphy; i.e. the basement basaltic sequence and the overlying sedimentary sequence. His terms and definitions of Kwaraae Mudstone and Alite Limestone are retained in the present paper. The basement sequence, which he termed the Fiu Lavas, has been renamed the Malaita Volcanic Group (MVG) (Petterson *et al.* 1997). Rickwood's 'Suaba Chalk' is broadly equivalent to the Haruta and Suafa Limestone Formations as defined in the present paper (see also Petterson *et al.* 1997).

Studies of the sedimentary characteristics and fossil content of the Malaitan pelagic sediment sequence were published by McTavish (1966), Coleman (1968) and Van Deventer & Postuma (1973). These studies confirmed the deep- to very-deep-water pelagic depositional environments of the older Malaitan sediments, and were able to define more precisely the sedimentary facies and diversity of fauna found within the sediments. Nannofossil ages ranging from Albian (Van Deventer & Postuma 1973) to Senonian–Maastrichtian (Coleman 1968) to Pliocene were determined. These stratigraphic studies helped to define the age of the Alite, Haruta and Suafa Limestone Formations.

Mapping progress continued with the publication of a whole-Malaita map based on aerial photograph interpretation by Allum (1967). Hackman (1968) described the folding style of Malaitan limestones. Hughes & Turner (1976) published the results of a comprehensive geological mapping survey of south Malaita and Maramasike (the island immediately south of Malaita, formerly also called Small Malaita). The work of Hughes & Turner represents a major advance in our understanding of the stratigraphy; they defined the stratigraphic column and stratigraphic age range of different lithological units more precisely than any previous study. This work resulted in serious comparisons (as opposed to rather speculative previous comparisons) being made between the stratigraphy of Malaita and that of the OJP (e.g. Hughes & Turner 1976). Like Rickwood's Fiu Lavas, Hughes & Turner's 'Older Volcanics' were formally defined as part of the MVG, and their 'Younger Volcanics' as the Maramasike Volcanic Formation by Petterson *et al.* (1997). The Apuloto Limestone of Hughes & Turner has been renamed the Alite Limestone Formation by Petterson *et al.* (1997), as Rickwood's original term has precedence. Similarly, the 'Hada Calcisiltites' as defined by Hughes & Turner are broadly equivalent to the Suafa Limestone Formation as used by Petterson *et al.* (1997) and in the present study.

Barron (1993) published a report on the geology of northern Malaita that included the results of a nannofossil study by Hine (1991). Many of the stratigraphic units defined in Barron's report are retained in the present paper, including a new local stratigraphic unit, the Lau Limestone Formation.

The stratigraphic age of the Malaitan sequence was further determined by a series of papers resulting from studies of the Malaitan alnöites (Allen & Deans 1965*a*, *b*, *c*; Davis 1977; Nixon & Coleman 1978; Nixon & Boyd 1979; Nixon 1980; Nixon *et al.* 1980; Neal 1983, 1985; Nixon & Neal 1987; Neal & Davidson 1989). Alnöites intruding the Alite and Haruta Limestone Formations have yielded an Oligocene U–Pb zircon age of 34 Ma (Davis 1977).

Several ^{40}Ar–^{39}Ar age determinations of MVG basalts have been published, together with age measurements of basalts from other

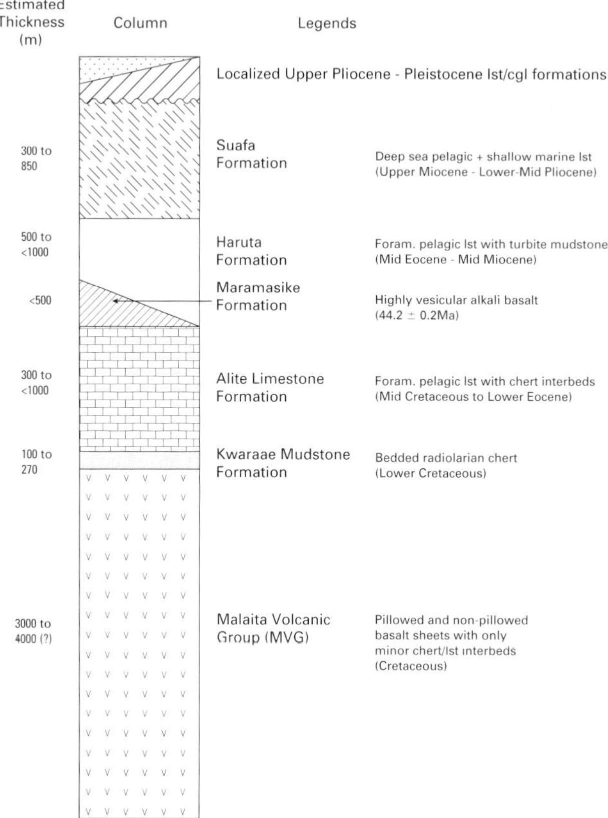

Fig. 3. Generalized stratigraphic column for Malaita (Petterson *et al.* 1999). Reproduced, courtesy of Elsevier Publications.

exposures of OJP basement in the Solomons and from the drill sites on the plateau. The MVG basalts are early Aptian in age, around 120–125 Ma. Readers are directed to Tejada *et al.* (1996, 2002) for details.

Generalized stratigraphy of Malaita

Figure 3 and Table 1 summarize the names of the most important lithostratigraphic units together with their respective stratigraphic age and their relationship to other named stratigraphic units in the literature. Figure 4 shows a simplified geological map of northern and central Malaita. A total of 10 stratigraphic units have been recognized. They are defined by a range of lithological criteria including grain size, composition, bedding characteristics, structural style, field and weathering appearance, and petrographic characteristics. A unit has only been defined if it proves to be useful from a mapping perspective; i.e. that its recognition increases understanding of the geology, allows further elucidation of structure and that it is possible to map the unit over a medium- to large-scale area. All previous work has been taken into account in proposing stratigraphic units and, where possible, names that have literature precedence have been used, although some have been redefined. There has been an attempt to use stratigraphic names that are compatible with current stratigraphic usage. Thus, stratigraphic units either have a group or formation status; no member status has been assigned to any unit.

The MVG has been given a group status, mainly because of its considerable thickness, although Tejada *et al.* (2002) have shown that the volcanic rocks can be subdivided into formations (e.g. the Singgalo and Kwaimbaita Formations) on the basis of geochemistry.

The overlying pelagic sedimentary units, which together with the basaltic basement have been extensively deformed, are named the Kwaraae Mudstone, Alite Limestone, Haruta Limestone and Suafa Limestone Formations, respectively. They are predominantly fine- to

Table 1. *Comparison between stratigraphic nomenclature used in the present study and stratigraphic nomenclature used by previous workers. See main text for further discussion*

Rickwood (1957) North Malaita	Hughes & Turner (1976, 1977) South Malaita	Barron (1993) North Malaita	Present study	Stratigraphic age
	Rokera Limestone	Rokera Limestone	Rokera Limestone Formation	Pleistocene
	Hauhui Congolomerate		Hauhui Conglomerate Formation	Pleistocene
Tomba Silts		Tomba Limestone	Tomba Limestone Formation	Pliocene
		Lau Limestone	Lau Limestone Formation	Pliocene
Suaba Chalk	Hada Calcisiltites	Suafa Calcisiltite Formation	Suafa Limestone Formation	Miocene–Pliocene
Suaba Chalk	Haruta Calcisiltites	Suafa Calcisiltite Formation	Haruta Limestone Formation	Eocene–Miocene
	Younger Volcanics	Younger Volcanics	Maramasike Volcanic Formation	Eocene
Alite Limestone	Apuloto Limestone	Alite Limestone Formation	Alite Limestone Formation	Cretaceous, (Albian)–Eocene
Kwaraae Mudstone	Kwaraae Mudstone	Kwaraae Mudstone Formation	Kwaraae Mudstone Formation	Cretacious, (Aptian–Albian)
Fiu Lavas & Fo'ondo Clastics	Malaita Volcanics (Older Volcanics)	Malaita Volcanics	Malaita Volcanic Group	Cretaceous, (Aptian) (125–120 Ma)

very-fine-grained calcilutites–calcisiltites ± chert ± mudstone, except for the Kwaraae Mudstone, which is chert-dominated. These sedimentary units are Cretaceous–Pliocene in age and, as noted above, have a combined present-day apparent thickness of 1–3.5 km. The Maramasike Volcanic Formation comprises basalts with a variable stratigraphic thickness that erupted during Eocene time.

Unconformably overlying the basement and pelagic sedimentary cover sequence are a number of local stratigraphic units, with individual thicknesses of tens of metres to as much as 200 m. These units comprise calcisiltites, uplifted coralline limestones, conglomerates and sandstones/siltstones of Pliocene–Pleistocene age.

Malaita Volcanic Group (MVG)

Location

In north and central Malaita, the MVG crops out as inliers within the cores of four main anticlinal to periclinal structures (Fig. 4). From south to north these structures are: (1) the Kwaio Anticline; (2) the Kwaraae Anticline; (3) the Fateleka Anticline; and (4) the Toambaita Anticline. The MVG also crops out within the cores of several smaller anticlines, and within thrusted sequences.

Thickness

The total thickness of the MVG is impossible to determine at present, as it represents the basement of Malaita and its lower contacts are not exposed. Balanced cross-sections indicate an *exposed* basement thickness on Malaita of 3–4 km (Petterson *et al.* 1997). However, it is possible that this thickness includes some tectonic thickening. The cover limestones are strongly deformed in places: evidence of thrust faults and tight, recumbent, asymmetrical folds abound in such places. Although the basement sequence appears much less deformed than the cover sequence, the absence of obvious marker units in the monotonous MVG stratigraphy makes it difficult to observe structural breaks. It is probable that the MVG accumulated quickly from large-scale eruptions (see below). If so, it is possible that initially high local geothermal gradients may have existed. However, the metamorphic grade at the deepest exposed level of the MVG does not rise above lower greenschist facies, suggesting that a portion of the exposed

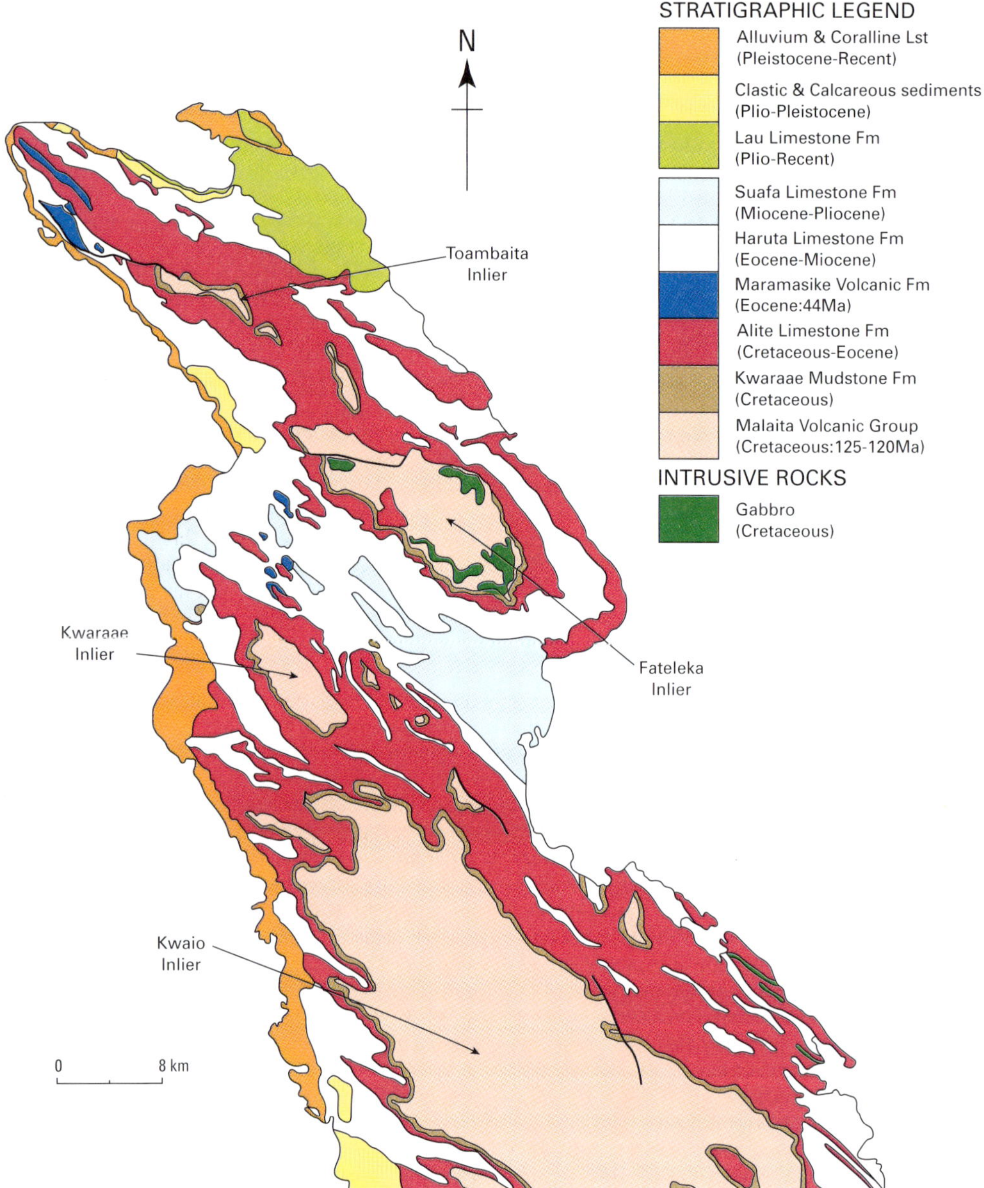

Fig. 4. Geological map of north-central Malaita. Note how the geology is dominated by a series of en echelon periclinal asymmetrical antiforms cored by the Malaita Volcanic Group.

3–4 km thickness could be a result of tectonic thickening in the form of 'cold' thrust stacking. Further research is required to assess the amount of tectonic thickening in the exposed MVG on Malaita.

Seismic studies of the OJP (e.g. Furumoto *et al.* 1976; Gladczenko *et al.* 1997) indicate that the crust of the plateau is as much as 35 km thick. One interesting observation from Malaita is that the great bulk of the exposed MVG comprises basaltic lava flows and some sills, and very little else; this result is consistent with a much thicker lava sequence than for normal ocean crust. Sheeted dykes have not been encountered. Gabbro bodies are present, especially within the northern Fateleka Inlier; however, they only form a minor portion of the total basement outcrop.

Lithological Characteristics

Basalts. The MVG is dominated by a sequence of parallel sheets of pillowed and non-pillowed submarine basalts with rare interbeds of chert and mudstone (calcareous and non-calcareous), and minor microgabbro and gabbro intrusive rocks. Individual basalt sheets vary in thickness between 60 and 80 m; >95% of measured basalt sheets are *c.* 25 m thick and about 50% are 5–10 m thick. Grain size in the basalts varies between very fine and coarse, with an average of fine to medium grained. Basalt textures vary and include glassy, fine-grained aphanitic, porphyritic, ophitic–subophitic and amygdaloidal/vesicular. Most basalts are dark to steel grey, typically with a melanocratic appearance; however, the finer-grained sheets are a deep coal-black, whereas the coarser-grained sheets are a lighter shade of grey and have a rather spotted appearance which is mainly a result of the weathering of subophitic textures. A number of sheets have a 'knobbly' appearance that is caused by tropical 'onion skin' weathering and/or poorly developed pillow structure. The mineralogy is dominated by augite and plagioclase, with a porphyritic or ophitic–subophitic texture being most common. Some sheets are amygdaloidal, particularly toward the outer zones of pillows, although amygdales are rare overall and many basalts are non-vesicular. Where present, amygdales are composed of zeolite (usually natrolite), calcite or microcrystalline quartz. Figure 5 illustrates some typical field exposures of MVG basalts.

Trap-like topography is particularly well-developed in areas where dips are relatively gentle (20°–45°), such as along the southern limb of the Fateleka Anticline (e.g. Longana River) or the western side of the Kwaio Anticline (e.g. the Singgalo River). These areas comprise a series of scarp and dip slopes, with the top surfaces of basaltic sheets forming benches (Fig. 5c) and the edges forming scarp slopes, where waterfalls are common.

Pillow lava flows are commonly almost 100% pillowed. Pillow diameters vary between 20 cm and 4–6 m, with a modal range of 60 cm–1 m. Joint surfaces within pillow structures may have a veneer of carbonate ± sulphide. Palagonitized basaltic glass, brecciated pillow skin, calcite, fine-grained basaltic fragments or simply voids occupy inter-pillow spaces. In some cases, malachite, chalcopyrite and/or quartz form minor components of the inter-pillow material.

Unpillowed lavas have a range of field characteristics. Many sheets are relatively featureless and massive, apart from ubiquitous jointing and, in some cases, pipe vesicles. In general, unpillowed sheets form the thicker, coarser-grained sheets; some are probably doleritic sills rather than lavas. In most unpillowed sheets, joint sets aligned parallel to and orthogonal to sheet boundaries are well developed. A minor proportion of sheets are columnar-jointed (Fig. 5b); one sheet exposed by the Kwaimanafu River exhibits radial columnar jointing over a lateral distance of about 10 m. Also exposed along the Kwaimanafu River is a terminal lobe of a thick sheet that exhibits a massive medium-grained (doleritic) core and an outer pillowed zone. Contacts between pillowed and unpillowed sheets are particularly clear in the field. These contacts show that some sheets are locally discordant and laterally thin or thicken. These characteristics may reflect lava flowing over locally irregular topography and are most consistent with lobes of compound flows, although some sheets may be sills. Exposure of basalt sheets is almost 100% along the upper courses of some of the larger Malaitan rivers (e.g. parts of the Longana River, the Kwaimbaita River system, and the Kwaiiafa, Fiu, Alaola, and Singgalo rivers). In such areas, it is possible to study many inter-sheet boundaries. One rather remarkable observation is that most basalt sheets rest directly on the underlying sheet with no intervening non-basalt material; approximately 80% of observed sheet boundaries are of this type, indicating very rapid effusion rates. Where observed, inter-sheet sediment is generally very thin (millimetres to 5 cm), and usually consists of very-fine-grained laminated to massive chert. Green, altered, soft mudstone and calcilutite material have also been observed as inter-sheet fill. These sedimentary interbeds are essentially barren of microfossils

Dolerite and gabbro/microgabbro. A small proportion of basalt sheets attain a considerable thickness; some are 40–60 m thick. The central massive portion of such thick sheets is medium- to coarse-grained dolerite with a predominant augite + plagioclase + opaque mineralogy, and subophitic–ophitic texture. These thick sheets are present within all major outcrop areas of the MVG.

A number of highly discordant basaltic–doleritic dykes have been mapped on the eastern and SE side of the Fateleka Anticline, and associated with the lavas and sills are (at least locally) discordant microgabbro–gabbro intrusions. The dykes are oriented

Fig. 5. (**a**) 'Spherulitic/orbicular' basalt with spherulites of coarse-grained ophitic anorthosite set within a finer groundmass. (**b**) Columnar-jointed sheet overlying pillowed basalt. (**c**) Typical 'trap' river scenery illustrating sheet-on-sheet contacts. (**d**) Intrusive contacts within gabbro. (**e**) Kwaraae Mudstone Formation: alternating calcareous and fissile mudstone layers with calcareous nodules. (**f**) Chevron folded alternating calcisiltites and cherts from the Alite Limestone Formation. (**g**) Alternating turbiditic mudstones and pelagic calcilutites within the Haruta Limestone Formation. Note the sharp bases and extensively bioturbated tops to mudstone units.

approximately N–S or E–W. A number of intrusive phases may be present within a single intrusion (Fig. 5d). One example from the Fateleka Anticline exhibits a 'background' or 'host' rock comprising a medium–coarse subophitic microgabbro–gabbro, which is cut by a banded gabbro that, in turn, is cut by late-stage pegmatitic gabbro and silica-rich veins.

Orbicular/spheruloidal basalt and gabbro. An unusual and striking facies observed within the MVG is termed here 'orbicular basalt', or 'spheruloidal basalt' (Fig. 5a) in reference to the superficial textural similarity it has with orbicular granite. Unfortunately, this lithology was only observed in the float. It comprises 1–6 cm-diameter spherulites of predominantly coarse–very coarse anorthosite set within a basaltic–doleritic host (Neal *et al.* 1997). The spherulites are well rounded, and display both radial and concentric jointing and an interlocking texture of augite and plagioclase crystals.

Petrography

The MVG have a rather simple mineral composition, which is dominated by approximately equal modal abundances of clinopyroxene and plagioclase, with around 5% opaque minerals and a variable proportion (5–30%, average 10–20%) of glass. Plagioclase has a maximum modal abundance of 50–58%, and clinopyroxene about 45%. Olivine pseudomorphs are present in 14% of the thin sections studied. A few thin sections contain trace amounts of apatite and/or sphene.

Plagioclase usually forms subhedral–euhedral crystals. In coarse-grained rocks it tends to have an ophitic or subophitic relationship with clinopyroxene. In the fine-grained rocks plagioclase tends to form variolitic to random aggregates of laths. Clinopyroxene forms large euhedral–subhedral crystals in coarser-grained rocks. In many sections, however, clinopyroxene crystals tend to be subhedral or even anhedral, particularly in the fine-grained rocks, and display an intergranular or intersertal texture. Where present, olivine usually forms small subhedral–euhedral crystals, with an average modal abundance of 5–7%. Olivine is almost invariably totally altered to iddingsite or smectite. Many thin sections are glass-rich, containing up to 30% modal abundance of glass, although 10–20% is usual. The glass is green-brown, has an interstitial texture, and varies from locally almost unaltered to (usually) totally devitrified and replaced by clays. Some thin sections contain a small number of vesicles or amygdales that are filled with zeolite, calcite or clay. The amygdales tend to be small and irregular to circular.

Eruptive environment

The key field evidence for the eruptive and final depositional environment of the MVG includes the following: (1) both pillowed and massive submarine basalt sheets are present; (2) there is very little inter-sheet sediment (basalt usually rests on top of basalt); where present, it is chert or fine-grained muddy limestone; (3) both thin (<1 m) and very thick (>50 m) basalt sheets are present within the volcanic pile; (4) overall, the MVG form a remarkably thick monolithological unit; and (5) some basaltic dykes and relatively small gabbroic intrusions are present.

These data suggest that MVG volcanism occurred within a deep oceanic setting, probably near or below the calcite compensation depth and far-removed from sources of terrigenous sediment. The great thickness of tholeiitic basalt coupled with the lack of inter-flow sediment implies high to very high effusion rates. The substantial range in lava flow thickness may imply contributions from both proximal and distal eruptive centres, although local thinning and ponding on an irregular topography also played a role. The intrusive rocks demonstrate that the lava pile was thickened to some extent by magmatic underplating. No volcanic centres, major dyke swarms or zones of significant hydrothermal activity have been found. Thus, the favoured eruptive environment is deep submarine, with multi-centred, large-volume and, probably, fissure-fed eruptions.

Cover sequence geology

Below, a summary of key lithological characteristics among the cover rocks that blanket the MVG is given. Readers are referred to Petterson *et al.* (1997) for more detailed descriptions.

Kwaraae Mudstone Formation (KMF)

The term Kwaraae Mustone Formation (KMF) is used to define a relatively thin sequence of Aptian–Albian siliceous–calcareous mudstones that overlie the MVG and are in turn overlain by the Alite Limestone Formation. The distribution of the KMF outcrop closely follows that of the MVG; the KMF is seen to 'wrap around' the major anticlinal structures of Malaita (Fig. 4), such as the Kwaio, Kwaraae, Fateleka and Toambaita anticlines. The KMF is also present within thrusted sequences, especially within the Auluta Thrust Belt (Petterson *et al.* 1997). The

KMF is the thinnest regional stratigraphic formation on Malaita. Estimated thickness is 100–270 m. Rickwood (1957) estimated a thickness of 400 ft (130 m) and Hughes & Turner (1976, 1977) estimated a thickness between 0 and 200 m in southern Malaita. The mudstones thin southwards within south Malaita until they are finally wedged out of the stratigraphy altogether in southernmost Malaita and Maramasike, where the Alite Limestone Formation directly overlies the MVG.

A typical outcrop of KMF (Fig. 5e) comprises parallel-laminated to thinly bedded mudstones that are siliceous in overall composition, but grade upwards to a more calcareous composition. The mudstones are pale-green–grey or grey-black, and fine to very fine grained. Very regular, parallel, sedimentary laminations are typical, indicating a generally low-energy depositional environment; individual laminations are usually millimetres to less than 1 mm thick. The more siliceous parts of the mudstones are hard and flint-like, exhibiting conchoidal–subconchoidal fracture; they may contain chert nodules and more persistent partings or laminations of chert. In addition, calcareous nodules are present.

The KMF also contains pyrite or marcasite nodules. They are generally spherical–subspherical, although more irregular shapes exist. The nodules vary in size between about 1 and 20 cm (longest axis), with fist-size nodules being common. In some rivers, pyrite–marcasite nodules in the float are the only evidence that the mudstones actually crop out; the mudstones themselves are not exposed. In thin section, the KMF typically comprises a very fine-grained, matrix-dominated rock, with the largest 'clasts' being radiolarian or foraminiferal tests. The KMF resulted from deep-water, low-energy, pelagic deposition during Aptian–Albian times.

Alite Limestone Formation (ALF)

The term Alite Limestone Formation (ALF) is used to define a thick sequence of alternating limestones and cherts that conformably overlies the KMF and is in turn conformably overlain by the Haruta Limestone Formation or, more locally, the Maramasike Volcanic Formation. Thickness estimates for the ALF vary between 300 and 1000 m, depending on location. These numbers compare well with thickness estimates of 2000–3000 ft (700–950 m) by Rickwood (1957) for the Alite Formation in north Malaita, and 200–550 m for the Apuloto Limestone in south Malaita by Hughes & Turner (1976, 1977). Tectonic thickening is important in the ALF. In nearly all of north-central Malaita (>95%), the ALF is conformably overlain by the Haruta Limestone Formation. However, locally the ALF is overlain by, and in part interbeds with, the Maramasike Volcanic Formation.

Most commonly, the ALF comprises an alternating sequence of hard, porcelaneous, well-bedded foraminiferal limestones with interbedded chert layers and pods. The limestone facies is very fine to fine grained and is most suitably classified as a calcilutite. It is generally thinly bedded, although massive units are not uncommon; parallel bedding and, rarely, lamination signify a relatively low-energy, current-free depositional environment. Rare, hard calcilutite nodules up to 50 cm long are situated within a very soft finer-grained calcilutite host. The chert facies forms layers that are parallel with the general bedding trends within the formation as a whole. Chert layers vary from red to pink to white and, rarely, green. The chert is very fine grained and usually massive, displaying a typical conchoidal fracture. Chert layers vary between 10 and 30 cm thick, comprising between 15 and 40% (average 20–25%) of the total formation, although this figure is variable and correlates inversely with stratigraphic height. The limestone interbeds between the chert layers are decimetres to 2 m thick. Sometimes the chert occurs in the form of nodules rather than continuous beds. Within some parts of the ALF the chert only forms a minor component of the formation or, more rarely, is absent.

The relative proportions of chert and limestone within the ALF vary systematically. Basal units are the most silica-rich, with siliceous limestones and continuous chert bands (as opposed to discrete pods) being particularly abundant. The number of chert beds decreases and the thickness of limestone interbeds between chert layers increases with stratigraphic height. The uppermost beds of the ALF display a transitional facies with the Haruta Formation, with the onset of rare, thin, wispy mud horizons.

The viscosity contrast between the relatively thin and brittle chert layers and the thicker, more ductile calcilutite layers produces a characteristic deformational style that is unique within the Malaitan stratigraphy. Pinch-and-swell and boudinage structures are very common. Fold scales are visible on an outcrop scale (metres to tens to hundreds of metres). Fold structures are generally geometrically predictable rather than chaotic (e.g. Fig. 5f). Chevron folds are most common, together with parallel folds with rounded fold noses. All shapes and varieties of minor 'Z', 'S' and 'M' folds are locally present, together with kink folds. Locally, fold structures are disharmonic. Thrust structures are commonly observed, sometimes with imbricate horse structures stacked in between sole and roof thrusts, or sheared out synforms in between adjacent antiforms. Many of the folds are recumbent and nappe-like; commonly, these structures 'break' to produce small thrusts as the shortening factor becomes too great to be accommodated by folding alone. Ductile shear zones up to 2 m wide were observed at a number of localities.

The most common petrographic facies within the ALF is a foraminifera-rich calcilutite. The foraminiferal content of individual thin sections is patchy and variable, although not to the same degree as for the Haruta Formation. Foraminiferal tests constitute between 10 and 60% (average 30–45%) of an individual section.

The matrix of the foraminiferal calcilutite facies is usually very fine grained. In most sections a trace modal abundance of plagioclase, strained quartz and vitric grains are present within the matrix. The chert layers within the ALF usually comprise a very fine grained silica mud with rare radiolaria, which can form up to 15% of the section.

The ALF contains a host of diagnostic microfauna, which have been used by a number of workers to date the formation. Van Deventer & Postuma (1973) determined a Cenomanian age for samples from the lower part of the ALF and inferred an Albian age for the base of the formation, whereas the youngest samples were Middle Eocene. Coleman (1968) determined a Senonian–Maastrichtian age for samples from near the base of the ALF. Hughes & Turner (1976) determined Early Cretaceous and Palaeocene–Eocene foraminiferal ages. Barron (1993) quoted ALF micropalaeontological ages of Maastrichtian–Middle Eocene. These data are in line with micropalaeontological ages determined from drill cores recovered during Ocean Drilling Program Leg 130 on the OJP (e.g. Sliter & Leckie 1993).

The ALF is dominated by a pelagic microfauna; few or no benthic species are present. There is no significant amount of terrigenous material; rare feldspar and vitric grains probably came from distal volcanic eruptions. There also is no evidence of high-energy currents. The origin of the chert layers is open to debate; they may have begun as deposits of radiolarian ooze, or they may be later chemical precipitates that formed as a result of diagenetic mobilization of any silica present in the original sediments. If the cherts represent primary deposits, then they could be used to infer that at least part of the ALF was deposited near or below the calcite compensation depth. Thus, it can be concluded that the ALF was deposited within a relatively deep oceanic setting far-removed from any source of terrigenous material, and as a result of slow deposition of pelagic faunal tests and lime mud within a low-energy, relatively current-free environment.

Maramasike Volcanic Formation (MVF)

The term Maramasike Volcanic Formation (MVF) is used here to define all alkaline basaltic lavas, sills and related intrusions that crop out throughout Malaita at the same stratigraphic level; i.e. in general terms, between the ALF and Haruta Limestone Formation. Tejada *et al.* (1996) reported an ^{40}Ar–^{39}Ar plateau age of 44.2 ± 0.2 Ma for an MVF lava, consistent with the Eocene biostratigraphic age of the upper ALF and lower Haruta. The MVF crops out locally within north-central Malaita but more extensively in southern Malaita. It does not exceed 500 m in thickness, and may be considerably thinner in some areas (<200 m). Key lithological observations in southern Malaita and Maramasike (Hughes & Turner 1976) include the following: peperitic sills intrude the ALF; and an average of 300 m of vesicular pillow lavas are sandwiched between the ALF and Haruta limestones. Barron (1993) described the MVF lavas as typically highly brecciated, highly weathered, pillowed, porphyritic and highly amygdaloidal. Tachylitic rocks and hyaloclastites are also present. These features imply rapid chilling of gas-charged lava with quench-fragmentation and autobrecciated volcaniclastic activity. The presence of peperite sills and hyaloclastite deposits demonstrates that magma–sediment and magma–water interaction occurred, which allows for the possibility that volcanism may have been relatively explosive, particularly compared to the MVG.

The highly amygdaloidal and autobrecciated nature of the Maramasike basalts contrasts greatly with the generally non-vesicular character and relatively quiescent eruptive nature of the MVG basalts. The differences are strongly suggestive of a shallower-water eruptive environment for the Maramasike basalts. Field evidence suggests that the Maramasike basalts issued from a number of local eruptive sites within Malaita, many of which were only small-volume centres. Thus, a multi-centred submarine volcanic field is envisaged.

Haruta Limestone Formation (HLF)

The Haruta Limestone Formation (HLF) is a thick sequence of calcilutites–calcisiltites with interbedded mudstones, which overlies the ALF or, locally, the MVF, and is conformably overlain by the Suafa Limestone Formation. The HLF crops out over large areas of north and central Malaita. Estimated thicknesses of the HLF are between 500 and 1000 m.

The most fundamental characteristic of the HLF is that it comprises a fine- to very-fine-grained well-bedded, foraminiferal limestone with regular interbeds of mudstone (Fig. 5g). The host limestone, where unbioturbated, is a reasonably pure lime mud composed of foraminiferal tests and material formed by chemical precipitation; the limestone is much softer and less compacted than limestones in the ALF. Within most outcrops, a significant proportion of the limestone component of the HLF is bioturbated. The bioturbation has produced an intimate mixture of lime and mud that gives the limestone a 'dirty' appearance. The mud occurs within the limestone matrix as particles, flakes, or broken up and disrupted thin laminae that give the rock a flaser-bedded appearance. The limestone is parallel-bedded, and thickly laminated to medium-bedded; rarely, massive units are present.

The limestone matrix of the HLF is host to a regular series of mudstone layers that typically give the HLF a banded appearance. The thickness of individual mudstone beds and the frequency at which they occur throughout the formation are variable. Mudstone layer thicknesses may vary between millimetres and 20 cm, with mudstone horizons occuring every 10 cm–1 m, averaging every 40–50 cm. They have sharp, sometimes erosional, bases, and highly diffuse and bioturbated tops; because of these characteristics they form excellent way-up criteria. The mud/lime ratio decreases with stratigraphic height within an individual mudstone layer and the colour changes correspondingly, from black within the lower parts to dark–medium grey in the upper parts.

The folding style of the HLF is quite distinct from that of the ALF. Larger, more open folds are typical within the HLF, in contrast to the more localized, small-scale, tight folds within the ALF. At many outcrop localities, only parts of fold structures are exposed, and the structure has to be pieced together, bit by bit, along several tens of metres of discrete river exposures. The HLF exhibits a moderate–high degree of intra-formational thrusting on a local scale. Where the HLF is highly sheared within these intra-formational thrust zones, it takes on a sigmoidal C–S style fabric, in which highly sheared bands separate domains/bands of limestone with a sigmoidal fabric.

In thin section, a typical well-preserved pelagic foraminiferal HLF limestone may contain foraminiferal tests showing a high level of internal and external morphology. The external surfaces of the tests are ornamented with a regular geometric arrangement of circular depressions or granules/pores. The tests range in diameter from 0.5–0.7 to <0.1 mm. The abundance of tests is highly variable and patchy, ranging from a maximum percentage of 60% to a minimum value of 5–10% in the more matrix-dominated areas. Thin sections contain a variable modal abundance of plagioclase and vitric fragments.

The turbiditic mudstone facies of the HLF typically shows a sharp contact zone between turbiditic mudstone and a mixed pelagic limestone and mudstone. The mudstones comprise a vitric-crystal-rich mudstone, with vitric clasts and crystals forming between 30 and 70% (average 40–50%) of the total mudstone.

The HLF contains abundant microfauna that are diagnostic of age. McTavish (1966) identified Upper Eocene–Oligocene microfossils, whereas Hughes & Turner (1976) quoted an age range of Upper Eocene–Middle Miocene. Hine (1991) determined the ages of 10 HLF samples as Lower Miocene. One further age constraint is the 34 Ma (Oligocene) U–Pb zircon age determined for the Malaitan alnöites, which intrude the HLF.

Many of the arguments used above to elucidate the depositional environment of the ALF can be used for the dominant lime mud of the HLF. Unlike the ALF, during the time the HLF was deposited, Malaita was clearly not situated at a great distance from a source of terrigenous material. The mudstone bands represent erosion and transportation by turbidity currents of largely volcaniclastic material from some island(s) or continent, and possibly include some primary pyroclastic ash input. The mudstones formed by rapid deposition of clastic material from turbidity currents; these events were essentially instantaneous relative to the very slow 'background' pelagic sedimentation. The time intervals between deposition of individual mudstone bands were large enough to allow for intense bioturbation to occur. Deposition of the mudstone units greatly accelerated sedimentation rates during HLF time. Hughes & Turner (1976, 1977) suggested that the mudstones are distal turbidites of material derived from the basement of some of the older Solomon Islands, which were subaerial at the time. They suggested the Cretaceous Mbirao Volcanics of Guadalcanal as a possible source for the mudstones. This proposal has two important implications. (1) The Palaeocene–Lower Miocene arc volcanism occurred above a S-directed (Vitiaz) subduction zone. A trench was thus situated between the arc and the OJP during the time of HLF deposition; yet, it could not have acted as a totally effective physical barrier to the transport (if this were the source area) of detrital material from the Vitiaz arc. (2) Malaita was within 'striking distance' of turbidites that probably originated in the area of the Vitiaz arc during HLF time.

Suafa Limestone Formation (SLF)

The Suafa Limestone Formation (SLF) comprises a sequence of soft, massively bedded, silty to coarse-grained, highly fossiliferous limestones that conformably overlie the HLF and are unconformably overlain by a number of localized Plio-Pleistocene stratigraphic units. The outcrop pattern of the SLF is domainal rather than contiguous, which contrasts with the ALF and HLF. The thickness of the SLF is highly variable and probably structurally controlled: current estimates based on balanced cross-sections put the thickness at 300–850 m. The basal parts of the SLF represent a transitional facies between typical HLF and SLF facies.

The main body of the SLF is composed of a soft, cream-brown–grey, medium to thickly bedded calcisiltite–calcarenite, which is moderately well cemented and contains an abundant pelagic (e.g. pelagic foraminifera) and benthic (e.g. benthonic foraminifera, mollusk shell debris) fauna. The limestones display a characteristic pitted karstic weathering surface. This main facies is interbedded with dark grey calcareous mudstones–siltstones, and medium- to coarse-grained, cross-bedded sandstones. The limestones are often intensely bioturbated with worm burrows up to 3 mm in diameter. The bioturbation tends to introduce coarser, more arenaceous material into the predominantly silt-grade host limestone in the form of burrow fillings and related sedimentary structures.

Thin sections of the SLF reveal a very-fine- to fine-grained carbonate matrix, with foraminiferal and other benthic shell material. Foraminiferal tests may form up to 60% of the rock. Volcanic and lithic grains may form as much as 20% of the rock, and include plagioclase, glass shards, opaques, pyroxene and hornblende. Some thin sections from the Lower Fiu River Valley are highly fossiliferous and contain a much more diverse fauna than is present in the older formations (i.e. HLF and older). Many foraminferal morphotypes are present, and mollusk shell fragments are common. Thus, both pelagic and benthonic fauna are present and indicate a much shallower-water environment than those of the older formations.

McTavish (1966) was the first worker to positively identify Upper Miocene–Pliocene fauna from probable Suafa deposits. Van Deventer & Postuma (1973) and Hughes & Turner (1976) also determined Upper Miocene–Pliocene micropalaeontological ages. Hine (1991) determined a Lower–Middle Pliocene palaeontological age.

Several lithological, palaeontological and petrographical characteristics present within the SLF indicate that the sedimentary environment of Malaita changed considerably during the Upper Miocene–Pliocene. The SLF shows clear evidence of current activity (cross-bedding), soft-sediment deformation (convolute bedding), input of terrigenous material, including calc-alkaline arc volcanic material (presence of reasonably abundant igneous crystals such as biotite and hornblende, and lithic fragments) erosion of OJP cover material by submarine erosion or perhaps subaerial erosion (presence of likely ALF lithic clasts) and sediment deposition within shallower water than had occurred previously (many of the above points plus the presence of benthic fauna, including mollusk shell fragments). Thus, the SLF records sedimentation on the Malaitan segment of the OJP as it began its journey from the deep waters of ALF and HLF times to medium water depths (?2000–1000 m) during early SLF time, to subsequent emergence above sea level in the Late Pliocene or Pleistocene (e.g. Table 2).

Local stratigraphic formations above the Middle Pliocene unconformity

The formations described above are relatively thick stratigraphic units of at least island-wide, regional extent. All of these formations (MVG through to the SLF) underwent a period of deformation, uplift and erosion that produced an island-wide middle Pliocene unconformity. From the Upper Pliocene to the Pleistocene, a number of relatively thin and localized formations were deposited onto the surface of the unconformity. Four such formations have been identified and given separate formational status names as a result of the 1990–1995 surveys. These formations are: (1) Lau Limestone Formation; (2) Tomba Limestone Formation; (3) Hauhui Conglomerate Formation; and (4) Rokera Limestone Formation. The formations are a series of comparatively deep-water

Table 2. *Tectonic and chronological summary of Malaita and the OJP (Petterson* et al. *1999a). Reproduced courtesy of Elsevier Publications*

Tectonic event	Associated structures	Volcanism/ sedimentation	Timing	References
Formation of OJP	Extensional structures related to oceanic plateau formation	Massive ?fissure or ?rift-centred, plume-related, basaltic magmatism	122 Ma and *c*.90 Ma	Mahoney *et al.* (1993), Tejada *et al.* (1996)
Passive movement of OJP southwards, punctuated by ?plume-related magmatism	Extensional structures related to 90 Ma plateau-building event. Local (e.g. Faufaumela) basin formation from the Eocene	Deep-sea pelagic sedimentation punctuated by 90 Ma plateau-building basalt, 44 Ma alkaline basalt and 34 Ma Alnöite magmatism	From initial plateau formation to Miocene	Hughes & Turner (1976, 1977), Petterson (1995)
Stage 1 Solomon arc. South-directed subduction at Vitiaz trench	Formation of Vitiaz trench and Solomon arc-block	Arc-related magmatism and sedimentation on numerous Solomon islands	Paleocene/Eocene–Lower Miocene	Kroenke (1984), Coulson & Vedder (1986)
First contact: OJP and Solomon arc-block. Cessation of Stage-1 Solomon arc; temporary cessation of S-directed subduction at Vitiaz trench	Arching of OJP on approach to Vitiaz trench with associated extensional faults. Thrust faulting in trench area	Final stage of Stage-1 Solomon arc magmatism. Poha diorite age of 24 ± 0.4 Ma. Continued deep-sea sedimentation on Malaita	22–22 Ma	Kroenke (1972, 1984), Chivas (1981), Yan & Kroenke (1993)
'Soft-docking' collisional stage between OJP and Solomon block	Possible commencement of thrust and strike-slip deformation within collisional zone of Malaita anticlinorium	Continued deep-sea sedimentation in Malaita area	25–22/8–7 Ma	

calcisiltites and calcilutites (e.g. Lau and Tomba limestones), shallow-water proximal deltaic-fan clastic rocks (Hauhui Conglomerate) and reef-forming limestones (Rokera Limestone).

Alnöite intrusions

A number of alnöite intrusions intrude the MVG and cover sequence up to the stratigraphic level of the HLF. As noted earlier, Davis (1977) dated zircons from one of these alnöites at 34 Ma. The alnöites have been studied extensively. Readers are referred, for example, to Rickwood (1957), Allen & Deans (1965*a*, *b*, *c*), Gerryts (1965), Kroenke (1972), Nixon & Coleman (1978), Nixon & Boyd (1979), Nixon (1980), Neal (1985), Nixon & Neal (1987), Neal & Davidson (1989) and Mahoa & Petterson (1995).

Malaita: an 'accessible' laboratory for studying the OJP

The objective of this paper is to provide a basic factual description-oriented background on exposed sections of the OJP on Malaita. Although Malaita is not one of the most readily accessible places on Earth, it is, and will remain, far more accessible than the submerged body of the OJP. It is my belief that Malaita will become an important natural laboratory in the future for OJP and other plateau-oriented studies. On Malaita, it is possible to test ideas and hypotheses and produce extensive data sets on a range of materials from a range of depths. Future work will refine the geological interpretations presented in this paper and allow more detailed comparisons between on-land and offshore data sets. The author hopes that this paper will form a modest contribution to these predicted future studies.

The author acknowledges the assistance of Solomon Island Geological Survey geologists without whom it would not have been possible to write this paper. Whilst it is not possible to thanks all geologists in person the author acknowledges the help of D. Tolia, A. Mason, W. Satokana, P. Diau, H. Mahoa, A. Ramo, P. Amusae, D. Billy, P. Auga, D. Natogga and C. Qopoto. The author also thanks the people of Malaita and the Solomon Islands who allowed geological surveys to be undertaken on their land. J. Mahoney and A. Saunders are thanked for their reviews of this paper, which significantly improved the quality of the manuscript. J. Mahoney is particularly thanked for his patience. D. Rayner is thanked for assistance with the diagrams. M. G. Petterson publishes by permission of the Director of BGS.

References

ALLEN, J.B. & DEANS, T. 1965*a*. An alnoite breccia associated with the ilmenite–pyrope gravels of Malaita, 1962. Report No. 49A. *British Solomon Islands Geological Record*, **2**, 1959–1962, 136–138.

ALLEN, J.B. & DEANS, T. 1965*b*. An alnoite breccia and ankaratrite associated with the ilmenite–pyrope gravels of Malaita, 1962. Report No. 49B. *British Solomon Islands Geological Record*, **2**, 1959–1962, 138–141.

ALLEN, J.B. & DEANS, T. 1965*c*. Ultrabasic eruptives with alnoitic–kimberlitic affinities from Malaita, Solomon Islands. *Mineralogical Magazine*, **34**, 16–34.

ALLUM, J.A.E. 1967. Regional photogeological interpretation of the British Solomon Islands. Report A6, Geology Division, Ministry of Natural Resources, Honiava, Solomon Islands.

AUZENDE, J.M., COLLOT, J.-Y., LAFOY, Y., GRACIA, E, GELI, L., ONDREAS, H., EISSEN, J-P., LARUE, M.B., OLISUKULU, C., TOLIA, D. & BILIKI, N. 1994. Evidence for sinistral strike slip deformation in the Solomon Islands arc. *Geo-Marine Letters*, **14**, 232–237.

BARRON, A.J.M. 1993. *The Geology of Northernmost Malaita. A Description of Sheets 8/160/7 and 8/160/8*. Publication of the Geological Survey Division, Ministry of Natural Resources, Honiara, **TR 5/93**.

CHIVAS, A.R. 1981. Gechemical evidence for magmatic fluids in porphyry copper mineralization, Part 1. Mafic silicates from the Koloula Igneous Complex. *Contributions to Mineralogy and Petrology*, **78**, 389–403.

COLEMAN, P.J. 1968. Upper Cretaceous deep water pelagic sediments from north Malaita. Report No. 79. *British Solomon Islands Geological Record*, **3**, 1963–1967, 53–57.

COULSON, F.I. & VEDDER, J.G. 1986. Geology of the central and western Solomon Islands. *In*: VEDDER, J.G., POUND, K.S. & BOUNDY, S.Q. (eds) *Geology and Offshore Resources of Pacific Island Arcs–Central and Western Solomon Islands*. Circum-Pacific Council for Energy and Mineral Resources. Earth Science Series, **4**, 59–87.

CROOK, K.A.W. & TAYLOR, B. 1994. Structural and Quaternary tectonic history of the Woodlark Triple Junction region, Solomon Islands. *Marine Geophysical Researches*, **16**, 65–89.

DAVIS, G.L. 1977. The ages and uranium contents of zircons from kimberlites and associated rocks. *In*: *Extended Abstracts, 2nd International Kimberlite Conference, Santa Fe, New Mexico*.

FURUMOTO, A.S., WEBB, J.P., ODEGARD, M.E. & HUSSONG, D.M. 1976. Seismic studies on the Ontong Java Plateau, 1970. *Tectonophysics*, **34**, 71–90.

GERRYTS, E. 1965. A visit to Malaita Island, 1962. Report No. 50. *British Solomon Islands Geological Record*, **2**, 1959–1962, 141–143.

GLADCZENKO, T.P., COFFIN, M.F. & ELDHOLM, O. 1997. Crustal structure of the Ontong Java Plateau: Modeling of new gravity and existing seismic data. *Journal of Geophysical Research*, **102**, 22 711–22 729.

HACKMAN, B.D. 1968. Observations on folding in the Oligocene–Miocene limestones of central Kwara'ae, Malaita. Report No. 76. *British Solomon Islands Geological Record*, **3**, 1963–1967, 47–50.

HINE, N. 1991. Calcareous nannofossil study of Upper Cretaceous to Pliocene sediments from Northern Malaita. British Geological Survey Technical Report. Stratigraphic Series, Report, **PD/91/278**.

HUGHES, G.W. & TURNER, C.C. 1976. *Geology of Southern Malaita. Solomon Islands Geology Division, Bulletin No. 2*. Publication of the Geological Survey Division of the Ministry of Natural Resources, Honiara, Solomon Islands.

HUGHES, G.W. & TURNER, C.C. 1977. Upraised Pacific Ocean floor, southern Malaita, Solomon Islands. *Geological Society of America Bulletin*, **88**, 412–424.

JOHNSON, R.W. & TUNI, D. 1987. Kavachi, an active forearc volcano in the Western Solomon Islands. *In*: TAYLOR, B. & EXON, N.F. (eds) *Marine Geology, Geophysics, and Geochemistry of the Woodlark Basin–Solomon Islands*. Circum-Pacific Council for Energy and Mineral Resources, (CPCEMR) Earth Science Series, **7**, 89–112.

KROENKE, L.W. 1972. Geology of the Ontong Java Plateau. PhD thesis. Hawaiian Institute of Geophysics, University of Hawaii.

KROENKE, L.W. 1984. Cenozoic tectonic development of the southwest Pacific. *UN ESCAP, CCOP/SOPAC Technical Bulletin*, **6**.

KROENKE, L.W., WESSEL, P. & STERLING, A. 2004. Motion of the Ontong Java Plateau in the hot-spot frame of reference: 122 Ma–present. *In*: FITTON,

J.G., MAHONEY, J.J., WALLACE, P.J. & SAUNDERS, A.D. (eds) *Origin and Evolution of the Ontong Java Plateau*. Geological Society, London, Special Publications, **229**, 9–20.

MAHOA, H. & PETTERSON, M.G. 1995. *Stream Sediment Geochemistry of North–central Malaita: Implications for Mineral Reconnaissance and Geological Studies*. Publication of the Water & Mineral Resources Division, Ministry of Energy, Water & Mineral Resources, Honiara, Solomon Island, Memoir, **2/95**.

MAHONEY, J.J., STOREY, M., DUNCAN, R.A., SPENCER, K.J. & PRINGLE, M. 1993. Geochemistry and geochronology of the Ontong Java Plateau. *In*: PRINGLE, M., SAGER, W., SLITER, W. & STEIN, S. (eds) *The Mesozoic Pacific. Geology, Tectonics, and Volcanism*. American Geophysical Union, Geophysical Monograph, **77**, 233–261.

MAHONEY, J.J., FITTON, J.G., WALLACE, P.J. *et al.* 2001. *Proceedings of the Ocean Drilling Program, Initial Reports,* **192**.

MANN, P., TAYLOR, F.W., LAGOE, M.B., QUARLES, A. & BURR, G. 1998. Accelerating late Quaternary uplift of the New Georgia Island Group (Solomon island arc) in response to subduction of the recently active Woodlark spreading center and Coleman seamount. *Tectonophysics*, **295**, 259–306.

MCINNES, B.I.A. 2000. *Kavachi Eruption 2000*. Video publication of CSIRO, North Ryde, NSW, Australia.

MCTAVISH, R.A. 1966. Planktonic foraminifera from the Malaita Group, British Solomon Islands. *Micropalaeontology,* **12**, 1–36.

NEAL, C.R. 1985. *Mantle studies in the western Pacific and kimberlite-type intrusives*. PhD Thesis, University of Leeds.

NEAL, C.R. & DAVIDSON, J.P. 1989. An unmetasomatised source for the Malaitan alnoite (Solomon Islands): Petrogenesis involving zone refining, megacryst fractionation, and assimilation of oceanic lithosphere. *Geochimica et Cosmochimica Acta,* **53,** 1975–1990.

NEAL, C.R., MAHONEY, J.J., KROENKE, L.W., DUNCAN, R.A. & PETTERSON, M.G. 1997. The Ontong Java Plateau. *In*: MAHONEY, J.J. & COFFIN, M.F. (eds) *Large Igneous Provinces: Continental, Oceanic, and Planetary Flood Volcanism*. American Geophysical Union, Geophysical Monograph, **100**, 183–216.

NIXON, P.H. 1980. Kimberlites in the south-west Pacific. *Nature,* **287**, 718–720.

NIXON, P.H. 1981. The prospect of diamonds in the south-west Pacific. *Indiaqua,* **28**, 11–16.

NIXON, P.H. & BOYD, F.R. 1979. Garnet bearing lherzolites and discrete nodule suites from the Malaita alnoite, Solomon Islands, SW pacific, and their bearing on ocean mantle composition and geothermal gradients. *In*: BOYD, F.R. & MEYER, H.O.A. (eds) *The Mantle Sample: Inclusions in Kimberlites and Other Volcanics. Proceedings of the Second International Kimberlite Conference*. American Geophysical Union, **2**, 400–423.

NIXON, P.H. & COLEMAN, P.J. 1978. Garnet-bearing lherzolites and discrete nodule suites from the Malaita alnoite, Solomon Islands, and their bearing on the nature and origin of the Ontong Java Plateau. *Bulletin of the Australian Society of Exploration Geophysicists*, **9**, (3), 103–106.

NIXON, P.H. & NEAL, C.R. 1987. Ontong Java Plateau: deep seated xenoliths from thick oceanic lithosphere. *In*: NIXON, P.H. (ed.) *Mantle Xenoliths*. Wiley, Chichester, 335–346.

NIXON, P.H., MITCHELL, R.H. & ROGERS, N.H. 1980. Petrogenesis of alnoitic rocks from Malaita, Solomon Islands, Melanesia. *Mineralogical Magazine,* **43**, 587–596.

PETTERSON, M.G. 1995. *The Geology of North and Central Malaita, Solomon Islands (Including Implications of Geological Research on Makira, Savo, Isabel, Guadalcanal, and Choiseul Between 1992 and 1995)*. Water and Mineral Resources Division, Honiara, Solomon Islands, Memoir, **1/95**.

PETTERSON, M.G., NEAL, C.R., MAHONEY, J.J., KROENKE, L.W., SAUNDERS, A.D., BABBS, T.L., DUNCAN, R.A., TOLIA, D. & MCGRAIL, B. 1997. Structure and deformation of north and central Malaita, Solomon Islands: tectonic implications for the OJP–Solomon arc collision, and for the fate of ocean plateaus. *Tectonophysics*, **283**, 1–33.

PETTERSON, M. G., BABBS, T., NEAL, C.R., MAHONEY, J.J., SAUNDERS, A.D., DUNCAN, R.A., TOLIA, D., MAGU, R., QOPOTO, C., MAHOA, H. & NATOGGA, D. 1999*a*. Geological–tectonic framework of Solomon Islands, SW Pacific: crustal accretion and growth within an intra-oceanic setting. *Tectonophysics*, **301**, 35–60.

PETTERSON, M.G., WALLACE, S. & TOLIA, D. 1999*b*. Records of explosive Surtseyan eruptions from Kavachi, Solomon Islands in 1961, 1970, 1976, 1978, 1991, 1998 and 1999. Published abstracts of the SOPAC STAR Annual sessions. Suva, Fiji. *SOPAC Miscellaneous Report*, **355**.

PETTERSON, M.G., CRONIN., S., TAYLOR, P., TOLIA, D., PAPABATU, A., TOBA, T. & QOPOTO, C. 2003. The eruptive history and volcanic hazards of Savo, Solomon Islands. *Bulletin of Volcanology,* **65**, 165–181.

PHINNEY, E.J., MANN, P., COFFIN, M.F. & SHIPLEY, T.H. 1999. Sequence stratigraphy, structure and tectonic history of the SW Ontong Java Plateau adjacent to the North Solomon trench and Solomon Island arc. *Journal of Geophysical Research*, **104**, 20 449–20 466.

RICKWOOD, F.K. 1955. Interim report on the geology of Malaita. *In*: GROVER, J.C. (ed.) *Geology, Mineral Deposits and Prospects of Mining Development in the British Solomon Islands Protectorate*. British Solomon Islands Interim Geological Survey, Memoir, **1**, 38–39.

RICKWOOD, F.K. 1957. Geology of the island of Malaita. *Colonial Geology and Mineral Resources*, **6,** 300–306.

SLITER, W.V. & LECKIE, R.M. 1993. Cretaceous planktonic foraminifers and depositional environments from the Ontong Java Plateau with emphasis on Sites 803 and 807. *In*: BERGER, W.H., KROENKE, L.W., MAYER, L.A., *et al.* (eds) *Proceedings of the*

Ocean Drilling Program, Scientific Results, **130**, 63–84.

TAYLOR, B. & EXON, N.F. (eds). 1987. *Marine Geology, Geophysics, and Geochemistry of the Woodlark Basin-Solomon Islands*. Circum-Pacific Council for Energy and Mineral Resources, (CPCEMR) Earth Science Series, **7**.

TEJADA, M.L.G., MAHONEY, J.J., DUNCAN, R.A. & HAWKINS, M.P. 1996. Age and geochemistry of basement and alkalic rocks of Malaita and Santa Isabel, southern margin of the Ontong Java Plateau. *Journal of Petrology*, **37**, 361–394.

TEJADA, M.L.G., MAHONEY, J.J., NEAL, C.R., DUNCAN, R.A. & PETTERSON, M.G. 2002. Basement geochemistry and geochronology of Central Malaita, Solomon Islands with implications for the origin and evolution of the Ontong Java Plateau. *Journal of Petrology*, **43**, 449–484.

VAN DEVENTER, J. & POSTUMA, J.A. 1973. Early Cenomanian to Pliocene deep-marine sediments from North Malaita, Solomon Islands. *Journal of the Geological Society of Australia*, **20**, Pt 2, 145–150.

YAN, C.Y. & KROENKE, L.W. 1993. A plate tectonic reconstruction of the Southwest Pacific, 0–100 Ma. *In*: BERGER, W.H., KROENKE, L.W., MAYER, L.W. *et al.* (eds) *Proceedings of the Ocean Drilling Program, Scientific Results*, **130**, 697–709.

Lower Cretaceous planktonic foraminiferal and nannofossil biostratigraphy of Ontong Java Plateau sites from DSDP Leg 30 and ODP Leg 192

P. J. SIKORA[1] & J. A. BERGEN[2]

[1]*Energy and Geoscience Institute at the University of Utah, 423 Wakara Way, Ste 300, Salt Lake City, UT 84108, USA (e-mail: psikora@egi.utah.edu)*

[2]*BP Exploration; 501 Westlake Park Boulevard, P.O. Box 3092, Houston, TX 77253-3092, USA*

Abstract: The Lower Cretaceous sediments of the Ontong Java Plateau of the SW Pacific Ocean provide a depositional history for the period immediately following the termination of one of the largest extrusive igneous events of the Phanerozoic eon. A more complete stratigraphic record is formulated of this critical event in Earth's history than previously available by integration of previous data and new analyses from DSDP Leg 30 sites combined with shipboard and post-cruise analyses from ODP Leg 192. The oldest sediment occurs within the upper part of the *Leupoldina cabri* planktonic foraminiferal zone, indicating equivalence with the last half of Oceanic Anoxic Event (OAE) 1a of which the Ontong Java eruption is a postulated cause. The remainder of the Aptian section is marked by major disconformities, with little section in common between central and marginal plateau sites. The Aptian–Albian boundary is conformable at both Leg 192 Sites 1183 and 1186 based on integration of biostratigraphy and preliminary $\delta^{13}C$ data. However, the overall Albian interval is very incomplete, with regional distribution noted for only the lowermost and upper Albian sections.

A well-constrained age record for the sediments immediately overlying basaltic basement on the Ontong Java Plateau is based on new biostratigraphic analyses of the Aptian–Albian section. The results provide a minimum age for the termination of one of the largest volcanic eruptive events in Phanerozoic time (Mahoney 1987), as well as providing time constraints for the elucidation of the early post-depositional history of the plateau. New biostratigraphic data from the sites of ODP Leg 192 are combined with new analyses of the Lower Cretaceous sections from DSDP Leg 30 Sites 288 and 289, and then integrated with existing data for the locations; i.e. the original shipboard studies from DSDP Leg 30 (Michael 1975; Shafik 1975), ODP 192 (Mahoney *et al.* 2001), and the integrative studies of Sliter (1992) and Sliter & Leckie (1993). New foraminiferal analyses of a small number of samples from ODP Leg 130 Sites 803 and 807 did not yield additional stratigraphic insights. Nannofossil analyses for these Leg 130 localities are ongoing and will be incorporated with the current results in a later publication. The nannofossil and planktonic foraminiferal zonation used is that of Leckie *et al.* (2002), with modifications as illustrated in Figure 1. The Leg 192 sites are discussed in the order of the most complete recovered Lower Cretaceous sections; i.e. Sites 1183 and 1186, followed by Sites 1185 and 1187. Anomalous Site 1184 is considered last.

Methods

The nannofossil methodology is discussed in Bergen (2004). Nannofossil abundances are highly variable in the Leg 192 mid-Cretaceous sediments, although preservation is consistently poor. All analyses were carried out on smear slide preparations, and varied between 2 and 12 h per slide.

Free-specimen foraminiferal analyses are based on sample volumes of approximately 20 cm^3 that were washed in tap water over a 63 μm mesh. More indurated samples were first crushed using a mortar and pestle before washing. Before processing each sample, sieves were soaked in a solution of methylene blue in order to stain any contaminants remaining from previous washes. Washed residues were dried in an oven at approximately 50°C. Dried sample residues were examined under a binocular microscope with foraminiferal assemblage composition in qualitative terms based on an assessment of species observed in a random sample of

From: FITTON, J. G., MAHONEY, J. J., WALLACE, P. J. & SAUNDERS, A. D. (eds) 2004. *Origin and Evolution of the Ontong Java Plateau*. Geological Society, London, Special Publications, **229**, 83–111. 0305-8719/$15.00

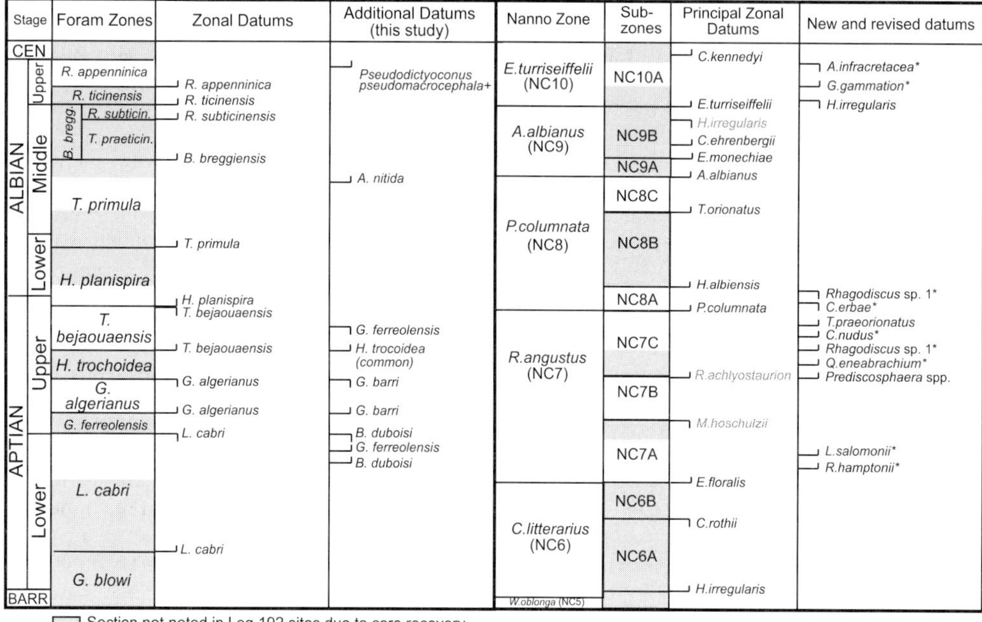

Fig. 1. Integrated Aptian–Albian planktonic foraminiferal and nannofossil zonation used for Ontong Java Sites, after Leckie *et al.* (2002); nannofossil datums in grey were found to be inapplicable to the Ontong Java section; +radiolarian datum; new and revised nannofossil datums are from this paper(*) or Kennedy *et al.* 2000.

200–400 specimens from the >63 μm-size fraction. Relative abundances were reported using the following categories:

A – abundant (>30%);
C – common (15%–30%);
F – few (3%–15%);
R – rare (2%–3%);
T – trace (<2%).

Preservation of planktonic foraminifer assemblages was recorded as follows:

G – good (<30% of specimens showing signs of dissolution or recrystallization/replacement);
M – moderate (30–80% of specimens showing signs of dissolution or recrystallization/replacement);
P – poor (>80% of specimens showing signs of dissolution or recrystallization/replacement).

For recrystallized and silicified samples, examination in thin section was required. Relative abundance in thin section was estimated at a magnification of ×100 using the following scale:

A – abundant (>30% of the total visible assemblage);
C – common (20–30%);
F – few (10–20%);
R – rare (3–10%);
T – trace (1–2%).

Biozonation

A modified version of the integrated calcareous plankton biozonation of Leckie *et al.* (2002) is used herein (Fig. 1) and the reader is referred to this article for more detailed discussions of mid-Cretaceous calcareous plankton stratigraphy. Supplemental nannofossil events (Fig. 1) were derived from ODP Holes 1183A and 1186A, as well as sections in SE France (Kennedy *et al.* 2000; Bergen 2004).

Biostratigraphic integration of the planktonic foraminifera and nannofossils from the ODP Leg 192 Ontong Java Plateau sites was difficult due to low core recoveries and overall poor preservation of nannofossil assemblages, especially in the Aptian interval. In addition, the authors found the stratigraphic calibration of some of the nannofossil datums in the zonation of Leckie *et al.* (2002) to be inaccurate or inapplicable to the Ontong Java section. This resulted in tentative recognition of nannofossil subzones NC7A–NC7C that often relied on proxy datums listed in Table 1. The highest stratigraphic occurrence of the nannofossil *Micrantholithus hoschulzii*, marking the base of NC7B in the Leckie *et al.* (2002) zonation, is noted in only one sample in the present study. The base of NC7C, marked by the lowest occurrence of *Rhagodiscus achlyostaurion*, is improperly defined because this species occurs throughout the lower Aptian into the upper Barremian (Bergen 1994, 1998). For planktonic foraminifera, all Aptian zones are either total or partial range zones (Fig. 1). In our material *Globigerinelloides algerianus* is sporadic in occurrence and the total range of *Globigerinelloides barri* has been used as a proxy to define the extent of the *Globigerinelloides algerianus* total range zone. Although the stratigraphic ranges of these two index species are largely congruent (e.g. Leckie *et al.* 2002), in the condensed Ontong Java section small differences in range may generate errors in estimation of both the thickness and presence of the underlying (*G. ferreolensis*) and overlying (*Hedbergella trocoidea*) partial range zones, as well as ties to the nannofossil biostratigraphy. This problem is further compounded by poor core recovery.

A summary of the principal nannofossil and foraminiferal biostratigraphic events for the two main Leg 192 sites (1183 and 1186) are presented in Table 1, showing few discrepancies in the relative correlation between the nannofossils and foraminifera. Detailed planktonic foraminiferal and nannofossil species distribution charts for the Leg 192 sites are available from the ODP *Initial Reports* volume (Mahoney *et al.* 2001), but condensed versions for the Aptian–Albian section are presented in Tables 2–5. Carbon isotope values for Sites 1183 and 1186 can be found in Tables 6 and 7, and planktonic foraminiferal species distribution charts for new sample analyses from DSDP Leg 30 Sites 288 and 289 are presented as Tables 8 and 9.

Carbon isotope stratigraphy

Carbon isotope analyses of the Aptian–Albian Leg 192 section were performed by Dr J. G. Ogg of Purdue University and will be described in detail in a forthcoming publication. However, values for the Aptian–Albian section of ODP Leg 192 Sites 1183 and 1186 are presented with permission for comparison to the biostratigraphic results (Tables 6 and 7). For Hole 1183A, $\delta^{13}C$ values range from 2.8 and 4.8‰ for Cores 50-2, 140–145 cm (1091.10 mbsf (metres below seafloor)) through to 54-3, 102–106 cm (1130.19 mbsf). For Hole 1186A, $\delta^{13}C$ values vary from 3.0 to 4.5‰ for Cores 26-CC 13–15 cm (931.36 mbsf) to 30-1, 28–31 cm (966.68 mbsf). Supported by biostratigraphic evidence, peak $\delta^{13}C$ values (4.5–4.8‰) indicate that portions of Oceanic Anoxic Event (OAE) 1a ('sellii') and OAE 1b events are recorded in the recovered section of Leg 192 Holes 1183A and 1186A. In particular, recognition of the 'Pacquier' horizon within OAE 1b is of critical importance because this datum has been proposed as the global stratigraphic datum defining the Aptian–Albian stage boundary (Kennedy *et al.* 2000).

OAE 1a has been calibrated to the late early Aptian and includes the carbonaceous 'selli' horizon in Italy and the 'Niveau Goguel' bed in SE France (Leckie *et al.* 2002). Bralower *et al.* (1994) calibrated the period of peak anoxia to the middle part of the lower Aptian (within the *G. blowi* foraminiferal zone and at the middle of the NC6A nannofossil zone). However, more recent evidence places peak anoxia at a younger age within the lower part of the *L. cabri* foraminiferal zone, below the base of the NC7A nannofossil zone (Larson & Erba 1999). In SE France, the dark beds of the 'Niveau Goguel' range from Beds 150 to 158 (Moullade pers. comm.) in the historical lower Aptian section at Cassis–La Bédoule, although the highest $\delta^{13}C$ values of around 4.5‰ are recorded above Bed 170 in the upper part of the *L. cabri* foraminiferal zone (Moullade *et al.* 1998), as constrained by the lowest occurrence of *Eprolithus floralis* in Bed 169A (Bergen 1998). These latter studies indicate that the positive excursion of $\delta^{13}C$ values associated with the 'sellii' and 'Niveau Goguel' beds continues through the remainder of the early Aptian as defined by the upper *L. cabri* zone.

Bralower *et al.* (1993) constrained OAE 1b in deep-sea sections within the basal Albian *H. planispira* foraminiferal zone and the lower half of nannofossil zone NC8B. OAE 1b has since been extended from the latest Aptian through to earliest Albian and recognized as including three carbonaceous horizons associated with three positive peaks in $\delta^{13}C$ values named for beds that crop out in European sections: the 'Jacob' in the upper Aptian, and the 'Pacquier' and 'Leenhardt' in the lower Albian (Leckie *et al.* 2002). The Jacob bed occurs just below the base of the potential Aptian–Albian boundary stratotype section of Col de Pré–Guittard (Kennedy *et al.* 2000), within the *Ticinella bejaouaensis* foraminiferal zone (Kennedy *et al.* 2000; Herrle & Mutterlose 2003) and immediately below nannofossil zone NC8A (Kennedy *et al.* 2000; Herrle and Mutterlose 2003; Bergen 2004).

Table 1. *Planktonic foraminiferal (F) and nannofossil (N) datums common to ODP Leg 192, Holes 1183A and 1186A, with OAE1b C isotope excursions in grey; principal datums in bold*

Datum				Hole 1183A		Hole 1186A	
	Event	Nanno Zone	Foram Zone	Core	Depth (mbsf)	Core	Depth (mbsf)
Assipetra infracretacea (N)	HO	NC10A		50-2, 78 cm	1090.48	26-3, 33 cm	931.33
R. appenninica (F)	LO	NC10A	appenninica	50-CC, 13 cm	1091.73	27-1, 66 cm	938.26
E. turriseiffelii (N)	LO	NC10A		52-1, 1 cm	1107.51	26CC, 7 cm	931.36
T. orionatus (N)	LO	NC8C	primula?	52-1, 24 cm	1107.74	27 CC	938.88
Rhagodicus sp. 1 (N)	HO	NC8A	planispira	52-1, 52 cm	1108.02	28-1, 52 cm	947.72
C. erbae (N)	HO	NC8A		52-1, 90 cm	1108.40	28-1, 52 cm	947.72
'Pacquier'				52-2,141 cm	1111.91	28-2,114 cm	949.84
H. trocoidea (F)	HO			52CC, 14 cm	1113.38	28 CC, 23 cm	950.81
T. bejaouaensis (F)	HO		bejaouaensis	52CC, 14 cm	1113.38	28 CC, 23 cm	950.81
P. columnata (N)	LO	NC8A		52CC, 14 cm	1113.38	28 CC, 23 cm	950.81
'Jacob'				53-1, 17 cm	1117.37	ABSENT	ABSENT
C. nudus (N)	LO	NC7C		53-2, 10 cm	1118.77	28 CC, 23 cm	950.81
Rhagodiscus sp. 1 (N)	LO	NC7C		53-2, 10 cm	1118.77	28 CC, 23 cm	950.81
G. ferreolensis (F)	HO			53-3, 21 cm	1120.58	29-1, 68 cm	957.48
T. bejaouaensis (F)	LO			53-3, 21 cm	1120.58	29-1, 10 cm	956.90
G. barri (F)	HO		algerianus	53-3, 95 cm	1121.12	29-1, 68 cm	957.48
G. algerianus (F)	HO			53-3, 95 cm	1121.12	29CC, 17 cm	959.88
Q. enebrachium (N)	LO	NC7C		53-3, 95 cm	1121.12	30-1, 1 cm	966.41
H.trochoidea (F)	LO			53-3, 123 cm	1121.40	29-1, 68 cm	957.48
G. algerianus (F)	LO			54-2, 28 cm	1128.58	29CC, 17 cm	959.88
G. barri (F)	LO			54-2, 54 cm	1128.84	29CC, 17 cm	959.88
L.cabri (F)	HO		cabri	54-2, 85 cm	1129.15	30-1, 2 cm	966.42
R.hamptonii (N)	LCO	NC7C		54-2, 88 cm	1129.18	30-1, 33 cm	966.73
G. ferreolensis (F)	LO			54-3, 116 cm	1130.33	30-1, 11 cm	966.51
E. floralis (N)	LO	NC7A		54-3, 119 cm	1130.99	30-1, 43 cm	966.83
R. gallagheri (N)	LO	NC6A		54-3, 119 cm	1130.99	30-1, 43 cm	966.83
H. irregularis (N)	LO	NC6A		54-3, 119 cm	1130.99	30-1, 43 cm	966.83
Basement				54-3, 107 cm	1130.87	30-1, 45 cm	966.85

HO, highest occurrence; LO, lowest occurrence; LCO, lowest consistent occurrence.

Table 2. *Stratigraphic distribution of planktonic foraminifera in Lower Cretaceous section of Hole 1183A*

Core	Section	Top (cm)	Depth (mbsf)	Group abundance	Group preservation	Stage		Zone	*Blefuscuiana daminiae*	*Praehedbergella sigali*	*Blefuscuiana hispaniae*	*Blefuscuiana rudis s.s.*	*Blefuscuiana kuznetsovae*	*Globigerinelloides ferreolensis*	*Praehedbergella ruka contritus*	*Praehedbergella tuschepsensis*	*Praehedbergella tatianae*	*Gubkinella graysonensis*	*Leupoldina cabri*	*Globigerinelloides barri*	*Globigerinelloides algerianus*	*Blefuscuiana aptiana orientalis*	*Blefuscuiana infracretacea*	*Blefuscuiana excelsa cumulus*	*Globigerinelloides aptiense*	*Schakoina bizonae*	*Blefuscuiana albiana*	*Ticinella bejaouaensis*	*Hedbergella trocoidea*	*Hedbergella delrioensis*	*Blefuscuiana gorbachikae*	*Hedbergella* cf. *rischi*	*Ascoliella nitida*	*Hedbergella planispira*	*Rotalipora appenninica*
50	CC	13	1091.73	C	P	Albian	upper	appen.																						F				R	T
51	3	13	1101.03	F	M	Albian	middle	upper primula																						A					
52	1	139	1108.89	T	P	Albian	middle	upper primula																						T			T	R	
52	2	105	1110.05	T	M	Albian	lower	planispira															F							R		T			
52	4	117	1113.17	F	M	Albian	lower	planispira															R							F		R			
52	CC	15	1113.38	A	M	Aptian	upper	bejaouaensis																				R	A	R		R			
53	1	86	1118.06	A	M	Aptian	upper	bejaouaensis															T						A	T					
53	2	57	1119.24	A	M	Aptian	upper	bejaouaensis																				T	A	T					
53	2	81	1119.48	A	G	Aptian	upper	bejaouaensis															R						F	A	T				
53	2	147	1120.14	A	G	Aptian	upper	bejaouaensis																				R	A	T					
53	3	21	1120.58	A	G	Aptian	upper	bejaouaensis						R									T						A	T					
53	3	95	1121.12	C	P	Aptian	upper	algerianus						R						F															
53	3	112	1121.29	A	P	Aptian	upper	algerianus	F		T			R											T										
53	3	123	1121.40	C	M	Aptian	upper	algerianus	T					R		R	T			T			R		T	T	T								
53	4	86	1122.30	C	M	Aptian	upper	algerianus	F									R					T	T											
53	CC	0	1122.58	F	P	Aptian	upper	algerianus												F															
54	2	28	1128.58	C	M	Aptian	upper	algerianus	F					T				F			T	R													
54	2	54	1128.84	C	P	Aptian	upper	algerianus												T															
54	3	59	1129.76	R	P	Aptian	lower	cabri	R	R	T								T																
54	3	96	1130.13	F	P	Aptian	lower	cabri	F		T							R	T																
54	3	105	1130.22	F	M	Aptian	lower	cabri								T	T																		
54	3	116	1130.33	F	M	Aptian	lower	cabri	F		T	R	T	T	R																				
54	4	7	1130.74	T	P	?			T	T																									
55	1	6	1136.56	T	P	?			T	T																									

* For abbreviations, see Methods.

Table 3. *Stratigraphic distribution of selected nannofossils in Lower Cretaceous section of Hole 1183A*

Core	Section	Depth (cm)	Depth (mbsf)	Abundance	Preservation	Stage		NC Zone	*Assipetra infracretacea*	*Axopodorhabdus albianus*	*Calcicalathina erbae*	*Cretarhabdus loriei*	*Cribrosphaerella ehrenbergii*	*Cylindralithus nudus*	*Eiffellithus monechiae*	*Eiffellithus turriseiffelii*	*Eprolithus floralis*	*Gartnerago gammation*	*Hayesites irregularis*	*Laguncula dorotheae*	*Loxolithus armilla*	*Lucianorhabdus salomonii*	*Metadoga* spp.	*Micrantholithus hoschulzii*	*Microstaurus chiastius*	*Nannoconus globulus*	*Perchnielseniella stradneri*	*Prediscosphaera columnata*	*Quadrum eneabrachium*	*Rhagodiscus achlyostaurion*	*Rhagodiscus angustus*	*Rhagodiscus gallagheri*	*Rhagodiscus hamptonii*	*Rhagodiscus* sp. 1	*Tranolithus orionatus*	*Tranolithus praeorionatus*	*Tranolithus pseudoangustus*	*Zebrashapka vanhintei*
50	2	78	1090.48	5.00	P	Albian	Upper	10A	R					R			1				1				1		F		R									
50	2	85	1090.55	6.00	P				R	1				F		1					R				1		F		R									
50	2	105	1090.75	10.00	P				1	1				F							1				R		F		R						2	1		
50	2	142	1091.12	3.00	P				1					R											1													
50	CC		1091.78	20.00	P				2	3		2		F		R	R				R				R			R	R						R			
51	1	28	1098.18	8.00	P				1					R		R					1				R				1		1							
51	2	23	1098.63	20.00	P				2	1			2	F		R		2			F				R										2	1		
51	3	28	1101.18	20.00	P				3	1		1		R		R		1							R				1						1			
51	CC		1101.59	5.00	P				1	1			1	R		R	R			1			1		R			1							1			
52	1	1	1107.51	15.00	P					1			1	1		1			1						R										1			
52	1	24	1107.74	20.00	P		Middle	8C	1					R	1		1		R						R					2					2			
52	1	52	1108.02	10.00	P		Lower	8A	R					2			1		2		1				R			1						2				
52	1	79	1108.29	8.00	P				R					R			1		1						R			1	1				1	2				
52	1	90	1108.40	20.00	P						3	1		R			1		R						R				1	R				R				
52	1	104	1108.54	4.00	P				R					R			1		3						R									1				
52	2	105	1110.05	8.00	P				1		1			R			1		R	1?					R				2	1		1		2				
52	3	107	1111.57	6.00	P				1		2			2			1		2						R				1	1			2	R				
52	4	107	1113.07	12.00	P				1		2			2					3						R					1								
52	CC		1113.38	10.00	P				1		1			3			1		1						R			1	2	1				1				
53	1	10	1117.30	15.00	P	Aptian	Upper	7C			R			2					2						R					R			2	3				
53	2	10	1118.77	5.00	P				1		2			2			1		1						R				1	2		1	1	2				
53	3	15	1020.32	4.00	P						2														R					R								
53	3	21	1020.38	2.00	P																																	
53	3	95	1021.12	8.00	P				1		2			1?			1		R						R				1			1						
53	4	3	1121.47	20.00	P			7B	1		1			1?					R						R					1	1	1						
53	4	29	1121,73	15.00	P						1								R						R					R								
53	4	67	1122.11	25.00	P				1		3								R						F					4		1	1					
53	CC		1122.58	20.00	M				1		1				1?				R						F					2	1	R	3					
54	1	9	1126.89	2.00	P																																	
54	1	53	1127.33	12.00	VP				2		1						1		5						R							2						
54	1	60	1127.40	8.00	P				1		5														R					2								
54	1	119	1127.99	8.00	P				1		1								1						F					2	1	1						
54	2	5	1128.35	12.00	P				1								2								F					1	1	R	2					2
54	2	34	1128.64	10.00	P				2								1		R			1			F	1					3	3					1?	1
54	2	88	1129.18	4.00	P				R																R					2	1	3	1					
54	3	3	1129.83	0.01	P		Lower	7A																														
54	3	60	1130.40	0.50	P				2																1							1						
54	3	72	1130.52	0.10	P				1																													
54	3	100	1130.80	2.00	P				1																2							1					1	
54	3	119	1130.99	3.00	P						2?						1		1			2?		1	1							1						
54	4	7	1131.37	0.00		?																																
55	1	10	1131.40	0.01	P																																	
55	1	83	1132.13	0.00																																		
55	2	84	1133.64	0.00																																		

The Pacquier horizon, also defined in SE France, has been suggested as the boundary datum in the proposed Aptian–Albian stage boundary stratotype section at Col de Pré–Guittard (Kennedy *et al.* 2000). However, relative to the most widely used microfossil biozonation, the Pacquier horizon lies within the lowermost Albian, within the upper *H. planispira* foraminiferal and NC8B nannofossil zones (Leckie *et al.* 2002; Herrle & Mutterlose 2003). This relative discrepancy in the stratigraphic placement of the stage boundary is lessened by other nannofossil data from the Col de Pré–Guittard section (Kennedy *et al.* 2000; Bergen 2004), indicating that the Pacquier bed is actually equivalent to an older age within nannofossil zone NC8A. Further supporting evidence for this older age for the Pacquier event is provided by nannofossil data from Col de Pré-Guittard (Bergen 2004) and by the results of the current study.

ODP Leg 192

Site 1183

Located near the crest of the main, or high portion of the Ontong Java Plateau, Site 1183 is physiographically the highest of any of the DSDP/ODP locales (Fig. 2). Fair to excellent recovery of nannofossils and planktonic foraminifera mark over 40 m of Aptian–Albian section drilled (Fig. 3). The Aptian interval is characterized by an alternation between section indicative of continuous pelagic deposition and condensed section and/or stratigraphic hiatuses. The Albian interval is marked by two disconformities associated with major hiatuses at the lower–middle substage and the Lower–Upper Cretaceous boundaries.

The oldest sediment in Hole 1183A is assigned to the uppermost lower Aptian *Leupoldina cabri* planktonic foraminiferal zone and the NC7A nannofossil zone (Fig. 1). Only about 1.4 m thick (1183A-54R-2, 84–86 cm–54R-3, 119 cm; 1129.14–1130.56 mbsf), the interval is largely composed of reworked lithic tuff and lapillistone interbedded with very thin, bioturbated limestone layers (Shipboard Scientific Party 2001*a*) and immediately overlies the top of basaltic basement at 1183A-54R-3, 120 cm (1130.57 mbsf). The lower–upper Aptian boundary, as defined by the top of the *L. cabri* zone (54R-2, 84 cm; 1129.14 mbsf), falls just above a lithofacies change (1183A-54R-3, 34 cm, 1129.51 mbsf) from the interbedded volcaniclastic–carbonate interval that characterizes the

Table 4. *$\delta^{13}C$ values for Hole 1183A, fine bulk limestone fraction (<62.5 μm) against PDB standard; data by J. G. Ogg*

Sample	Core depth (mbsf)	$\delta^{13}C$
192-1183A-50R-2, 140–145 cm	1091.10	3.4
192-1183A-50R-3, 20–24 cm	1091.40	2.8
192-1183A-50R-CC, 7–9 cm	1091.67	2.8
192-1183A-51R-1, 67–72 cm	1098.57	2.6
192-1183A-51R-2, 106–110 cm	1100.46	2.2
192-1183A-52R-1, 75–78 cm	1108.25	3.4
192-1183A-51R-2, 25–28 cm	1099.65	2.9
192-1183A-52R-2, 69–73 cm	1109.69	4.0
192-1183A-52R-3, 17–23 cm	1110.67	3.9
192-1183A-52R-3, 141–145 cm	1111.91	4.5
192-1183A-52R-4, 69–74 cm	1112.69	4.0
192-1183A-53R-1, 17–23 cm	1117.37	4.8
192-1183A-53R-1, 124–128 cm	1118.44	4.3
192-1183A-53R-2, 98–102 cm	1119.65	4.3
192-1183A-53R-3, 44–48 cm	1120.61	3.8
192-1183A-53R-3, 104–108 cm	1121.21	3.8
192-1183A-53R-4, 6–16 cm	1121.50	4.3
192-1183A-53R-4, 30–34 cm	1121.74	3.9
192-1183A-53R-4, 52–56 cm	1121.96	4.2
192-1183A-53R-4, 95–100 cm	1122.39	3.9
192-1183A-54R-1, 35–39 cm	1127.15	3.9
192-1183A-54R-1, 97–101 cm	1127.77	3.7
192-1183A-54R-2, 8–13 cm	1128.38	4.3
192-1183A-54R-2, 46–51 cm	1128.76	4.7
192-1183A-54R-2, 84–88 cm	1129.14	4.3
192-1183A-54R-3, 59–62 cm	1129.76	3.8
192-1183A-54R-3, 102–106 cm	1130.19	4.3

bulk of the *L. cabri* zone to the overlying continuous limestone section (Fig. 3). The *L. cabri* zone therefore records the first period of continuous sedimentation following termination of the major period of Ontong Java volcanic activity on the high plateau at Site 1183. More specifically, the basal sediment can be assigned to the upper part of the *L. cabri* zone based on the co-occurrence of the planktonic foraminiferal species *Leupoldina cabri* and *Globigerinelloides ferreolensis* with the nannofossil indexes *Eprolithus floralis* and *Rhagodiscus hamptonii* (Fig. 1). Stable isotope analysis indicates a strong positive $\delta^{13}C$ maximum at the top of the *L. cabri* zone at 54R-2, 84 cm (1129.14 mbsf) that probably marks the post-sellii maximum (Moullade *et al.* 1998; Larson & Erba 1999) of the positive isotopic excursion associated with OAE 1a (Fig. 3, Table 6) for which the Ontong Java eruption has been postulated as the cause (e.g. Leckie *et al.* 2002). The most recent absolute age calibration of the top of the *L. cabri* zone is 118.2 Ma (Erba *et al.* 1999). This age compares well with a minimum radiometric age from

Table 5. *Stratigraphic distribution of planktonic foraminifera in Lower Cretaceous section of Hole 1186A*

Core	Section	Top (cm)	Depth (mbsf)	Group abundance	Group preservation	Stage		Zone	*Praehedbergella ruka papillata*	*Praehedbergella pseudosigali*	*Praehedbergella sigali*	*Blefuscuiana daminiae*	*Blefuscuiana occulta*	*Blefuscuiana infracretacea*	*Claviblowiella sigali*	*Globigerinelloides blowi*	*Blefuscuiana aptiana*	*Schackoina cenomana*	*Leupoldina cabri*	*Blowiella duboisi*	*Lilliputianella globulifera*	*Blefuscuiana speetonensis*	*Globigerinelloides aptiense*	*Praehedbergella tuschepsensis*	*Blefuscuiana hispaniae*	*Blowiella moulladei*	*Claviblowiella minai*	*Globigerinelloides ferreolensis*	*Globigerinelloides maridalensis*	*Blowiella solida*	*Blowiella gottisi*	*Blefuscuiana hexacamerata*	*Praehedbergella ruka contritus*	*Lilliputianella longorii*	*Globigerinelloides algerianus*	*Globigerinelloides barri*	*Blefuscuiana gorbachikae*	*Hedbergella delrioensis*	*Blefuscuiana excelsa cumulus*	*Blefuscuiana praetrocoidea*	*Blefuscuiana excelsa*	*Hedbergella trocoidea*	*Ticinella bejaouaensis*	*Hedbergella planispira*	*Hedbergella rischi*	*Rotalipora appenninica*
27	1	66	938.26	R	P	Albian	upper	appenninica																														R						T		R
28	2	44	949.14	T	P		lower	planispira																														T						T	T	
28	CC	23	950.81	A	M	Aptian	upper	bejaouaensis						R																								R		F		A	O			
29	1	10	956.90	A	M																																					A	O			
29	1	68	957.48	A	P			algerianus																	R			R								F				A		F				
29	1	148	958.28	A	P																							R								A			R	F	F					
29	2	73	959.03	A	P																							F								A				F	F					
29	2	131	959.61	A	G							A		F											R							R	F						R	T						
29	CC	17	959.88	A	G				R			R		F			R								T		T					R	T	R	A	A	R	R	R							
30	1	11	966.51	A	VG		lower	upper cabri				R			R	T		T	R	A	R	R	R			R	F	T	R	F	F	R														
30	1	28	966.68	A	M				R			A		R									R	T	T																					
30	1	43	966.83	C	M				T			F		R	R	T	R	T	R	F	R	T																								
30	1	73	967.13	F	P				F			F																																		
32	3	79	979.96	C	P			lower cabri?	F		R	F	R																																	
32	4	89	981.55	R	P			?	R	R	R																																			

Table 6. *Stratigraphic distribution of selected nannofossils in Lower Cretaceous section of Hole 1186A*

Core	Section	Depth (cm)	Depth (mbsf)	Abundance	Preservation	Stage		NC Zone	*Assipetra infracretacea*	*Axopodorhabdus albianus*	*Calcicalathina erbae*	*Cretarhabdus loriei*	*Cribrosphaerella ehrenbergii*	*Cylindralithus nudus*	*Eiffellithus turriseiffelii*	*Eiffellithus paragogus*	*Eprolithus floralis*	*Hayesites irregularis*	*Isocrystallithus partitum*	*Loxolithus armilla*	*Microstaurus chiastius*	*Nannoconus globulus*	*Perchnielseniella stradneri*	*Prediscosphaera columnata*	*Prediscosphaera cretacea*	*Quadrum eneabrachium*	*Rhagodiscus achlyostaurion*	*Rhagodiscus angustus*	*Rhagodiscus gallagheri*	*Rhagodiscus hamptonii*	*Rhagodiscus* sp. 1	*Tranolithus orionatus*	*Tranolithus praeorionatus*
26	3	33	931.33	5.00	P	Albian	Upper	10A	4					R	1		R			F	R		F	1		1							
26	3	42	931.42	6.00	P				2	1				R			F			R	R		F	R	1	1							1
26	3	54	931.54	15.00	P				R	1				R	1		F			F	R		F	1	1							1	1
26	3	61	931.61	20.00	P				1								F			R	2		F	R	2								
26	3	68	931.68	10.00	P				R								R			R	R		F	1	1								R
26	CC	7	931.36	25.00	P				R	2		2	1		R		F			R	R		R	R	R	F							
27	1	33	937.93	1.00	P		Middle	8C	R					R				1?	1?		R				1								
27	CC	28	938.88	12.00	P				R				3	2			1	2			R			1			R	1				1	1
28	1	52	947.72	25.00	P		Lower	8A	R		2			R			1	R			R			1		3					3		
28	2	3	948.73	20.00	P				1		1			1				2			F					1	R				2		
28	3	22	950.42	10.00	P									1			1	2			R					2	R			2	2		
28	CC	28	950.81	8.00	P				1		3			1			1	R			F	1		1		2	F		1	1	R		
29	1	74	957.54	4.00	P	Aptian	Upper	7C	1		2						1	1			R					2	1						
29	2	60	958.90	15.00	P			7B			1							R			F						R			1			
29	CC		959.88	25.00	P				1		4						2	3			F						1		1	1			
30	1	1	966.41	25.00	P						3							R			C						R		2	3			
30	1	14	966.54	15.00	P				1		2						1	R			F						R		1	1			
30	1	24	966.64	10.00	P				2		1					1	1	R			F						R		2				
30	1	33	966.73	5.00	P				2		2						6	R			F						F		2	2			
30	1	41	966.81	0.30	P		Lower	7A																									
30	1	43	966.83	0.50	P				1								4	1			R								1	1			
30	1	44	966.84	0.01	P	?																											
32	1	1	976.21	0.00																													
32	3	80	979.97	0.02	P																												
32	3	82	979.99	0.01	P																												
32	4	90	981.56	0.00																													

Table 7. *$\delta^{13}C$ values for Hole 1186A, fine bulk limestone fraction (<62.5 µm) against PDB standard; data by J. G. Ogg*

Sample	Core depth (mbsf)	$\delta^{13}C$
192-1186 26R-CC, 13–15 cm	931.36	3.0
192-1186A-27R-1, 66–71 cm	938.26	3.3
192-1186 27R-CC, 28–30 cm	938.88	4.1
192-1186A-28R-1, 37–41 cm	947.57	4.2
192-1186A-28R-1, 134–138 cm	948.54	4.3
192-1186A-28R-2, 44–48 cm	949.14	3.9
192-1186A-28R-2, 114–118 cm	949.84	4.5
192-1186 28R-CC, 23–27 cm	950.81	4.5
192-1186A-29R-1, 10–14 cm	956.90	4.0
192-1186A-29R-1, 43–48 cm	957.23	4.2
192-1186A-29R-1, 127–131 cm	958.07	3.6
192-1186A-29R-2, 55–60 cm	958.85	4.0
192-1186A-29R-2, 135–139 cm	959.65	4.1
192-1186 29R-CC, 17–22 cm	959.88	4.1
192-1186A-30R-1, 28–31 cm	966.68	4.1

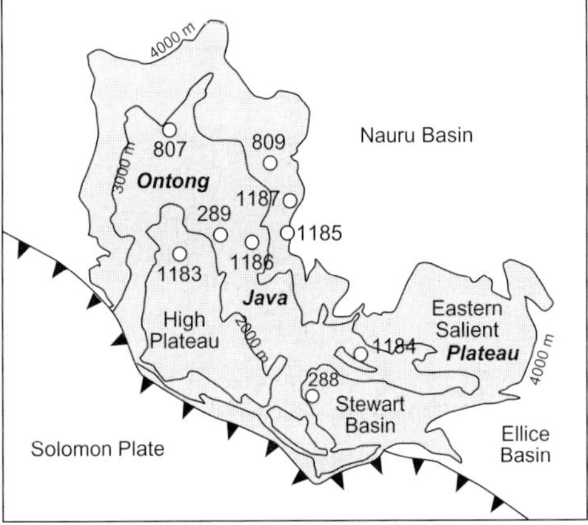

Fig. 2. Location map of ODP Leg 192 sites as well as previously drilled sites from ODP Leg 130 and DSDP Leg 30.

the basaltic basement at Site 1183 of 118.6 Ma (Chambers *et al.* 2002, unpublished data) based on a sample at 1183A-66R-2, 92–102 cm (1202.42 mbsf), approximately 72 m below the base of the continuous sedimentary interval. In addition, in the upper portion of the drilled basement section (1183A-54R-3, 120 cm–56R-1, 3 cm; 1130.57–1146.1 mbsf) occur rare, thin and sporadic limestone beds and/or nodules intercalated with pillow basalts (Shipboard Scientific Party 2001*a*). These limestone intervals represent thermally altered chalk and, although ghosts of planktonic foraminifera and/or radiolaria are often distinguishable, preservation is too poor to allow definite species identification. Nevertheless, no evidence is noted to indicate any section older than the *L. cabri* zone.

The great majority of the Aptian section (1183A-53R-3, 95–98 cm–54R-3, 116 cm) is assigned to the uppermost Aptian (foraminiferal *Ticinella bejaouaensis* and nannofossil NC7C zones) and the middle upper Aptian (*Globigerinelloides algerianus* and NC7B zones; Fig. 3). The first stratigraphic occurrence of the planktonic foraminiferal index *Globigerinelloides barri* (54R-2, 31 cm; 1128.60 mbsf) is used as a proxy for the base of the *G. algerianus* zone. The near coincidence of this datum and the NC7A first occurrence of the nannofossil *Rhagodiscus hamptonii* (54–2, 88 cm; 1129.18 mbsf) indicates the absence of the *Globigerinelloides ferreolensis* foraminiferal zone (Table 1, Fig. 3). The *G. algerianus* zone extends up-section to 54R-2, 31 cm (1128.60 mbsf) based on the continued occurrence of *G. barri*, as well as a more sporadic occurrence of *G. algerianus*. According to the integrated zonation of Leckie *et al.* (2002), the *G. algerianus* planktonic foraminiferal zone is coincident with nannofossil zone NC7B. However, only by negative evidence can this section be tentatively assigned to zone NC7B, based on the absence of several nannofossil taxa with NC7C first stratigraphic occurrences such as *Quadrum eneabrachium* (Figs 1 and 3). The first stratigraphic occurrence of the nannofossil *Rhagodiscus achlyostaurion*, which defines the base of nannofossil zone NC7C in the integrated zonation of Leckie *et al.* (2002), occurs at 54R-2, 5 cm (1128.35 mbsf) near the base of the *G. algerianus* planktonic foraminiferal zone. As noted in the Biozonation section, we believe that *R. achlyostaurion* has a much older stratigraphic first appearance than indicated by the published zonation.

One of the most conspicuous characteristics of the *G. algerianus* zonal interval in Hole 1183A is the best developed carbonaceous bed in the Lower Cretaceous section noted on Leg 192. This nearly 1 m-thick bed occurs in the upper part of the zone from 1183A-53R-3, 115 cm to 53R-4, 85 cm (1121.33–1122.29 mbsf) and is a reddish-brown to black, ferruginous nannofossil claystone containing scattered horizontal stringers and lenses of packed foraminifer tests and fine sand-size clasts (Shipboard Scientific Party 2001*a*). Unlike the heavily bioturbated limestone with which it is interbedded, the claystone exhibits no sign of bioturbation. Planktonic foraminifera occur frequently through the claystone interval, but benthic foraminifera are nearly absent. This interval probably marks a

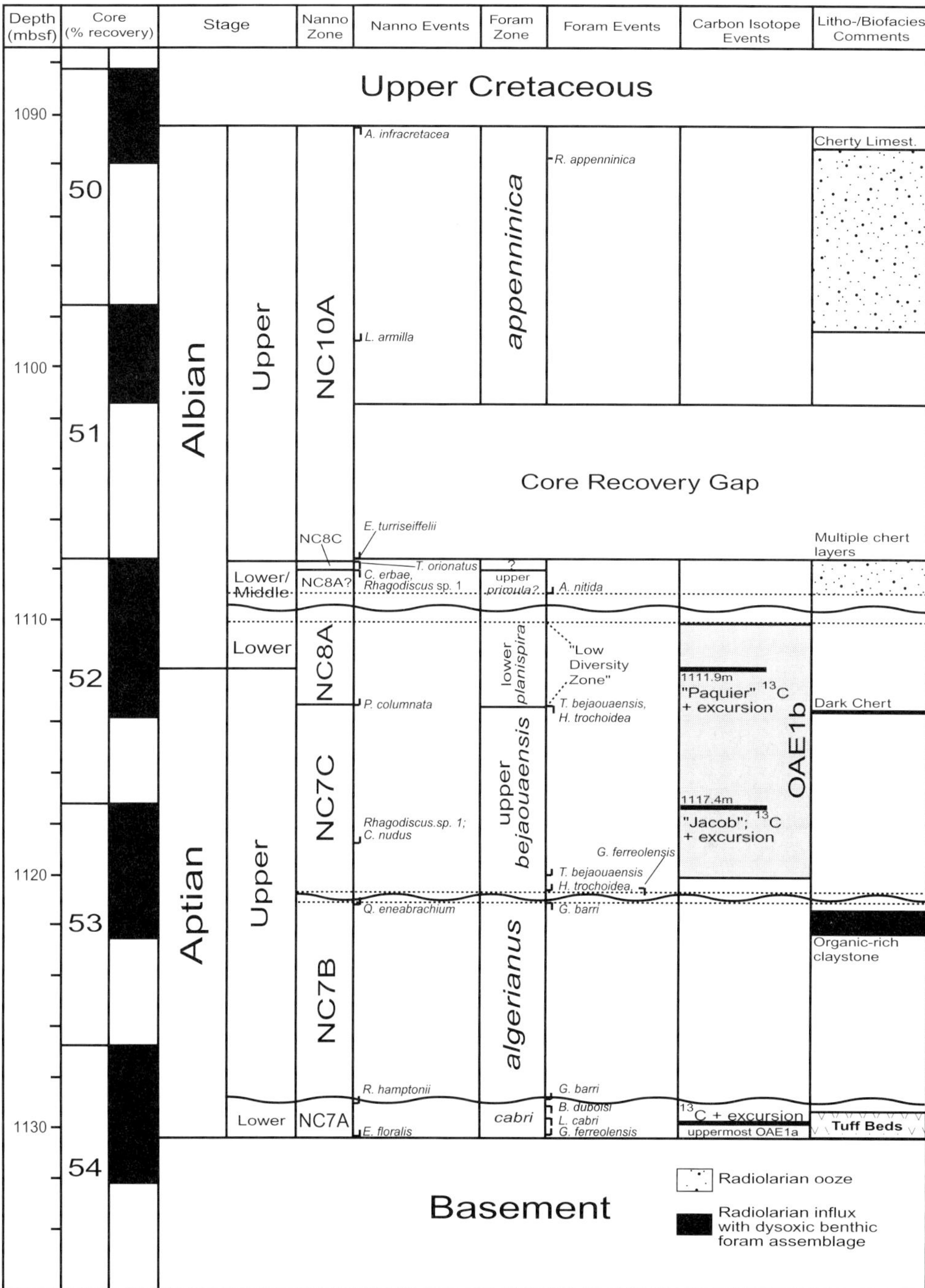

Fig. 3. Stratigraphic summary for ODP Leg 192 Site 1183; isotope summary based on data of J. G. Ogg (see Table 6). Dashed lines indicate the range of uncertainty on the placement of disconformities.

period of low-oxygen deposition on the central Ontong Java Plateau, possibly a local event. Its regional extent is unknown because the unit was not observed in other Leg 192 sites. However, another possibility is that this bed reflects a more global event, equivalent to similar carbonaceous beds within the *G. algerianus* zone noted in northern California and NE Mexico (Premoli Silva & Sliter 1999; Sliter 1999). If so, the Site 1183A claystone would mark another global OAE intermediate in chronology to OAE 1a and 1b.

Occurring immediately above the *G. algerianus* zone, the *T. bejaouaensis* zone is partially defined by sporadic occurrences of the planktonic foraminifer *Ticinella bejaouaensis* (Table 2). However, the complete zonal range at Site 1183 (1183A-52R-CC, 10 cm–53R-3, 21 cm; 1113.38–1120.58 mbsf) is best defined by common occurrence of the planktonic foraminifer *Hedbergella trocoidea*, an event apparently restricted to the *T. bejaouaensis* zone across the Ontong Java Plateau. This association is best represented in deeper-water localities, such as Site 1186, that exhibit a more continuous stratigraphic distribution of *T. bejaouaensis* (see below).

The coincidence of the NC7C first occurrence of the nannofossil *Quadrum eneabrachium* and the last stratigraphic occurrence of the *Globigerinelloides barri*, marking the top of the *G. algerianus* foraminiferal zone (Fig. 3), indicates a major intra-upper Aptian disconformity at 1183A-53R-3, 95 cm (1121.12 mbsf) and the absence of the *Hedbergella trocoidea* foraminiferal zone and the lower portion of nannofossil zone NC7C. Furthermore, the close association of the first occurrence of the foraminifer *Ticinella bejaouaensis* and the mid NC7C first occurrence of the nannofossil *Cylindralithus nudus* and last stratigraphic occurrence of the planktonic foraminifer *Globigerinelloides ferreolensis* (Fig. 3) indicates that much of the lower *T. bejaouaensis* zone is missing as well. This interpretation is further supported by a strong positive maximum in $\delta^{13}C$ values (Fig. 3, Table 6) at 1183A-53R-1, 17–33 cm (1117.37 mbsf) that occurs just above the first stratigraphic appearance of *T. bejaouaensis* in the Site 1183A section. The $\delta^{13}C$ excursion is probably indicative of the late Aptian Jacob event of OAE 1b, which occurs well above the base of the *T. bejaouaensis* zone (Leckie *et al.* 2002).

The last stratigraphic occurrence of the planktonic foraminifer *Ticinella bejaouaensis* marks the top of the *T. bejaouaensis* zone at 1183A-52R-CC, 10–14 cm (1113.38 mbsf). Immediately above occurs the *Hedbergella planispira* foraminiferal zone (52R-3, 17 cm–52R-4, 117 cm; 1110.67–1113.17 mbsf), marked by a low-diversity assemblage of very small planktonic foraminifera that includes the nominate taxon. The base of this zone at Site 1183A is also coincident with the first stratigraphic occurrence of the nannofossil *Prediscosphaera columnata*, marking the base of nannofossil zone NC8A. In addition, within the *H. planispira* zone at 1183A-52R-2, 141–145 cm (1111.91 mbsf), another maximum in $\delta^{13}C$ values occurs (Fig. 3, Table 6) that indicates the Pacquier or middle positive excursion of OAE 1b. The zonation of Leckie *et al.* (2002) indicates that the *T. bejaouaensis* zone overlaps all of nannofossil zone NC8A, as well as the lower portion of zone NC8B in which they place the Pacquier event. Based upon this zonation, the near coincidence in Hole 1183A of the last stratigraphic occurrence of *T. bejaouaensis* and the first stratigraphic occurrence of the nannofossil *P. columnata* followed shortly upsection by the Pacquier isotope event (Fig. 3) indicates a major unconformity in which the upper half of the foraminifer *T. bejaouaensis* and lower half of the nannofossil NC8A zones are missing. It also indicates that most of the *H. planispira* zone in Hole 1183A would be coincident with nannofossil zone NC8B. However, the nannofossil *Calcicalathina erbae* occurs at the Pacquier level and throughout the *H. planispira* zone at Site 1183A, indicating that the section is no younger than zone NC8A (Figs 1 and 3). Also, as already noted, biostratigraphic studies of the Paquier section in SE France have indicated that the Pacquier bed actually falls within nannofossil zone NC8A (Kennedy *et al.* 2000; Bergen 2004). Finally, the current study finds no evidence of overlap between the foraminiferal *T. bejaouaensis* zone and nannofossil zone NC8A or NC8B. The authors therefore believe that the *T. bejaouaensis–H. planispira* zonal boundary (1183A-52R-3, 17 cm; 1113.17 mbsf) is probably conformable and the overlying *H. planispira* interval is coincident *in toto* with the NC8A nannofossil zone. In accordance with the recommendation of Kennedy *et al.* (2000), a conformable Aptian–Albian stage boundary is set at the Pacquier event at 1183A-52R-2, 141–145 cm (1111.91 mbsf) within the *H. planispira* and NC8A zones (Fig. 3).

The only major discrepancy between planktonic foraminiferal and nannofossil biostratigraphy noted in the Lower Cretaceous section of Hole 1183A occurs in a thin interval overlying the *H. planispira* zone (1183A 52R-1, 90 cm–52R-1, 139 cm; 1108.40–1108.89 mbsf). Occurrence of the nannofossils *Calcicalathina erbae* and the informal taxon *Rhagodiscus* sp. 1

indicates a continuation of lowermost Albian zone NC8A. However, thin-section analysis indicates a major biofacies change from the underlying low-diversity *H. planispira* zone to an interval with common, large hedbergellid planktonic foraminifera, mainly *Hedbergella delrioensis*. Furthermore, recovery of a free specimen of the favusellid foraminifer *Ascoliella nitida* indicates a stratigraphy no lower than the upper middle Albian (Leckie *et al.* 2002). The section is probably middle Albian in age, comparable to a similarly thin section defined by nannofossils at Site 1186 (see below). However, in the Hole 1183A section, some reworking of lower Albian sediment occurred in the basal sediments overlying a major unconformity that marks the absence of the upper lower and most of the middle Albian (Fig. 3). Supporting this hypothesis, the first stratigraphic occurrence of the nannofossil *Tranolithus orionatus* is noted at 1183A-52-1, 24 cm (1107.74 mbsf), marking the base of zone NC8C within the upper middle Albian. This thin middle Albian section is overlain, perhaps conformably, by the upper Albian, marked by the first stratigraphic occurrence of the nannofossil *Eiffellithus turriseiffelii* that indicates the base of nannofossil zone NC10A (1183A-52R-1, 1 cm; 1107.51mbsf). Thin-section analysis identified another major biofacies change at this level to siliceous radiolarian wackestone nearly devoid of planktonic foraminifera, a facies typical of the upper Albian across the Ontong Java Plateau.

The upper Albian at Site 1183 (nannofossil zone NC10A) continues up-section to the last stratigraphic occurrence of the nannofossil *Assipetra infracretacea* (Bergen 2004) at 1183A-50R-2, 78 cm (1090.48 mbsf). Planktonic foraminifera are nearly absent from this section, which is dominated by spherical radiolaria. However, a single specimen of the species *Rotalipora appenninica* at 50R-CC, 13–16 cm (1091.73 mbsf) indicates that at least the upper NC10A interval in Hole 1183A falls within the upper half of the upper Albian and the planktonic foraminiferal *R. appenninica* zone. The upper Albian section is truncated by the largest disconformity noted in Hole 1183A and is overlain by condensed Coniacian limestone (Shipboard Scientific Party 2001*a*).

Site 1186

Site 1186 is located on the eastern slope of the high Ontong Java Plateau, 319 km east of Site 1183 (Fig. 2). As at Site 1183, the greatest portion of the Aptian section falls within the upper Aptian *Ticinella bejaouaensis* zone and the middle upper Aptian *Globigerinelloides algerianus* zone (Fig. 1). Together these two zones constitute 15.8 m of a total of 16.2 m of Aptian section (1186A-28R-CC, 23 cm–30R-1, 44 cm; 950.60–966.83 mbsf). The Site 1186 section is more fossiliferous than the Aptian interval at Site 1183, with more diverse and abundant planktonic foraminiferal and nannofossil assemblages. The better microfossil recovery probably reflects deeper Aptian palaeobathymetry on the plateau slope at Site 1186 than on the plateau high at Site 1183. Unfortunately, this advantage for biostratigraphic analysis is mitigated by poor core recovery throughout the Site 1186 sedimentary section. In the Aptian–Albian interval, core recovery averages only 29.8% (Shipboard Scientific Party 2001*b*), which makes the true thickness of some biozones unknown.

The thickness of the recovered *Leupoldina cabri* zone is 43 cm (1186A-30R-1, 2 cm–30R-1, 45 cm; 966.42–966.85 mbsf). The interval shows many similarities to the comparable section at Site 1183 (Figs 3–4): only the upper part of the zone is present as indicated by the co-occurrence of the planktonic foraminifera *L. cabri* and *Globigerinelloides ferreolensis* together with the nannofossil *Eprolithus floralis*; the first stratigraphic occurrence of the nannofossil *Rhagodiscus achlyostaurion* occurs near the top of the *L. cabri* zone at 1186A-30R-1, 33–36 cm (966.74 mbsf), again demonstrating a more extensive stratigraphic range for this taxon than indicated by previously published zonations (Leckie *et al.* 2002); the basal sedimentary section is volcaniclastic in origin, composed of a very thin interval of ferruginous claystone with a basal breccia containing angular basaltic glass fragments (1186A-30R-1, 38 cm–30R-1, 45 cm; 966.79–966.85 mbsf; Shipboard Scientific Party 2001*b*); but the *L. cabri* zone extends above the volcaniclastic layer into continuous carbonate section (1186A-30R-1, 2 cm–30R-1, 38 cm; 966.42–966.79 mbsf). However, unlike the Site 1183 section, the upper part of basaltic basement at Site 1186 lacks any intercalated carbonate layers; instead sedimentary interbeds are composed of hyaloclastite and volcanogenic sandstone. Nevertheless, carbonate intraclasts are noted within a reddish brown conglomerate at 1186A-32R-3, 76–85 cm (979.93–980.02 mbsf). Thin-section analysis of some of the limestone intraclasts reveals a relatively fossiliferous wackestone with common hedbergellid planktonic foraminifera. Species restricted to the upper *L. cabri* zone that occur frequently in Section 30R-1, such as *Claviblowiella sigali* and *Lilliputianella globulifera*, are absent. *Leupoldina cabri* is absent as well.

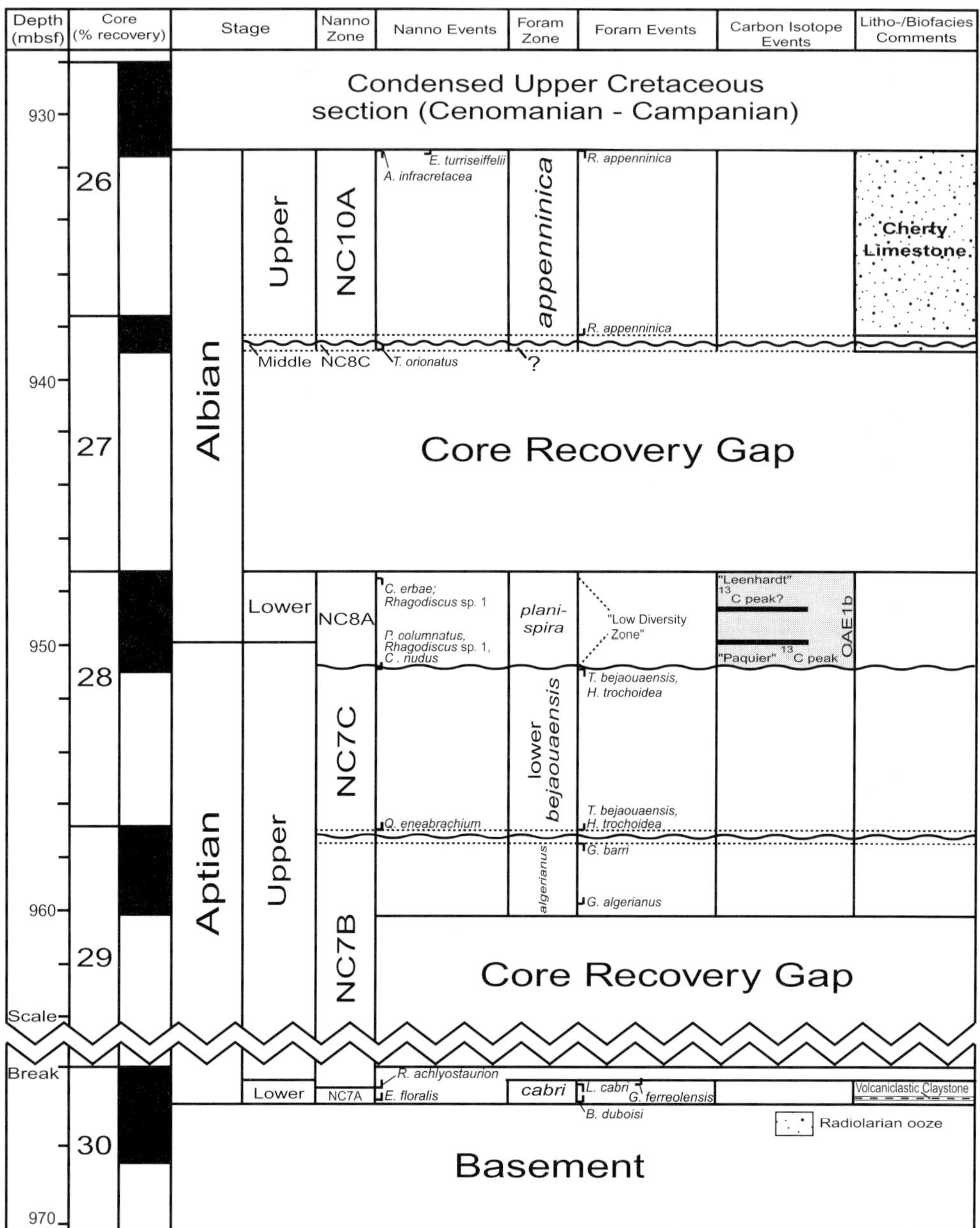

Fig. 4. Stratigraphic summary for ODP Leg 192 Site 1186; isotope summary based on data of J. G. Ogg (see Table 7). Dashed lines indicate the range of uncertainty on the placement of disconformities.

However, rare specimens do occur that are similar to the planktonic foraminifer *Blefuscuiana occulta*, a taxon that does not range below the *L. cabri* zone (Leckie *et al.* 2002). It is therefore possible that these limestone intraclasts were derived from a carbonate bed of the lower *L. cabri* zone.

At Site 1183, an unconformable lower–upper Aptian substage boundary is indicated by the absence of the *G. ferreolensis* foraminiferal zone. However, the presence of a similar stratigraphic discontinuity in the Hole 1186A section cannot be definitely determined because of a recovery gap between Cores 1186A-29R and 30R from 959.88 to 966.42 mbsf (Shipboard Scientific Party 2001*b*). Most of the recovered

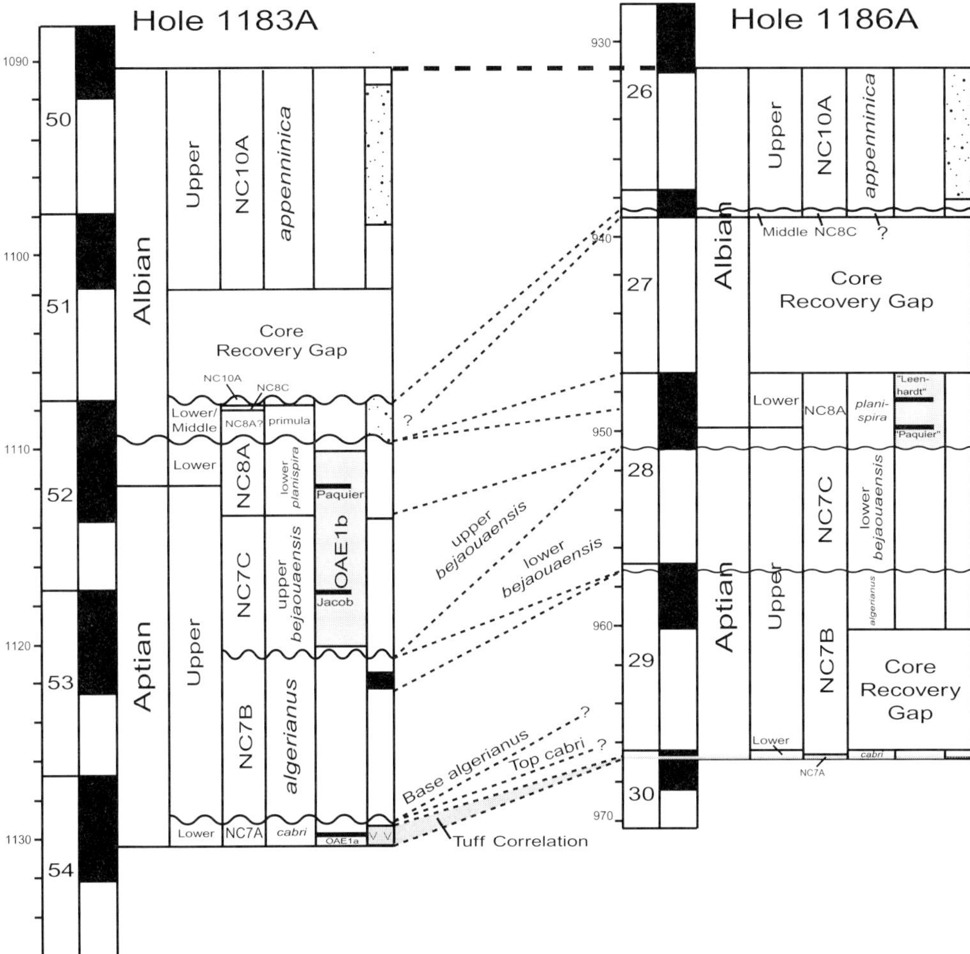

Fig. 5. Correlation of the Lower Cretaceous sedimentary section at ODP Leg 192 Sites 1183 and 1186; datum, top Albian.

section from Core 29R falls within the middle upper Aptian planktonic foraminiferal *G. algerianus* zone (Fig. 4), from the first stratigraphic occurrence of the nominate species at 29R-CC, 17–20 cm (959.88 mbsf) to the last stratigraphic occurrence of *Globigerinelloides barri* at 29R-1, 68–71 cm (957.48 mbsf). The absence of any nannofossil with a first stratigraphic occurrence in zone NC7C provides negative evidence that this interval falls within zone NC7B.

The near coincidence of the first and last stratigraphic occurrences of the planktonic foraminifera *Ticinella bejaouaensis* and *Globigerinelloides barri*, respectively (Fig. 4) indicates the absence of the *Hedbergella trocoidea* zone and an intra-upper Aptian disconformity between these datums at 1186A-29R-1, 10–13 cm (956.90 mbsf) and 29R-1, 68–71 cm (957.48 mbsf). In the overlying section, the upper Aptian *T. bejaouaensis* zone extends from the first to last stratigraphic occurrence of the nominate species (1186A-28R-CC, 23–25 cm–29R-1, 10–13 cm; 956.90–950.81 mbsf), although most of the zone is lost in a recovery gap of over 6 m between Cores 28R and 29R (Fig. 4). However, unlike the Site 1183 section, co-occurrence between *T. bejaouaensis* and the nannofossil *Quadrum eneabrachium*, together with the absence of nannofossils with mid-NC7C first occurrences (i.e. *Cylindralithus nudus* and *Rhagodiscus* sp. 1; Figs 1 and 4), indicate that only the lower portion of the *T. bejaouaensis* zone is present and is not correlative with the younger, upper *T. bejaouaensis* section at Site 1183 (Fig. 5). In addition, the absence of the thick carbonaceous bed noted in the *Globigerinelloides algerianus* zonal interval at Site 1183 (1183A-53R-3, 115 cm–53R-4, 85 cm; 1121.33–1122.29 mbsf) indicates that the uppermost *G. algerianus* zone is also absent at Site 1186 (Fig. 5).

Another major intra-upper Aptian disconformity truncates the top of the *T. bejaouaensis* and NC7C zones at 1186A-28R-CC, 23 cm (950.60 mbsf). The discontinuity is sharply marked by the co-occurrence of the last stratigraphic occurrence of *T. bejaouaensis* and the first stratigraphic occurrences of the mid-NC7C nannofossils *Cylindralithus nudus* and *Rhagodiscus* sp. 1, as well as the NC8A first occurrence of the nannofossil *Prediscosphaera columnata* (Fig. 4). However, similar to Site 1183, carbon isotope stratigraphy indicates a conformable Aptian–Albian stage boundary at Site 1186. A positive maximum in $\delta^{13}C$ values occurs from 1186A-28R-2, 114 cm–28R-CC, 23 cm (949.84–950.81 mbsf) just above the first occurrence of *P. columnata*, marking the Pacquier maximum of OAE 1b and the Aptian–Albian boundary (Fig. 4, Table 7). The absence of the earlier Jacob excursion in the Site 1186 section provides further evidence of the major discontinuity at the top of the *T. bejaouaensis* zone. Although the earlier record of OAE 1b is absent at Site 1186, the section may contain evidence of the later portion of the event missing at Site 1183; i.e. a broader $\delta^{13}C$ maximum occurs further up-section from 1186A-28R-1, 37 cm to 28R-1, 134 cm (947.57–948.54 mbsf) that may be indicative of the youngest OAE 1b excursion, that of the 'Leenhardt' event (Fig. 4, Table 7).

The major lower–middle Albian unconformity that characterizes the Site 1183 section cannot be definitely determined to occur at Site 1186 because of a recovery gap of 8.32 m between Cores 1186A-27R and 28R (938.88–947.20 mbsf; Shipboard Scientific Party 2001*b*). However, the middle–upper Albian stage boundary is recovered in the 1186A-27R section (Fig. 4). Upper middle Albian is indicated by the NC8C first stratigraphic occurrence of the nannofossil *Tranolithus orionatus* at 1186A-27R-CC, 28 cm (938.88 mbsf). As at Site 1183, this middle Albian datum is followed a short distance up-section by the upper Albian. However, unlike Site 1183 where no definite evidence of a disconformity characterized this substage boundary, at Site 1186 the first stratigraphic occurrence of the planktonic foraminifer *Rotalipora appenninica* at 1186A-27R-1, 66 cm (938.26 mbsf) indicates uppermost Albian directly overlying nannofossil zone NC8C. This implies a stratigraphic hiatus marking the absence of the uppermost middle and lower upper Albian (Fig. 4).

The upper Albian continues up-section to 1186A-26R-3, 33–36 cm (931.33 mbsf) as indicated by the co-occurrence of the nannofossils *Assipetra infracretacea* and *Eiffellithus turriseiffelii* marking zone NC10A. As is typical of upper Albian facies across the Ontong Java Plateau, this interval is composed of siliceous wackestone rich in radiolaria. In addition to sporadic occurrence of the planktonic foraminifer *Rotalipora appenninica*, a late Albian age for this section is also supported by the occurrence of the nassellarian radiolarian *Pseudodictyometra pseudomacrocephala*. As at Site 1183, the upper Albian section at Site 1186 is truncated by a large disconformity and is overlain by condensed Coniacian limestone (Shipboard Scientific Party 2001*b*).

Site 1185

Site 1185 is on the eastern edge of the slope of the high Ontong Java Plateau at the northern side of an enormous submarine canyon system (provisionally named Kroenke Canyon) that extends from Ontong Java Atoll and the Nukumanu Islands into the Nauru Basin (Fig. 2). The closest previously drilled location is ODP Leg 130 Site 803, 334 km to the NNW. Together with Sites 1183 and 1186 at the crest and the upper slope of the high plateau, respectively, Site 1185 forms a physiographical transect of the eastern slope of the high Ontong Java Plateau. The site is midway between two seafloor scarps that are probably related to erosion in the canyon system, such as the heads of large slumps. The seismic profile of the seafloor often displays a rough texture, suggesting a high-energy depositional environment with frequent sediment displacement via sediment flows and mass wasting (Shipboard Scientific Party 2001*c*).

With such a depositional setting, it was thought that the sedimentary record at Site 1185 might be discontinuous. This hypothesis was verified when chalk of Eocene age, as defined by both nannofossils and planktonic foraminifera, was found to lie immediately above basaltic basement (Fig. 6). However, radiometric dating of the underlying basalt yields an Aptian age (Chambers *et al.* 2002, unpublished data). Rather than indicative of a surface of non-deposition with a hiatus of the order of 70–75 Ma, it is more likely that sediment at Site 1185 has been periodically swept into the basin by sediment flows and/or mass-wasting events throughout middle and Late Cretaceous, as well as older Palaeogene time.

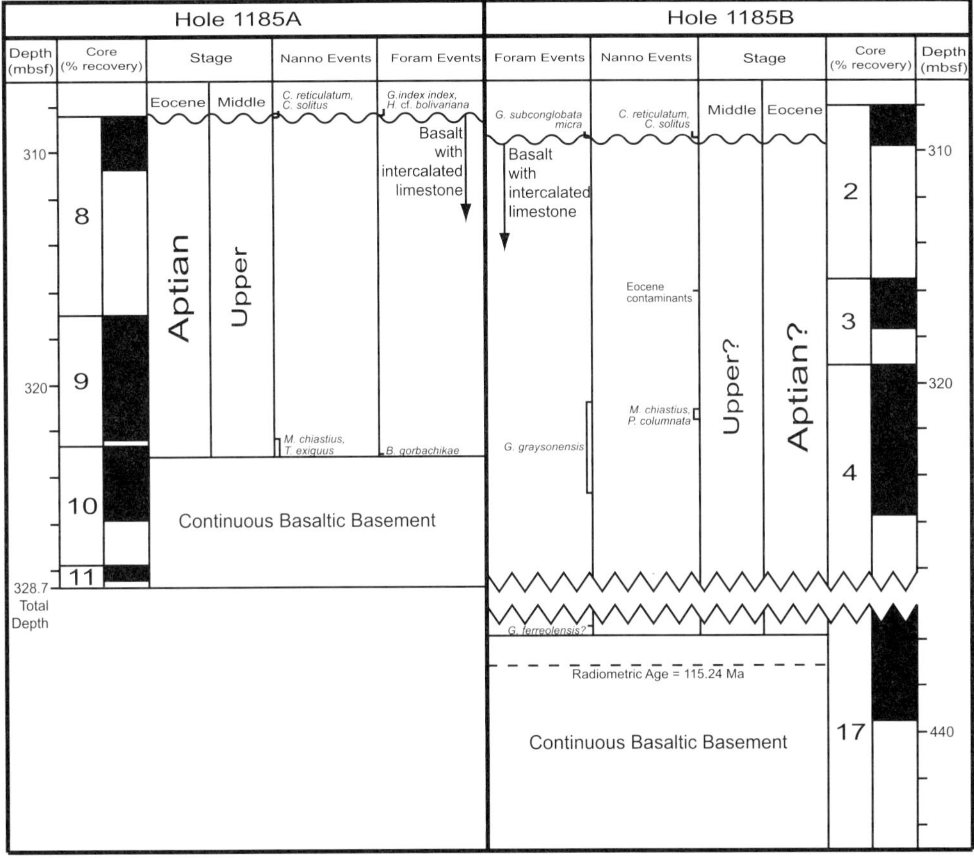

Fig. 6. Stratigraphic summary for ODP Leg 192 Site 1185, Holes 1185A and 1185B.

Nannofossil evidence from Hole 1185A clearly correlates the basal sedimentary section to middle Eocene Zone NP16 of Martini (1971), and most probably to the upper part of Zone NP16. Using the absolute age calibration of Berggren *et al.* (1995), nannofossils indicate an age range of 42.0–43.7 Ma. *Reticulofenestra umbilicus* occurs at the base of the sedimentary section (1185A-8R-1, 10 cm; 308.50 mbsf) and its appearance has been assigned an age of 43.7 Ma by Berggren *et al.* (1995). The lowest occurrence of *Cribrocentrum reticulatum* is immediately above (1185A-8R-1, 8 cm; 308.48 mbsf) and its appearance has been calibrated at 42.0 Ma by Berggren *et al.* (1995). Planktonic foraminifera recovered from 1185A-7R-CC, 48–50 cm (301.82 mbsf) to 8R-8R-1, 10 cm (308.50 mbsf) are indicative of Zone P12 (Blow 1969), which is in agreement with the middle Eocene nannofossil age. Planktonic foraminiferal indexes include *Globigerinatheka index index* and *'Hastigerina'* cf. *bolivariana* (*sensu* Tourmarkine & Luterbacher 1985). The first stratigraphic appearance of *Globigerinatheka index*, which occurs at 1185A-8R-1, 10 cm (308.50 mbsf), has been assigned an age of 42.9 Ma by Berggren *et al.* (1995). Thus, the combined nannofossil and foraminiferal stratigraphy is indicative of an age of 42.9–42.0 Ma for the basal portion of the continuous sedimentary section of Hole 1185A.

The basaltic section drilled in both Hole 1185A and Hole 1185B (i.e. 1185A-8R-1, 14 cm–11R-1, 92 cm; 308.54–328.7 mbsf and 1185B-2R-2, 0 cm–28R-1, 104 cm; 309.51–526.1 mbsf) is characterized by rare limestone that occurs as intercalated beds, as sediment-filled fissures within basalt and as the matrix of hyaloclastite breccia (Shipboard Scientific Party 2001*c*). The limestone is composed of finely recrystallized mudstone that contains poorly preserved foraminifera, nannofossils and radiolaria.

Very rare nannofossils were recovered from six of nine samples taken from intercalated limestone within the upper basaltic basement section of Hole 1185A. One of these, 1185A-8R-1, 14 cm (308.54 mbsf) yields only an undifferentiated Eocene assemblage, possibly indicative of drilling contamination. The other five nannofossil-bearing samples (from Sections 9R-4 to 10R-1) yield Cretaceous taxa. Samples 9R-4, 47 cm (318.97 mbsf) and 10R-1, 49 cm (323.07 mbsf) contain nannofossils indicative of a latest Aptian–Cenomanian age; i.e. *Microstaurus chiastius*, which had a terminal Cenomanian extinction, together with *Tranolithus exiguus*, whose first occurrence is in the latest Aptian (*jacobi* ammonite zone) near the base of zone NC8A.

Thin-section analysis of some of the limestone intercalated in the basalt section of Hole 1185A reveal frequent, small hedbergellid planktonic foraminifera (*Blefuscuiana*? species) and spherical radiolaria, a biofacies noted from intercalated limestone or limestone intraclasts of the upper basement section at both Sites 1183 and 1186. As is typical of this biofacies, definite identification of species is not possible due to recrystallization of the limestone caused by thermal alteration. However, a specimen strongly resembling the planktonic foraminifer *Blefuscuiana gorbachikae* is noted in the sample at 1185A-10R-1, 38 cm (322.98 mbsf). This taxon is known to range from the middle upper Aptian to at least the lowermost Albian (Leckie *et al.* 2002) and, if actually present and together with the nannofossils noted above, would correlate the sample to the uppermost Aptian–lowermost Albian (lower *H. planispira* and NC8A–lower NC8B zones).

Nannofossil evidence from the basal sedimentary section in Hole 1185B is nearly identical to that of Hole 1185A. Nannofossils again clearly indicate a middle Eocene age for the basal sediments in Hole 1185B, equivalent to Zone NP16 of Martini (1971), and most probably from the upper part of Zone NP16. The nannofossil age calibration of Berggren *et al.* (1995) indicates an age range of 43.7–42.0 Ma for the Hole 1185B section. *Reticulofenestra umbilicus* is again present at the base of the sedimentary section at 1185B-2R-1, 149 cm (309.43 mbsf), and the lowest occurrence of *Cribrocentrum reticulatum* occurs immediately above at 2R-1, 133 cm (309.27 mbsf). A middle Eocene planktonic foraminiferal assemblage from 1185B-2R-1, 146–149 cm (309.46 mbsf) includes the subspecies *Globigerinatheka subconglobata micra*, again indicating Zone P12 and supporting the nannofossil age determination.

In the upper basaltic section from Hole 1185B very rare nannofossils are recovered from six of the 26 samples taken from intercalated limestone between Cores 1185B-3R and 17R. At 1185B-3R-1, 32 cm (316.02 mbsf) a single specimen of *Cyclicargolithus floridanus* was recovered, indicative of an Eocene age with reference to the basal portion of the overlying continuous sedimentary section. An Eocene age conflicts with the radiometric date for the basaltic section (Chambers *et al.* 2002, unpublished data) and may therefore be indicative of drilling contamination. Three samples from 1185B-4R-2 to 4R-4 yield very rare Cretaceous nannofossils and a broad late Albian–Cenomanian age. Each of these three samples (4R-2, 42 cm, 4R-3, 12.5 cm and 4R-4, 43 cm) contains the terminal Cenomanian index *Microstaurus chiastius*. Single specimens of *Prediscosphaera columnata,* whose first occurrence datum defines the base of nannofossil zone NC8A, are found in samples at 1185B-4R-2, 42 cm and 4R-3, 12.5 cm. Finally, 1185B-6R-1, 53 cm and 6R-1, 59 cm yield only *Watznaueria barnesae* and *Microstaurus chiastius*, taxa that provide only a very broad Tithonian–Cenomanian age. For planktonic foraminifera, preservation of the hedbergellid–radiolarian biofacies is slightly better in the 1185B section. The planktonic foraminifer *Guembelitriella graysonensis* occurs from 1185B-4R-2, 9 cm to 4R-5, 70 cm (320.79–324.7 mbsf), indicating an age of late Aptian–late Cenomanian, in accordance with the nannofossil biostratigraphy.

A more specific age for these limestone beds may be indicated by foraminifera noted from Sample 1185B-17R-1, 80 cm (435.4 mbsf), in which occurs a large specimen of *Globigerinelloides* that resembles *G. ferreolensis.* If actually present, this species would indicate an age of latest early Aptian–middle late Aptian (upper *L. cabri*–lower *T. bejaouaensis* zones). Radiometric dating of the basalt from the 1185B-17R-3, 4–16 cm interval (437.12–437.24 mbsf) yields an $^{40}Ar/^{39}Ar$ age of 115.24 Ma (Chambers *et al.* 2002, unpublished data). The latest absolute age calibration for the last stratigraphic appearance of the planktonic foraminifer *Globigerinelloides ferreolensis* that may occur just above this level is 114.0 Ma (Leckie *et al.* 2002), consistent with the radiometric estimate. The nannofossils that occur within the limestone beds intercalated with basalt from higher in the basement sections in Holes 1185A and 1185B indicate an age of latest Aptian–terminal Cenomanian. However, no Cenomanian section has been noted in any of the sections drilled from the high Ontong Java Plateau and, although Albian-restricted nannofossil species frequently occur in the Albian section of Sites 1183 and 1186, they are not present within the Cretaceous limestone of the

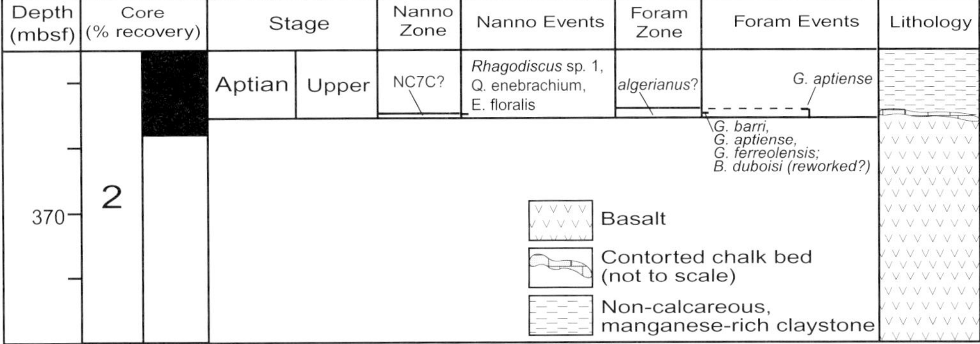

Fig. 7. Stratigraphic summary for ODP Leg 192 Site 1187.

Site 1185 holes. Therefore, a latest Aptian–very earliest Albian age (zone NC8A) is probable for these Site 1185 limestone intervals. This age is only slightly younger than the estimate indicated by foraminifera and radiometric analyses for the deeper section from Core 1185B-17R (Fig. 6).

Site 1187

Site 1187 was chosen to further elucidate the distributional pattern of different basalt types noted from the high Ontong Java Plateau. Specifically, more information was sought regarding the apparent pattern of chemically homogeneous flows dominating on the central high plateau (e.g. at Sites 1183 and 1186) and basalts characterized by more variable composition occurring on the eastern edge of the plateau (e.g. at Site 1185). As such, Site 1187 was drilled on the eastern edge of the high plateau in a physiographical position similar to that of Site 1185. It is located 194 km SE of ODP Leg 130 Site 803 and 146 km north of Site 1185. It was hoped that Site 1187 would also yield a more complete stratigraphic section than the depositionally active Site 1185. Unfortunately, while drilling ahead to about 20 m above the predicted basement level, basaltic section was encountered nearly 40 m higher in the section than anticipated, so that a total of only 1.47 m of sedimentary section was recovered.

The other sites drilled on Leg 192 that penetrated Cretaceous sedimentary section (i.e. Sites 1183, 1185 and 1186) all indicated carbonate deposition above the calcite compensation depth (CCD) during the very latest Ontong Java eruptive event and/or immediately after its termination. However, the sediment recovered at Site 1187 presents a more ambiguous depositional palaeoenvironment (Fig. 7). The 1.47 m-interval overlying basaltic basement is composed of dark brown-red–red-brown, ferromanganese-rich claystone in the upper part, grading down-section into a very dark brown, burrow-mottled–laminated claystone that in turn overlies a contorted 2 cm-thick chalk layer just above basement (Shipboard Scientific Party 2001*d*). The top of basaltic basement occurs at 1187A-2R-2, 30 cm (366.98 mbsf)

The thin chalk layer at 1187A-2R-2, 28 cm (366.96 mbsf) contains common, poorly preserved nannofossils. These include the species *Eprolithus floralis* and *Hayesites irregularis*, indicating an age range of late Aptian–late Albian. However, based on the absence of Albian-restricted nannofossil species observed in cored intervals from Holes 1183A and 1186A, a late Aptian age is probable.

Planktonic foraminifera exhibiting fair preservation are also frequent in the chalk layer. The occurrence of the species *Globigerinelloides barri, G. ferreolensis* and *G. aptiense* indicate the *Globigerinelloides algerianus* zone and a middle late Aptian age. One discrepancy to this assemblage, however, is the presence of a single specimen of the planktonic foraminifer *Blowiella duboisi*, a taxon that does not range above the upper lower Aptian *Leupoldina cabri* zone (Leckie *et al.* 2002). Presumably, the specimen is reworked from older chalk. However, such reworking, combined with the contorted appearance of the chalk bed and its anomalous occurrence in an otherwise non-calcareous sedimentary section, calls into question whether or not the entire layer is *in situ*. If it is allochthonous, presumably indicative of transport from more central regions of the high plateau, then the entire microfossil assemblage noted could have been reworked and deposited at Site 1187 at an indeterminate younger time. This hypothesis is supported by a radiometric age from the underlying basalt section at 1187A-13R-2, 38 cm

(472.37 mbsf) of 106.63 Ma that would place the basalt in the middle Albian. However, this age may be only a minimum estimate (Chambers *et al.* 2002, unpublished data).

There are two difficulties with this younger age hypothesis. The thickness of the chalk bed is only 2 cm. If the chalk is allochthonous, the overlying ferro-manganese-rich claystone would presumably represent the post-eruptive depositional palaeoenvironment at Site 1187. This claystone is a typical condensed basinal 'red clay' deposited well below the CCD. It is unlikely that such a low-volume sediment flow, as represented by the 2 cm-thick chalk layer, would be able to travel for what must have been a considerable distance from an environment above the CCD to Site 1187. Even if such an event did occur, it is unlikely that such a thin calcareous deposit would survive dissolution in the intensely calcite-depleted bottom water that is indicated by the claystone deposit. Washed residues from the claystone are rich in ferromanganese nodules and fish-bone debris, indicating a slow sedimentation rate that would have left any allochthonous chalk deposit exposed to corrosive bottom water for a considerable period. The other problem with the chalk layer being an allochthonous unit deposited at a much younger period is the very rare occurrence of the planktonic foraminifer *Globigerinelloides aptiense* in the lower part of the overlying claystone section at 1187A-2R-1, 90–92 cm and 2R-2, 21–24 cm (366.40 m and 366.89 mbsf). This species does not occur above the top of the Aptian (Leckie *et al.* 2002) and thus indicates that the basal portion of the claystone unit, indicative of very slow depositional rates, cannot be much younger than the bulk of the microfossil assemblage occurring in the basal chalk.

Site 1184

Site 1184 was drilled to obtain information on the poorly understood eastern lobe of the Ontong Java Plateau (Fig. 2). Located near the crest of the northern ridge adjoining the Stewart Basin, it was also hoped that the site would yield shallower palaeobathymetry for the time of the terminal eruption than the sites of the high plateau. Unfortunately, the recovered section from Site 1184 is composed of Neogene (lowermost Miocene) pelagic ooze overlying a thick volcaniclastic interval made up of tuff, lapilli tuff and lapillistone (Shipboard Scientific Party 2001*e*). After drilling over 330 m of volcaniclastic section, the site was abandoned.

The volcaniclastic interval is barren of foraminifera, but rare and poorly preserved nannofossils were recovered from much of the drilled interval. Recovered from pods and stringers of sediment in 32 of 58 samples taken between 1184A-9R-1, 13 cm (201.23 mbsf) and 45R-5, 12 cm (524.94 mbsf), very rare nannofossils indicate a middle Eocene age essentially coeval with the basal sedimentary section at Site 1185. However, radiometric dating of plagioclase crystals from this volcaniclastic sequence yields an age estimate of 123.5±1.8 Ma (Chambers *et al.* 2004). Although this volcaniclastic sequence contains extremely rare Late Cretaceous and Palaeocene nannofossils, it is very difficult to explain the co-occurrence of the 123.5 Ma radiometric age estimate and the Eocene nannofossil recovery from this section.

This discrepancy is one of many enigmas presented by the volcaniclastic section from Cores 1184A-9R–1184A-46R. The sediments are quite different from the much thinner volcaniclastic deposits of the high Ontong Java Plateau noted at Sites 1183 and 1186. The Site 1184 volcaniclastic material was derived from a volcanically active island. Most eruptions were, however, submarine, as demonstrated by the abundance of vitric clasts, but the eruptive column often reached the atmosphere, as indicated by the occurrence of accretionary lapilli and armoured lapilli. The presence of wood and other organic material at several intervals suggest that the volcano grew to, and remained above, sea level long enough to become vegetated (Shipboard Scientific Party 2001*e*).

The vast majority of recovered nannofossils from this section indicate a middle Eocene age (Zone NP16 of Martini 1971), coeval to the basal sedimentary section in both Holes 1185A and 1185B. Nannofossil taxa include *Reticulofenestra umbilicus*, present in 18 of the 33 analysed samples from the volcaniclastic section in Hole 1184A, as well as in the basal sedimentary section of Holes 1185A and 1185B. *Sphenolithus furcatolithoides* and *Sphenolithus spiniger* evidence last stratigraphic occurrences in 1184A-9R-1, 13 cm and 1184A-9R-1, 28 cm. Both these last occurrences are observed at or near the top of the nannofossil Zone NP16 at other Leg 192 sites (1183, 1185 and 1186), at which they are associated with the NP16 zonal datum, the highest occurrence of *Chiasmolithus solitus* (age of 40.4 Ma in Berggren *et al.* 1995). In addition to the more pervasive middle Eocene nannofossil assemblages from the volcaniclastic sequence in Hole 1184A, a total of seven single Cretaceous nannofossil specimens occur in six of 33 samples analysed. These include single specimens of the Late Cretaceous species *Micula staurophora* (1184A-9R-1R, 13 cm) and

Arkhangelskiella cymbiformis (1184A-35R, 57 cm). These two species had respective first appearances in the early Coniacian and late Santonian. In addition, two specimens of *Fasciculithus tympaniformis* occur in Sample 1184A-14–4R, 47 cm. This species is restricted to the late Palaeocene to possibly earliest Eocene. Therefore, no nannofossils recovered from the volcaniclastic section of Hole 1184A are supportive of the radiometric age estimate of 123.5 Ma obtained by Chambers *et al.* (2004). The plagioclase crystals on which the age is based may be xenocrysts derived from older Aptian basalt ejected as part of a much later Eocene volcanic eruption.

DSDP Leg 30

Site 289

Located on the upper part of the NE flank of the high Ontong Java Plateau, Site 289 was drilled to obtain a relatively continuous sequence of tropical biogenic sediments (Andrews *et al.* 1975*b*). The location lies 149 km WNW of ODP Leg 192 Site 1186, and physiographically falls about midway between that eastern slope site and high crestal Site 1183 (Fig. 2). The top of basaltic basement in Hole 289 was encountered at 289-132R-2, 92 cm (1261.82 mbsf). Palaeontological data originally reported for the sediment section immediately overlying basement is quite sparse. Shafik (1975) indicated that the section from 289-131R-1, 149 cm to 132R-2, 68 cm (1232.49–1261.68 mbsf) is Aptian in age, but failed to list or mention any species that supported this determination. No results for foraminifera were reported below the Campanian–Maastrichtian section in Core 289-129R (Michael 1975). The only subsequent biostratigraphic data come from the regional biostratigraphic synthesis of the Ontong Java region by Sliter (1992), in which he examined thin sections from two samples from Site 289 (131R-2, 16 cm and 132R-2, 37 cm), and from selected samples examined in washed residue and thin section by Sliter & Leckie (1993). For this study, the existing data set is combined with results of a reanalysis for foraminifera for the interval 289-131R-1, 57 cm–289-132-2, 70 cm (1231.57–1261.7 mbsf).

Reanalysis of the section for this study indicates that the Aptian–Albian interval at Site 289 is similar in many ways to that analysed at ODP Leg 192 Sites 1183 and 1186. However, the Site 289 section differs by a greater degree of dissolution of the Aptian foraminiferal assemblages and greater overall diagenetic alteration of the carbonate section, mainly recrystallization but also patchy silicification. This required foraminiferal examination almost entirely in thin section. The lack of free microfossil specimens, combined with generally poor preservation and the lack of nannofossil data, results in more sporadic biostratigraphic control relative to equivalent sections from Sites 1183 and 1186. The following interpretation is therefore based on a combination of biostratigraphy and a biofacies succession similar to that which occurs in the Aptian–Albian sections of Sites 1183 and 1186. This biofacies succession proved to be largely isochronous between the two ODP Leg 192 sites.

The sediment immediately overlying basement at Site 289 (132R-2, 30 cm–132R-2, 92 cm; 1261.30–1261.82 mbsf) is composed of a reworked volcaniclastic deposit composed of basaltic lithic sandstone and finer-grained tuff, interbedded with thin (approximately 2 cm thick) limestone beds (Andrews *et al.* 1975*b*). A very similar unit occurs in Hole 1183A (Shipboard Scientific Party 2001*a*) that falls entirely within the uppermost lower Aptian *Leupoldina cabri* zone (see above). Unfortunately the Site 289 tuff layer at 289-132R-2, 70 cm (1261.70 mbsf) yields only very rare, poorly preserved, tiny planktonic foraminifera resembling the Aptian–lower Albian taxon *Blefuscuiana daminae*. Thin-section analysis of the very thin carbonate bed at 289-132R-2, 66–68 cm (1261.66 mbsf) reveals abundant, tiny planktonic foraminifera (*Blefuscuiana* species) that have been subjected to extensive dissolution and recrystallization, together with abundant spherical radiolaria that have been calcified to finely crystalline spar. A similar radiolarian influx marks the uppermost part of the *L. cabri* interval at Site 1183 (1183A-54R-3, 59 cm–54R-3, 96 cm; 1129.76–1130.13 mbsf). The upper tuff bed at 289-132R-2, 53 cm is barren of foraminifera. Nevertheless, based on similarity of both biofacies and lithofacies to the basal sediment section in Hole 1183A, this interval is tentatively correlated to the *L. cabri* zone (Fig. 8). The base of the overlying continuous carbonate interval (289-132R-2, 37 cm; 1261.37 mbsf) is composed of facies very similar to the thin limestone beds within the tuff section; i.e. abundant, tiny planktonic foraminifera resembling species of *Blefuscuiana*. This interval is therefore also included within the *L. cabri* zone, which is consistent with the continuation of this zone above the tuff beds in both basal sedimentary sections from Sites 1183 and 1186. Although not observed in the present study, Sliter (1992)

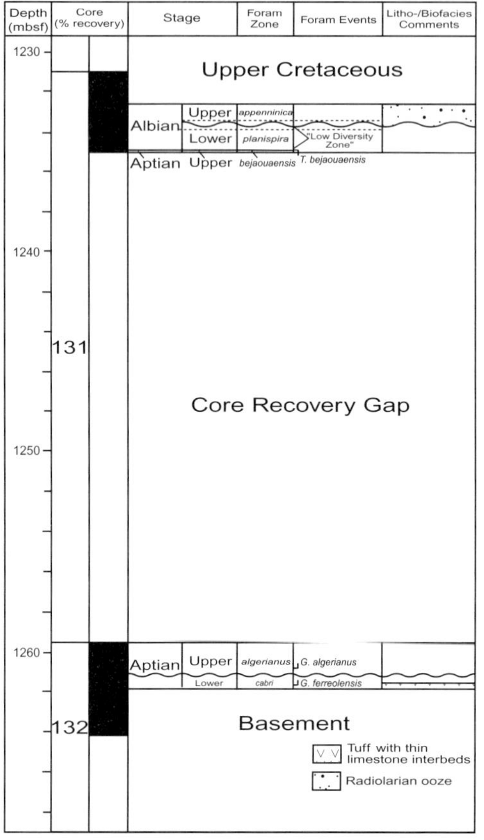

Fig. 8. Stratigraphic summary for DSDP Leg 30 Site 289. Dashed lines indicate the range of uncertainty on the placement of disconformities.

described *Globigerinelloides ferreolensis* and Sliter & Leckie (1993) indicated the presence of *Leupoldina reicheli* in the 289-132R-2, 37 cm (1261.37 mbsf) sample. *Globigerinelloides ferreolensis* also occurs in the upper *L. cabri* zone in Hole 1183A, and both it and *L. reicheli* have first stratigraphic occurrences in the upper *L. cabri* zone (Leckie *et al.* 2002).

The consistent occurrence of large specimens of *Globigerinelloides* (mainly *G. ferreolensis*) in the section from 289-132R-1, 129 cm (1260.79 mbsf) to 132R-2, 16 cm (1261.20 mbsf) indicates the middle upper Aptian *G. algerianus* zone (the nominate taxon occurs very rarely in the top sample). Sliter & Leckie (1993) assigned the 289-132R-1, 21 cm sample to the *G. ferreolensis* zone based on an assemblage of the planktonic foraminifera *Blowiella blowi, B. gottisi, Blefuscuiana hispaniae* and *B. praetrocoidea.* However, these species have last stratigraphic occurrences that range from the *H. trocoidea* zone (*B. hispaniae*) to at least the top of the *T. bejaouaensis* zone (*B. blowi*) and therefore their presence would not preclude a stratigraphic assignment to the *G. algerianus* zone (Leckie *et al.* 2002). Based on the absence of the *G. ferreolensis* zone in both Holes 1183A and 1186A, a disconformity is tentatively set between 289-132R-2, 16 cm (1261.20 mbsf) and 132R-2, 37 cm (1261.37 mbsf).

Interpretation of the remaining Lower Cretaceous section in Hole 289 is complicated by a large core recovery gap of approximately 15 m between Cores 289-131R and 132R (Fig. 8). The base of the recovered section in Core 131R from 131R-3, 89 cm (1234.89 mbsf) to 131R-3, 114 cm (1235.14 mbsf) is marked by common planktonic foraminifera, including rare specimens closely resembling *Hedbergella trocoidea* and *T. bejaouaensis* indicative of the upper Aptian *T. bejaouaensis* zone. Other planktonic species noted include *G. ferreolensis* and *Blefuscuiana gorbachikae*. The former does not range above the lower *T. bejaouaensis* zone (Leckie *et al.* 2002), indicating section correlative to the Hole 1186A *T. bejaouaensis* zonal interval rather than that of Hole 1183A. This correlation is supported by Sliter & Leckie (1993) who report both *T. bejaouaensis* and *H. trocoidea* at 289-131R-3, 80 cm (1234.80 mbsf). This section is also marked by an influx of relatively deep-water, fine-grained agglutinated foraminifera (species of *Dorothia* and *Verneuilinoides*) that indicate much deeper-water facies than evident in the *G. algerianus* section of Core 289-132R, a palaeobathymetric trend also noted in a comparable section at Site 1183. Thus, there is compelling evidence for assigning this interval to the *T. bejaouaensis* zone. This correlation indicates over 28 m of Aptian section in Hole 289 (taking into account the core recovery gap), whereas less than 20 m of Aptian occurs at both Sites 1183 and 1186.

A very poorly fossiliferous interval from 289-131R-2, 126 cm to 131R-3, 17 cm (1233.76–1234.17 mbsf) contains only very rare planktonic foraminifera, including the zonal index *Hedbergella planispira*. Benthic foraminifera are relatively frequent in the interval, mainly composed of rectilinear nodosariids and compressed *Gavelinella* species, and include specimens that strongly resemble the Albian–Cenomanian European species *Nodosaria prismatica* and *Gavelinella berthelini*. These taxa, together with the low-diversity planktonic biofacies, suggest a correlation to the *H. planispira* zone (Fig. 8).

A major change in biofacies occurs in the overlying section of Hole 289 from 289-131R-2, 1 cm to 131R-2, 96 cm (1232.77–1233.46 mbsf) to a densely fossiliferous, radiolarian wackestone. Preservation is poor, with the interior of most radiolarians recrystallized to radial chert. Spherical radiolarians are dominant, but, unlike most of the radiolarian-rich facies observed in the Leg 192 sites, nassellarian species are frequent and include the latest Albian–Cenomanian index *Pseudodictyomitra pseudomacrocephala*. Foraminifera are limited to rare ataxophragmiid benthic agglutinated species. The occurrence of *P. pseudomacrocephala* clearly indicates a correlation between this interval and the prominent radiolarian influx associated with the first stratigraphic occurrence of the nannofossil *Eiffellithus turriseiffelii* in both upper Albian sections of ODP Sites 1183 and 1186. Sliter (1992) reported an occurrence of the Aptian planktonic foraminifer *Globigerinelloides ferreolensis* in this interval at 289-131R-2, 16 cm (1231.23 mbsf). However, taking into account the very poor microfossil preservation characteristic of this section coupled with examination in thin section, it is assumed that this report confuses *G. ferreolensis* with a similar species from the Albian section, such as *Globigerinelloides caseyi*.

The radiolarian-rich interval probably marks the top of the Lower Cretaceous section at Site 289. In contrast to the Aptian, the Albian section of Site 289 is much thinner than comparable sections at Sites 1183 and 1186. The Site 289 Albian interval ranges only from 289-131R-2, 27 cm (1232.77 mbsf) to a maximum lowest occurrence, assuming the *H. planispira* zone falls wholly within the Albian, just above the top of the *T. bejaouaensis* zone at 289-131R-3, 80 cm (1234.80 mbsf; Sliter & Leckie 1993). This indicates no more than 2 m of Albian section, compared to approximately 11 and 17 m of section at Sites 1183 and 1186, respectively.

The thin Site 289 Albian section is overlain by a siliceous, manganese-rich, non-calcareous claystone that is barren of microfossils other than fine-grained fish-bone debris (289-131R-1, 84 cm–131R-1, 145 cm; 1231.84–1232.45 mbsf) that, in turn, is overlain by a red-brown mottled limestone from 289-131R-1, 57 cm to 131R-1, 67 cm (1231.57–1231.67 mbsf). The limestone is nearly barren of microfossils as well, but does contain very small heterohelicid planktonic foraminifera and very rare hyaline benthic taxa such as *Globorotalites micheliana*. As such, the section is probably indicative of the lengthy period the high Ontong Java Plateau spent below the foraminiferal lysocline and/or CCD from the Cenomanian to early Campanian (Shipboard Scientific Party 2001*a*, *b*).

Site 288

Located on the lower part of the northern slope of the Stewart Arc near the head of Stewart Basin, 286 km SW of ODP Leg 192 Site 1184, Site 288 was chosen for what was thought to be a relatively thin sediment cover that would increase the probability of successful drilling to oceanic basement. However, Hole 288A penetrated nearly 990 m of sedimentary section before the site was abandoned without reaching basement.

As with Site 289, the original biostratigraphy for the oldest sediments is very sparse. The section in Hole 288A is not addressed in the nannofossil chapter of the *Initial Report* (Shafik 1975) and foraminiferal descriptions for the oldest section simply note heavily recrystallized *Hedbergella* species (Michael 1975). The site chapter makes additional brief reference to the palaeontology of the oldest section (Andrews *et al.* 1975*a*): an age for Cores 288A-29 and 288A-30 said to be 'probably Aptian to Albian age' based on nannofossils, but without further details; and an Aptian age for lower Core 288A-30R based on a foraminiferal assemblage of '*Hedbergella infracretacea* and several other species of primitive *Hedbergella* and *Praeglobotruncana*', although the latter genus does not occur below the upper Albian. Therefore, no convincing evidence of Aptian section is provided by the original report. Reanalyses of the older sedimentary section from Hole 288A were undertaken to clarify the stratigraphic relationship between this section and that of other Ontong Java localities.

The oldest interval examined from samples 288A-30R-1, 53 cm and 288A-30R-1, 128 cm (980.03 m and 980.78 mbsf) is composed of silicified, tuffaceous limestone with abundant radiolaria and sponge spicules. Also common are hedbergellid planktonic foraminifera, some similar to the species *Hedbergella planispira* and *Blefuscuiana infracretacea*. Rare bathyal agglutinated benthic foraminifera are also present, including a specimen resembling *Spiroplectinata annectens*. The biofacies is similar to the section at Site 289 from 131R-3, 89 cm to 131R-3, 114 cm (1234.89–1235.14 mbsf) that is assigned to the upper Aptian *T. bejaouaensis* zone. Supporting this hypothesis, Sliter & Leckie (1993) assigned the sample at 288A-30R-1, 81cm (980.31 mbsf) to the *T. bejaouaensis* zone, but do not provide a justification for this conclusion. Although this section is probably equivalent to the *T. bejaouaensis* zone, there is also the possibility that it may be indicative of the lower Albian *H. planispira* zone.

Nearly 7 m of unrecovered section separate these samples from considerably younger sediments in Core 288A-29R (Andrews *et al.* 1975*a*). The interval from 288A-28R-1, 140 cm to 29R-1, 100 cm (952.40–971.00 mbsf) continues to be dominated by radiolaria, but, at least in the lower part of the section, is also characterized by abundant planktonic foraminifera. Among the latter, few to common *Planomalina buxtorfi* indicate an uppermost Albian section (upper *R. appenninica* zone) that is not observed at the high Ontong Java Plateau sites. Sliter & Leckie (1993) also assign Core 30–288A-29R to the *R. appenninica* zone, but do not provide their rationale for this conclusion. Radiolaria become very dominant in the upper part of the interval (28R-1, 140 cm) in which planktonic foraminifera are rare and partially silicified. A rugose hedbergellid observed in thin section from this interval resembles the taxon *Hedbergella libyca*, whose co-occurrence with *P. buxtorfi* would indicate a very latest Albian age for the section.

Radiolaria continue to dominate the section from 288A-27R-2, 135 cm to 28R-1, 50 cm (934.85–951.5 mbsf). Planktonic foraminifera are frequent, however, and include *Praeglobotruncana stefani, Rotalipora appenninica, Globigerinelloides bentonensis* and *Whiteinella brittonensis/paradubia*. Together with the absence of *Planomalina buxtorfi*, the assemblage indicates an age of early–early middle Cenomanian (*R. globotruncanoides*–lower *R. reicheli* zones).

The thick, overlying section from Cores 288A-20R–288A-27R has long been perplexing to stratigraphers interested in the Ontong Java region. Whereas stratigraphic evidence of upper Cenomanian to Turonian section is absent from all other sites drilled on the Ontong Java Plateau, including those from ODP Leg 192, the original biostratigraphic report implies nearly 70 m of upper Cenomanian–Turonian in this portion of the Hole 288A section (Andrews *et al.* 1975*a*). The original evidence is unconvincing, however, because the results described from the foraminiferal chapter of the *Initial Report* (Michael 1975) contradict the results described in the site chapter (Andrews *et al.* 1975*a*). For example, the site chapter assigns 288A-23R-CC (882.30 mbsf) to the lower Turonian based on the occurrence of the planktonic foraminifer *Helvetoglobotruncana helvetica*, whereas the foraminiferal chapter lists the upper Cenomanian indexes *Rotalipora cushmani* and *Rotalipora greenhornensis* as high as 288A-20R-1, 38–40 cm (846.88 mbsf).

Re-examination of this section reveals a depositionally complex interval in which samples containing abundant and diverse planktonic foraminifera alternate abruptly and frequently with sections highly dominated by radiolaria in which foraminifera are nearly absent. Silicification is common, resulting in indurated lithology requiring examination by thin section. Surprisingly, the re-examination finds that the original age contradictions accurately described the biostratigraphy for this section. For example, 288A-27R-2, 41 cm (933.91 mbsf) is marked by abundant planktonic foraminifera that include *Dicarinella algeriana, D. imbricata, Praeglobotruncana gibba* and *Helvetoglobotruncana praehelvetica*, along with abundant hedbergellids/whiteinellids. Together with the absence of *Rotalipora* species and *H. helvetica*, the assemblage is indicative of the latest Cenomanian–earliest Turonian *Whiteinella archaeocretacea* zone. However, the overlying Sample 288A-25R-1, 129–132 cm (895.29 mbsf) contains the planktonic foraminifera *Rotalipora greenhornensis* and *R. micheli*, indicating an upper middle–lower upper Cenomanian section (upper part of the *Rotalipora reicheli* zone–lower *R. cushmani* zone). Similar apparent age discrepancies characterize the entire section from Cores 288A-20R–288A-27R, marked by alternations of middle Cenomanian, upper Cenomanian or uppermost Cenomanian–lower Turonian planktonic foraminiferal assemblages with no apparent stratigraphic order (Table 9).

The interval is very probably indicative of a repeated section due to sediment flow deposition at Site 288 that resulted in interbedding of reworked chalk of various ages. The latest Cenomanian–earliest Turonian assemblage is the youngest noted in the interval, and may be indicative of *in situ* deposition. Alternatively, all the beds rich in foraminifera may be allochthonous with the time of deposition unknown and the radiolarian-rich intervals representative of autochthonous deposition. The source area for this reworked chalk is also unknown, but presumably lay farther up-gradient on the Stewart Arc (Fig. 2). This would imply that the stratigraphic section that drapes the Stewart Arc may be radically different from that of the high Ontong Java Plateau. The top of this expanded allochthonous section occurs between samples 288A-20R-1, 78–83 cm (847.28 mbsf) and 20R-2, 64–68 cm (848.64 mbsf). The overlying section is composed of mudstone nearly barren of microfossils except for very tiny hedbergellid and heterohelicid planktonic foraminifera, probably

Table 8. *Stratigraphic distribution of planktonic foraminifera in Lower Cretaceous sediment of DSDP Leg 30, Site 289*

Core	Section	Top (cm)	Depth (mbsf)	Group abundance	Group preservation	Stage		Zone	*Blefuscuiana daminiae*	*Praehedbergella sigali*	*Blefuscuiana* spp.	*Globigerinelloides ferreolensis*	*Blefuscuiana infracretacea*	*Globigerinelloides algerianus*	*Ticinella bejaouaensis*	*Blefuscuiana kuznetzovae*	*Blefuscuiana gorbachikae*	?*Hedbergella* spp.	*Hedbergella planispira*	*Pseudodictyometra pseudomacrocephala**
131	2	27	1232.51	barren		Albian	upper	appenninica?												F
131	2	96	1233.46																	R
131	2	126	1233.76	F	P		lower	planispira					P					F	R	
131	3	17	1234.17	T	P													R		
131	3	89	1234.89	R	P				P									F		
131	3	114	1235.14	C	M	Aptian	upper	bejaouaensis	C		C	R	R		P	R	P			
132	1	129	1260.79	A	P			algerianus	C		A		R	P						
132	2	16	1261.20	C	P				R		C	R								
132	2	37	1261.37	C	P		lower	cabri?	F		A									
132	2	47	1261.47	barren				?												
132	2	53	1261.53																	
132	2	62	1261.62	A	P			cabri	F	P	A	P	P							
132	2	70	1261.66	A	M				O	P	C									

* Radiolarian with upper Albian first occurrence.

indicative of the Coniacian–Campanian condensed section deposited below the foraminifer lysocline that is also noted from DSDP Leg 30 Site 289 and ODP Leg 192 Sites 1183 and 1186.

Conclusions

The Lower Cretaceous section of the Ontong Java Plateau is quite variable in the thickness and age of the sediment immediately overlying basaltic basement. Stratigraphic correlation is hampered by the effects of dissolution and diagenesis, frequent disconformities and poor core recovery in the section available from DSDP and ODP sites. Therefore, high-resolution chronostratigraphy requires a multi-disciplinary approach utilizing nannofossil, foraminiferal (using both free specimen and thin section), and stable and radiometric isotope analyses. Despite the variability of the section, some regional stratigraphic conclusions are possible.

- Stable carbon isotopic evidence indicates that the main Ontong Java eruption continued through much of OAE 1a, for which it is postulated to have been the cause (Bralower *et al.* 1994; Erba 1994; Larson & Erba 1999). This is based on a prominent $\delta^{13}C$ positive maximum in the basal sedimentary section at Site 1183 near the top of the *L. cabri* zone, indicating that sediment deposition did not begin until near the end of OAE 1a relative to the current calibration of the event (Moullade *et al.* 1998; Larson & Erba 1999).
- Pyroclastic eruptions immediately following the main extrusive phase of the Ontong Java event resulted in widespread deposition of reworked tuff deposits immediately overlying basement. The volcaniclastic deposits are indicative of multiple events, interbedded with thin limestone of the *L. cabri* zone. The tuffaceous unit varies in thickness from about 1 m in the crestal region of the plateau (Site 1183) to 7 cm on the eastern slope (Site 1186).
- No convincing evidence for the presence of the *G. ferreolensis* planktonic foraminiferal zone is noted from any of the study sections, indicating a regional disconformity. Together with the thinness of the limestone beds associated with the underlying tuffaceous interval, this suggests that the period immediately following the termination of the main Ontong Java eruptive phase was one of low pelagic productivity and sediment starvation across

Table 9. *Stratigraphic distribution of planktonic foraminifera in Lower and Lower Upper Cretaceous sediment of DSDP Leg 30, Hole 288A. Questionable intervals are generally composed of siliceous radiolarian limestone/claystone*

Core	Section	Top (cm)	Depth (mbsf)	Group abundance	Group preservation		Stage	Zone	*Blefuscuiana/Hedbergella* spp.	*Hedbergella planispira*	*Blefuscuiana infracretacea*	*Planomalina buxtorfi*	*Hedbergella? libyca*	*Praeglobotruncana stefani*	*Whiteinella paradubia/brittonensis*	*Rotalipora* spp.	*Globigerinelloides bentonensis*	*Whiteinella archaeocretacea*	*Dicarinella algeriana*	*Dicarinella imbricata*	*Whiteinella aumalensis*	*Marginotruncana pseudolinneiana*	*Heterohelix* spp.	*Praeglobotruncana gibba*	*Helvetoglobotruncana praehelvetica*	*Hedbergella delrioensis*	*Rotalipora micheli*	*Rotalipora greenhornensis*	*Rotalipora cushmani*	*Rotalipora deecki*	*Rotalipora reicheli*	*Dicarinella canaliculata*	*Pessaginella oviformis*	*Praeglobotruncana turbinata*
20	2	64	848.64	O	P			?	O														R											
20	2	97	848.97	A	P			cushmani	A					R		R							F	R				R	R	T				T
20	3	50	850.00	barren																														
21	1	126	857.26	O	P		upper	?	O																									
21	2	47	857.97	A	M			upper reicheli	O					O	F	C				T	T?		F	T				F	T		R			
21	3	96	859.96	A	P				A																T?									
22	1	125	866.75	F	P			?	F																									
22	2	63	867.63	C	P				C														T											
23	1	109	876.09	A	P		middle	upper reicheli	O					O		F							R					T			R			
23	2	22	876.72	A	M				O					O	R	C				T			F	T				F	T	R			R	
23	2	110	877.60	A	P		uppermost	archaeocretacea	A											T				T										
23	3	1	878.01	A	P				A					T									R									T	T	
23	3	116	879.16	O	G	Cenomanian			F					R	R	R							R					T	R					
24	1	1	884.51	C	P		upper	cushmani	O					T									O							T?				
24	1	143	885.93	A	G				C					O	O	F		O					O					R	T	O	T			
25	1	48	894.48	C	P			?	C														T											
25	1	129	895.29	O	P		middle	upper reicheli	O	T					T												T	T						
26	1	11	913.11	C	P			?	C														F											
26	1	116	914.16	R	P				R				T										R		T	T								
27	1	118	933.18	F	P		uppermost	archaeocretacea	F						F																			
27	2	41	933.91	A	M				C					R				T?	T	T	T?	T	C	R										
27	2	135	934.85	C	P				C									R?																
28	1	50	951.50	O	P		lower?	globotruncanoides?	O					R	R	R	R																	
28	1	140	952.40	F	P				F			R	T	T																				
29	1	53	970.53	O	P	Albian	upper	buxtorfi	F	R		R																						
29	1	100	971.00	C	P				C			F																						
30	1	53	980.03	C	P	Aptian	upper?	bejaouaensis?	C		R																							
30	1	128	980.78	O	P				O	O?																								

the high plateau. An alternative cause may have been a eustatic sea-level fall resulting in current erosion across the central plateau with concomitant erosion on the plateau slope by sediment flows.

- Pelagic sedimentation first increased during the middle part of the late Aptian, as indicated by the greater thickness and wide distribution of *G. algerianus* zone. A carbonaceous, radiolarian-rich claystone bed in this section at Site 1183 (Fig. 3) indicates dysoxia possibly related to a local cause unassociated with any OAE. However, the unit may also be equivalent to similar carbonaceous beds within the *G. algerianus* zone noted in northern California and NE Mexico (Premoli Silva & Sliter 1999; Sliter 1999) and therefore indicative of another global OAE intermediate in chronology to OAE 1a and 1b.
- Another regional unconformity overlies the *G. algerianus* zone, marked by the absence of the *H. trocoidea* zone and indicating another period of sediment starvation or submarine erosion.
- Pelagic sedimentation increased again during deposition of the *T. bejaouaensis* zone (nannofossil zone NC7C), but the resumption was diachronous, beginning first on the eastern flank of the high plateau (DSDP Site 289 and ODP Site 1186) and later in the central crestal region (Site 1183).
- A partial record of OAE 1b is noted across the high plateau. The record varies with completeness of the Aptian–Albian boundary section across the region. In the central crestal region of Site 1183, $\delta^{13}C$ values indicate a record of the early portion of the event and the Jacob and Paquier positive maxima. Conversely, only the younger portion of the event was recorded on the eastern plateau slope (the Paquier and, possibly, Leenhardt maxima). Nevertheless, a conformable Aptian–Albian stage boundary (as defined by the Paquier datum) appears to have regional extent across the Ontong Java Plateau.
- The largest disconformity of the Lower Cretaceous section occurs at the lower–middle Albian substage boundary and is noted from all studied sections.
- Another regional disconformity may mark the middle–upper Albian substage boundary, underlain by a thin middle Albian section correlative to nannofossil zone NC8C.
- An upper Albian section occurs at all high plateau sites except Site 1185 and, possibly, Site 1187. This interval is always associated with radiolarian-rich cherty limestone indicative of an increase in primary productivity at a time equivalent to the middle portion of nannofossil zone NC10A.
- Biostratigraphic and radiometric results indicate that the uppermost basalt from Sites 1185 and 1187 are younger (upper Aptian) than the upper basaltic section at Sites 1183 and 1186 that are more centrally located on the Ontong Java Plateau. There is no definitive radiometric evidence of extrusive rock younger than Aptian from any Leg 192 material. Nevertheless, the presence of Eocene nannofossils in the uppermost intercalated limestone of the basaltic section in Holes 1185A and 1185B, if not drilling contaminants, would indicate a period of Palaeogene volcanism. More persuasive evidence of post-Aptian volcanism is provided by the sporadic occurrence of Palaeogene nannofossils in stratigraphic order throughout the volcaniclastic section of Site 1184. Although radiometric analysis of plagioclase crystals suggests an Aptian age for the deposit, these crystals could be xenocrysts derived from older extrusive rock during violent Palaeogene volcanic eruptions. The authors believe that further research is needed to resolve whether a period of such Palaeogene volcanism occurred.
- Despite two attempts (DSDP Leg 30 and ODP Leg 192), the basaltic basement of the eastern salient of the Ontong Java Plateau remains unsampled. Reanalysis for foraminifera from the Lower Cretaceous sedimentary section recovered from DSDP Leg 30 Site 288 indicates a similar late Aptian–early late Albian depositional history between this northern Stewart Arc slope site and the main Ontong Java Plateau. However, the uppermost Albian–Cenomanian interval at Site 288 indicates that portions of the Stewart Arc remained well above the CCD during a period when the main portion of the Ontong Java Plateau dropped well below this oceanographic boundary. Therefore, the authors believe further research is needed to fully relate both the tectonic and depositional history of the eastern salient to the main plateau.

This research used samples (DSDP Leg 30 Sites 288 and 289, and ODP Leg 192 Sites 1183 and 1186) provided by the Ocean Drilling Program (ODP). ODP is sponsored by the US National Science Foundation (NSF) and participating countries under management of Joint Oceanographic Institutions (JOI), Inc. Funding for this research was provided by a post-cruise science grant from JOI. The authors are grateful to Dr J. G. Ogg of Purdue University for permission to publish the carbon isotopic data for the Lower Cretaceous section of Leg 192 Sites 1183 and 1186.

Finally, the authors also wish to thank Dr R. M. Leckie of the University of Massachusetts at Amherst and Dr E. Erba of the University of Milan for their thoughtful reviews, which resulted in significant improvement of this paper.

References

ANDREWS, J.E., PACKHAM, G. *et al.* 1975*a*. Site 288. *Initial Reports of the Deep Sea Drilling Project*, **30**, 175–230.

ANDREWS, J.E., PACKHAM, G. *et al.* 1975*b*. Site 289. *Initial Reports of the Deep Sea Drilling Project*, **30**, 231–398.

BERGEN, J.A. 1994. Berriasian to Early Aptian calcareous nannofossils from the Vocontian Trough (SE France) and Deep Sea Drilling Site 534: New nannofossil taxa and a summary of low-latitude biostratigraphic events. *Journal of Nannoplankton Research*, **16**, 59–69.

BERGEN, J.A. 1998. Calcareous nannofossils from the lower Aptian historical stratotype at Cassis–La Bedoule (SE France). *Geologie Mediterraneene*, **25**, 227–255.

BERGEN, J.A. 2004. Calcareous nannofossils from ODP Leg 192, Ontong Java Plateau. *In*: FITTON, J.G., MAHONEY, J.J., WALLACE, P.J. & SAUNDERS, A.D. (eds) *Origin and Evolution of the Ontong Java Plateau*. Geological Society, London, Special Publications, **229**, 113–132.

BERGGREN, W.A., KENT, D.V., SWISHER, C.C., III. & AUBRY, M.-P. 1995. A revised Cenozoic geochronology and chronostratigraphy. *In*: BERGGREN, W.A., KENT, D.V., AUBRY, M.-P. & HARDENBOL, J. (eds) *Geochronology, Time Scales and Global Stratigraphic Correlation*. SEPM, Special Publication, **54**, pp. 129–212.

BLOW, W.H. 1969. Late middle Eocene to Recent planktonic foraminiferal biostratigraphy. *Proceedings First International Conference on Planktonic Microfossils*, Geneva, E.J. Brill, Leiden 1967, **1**, 199–422.

BRALOWER, T.J., ARTHUR, M.A., LECKIE, R.M., SLITER, W.V, ALLARD, D. & SCHLANGER, S.O. 1994. Timing and paleooceanography of oceanic dysoxia/anoxia in the late Barremian to early Aptian. *Palaios*, **9**, 35–369.

BRALOWER, T.J., SLITER, W.V., ARTHUR, M.A., LECKIE, R.M., ALLARD, D. & SCHLANGER, S.O. 1993. *Dysoxic/Anoxic Episodes in the Aptian–Albian (Early Cretaceous). Schlanger Memorial Volume.* American Geophysical Union, Geophysical Monograph, **73**, 5–37.

CHAMBERS, L.M., PRINGLE, M.S. & FITTON, J.G. 2002. Age and duration of magmatism on the Ontong Java Plateau: $^{40}Ar/^{39}Ar$ results from ODP Leg 192. *Eos, Transactions of the American Geophysical Union*, **83**, F47, abstract V71B-1271.

CHAMBERS, L.M., PRINGLE, M.S. & FITTON, J.G. 2004. Phreatomagmatic eruptions on the Ontong Java Plateau: an Aptian $^{40}Ar/^{39}Ar$ age for volcaniclastic rocks at ODP Site 1184. *In*: FITTON, J.G., MAHONEY, J.J., WALLACE, P.J. & SAUNDERS, A.D. (eds) *Origin and Evolution of the Ontong Java Plateau*. Geological Society, London, Special Publications, **229**, 325–331.

ERBA, E. 1994. Nannofossils and superplumes: The early Aptian 'nannoconid crisis'. *Paleoceanography*, **9**, 483–501.

ERBA, E., CHANNELL, E.T., CLAPS, M., JONES, C., LARSON, R., OPDYKE, B., PREMOLI SILVA, I., RIKVA, A., SALVINI, G. & TORRICELLI, S. 1999. Integrated stratigraphy of the Cismon Apticore (southern Alps, Italy): A 'reference section' for the Barremian–Aptian interval at low latitudes. *Journal of Foraminiferal Research*, **29**, 371–391.

HERRLE, J.O. & MUTTERLOSE, J. 2003. Calcareous nannofossils from the Aptian–lower Albian of southeast France: palaeoecological and biostratigraphic implications. *Cretaceous Research*, **24**, 1–22.

KENNEDY, W.J., GALE, A.S., BOWN, P.R., CARON, M., DAVEY, R.J., GROCKE, D. & WRAY, D.S. 2000. Integrated stratigraphy across the Aptian–Albian boundary in the Marnes Bleues, at the Col de Pré-Guittard, Arnayon (Drôme), and at Tartonne (Alpes-de-Hautes–Provence), France: a candidate Global Boundary Stratotype Section and Boundary Point for the base of the Albian Stage. *Cretaceous Research*, **21**, 591–720.

LARSON, R.L. & ERBA, E. 1999. Onset of the mid-Cretaceous greenhouse in the Barremian–Aptian: Igneous events and the biological, sedimentary, and geochemical responses. *Paleoceanography*, **14**, 663–678.

LECKIE, R.M., BRALOWER, T.J. & CASHMAN, R. 2002. Oceanic anoxic events and plankton evolution: Biotic response to tectonic forcing during the mid-Cretaceous. *Paleoceanography*, **17**, 13-1–13-29.

MAHONEY, J.J. 1987. An isotopic survey of Pacific oceanic plateaus: implications for their nature and origin. *In*: KEATING, B.H., FRYER, P., BATIZA, R. & BOEHLERT, G.W. (eds) *Seamounts, Islands, and Atolls*. American Geophysical Union, Geophysical Monograph, **43**, 207–220.

MAHONEY, J.J., FITTON, J.G., WALLACE, P.J. *et al.* 2001. *Proceedings of the Ocean Drilling Program, Initial Reports*, **192** (CD-ROM). Available from Ocean Drilling Program, Texas A&M University, College Station TX 77845-9547, USA.

MARTINI, E. 1971. Standard Tertiary and Quaternary calcareous nannoplankton zonation. *In*: FARINACCI, A. (ed.) *Proceedings II Planktonic Conference, Roma, 1970*, **2**, 729–785.

MICHAEL, F. 1975. Mesozoic Foraminifera, Leg 30. *In*: ANDREWS, J.E., PACKHAM, G. *et al.* (eds) Site 289. *Initial Reports of the Deep Sea Drilling Project*, **30**, 599–602.

MOULLADE, M., MASSE, J.P., TRONCHETTI, G., KUHNT, W., ROPOLO, P., BERGEN, J.A., MASURE, E. & RENARD, M. 1998. Le stratotype historique de l' Aptien inférieur (region de Cassis–La Bédoule, SE France): synthèse stratigraphique, *Geologie Mediterraneene*, **25**, 289–298.

PREMOLI SILVA, I. & SLITER, W.V. 1999. Cretaceous paleoceanography: Evidence from planktonic foraminiferal evolution. *In*: BARRERA, E. & JOHNSON, C.C. (eds) *Evolution of the Cretaceous Ocean–Climate System*. Geological Society of America, Special Paper, **332**, 301–328.

SHAFIK, S. 1975. Nannofossil biostratigraphy of the Deep Sea Drilling Project, Leg 30. *In*: ANDREWS,

J.E., PACKHAM, G. *et al.* (eds) Site 289. *Initial Reports of the Deep Sea Drilling Project*, **30**, 549–598.

SHIPBOARD SCIENTIFIC PARTY. 2001*a*. Site 1183. *In*: MAHONEY, J.J., FITTON, J.G., WALLACE, P.J. *et al.* (eds) *Proceedings of the Ocean Drilling Program, Initial Reports*, **192**, 169 pp. (CD-ROM).

SHIPBOARD SCIENTIFIC PARTY. 2001*b*. Site 1186. *In*: MAHONEY, J.J., FITTON, J.G., WALLACE, P.J. *et al.* (eds) *Proceedings of the Ocean Drilling Program, Initial Reports*, **192**, 117 pp. (CD-ROM).

SHIPBOARD SCIENTIFIC PARTY. 2001*c*. Site 1185. *In*: MAHONEY, J.J., FITTON, J.G., WALLACE, P.J. *et al.* (eds) *Proceedings of the Ocean Drilling Program, Initial Reports*, **192**, 124 pp. (CD-ROM).

SHIPBOARD SCIENTIFIC PARTY. 2001*d*. Site 1187. *In*: MAHONEY, J.J., FITTON, J.G., WALLACE, P.J. *et al.* (eds) *Proceedings of the Ocean Drilling Program, Initial Reports*, **192**, 66 pp. (CD-ROM).

SHIPBOARD SCIENTIFIC PARTY. 2001*e*. Site 1184. *In*: MAHONEY, J.J., FITTON, J.G., WALLACE, P.J. *et al.* (eds) *Proceedings of the Ocean Drilling Program, Initial Reports*, **192**, 131 pp. (CD-ROM).

SLITER, W.V. 1992. Cretaceous planktonic foraminiferal biostratigraphy and paleoceanographic events in the Pacific Ocean with emphasis on indurated sediment. *In*: ISHIZAKI, K. & SAITO, T. (eds) *Centenary of Japanese Micropaleontology*. Terra, Tokyo, 281–289.

SLITER, W.V. 1999. Cretaceous planktic foraminiferal biostratigraphy of the Calera Limestone, northern California, USA. *Journal of Foraminiferal Research*, **29**, 318–339.

SLITER, W.V. & LECKIE, R.M. 1993. Cretaceous planktonic foraminifers and depositional environments from the Ontong Java Plateau with emphasis on Sites 803 and 807. *In*: BERGER, W.H., KROENKE, L.W., MAYER, L.A. *et al. Proceedings of the Ocean Drilling Program, Results*, **130**, 63–84.

TOURMARKINE, M. & LUTERBACHER, H. 1985. Paleocene and Eocene planktic foraminifera. *In*: BOLLI, H.M., SAUNDERS, J.B. & PERCH-NIELSEN, K. (eds) *Plankton Stratigraphy*, Volume 1. Cambridge University Press, Cambridge, 87–154.

Calcareous nannofossils from ODP Leg 192, Ontong Java Plateau

J. A. BERGEN
BP America, 501 Westlake Park Boulevard, Houston, TX 77079, USA
(e-mail: bergenj2@bp.com)

Abstract: Middle Miocene–upper lower Aptian calcareous nannofossils were recovered from Sites 1183–1187 drilled by Ocean Drilling Project Leg 192 on the Ontong Java Plateau. Nannofossil biostratigraphy indicates the presence of six unconformities among the five Leg 192 sites. These are: (1) between the lowermost Albian and upper middle Albian; (2) between the upper Albian and middle Coniacian; (3) within the lower Maastrichtian; (4) between the lower upper Maastrichtian and basal Danian; (5) within the upper Palaeocene; and (6) between the Oligocene and Miocene.

Previous drilling before Leg 192 on the Ontong Java Plateau and in the SW Pacific (Legs 30 and 130) indicated two episodes for major emplacement of basement during the earliest Aptian and Turonian. The Leg 192 drilling was not able to confirm either of these episodes, but instead indicated that the emplacement of basement on the Ontong Java Plateau was relatively continuous from the latest early Aptian to latest Aptian. Results from Sites 1184 and 1185 indicate the possibility of another magmatic episode during the middle Eocene (Zone NP16).

Nannofossil biostratigraphy

The nannofossil biostratigraphy of the sedimentary sections drilled at Sites 1183–1187 is discussed in the following sections. A summary of Cretaceous–Eocene nannofossil events for Leg 192 is presented in Tables 1–3. Fourteen species distribution charts are posted in the ODP *Scientific Results* volume (available as CD-ROM and on-line).

The nannofossil zones applied to the Leg 192 sediments are the NC Zones (Roth 1978; Bralower *et al.* 1993) for the Lower Cretaceous, CC Zones (Sissingh 1977, 1978; Perch-Nielsen 1985*a*) for the Upper Cretaceous and NP Zones (Martini 1971; Perch-Nielsen 1985*b*; Aubry 1991) for the Palaeogene. The NN zonation of Martini (1971) is applied to lower–middle Miocene sediments recovered at Sites 1183 and 1184.

The aforementioned nannofossil zonations were founded mainly on circum-Atlantic outcrops or deep-sea material. Thus, it is necessary to use a number of 'unconventional' nannofossil events due to both the provenance and poor preservation of the Leg 192 material. For the Cretaceous, a number of ancillary events were used to modify and enhance existing zonations. These include both published (Bergen 1994; Bralower & Bergen 1998; Burnett 1998; Kennedy *et al.* 2000; Bergen & Sikora 2000) and unpublished data. The unpublished data sources are from the author's Amoco research projects from 1988 to 1999 on outcrops in Tunisia (El Kef and Hammam Mellegue areas), southern England (Isle of Wight), SE France (Vocontian Trough) and North America (eastern Gulf of Mexico and Western Interior).

Cretaceous

Lower Cretaceous

An Upper lower Aptian–upper Albian section was drilled at Sites 1183 and 1186 (Table 1). At Site 1187, a single core of upper Aptian sediment was recovered above basement (Table 1). Latest Aptian (to possibly Cenomanian) nannofossils were recovered from intercalated sediments within basement at Site 1185 (Tables 1; see also the section on Basement ages).

A modified version of the Bralower *et al.* (1993) biozonation, last updated by Leckie *et al.* (2002), was used to date the Leg 192 Aptian–Albian sediments (Fig. 1). This integrated foraminifera–nannofossil biozonation is based on study of numerous deep-sea sections and provides additional resolution beyond the original 'NC' Zone scheme of Roth (1978). The three subzones of NC7C could not be recognized in the Leg 192 material. Only a single specimen of *Micrantholithus hoschulzii*, whose extinction marks the base of NC7B, was observed in all samples. *Rhagodiscus achlyostaurion*, whose appearance marks the base of Zone NC7C, actually ranges into the upper Barremian in the Barremian stratotype in

From: FITTON, J. G., MAHONEY, J. J., WALLACE, P. J. & SAUNDERS, A. D. (eds) 2004. *Origin and Evolution of the Ontong Java Plateau*. Geological Society, London, Special Publications, **229**, 113–132. 0305-8719/$15.00

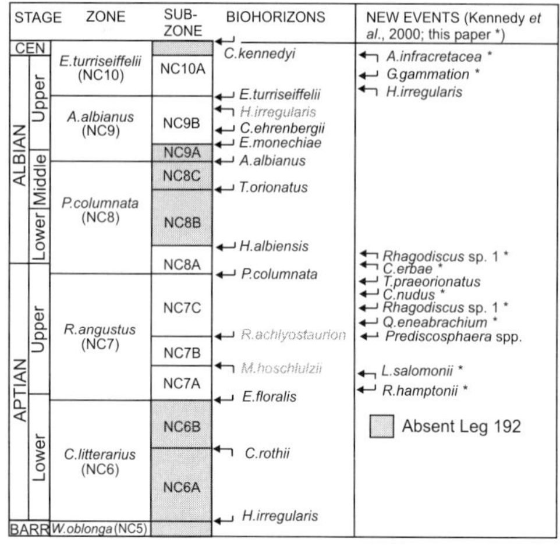

Fig. 1. Aptian–Albian biozonation after Bralower *et al.* (1993). Events in grey are not applicable to Leg 192.

SE France and at DSDP Site 534 (Bergen 1994). The lowest occurrence of the genus *Prediscosphaera* (small, elliptical forms having an axial cross) is used as a proxy for the base of NC7C. In addition, *Hayesites irregularis* and *Eiffellithus turriseiffelii* co-occur in the Leg 192 material, as well as in SE France (Thierstein 1973). Bralower *et al.* (1993) placed the highest occurrence of *Hayesites irregularis* within the uppermost portion of Zone NC9B, immediately below the lowest occurrence of *Eiffellithus turriseiffelii* (the marker for the base of Zone NC10A).

More recent data from potential boundary stratotypes for the base of the Cenomanian and the base Albian in SE France not only provide additional information on the location of these stage boundaries, but also additional nannofossil events. The recommended boundary stratotype section for the base of the Cenomanian is the Mont Risou section near Rosans (Gale *et al.* 1996). Most of this section is within the *Eiffellithus turriseiffelii* Zone (NC10A). The basal Cenomanian in this section is defined on the appearance of the planktonic foraminifera *Rotalipora globotruncanoides*. Nannofossil analyses from sample sets provided by Gale to the author have yielded two new events pertinent to the Leg 192 material. The highest occurrence of *Assipetra infracretacea* lies about 47 m below the basal Cenomanian and the lowest occurrence of *Gartnerago gammation* is 67 m below the basal Cenomanian, thus co-occurring in approximately 20 m of section. These two species co-occur in Hole 1183A between samples 51-3, 28 cm and 50-2, 78 cm within Zone NC10A. For the base of the Albian, there are two potential boundary reference sections (Col de Pré-Guittard and Tartonne) in SE France (Kennedy *et al.* 2000). Published nannofossil data (Fig. 2) from these sections by Bown (in Kennedy *et al.* 2000) indicate that: (1) Zone NC8A actually straddles the Aptian–Albian boundary, as *Prediscosphaera columnata* ranges into the uppermost Aptian; and (2) that *Tranolithus praeorionatus* also ranges into the uppermost Aptian. My examination (see Fig. 2) of these same two boundary reference sections also show that the highest occurrences of both *Calcicalthina erbae* and *Rhagodiscus* sp. 1 range into the basal Albian, whose base is defined by the base of the 'Pacquier' horizon. In the Leg 192 material *Prediscosphaera columnata* co-occurs with *Calcicalathina erba* and *Rhagodiscus* sp. 1 in both Holes 1186A and 1183A (Table 1), thus bracketing the Aptian–Albian boundary. Three additional nannofossil first occurrences supplement the Bralower *et al.* (1993) scheme (Figs 1 and 2, Table 1) based on my examination of an upper Aptian section near Angles in SE France and the Leg 192 material: *Quadrum eneabrachium*, *Rhagodiscus* sp. 1 and *Cylindralithus nudus*.

There is a major unconformity between the upper Albian and middle Coniacian that is well constrained by detailed nannofossil sampling in Holes 1183A and 1186A (Table 1). Recovered nannofossils also indicate an unconformity between the lowermost Albian and upper middle Albian in the same two holes (Fig. 1, Table 1). Again, zone NC8A, which straddles the Aptian-Albian boundary, is constrained between the tops of *Calcicalathina erbae* and *Rhagodiscus* sp. 1 and the base of *Prediscosphaera columnata*. These events are followed immediately up-section in both Holes 1183A and 1118A by events (Table 1) that mark the base of the upper Albian (Zone NC9B).

Upper Cretaceous

Upper Cretaceous nannofossils were recovered from Holes 1183A and 1186A. Nannofossils indicate that there are two discrete Upper Cretaceous sedimentary packages in the Leg 192 material. Extremely condensed middle Coniacian–lower middle Campanian sediments (Zones CC-19–CC-20), ranging from 0.3 m to possibly 1.5 m, overlie a major unconformity with the upper Albian (Table 1). Higher abundance, upper middle Campanian–upper lower Maastrichtian assemblages overlie this condensed section, but are also unconformably

Table 1. *Summary of Cretaceous nannofossil events*

Species	Event	Zone	Hole 1183A Core	Depth (m)	Hole 1186A Core	Depth (m)	Hole 1187A Core	Depth (m)
Base Cenozoic			39–4, 86 cm	987.36	13 CC	804.99		
Top K			39–4, 95 cm	987.40	14 CC	812.88		
L. quadratus	LO	CC25C	40 CC	995.36	14 CC	812.88		
C. self-trailiae	LO	CC25B	40 CC	995.36	14 CC	812.88		
L. praequadratus	LO	CC25B	41–1, 55 cm	1001.95	16–1, 72 cm	832.72		
C. gallica	LO	CC25B	41 CC	1004.26	16–1, 72 cm	832.72		
C. indiensis	HO	CC25B	42–1, 54 cm	1011.64	16 CC	833.16		
S. biarcus	HO	CC25B	42–3, 30 cm	1014.40	18 CC	851.10		
M. praemura	LO	CC25B	42–3, 30 cm	1014.40	18 CC	851.10		
Z. bicrescenticus	HO	CC25B	43 CC	1022.08	19–1, 51 cm	861.11		
Nannoconus sp. 1	HO	CC25B	43 CC	1022.08	19–1, 51 cm	861.11		
C. pricei	LO	CC25B	43 CC	1022.08	19–1, 51 cm	861.11		
A. cymbiformis var. W	LO	CC25B	43 CC	1022.08	19–1, 51 cm	861.11		
S. biarcus (elliptical)	LO		43 CC	1022.08	19 CC	861.42		
T. orionatus	HO	CC23B	44 CC	1031.28				
Nannoconus sp. 1	LO		44 CC	1031.28	19 CC	861.42		
A. parcus constrictus	HO	CC23A	45–1, 56 cm	1040.56	20 CC	870.78		
Q. gartneri	HO	CC23A	45–1, 56 cm	1040.56	20 CC	870.78		
U. trifidus	HO	CC23A	45–1, 56 cm	1040.56	20 CC	870.78		
U. gothicus	HO	CC23A	45–1, 56 cm	1040.56	20 CC	870.78		
A. cymbiformis var. N/T	HO	CC22	46–2, 22 cm	1051.20	21 CC	888.03		
W. fossacincta	HO	CC22	46–2, 22 cm	1051.20	22–1, 25 cm	889.75		
A. parcus parcus	HO	CC22	46–2, 22 cm	1051.20	22–1, 25 cm	889.75		
U. trifidus	LO	CC22	48–1, 49 cm	1069.29	24–1, 13 cm	908.83		
A. parcus parcus	HCO	CC21	48–2, 49 cm	1070.79	24 CC	911.81		
W. fossacincta	HCO	CC21	48–2, 49 cm	1070.79	24 CC	911.81		
H. circumradiatus	HO	CC21	48 CC	1071.79	24 CC	911.81		
E. eximius	HO	CC21	49 CC	1078.66	25–1, 76 cm	919.06		
B. magnus	LO	CC21	49 CC	1078.66	25 CC	925.51		
U. gothicus	LO	CC21	49 CC	1078.66	26–1, 3 cm	928.03		
C. self-trailiae	LO	CC20	49 CC	1078.66	26–1, 80 cm	928.80		
A. terebrodentarius	HO	CC20	50–1, 20 cm	1088.40	26–2, 62 cm	930.15		
L. grillii	HO	CC20	50–1, 20 cm	1088.40	26–3, 1 cm	931.01		
C. indiensis	LO	CC20	50–1, 30 cm	1088.50	26–3, 1 cm	931.01		
C. longissiimus	LO	CC20	50–1, 43 cm	1088.63	26–3, 6 cm	931.06		
B. hayii	HO	CC19A	50–1, 88 cm	1089.07	absent			
C. crassus	HO	CC19A	50–1, 112 cm	1089,31	absent			
M. furcatus	HO	CC18B	50–1, 112 cm	1089,31	26–3, 7 cm	931.07		
A. parcus constrictus	LO	CC18B	50–2, 15 cm	1089.85	26–3, 7 cm	931.07		
B. hayii	LO	CC18B	50–2, 15 cm	1089.85	absent			
A. parcus parcus	LO	CC18A	50–2, 15 cm	1089.85	26–3, 7 cm	931.07		
C. crassus	LO	CC16	50–2, 35 cm	1090.05	absent			
Q. enebrachium	HO	CC14/15	50–2, 60 cm	1090.30	26–3, 10 cm	931.10		
M. furcatus	HAbO		50–2, 60 cm	1090.30	26–3, 10 cm	931.10		
E. floralis	HO	CC14	50–2, 68 cm	1090.38	26–3, 13 cm	931.13		
L. grillii	LO	CC14	50–2, 68 cm	1090.38				
M. cubiformis	LO	CC14	50–2, 74 cm	1090.44	26–3, 20 cm	931.20		
E. orbiculatus	HO	CC14	absent		26–3, 28 cm	931.28		
M. staurophora	LO	CC14	50–2, 74 cm	1090.44	26–3, 32 cm	931.32		
M. furcatus	LO	CC13	50–2, 74 cm	1090.44	26–3, 32 cm	931.32		
A. infracreactea	HO	NC10A	50–2, 78 cm	1090.48	26–3, 33 cm	931.33		
P. stradneri	LO	NC10A	50–2, 105 cm	1090.75	26 CC	931.36		
L. armilla	LO	NC10A	51–2, 23 cm	1098.63	26 CC	931.36		
G. gammation	LO	NC10A	51–3, 28 cm	1101.18	absent			
Metadoga spp.	LO	NC10A	51 CC	1101.59	absent			
L. dorothae	LO	NC10A	51 CC	1101.59	absent			
H. irregularis	HO	NC10A	52–1, 1 cm	1107.51	27–1, 33 cm	937.93		
E. turriseifellii	LO	NC10A	52–1, 1 cm	1107.51	26 CC	931.36		
A. albianus	LO	*NC9A*	52–1, 1 cm	1107.51	26 CC	931.36		
C. ehrenbergii	LO	NC9B	52–1, 1 cm	1107.51	27 CC	938.88		
E. monechiae	LO	NC9B	52–1, 24 cm	1107.74	absent			
T. praeorionatus	LO	NC8C	52–1, 24 cm	*1107.74*	27 CC	938.88		
Rhagodiscus sp. 1	HO	NC8A	52–1, 52 cm	1108.02	28–1, 52 cm	947.72		
C. erbae	HO	NC8A	52–1, 90 cm	1108.40	28–1, 52 cm	947.72		

Table 1. *continued*

Species	Event	Zone	Hole 1183A Core	Depth (m)	Hole 1186A Core	Depth (m)	Hole 1187A Core	Depth (m)
P. columnata	LO	NC8A	52 CC	1113.38	28CC	950,81		
C. nudus	LO	NC7C	53–2, 10 cm	1118.77	28 CC	950.81		
Rhagodiscus sp. 1	LO	NC7C	53–2, 10 cm	1118.77	28 CC	950.81	2–2, 29 cm	367.29
Q. enebrachium	LO	NC7C	53–3, 95 cm	1021.12	29–1, 74cm	957.54	2–2, 29 cm	367.29
L. salomonii	HO	NC7A	54–2, 34 cm	1128.64	absent			
R. hamptonii	LO	NC7A	54–2, 88 cm	1129.18	30–1, 43 cm	966.83		
E. floralis	LO	NC7A	54–3, 119 cm	1130.99	30–1, 43 cm	966.83	2–2, 29 cm	367.29
R. gallagheri	LO	NC6A	54–3, 119 cm	1130.99	30–1, 43 cm	966.83		
H. irregularis	LO	NC6A	54–3, 119 cm	1130.99	30–1, 43 cm	966.83	2–2, 29 cm	367.29
BASEMENT			54–3, 107 cm	1130.87	30–1, 45 cm	966.85	2–2, 30 cm	367.30

HO and LO, highest and lowest occurrence; HAbO, highest abundant occurrence.

overlain by the Palaeocene (Tables 1 and 2), as the terminal Maastrichtian (Zone CC-26) is not present in either Holes 1183A or 1186A. Both poor preservation and the resulting low diversity limit biostratigraphic resolution within the Leg 192 Upper Cretaceous. Low core recoveries, due to the presence of cherts, further limit dating and correlation of the richer middle Campanian–Maastrichtian between Holes 1183A and 1186A.

The condensed middle Coniacian–lower Campanian section is characterized by dissolution assemblages dominated by the nannolith species *Marthasterites furcatus* and *Assipetra terebrodentarius*. Although *Marthasterites furcatus* can be very common in well-preserved and diverse shelf and epicontinental Coniacian–Santonian marine deposits, the high percentages of *Assipetra terebrodentarius* in the Leg 192 material is unusual. The combination of low sedimentation rates, dissolution and provenance (i.e. low latitudes) in the Leg 192 material have all probably contributed to these high concentrations of *A. terebrodentarius*. The base of Zone CC-14 is defined on the first occurrence of *Micula staurophora*, which helps delineate the upper Albian–middle Coniacian unconformity that must lie between Samples 50–2, 74 cm and 50-2-78 cm in Hole 1183A, and between Samples 26–2, 32 cm and 26–2, 33 cm in Hole 1186A (Table 1, Figs 1 & 3).

The Santonian portion of the Sissingh (1997, 1978) CC zonation, which was based on Tunisian outcrop sections, could not be applied to the Leg 192 material. The three species whose appearances mark the base of Zones CC-15–CC-17 (see Fig. 3) in the Sissingh (1977, 1978) zonation all actually range down into in the lower Coniacian–Turonian in the Western Interior Basin of the United States (see Bralower & Bergen 1998). This is further validated by more recent examination of Tunisian outcrop sections (unpublished data), as well as by cores in the North Sea and outcrops in southern England (Bergen & Sikora 2000). Proxies for the bases of Zones CC-15–CC-17 are presented in Figure 3. As far as the remaining condensed lower Campanian Leg 192 section, the bases of Zones CC-18 and CC-19 (Sissingh 1997, 1978) are generally applicable, but somewhat compromised by both the low core recovery and condensation.

Nannofossil abundance from Holes 1183A and 1186A dramatically increase near the base of Zone CC-20 (Sissingh 1977, 1978). However, dating and correlation of the nannofossil-rich upper middle Campanian–lower upper Maastrichtian sediments are severely compromised by low core recoveries. In addition, the genus *Reinhardtites* is extremely rare in the Leg 192 material, making it impossible to precisely delineate the upper Campanian Zone CC-22B and the base of Zone CC-25A in the lower Maastrichtian. The extinctions of three species (Fig. 3) were utilized as proxies to approximate Zone CC-22B based on research from circum-Atlantic northern hemisphere sections (Bergen & Sikora 2000), including Tunisia (unpublished data). It was not possible to constrain the boundary between Zones CC-24 and CC25A in the Leg 192 material based on any published or unpublished data due to the scarcity of *Reinhardtites levis* in the Leg 192 material. Sissingh (1977, 1978) used the appearance of *Arkhangelskiella cymbiformis* to define the base of Subzone CC25B in the upper Maastrichtian. This event is problematic, as subsequent published zonations (e.g. Perch-Nielsen 1985*a*; Burnett, 1998) show this species ranging into the Campanian, but a broader species concept would extend its appearance into the late Santonian based on outcrops in southern England, Tunisia and the Gulf of Mexico (unpublished

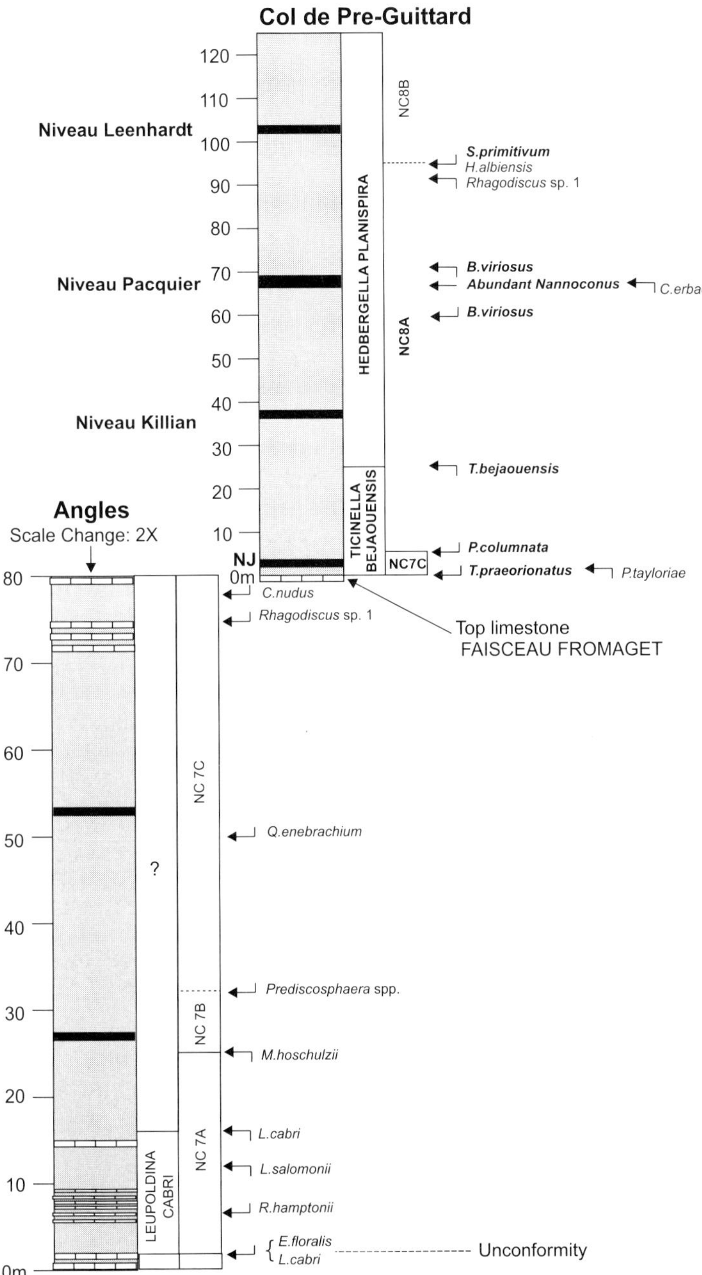

Fig. 2. Calcareous microfossil biostratigraphy from SE France. Col de Pré–Guittard from Kennedy *et al.* (2000); their events are in bold type, new events are in normal type. Levels of organic-rich shales also shown and Niveau Pacquier is used as base of the Albian. The Angles section was collected by Amoco; all events are nannofossils, except for the planktonic foraminifera *Leupoldina cabri*.

data). Varol (1989) provides potential resolution of this problematic zonal event. Varol (1989) defined three varieties of *A. cymbiformis* based on central area to rim width ratios. *Arkhangelskiella* cymbiformis var. W (Varol 1989) has a broad rim and relatively narrow central area and

Table 2. *Summary of Palaeocene nannofossil events*

Species	Event	Zone	Age*	Hole 1183A Core	Depth (m)	Hole 1186A Core	Depth (m)
T. bramlettei	**LO**	**NP10**	55.0	35–1, 3 cm	943.53		
C. eograndis	LO	NP9/10		35–1, 24 cm	943.74	7 CC	745.40
C. dela	LO	NP9/10		35–1, 41 cm	943.93	7 CC	745.40
F. tympaniformis	HO	NP9	55.3	35–2, 1 cm	945.01	8 CC	755.14
N. bukryi	HO	NP9		35–2, 1 cm	945.01	9–1, 10 cm	764.70
T. eminens	HO	NP9		35–2, 48 cm	945.48	9–1, 10 cm	764.70
C. frequens	HO	NP9		36–1, 0 cm	953.10	9–1, 10 cm	764.70
D. megastypus	HO	NP9		36–1, 0 cm	953.10	9–1, 10 cm	764.70
E. robusta	HO	NP9		36–1, 0 cm	953.10	9–1, 10 cm	764.70
H. universus	HO	NP9		36–1, 85 cm	953.95	9–1, 10 cm	764.70
P. martinii	HO	NP9		36–2, 0 cm	954.60	9–1, 102 cm	765.72
P. sigmoides	HO	NP9		*36–4, 0 cm*	*957.60*	9–1, 102 cm	765.72
C. eodela	LO	NP9	55.5	36–3, 71cm	956.71	9 CC	765.92
T. occultatus	LO	NP9		36–4, 0 cm	957.60	9 CC	765.92
H. kleinpelii	HO	NP9		37–1, 0 cm	962.70	10–1, 0 cm	774.20
B. elegans	HO	NP9		37–1, 73 cm	963.43	10 CC	775.53
H. cantabriae	HO	NP9		37–1, 73 cm	963.43	10 CC	775.53
C. primus	HO	NP9		37–2, 0 cm	964.20	9 CC	765.92
D. multiradiatus	**LO**	**NP9**	56.2	37–2, 72 cm	964.92	10 CC	775.53
H. riedelii	HO	NP8		37–3, 0 cm	965.70	11–1, 0 cm	783.90
C. edwardsii	HO	NP8		37–3, 0 cm	965.70	11–1, 0 cm	783.90
D. megastypus	LO	NP8		*37–2, 0 cm*	*964.20*	11–1, 0 cm	783.90
H. kleinpelii	LO	NP8	*58.4*	37–3, 19 cm	965.89	11–1, 0 cm	783.90
N. concinnus	HO	NP8		37–3, 74 cm	966.44	11–1, 90 cm	784.80
H. riedelii	**LO**	**NP8**	57.3	37–3, 74 cm	966.44	*11–1, 0 cm*	*783.90*
N. bukryi	LO	NP8		37–3, 74 cm	966.44	11–1, 90 cm	784.80
S. anarhopus	LO	NP8	58.4	37–3, 74 cm	966.44	11–1, 90 cm	784.80
H. universa	LO	NP8		37–3, 74 cm	966.44	11–1, 90 cm	784.80
T. oztunalii	LO	NP8		37–3, 74 cm	966.44	11–1, 90 cm	784.80
C. danicus	HO	NP5		37–4, 16 cm	967.36	11 CC	785.00
N. distentus	LO	NP5		37–4, 56 cm	967.76	*11–1, 90 cm*	*784.80*
H. cantabriae	LO	NP5		37–4, 56 cm	967.76	*11–1, 90 cm*	*784.80*
C. tenuis	HO	NP5		37–4, 96 cm	968.16	11 CC	785.00
F. billii	HO	NP5		37–4, 96 cm	968.16	absent	
F. ulii	HO	NP5		37–4, 96 cm	968.16	11 CC	785.00
N. modestus	HO	NP5		37–4, 96 cm	968.16	11 CC	785.00
F. pileatus	HO	NP5		38–1, 0 cm	972.40	12 CC	793.55
T. eminens	LO	NP5		*37 CC*	*968.71*	12 CC	793.55
C. bidens	LO	NP5	*60.7*	38–2, 0 cm	973.70	12 CC	793.55
F. pileatus	LO	NP5		38–2, 74 cm	974.44	12 CC	793.55
C. consuetus	LO	NP5	59.7	38–3, 59 cm	975.99	12 CC	793.55
N. concinnus	LO	NP5		38–3, 147 cm	976.87	12 CC	793.55
F. tympaniformis	**LO**	**NP5**	59.7	38–3, 147 cm	976.87	12 CC	793.55
C. frequens	LO	NP4		38–3, 147 cm	976.87	12 CC	793.55
F. bitectus	HO	NP4		38–4, 75 cm	977.65	absent	
F. chowii	HO	NP4		38–4, 75 cm	977.65	absent	
F. chowii	LO	NP4		38–4, 75 cm	977.65	absent	
F. bitectus	LO	NP4		38–4, 75 cm	977.65	absent	
F. ulii	LO	NP4	59.9	38–4, 75 cm	977.65	12 CC	793.55
S. primus	LO	NP4	60.6	38–4, 75 cm	977.65	12 CC	793.55
F. magnicordis	HO	NP4		38–5, 0 cm	978.40	absent	
F. magnicordis	LO	NP4		38–5, 0 cm	978.40	absent	
C. inseadus	HO	NP4		38–5, 59 cm	978.99	absent	
N. perfectus	LO	NP4		38–5, 59 cm	978.99	12 CC	793.55
P. dimorphosus	HO	NP4		38–5, 144 cm	979.84	13–1, 22 cm	803.32
T. pertusus	LO	NP4		38–6, 15 cm	980.05	*12 CC*	*793.55*
P. tenuiculus	HO	NP4		38 CC	980.42	13–1, 22 cm	803.32
E. macellus	**LO**	**NP4**	62.2	38 CC	980.42		

Table 2. *continued*

Species	Event	Zone	Age*	Hole 1183A Core	Depth (m)	Hole 1186A Core	Depth (m)
P. bisulcus	LO	NP4		39–1, 4 cm	982.04	*12 CC*	*793.55*
E. robusta	LO	NP4		39–1, 4 cm	982.04	*12 CC*	*793.55*
C. tenuis	LO	*NP2*	*64.5*	39–1, 4 cm	982.04	*12 CC*	*793.55*
P. martinii	LO	NP3		39–1, 49 cm	982.49	*12 CC*	*793.55*
C. alta	HO	NP3		39–2, 25 cm	983.75	absent	
E. alternans	LO	NP3		39–2, 25 cm	983.75	13–1, 22 cm	803.32
E. ovalis	LO			39–2, 75 cm	984.25	13–1, 22 cm	803.32
C. danicus	**LO**	**NP3**	63.8	39–2, 124 cm	984.74	13–1, 70 cm	803.32
C. pelagicus	LO	NP2		39–2, 124 cm	984.74	13–1, 70 cm	803.32
E. supertusa	LO			39–3, 24 cm	985.24	13–1, 70 cm	803.32
E. cava	LO	NP2		39–3, 83 cm	985.83	13 CC	804.99
C. intermedius	**LO**	**NP2**		39–3, 83 cm	985.83	13 CC	804.99
C. primus	LO	NP1	64.8	39–2, 125 cm	986.25	13 CC	804.99
Thoracopshaera spp.	LO	NP1		39–4, 14 cm	986.66	13 CC	804.99
base Cenozoic (forams)	LO			39–4, 86 cm	987.36	13 CC	804.99

*Age from Berggren *et al.* (1995). Values in italics indicate discrepancies. Bold type indicates zonal markers.

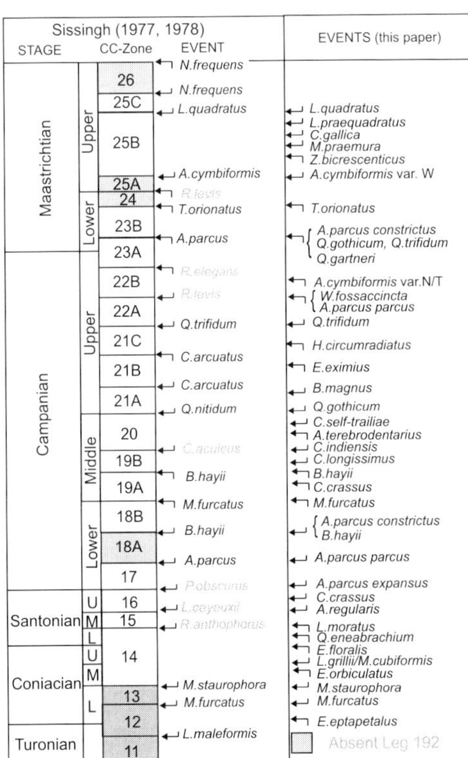

Fig. 3. Upper Cretaceous Zonation.

is the youngest morphotype in this progression. Its rim also has a first-order yellow–orange birefringence compared to his other two morphotypes with narrower rims that exhibit a first-order white birefringence. In the Leg 192 material the lowest occurrence *Arkhangelskiella cymbiformis* var. W can be used to constrain the base of Zone CC-25B, whereas as the other two morphotypes range down into much older section. As defined, other nannofossil events from Holes 1183A and 1186A define an unconformity between Zones CC-23B and Zone CC-25B (Table 1). The appearance of *Lithraphidites quadratus*, which marks the base of Zone CC-25C, occurs near the terminal Cretaceous in Holes 1183A and 1186A. Zone CC-26 was not recognized in the Leg 192 material, indicating another unconformity with the basal Cenozoic.

Cenozoic

Palaeocene–Eocene

The 'NP' zonation of Martini (1971) was applied to the Leg 192 material with additional events (Tables 2 and 3) mostly derived from Perch-Nielsen (1985*b*). The lower–middle Eocene Zones NP12–NP15 have since been subdivided to subzones (Valentine 1987; Aubry 1991), mainly to incorporate elements from the enumerated Cenozoic 'CP' Zonation of Okada & Bukry (1980). For this paper, only the subdivisions of Zones NP14 and NP15 from Aubry (1991) were applicable (Fig. 4). Additional Palaeocene nannofossil events are based on the Leg 192 material, as well as unpublished Palaeocene data (de Kaenel) from Tunisia. Berggren *et al.* (1995) provided estimations of

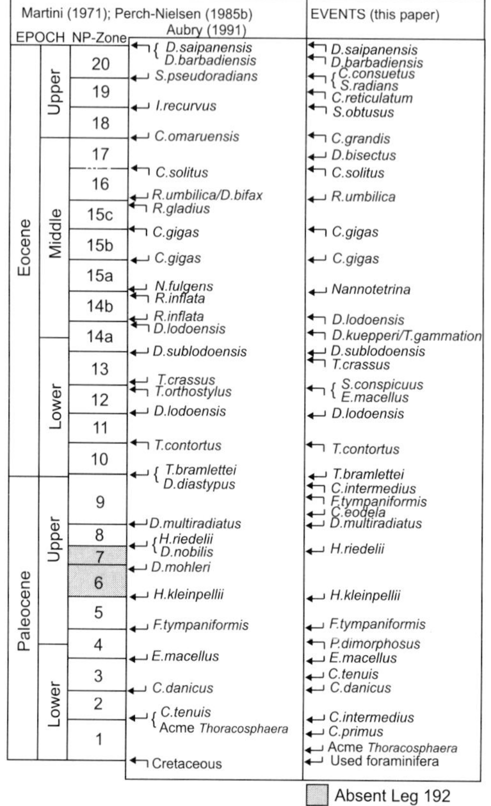

Fig. 4. Palaeocene–Eocene Zonation.

Palaeocene–Eocene nannofossil events, and these ages are also included in Tables 2 and 3.

Holes 1183A and 1186A (Fig. 4, Tables 2 and 3) are the only holes that recovered Zone NP15 and older Cenozoic strata. Both these holes are characterized by both low core recoveries and poor nannofossil preservation. However, Hole 1183A provides the basic framework because of its higher core recovery. Zones NP16 and NP17 were also recovered from section immediately above basaltic basement in both Holes 1185A and 1185B (Table 3), in addition to NP16 nannofossil assemblages recovered from the majority of the volcaniclastic sequence in Hole 1184A (Table 3; see also the Basement ages section).

The Cretaceous–Tertiary boundary was drilled in both Holes 1183A and 1186A. In Hole 1183A recovered foraminifera and nannofossils constrain this boundary between Samples 1183A-39-4, 86 cm and 1183A-39-4, 95 cm. However, the basal Palaeocene section in Hole 1183A is barren of calcareous nannofossils and the lowest occurrence of Palaeocene nannofossils (Zone NP1) is from Sample 1183A-39-4, 16 cm. In Hole 1186A the Cretaceous–Tertiary boundary is typified by very low core recoveries, and lies between Samples 1186A-13CC (Zone NP2) and 1186A-14CC. Recovered nannofossils from both Holes 1183A and 1186A (Table 2) also indicate an unconformity between Zones NP5 and NP8 (Fig. 4, Table 2). Above that, the remainder of the upper Palaeocene–Eocene section appears relatively complete among the Leg 192 sites (Tables 2 and 3).

Oligocene–Miocene

Oligocene section was drilled in Hole 1183A. Miocene section was penetrated in both Holes 1183A and 1184A. Dating of the Oligocene–Miocene section in both Holes 1183A and 1184A is based only on shipboard examination of samples. Cursory results from examination of core-catcher samples from Hole 1183A indicate an unconformity somewhere between the basal Miocene in Sample 1183A-15-CC (Zone NN1) and the upper Oligocene in Sample 1183A-16-CC (Zone NP24). In Hole 1183A, core-catcher samples also indicate a potential missing lower Miocene section (Zones NN2 and NN3) between Sample 1183A-15-CC and Sample 1184A-14-CC, but a relatively complete section up into Zone NN7 (Core 2) in the middle Miocene. In contrast, lower Miocene nannofossils dated as Zone NN2 were recovered from Samples 2-1, 20 cm–9-1, 5 cm in Hole 1184A.

Basement ages

Four of the five Leg 192 sites (1183, 1185, 1186 and 1187) penetrated basaltic basement. Hole 1184A drilled a thick volcaniclastic sequence in Cores 9–46. Nannofossils recovered among these five Leg 192 sites indicate two periods of basement emplacement, which are discussed in the following two sections. The number of samples taken and the time for sample analyses were critical to achieving the results from the Leg 192 material both in sediments immediately above basaltic basement and especially from intercalated sediments within basaltic basement. Sample analyses on such samples varied between 2 and 12 hours.

Aptian ages

Poorly preserved Aptian nannofossil assemblages were recovered from the basal sedimentary section in Holes 1183A, 1186A and 1187A (Fig. 5, Table 1). Extremely rare and very

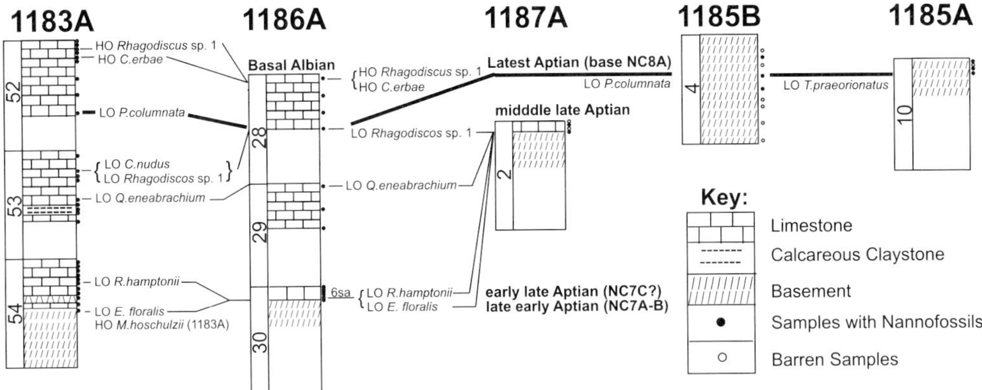

Fig. 5. Aptian nannofossil correlations among Leg 192 basement holes. Note: lowest occurrences of *P. columnata* and *T. praeorionatus* are roughly coeval.

poorly preserved mid-Cretaceous nannofossil assemblages were also recovered from intercalated sediments within basaltic basement in Holes 1185A and 1185B (Table 1). Recovered nannofossil assemblages among these four Leg 192 sites indicate a general age progression of basaltic basement emplacement (Fig. 5) from the late early Aptian (Hole 1183A) to the latest Aptian (Holes 1185A and 1185B), as detailed below.

In Hole 1183A, single specimens of *Eprolithus floralis* and *Micrantholithus hoschulzii* were recovered from intercalated sediments (Sample 54-3, 119 cm) immediately below the basement–sedimentary contact at 54-3, 107 cm (Fig. 5, Table 1). This defines Zone NC7A (Fig. 1), which straddles the lower–upper Aptian boundary (Bralower *et al.* 1993). The lowest occurrence of *Rhagodiscus hamptonii* occurs in Sample 54–2, 88 cm in Hole 1183A (Table 1). In Hole 1186A both *Eprolithus floralis* and *Rhagodiscus hamptonii* occur at the base of the sedimentary section in Sample 30–1, 43 cm, without *Micrantholithus hoschulzii* (Table 1). This indicates a younger early late Aptian age possibly within Zone NC7C for the base of the sedimentary section in Hole 1186A (Fig. 1 and 5). In Hole 1187A both *Rhagodiscus* sp. 1 and *Quadrum eneabrachium* are present in the basal sedimentary section in Sample 2-2, 29 cm (Table 1). The lowest occurrences of these two species (Fig. 5) are well above basaltic basement in both Hole 1183A (Samples 53-2, 10 cm and 53-3, 95 cm, respectively) and Hole 1186A (28CC and 29-1, 74 cm, respectively), which indicate that the basal sedimentary section in Hole 1187A is younger than that recovered in Hole 1186A.

In Holes 1185A and 1185B recovered nannofossils clearly indicate that the basal sedimentary section is middle Eocene (Zone NP 16; see below). However, extremely rare and poorly preserved nannofossil assemblages were recovered from intercalated sediments within basaltic basement in both Hole 1185A (Cores 9 and 10) and Hole 1185B (Core 4). *Prediscosphaera columnata* was recovered from two samples (4-2, 42 cm and 4-3, 58 cm) within basaltic basement in Hole 1185B, and *Tranolithus praeorionatus* from two samples (9-4, 97 cm and 10-1, 49 cm) in Hole 1185A. The appearance of *Prediscosphaera columnata* defines the base of Zone NC8A in the uppermost Aptian (Bralower *et al.* 1993; Kennedy *et al.* 2000). *Tranolithus praeorionatus* also ranges into the uppermost Aptian (Kennedy *et al.* 2000) and its appearance is close to that of *Prediscosphaera columnata* (Kennedy *et al.* 2000; unpublished data from SE France). The highest occurrence of *Microstaurus chiastius* can only constrain the upper age limit ages of nannofossil assemblages recovered in Cores 1185B-9, 1185B-10 and 1185A-4 as no younger than latest Cenomanian. However, exhaustive analyses of each of these samples (10–12 h) failed to produce any first occurrence events younger than latest Aptian. There is no question that the nannofossil assemblages recovered from intercalated sediments within basaltic basement in Cores 1185B-9, 1185B-10 and 1185A-4 are younger than any basal sediments drilled in Hole 1187A.

Eocene ages

In both Holes 1185A and 1185B the basal sedimentary section is middle Eocene and dated

Table 3. *Summary of Eocene nannofossil events*

Species	Event	Zone	Age*	Hole 1183A Core	Depth (m)	Hole 1186A Core	Depth (m)	Hole 1185A Core	Depth (m)	Hole 1185B Core	Depth (m)	Hole 1184A Core	Depth (m)
D. saipanensis	**HO**	**NP20**	34.2	24–2, 25 cm	839.95	2 CC	698.44	2–1-, 1 cm	250.61	2–1, 1 cm	308.01		
D. barbadiensis	**HO**	**NP20**	34.3	24–2, 114 cm	840.84	2 CC	698.44	2–1-, 1 cm	250.61	2–1, 1 cm	308.01		
C. consutetus	HO	NP19		24–2, 114 cm	840.84	2 CC	698.44	2–1-, 1 cm	250.61	2–1, 1 cm	308.01	9–1, 13 cm	201.23
S. radians	HO	NP19		24–2, 114 cm	840.84	2 CC	698.44	2–1-, 1 cm	250.61				
C. reticulatum	HO	NP19	35.0	25–1, 56 cm	847.96	2 CC	698.44	3–1, 30 cm	260.50				
G. bramlettei	HCO	NP19		25–1, 56 cm	847.96	2 CC	698.44	3CC	262.26				
S. obtusus	**HO**	**NP18**		25–1, 74 cm	848.14	2 CC	698.44	4–1, 25 cm	270.05				
S. intercalaris	HO	NP18		25–2, 1 cm	848.91	2 CC	698.44	4–1, 25 cm	270.05				
S. aff. spiniger	HO	NP18		26 CC	857.37			4 CC	271.76	2–1, 1 cm	308.01		
H. lophota	HO	NP18		26 CC	857.37	2 CC	698.44	5–1, 26 cm	279.76				
C. oamaruensis	LO	NP18	37.0	26 CC	857.37	absent		6–1, 53 cm	289.63	absent			
C. grandis	**HO**	**NP17**	37.1	27 CC	866.97	2 CC	698.44	6–1, 108 cm	290.18	2–1, 1 cm	308.01		
S. intercalaris	LO	NP17		27 CC	*857.37*	absent		6–2, 6 cm	289.66	absent		absent	
H. compacta	LO	NP17		27 CC	866.97	absent		6–2, 6 cm	289.66	absent		absent	
C. dela	HO	*NP16*		*28 CC*	876.39	2 CC	698.44	7–1, 91 cm	299.71	2–1, 1 cm	308.01		
D. bisectus	LO	NP17	38.0	27 CC	866.97	2 CC	698.44	7 CC	301.84	absent		absent	
D. tani ornatus	LO	NP17		27 CC	866.97	2 CC	698.44	7 CC	301.84	2–1, 3 cm	308.03	absent	
C. solitus	**HO**	**NP16**	40.4	28 CC	876.39	3–1, 0 cm	706.90	8–1, 1 cm	308.41	2–1, 40 cm	308.4		
S. furcatolithoides	HO	NP16		29 CC	886.03	3–1, 0 cm	706.90	8–1, 1 cm	308.41	2–1, 40 cm	308.4	9–1, 13 cm	201.23
S. spiniger	HO	*NP15*		29 CC	886.03	3–1, 0 cm	706.90	8–1, 1 cm	308.41	2–1, 40 cm	308.4	9–1, 28 cm	201.38
B. serraculoides	LO	NP16		29 CC	886.03	2 CC	698.44	7 CC	301.84	2–1, 40 cm	308.4	absent	
C. reticulatum	LO	NP16	42.0	29 CC	886.03	2 CC	698.44	8–1, 8 cm	308.48	2–1, 133 cm	309.33	absent	
S. obtusus	LO	NP16		29 CC	886.03	2 CC	698.44	8–1, 8 cm	308.48	2–1, 133 cm	309.33	absent	
R. hillae	LO	NP16		29 CC	886.03	2 CC	698.44	8–1, 8 cm	308.48	2–1, 145 cm	309.45	absent	
D. tani nodifer	LO	*NP17*		29 CC	886.03	*2 CC*	698.44	8–1, 10 cm	308.50	2–1, 149 cm	309.49	in 11–1, 82 cm	207.82
D. bifax	TR	NP16		absent		absent		absent		absent		in 13–1, 89 cm	221.29
R. umbilica	**LO**	**NP16**	43.7	29 CC	886.03	2 CC	698.44	8–1, 10 cm	308.50	2–1, 149 cm	309.49	in 45–5, 12 cm	525.62
N. alata	HO	NP16	43.1	30–1, 17 cm	895.67	3–1, 0 cm	706.90	absent		absent		absent	
D. sublodoensis	HO	NP15C		30–2, 1 cm	897.01	3–1, 0 cm	706.90						
C. gigas	**HO**	**NP15B**	44.5	30–2, 1 cm	897.01	3–1, 0 cm	706.90						
T. carinatus	HComO	NP15B		31–1, 23 cm	905.33	3–1, 0 cm	706.90						
B. staurion	HO	NP15B		31 CC	905.90	3–1, 0 cm	706.90						
S.orphanknollensis	HO	NP15B		31 CC	905.90	3–1, 28 cm	707.18						
E. alternans	HO	NP15B		31 CC	905.90	4–1, 6 cm	716.56						
E. cava	HO	NP15B		32–1, 19 cm	914.89	4–1, 6 cm	716.56						
C. gigas	**LO**	**NP15B**	46.1	32–1, 70 cm	915.40	3 CC	707.62						
R. coenurum	LO	NP15A		32 CC	916.42	3 CC	707.62						
S. furcatolithoides	LO	NP15A		32 CC	916.42	3 CC	707.62						
Nannotetrina	**LO**	**NP15A**	47.3	32 CC	916.42	3 CC	707.62						
D. lodoensis	HO	NP14A		33–1, 10 cm	924.40	4–1, 6 cm	716.56						

S. peudoradians	LO	*NP15*		32 CC	916.42	4–1, 6 cm	716.56
T. carinatus	LComO			32 CC	916.42	4–1, 6 cm	716.56
S. spiniger	LO	NP14A		32 CC	916.42	4–1, 6 cm	716.56
T. gammation	HO	NP14A		33–1, 10 cm	924.40	4 CC	717.70
D. kuepperi	HO	NP14A		33–1, 10 cm	924.40	4 CC	717.70
E. cf. E. robusta	HO	NP14A		33–1, 10 cm	924.40	5–1, 17 cm	726.27
D. wemmelensis	LO	NP14A		33–1, 10 cm	924.40		
E. ovalis	HO	NP14A		33–1, 71 cm	925.01	5–1, 17 cm	726.27
D. sublodoensis	**LO**	**NP14A**	49.7	33–1, 71 cm	925.01	4 CC	717.70
T. crassus	HO	NP13		33–2, 77 cm	926.57	5–1, 17 cm	726.27
T. pertusus	HO	NP13		33–2, 77 cm	926.57	5–1, 17 cm	726.27
S. conspicuus	HO	NP12		33–2, 77 cm	926.57	5–1, 17 cm	726.27
E. macellus	HO	NP12		33–2, 77 cm	926.57		
C. eodela	HO	*NP10*		33–2, 77 cm	926.57	*5 CC*	*726.97*
D. lodoensis	**LO**	**NP12**	52.9	33–3, 10 cm	927.40	4 CC	717.70
S. anarhopus	HO	NP11		33–3, 130 cm	928.60	5–1, 17 cm	726.27
S. editus	HO	NP11		33–3, 130 cm	928.60	5–1, 17 cm	726.27
D. diastypus	HO	NP11		33–3, 130 cm	928.60		
S. conspicuus	LO	NP11		33–3, 130 cm	928.60	5–1, 17 cm	726.27
S. radians	LO	NP11		33 CC	928.67	5 CC	726.97
D. multiradiatus	HO	NP11		34–1, 24 cm	934.14	6 CC	735.98
T. occulatatus	HO	NP11		34–1, 24 cm	934.14	6 CC	735.98
N. distentus	HO	NP11		34–1, 24 cm	934.14	6 CC	735.98
C. bidens	HO	NP11		34–1, 24 cm	934.14	6 CC	735.98
S. editus	LO	NP11		34–1, 24 cm	934.14	5 CC	726.97
D. barbadiensis	LO	NP11		34–1, 73 cm	934.63	5 CC	726.97
T. contortus	**HO**	**NP10**	53.6	34–2, 0 cm	944.40	6 CC	735.98
T. bramlettei	HO	NP10	53.9	34–2, 89 cm	945.29		
S. primus	HO	NP10		34–2, 89 cm	945.29	6 CC	735.98
N. junctus	HO	NP10		34–2, 89 cm	945.29	6 CC	735.98
S. moriformis	LO	NP10		34–3, 17 cm	937.09	5 CC	726.97
D. mohleri	HO	NP10		34–3, 56 cm	937.46	6 CC	735.98
P. bisulcus	HO	NP10		34–3, 56 cm	937.46	6 CC	735.98
C. intermedius	HO	NP10		34–3, 56 cm	937.46	6 CC	735.98
T. orthostylus	LO	NP10	*53.6*	34–3, 56 cm	937.46		
E. supertusa	HO	NP10		34–3, 58 cm	937.48	6 CC	735.98
T. contortus	LO	NP10	54.4	34–3, 58 cm	937.48	7 CC	745.40
T. bramlettei	**LO**	**NP10**	55.0	35–1, 3 cm	943.53		
C. eograndis	LO	NP9/10		35–1, 24 cm	943.74	7 CC	745.40
C. dela	LO	NP9/10		35–1, 41 cm	943.93	7 CC	745.40
F. tympaniformis	HO	NP9	55.3	35–2, 1 cm	945.01	8 CC	755.14

* Age: from Berggren *et al.* (1995). Values in italics indicate discrepancies. Bold type indicates zonal marker events.

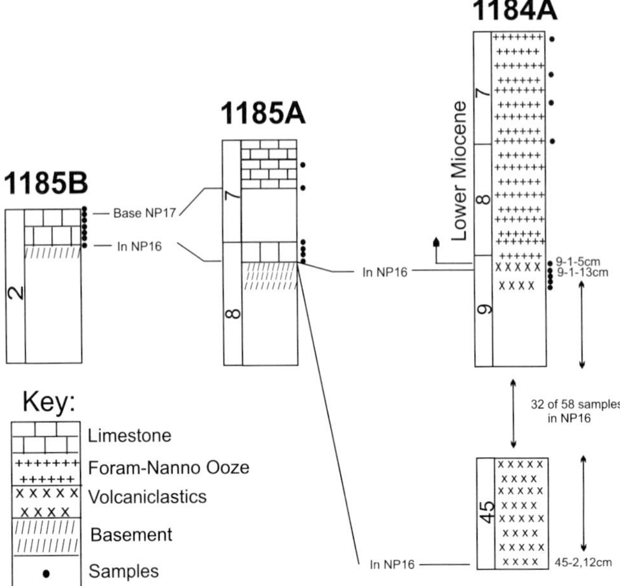

Fig. 6. Middle Eocene nannofossil correlations of Leg 192 basement holes.

within Zone NP16 (Fig. 6, Table 3). The base of Zone NP17 is constrained in Hole 1185B between 109 and 146 cm above basaltic basement (Table 3) based on the highest occurrence of *Chiasmolithus solitus*. In Hole 1185A the boundary between NP16 and NP17 occurs within a gap in recovery between Cores 7 and 8 (Table 3). Dating of basal sediments at Site 1185 are based on the following ages from Berggren *et al.* (1995), which include:

- first occurrence of *Reticulofenestra umbilica* at 43.7 Ma (base NP16 proxy);
- last occurrence of *Nannotetrina fulgens* at 43.1 Ma (NP16 event);
- first occurrence of *Cribrocentrum reticulatum* at 42.0 Ma (NP16 event);
- last occurrence of *Chiasmolithus solitus* at 40.4 Ma (top NP16);
- first occurrence of *Dictyococcites bisectus* at 38.0 Ma (NP17 event).

Results from Site 1185 indicate that these basal sediments are condensed. *Reticulofenestra umbilica* occurs at the base of the sedimentary section in both Holes 1185A and 1185B, but *Nannotetrina fulgens* was not observed within the basal sedimentary section in either Hole 1185A or 1185B. This gives a maximum age constraint on basal sediments at Site 1185 of 43.1 Ma. The lowest occurrence of *Cribrocentrum reticulatum* occurs between 6 (Section 1185A-8-1) and 16 cm (Section 1185B-2–1) above basaltic basement at Site 1185, which provides a minimum age of 42.0 Ma. The top of Zone NP16 is constrained between 109 and 143 cm above basaltic basement in Hole 1185B and at 40.4 Ma based on the highest occurrence of *Chiasmolithus solitus*. *Dictyococcites bisectus* (first occurrence (FO) at 38.0 Ma) was recovered from Sample 1185A-7CC, but there is a 6.56 m core gap with Core 1185A-8. *Dictyococcites bisectus* was not recovered from Hole 1185B.

In Hole 1184A calcareous oozes were recovered in Cores 2–8. Approximately 330 m of volcaniclastics were drilled from Cores 9 through to 46. Nannofossils recovered from Cores 1184A-2 to 1184-7 are all lower Miocene (Fig. 6) and dated within Zone NN2. Core 1184A-8 was not sampled. However, a lower Miocene nannofossil assemblage (also Zone NN2) was recovered from a stringer of sediment (Sample 1184A-9–1, 5 cm) just within the volcaniclastics.

Below this, 59 samples were taken from pods and stringers of sediments within this volcaniclastic sequence from Cores 1184A-9 to 1184A-46. Forty-five of these samples yielded extremely rare nannofossils. The vast majority of nannofossils recovered from these 45 samples indicate a middle Eocene age (Zone

NP16), as constrained by the lowest occurrence of *Reticulofenestra umbilica* in Sample 1184A-45–5, 12 cm and the highest occurrence of *Sphenolithus furcatolithoides* in Sample 1184A-9–1, 13 cm (Fig. 6, Table 3). A total of two Palaeocene specimens and seven Upper Cretaceous specimens were also recovered from these 45 samples.

In summary, the basal sedimentary section in Hole 1184A is lower Miocene, possibly the youngest of the Leg 192 sites relative to any 'volcaniclastics' or emplacement of basaltic basement. Nannofossils recovered from numerous stringers and pods of sediment within the volcaniclastic sequence in Hole 1184 indicate a middle Eocene age (Zone NP16); these are possibly coeval with basal sediments at Site 1185 and, as such, would indicate the oldest possible age for the Site 1184A volcaniclastics.

Systematic palaeontology

The systematic palaeontology includes six Cretaceous species that are re-assigned to new genera or are not formally described. New species are not described from the Leg 192 material because of its general poor preservation.

All the binomial designations and associated author citations for all remaining Cenozoic and Mesozoic species are listed following the Systematic Palaeontology. The reader is referred to Perch-Nielsen (1985*a*, *b*), de Kaenel & Villa (1996), Bergen (1998) and Bown (1998) for references relative to authors cited on all species.

Bukryaster magnus *(Bukry 1975) new combination*

1975 *Rucinolithus* ? *magnus* Bukry, p. 690, plate 3, figs 12–14.

Discussion. The elements of this six-rayed, late Campanian species have the characteristic shape and birefringence of *Bukryaster*. *Rucinolithus* is not appropriate because this is an Early Cretaceous genus and its morphology is much different (size and element shape).

Occurrence. Bukryaster magnus ranges from the Zone CC-21 to CC-23A in low–middle latitude (Gulf of Mexico, Tunisia and southern France) upper Campanian sections. Authors who have reported a late Campanian extinction for *Lithastrinus grillii* (*e.g.* Perch-Nielsen 1985*a*) have probably confused that species with *Bukrylithus magnus*.

Gartnerago gammation *(Hill 1976), new combination*

1976 *Broinsonia gammation* Hill, p. 126, plate 2, figs 32–43.

Discussion. The specimens illustrated by Hill (1976) show that this species belongs to the genus *Gartnerago* (i.e. a placolith with a central plate). It may be the ancestral species of *Gartnerago*. In this context, it is critical to realize that other species with identical rim birefringence patterns, but with central cross-structures (e.g. *Gartnerago theta* and *Gartnerago nanum*), may actually be muroliths and unrelated.

Occurrence. Gartnerago gammation was observed in two samples from the upper Albian (Zone NC10A) in Hole 1183A. The author has observed this species in the upper Albian–lower Cenomanian from outcrops in southern England, SE France and Tunisia (unpublished data). In the potential base Cenomanian boundary section at Mont Risou near Rosans (SE France), the lowest occurrence of this species is approximately 86.2 m below the base of the Cenomanian within the upper Albian in Zone NC10A.

Gorkaea pomerolii *(Perch-Nielsen 1973), new combination*

1973 *Zygodiscus pomerolii* Perch-Nielsen, p. 327, plate 6, figs 3 and 5; plate 10, figs 41 and 42.

Discussion. Varol & Girgis (1994) erected the genus *Gorkaea* for muroliths having a transverse central bar and inner wall (proximal rim in distal view) constructed of numerous cycles of tangentially arranged elements. Both the inner wall and transverse bar of these large muroliths also display a high first-order yellow–orange birefringence. Varol & Girgis (1994) correctly stated that many authors had incorrectly placed the type species of *Gorkaea*, *G. pseudanthophorus*, within *Zeugrhabdotus embergeri*. Perch-Nielsen (1973) described *Zygodiscus pomerolii*, but she placed it within the Palaeogene genus *Zygodiscus*. This Upper Cretaceous species possesses the same rim ultra-structure and birefringence characteristics of *Gorkaea pseudanthophorus* and is transferred into that genus herein.

Nannoconus *sp. 1*

Discussion. This medium-sized *Nannoconus* is round and has a large central cavity. It most closely resembles *Nannoconus globulus*, but that species is slightly larger and had a latest Aptian extinction.

Occurrence. Nannoconus sp. 1 was observed in the lower–mid Maastrichtian (Zones CC23B–CC25B) in both Holes 1183A and 1186A (see Table 1).

Rhagodiscus *sp. 1*

Description. Rhagodiscus sp. 1 is a large, normally elliptical murolith. It has a very broad rim (central area about one-half the coccolith width) that displays a first-order orange birefringence. It possesses a granular central plate.

Discussion. Three other large muroliths occur in the Aptian Leg 192 material. *Calcicalathina erbae* and *Rhagodiscus hamptonii* both have narrow rims that are less birefringent than *Rhagodiscus* sp. 1. *Calcicalathina erbae* is further distinguished by its higher central area (out of the plane of focus) and *Rhagodiscus hamptonii* by its bicyclic rim extinction. Large specimens of *Rhagodiscus asper* are also present in the Aptian; these have narrower rims than *Rhagodiscus* sp. 1 and display a first-order grey–faint yellow birefringence.

Occurrence. The lowest occurrence of *Rhagodiscus* sp. 1 is critical to establishing the relative age of basal sediments between Holes 1183A, 1186A and 1187A (Fig. 5). This short-ranging species has also been observed in the upper Aptian–lowermost Albian (upper NC7C–lower NC8A) in sections in SE France (unpublished data).

Rhagodiscus thiersteinii *(Roth 1973), new combination*

1973 *Crepidolithus thiersteinii* Roth, p. 725, plate 21, fig. 1; plate 22, figs 5a–d and 6a–d.

Discussion. Roth (1973) described *Crepidolithus thiersteinii* from the lower to middle Campanian of DSDP Leg 17 and reported a Turonian–Maastrichtian range for this species. *Crepidolithus* is a Jurassic genus. Roth (1973) illustrated this species that clearly belongs within the Cretaceous genus *Rhagodiscus*. *Rhagodiscus thiersteinii* most closely resembles *Rhagodiscus asper*, but is distinguished from that species by its more coarsely granular central area and more prominent inner rim cycle in cross-polarized light.

Occurrence. Rhagodiscus thiersteinii was observed in only two mid-Campanian sample from Hole 1183A. *Rhagodiscus asper* had an early Turonian extinction, whereas *Rhagodiscus thiersteinii* appeared in the late Cenomanian and ranged at least into the Campanian (unpublished data).

Cenozoic species

Bomolithus elegans Roth 1973
Bramletteius serraculoides Gartner 1969
Calcidiscus fuscus (Backman 1980) Janin 1987
Calcidiscus leptoporus (Murray & Blackman 1898) Loeblich & Tappen 1978
Calcidiscus premacintyrei Theodoridis 1984
Calcidiscus radiatus (Kamptner 1955) Martin Perez & Aguado 1990
Calcidiscus tropicus (Kamptner 1955) Varol 1989
Campylosphaera dela (Bramlette & Sullivan 1961) Hay & Mohler 1967
Campylosphaera eodela Bukry & Percival 1971
Camuralithus pelliculathus de Kaenel & Villa 1996
Chiasmolithus bidens (Bramlette & Sullivan 1961) Hay & Mohler 1967
Chiasmolithus californicus (Sullivan 1964) Hay & Mohler 1967
Chiasmolithus consuetus (Bramlette & Sullivan 1961) Hay & Mohler 1967
Chiasmolithus danicus (Brotzen 1959) Hay & Mohler 1967
Chiasmolithus edentulus van Heck & Prins 1987
Chiasmolithus eograndis Perch-Nielsen 1971
Chiasmolithus expansus (Bramlette & Sullivan 1961) Gartner 1970
Chiasmolithus gigas (Bramlette & Sullivan 1961) Radomski 1968
Chiasmolithus grandis (Bramlette & Riedel 1965) Radomski 1968
Chiasmolithus oamaruensis (Deflandre 1954) Hay *et al.* 1966
Chiasmolithus solitus (Bramlette & Sullivan 1961) Locker 1968
Clausicoccus cribellus (Bramlette & Sullivan 1961) Prins 1979
Clausicoccus eroskayi Varol 1989
Clausicoccus fenestratus (Deflandre & Fert 1954) Prins 1979
Clausicoccus obrutus (Perch-Nielsen 1971) Prins 1979
Clausicoccus subdistichus (Roth & Hay, in Hay *et al.* 1967) Prins 1979
Clausicoccus vanheckiae (Perch-Nielsen 1986) de Kaenel & Villa 1996
Coccolithus miopelagicus Bukry 1971*a*
Coccolithus pelagicus (Wallich 1877) Schiller 1930

Coronocyclus nitescens (Kamptner 1963) Bramlette & Wilcoxon 1967
Coronocyclus prionion (Deflandre & Fert 1954) Stradner, in Stradner & Edwards 1968
Cribrocentrum reticulatum (Gartner & Smith 1967) Perch-Nielsen 1971*b*
Cruciplacolithus asymmetricus van Heck & Prins 1987
Cruciplacolithus edwardsii Romein 1979
Cruciplacolithus frequens (Perch-Nielsen 1977) Romein 1979
Cruciplacolithus staurion (Bramlette & Sullivan 1961) Gartner 1971
Cruciplacolithus inseadus Perch-Nielsen 1969
Cruciplacolithus intermedius van Heck & Prins 1987
Cruciplacolithus primus Perch-Nielsen 1977
Cruciplacolithus subrotundus Perch-Nielsen 1969
Cruciplacolithus tenuis (Stradner 1961) Hay & Mohler, in Hay *et al.* 1967
Cyclagelosphaera alta Perch-Nielsen 1979
Cyclicargolithus abisectus (Müller 1970) Wise 1973
Cyclicargolithus floridanus (Roth & Hay, in Hay *et al.* 1967) Bukry 1971
Dictyococcites bisectus (Hay *et al.* 1966) Bukry & Percival 1971
Dictyococcites scrippsae Bukry & Percival 1971
Discoaster adamanteus Bramlette & Wilcoxon 1967
Discoaster barbadiensis Tan 1927
Discoaster bifax Bukry 1971
Discoaster calculosus Bukry 1971
Discoaster deflandrei Bramlette & Riedel 1954
Discoaster diastypus Bramlette & Sullivan 1961
Discoaster druggii Bramlette & Wilcoxon 1967
Discoaster gemmeus Stradner 1959
Discoaster kuepperi Stradner 1959
Discoaster kugleri Martini & Bramlette 1963
Discoaster lenticularis Bramlette & Sullivan 1961
Discoaster lodoensis Bramlette & Riedel 1954
Discoaster megastypus (Bramlette & Sullivan 1961) Perch-Nielsen 1984
Discoaster mohleri Bukry & Percival 1971
Discoaster multiradiatus Bramlette & Riedel 1954
Discoaster nobilis Martini 1961
Discoaster saipanensis Bramlette & Riedel 1954
Discoaster salisburgensis Stradner 1961
Discoaster sublodoensis Bramlette & Sullivan 1961
Discoaster tanii Bramlette & Riedel 1954
Discoaster tanii nodifer Bramlette & Riedel 1954
Discoaster wemmelensis Achuthan & Stradner 1969
Ellipsolithus macellus (Bramlette & Sullivan 1961) Sullivan 1964
Ericsonia alternans Black 1964
Ericsonia cava (Hay & Mohler 1967) Perch-Nielsen 1969
Ericsonia detecta de Kaenel & Villa 1996
Ericsonia formosa (Kamptner 1963) Haq 1971
Ericsonia ovalis Black 1964
Ericsonia robusta (Bramlette & Sullivan 1961) Perch-Nielsen 1977
Ericsonia subpertusa Hay & Mohler 1967
Fasciculithus billii Perch-Nielsen 1971
Fasciculithus bitectus Romein 1979
Fasciculithus chowii Varol 1989
Fasciculithus clinatus Bukry 1971
Fasciculithus hayi Haq 1971
Fasciculithus magnicordis Romein 1979
Fasciculithus pileatus Bukry 1973
Fasciculithus tympaniformis Hay & Mohler, in Hay *et al.* 1967
Fasciculithus ulii Perch-Nielsen 1971
Geminilithella bramlettei (Hay & Towe 1962) Varol 1989
Hayaster perplexus (Bramlette & Riedel 1954) Bukry 1973
Hayella situliformis Gartner 1969
Helicosphaera ampliaperta Bramlette & Wilcoxon 1967
Helicosphaera carteri (Wallich 1877) Kamptner 1954
Helicosphaera compacta Bramlette & Wilcoxon 1967
Helicosphaera euphratis Haq 1966
Helicosphaera gertae Bukry 1981
Helicosphaera granulata (Bukry & Percival 1971) Jafar & Martini 1975
Helicosphaera intermedia Martini 1965
Helicosphaera lophota Bramlette & Sullivan 1961
Helicosphaera obliqua Bramlette & Wilcoxon 1967
Helicosphaera perch-nielseniae (Haq 1971) Jafar & Martini 1975
Helicosphaera recta (Haq 1966) Jafar & Martini 1975
Helicosphaera reticulata Bramlette & Wilcoxon 1967
Helicosphaera salebrosa Perch-Nielsen 1971
Helicosphaera waltrans Theodoridis 1984
Heliolithus cantabriae Perch-Nielsen 1971
Heliolithus kleinpellii Sullivan 1964
Heliolithus riedelii Bramlette & Sullivan 1961
Heliolithus universus Wind & Wise, in Wise & Wind 1977
Hughesius gizoensis Varol 1989
Hughesius tasmaniae (Edwards & Perch-Nielsen 1975) de Kaenel & Villa 1996
Ilselithina fusa Roth 1970

Lanternithus minutus Stradner 1962
Lophodolithus nascens Bramlette & Sullivan 1961
Markalius inversus (Deflandre, in Deflandre & Fert 1954) Bramlette & Martini 1964
Nannotetrina alata (Martini 1960) Haq & Lohmann 1976
Nannotetrina fulgens (Stradner 1960) Achuthan & Stradner 1969
Neochiastozygus chiastus (Bramlette & Sullivan 1961) Perch-Nielsen 1971
Neochiastozygus concinnus (Martini 1961) Perch-Nielsen 1971
Neochiastozygus distentus (Bramlette & Sullivan 1961) Perch-Nielsen 1971
Neochiastozygus junctus (Bramlette & Sullivan 1961) Perch-Nielsen 1971
Neochiastozygus modestus Perch-Nielsen 1971
Neochiastozygus perfectus Perch-Nielsen 1981
Neocrepidolithus bukryi Perch-Nielsen 1981
Neocrepidolithus rimosus (Bramlette & Sullivan 1961) Varol 1989
Orthozygus aureus (Stradner 1962) Bramlette & Wilcoxon 1967)
Pedinocyclus larvalis (Bukry & Bramlette 1969) Loeblich & Tappen 1973
Placozygus sigmoides (Bramlette & Sullivan 1961) Romein 1979
Pontosphaera anisotrema (Kamptner 1956) Backman 1980
Pontosphaera longiforaminis (Baldi-Beke 1964) de Kaenel & Villa 1996
Prinsius bisulcus (Stradner 1963) Hay & Mohler 1967
Prinsius dimorphosus (Perch-Nielsen 1969) Perch-Nielsen 1977
Prinsius martinii (Perch-Nielsen 1969) Haq 1971
Prinsius tenuiculum (Okada & Thierstein 1979) Perch-Nielsen 1984
Reticulofenestra circus de Kaenel & Villa 1996
Reticulofenestra coenura (Reinhardt 1966) Roth 1970
Reticulofenestra daviesii (Haq 1968) Haq 1971
Reticulofenestra dictyoda (Deflandre, in Deflandre & Fert 1954) Stradner, in Stradner & Edwards 1968
Reticulofenestra hillae Bukry & Percival 1971
Reticulofenestra pseudoumbilica (Gartner 1967) Gartner 1969
Reticulofenestra umbilica (Levin 1965) Martini & Ritzkowski 1968
Reticulofenestra spp. (small)
Rhomboaster cuspis Bramlette & Sullivan 1961
Scyphosphaera amphora Deflandre 1942
Scyphosphaera martinii Jafar 1975
Sphenolithus abies Deflandre, in Deflandre & Fert 1954
Sphenolithus anarrhopus Bukry & Bramlette 1969
Sphenolithus calyculus Bukry 1985
Sphenolithus capricornutus Bukry & Percival 1971
Sphenolithus ciperoensis Bramlette & Wilcoxon 1967
Sphenolithus cometa de Kaenel & Villa 1996
Sphenolithus conicus Bukry 1971
Sphenolithus conspicuus Martini 1976
Sphenolithus delphix Bukry 1973
Sphenolithus disbelemnos Foruaciari & Rio 1996
Sphenolithus dissimilis Bukry & Percival 1971
Sphenolithus distentus (Martini 1965) Bramlette & Wilcoxon 1967
Sphenolithus editus Perch-Nielsen, in Perch-Nielsen *et al.* 1978
Sphenolithus furcatolithoides Locker 1967
Sphenolithus heteromorphus Deflandre 1953
Sphenolithus intercalcaris Martini 1976
Sphenolithus moriformis (Brönnimann & Stradner 1960) Bramlette & Wilcoxon 1967
Sphenolithus neoabies Bukry & Bramlette 1969
Sphenolithus obtusus Bukry 1971
Sphenolithus orphanknollensis Perch-Nielsen 1971
Sphenolithus predistentus Bramlette & Wilcoxon 1967
Sphenolithus primus Perch-Nielsen 1971
Sphenolithus pseudoradians Bramlette & Wilcoxon 1967
Sphenolithus radians Deflandre, in Grassé 1952
Sphenolithus spiniger Bukry 1971
Sphenolithus aff. *S. spiniger* Bukry 1971
Sphenolithus tintinnabulum Maiorano & Monechi 1997
Syracosphaera lamina de Kaenel & Villa 1996
Tetralithoides symeonidesii Theodoridis 1984
Toweius crassus (Bramlette & Sullivan 1961) Perch-Nielsen 1984
Toweius craticulus Hay & Mohler 1967
Toweius eminens (Bramlette & Sullivan 1961) Perch-Nielsen 1971
Toweius gammation (Bramlette & Sullivan 1961) Romein 1979
Toweius magnicrassus (Bukry 1971) Romein 1979
Toweius occultatus (Locker 1967) Perch-Nielsen 1971
Toweius oztunalii Varol 1989
Toweius pertusus (Sullivan 1965) Romein 1979
Thoracosphaera spp.
Tribrachiatus bramlettei (Brönnimann & Stradner 1960) Proto Decima *et al.* 1975
Tribrachiatus contortus (Stradner 1958) Bukry 1972
Tribrachiatus orthostylus Shamrai 1963
Triquetrorhabdulus carinatus Martini 1965

Triquetrorhabdulus milowii Bukry 1971
Triquetrorhabdulus rioi Olafsson 1989
Triquetrorhabdulus rugosus Bramlette & Wilcoxon 1967
Triquetrorhabdulus serratus (Bramlette & Wilcoxon 1967) Olafsson 1989
Umbilicosphaera rotula (Kamptner 1956) Varol 1982
Umbilicosphaera sibogae (Weber van Bosse 1901) Gaarder 1970
Zygrhablithus bijugatus (Deflandre, in Deflandre & Fert 1954) Deflandre 1959

Cretaceous species

Ahmuellerella regularis (Górka 1957) Reinhardt & Górka 1967
Arkhangelskiella cymbiformis Vekshina 1959
Arkhangelskiella cymbiformis Vekshina 1959 var. N Varol 1989
Arkhangelskiella cymbiformis Vekshina 1959 var. NT Varol 1989
Arkhangelskiella cymbiformis Vekshina 1959 var. W Varol 1989
Aspidolithus parcus constrictus (Hattner *et al.* 1980) Perch-Nielsen 1984
Aspidolithus parcus expansus (Wise & Watkins, in Wise 1983) Perch-Nielsen 1984
Aspidolithus parcus parcus (Stradner 1963) Noel 1969
Assipetra infracretacea (Thierstein 1973) Roth 1973
Assipetra terebrodentarius (Applegate *et al.*, in Covington & Wise 1987) Rutledge & Bergen, in Bergen 1994
Axopodorhabdus albianus (Black 1967) Wind & Wise, in Wise & Wind 1977
Biscutum ellipticum (Górka 1957) Grün, in Grün & Allemann 1975
Bisdiscus rotatorius Bukry 1969
Broinsonia bevieri Bukry 1969
Broinsonia matalosa (Stover 1966) Burnett, in Gale *et al.* 1996
Bukryaster hayii (Bukry 1969) Prins & Sissingh, in Sissingh 1977
Bukryaster magnus (Bukry 1975) new combination
Bukrylithus ambiguus Black 1971
Calcicalathina erbae Bergen 1998
Calculites obscurus (Deflandre 1959) Prins & Sissingh, in Sissingh 1977
Calculites ovalis (Stradner 1963) Prins & Sissingh, in Sissingh 1977
Ceratolithoides brevicorniculans Burnett 1998
Ceratolithoides indiensis Burnett 1998
Ceratolithoides longissimus Burnett 1998
Ceratolithoides pricei Burnett 1998
Ceratolithoides self-trailiae Burnett 1998
Ceratolithoides ultimus Burnett 1998
Ceratolithoides verbeekii Perch-Nielsen 1979
Chiastozygus fessus (Stover 1966) Shafik 1979
Chiastozygus tenuis Black 1971
Corollithion completum Perch-Nielsen 1973
Cretarhabdus conicus Bramlette & Martini 1964
Cretarhabdus loriei Gartner 1968
Cribrocorona gallica (Stradner 1963) Perch-Nielsen 1973
Cribrosphaerella ehrenbergii (Arkhangelsky 1912) Deflandre, in Piveteau 1952
Cribrosphaerella romanica Reinhardt 1964
Cyclagelosphaera margerelii Noël 1965
Cylindralithus crassus Stover 1966
Cylindralithus nudus Bukry 1969
Diazomatolithus lehmanii Noël 1965
Eiffellithus eximius (Stover 1966) Perch-Nielsen 1968
Eiffellithus hancockii Burnett 1998
Eiffellithus monechiae Crux 1991
Eiffellithus paragogus Gartner 1993
Eiffellithus turriseiffelii (Deflandre, in Deflandre & Fert 1954) Reinhardt 1965
Ellipsagelosphaera britannica (Stradner 1963) Perch-Nielsen 1968
Eprolithus eptapetalus Varol 1992
Eprolithus floralis (Stradner 1962) Stover 1966
Eprolithus orbiculatus (Forchheimer 1972) Crux, in Crux *et al.* 1982
Flabellites biforaminis Thierstein 1973
Gartnerago gammation (Hill 1976) new combination
Gartnerago nanum Thierstein 1974
Gartnerago obliquum (Stradner 1963) Noël 1970
Gartnerago theta (Black, in Black & Barnes 1959) Jakubowski 1986
Gorkaea pomerolii (Perch-Nielsen 1973) new combination
Gorkaea pseudanthophorus (Bramlette & Martini 1964) Varol & Girgis 1994
Haqius circumradiatus (Stover 1966) Roth 1978
Hayesites irregularis (Thierstein, in Roth & Thierstein 1972) Applegate *et al.*, in Covington & Wise 1987
Helicolithus trabeculatus (Gorka 1957) Verbeek 1977
Hemipodorhabdus biforatus Black 1972
Heterorhabdus primitivus Perch-Nielsen 1973
Isocrystallithus partitum (Varol, in Al-Rifaiy *et al.* 1990) Bergen 1998
Laguncula dorotheae Black 1971
Lithastrinus grillii Stradner 1962
Lithastrinus moratus Stover 1966
Lithraphidites carniolensis Deflandre 1963
Lithraphidites praequadratus Roth 1978
Lithraphidites quadratus Bramlette & Martini 1964

Loxolithus armilla (Black, in Black & Barnes 1959) Noël 1965
Lucianorhabdus salomonii Bergen 1994
Manivitella pemmatoidea (Deflandre, in Manivit 1965) Thierstein 1971
Marthasterites furcatus (Deflandre, in Deflandre & Fert 1954) Deflandre 1959
Metadoga spp.
Micrantholithus hoschulzii (Reinhardt 1966) Thierstein 1971
Microrhabdulus decoratus Deflandre 1959
Microstaurus chiastius (Worsely 1971) Grün, in Grün & Allemann 1975
Microstaurus quadratus Black 1971
Micula concava (Stradner, in Martini & Stradner 1960) Verbeek 1976
Micula cubiformis Forchheimer 1972
Micula praemurus (Bukry 1973) Stradner & Steinmetz 1984
Micula quadrata (Stradner 1961) Perch-Nielsen 1984
Micula staurophora (Gardet 1955) Stradner 1963
Micula swastica Stradner & Steinmetz 1984
Nannoconus elongatus Brönnimann 1955
Nannoconus globulus Brönnimann 1955
Nannoconus sp. 1
Neocrepidolithus watkinsii Pospichal & Wise 1990
Perchnielseniella stradneri (Perch-Nielsen 1973) Watkins 1984
Percivalia fenestrata (Worsley 1971) Wise 1983
Petrarhabdus copulatus (Deflandre 1959) Wind & Wise, in Wise 1983
Pickelhaube furtiva (Roth 1983) Applegate *et al.*, in Covington & Wise 1987
Prediscosphaera columnata (Stover 1966) Perch-Nielsen 1984
Prediscosphaera cretacea (Arkhangelsky 1912) Gartner 1968
Prediscosphaera grandis Perch-Nielsen 1979
Prediscosphaera intercisa (Deflandre, in Deflandre & Fert 1954) Shumenko 1976
Prediscosphaera majungae Perch-Nielsen 1973
Quadrum eneabrachium Varol 1992
Quadrum gartneri Prins & Perch-Nielsen, in Manivit *et al.* 1977
Quadrum gothicum (Deflandre 1959) Prins & Perch-Nielsen, in Manivit *et al.* 1977
Quadrum octobrachium Varol 1992
Quadrum svabenickae Burnett 1998
Quadrum trifidum (Stradner, in Stradner & Papp 1961) Prins & Perch-Nielsen, in Manivit *et al.* 1977
Reinhardtites anthophorus (Deflandre 1959) Perch-Nielsen 1968
Reinhardtites elegans (Gartner 1969) Wise 1983
Reinhardtites levis Prins & Sissingh, in Sissingh 1977
Retecapsa crenulata (Bramlette & Martini 1964) Grün, in Grün & Allemann 1975
Retecapsa octofenestrata (Bralower, in Bralower *et al.* 1989) Bown 1998
Retecapsa schizobrachiata (Gartner 1968) Grün, in Grün & Allemann 1975
Rhagodiscus achlyostaurion (Hill 1976) Doeven 1983
Rhagodiscus angustus (Stradner 1963) Reinhardt 1971
Rhagodiscus asper (Stradner 1963) Reinhardt 1967
Rhagodiscus gallagheri Rutledge & Bown 1996
Rhagodiscus hamptonii Bown, in Kennedy *et al.* 2000
Rhagodiscus indistinctus Burnett 1998
Rhagodiscus reniformis Perch-Nielsen 1973
Rhagodiscus thiersteinii (Roth 1973) new combination
Rhagodiscus sp. 1
Rhomboaster svabenickiae Bergen, in Bralower & Bergen 1998
Rotelapillus crenulatus (Stover 1966) Perch-Nielsen 1984
Seribiscutum primitivum (Thierstein 1974) Filewicz *et al.*, in Wise & Wind 1977
Stoverius biarcus (Bukry 1969) Perch-Nielsen 1984
Stoverius coangustatus Bergen & Howe, in Howe *et al.* 2003
Tegumentum stradneri Thierstein, in Roth & Thierstein 1972
Tetrapodorhabdus coptensis Black 1971
Tetrapodorhabdus decorus (Deflandre, in Deflandre & Fert 1954) Wind & Wise, in Wise & Wind 1977
Tranolithus exiguus Stover 1966
Tranolithus orionatus (Reinhardt 1966) Perch-Nielsen 1968
Tranolithus praeorionatus Bown, in Kennedy *et al.* 2000
Tranolithus pseudoangustus Crux 1987
Vagalapilla elliptica (Gartner 1968) Bukry 1969
Watznaueria barnesae (Black, in Black & Barnes 1959) Perch-Nielsen 1968
Watznaueria biporta Bukry 1969
Watznaueria fossacincta (Black 1971) Bown, in Bown & Cooper 1989
Zebrashapka vanhintei Covington & Wise 1987
Zeugrhabdotus bicrescenticus (Stover 1966) Burnett, in Gale *et al.* 1996
Zeugrhabdotus diplogrammus (Deflandre, in Deflandre & Fert 1954) Burnett, in Gale *et al.* 1996
Zeugrhabdotus howei Bown 2000, in Kennedy *et al.* 2000
Zeugrhabdotus moulladei Bergen 1998

Zeugrhabdotus praesigmoides Burnett 1998
Zeugrhabdotus spiralis (Bramlette & Martini 1964) Burnett 1998
Zeugrhabdotus xenotus (Stover 1966) Burnett, in Gale *et al.* 1996

This research used samples provided by the Ocean Drilling Program (ODP). ODP is sponsored by the US National Science Foundation (NSF) and participating countries under management of Joint Oceanographic Institutions (JOI), Inc. I would like to thank Dr D. K. Watkins, Dr S. W. Wise and Dr E. de Kaenel for their many helpful suggestions and careful review of the original manuscript.

References

AUBRY, M.-P. 1991. Sequence stratigraphy: Eustasy or tectonic imprint. *Journal of Geophysical Research*, **96**, 6641–6679.

BERGEN, J.A. 1994. Berriasian to early Aptian calcareous nannofossils from the Vocontian Trough (SE France) and Deep Sea Drilling Site 534: new nannofossil taxa and a summary of low-latitude biostratigraphic events. *Journal of Nannoplankton Research*, **16**, 59–69.

BERGEN, J.A. 1998. Calcareous nannofossils from the lower Aptian historical stratotype at Cassis–La Bedoule (SE France). *Geologie Mediterraneene*, **25**, 227–255.

BERGEN, J.A. & SIKORA, P.J. 2000. Microfossil diachronism in southern Norwegian Chalks: Valhall and Hod Fields. *In*: JONES, R.W. & SIMMONS, M.D. (eds) *Biostratigraphy in Production and Development Geology*. Geological Society, London, Special Publications, **152**, 85–111.

BERGGREN, W.A., KENT, D.V., SWISHER, C.C., III & AUBRY, M.-P. 1995. A revised Cenozoic geochronology and chronostratigraphy. *In*: BERGGREN, W.A., KENT, D.V., AUBRY, M.-P. & HARDENPOL, J. (eds) *Geochronology, Time Scales and Global Correlation*. SEPM, Special Publication, **54**, 129–212.

BOWN, P.R. (ed.) 1998. *Calcareous Nannofossil Biostratigraphy*. British Micropalaeontological Society, Publication Series. Chapman & Hall, 314 pp.

BRALOWER, T.J. & BERGEN, J.A. 1998. Cenomanian–Santonian calcareous nannofossil biostratigraphy of a transect of cores drilled across the Western Interior Seaway. *In*: *Stratigraphy and Paleoenvironments of the Cretaceous Western Interior Seaway*. SEPM, Concepts in Sedimentology and Paleontology, **6**, 59–77.

BRALOWER, T.J., SLITER, W.V., ARTHUR, M.A., LECKIE, R.M., ALLARD, D. & SCHLANGER, S.O. 1993. *Dysoxic/Anoxic Episodes in the Aptian-Albian (Early Cretaceous)*. Schlanger Memorial Volume. American Geophysical Union, Geophysical Monograph, **73**, 5–37.

BUKRY, D. 1975. Coccolith and silicoflagellate stratigraphy, northwestern Pacific Ocean, DSDP Leg 32. *In*: LARSON, R.L., MOBERLY, R. *et al.* *Initial Reports of the Deep Sea Drilling Project*, **132**, 677–701.

BURNETT, J.A. 1998. Upper Cretaceous. *In*: BOWN, P.R. (ed.) *Calcareous Nannofossil Biostratigraphy*. British Micropalaeontoloogical Society, Publication Series, 132–199, Chapman & Hall.

DE KAENEL, E. & VILLA, G. 1996. Oligocene–Miocene calcareous nannofossil biostratigraphy and paleoecology from the Iberia Abyssal Plain. *Proceedings of the Ocean Drilling Program, Scientific Results*, **149**, 79–145.

GALE, A.S., KENNEDY, W.J., BURNETT, J.A., CARON, M. & KIDD, B.E. 1996. The Late Albian to Early Cenomanian succession at Mont Risou, near Rosans (Drôme, SE France): an integrated study (ammonites, inoceramids, planktonic foraminifera, nannofossils, oxygen and carbon isotopes). *Cretaceous Research*, **17**, 515–606.

HILL, M.E. 1976. Lower Cretaceous calcareous nannofossils from Texas and Oklahoma. *Palaoontographica Abteilung, B*, **56**, 103–179.

KENNEDY, W.J., GALE, A.S., BOWN, P.R., CARON, M., DAVEY, R.J., GROCKE, D. & WRAY, D.S. 2000. Integrated stratigraphy across the Aptian–Albian boundary in the Marnes Bleues, at the Col de Pre–Guittard, Arnayon (Drome), and at Tartonne (Alpes-de-Hautes–Provence), France: a candidate global boundary stratotype section and boundary point for the base of the Albian stage. *Cretaceous Research*, **21**, 591–720.

LECKIE, R.M., BRALOWER, T.J. & CASHMAN, R. 2002. Oceanic anoxic events and plankton evolution during the mid-Cretaceous. *Paleooceanography*, **17/3**, 13, 1–29.

MARTINI, E. 1971. Standard Tertiary and Quaternary calcareous nannoplankton zonation. *In*: FARINACCI, A. (ed.) *Proceedings of the Second Planktonic Conference Roma 1970*, Volume 2. Edizioni Tecnoscienza, Rome, 739–785.

OKADA, H. & BUKRY, D. 1980. Supplementary modification and introduction of code numbers to the low-latitude coccolith biostratigraphic zonation (Bukry 1973, 1975). *Marine Micropaleontology*, **5**, 321–325.

PERCH-NIELSEN, K. 1973. Neue Coccolithen aus dem Maastrichtien von Dänemark, Madagaskar und Ägypten. *Bulletin of Geological Society of Denmark*, **22**, 306–335.

PERCH-NIELSEN, K. 1985*a*. Mesozoic calcareous nannofossils. *In*: BOLLI, H.M., SAUNDERS, J.B. & PERCH-NIELSEN, K. (eds) *Plankton Stratigraphy*. Cambridge University Press, Cambridge, 329–426.

PERCH-NIELSEN, K. 1985*b*. Cenozoic calcareous nannofossils. *In*: BOLLI, H.M., SAUNDERS, J.B. & PERCH-NIELSEN, K. (eds) *Plankton Stratigraphy*. Cambridge University Press, Cambridge, 427–554.

ROTH, P.H. 1973. Calcareous nannofossils – Leg 17. *Initial Reports of the Deep Sea Drilling Project*, **17**, 695–795.

ROTH, P.H. 1978. Cretaceous nannoplankton biostratigraphy and oceanography of the northwestern Atlantic Ocean. *Initial Reports of the Deep Sea Drilling Project*, **44**, 731–760.

SISSINGH, W. 1977. Biostratigraphy of Cretaceous calcareous nannoplankton. *Geologie en Mijnbouw*, **56**, 37–65.

SISSINGH, W. 1978. Microfossil biostratigraphy and stage-stratotypes of the Cretaceous. *Geologie en Mijnbouw*, 57, 433–440.

THIERSTEIN, H.R. 1973. Lower Cretaceous calcareous nannoplankton biostratigraphy. *Abhandlungen der Geologischen Bundesanstalt*, **29**, 1–52.

VALENTINE, P.C. 1987. Lower Eocene Calcareous nannofossil biostratigraphy beneath the Atlantic Slope and Upper Rise off New Jersey – New Zonation based on Deep Sea Drilling Project Sites 612 and 613. *In*: POAG, C.W., WATTS, B. *et al. Initial Reports of the Deep Sea Drilling Project*, **95**, 359–394.

VAROL, O. 1989. Quantitative analysis of the *Arkhangelskiella cymbiformis* group and biostratigraphic usefulness in the North Sea area. *Journal of Micropalaeontology*, **8**, 131–134.

VAROL, O. & GIRGIS, M. 1994. New taxa and taxonomy of some Jurassic to Cretaceous calcareous nannofossils. *Neues Jahrbuch für Geologie und Paläontologie, Abhandlungen*, 192, 221–253.

Pin-pricking the elephant: evidence on the origin of the Ontong Java Plateau from Pb–Sr–Hf–Nd isotopic characteristics of ODP Leg 192 basalts

M. L. G. TEJADA[1], J. J. MAHONEY[2], P. R. CASTILLO[3], S. P. INGLE[4,5], H. C. SHETH[2,6] & D. WEIS[4,7]

[1]*National Institute of Geological Sciences, University of the Philippines, Diliman, Quezon City, 1101 Philippines*

[2]*School of Ocean and Earth Science and Technology, University of Hawaii, Honolulu, HI 96822, USA (e-mail: jmahoney@hawaii.edu)*

[3]*Scripps Institution of Oceanography, University of California, San Diego, La Jolla, CA 92093-0220, USA*

[4]*Département des Sciences de la Terre et de l'Environnement, Université Libre de Bruxelles, CP 160/02, Avenue F.D. Roosevelt, 50B-1050 Brussels, Belgium*

[5]*Present address: Earth and Planetary Sciences, Tokyo Institute of Technology, 2-12-1 Ookayama, Meguroku Tokyo 152–8551, Japan*

[6]*Present address: Department of Earth Sciences, Indian Institute of Technology, Powai, Bombay 400 076, India*

[7]*Present address: Department of Earth and Ocean Sciences, University of British Columbia, Vancouver, B.C., Canada V6T 1Z4*

Abstract: Age-corrected Pb, Sr and Nd isotope ratios for early Aptian basalt from four widely separated sites on the Ontong Java Plateau that were sampled during Ocean Drilling Program Leg 192 cluster within the small range reported for three earlier drill sites, for outcrops in the Solomon Islands, and for the Nauru and East Mariana basins. Hf isotope ratios also display only a small spread of values. A vitric tuff with $\varepsilon_{Nd}(t)$ = +4.5 that lies immediately above basement at Site 1183 represents the only probable example from Leg 192 of the Singgalo magma type, flows of which comprise the upper 46–750 m of sections in the Solomon Islands and at Leg 130 Site 807 on the northern flank of the plateau. All of the Leg 192 lavas, including the high-MgO (8–10 wt%) Kroenke-type basalts found at Sites 1185 and 1187, have $\varepsilon_{Nd}(t)$ between +5.8 and +6.5. They are isotopically indistinguishable from the abundant Kwaimbaita basalt type in the Solomon Islands, and at previous plateau, Nauru Basin and East Mariana Basin drill sites. The little-fractionated Kroenke-type flows thus indicate that the uniform isotopic signature of the more evolved Kwaimbaita-type basalt (with 5–8 wt% MgO) is not simply a result of homogenization of isotopically variable magmas in extensive magma chambers, but instead must reflect the signature of an inherently rather homogeneous (relative to the scale of melting) mantle source. In the context of a plume-head model, the Kwaimbaita-type magmas previously have been inferred to represent mantle derived largely from the plume source region. Our isotopic modelling suggests that such mantle could correspond to originally primitive mantle that experienced a rather minor fractionation event (e.g. a small amount of partial melting) approximately 3 Ga or earlier, and subsequently evolved in nearly closed-system fashion until being tapped by plateau magmatism in the early Aptian. These results are consistent with current models of a compositionally distinct lower mantle and a plume-head origin for the plateau. However, several other key aspects of the plateau are not easily explained by the plume-head model. The plateau also poses significant challenges for asteroid impact, Icelandic-type and plate separation (perisphere) models. At present, no simple model appears to account satisfactorily for all of the observed first-order features of the Ontong Java Plateau.

Several massive volcanic plateaus appeared at equatorial to mid-southern latitudes in the Pacific Basin between the latest Jurassic and the middle Cretaceous. Of these, the Ontong Java Plateau (OJP; Fig. 1) in the western Pacific is the world's largest (the 'elephant' in our title), with

From: FITTON, J. G., MAHONEY, J. J., WALLACE, P. J. & SAUNDERS, A. D. (eds) 2004. *Origin and Evolution of the Ontong Java Plateau*. Geological Society, London, Special Publications, **229**, 133–150. 0305-8719/$15.00

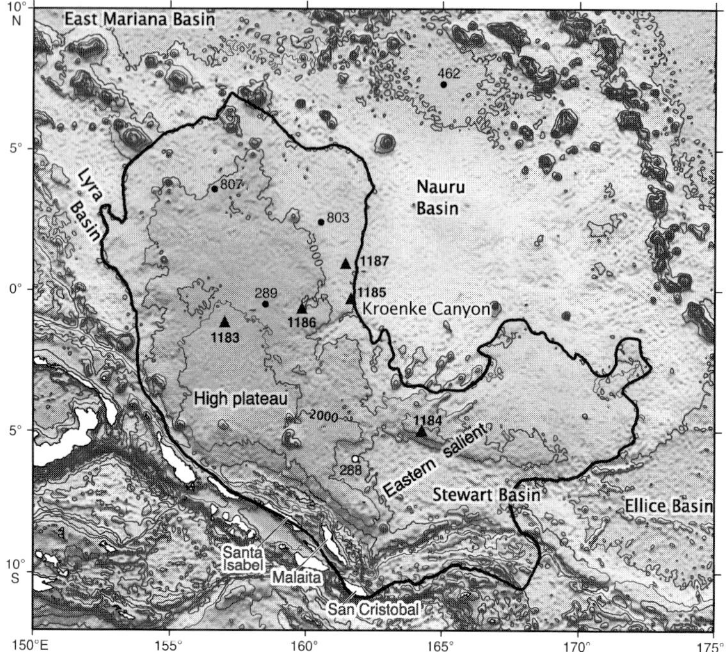

Fig. 1. Map of the Ontong Java Plateau (outlined) showing the locations of sites drilled during Leg 192 (triangles). Dots are previous drill sites that reached basement. The white dot represents Site 288, which did not reach basement but bottomed in Aptian limestone. The bathymetric contour interval is 1000 m (predicted bathymetry from Smith & Sandwell 1997).

a Greenland-size area of approximately 2×10^6 km^2 and an average crustal thickness of about 32 km (e.g. Gladczenko *et al.* 1997; Richardson *et al.* 2000; J.G. Fitton & M.F. Coffin pers. comm. 2003; Miura *et al.* 2004). Despite their great size, the origin of the Pacific plateaus is only poorly understood, having been attributed variously to: (1) cataclysmic melting in the inflated heads of newly risen mantle plumes (e.g. Richards *et al.* 1989) or even a single 'super' plume (Larson 1991); (2) formation above near-ridge plume tails over much longer periods of time (e.g. Mahoney & Spencer 1991; Ito & Clift 1998); (3) plate separation above extensive, near-solidus, but non-plume regions of the shallow asthenosphere (e.g. Anderson *et al.* 1992; Smith & Lewis 1999; Hamilton 2003); and (4) asteroid impact (Rogers 1982). The variety of models that have been applied in part reflects a lack of detailed knowledge of Pacific plateau composition and age, which in turn is a result of the very sparse sampling of crustal basement. Although it is by far the largest, the OJP is also presently the best sampled of any Pacific plateau.

Along its southern margin the plateau has collided with the Solomon island arc, where fragments of OJP crustal basement have been uplifted and exposed in several places, particularly in the islands of Santa Isabel, Malaita and San Cristobal (also known as Makira) (see Petterson *et al.* 1999). Away from the collision zone, however, basement on the plateau is buried under a thick marine sedimentary section, itself submerged approximately 1.7–4 km below sea level. Thus, drilling is the only effective means of sampling volcanic basement, in general. Until recently, it had been reached at only three drill sites; penetration of 149 m into the volcanic section was achieved at one site (Site 807), but the other two holes penetrated only 9 (Site 289) and 26 m (Site 803) into basement (Fig. 2) (Andrews *et al.* 1975; Kroenke *et al.* 1991). Sampling of basement was augmented considerably in September and October of 2000, when Ocean Drilling Program (ODP) Leg 192 cored sections at four sites on the OJP's main or high plateau (Sites 1183, 1185, 1186 and 1187) to subbasement depths ranging from 65 to 217 m (Mahoney *et al.* 2001). A fifth site, Site 1184, cored 338 m of a basaltic volcaniclastic sequence on the eastern lobe or salient of the plateau.

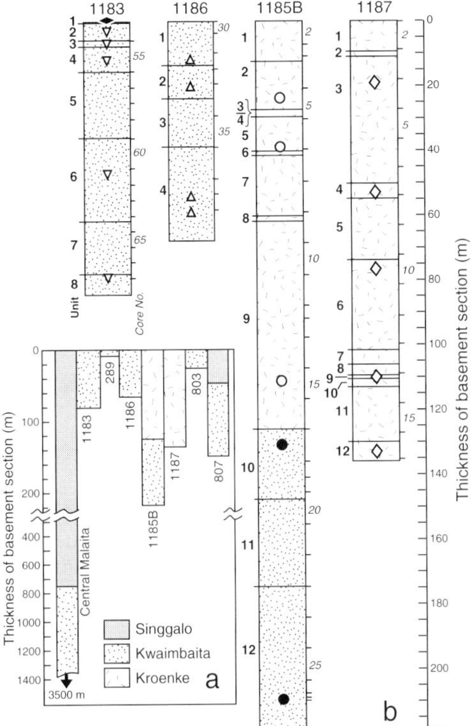

Fig. 2. (**a**) Basement thickness and magma types in OJP drill sites and central Malaita. (**b**) Basement sections of Leg 192 sites on the high plateau, showing drill-core number, unit boundaries and magma type. Symbols indicate sample locations.

Prior to Leg 192, study of samples from the Solomon Islands and the three previous drill sites had established that basement at all of these locations is composed of massive and pillowed submarine lava flows. The rock is low-K tholeiitic basalt with only a small range of major-element, trace-element and Nd–Pb–Sr isotopic composition, a surprising result in view of the immensity of the plateau (e.g. Mahoney *et al.* 1993; Tejada *et al.* 1996, 2002; Babbs 1997; Neal *et al.* 1997). Dating by ^{40}Ar–^{39}Ar revealed that a major plateau-forming event occurred in the early Aptian, with ages clustering around 122 Ma; however, ages near 90 Ma were obtained for Site 803 and parts of Santa Isabel and San Cristobal (Mahoney *et al.* 1993; Birkhold-VanDyke *et al.* 1996; Parkinson *et al.* 1996; Tejada *et al.* 1996, 2002). All of the basalts were found to be distinct from both N-MORB (normal mid-ocean ridge basalt) and OIB (ocean island basalt). They have low, MORB-like concentrations of many incompatible elements, but OIB-like isotopic characteristics rather similar to those of the Hawaiian shield volcanoes of Kilauea and Mauna Loa; moreover, unlike either N-MORB or OIB, their primitive-mantle-normalized incompatible-element patterns and chondrite-normalized rare-earth patterns are relatively flat. Despite the limited compositional variability, two isotopically distinct, stratigraphically separate groups of basalt, termed the Kwaimbaita and Singgalo types by Tejada *et al.* (2002), were identified at Site 807, in Santa Isabel and in Malaita. The stratigraphically lower Kwaimbaita type is characterized by higher age-corrected $\varepsilon_{Nd}(t)$ (+5.4–+6.4), higher $(^{206}Pb/^{204}Pb)_t$ (18.21–18.42) and lower $(^{87}Sr/^{87}Sr)_t$ (0.7034–0.7039) than the overlying Singgalo type (with +3.7–+5.3, 17.71–17.99 and 0.7039–0.7044, respectively). Kwaimbaita-type basalts also tend to have slightly lower ratios of highly incompatible elements to moderately incompatible elements. The thickest basement section (3.5 km) is found in central Malaita (Fig. 2), where the two groups are defined as formations; the lower group, the Kwaimbaita Formation, is >2.7 km thick and the upper group, the Singgalo Formation, reaches a thickness of 750 m (Tejada *et al.* 2002). At Site 807, 1600 km to the north, the thickness of Singgalo-type flows is only 46 m. At Site 289, located between Site 807 and the Solomons, the single flow sampled at the top of basement is isotopically Kwaimbaita type. Isotopic data for glasses from the 640 m-thick basalt sequence drilled at Site 462A in the Nauru Basin to the NE of the OJP proper (Mahoney 1987; Castillo *et al.* 1994) show that they, too, are of the Kwaimbaita type. North of the OJP, 51 m of Kwaimbaita-type flows were drilled at Site 802 in the East Mariana Basin (Castillo *et al.* 1994). Both Singgalo and Kwaimbaita magma types appear to be the products of high amounts of partial melting; pre-Leg 192 estimates yielded values in the 20–30% range (assuming peridotite source rock), with the Kwaimbaita representing the upper end of this range (Mahoney *et al.* 1993; Neal *et al.* 1997; Tejada 1998).

Shipboard analysis of TiO_2, Zr and several other elements during Leg 192 suggested that basalt recovered from Sites 1183, 1186 and the lower 92 m of the 217 m-thick lava section at Site 1185 was of the Kwaimbaita type (Fig. 2), whereas biostratigraphic evidence indicated an early Aptian basement age (Shipboard Scientific Party 2001). The volcaniclastic deposits at Site 1184 also appeared chemically Kwaimbaita-like. No obvious Singgalo-type compositions were found at any of the sites. In contrast, the 136 m-thick basement section drilled at Site 1187 and the upper 125 m of flows at Site 1185, 146 km south of Site 1187, were discovered to be a low-TiO_2, high-MgO type of basalt not seen previously on the OJP. We term this magma type, the least differentiated of any found thus far, the Kroenke type, after the location of Site 1185 adjacent to Kroenke Canyon, a large submarine canyon just south of the site. Shipboard measurements showed it to be characterized by approximately 0.75 wt% TiO_2, 8–10 wt% MgO, *c.* 200 ppm Ni and *c.* 500 ppm Cr; in contrast, Kwaimbaita-type basalts average around 1 wt% TiO_2, 7 wt% MgO, and have <120 ppm Ni and <250 ppm Cr.

In this chapter we present Sr, Pb, Hf and Nd isotopic data for samples from Sites 1183, 1185, 1186 and 1187, and, in combination with previous data, discuss their implications for the source and origin of the OJP. Complementary to our work, Fitton & Godard (2004) have carried out a detailed major- and trace-element study of basalt from these sites, and mineral and glass compositions have been determined by Sano & Yamashita (2004). Elemental and isotopic data for the volcaniclastic rocks of Site 1184 are reported by Shafer *et al.* (2004) and White *et al.* (2004).

Analytical methods

During the cruise, we obtained small slabs from the least altered portions of crystalline basalt and, in some cases, glassy flow margins recovered from a representative number of basement units from each site (see Fig. 2) (Mahoney *et al.* 2001). Both massive and pillowed lava units were sampled, as was one of two vitric tuff beds lying immediately above basement at Site 1183. Preparation and analysis onshore followed our standard procedures for glass and for moderately altered bulk-rock samples (e.g. see descriptions and references of Castillo *et al.* 1994; Mahoney *et al.* 1998; Ingle *et al.* 2004). Basalts from Sites 1183 and 1186 were processed for Pb, Nd and Sr isotope analyses at the Scripps Institution of Oceanography, whereas those from Sites 1185 and 1187, and the tuff from Site 1183, were processed and analysed at the University of Hawaii. In both laboratories, parent and daughter element abundances were obtained on the same dissolution of sample analysed for isotope ratios. Concentration measurements at Scripps employed single-collector, high-resolution, inductively coupled plasma-mass spectrometry (ICP-MS), whereas at Hawaii parent and daughter concentrations were determined by isotope dilution using a multi-collector thermal-ionization mass spectrometer. Analysis of Hf isotopes was carried out at the Université Libre de Bruxelles using multi-collector ICP-MS (Nu Plasma). In addition to the Leg 192 samples, Hf isotopes also were measured for four samples of Singgalo Formation basalt and one sample of Kwaimbaita Formation basalt from Malaita previously analysed by Tejada *et al.* (2002) for Nd–Pb–Sr isotopes, and for major and trace elements. Concentrations of Lu and Hf for the Leg 192 samples were determined by high-resolution ICP-MS at the Université de Montpellier II on separate splits of sample from those used for isotope analysis (see Fitton & Goddard 2004). Our results are presented in Tables 1 and 2, along with estimated analytical uncertainties, isotopic fractionation corrections and standard values.

Results

Pb, Nd and Sr isotopes

An important, and unexpected, result of our study is that the high-MgO Kroenke-type basalt flows at Site 1187 and comprising the upper 125 m of basement units at Site 1185 are isotopically indistinguishable from previous data for the Kwaimbaita magma type (Figs 3 and 4). Recent $^{187}Re–^{187}Os$ dating indicates that basement at these sites and at Sites 1183 and 1186 is early Aptian, 121.7 ± 1.4 Ma (Parkinson *et al.* 2001), although $^{40}Ar–^{39}Ar$ and biostratigraphic data suggest a slight progression of ages within the Aptian (L. Chambers pers. comm. 2003). For age-correcting our isotopic data we have assumed an age (t) of 120 Ma. The Kroenke-type lavas have age-corrected $\varepsilon_{Nd}(t)$ = +6.1–+6.4, $(^{87}Sr/^{87}Sr)_t$ = 0.70374–0.70381, $(^{206}Pb/^{204}Pb)_t$ = 18.317–18.360, $(^{207}Pb/^{204}Pb)_t$ = 15.514–15.522 and $(^{208}Pb/^{204}Pb)_t$ = 38.169–38.220 (where t = 120 Ma). This range is within, or only slightly greater than, the propagated analytical errors for age-corrected Nd and Sr isotope ratios and double-spike $^{207}Pb/^{204}Pb$ (see the footnotes to Tables 1 and 2), which is remarkable considering the distance of 146 km between these two sites. Although relative variability in $(^{206}Pb/^{204}Pb)_t$ and $(^{208}Pb/^{204}Pb)_t$ is slightly greater, among the glasses it is only 18.320–18.340 and 38.180–38.194, respectively.

The age-corrected Pb, Nd and Sr isotope ratios for basalts from Sites 1183, 1186 and the lower 92 m of Site 1185 also fall within, or very close to, the small range measured for Kwaimbaita-type basalts from Malaita, Santa Isabel, Sites 289 and 807, and glasses from the Nauru and East Mariana basins. The Kwaimbaita-type nature of

Table 1. *Sr, Nd and Hf isotope data*

Sample			$(^{87}Sr/^{86}Sr)_t$	$(^{143}Nd/^{144}Nd)_t$	$\varepsilon_{Nd}(t)$	$(^{176}Hf/^{177}Hf)_t$	$\varepsilon_{Hf}(t)$	Rb	Sr	Sm	Nd	Lu	Hf
Site 1183A													
54–3 (28–34) tuff	L	Sg	0.70254	0.512714	+4.5			55.39	28.95	1.106	3.599		
54–4 (2–6) Unit 2B	L	Kw	0.70329	0.512812	+6.4	0.28298	+10.1	8.46	145.0	2.10	5.17	0.33	1.72
55–1 (77–78) Unit 3B	L	Kw	0.70347	0.512820	+6.5	0.28301	+11.0	0.91	104.1	0.84	1.83	0.37	1.83
55–3 (120–122) Unit 4B	L	Kw	0.70337	0.512786	+5.9	0.28301	+11.2	3.72	95.60	1.19	2.27	0.33	1.75
61–2 (6–7) Unit 6	L	Kw	0.70345	0.512789	+5.9	0.28300	+10.7	0.39	90.89	0.93	2.02	0.35	1.71
	R		0.70348										
	R		0.70347										
67–3 (93–95) Unit 8	L	Kw	0.70349	0.512802	+6.2	0.28301	+11.0	0.49	85.88	1.15	2.69	0.33	1.67
Site 1185B													
5–5 (142–143) Unit 2		Kr	0.70381	0.512799	+6.1	0.28301	+11.0	1.290	77.37	1.518	4.393	0.28	1.07
6–4 (95–96) Unit 5		Kr	0.70380	0.512803	+6.2	0.28307	+13.3	1.035	63.56	1.273	3.678	0.24	1.05
15–2 (113–114) Unit 9	L	Kr	0.70375	0.512813	+6.4	0.28304	+12.2	0.5998	82.15	1.485	4.333	0.28	1.10
17–3 (33–34) Unit 10		Kw	0.70352	0.512788	+5.9	0.28304	+12.1	0.8742	104.9	2.290	6.855	0.38	1.82
28–1 (56–57) Unit 12		Kw	0.70354	0.512785	+5.8	0.28303	+11.8	0.7935	104.1	2.163	6.424	0.33	1.62
Site 1186A													
32–2 (100–101) Unit 1	L	Kw	0.70343	0.512798	+6.1	0.28301	+11.2	3.70	124.0	1.82	3.84	0.36	1.79
34–1 (84–86) Unit 2B	L	Kw	0.70346	0.512801	+6.1	0.28301	+11.0	0.26	87.81	0.91	1.99	0.33	1.52
38–1 (38–40) Unit 4	L	Kw	0.70342	0.512820	+6.5	0.28301	+11.0	1.14	88.87	0.90	2.28	0.34	1.58
GRS-39–1 Unit 4	L	Kw	0.70345	0.512792	+6.0	0.28301	+11.1	0.19	71.69	0.95	2.06	0.34	1.58
Site 1187A													
3–4 (93–95) Unit 3	G	Kr	0.70376	0.512803	+6.2	0.28302	+11.4	1.198	83.47	1.556	4.571	0.29	1.17
7–7 (97–98) Unit 4	L	Kr	0.70377	0.512799	+6.1	0.28302	+11.6	0.4247	79.82	1.498	4.395	0.28	1.17
10–1 (124–126) Unit 6	G	Kr	0.70375	0.512807	+6.3	0.28300	+10.8	1.281	88.88	1.663	4.892	0.29	1.18
14–2 (4–6) Unit 9	G	Kr	0.70381	0.512799	+6.1	0.28301	+11.2	1.347	84.73	1.572	4.630	0.29	1.08
16–2 (87–89) Unit 12	G	Kr	0.70374	0.512802	+6.2			2.445	86.21	1.538	4.511		
Central Malaita													
SG-1		Sg	0.70405	0.512694	+4.1	0.28298	+10.0	1.277	107.8	1.689	4.607	0.43	2.50
SGB-17	L	Sg	0.70414	0.512690	+4.0	0.28299	+10.3	0.9675	93.10	1.540	3.425	0.38	2.47
SGB-18	L	Kw	0.70379	0.512783	+5.8	0.28305	+12.4	0.6878	112.5	1.439	3.253	0.32	2.00
ML-421	L	Sg	0.70416	0.512691	+4.1	0.28298	+10.1	0.0990	109.6	1.326	2.885	0.43	2.94
KF-4	L	Sg	0.70416	0.512687	+4.0	0.28298	+10.2	0.7995	81.30	1.558	3.673	0.47	3.02

Notes: All elemental concentrations are in ppm. Sg, Singgalo type; Kw, Kwaimbaita type; Kr, Kroenke type. G, glass; R, replicate analysis; L, strongly acid-leached powder for Sr and Nd isotope analysis, except for the two samples from Sites 1185B and 1187A, for which only Sr isotopes and Rb and Sr concentrations were measured on strongly leached powder. Isotopic fractionation corrections are $^{148}NdO/^{144}NdO = 0.242436$ ($^{148}Nd/^{144}Nd = 0.241572$), $^{86}Sr/^{88}Sr = 0.1194$ and $^{179}Hf/^{177}Hf = 0.7325$. Sm, Nd, Lu and Hf concentrations, and Nd isotope data for Central Malaitan samples are from Tejada *et al.* (2002). Nd and Sr isotope data are reported relative to $^{143}Nd/^{144}Nd = 0.511850$ for La Jolla Nd and $^{87}Sr/^{86}Sr = 0.71025$ for NBS 987 Sr. At the University of Hawaii, the total range measured for La Jolla Nd is $\pm$ 0.000008 (0.2 ε units); for NBS 987 it is $\pm$ 0.000020 over a 2 year-period; at Scripps, it is $\pm$ 0.000014 (0.3 e units) and 0.000018, respectively. At the Free University of Brussels, $^{176}Hf/^{177}Hf = 0.282160 \pm 0.000020$ (0.7 ε units) for the JMC-475 standard. Within-run errors on the isotopic data above are less than or equal to the external uncertainties on these standards. Estimated relative uncertainties on Sm and Nd concentrations are <0.2% and *c.*1%, respectively, at Hawaii and Scripps; for Sr the respective uncertainties are 0.5% and *c.*2%, and for Rb, 1% and *c.*2%. Total procedural blanks are negligible: for Hf, <20 pg; for Nd and Sr, <10 pg and <35 pg. $\varepsilon_{Nd} = 0$ today corresponds to $^{143}Nd/^{144}Nd = 0.51264$; for $^{147}Sm/^{144}Nd = 0.1967$, $\varepsilon_{Nd}(t) = 0$ corresponds to $^{143}Nd/^{144}Nd = 0.512486$ at 120 Ma. $\varepsilon_{Hf} = 0$ today corresponds to $^{176}Hf/^{177}Hf = 0.28277$; for $^{176}Lu/^{177}Hf = 0.0332$, $\varepsilon_{Hf}(t) = 0$ corresponds to 0.282696 at 120 Ma.

Table 2. *Pb isotope data*

Sample		$(^{206}Pb/^{204}Pb)_t$	$(^{207}Pb/^{204}Pb)_t$	$(^{208}Pb/^{204}Pb)_t$	Th	U	Pb	$(^{206}Pb/^{204}Pb)_0$	$(^{207}Pb/^{204}Pb)_0$	$(^{208}Pb/^{204}Pb)_0$
Site 1183A										
54–3 (28–34) tuff	Sg	18.400	15.556	38.436	0.2877	0.5119	1.1750	18.915	15.581	38.531
54–4 (2–6) Unit 2B	Kw	18.517	15.528	38.256	0.1701	0.0825	0.3122	18.829	15.543	38.467
55–1 (77–78) Unit 3B	Kw	18.495	15.514	38.286	0.1971	0.0778	0.6595	18.634	15.521	38.401
55–3 (120–122) Unit 4B	Kw	18.429	15.509	38.161	0.1788	0.0718	0.2562	18.759	15.525	38.431
61–2 (6–7) Unit 6	Kw	18.369	15.511	38.138	0.2175	0.0690	0.2463	18.700	15.528	38.479
67–3 (93–95) Unit 8	Kw	18.510	15.525	38.248	0.4968	0.1135	0.9453	18.651	15.532	38.451
Site 1185B										
5–5 (142–143) Unit 2	Kr	18.320	15.518	38.185	0.1526	0.0471	0.4057	18.456	15.525	38.329
6–4 (95–96) Unit 5	Kr	18.317	15.514	38.169	0.1416	0.0433	0.1012	18.823	15.539	38.712
15–2 (113–114) Unit 9	Kr	18.360	15.515	38.220	0.1496	0.0450	0.2470	18.574	15.537	38.453
17–3 (33–34) Unit 10	Kw	18.398	15.516	38.201	0.2411	0.0711	0.4350	18.590	15.525	38.414
28–1 (56–57) Unit 12	Kw	18.390	15.515	38.219	0.2292	0.0690	0.1674	18.879	15.539	38.751
Site 1186A										
32–2 (100–101) Unit 1	Kw	18.279	15.512	38.190	0.1936	0.1716	0.2278	19.173	15.556	38.520
34–1 (84–86) Unit 2B	Kw	18.465	15.509	38.240	0.1958	0.0685	0.4061	18.663	15.519	38.426
38–1 (38–40) Unit 4	Kw	18.427	15.521	38.244	0.2110	0.0686	0.6348	18.554	15.528	38.372
GRS-39–1 Unit 4	Kw	18.376	15.518	38.180	0.1658	0.0526	0.1496	18.792	15.538	38.609
Site 1187A										
3–4 (93–95) Unit 3 G	Kr	18.336	15.521	38.194	0.1564	0.0485	0.2402	18.573	15.533	38.445
7–7 (97–98) Unit 4	Kr	18.357	15.514	38.201	0.1376	0.0445	0.2494	18.567	15.524	38.413
10–1 (124–126) Unit 6 G	Kr	18.320	15.522	38.184	0.1978	0.0606	0.2595	18.595	15.540	38.478
14–2 (4–6) Unit 9 G	Kr	18.333	15.517	38.180	0.1634	0.0495	0.2427	18.573	15.529	38.439
16–2 (87–89) Unit 12 G	Kr	18.340	15.517	38.192	0.1551	0.0478	0.2402	18.574	15.528	38.441

Notes: G, glass; Sg, Singgalo type; Kw, Kwaimbaita type; Kr, Kroenke type. All elemental abundances are in ppm. For Scripps data, measured Pb isotopic ratios are corrected for fractionation using the NBS 981 standard values of Todt *et al.* (1996); the long-term errors measured for this standard are ±0.008 for $^{206}Pb/^{204}Pb$ and $^{207}Pb/^{204}Pb$, and ± 0.024 for $^{208}Pb/^{204}Pb$. For Hawaii data, a double-spike method (Galer 1999) was employed; the total range on approximately 5 ng loads of NBS 981 Pb in the last 3 years is 230 ppm for each ratio, and mean ratios measured are 16.937, 15.492 and 36.710. For both Scripps and Hawaii data, the within-run errors on measurements above are less than or equal to the external uncertainties on the standard. Estimated uncertainties on concentrations are <2% on Th, ≈1% on U and ≈0.5% on Pb for the Hawaii isotope-dilution data, and <2% for these elements for the Scripps data. Total procedural blanks are negligible: <3 pg for Th, <5 pg for U and <30 pg and <60 pg for Pb at Hawaii and Scripps, respectively.

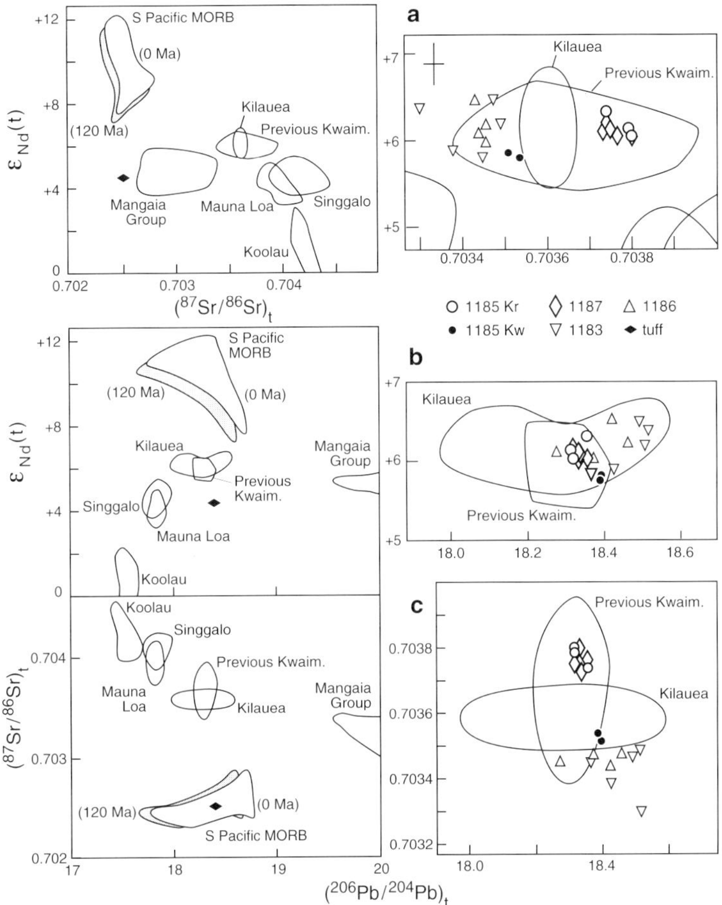

Fig. 3. Age-corrected (**a**) Sr v. Nd, (**b**), $^{206}Pb/^{204}Pb$ v. Nd and (**c**) $^{206}Pb/^{204}Pb$ v. Sr isotopic data for the Leg 192 lavas and tuff. Panels on the right are expanded portions of those on the left. Kw, Kwaimbaita type; Kr, Kroenke type. Fields are shown for previous Kwaimbaita- (Kwaim.) and Singgalo-type basalt from the pre-Leg 192 drill sites, Malaita and Santa Isabel, and glass from the Nauru and East Mariana basins (data sources: Mahoney 1987; Castillo *et al.* 1991, 1994; Mahoney & Spencer 1991; Mahoney *et al.* 1993; Tejada *et al.* 1996, 2002). Pb isotope data for two previously analysed samples with suspected analytical problems were not used. See Tejada *et al.* (2002) for data sources used for the fields of South (S) Pacific MORB, Kilauea, Mauna Loa (subaerial portion), Koolau (subaerial portion) and the Mangaia Group islands. The shaded 120 Ma field is for estimated South Pacific MORB source mantle (see Tejada *et al.* 2002). Error bars in (**a**) are for data in this paper (see Fig. 4 for Pb error bars). Note that symbols are the same as in Figure 2.

these Leg 192 lavas is confirmed by the incompatible-element data of Fitton & Godard (2004). For the Site 1183 basalts, the isotopic range is $\varepsilon_{Nd}(t)$ = +5.9–+6.5, $(^{87}Sr/^{87}Sr)_t$ = 0.70329–0.70349, $(^{206}Pb/^{204}Pb)_t$ = 18.369–18.517, $(^{207}Pb/^{204}Pb)_t$ = 15.509–15.528 and $(^{208}Pb/^{204}Pb)_t$ = 38.138–38.286. Samples from Site 1186 have $\varepsilon_{Nd}(t)$ = +6.0–+6.5, $(^{87}Sr/^{87}Sr)_t$ = 0.70342–0.70346, $(^{206}Pb/^{204}Pb)_t$ = 18.279–18.465, $(^{207}Pb/^{204}Pb)_t$ = 15.509–15.521 and $(^{208}Pb/^{204}Pb)_t$ = 38.180–38.244. Values for the lower basalt units at Site 1185 are +5.8–+5.9; 0.70352–0.70354; 18.390–18.398; 15.515–15.516 and 38.201–38.219, respectively.

Although the range of age-corrected isotopic values for all the Leg 192 and other Kwaimbaita-type basalt flows is small, subtle variations are apparent between and within sites, particularly in Sr isotopes (Fig. 3a). The $(^{87}Sr/^{86}Sr)_t$ values of lavas from Sites 1183, 1186 and the lower portion of Site 1185, together with the Kwaimbaita-type units at Site 807 and Site 289, are less than 0.7035. In contrast, Kwaimbaita-type lavas in

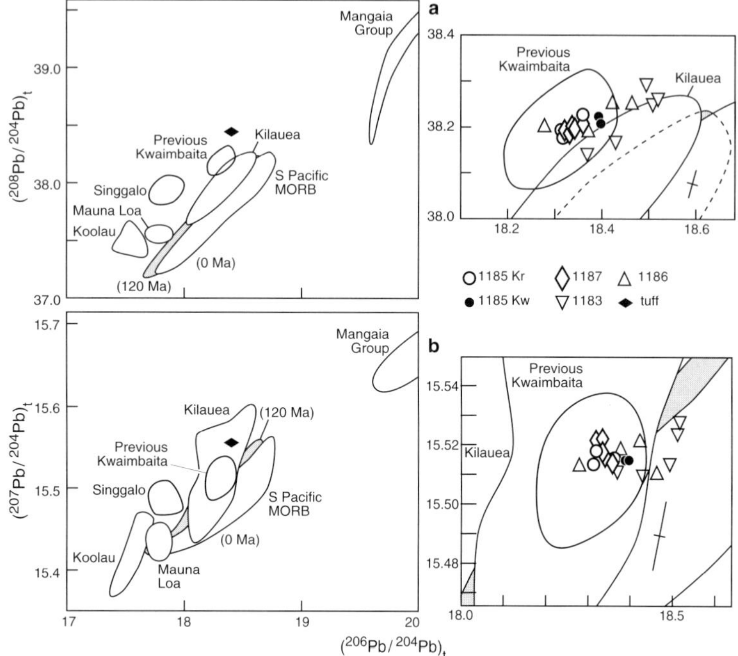

Fig. 4. (**a**) $(^{206}Pb/^{204}Pb)_t$ v. $(^{208}Pb/^{204}Pb)_t$ and (**b**) $(^{207}Pb/^{204}Pb)_t$ for the Leg 192 samples. Kw, Kwaimbaita type; Kr, Kroenke type. Panels on the right are expanded portions of those on the left. Fields for previous Kwaimbaita- and Singgalo-type basalt are for the pre-Leg 192 drill sites, Malaita and Santa Isabel, and glass from the Nauru and East Mariana basins. Note that for some of the previous data and for the Site 1183 and 1186 results, some of the range in age-corrected Pb isotope values probably reflects the determination of parent–daughter ratios by methods other than isotope dilution. In addition, for some previous samples and several in our data set, variable alteration of U/Pb ratios, in particular, has probably caused over- or under-corrections in initial Pb isotope ratios. See Figure 3 and Tejada *et al.* (2002) for data sources. Error bars are for measured values of Site 1183 and 1186 samples.

Malaita and Santa Isabel, the Site 803 basalts, and the Kroenke-type flows at Sites 1185 and 1187 have $(^{87}Sr/^{86}Sr)_t$ between 0.7036 and 0.7039. The Nauru Basin and East Mariana Basin data set straddles both ranges. Some of the higher values for these basins and the plateau proper no doubt reflect the effects of sea-water alteration not removed by acid-leaching, but values as high as 0.7038 are found in fresh glasses, indicating a small amount of real, pre-eruptive variability in Sr isotopes. The lowest value, 0.70329, is for sample 1183A-54-4 (2–6). This sample has a much higher Rb concentration (8.47 ppm) and $^{87}Rb/^{86}Sr$ value (0.169) than unaltered or slightly altered Kwaimbaita-type basalts (generally <2 ppm and <0.06, respectively), suggesting that alteration- or leaching-related disturbance of the Rb–Sr system may have caused its age-adjusted Sr isotope ratio to be over-corrected slightly. Three Site 1183 samples and one Site 1186 sample have slightly higher age-corrected $(^{206}Pb/^{204}Pb)_t$ (by up to 0.1) than seen previously, although their $(^{207}Pb/^{204}Pb)_t$ and $(^{208}Pb/^{204}Pb)_t$ values are within the previous range (Fig. 4). The reason for these small differences is not clear. They could represent pre-eruptive variability; however, disturbance of U/Pb ratios and, for these particular samples (and some of the pre-Leg 192 samples), measurement of U and Pb concentrations by methods other than isotope dilution probably account for a significant portion of the range in age-corrected $(^{206}Pb/^{204}Pb)_t$ values. Mis-corrections also may result when different splits of sample are used for concentration and Pb isotope measurements, as was the case for several pre-Leg 192 samples.

The only evidence of Singgalo-type magmatism at any of the Leg 192 sites is provided by the vitric tuff of Site 1183. The Singgalo magma type is characterized by higher $(^{87}Sr/^{86}Sr)_t$ and lower $\varepsilon_{Nd}(t)$ and $(^{206}Pb/^{204}Pb)_t$ than the Kwaimbaita type (Figs 3 and 4), and by slight relative enrichment in highly incompatible elements. The tuff's $\varepsilon_{Nd}(t)$ = +4.5, which is well within the Singgalo range. Although it is possible that the tuff could represent a volcanic source unrelated to the OJP, alteration-resistant incompatible elements also indicate that it belongs to the Singgalo magma type (Fitton & Godard 2004). However, its age-corrected ($^{87}Sr/^{87}Sr)_t$ is only 0.70254, much lower than values for either Singgalo or Kwaimbaita types. Likewise, its ($^{206}Pb/^{204}Pb)_t$ ratio (18.400) is well above the Singgalo range, and its ($^{207}Pb/^{204}Pb)_t$ (15.556) and ($^{208}Pb/^{204}Pb)_t$ (38.436) values are higher than for Singgalo- or Kwaimbaita-type basalts. The tuff is much more altered than any of the basalts, and these isotopic differences are probably a combination of the high level of alteration coupled with disturbance of the Rb–Sr system, and possibly the U–Pb and Th–Pb systems, by acid leaching during sample preparation. For example, relatively small changes in the high $^{87}Rb/^{86}Sr$ ratio (5.535) caused by leaching would lead to large under- or over- (as appears likely in this case) corrections in the calculated initial Sr isotope ratio of this sample.

Hf isotopes

Hafnium isotope ratios for all the Leg 192 basalts and the Kwaimbaita Formation sample from Malaita (SGB-18) fall within a limited range between $\varepsilon_{Hf}(t)$ = +10.1 and +13.3; however, values for all but two Leg 192 samples are within a much narrower range, from +10.7 to +12.2 (Table 1). As with Nd, Pb and Sr isotopes, the Kroenke-type lavas are indistinguishable from the Kwaimbaita type in their Hf isotope characteristics. The Singgalo Formation samples from Malaita possess slightly lower values of $\varepsilon_{Hf}(t)$, from +10.0 to +10.3.

The combined Hf–Nd isotope results place the OJP lavas within the field of OIB in Figure 5. However, relative to their $\varepsilon_{Nd}(t)$ values, the $\varepsilon_{Hf}(t)$ values of the Singgalo Formation samples are slightly higher (by approximately 1–2 ε units) than the corresponding values on a regression line fitted through the global Nd–Hf isotope array for oceanic basalts (Vervoort *et al.* 1999). Thus, the combined OJP data form an array with a slightly shallower slope than the mean slope for OIB globally, and in this respect, as in Figure 3, again appear broadly similar to the Kilauea and Mauna Loa shield volcanoes of Hawaii.

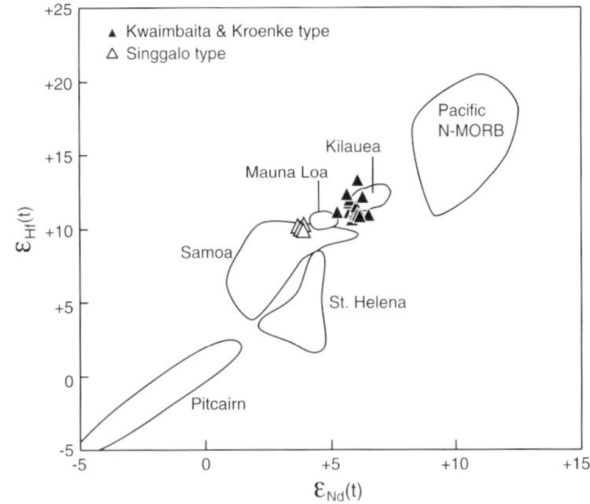

Fig. 5. $\varepsilon_{Hf}(t)$ v. $\varepsilon_{Nd}(t)$ for Ontong Java basalts. Fields for some Pacific and Atlantic oceanic island volcanoes and for high-precision data for Pacific N-MORB are shown for comparison. Data sources are Nowell *et al.* (1998), Salters & White (1998) and Chauvel & Blichert-Toft (2001) for the MORB field; Stracke *et al.* (1999), Blichert-Toft *et al.* (1999) and references therein for the Kilauea and Mauna Loa fields; Patchett & Tatsumoto (1980) and White & Hofmann (1982) for Samoa; Salters & White (1998) for St Helena; and Eisele *et al.* (2002) for Pitcairn. Note that, as in Figure 3, the Kwaimbaita- and Kroenke-type rocks are similar in their Hf–Nd isotope compositions to values reported for Kilauea, Hawaii, whereas the Singgalo-type compositions are closer to those of Mauna Loa, Hawaii.

Discussion

Was the OJP derived from a huge (almost) primitive mantle reservoir?

The small isotopic range defined by lavas from the OJP and surrounding basins is remarkable, given the great distances between sampling locations. In previous studies it was not clear to what extent the range for each magma type represents homogenization of more variable primary magma compositions by efficient mixing in extensive open-system magma chambers, or reflects an enormous, isotopically near-uniform mantle source (Tejada *et al.* 1996, 2002; Neal *et al.* 1997). The Singgalo magma type, although widespread and of considerable thickness in some OJP basement sections, now appears to have been volumetrically very minor relative to the Kwaimbaita type, at least during formation of the upper levels of basement crust. In considering the origin of the OJP, we therefore focus here on the Kwaimbaita mantle source.

To what extent was the Kwaimbaita source homogeneous? The Kroenke-type basalt, having lost only olivine by crystal fractionation (Fitton & Godard 2004; Sano & Yamashita 2004), is much closer to a primary magma composition than the Kwaimbaita type, which has lost substantial amounts of olivine, augite and plagioclase (e.g. Neal *et al.* 1997). Thus, the Kroenke-type isotopic signature cannot be a product of homogenization via open-system magmatic differentiation to any significant extent. That it is identical to the Kwaimbaita isotopic signature indicates that both groups of basalt were derived from the same mantle source and, although magma mixing may have played some role in damping isotopic variability, this source must indeed have been isotopically quite homogeneous relative to the scale of melting. The small amount of non-alteration-related site-to-site variability observed probably reflects small-scale heterogeneity present in the source region that, although largely averaged out at the high extents of partial melting involved, was not distributed perfectly uniformly.

Fitton & Godard (2004) show that Kwaimbaita-type chemical compositions can be derived by fractionation of olivine, plagioclase and augite from Kroenke-type magmas. For the Kroenke magma type, these authors estimate the percentage of partial melting to be approximately 30%. This value, which is at the high end of previous estimates for the Kwaimbaita magma type, is based on Zr content in primary Kroenke-type magma, the composition of which is calculated by incremental addition of equilibrium olivine to a Kroenke-type glass composition. In good agreement, a nearly identical value is obtained by Herzberg (2004) using a completely different method that relies on phase equilibria and major elements.

What might be the explanation for the particular combination of isotopic and incompatible-element characteristics in this mantle source? Specifically, the flat chondrite-normalized rare-earth patterns and, for all but the most incompatible elements, flat primitive-mantle-normalized element patterns of the Kwaimbaita- and Kroenke-type basalts (see Fitton & Godard 2004) point to a mantle source not too different from estimated primitive mantle in most of its inter-element ratios. However, the observed isotopic values (e.g. $\varepsilon_{Nd}(t)$ *c.* +6) are clearly far removed from those estimated for primitive mantle ($\varepsilon_{Nd} = 0$). Qualitatively similar characteristics are seen in some basalts from some other oceanic plateaus, but in those plateaus the range of isotopic and chemical variation is substantially greater than found for the OJP; the general explanation given is in terms of variable mixing of OIB mantle end-member types, with or without involvement of MORB-type mantle (e.g. Kerr *et al.* 2002). Contrary to early predictions (Mahoney 1987), no evidence for involvement of MORB-type mantle has yet been found in the OJP, including the Leg 192 basalts. Although mixing involving, for example, an EM-1-like (low $^{206}Pb/^{204}Pb$) end-member and anciently recycled oceanic lithosphere can explain the observed OJP isotopic ratios, it is not particularly supported by the lack of any discernible mixing arrays in the data, or by trace- or major-element modelling (Tejada *et al.* 2002).

An alternative possibility is that the Kwaimbaita source represents originally primitive mantle (i.e. originally of bulk silicate earth composition) that underwent minor fractionation long ago, after which it evolved isotopically in an essentially closed-system manner until the early Aptian. Indeed, simple two-stage evolution models assuming a fractionation event in the 3–4 Ga range can reproduce the Sr, Pb, Hf and Nd isotopic characteristics of the Kwaimbaita- and Kroenke-type basalts. An example is summarized in Table 3 and illustrated in Figure 6. In this case, the first stage of isotopic evolution occurs in a reservoir of primitive-mantle chemical and isotopic composition until about 3 Ga, when a fractionation event occurs that modifies the reservoir's chemical composition slightly. Subsequent closed-system isotopic evolution transpires until 120 Ma in this slightly modified mantle reservoir, which possesses parent–daughter ratios similar to those measured in the OJP basalts (i.e. we assume that because the OJP basalts represent high-degree partial melts, their parent–daughter ratios are not too different from those of their source). The *c.* 3 Ga fractionation event in this model is assumed to occur by removal of a small (1%) partial melt under upper-mantle conditions, leaving residual mantle (the future Kwaimbaita/Kroenke source) with the same normative phase proportions as those estimated for the Kwaimbaita-type source by Neal *et al.* (1997). In general agreement, Fitton & Godard (2004) show that the incompatible-element patterns of Kroenke- and Kwaimbaita-type basalts can be explained quite well by past removal of a similarly small amount of melt from primitive mantle. Such a modified primitive-mantle source also would be predicted to be depleted in volatiles and to be slightly enriched in highly compatible elements. Few data on volatiles in OJP basalt are available as yet, but those that exist show that water contents are indeed low (MORB-like or lower; Michael 1999; Roberge *et al.* 2004). Likewise, data for platinum-group elements suggest a source slightly enriched in highly siderophilic elements relative to most estimates of primitive-mantle composition (Chazey & Neal 2004).

Arguably, the most likely part of the planet in which a large volume of ancient, chemically near-primitive material might survive would be the lower mantle. (Note that the model

Table 3. *Parameters and results for two-stage model evolution of Kwaimbaita-type mantle*

	Start, stage 1	Start, stage 2	OJP eruption
Time (Ma)	4450	3050	120
Nd isotopes			
$^{147}Sm/^{144}Nd$	0.1967	0.2126	0.2126
OJP source $^{143}Nd/^{144}Nd$	0.506829	0.508675	0.512791
$\varepsilon_{Nd}(t)$ for OJP source	0.0	0.0	+6.0
Hf isotopes			
$^{176}Lu/^{176}Hf$	0.0332	0.0388	0.0388
OJP source $^{176}Hf/^{177}Hf$	0.279795	0.280759	0.283022
Sr isotopes			
$^{87}Rb/^{86}Sr$	0.0850	0.0538	0.0538
OJP source $^{87}Sr/^{86}Sr$	0.69956	0.70134	0.70363
Pb isotopes			
$^{238}U/^{204}Pb$	9	9.5	9.5
$^{232}Th/^{204}Pb$	36	36.575	36.575
OJP source $^{206}Pb/^{204}Pb$	9.311	12.792	18.361
OJP source $^{207}Pb/^{204}Pb$	10.301	14.210	15.522
OJP source $^{208}Pb/^{204}Pb$	29.476	32.478	38.218

Notes: Stage 1 source is estimated primitive mantle, and evolution to the beginning of Stage 2 is closed-system. Changes in parent–daughter ratios at 3.05 Ga (the start of Stage 2) are assumed to result from partial melting. Stage 2 evolution is also closed-system. Partial melting event at 3.05 Ga involves 1% batch melting, leaving behind a residue with normative phase proportions inferred for the OJP source (0.6 ol: 0.2 opx: 0.1 cpx: 0.1 gt; Neal *et al.* 1997) (ol, olivine; opx, orthopyroxene; cpx, clinopyroxene; gt, garnet). Phase proportions assumed entering the melt are 0.15 ol: 0.3 opx: 0.25 cpx: 0.3 gt (note that a sulphide-bearing residue would change the above Pb isotope results somewhat). Solid–liquid distribution coefficients used are from Kennedy *et al.* (1993) and the compilation of Green (1994). Starting isotope compositions are determined from meteorite (for Nd, Jacobsen & Wasserburg 1980; for Hf, Blichert-Toft & Albarède 1997; for Pb, Chen & Wasserburg 1983) and estimated original bulk earth (for Sr; McCulloch 1994) values.

calculations in Table 3 do not specify how such material would come to reside in the lower mantle, and the assumption of *c.* 3 Ga melting at upper-mantle depths is made simply because very little is known of solid–liquid distribution coefficients for lower-mantle minerals and conditions.) Recent seismological results, indeed, suggest the bottom *c.* 1000 km of today's mantle is chemically distinct, perhaps relatively primitive, separated from the rest of the mantle by a thermochemical interface with large undulations, and that most of the time little mixing occurs across the interface (e.g. Kellogg *et al.* 1999). Hypotheses involving an ultimately lower-mantle origin for the plateau are, of course, inevitably coupled to plume-head or mantle-overturn models.

Results of recent models of plume formation in the lower mantle are broadly consistent with the isotopic homogeneity exhibited by the voluminous Kwaimbaita (and Kroenke) magma type in that they suggest large plume heads should be well mixed, should entrain little non-plume mantle during their ascent and thus should be significantly more homogeneous than plume tails (e.g. Van Keken 1997; Farnetani *et al.* 2002). More generally, the OIB-like isotopic signature and the apparently rapid formation of the bulk of the OJP by high-degree partial melting are consistent with predictions of plume-head (plume-impact) models (e.g. Richards *et al.* 1989; Campbell & Griffiths 1990; Saunders *et al.* 1992); thus, this sort of model has been the type explored most commonly in previous attempts to understand the origin of the plateau (Mahoney *et al.* 1993; Tejada *et al.* 1996, 2002; Babbs 1997; Gladczenko *et al.* 1997; Neal *et al.* 1997).

However, these studies also noted significant discrepancies between observation and model. Tejada *et al.* (2002) recently emphasized several of the most important discrepancies, which include the following: (1) after an apparent eruptive hiatus of *c.* 30 Ma, the puzzling *c.* 90 Ma volcanic episode, which in several widespread locations produced tholeiitic basalts with

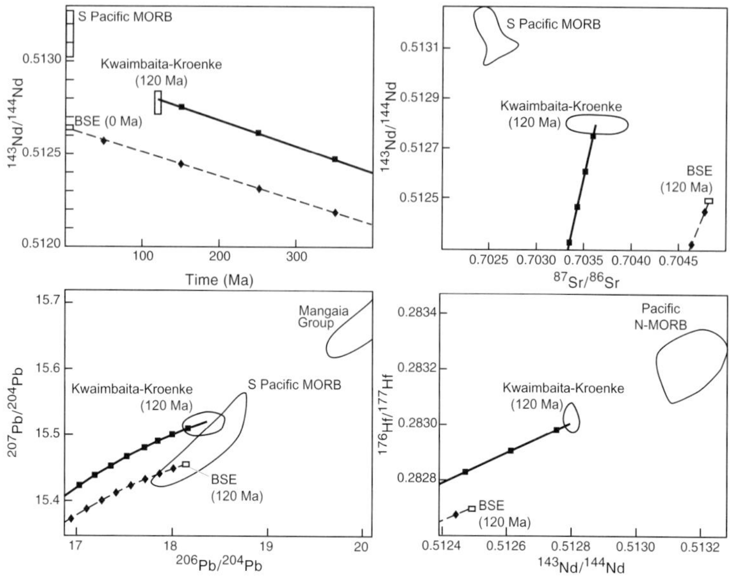

Fig. 6. Upper left panel: evolution of $^{143}Nd/^{144}Nd$ during the last several hundred million years in the two-stage model of Table 3. Other panels illustrate model evolution in Nd v. Sr isotope, $(^{206}Pb/^{204}Pb)_t$ v. $(^{207}Pb/^{204}Pb)_t$ and Nd v. Hf isotope space. Squares and diamonds on lines indicate successive 100 Ma increments since a fractionation event at 3.05 Ga. BSE is model bulk silicate Earth or primitive mantle.

isotopic and incompatible-element compositions closely similar to those of the *c.* 122 Ma Kwaimbaita-type magmas; (2) the lack of a post-plateau seamount chain corresponding to the plume-tail stage of hot-spot development theorized to follow the plume-head stage; (3) the lack of any presently active hot spot that can be linked unambiguously to the plateau; (4) post-eruptive subsidence of the plateau appears to have been much less than expected for oceanic lithosphere; and (5) all OJP lavas sampled thus far were erupted well below sea level, yet the standard form of the plume-head model (e.g. Richards *et al.* 1989; Campbell & Griffiths 1990) predicts that much of the surface should originally have been shallow or subaerial.

Some of these discrepancies may provisionally be accommodated by assuming a number of case-specific modifications to the standard plume-head model. (1) Following Leg 192, it now appears that the *c.* 90 Ma event was volumetrically minor (Shipboard Scientific Party 2001; L. Chambers pers. comm. 2003). It could represent plume-tail-related volcanism, assuming that the plateau did not drift much between 122 and 90 Ma (Neal *et al.* 1997), but additional assumptions are required to account for the apparent lack of any eruptive activity between 122 and 90 Ma. (2) The absence of a post-plateau chain of seamounts perhaps may be explained by appealing to a 'mid-mantle' plume of the type modelled by Cserepes & Yuen (2000). Alternatively, it can be explained by assuming a hypothetical jump or migration of a spreading centre near the OJP after the 90 Ma event, placing a different plate, now largely subducted, above the hot spot (Neal *et al.* 1997). Unfortunately, the precise relationship of the OJP to spreading centres in its vicinity at approximately 90 Ma is still largely conjectural (Gladczenko *et al.* 1997; Kroenke *et al.* 2004). (3) That the OJP cannot be linked clearly to a present-day hot spot could mean that the plume that produced the OJP no longer exists (e.g. Neal *et al.* 1997); alternatively, the problem could arise from large cumulative uncertainties in the amount of 120–0 Ma plume motion and true polar wander (Antretter *et al.* 2004). (4) Large amounts of intrusion and underplating during the 122–90 Ma period, and perhaps subsequently, have been postulated to explain the anomalous subsidence record of the plateau (Ito & Clift 1998; Ito & Taira 2000), but persistent lithospheric stress conditions that would prevent any significant accompanying volcanism during this *c.* 30 Ma period also must be assumed. (5) The lack of a large area initially at shallow depths may partly be explained by appealing to an eclogite-rich (Tejada *et al.* 2002) or, to a lesser extent, a volatile-rich plume head. Both would allow large-scale melting at lower plume temperatures than required for a dry or purely peridotitic plume head, and thus might lead to significantly less lithospheric uplift. Both seem unlikely, however, as OJP glasses point to a volatile-poor source (Michael 1999; Roberge *et al.* 2004) and chemical characteristics of OJP lavas are not matched well by assuming an eclogite-rich head, which additionally would require truly enormous amounts of eclogite to be concentrated in one deep-mantle location and then entrained within the plume head (Tejada *et al.* 2002). The amount of eclogite that plumes can carry is debated, but is likely to be rather small, in general (e.g. Gibson 2002). In any case, the high density of eclogite implies that eclogite-rich plume heads would have to be hotter, not cooler, than purely peridotitic ones in order to be sufficiently buoyant to rise through the mantle. For the OJP, Fitton & Godard (2004) point out that to produce the high-MgO Kroenke magma type from an eclogite-rich source would require approximately 100% fusion of the eclogite component and thus very high potential temperatures.

Plate separation and ridge-centred hot-spot hypotheses for the origin of the OJP

Proposed alternatives to the plume-head model also are problematic in the case of the OJP. The plate separation hypothesis (e.g. Anderson *et al.* 1992; Smith & Lewis 1999) posits an extensive layer of shallow, volatile-rich, near-solidus, OIB-like (but not plume-derived) asthenosphere ('perisphere') to have been residing beneath a region of the Pacific lithosphere that was rifted suddenly by a ridge jump around 122 Ma, causing cataclysmic melting. However, the pre-122 Ma seafloor within several hundred kilometres of the plateau is not much older than the plateau itself, having been formed only *c.* 2–35 Ma earlier (e.g. Taylor 1978; Nakanishi *et al.* 1992); thus, during this period, a spreading system was not too distant from the (future) location of the OJP. It is difficult to understand why such perisphere, assuming it existed, was not drained earlier by the nearby ridge. Also, the hypothesis predicts that normal MORB-type mantle lying just beneath the perisphere should have been tapped progressively more as OJP volcanism proceeded; yet, as noted above, no evidence of MORB-type mantle has been found thus far in OJP basalts. Indeed, the topmost part of the lava pile in several widespread locations is composed of the Singgalo-type basalts, which are even less MORB-like than the Kwaimbaita and Kroenke types. Also, as noted above, present indications are that the OJP's source mantle was poor in volatiles.

Similarly, the 'Icelandic' hypothesis that the OJP formed gradually, over several tens of millions of years, as the product of a large, approximately steady-state, ridge-centred plume (e.g. Mahoney & Spencer 1991; Ito & Clift 1998) is not supported by the isotopic and elemental homogeneity of OJP basalts over great distances, by the extensive Kwaimbaita-type basalts filling the Nauru and West Mariana basins, or by evidence suggesting that much of the plateau itself may have formed in an off-axis position (Coffin & Gahagan 1995). Also, in contrast to earlier plate-motion models, recent modelling suggests that the OJP drifted WNW as much as 2000 km in the 122–90 Ma period (Kroenke *et al.* 2004), implying that it could not have been situated above one hot spot the whole time. Furthermore, it now appears that the great bulk of the OJP formed rapidly in the Aptian (e.g. Tejada *et al.* 1996, 2002; Parkinson *et al.* 2001; L. Chambers pers. comm. 2003). A super-fast spreading rate would shorten the time needed for a ridge-centred hot spot to form the plateau, but the Aptian spreading rate in the vicinity of the OJP is unknown. However, for the Pacific–Phoenix ridge segments east of the plateau, a super-fast (*c.* 150 km Ma^{-1}) Barremian–late Hauterivian (*c.* 122–129 Ma) rate is indeed indicated by magnetic lineations M0–M7 (Larson 1997). What we can say at present is that the lack of any evidence for a long period of significant constructional volcanism appears to rule out the hypothesis as originally presented.

Meteorite impact instead of plume impact?

Although the isotopic and elemental characteristics of OJP lavas can be accommodated relatively well by plume-head models, we are impressed by the number of characteristics that are not explained satisfactorily by such models without the *ad hoc* postulation of special conditions. In the light of presently available evidence, we revisit a proposed alternative mechanism for the formation of oceanic plateaus by meteorite impact (Rogers 1982).

Noting that impact sites should be more numerous in the ocean than on land, Rogers (1982) proposed that the major Pacific plateaus represent massive outpourings of basalt formed by the cataclysmic excavation of asthenosphere by large, but rare, oceanic impacts. This hypothesis is attractive in that it can explain, without the special pleading required in the plume-head model, the absence of a post-plateau seamount chain and any obvious present-day hot spot that can be linked with the formation of the OJP (the same applies to several of the other Pacific plateaus). Also, the apparent lack of large areas at shallow-water depths during the construction of the OJP is not necessarily a problem, because inherently hotter-than-normal mantle is not required for widespread magmatism. Nor are huge amounts of eclogite- or volatile-rich mantle necessary. Post-volcanic subsidence likewise might be less than otherwise expected (although we question whether the pervasive serpentinization of shallow mantle suggested by Rogers (1982) is a viable mechanism for a feature the size of the OJP). Moreover, the limited range of elemental and isotopic variation among both the Singgalo- and Kwaimbaita-type basalts might be attributable in part to melt homogenization in extensive magma pools created by the impact. For additional discussion of the potential advantages of the impact hypothesis in explaining the OJP, we refer the reader to Ingle & Coffin (2004).

However, just as simple versions of the plume-head and other models for the OJP have significant shortcomings, the same appears to be true of the asteroid (or comet) impact hypothesis. Large-volume, high-degree partial melting of the upper few hundred kilometres of subseafloor mantle resulting from a large impact would ordinarily be expected to produce basalt with essentially N-MORB-type isotopic and incompatible-element characteristics. Although sampled in relatively few places, pre-OJP Pacific MORB are isotopically and chemically indistinguishable from modern Pacific MORB (Janney & Castillo 1997; Mahoney *et al.* 1998, J. Mahoney unpublished data). In contrast, the OJP is characterized by enormous volumes of basalt with OIB-like isotopic signatures and rather flat incompatible-element patterns, and there is as yet no evidence for any involvement of MORB-type mantle in the plateau (see above, and Tejada *et al.* 2002). Note also that neither chondritic, achondritic nor iron meteorites have a suitable combination of Sr–Pb–Nd isotopic characteristics (e.g. Kerridge & Matthews 1988 and references therein) to explain the isotopic signature of the OJP basalts via contamination of MORB-type mantle with meteoritic material. (Moreover, the isotopic and incompatible element contribution of an impacting body would be quite small in most plateau magmas, as most of the body's mass would be expelled in the impact ejecta, the volume of the object would be miniscule compared to that of the OJP, and meteoritic concentrations of most of these elements are low (e.g. Wolf *et al.* 1980; Schuraytze *et al.* 1996).)

Of course, just as with plume-head models, it is possible to appeal to special circumstances. For example, the impact site might have been above a geochemically anomalous region of asthenosphere containing a substantial amount of OIB-type mantle. Suitable isotopic compositions are notably rare among modern South Pacific hot spots and non-hot-spot volcanic areas, but it is conceivable that an impact fortuitously occurred near an area dominated by such material (cf. Ingle & Coffin 2004). Alternatively, as some authors have speculated, a large impact may actually trigger a deep-mantle plume beneath the impact site (e.g. Alt *et al.* 1988; Glikson 1999). In such a 'hybrid' scenario, it is not clear whether significant initial uplift of a plateau's surface would result or not.

Another potential difficulty for the impact hypothesis includes the reportedly systematic patterns in the gravity field and bathymetry of the OJP that Winterer & Nakanishi (1995), Neal *et al.* (1997) and Kroenke *et al.* (2004) have suggested represent a seafloor-spreading fabric. This interpretation remains to be evaluated rigorously but, if correct, is very difficult to reconcile with the expected widespread destruction and disruption of pre-existing oceanic lithosphere by a large impact, or with the short-lived outpouring of magma following an impact, which would be too rapid to allow formation of significant amounts of new seafloor.

To our knowledge, key features diagnostic of other large impact events, such as microspherules and siderophile element anomalies, have not been found in terrestrial or marine sediments around the Barremian–Aptian boundary. In contrast to the impact-linked Cretaceous–Tertiary boundary, no mass extinction occurred at the time of OJP formation. The statistical likelihood of an impact large enough to form the OJP can be estimated from cratering rates (e.g. Glikson 1999), but significant extrapolation is required. It is not clear that a sufficiently large object has been available in the Earth's vicinity in the last few hundred million years. The largest known Phanerozoic impact crater, the 65 Ma, approximately 200–250 km-wide crater at Chicxulub, Mexico, is thought to have been created by an object about 10 km in diameter (Grieve & Therriault 2000 and references therein). Although a large impact in relatively young, thin oceanic lithosphere would have different consequences than one on a continent (Rogers 1982; Glikson 1999; Jones *et al.* 2002), any object capable of creating the OJP must have been several times larger. However, no near-earth objects of such size are observed today (Binzel *et al.* 2002). Moreover, Venus, with a surface age estimated at approximately 600 Ma (e.g. Nimmo & McKenzie 1998), lacks any impact craters larger than Chicxulub (Schaber *et al.* 1992), whereas all lunar craters larger than about 100 km in diameter appear to be older than approximately 800 Ma (e.g. Eberhardt *et al.* 1973; Grier *et al.* 2001). Nevertheless, despite these potentially serious problems with the impact hypothesis, we regard it as deserving of further study in view of the difficulties encountered with any simple form of the plume-head or other proposed models for the OJP.

Conclusions

Age-corrected Nd, Pb and Sr isotope ratios of early Aptian basalt flows cored in four widely separated sites on the OJP during Leg 192, including the primitive Kroenke type, display only a small range of variation (e.g. $\varepsilon_{Nd}(t)$ = +5.8–+6.5, $(^{206}Pb/^{204}Pb)_t$ = 18.28–18.52, $(^{87}Sr/^{87}Sr)_t$ = 0.7033–0.7038). Moreover, all of the values fall within, or very near, the small field defined by the Kwaimbaita basalt type of the eastern Solomon Islands, previous OJP basement drill sites, and the adjacent Nauru and East Mariana basins. Age-corrected Hf isotope ratios display a correspondingly small range of variation ($\varepsilon_{Hf}(t)$ = +10.7–+12.2 for all but two samples), and the combined data indicate the Kwaimbaita/Kroenke-type mantle source was both immense and quite homogeneous relative to the scale of melting. Among the Leg 192 sites evidence for the Singgalo magma type, which forms lava piles lying stratigraphically above sections of Kwaimbaita-type basalt in the Solomon Islands and at Leg 130 Site 807, is confined to a thin interval of vitric tuff atop basement at Site 1183. It now seems apparent that the Singgalo magma type was a relatively minor component in the upper portions of crustal basement over much of the high plateau.

The isotopic characteristics of the Kwaimbaita/Kroenke mantle source can be modelled by simple two-stage evolution involving originally primitive mantle that underwent minor fractionation in the 3–4 Ga period, followed by closed-system radiogenic ingrowth until being tapped by plateau magmatism in the early Aptian. In the context of a plume-head model, such a source is compatible with recent geophysical evidence for a compositionally distinct lower-mantle layer, assuming this layer consists of chemically slightly modified primitive mantle. Although such a model can account for the isotopic and chemical characteristics of the OJP basalts and – with a large enough plume head – the sheer size of the plateau and the rapid formation of most of it in the early Aptian, a number of other first-order features require the assumption of *ad hoc* adjustments to the plume-head model. At least some of these appear unlikely or untenable on the basis of presently available evidence. However, alternative hypotheses, including an origin by asteroid impact, formation by plate separation, and gradual formation above a ridge-centred plume tail, also appear inadequate or require significant *ad hoc* modifications.

We thank S. Gibson and A. Saunders for helpful critical reviews, and R. Carmody, B. Cohen, N. Hulbirt, J. DeJong, C. MacIsaac, E. Scott, R. Solidum and K. Walda for help with various aspects of the work. We are grateful to our shipboard colleagues and the ODP and Transocean/Sedco-Forex staff of the *JOIDES Resolution* for making Leg 192 a success. Funding for the analyses at Scripps and SOEST was through USSSP grants and a SOEST YI award; the work in Brussels was funded by the Belgian FNRS and the Communauté Française de Belgique (ARC 98/03–233). This study used samples provided by ODP. ODP is sponsored by the National Science Foundation and participating countries under management of Joint Oceanographic Institutions, Inc. IL ms.3.03/7.03.

References

ALT, D., SEARS, J.W. & HYNDMAN, D.W. 1988. Terrestrial maria: the origins of large basalt plateaus, hotspot tracks, and spreading ridges. *Journal of Geology*, **96**, 647–662.

ANDERSON, D.L., ZHANG, Y.-S. & TANIMOTO, T. 1992. Plume heads, continental lithosphere, flood basalts and tomography. *In*: STOREY, B.C., ALABASTER, T. & PANKHURST, R.J. (eds) *Magmatism and the Causes of Continental Break-up*. Geological Society, London, Special Publications, **68**, 99–124.

ANDREWS, J.E., PACKHAM, G.H. *et al.* 1975. *Proceedings of the Deep Sea Drilling Project, Initial Reports*, **30**.

ANTRETTER, M., RIISAGER, P., HALL, S., ZHAO, X. & STEINBERGER, B. 2004. Modelled palaeolatitudes for the Louisville hot spot and the Ontong Java Plateau. *In*: FITTON, J.G., MAHONEY, J.J., WALLACE, P.J. & SAUNDERS, A.D. (eds) *Origin and Evolution of the Ontong Java Plateau*. Geological Society, London, Special Publications, **229**, 21–30.

BABBS, T.L. 1997. Geochemical and petrological investigations of the deeper portions of the Ontong Java Plateau: Malaita, Solomon Islands. PhD Thesis, University of Leicester.

BINZEL, R.P., LUPISHKO, D.F., DI MARTINO, M., WHITELEY, R.J. & HAHN, G.J. 2002. Physical properties of near-earth objects. *In*: BOTTKE, W., CELLINO, A., PAOLICCHOI, P. & BINZEL, R.P. (eds) *Asteroids III*. University of Arizona Press, Tucson, AZ, 255–271.

BIRKHOLD-VANDYKE, A.L., NEAL, C.R., JAIN, J.C., MAHONEY, J.J. & DUNCAN, R.A. 1996. Multi-stage growth for the Ontong Java Plateau? A progress report from San Cristobal (Makira). Abstract. *Eos, Transactions of the American Geophysical Union*, **77**, F714.

BLICHERT-TOFT, J. & ALBARÈDE, F. 1997. The Lu–Hf isotope geochemistry of chondrites and the evolution of the mantle–crust system. *Earth and Planetary Science Letters,* **148**, 243–258.

BLICHERT-TOFT, J., FREY, F.A. & ALBARÈDE, F. 1999. Hf isotope evidence for pelagic sediments in the source of Hawaiian basalts. *Science,* **285**, 879–882.

CAMPBELL, I.H. & GRIFFITHS, R.W. 1990. Implications of mantle plume structure for the evolution of flood basalts. *Earth and Planetary Science Letters,* **99**, 79–93.

CASTILLO, P.R., CARLSON, R.W. & BATIZA, R. 1991. Origin of Nauru Basin igneous complex: Sr, Nd, and Pb isotopes and REE constraints. *Earth and Planetary Science Letters,* **103**, 200–213.

CASTILLO, P.R., PRINGLE, M.S. & CARLSON, R.W. 1994. East Mariana Basin tholeiites: Jurassic ocean crust or Cretaceous rift basalts related to the Ontong Java plume? *Earth and Planetary Science Letters,* **123**, 139–154.

CHAUVEL, C. & BLICHERT-TOFT, J. 2001. A Hf isotope and trace element perspective on melting of the depleted mantle. *Earth and Planetary Science Letters,* **190**, 137–151.

CHAZEY, W.J., III & NEAL, C.R. 2004. Large igneous province magma petrogenesis from source to surface: platinum-group element evidence from Ontong Java Plateau basalts recovered during ODP Legs 130 and 192. *In*: FITTON, J.G., MAHONEY, J.J., WALLACE, P.J. & SAUNDERS, A.D. (eds) *Origin and Evolution of the Ontong Java Plateau.* Geological Society, London, Special Publications, **229**, 219–238.

CHEN, J.H. & WASSERBURG, G.J. 1983. The least radiogenic Pb in iron meteorites. *In*: *Fourteenth Lunar and Planetary Science Conference, Abstracts, Pt. 1.* Lunar and Planetary Institute, Houston, TX, 103–104.

COFFIN, M.F. & GAHAGAN, L.M. 1995. Ontong Java and Kerguelen plateaux: Cretaceous Icelands? *Journal of the Geological Society, London,* **152**, 1047–1052.

CSEREPES, L. & YUEN, D.A. 2000. On the possibility of a second kind of mantle plume. *Earth and Planetary Science Letters,* **183**, 61–71.

EBERHARDT, P., STETLER, A., GEISS, J. & GROSLER, N. 1973. How old is the lunar crater Copernicus? *Moon,* **8**, 1–04–114.

EISELE, J., SHARMA, M., GALER, S.J.G., BLICHERT-TOFT, J., DEVEY, C.W. & HOFMANN, A.W. 2002. The role of sediment recycling in EM-1 inferred from Os, Pb, Hf, Nd, Sr isotope and trace element systematics of the Pitcairn hotspot. *Earth and Planetary Science Letters,* **196**, 197–212.

FARNETANI, C.G., LEGRAS, B. & TACKLEY, P.J. 2002. Mixing and deformations in mantle plumes. *Earth and Planetary Science Letters,* **196**, 1–15.

FITTON, J.G. & GODARD, M. 2004. Origin and evolution of magmas on the Ontong Java Plateau. *In*: FITTON, J.G., MAHONEY, J.J., WALLACE, P.J. & SAUNDERS, A.D. (eds) *Origin and Evolution of the Ontong Java Plateau.* Geological Society, London, Special Publications, **229**, 151–178.

GALER, S.J.G. 1999. Optimal double and triple spiking for high precision lead isotopic measurement. *Chemical Geology,* **157**, 255–274.

GIBSON, S.A. 2002. Major element heterogeneity in Archean to Recent mantle plume starting-heads. *Earth and Planetary Science Letters,* **195**, 59–74.

GLADCZENKO, T.P., COFFIN, M.F. & ELDHOLM, O. 1997. Crustal structure of the Ontong Java Plateau: modeling of new gravity and existing seismic data. *Journal of Geophysical Research,* **102**, 22 711–22 729.

GLIKSON, A.Y. 1999. Oceanic mega-impacts and crustal evolution. *Geology,* **27**, 387–390.

GREEN, T.H. 1994. Experimental studies of trace-element partitioning applicable to igneous petrogenesis – Sedona 16 years later. *Chemical Geology,* **117**, 1–36.

GRIER, J.A., MCEWAN, A.S., LUCEY, P.G., MILAZZO, M. & STROM, R.G. 2001. Optical maturity of ejecta from large rayed lunar craters. *Journal of Geophysical Research,* **106**, E12, 32 847–32 862.

GRIEVE, R. & THERRIAULT, A. 2000. Vredefort, Sudbury, Chicxulub: three of a kind? *Annual Review of Earth and Planetary Sciences,* **28**, 305–338.

HAMILTON, W.B. 2003. The closed upper-mantle circulation of plate tectonics. *In*: STEIN, S. & FREYMUELLER, J.T. (eds) *Plate Boundary Zones.* American Geophysical Union, Geodynamics Monograph, **30**, 359–410.

HERZBERG, C. 2004. Partial melting below the Ontong Java Plateau. *In*: FITTON, J.G., MAHONEY, J.J., WALLACE, P.J. & SAUNDERS, A.D. (eds) *Origin and Evolution of the Ontong Java Plateau.* Geological Society, London, Special Publications, **229**, 179–183.

INGLE, S. & COFFIN, M.F. 2004. Impact origin for the greater Ontong Java Plateau? *Earth & Planetary Science Letters*, **218**, 123–134.

INGLE, S., WEIS, D., DOUCET, S. & MATTIELLI, N. 2004. Hf isotope constraints on mantle sources and shallow-level contaminants during Kerguelen hotspot activity since approximately 120 Ma. *Geochemistry, Geophysics, Geosystems*, in press.

ITO, G.T. & CLIFT, P.D. 1998. Subsidence and growth of Pacific Cretaceous plateaus. *Earth and Planetary Science Letters,* **161**, 85–100.

ITO, G.T. & TAIRA, A. 2000. Compensation of the Ontong Java Plateau by surface and subsurface loading. *Journal of Geophysical Research,* **105**, 11 171–11 183.

JACOBSEN, S.B. & WASSERBURG, G.J. 1980. Sm-Nd isotopic evolution of chondrites. *Earth and Planetary Science Letters*, **50**, 139–155.

JANNEY, P.E. & CASTILLO, P.R. 1997. Geochemistry of Mesozoic Pacific MORB: constraints on melt generation and the evolution of the Pacific upper mantle. *Journal of Geophysical Research,* **102**, 5207–5229.

JONES, A.P., PRICE, G.D., PRICE, N.J., DECARLI, P.S. & CLEGG, R.A. 2002. Impact induced melting and the development of large igneous provinces. *Earth and Planetary Science Letters*, **202**, 551–561.

KELLOGG, L.H., HAGER, B.H. & VAN DER HILST, R.D. 1999. Compositional stratification in the deep mantle. *Science,* **283**, 1881–1884.

KENNEDY, A.K., LOFGREN, G.E. & WASSERBURG, G.J.

1993. An experimental study of trace element partitioning between olivine, orthopyroxene, and melt in chondrules: equilibrium values and kinetic effects. *Earth and Planetary Science Letters,* **115**, 177–195.

KERR, A.C., TARNEY, J., KEMPTON, P.D., SPADEA, P., NIVIA, A., MARRINER, G.F. & DUNCAN, R.A. 2002. Pervasive mantle plume head heterogeneity: evidence from the Late Cretaceous Caribbean–Colombian oceanic plateau. *Journal of Geophysical Research,* **107**, 10 1029–10 1041.

KERRIDGE, J.F. & MATTHEWS, M.S. (eds). 1988. *Meteorites and the Early Solar System.* University of Arizona Press, Tucson, AZ.

KROENKE, L.W., BERGER, W., JANECEK, T.R. *et al.* 1991. *Proceedings of the Ocean Drilling Program, Initial Reports*, **130**.

KROENKE, L.W., WESSEL, P. & STERLING, A. 2004. Motion of the Ontong Java Plateau in the hot-spot frame of reference: 120 Ma–present. *In*: FITTON, J.G., MAHONEY, J.J., WALLACE, P.J. & SAUNDERS, A.D. (eds) *Origin and Evolution of the Ontong Java Plateau.* Geological Society, London, Special Publications, **229**, 9–20.

LARSON, R.L. 1991. Latest pulse of the earth: evidence for a mid-Cretaceous superplume. *Geology*, **19**, 547–550.

LARSON, R.L. 1997. Superplumes and ridge interactions between Ontong Java and Manihiki plateaus and the Nova-Canton Trough. *Geology,* **25**, 779–782.

MAHONEY, J.J. 1987. An isotopic survey of Pacific oceanic plateaus: implications for their nature and origin. *In*: KEATING, B., FRYER, P., BATIZA, R. & BOEHLERT, G. (eds) *Seamounts, Islands, and Atolls.* American Geophysical Union, Geophysical Monograph, **43**, 207–220.

MAHONEY, J.J. & SPENCER, K.J. 1991. Isotopic evidence for the origin of the Manihiki and Ontong Java oceanic plateaus. *Earth and Planetary Science Letters,* **104**, 196–210.

MAHONEY, J.J., STOREY, M., DUNCAN, R.A., SPENCER, K.J. & PRINGLE, M. 1993. Geochemistry and geochronology of the Ontong Java Plateau. *In*: PRINGLE, M., SAGER, W., SLITER, W. & STEIN, S. (eds) *The Mesozoic Pacific. Geology, Tectonics, and Volcanism.* American Geophysical Union, Geophysical Monograph, **77**, 233–261.

MAHONEY, J.J., FREI, R., TEJADA, M.L.G., MO, X.X., LEAT, P.T. & NÄGLER, T.F. 1998. Tracing the Indian Ocean mantle domain through time: isotopic results from old West Indian, East Tethyan, and South Pacific seafloor. *Journal of Petrology*, **39**, 1285–1306.

MAHONEY, J.J., FITTON, J.G., WALLACE, P.J. *et al.* 2001. *Proceedings of the Ocean Drilling Program, Initial Reports,* **192**. Available from World Wide Web: http://www-odp.tamu.edu/publications/192_IR/192ir.htm

MCCULLOCH, M.T. 1994. Primitive $^{87}Sr/^{86}Sr$ from an Archean barite and conjecture on the earth's age and origin. *Earth and Planetary Science Letters,* **126**, 1–13.

MICHAEL, P.J. 1999. Implications for magmatic processes at Ontong Java Plateau from volatile and major element contents of Cretaceous basalt glasses. *Geochemistry, Geophysics, Geosystems,* **1**, 1999GC000025.

MIURA, S., SUYEHIRO, K., SHINOHARA, M., TAKAHASHI, N., ARAKI, E. & TAIRA, A. 2004. Seismological structure and implications of double convergence and oceanic plateau collision of Ontong Java Plateau and Solomon Island arc from ocean bottom seismometer-airgun data. *Tectonophysics,* in press.

NAKANISHI, M., TAMAKI, K. & KOBAYASHI, K. 1992. Magnetic anomaly lineations from Late Jurassic to Early Cretaceous in the west-central Pacific Ocean. *Geophysical Journal International,* **109**, 701–719.

NEAL, C.R., MAHONEY, J.J., KROENKE, L.W., DUNCAN, R.A. & PETTERSON, M.G. 1997. The Ontong Java Plateau. *In*: MAHONEY, J.J. & COFFIN, M.F. (eds) *Large Igneous Provinces: Continental, Oceanic, and Planetary Flood Volcanism.* American Geophysical Union, Geophysical Monograph, **100**, 183–216.

NIMMO, F. & MCKENZIE, D. 1998. Volcanism and tectonics on Venus. *Annual Review of Earth and Planetary Sciences,* **26**, 23–51.

NOWELL, G.M., KEMPTON, P.D., NOBLE, S.R., FITTON, J.G., SAUNDERS, A.D., MAHONEY, J.J. & TAYLOR, R.N. 1998. High precision Hf isotope measurements of MORB and OIB by thermal ionisation mass spectrometry: insights into the depleted mantle. *Chemical Geology,* **149**, 211–233.

PARKINSON, I.J., ARCULUS, R.J. & DUNCAN, R.A. 1996. Geochemistry of Ontong Java Plateau basalt and gabbro sequences, Santa Isabel, Solomon Islands. *Eos, Transactions of the American Geophysical Union,* **77**, Abstract F715.

PARKINSON, I.J., SCHAEFER, B.F. & ODP LEG 192 SHIPBOARD SCIENTIFIC PARTY. 2001. A lower mantle origin for the world's biggest LIP? A high precision Os isotope isochron from Ontong Java Plateau basalts drilled on ODP Leg 192. Abstract V51C-1030. *Eos, Transactions of the American Geophysical Union,* **82**, F47.

PATCHETT, P.J. & TATSUMOTO, M. 1980. Hafnium isotope variations in oceanic basalts. *Geophysical Research Letters*, **7**, 1077–1080.

PETTERSON, M.G., BABBS, T., NEAL, C.R., MAHONEY, J.J., SAUNDERS, A.D., DUNCAN, R.A., TOLIA, D., MAGU, R., QOPOTO, C., MAHOA, H. & NATOGGA, D. 1999. Geological–tectonic framework of Solomon Islands, SW Pacific: crustal accretion and growth within an intra-oceanic setting. *Tectonophysics,* **301**, 35–60.

RICHARDS, M.A., DUNCAN, R.A. & COURTILLOT, V. 1989. Flood basalts and hot-spot tracks: plume heads and tails. *Science,* **246**, 103–107.

RICHARDSON, W.P., OKAL, E.A. & VAN DER LEE, S. 2000. Rayleigh-wave tomography of the Ontong Java Plateau. *Physics of the Earth and Planetary Interiors,* **118**, 29–61.

ROBERGE, J., WHITE, R.V. & WALLACE, P.J. 2004. Volatiles in submarine basaltic glasses from the Ontong Java Plateau (ODP Leg 192): implications for magmatic processes and source region compositions. *In*: FITTON, J.G., MAHONEY, J.J., WALLACE,

P.J. & SAUNDERS, A.D. (eds) *Origin and Evolution of the Ontong Java Plateau*. Geological Society, London, Special Publications, **229**, 239–257.

ROGERS, G.C. 1982. Oceanic plateaus as meteorite impact signatures. *Nature,* **299**, 341–342.

SALTERS, V.J.M. & WHITE, W.M. 1998. Hf isotope constraints on mantle evolution. *Chemical Geology,* **145**, 447–460.

SANO, T. & YAMASHITA, S. 2004. Experimental petrology of basement lavas from Ocean Drilling Program Leg 192: implications for differentiation processes of Ontong Java Plateau magmas. *In*: FITTON, J.G., MAHONEY, J.J., WALLACE, P.J. & SAUNDERS, A.D. (eds) *Origin and Evolution of the Ontong Java Plateau*. Geological Society, London, Special Publications, **229**, 185–218.

SAUNDERS, A.D., STOREY, M., KENT, R.W. & NORRY, M.J. 1992. Consequences of plume–lithosphere interactions. *In*: STOREY, B.C., ALABASTER, T. & PANKHURST, R.J. (eds) *Magmatism and the Causes of Continental Break-up*. Geological Society, London, Special Publications, **68**, 41–59.

SCHABER, G.G., STROM, R.G., MOORE, H.J., SODERBLOM, L.A., KIRK, R.L. *et al.* 1992. Geology and distribution of impact craters on Venus: what are they telling us? *Journal of Geophysical Research,* **97**, E8, 13 257–13 301.

SCHURAYTZE, B.C., LINDSTROM, D.J., MARIN, L.E., MARTINEZ, R.R., MITTLEFEHLDT, D.W. *et al.* 1996. Iridium metal in Chicxulub impact melt; forensic chemistry on the K–T smoking gun. *Science,* **276**, 1573–1576.

SHAFER, J., NEAL, C.R. & CASTILLO, P.C. 2004. Compositional variability in lavas from the Ontong Java Plateau: results from basalt clasts within the volcaniclastic succession at Ocean Drilling Program Leg 192 Site 1184. *In*: FITTON, J.G., MAHONEY, J.J., WALLACE, P.J. & SAUNDERS, A.D. (eds) *Origin and Evolution of the Ontong Java Plateau.* Geological Society, London, Special Publications, **229**, 333–351.

SHIPBOARD SCIENTIFC PARTY. 2001. Leg 192 summary. *In*: MAHONEY, J.J., FITTON, J.G., WALLACE, P.J. *et al. Proceedings of the Ocean Drilling Program, Initial Reports,* **192**, 1–75.

SMITH, A.D. & LEWIS, C. 1999. The planet beyond the plume hypothesis. *Earth Science Reviews,* **48**, 135–182.

SMITH, W.H.F. & SANDWELL, D.T. 1997. Global seafloor topography from satellite altimetry and ship depth soundings. *Science,* **277**, 1956–1962.

STRACKE, A., SALTERS, V.J.M. & SIMS, K.W.W. 1999. Assessing the presence of garnet–pyroxenite in the mantle sources of basalts through combined hafnium–neodymium–thorium isotope systematics. *Geochemistry, Geophysics, Geosystems,* **1**, 1999GC000013.

TAYLOR, B. 1978. Mesozoic magnetic anomalies in the Lyra Basin. *Eos, Transactions of the American Geophysical Union,* **59**, 320.

TEJADA, M.L.G. 1998. *Geochemical studies of Pacific oceanic plateaus: the Ontong Java Plateau and Shatsky Rise*. PhD Thesis, University of Hawaii.

TEJADA, M.L.G., MAHONEY, J.J., DUNCAN, R.A. & HAWKINS, M.P. 1996. Age and geochemistry of basement and alkalic rocks of Malaita and Santa Isabel, Solomon Islands, southern margin of Ontong Java Plateau. *Journal of Petrology,* **37**, 361–394.

TEJADA, M.L.G., MAHONEY, J.J., NEAL, C.R., DUNCAN, R.A. & PETTERSON, M.G. 2002. Basement geochemistry and geochronology of Central Malaita, Solomon Islands, with implications for the origin and evolution of the Ontong Java Plateau. *Journal of Petrology,* **43**, 449–484.

TODT, W., CLIFF, R.A., HANSER, A. & HOFFMAN, A.W. 1996. Evaluation of a ^{202}Pb–^{205}Pb double spike for high-precision lead isotope analysis. *In*: BASU, A. & HART, S. (eds) *Earth Processes: Reading the Isotopic Code.* American Geophysical Union, Geophysical Monograph, **95**, 429–437.

VAN KEKEN, P. 1997. Evolution of starting mantle plumes: a comparison between numerical and laboratory models. *Earth and Planetary Science Letters,* **148**, 1–11.

VERVOORT, J.D., PATCHETT, P.J., BLICHERT-TOFT, J. & ALBARÈDE, F. 1999. Relationships between Lu–Hf and Sm–Nd isotopic systems in the global sedimentary system. *Earth and Planetary Science Letters,* **168**, 79–99.

WHITE, R.V., CASTILLO, P.R., NEAL, C.R., FITTON, J.G. & GODARD, M.M. 2004. Phreatomagmatic eruptions on the Ontong Java Plateau: chemical and isotopic relationship to Ontong Java Plateau basalts. *In*: FITTON, J.G., MAHONEY, J.J., WALLACE, P.J. & SAUNDERS, A.D. (eds) *Origin and Evolution of the Ontong Java Plateau.* Geological Society, London, Special Publications, **229**, 307–323.

WHITE, W.M. & HOFMANN, A.W. 1982. Sr and Nd isotope geochemistry of oceanic basalts and mantle evolution. *Nature,* **296**, 821–825.

WINTERER, E.L. & NAKANISHI, M. 1995. Evidence for a plume-augmented, abandoned spreading center on Ontong Java Plateau. Abstract. *Eos, Transactions of the American Geophysical Union,* **76**, F617.

WOLF, R., WOODROW, A.B. & GRIEVE, R.A.F. 1980. Meteoritic material at four Canadian impact craters. *Geochimica et Cosmochimica Acta,* **44**, 1015–1022.

Origin and evolution of magmas on the Ontong Java Plateau

J. GODFREY FITTON[1] & MARGUERITE GODARD[2]

[1]*School of GeoSciences, University of Edinburgh, Grant Institute, West Mains Road, Edinburgh EH9 3JW, UK (e-mail: Godfrey. Fitton@ed.ac.uk)*

[2]*Laboratoire de Tectonophysique – CNRS UMR 5568, ISTEEM, Université de Montpellier 2, Place Eugéne Bataillon, F-34095 Montpellier Cedex 5, France (e-mail: Marguerite.Godard@dstu.univ-montp2.fr)*

Abstract: The Early Cretaceous Ontong Java Plateau (OJP) represents by far the largest igneous event on Earth in the last 200 Ma and yet, despite its size, the OJP's basaltic crust appears to be remarkably homogeneous in composition. The most abundant rock type is a uniform low-K tholeiite, represented by the Kwaimbaita Formation on Malaita and found at all but one of the Deep Sea Drilling Project (DSDP) and Ocean Drilling Program (ODP) drill sites on the plateau and in the adjacent basins. This is capped by a thin and geographically restricted veneer of a slightly more incompatible-element-rich tholeiite (the Singgalo Formation on Malaita and the upper flow unit at ODP Site 807), distinguished from Kwaimbaita-type basalt by small but significant differences in Sr-, Nd- and Pb-isotope ratios. A third magma type is represented by high-Mg (Kroenke-type) basalt found in thick (>100 m) successions of lava flows at two drill sites (ODP Sites 1185 and 1187) 146 km apart on the eastern flank of the plateau. The high-Mg basalt is isotopically indistinguishable from Kwaimbaita-type basalt and may therefore represent the parental magma for the bulk of the OJP. Low-pressure fractional crystallization of olivine followed by olivine+augite+plagioclase can explain the compositional range from high-Mg Kroenke-type to Kwaimbaita-type basalt. The Singgalo-type basalt probably represents slightly smaller-degree, late-stage melting of an isotopically distinct component in the mantle source. Primary magma compositions, calculated by incremental addition of equilibrium olivine to aphyric Kroenke-type basalt glass, contain between 15.6% (in equilibrium with Fo_{90}) and 20.4% (Fo_{92}) MgO. Incompatible-element abundances in the primary OJP magma can be modelled by around 30% melting of a peridotitic primitive-mantle source from which about 1% by mass of average continental crust had previously been extracted. This large degree of melting implies decompression of very hot (potential temperature >1500°C) mantle beneath very thin lithosphere. The initiation of an exceptionally large and hot plume head close to a mid-ocean ridge provides the best explanation for the size, homogeneity and composition of the OJP, but is difficult to reconcile with the submarine eruption of virtually all of the basalt so far sampled.

The Ontong Java Plateau (OJP) is the largest of the Earth's large igneous provinces (Coffin & Eldholm 1994). The plateau, defined mostly by the 4000-m bathymetric contour (Fig. 1), covers an area of 2.0×10^6 km^2 (comparable in size with western Europe), but OJP-related volcanism extends over a considerably larger area into the adjacent Nauru and East Mariana basins. With an average thickness of crust beneath the plateau of 30–35 km (Gladczenko *et al.* 1997; Richardson *et al.* 2000), the volume of igneous rock forming the plateau may be as high as 6×10^7 km^3 (Coffin & Eldholm 1994).

Collision with the Solomon arc has resulted in folding and uplift of the southern margin of the OJP in the last 6 Ma (e.g. Kroenke *et al.* 2004). Thick (up to about 3.5 km) sections of basaltic rocks are exposed on land in the Solomon Islands, notably in Malaita, Santa Isabel and San Cristobal (e.g. Petterson *et al.* 1999; Petterson 2004). Prior to ODP Leg 192 (September–November 2000) only three drill sites on the plateau (Deep Sea Drilling Project (DSDP) Site 289 and Ocean Drilling Program (ODP) Sites 803 and 807) had penetrated basaltic basement. Five more basement sites were drilled during ODP Leg 192; basaltic lava flows were sampled at four of these (Sites 1183, 1185, 1186 and 1187) and a fifth (Site 1184) penetrated 338 m into a volcaniclastic sequence. In addition to these eight plateau sites, basaltic basement with the same composition as basalt from the OJP has been penetrated at DSDP Site 462 in the Nauru Basin and at ODP Site 802 in the East Mariana Basin. The locations of all DSDP and ODP drill sites on and around the OJP are shown in Figure 1, and stratigraphic sections are given in Figure 2.

From: FITTON, J. G., MAHONEY, J. J., WALLACE, P. J. & SAUNDERS, A. D. (eds) 2004. *Origin and Evolution of the Ontong Java Plateau*. Geological Society, London, Special Publications, **229**, 151–178. 0305-8719/$15.00

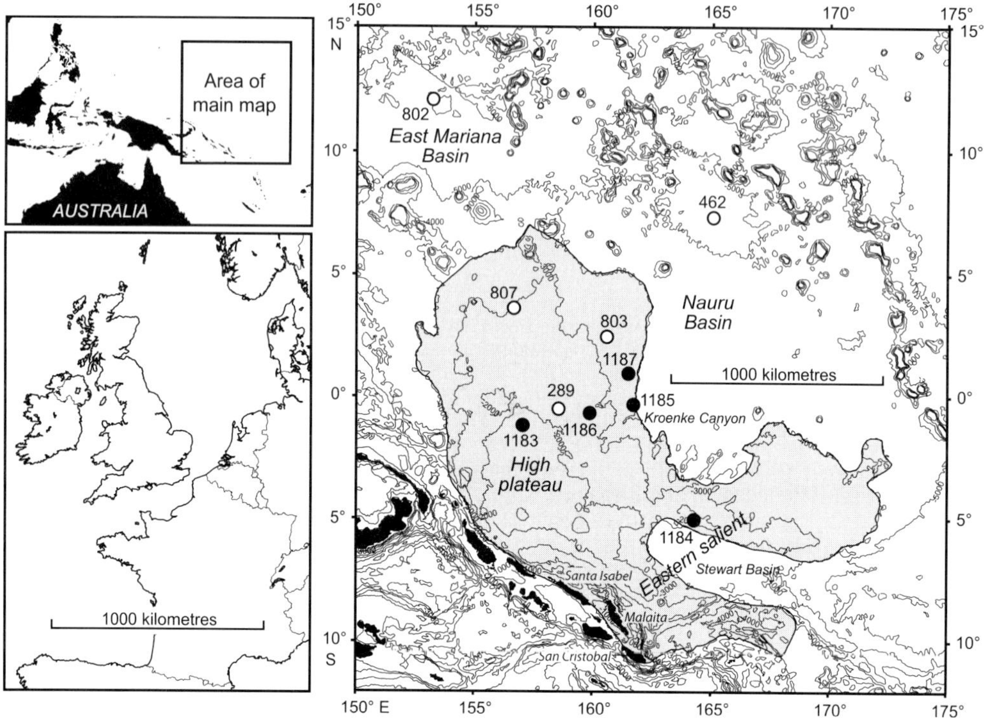

Fig. 1. Predicted bathymetry (after Smith & Sandwell 1997) of the OJP and surrounding areas showing the location of DSDP and ODP basement drill sites. Leg 192 drill sites are marked by black circles; open circles represent pre-Leg 192 drill sites. The edge of the plateau is defined by the –4000 m-contour, except in the SE part where it has been uplifted through collision with the Solomon arc. A map of part of western Europe, drawn to the same scale and on the same projection, is shown for comparison.

Published $^{40}Ar/^{39}Ar$ data (Mahoney *et al.* 1993; Tejada *et al.* 1996, 2002) suggest a major episode of OJP volcanism at approximaely 122 Ma and a minor episode at about 90 Ma. $^{40}Ar/^{39}Ar$ analysis of samples from ODP Leg 192 Sites 1186 and 1187 (Chambers *et al.* 2002; L. M. Chambers pers. comm) gives ages ranging from 105 to 122 Ma. These authors argue that the younger apparent ages (and, by implication, the data on which the 90 Ma episode is based) are the result of argon-recoil and therefore represent minimum ages. Biostratigraphic dating (Bergen 2004; Sikora & Bergen 2004) of sediment intercalated with lava flows at ODP Sites 1183, 1185, 1186 and 1187 suggests that magmatism on the high plateau extended from latest Early Aptian to latest Aptian. This corresponds to age ranges of 122–112 Ma (Harland *et al.* 1990) or 118–112 Ma (Gradstein *et al.* 1995). However, Re–Os isotopic data on basalt samples from these same four drill sites define a single isochron with an age of 121.5 ± 1.7 Ma (Parkinson *et al.* 2002). The eruption age of the volcaniclastic succession at Site 1184, on the eastern salient of the OJP, has proved equally enigmatic. Rare nannofossils suggest an Eocene age (Bergen 2004), but $^{40}Ar/^{39}Ar$ analysis of plagioclase crystals (Chambers *et al.* 2004) suggests an age of 123.5 ± 1.8 Ma, consistent with the steep (–54°) palaeomagnetic inclination (Riisager *et al.* 2004).

Because of the uncertainties in the various dating methods, there is no clear consensus on the age and duration of OJP magmatism. The plateau could have formed in a single, short-lived event at about 122 Ma, or over a period of 10 Ma or longer. With magmatism extending over a 10 Ma interval the average rate of magma production would have been 6 km^3 $year^{-1}$. If, however, the bulk of the plateau was emplaced over a significantly shorter interval than this then the peak OJP magma production rate may well have exceeded the global mid-ocean ridge magma production rate at the time (e.g. Coffin & Eldholm 1994).

With the exception of the volcaniclastic succession at Site 1184, the basaltic basement

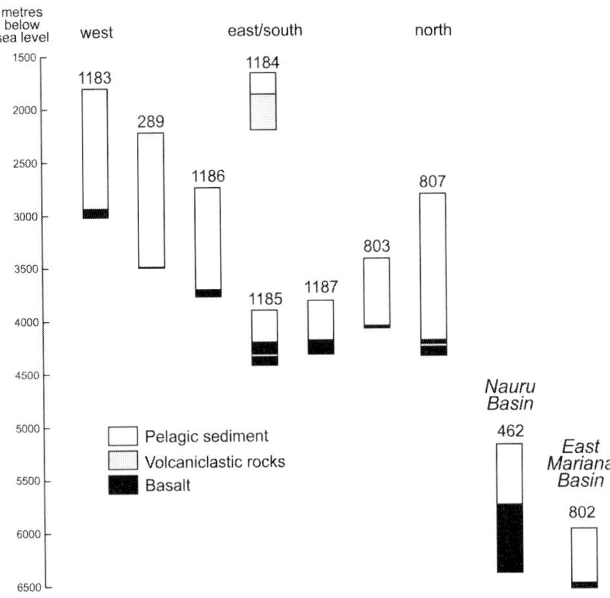

Fig. 2. Stratigraphic sections drilled at the 10 DSDP and ODP drill sites marked on Figure 1. Seven of the OJP sites are arranged on a transect from the crest of the plateau (Site 1183) eastward to the plateau rim (Site 1185), and then north and NW to Site 807 on the northern flank. Site 1184 lies off the transect, 586 km to the SE of Site 1185 on the eastern salient of the OJP. The white lines in the basement at Sites 807 and 1185 represent compositional breaks in the basaltic successions at these two sites. Basement penetration and data sources: DSDP Site 289 (9 m), Andrews *et al.* (1975); Site 462 (640 m), Larson *et al.* (1981) and Moberley *et al.* (1986). ODP Site 802 (51 m), Lancelot *et al.* (1990); ODP Sites 803 (26 m) and 807 (149 m), Kroenke *et al.* (1991); ODP Sites 1183 (81 m), 1184 (338 m of volcaniclastic rocks), 1185 (217 m), 1186 (65 m) and 1187 (136 m), Mahoney *et al.* (2001).

drilled on the OJP and in the adjacent basins, and exposed in the Solomon Islands, consists almost entirely of pillow lavas and submarine sheet flows. Basaltic units in the upper part of the Nauru Basin succession drilled at Site 462 were originally described as sills (Larson *et al.* 1981), but Saunders (1986) has argued that these are thick sheet flows. The only evidence for subaerial or shallow-water volcanism during emplacement of the OJP is provided by the thick volcaniclastic succession at Site 1184 (Thordarson 2004; White *et al.* 2004), two thin intervals of vitric tuff in the Aptian limestone immediately above basement at Site 1183 (Mahoney *et al.* 2001) and a vitric tuff just above basement at Site 289. There is no evidence for subaerial eruption of lava flows, and the CO_2 contents of basaltic glass (Michael 1999; Roberge *et al.* 2004) provide convincing evidence for relatively deep submarine eruption at all the drill sites on and around the OJP. For example, Roberge *et al.* (2004) estimate eruption depths of approximately 1000 m at Site 1183 on the crest of the plateau and approximately 2500 m at Site 1187 on its eastern margin (Fig. 1).

Several previous petrological and geochemical studies (notably those of Mahoney *et al.* 1993; Tejada *et al.* 1996, 2002; Neal *et al.* 1997) have highlighted the remarkable uniformity of the tholeiitic basalt forming the OJP. These studies were based on rock samples collected from the Solomon Islands and recovered from the three pre-Leg 192 drill sites. In this paper, we present a comprehensive major- and trace-element data set for basaltic rocks recovered during ODP Leg 192, from previous ODP and DSDP drill sites on the OJP, and from Site 462 in the Nauru Basin. We will use these data, in conjunction with isotopic data presented by Tejada *et al.* (2004), to deduce the nature of the OJP mantle source, the temperature and degree of melting of the source, and the extent of subsequent fractional crystallization of the primitive OJP magmas.

Ontong Java Plateau magma types

Mahoney *et al.* (1993) showed that the basement section drilled at ODP Site 807 consisted of two chemically and isotopically distinct parts; an upper part (unit A) and a lower part (units C–G) separated by a thin sediment layer (unit B). The basaltic lava flows forming unit A have a small relative enrichment in the more incompatible elements compared with those from units C–G, and also have slightly lower $^{206}Pb/^{204}Pb$ and $^{143}Nd/^{144}Nd$, and higher $^{87}Sr/^{86}Sr$. Basalt from Site 803 is indistinguishable from units C–G at Site 807. Two similarly distinct groups of basalt have been recognized by Tejada *et al.* (1996) and Neal *et al.* (1997) in the uplifted SE edge of the OJP exposed on Malaita and Santa Isabel in the Solomon Islands (Fig. 1). OJP basalt on Malaita can be divided into two compositionally distinct stratigraphic units: the Kwaimbaita Formation (>2.7 km thick) and the overlying Singgalo Formation (*c.* 750 m thick) (Tejada *et al.* 2002). Basalt of the Kwaimbaita Formation is chemically and isotopically similar to basalt at Site 807 (units C–G) and Site 803, while the Singgalo Formation is similar to unit A at Site 807. Thus, Kwaimbaita- and Singgalo-type basalt flows with the same stratigraphic relationship are found at sites 1500 km apart on the plateau (Tejada *et al.* 2002). Kwaimbaita-type basalt appears to represent the dominant OJP magma type because it has been found at all but one of the OJP drill sites (Tejada *et al.* 2004). Singgalo-type basalt, on the other hand, appears to be volumetrically minor.

A third OJP basalt type was recognized during ODP Leg 192. The 136 m-basement section at Site 1187 and the upper 125 m at Site 1185 are composed of basalt with higher MgO and lower concentrations of incompatible elements than any previously reported from the OJP (Mahoney *et al.* 2001). This basalt is isotopically identical to Kwaimbaita-type basalt (Tejada *et al.* 2004) and may represent the parental magma for the bulk of the OJP. We propose the term *Kroenke-type basalt* because it was discovered on the flanks of the submarine Kroenke Canyon at Site 1185 (Fig. 1).

To summarize, we now recognize three basalt types on the OJP.

- *Kwaimbaita-type* basalt is named after Kwaimbaita River in Malaita. It is by far the most abundant basalt type on the OJP; found in Malaita and Santa Isabel, and at DSDP Sites 289 and 462, and ODP Sites 803, 807, 1183, 1185 and 1186. It has typically 6–8 wt% MgO.
- *Kroenke-type* basalt is named after the submarine Kroenke Canyon on the east side of the OJP. It is found at ODP Site 1187 and overlying Kwaimbaita-type basalt at ODP Site 1185. It is magnesian (up to 11 wt% MgO) and isotopically indistinguishable from Kwaimbaita-type basalt. Kroenke-type magma may have been parental to the dominant Kwaimbaita type.
- *Singgalo-type* basalt is named after the Singgalo River in Malaita. It forms a thin late-stage veneer over the western and southern parts of the plateau; overlies Kwaimbaita-type basalt in Malaita, Santa Isabel and at ODP Site 807 (unit A). It has a similar range of MgO to Kwaimbaita-type basalt, but a small relative enrichment in the more incompatible elements. Its $^{206}Pb/^{204}Pb$ and $^{143}Nd/^{144}Nd$ are lower, and $^{87}Sr/^{86}Sr$ higher, than Kwaimbaita-type basalt.

Analytical methods

A total of 102 samples were analysed in the present study. Of these, 72 were from samples collected during ODP Leg 192 (Sites 1183–1187), 18 from previous drill sites on the plateau (DSDP Site 289; ODP Sites 803 and 807) and 12 from the Nauru Basin (DSDP Site 462). The samples were trimmed with a diamond-impregnated saw to remove veins and altered zones, washed, dried and then ground in an agate-lined barrel on a Tema mill. The powders were analysed by X-ray fluorescence (XRF) spectrometry in Edinburgh for major elements and some trace elements, and by inductively coupled plasma-source mass spectrometry (ICP-MS) at ISTEEM, Montpellier for a larger range of trace elements.

The XRF techniques used are essentially similar to those described by Fitton *et al.* (1998), except that a Philips PW2404 automatic X-ray spectrometer was used, and the analytical conditions for the determination of Nb and Zr were different. Because the background around the wavelength of NbKα is curved, linear interpolation between two background positions can introduce significant error at low concentrations. Instead, the background correction was calculated by fitting a third-order polynomial to background count rates measured at four positions, two either side of the peak. Long counting times (500 s at the peak and 500 s in total at the background positions) were used in order to improve precision in the determination of Nb. Each sample was analysed three times and the average value taken. A LiF220 analysing crystal was used in the determination of Zr to give better wavelength dispersion and hence reduce overlap of the SrKβ peak on ZrKα.

Table 1. *Average values obtained for international standards BIR-1 and BHVO–1*

	BIR-1			BHVO–1		
	This study			This study		
	Mean (ppm)	RSD %	GSNL (ppm)	Mean (ppm)	RSD %	GSNL (ppm)
XRF (ppm)*						
n	*6*			*6*		
Rb	0.25	32.1	0.25	9.1	1.6	11.0
Sr	106.8	0.2	108.0	398.5	0.3	403.0
Y	16.1	0.5	16.0	27.5	0.8	27.6
Zr	16.1	0.9	15.5	175.5	0.3	179.0
Nb	0.60	16.2	0.6	19.8	0.7	19.0
ICP-MS (ppm)						
n	*21*			*17*		
Li	2.8	7.3	3.4	4.2	7.5	4.6
Sc	41.3	4.3	44.0	29.8	6.3	31.8
Ti	5539.0	13.0	5755.0	15 271.0	15.3	16 246.0
Co	51.2	2.2	51.4	45.0	5.1	45.0
Cu	120.9	3.7	126.0	144.4	5.9	136.0
Rb	0.2	6.2	0.25	9.19	3.9	11.0
Sr	99.8	4.8	108.0	366.0	7.3	403.0
Y	16.3	5.0	16.0	27.9	3.4	27.6
Zr	14.5	4.2	15.5	171.8	4.6	179.0
Nb	0.58	3.5	0.6	19.15	3.0	19.0
Cs	0.007	31.5		0.098	6.3	0.13
Ba	6.78	8.5	7.0	135.7	4.2	139.0
La	0.63	7.6	0.62	15.63	4.0	15.8
Ce	1.91	4.5	1.95	39.08	3.8	39.0
Pr	0.37	5.4	0.38	5.37	4.1	5.7
Nd	2.41	4.7	2.5	25.35	4.0	25.2
Sm	1.07	4.6	1.1	6.03	4.0	6.2
Eu	0.53	4.6	0.54	2.15	4.6	2.06
Gd	1.86	6.0	1.85	6.22	6.0	6.4
Tb	0.36	4.8	0.36	0.94	4.5	0.96
Dy	2.64	6.6	2.5	5.43	7.3	5.2
Ho	0.59	3.8	0.57	1.02	3.6	0.99
Er	1.71	5.1	1.7	2.55	4.5	2.4
Tm	0.25	4.7	0.26	0.34	4.6	0.33
Yb	1.63	4.9	1.65	1.95	4.9	2.02
Lu	0.27	5.4	0.26	0.29	4.4	0.29
Hf	0.6	5.2	0.6	4.57	4.7	4.38
Ta	0.041	6.6	0.04	1.27	6.4	1.23
Th	0.031	6.9	0.03	1.222	4.3	1.08
U	0.011	13.0	0.01	0.402	5.6	0.42

* Precision estimates for other elements by XRF are given in Fitton *et al.* (1998). GSNL, values given in Govindaraju (1994); RSD (%), per cent relative standard deviation; *n*, number of analyses.

ICP-MS trace-element concentrations were determined on a VG-PQ2 Turbo+ spectrometer. Rare-earth elements, Cs, Rb, Ba, Th, U, Zr, Hf and Y were determined following the method described by Ionov *et al.* (1992), and Nb and Ta by surrogate calibration following the procedure described in Godard *et al.* (2000).

Precision estimates obtained during this study are given in Table 1, and additional XRF precision data are given in Fitton *et al.* (1998). Analytical data for the suite of OJP and Nauru Basin basalt samples are given in Tables 2–4. Two sets of data are given for elements determined by both ICP-MS and XRF, and the set used is noted in appropriate figure captions.

Results

For clarity when displaying the large amount of data in Tables 2–4, we have divided the data into

Table 2. *Analyses of basalt from the Ontong Java Plateau (OPD Leg 192)*

Sample	1183-1	1183-2	1183-3	1183-4	1183-5	1183-6	1183-7	1183-8	1183-9	1183-10	1186-1	1186-2	1186-3	1186-4	1186-5
Site, hole	1183A	1183A	1183A	1183A	1183A	1183A	1183A	1183A	1183A	1183A	1186A	1186A	1186A	1186A	1186A
Core-section	54R-3	54R-3	54R-3	54R-4	55R-1	55R-3	59R-1	61R-2	66R-2	67R-3	32R-2	34R-1	34R-4	38R-1	39R-5
Interval (cm)	28–34	68–72	145–149	81–85	78–82	124–129	30–35	7–12	79–84	76–81	85–90	78–84	51–55	40–46	45–50
Depth (mbsf)	1129.45	1129.85	1130.62	1131.48	1137.28	1140.38	1160.90	1177.55	1202.29	1208.60	978.52	986.58	990.70	1019.80	1029.86
Subunit	IIIB (tuff)	IIIB (tuff)	1	2B	3B	4B	5B	6	7	8	1	2	3	4	4
Magma type	Sg	Sg	Kw	Kw	Kw	Kw	Kw	Kw	Kw	Kw	Kw	Kw	Kw	Kw	Kw
XRF (wt%)															
SiO_2	55.14	57.24	50.16	50.08	50.38	50.10	49.92	50.21	49.47	49.71	49.78	49.94	49.83	49.56	49.49
Al_2O_3	12.07	12.50	15.34	14.64	15.13	15.14	14.17	14.13	14.05	14.17	14.34	14.20	15.45	14.12	14.01
Fe_2O_3T	11.51	10.09	11.19	11.15	9.56	10.15	12.07	12.36	12.26	12.25	12.36	12.36	9.60	12.21	12.23
MnO	0.069	0.033	0.136	0.165	0.161	0.187	0.192	0.184	0.193	0.199	0.207	0.196	0.224	0.198	0.191
MgO	4.94	3.59	6.83	7.65	8.56	8.04	7.59	7.43	7.64	7.76	7.34	7.63	8.26	7.80	7.96
CaO	1.73	1.18	10.53	11.80	11.95	11.86	12.05	12.07	12.34	12.35	11.91	12.22	11.81	12.25	12.07
Na_2O	1.44	1.07	2.47	2.18	2.33	2.23	2.14	2.09	2.08	2.09	2.14	2.06	2.46	2.00	2.11
K_2O	5.537	7.713	0.717	0.318	0.168	0.103	0.134	0.081	0.045	0.062	0.416	0.065	0.142	0.091	0.059
TiO_2	1.922	1.405	1.184	1.120	1.163	1.170	1.099	1.100	1.075	1.087	1.094	1.037	1.155	1.012	1.042
P_2O_5	0.058	0.037	0.111	0.091	0.092	0.095	0.091	0.091	0.086	0.086	0.107	0.082	0.091	0.078	0.085
LOI	5.10	4.64	1.05	0.45	0.55	0.33	–0.06	–0.03	0.34	0.00	–0.02	0.20	0.65	0.14	0.17
Total	99.52	99.50	99.72	99.64	100.04	99.41	99.40	99.72	99.58	99.76	99.67	99.99	99.67	99.46	99.42
XRF (ppm)															
Sc	56.0	55.9	56.5	48.6	49.7	50.4	47.1	43.5	40.9	41.0	50.6	41.2	53.7	40.3	41.4
Ni	151.3	78.4	105.4	137.6	163.5	125.8	107.7	107.3	104.5	103.3	97.9	106.2	122.4	107.5	98.8
Cu	162.6	105.3	174.4	131.2	93.6	165.8	158.4	154.9	158.0	157.8	159.9	137.7	163.9	128.4	131.0
Zn	208.9	117.7	110.7	108.8	104.6	104.1	94.3	88.6	84.9	89.3	95.5	84.8	103.3	75.7	77.9
Cr	102.1	147.7	244.2	214.8	205.3	220.1	212.1	199.4	189.1	191.5	218.4	188.8	204.6	214.0	169.6
V	262.4	124.1	385.0	354.6	365.0	373.4	341.2	319.0	314.2	324.2	338.2	306.8	380.0	294.1	316.4
Rb	56.9	62.9	11.6	6.6	1.1	0.8	2.9	0.7	0.3	0.7	9.1	0.6	0.3	1.2	1.0
Sr	174.4	138.5	149.9	131.2	136.6	135.5	118.4	116.9	118.0	116.1	116.8	111.3	126.9	109.2	110.6
Y	20.2	11.3	24.5	23.0	24.5	24.1	22.5	24.3	24.1	24.1	23.7	23.0	24.8	22.6	22.4
Zr	112.7	78.7	66.1	62.8	64.0	64.3	60.9	61.2	60.8	61.6	61.0	58.5	64.6	56.3	58.3
Nb	6.8	4.5	3.5	3.4	3.5	3.5	3.3	3.3	3.3	3.3	3.4	3.1	3.7	3.0	3.0
Ba	8.5	0.3	10.2	6.7	9.3	10.5	7.4	12.4	7.1	5.2	2.3	14.5	10.5	13.7	14.1
ICP-MS (ppm)															
Li	78.5	70.1	11.0	10.9	12.2	9.8	7.1	5.1	6.1	4.6	7.4	7.8	9.6	6.1	7.3
Sc	34.6	36.7	49.6	47.7	49.0	49.3	47.0	46.5	45.4	45.7	46.7	45.0	48.9	44.8	43.2
Ti	8800	6900	6000	5800	5800	5820	5530	5480	5360	5450	5360	4970	5510	4910	4760
Co	54.5	26.3	65.4	61.7	59.1	53.8	50.6	51.4	49.8	49.2	45.3	49.0	54.3	49.0	47.1
Cu	142.9	105.0	182.0	160.2	106.9	178.7	170.6	168.6	164.6	160.2	171.9	152.8	171.1	151.4	146.6
Rb	51.59	61.51	11.51	6.56	1.15	0.85	2.88	0.72	0.32	0.62	8.40	0.57	0.37	1.12	0.84
Sr	141.7	119.6	133.5	120.3	123.8	124.6	106.5	103.9	104.6	101.9	107.2	99.6	117.3	98.7	93.5
Y	17.7	11.0	22.7	22.2	23.4	23.6	21.2	22.6	22.7	22.5	22.8	22.1	24.0	22.0	20.3
Zr	103.34	72.40	64.10	62.17	64.62	66.00	61.01	60.20	59.98	58.64	58.71	56.08	63.04	55.25	53.14
Nb	5.87	3.61	3.37	3.27	3.37	3.49	3.22	3.18	3.21	3.02	3.08	2.91	3.25	2.84	2.72
Cs	1.120	0.756	0.206	0.107	0.010	0.009	0.046	0.038	0.006	0.018	0.206	0.006	0.002	0.010	0.024
Ba	10.06	9.42	14.62	9.92	10.51	9.33	10.60	14.71	10.82	11.74	9.56	15.53	9.64	14.32	11.00
La	4.59	4.20	4.29	2.86	3.01	3.20	2.76	2.94	2.92	3.03	3.03	2.92	3.31	2.91	2.78
Ce	11.34	8.73	9.65	8.30	8.71	9.08	8.04	8.21	8.14	8.56	8.40	7.95	9.08	7.84	7.66
Pr	1.82	1.37	1.44	1.28	1.37	1.40	1.25	1.28	1.25	1.31	1.31	1.24	1.40	1.22	1.18
Nd	9.38	6.96	7.68	7.10	7.36	7.66	6.81	6.97	6.91	6.95	6.98	6.72	7.56	6.61	6.31
Sm	2.84	2.07	2.41	2.34	2.40	2.44	2.21	2.26	2.19	2.21	2.23	2.13	2.40	2.13	1.99
Eu	1.03	0.76	0.99	0.96	0.99	1.01	0.92	0.92	0.90	0.93	0.95	0.90	1.00	0.90	0.86
Gd	3.58	2.57	3.42	3.32	3.45	3.47	3.14	3.25	3.22	3.35	3.31	3.15	3.53	3.10	3.00
Tb	0.61	0.44	0.60	0.58	0.61	0.61	0.56	0.58	0.56	0.59	0.58	0.55	0.61	0.54	0.52
Dy	4.01	2.91	4.24	4.10	4.34	4.36	3.91	4.03	4.01	4.17	4.10	3.95	4.32	3.79	3.73
Ho	0.78	0.56	0.88	0.86	0.92	0.91	0.82	0.86	0.85	0.87	0.85	0.82	0.90	0.79	0.76
Er	2.05	1.52	2.41	2.32	2.53	2.49	2.24	2.37	2.35	2.38	2.36	2.27	2.45	2.23	2.15
Tm	0.29	0.22	0.36	0.35	0.37	0.37	0.34	0.35	0.35	0.36	0.35	0.33	0.36	0.33	0.31
Yb	1.70	1.42	2.34	2.23	2.40	2.36	2.15	2.28	2.20	2.19	2.27	2.19	2.30	2.14	2.06
Lu	0.25	0.22	0.37	0.36	0.37	0.38	0.35	0.36	0.36	0.37	0.36	0.34	0.37	0.34	0.33
Hf	2.81	2.24	1.85	1.80	1.83	1.89	1.74	1.71	1.70	1.71	1.70	1.63	1.83	1.58	1.51
Ta	0.363	0.295	0.218	0.205	0.210	0.217	0.196	0.203	0.197	0.197	0.207	0.193	0.220	0.187	0.182
Th	0.463	0.353	0.288	0.265	0.271	0.275	0.255	0.252	0.247	0.248	0.256	0.246	0.274	0.240	0.225
U	0.467	0.326	0.148	0.093	0.100	0.132	0.107	0.072	0.086	0.075	0.222	0.066	0.167	0.063	0.060

mbsf is metres below sea floor; Fe_2O_3T is total Fe expressed as Fe_2O_3 ; LOI is weight loss on ignition at 1100ºC.
Magma type: Kr, Kroenke type; Kw, Kwaimbaita type; Sg, Singgalo type.

Table 2. *continued*

1185-1	1185-2	1185-3	1185-4	1185-5	1185-6	1185-7	1185-8	1185-9	1185-10	1185-11	1185-12	1185-13	1185-14	1185-15	1185-16	1185-17	1185-18
1185A	1185A	1185A	1185B	1185B	1185B	1185B	1185B	1185B	1185B	1185B	1185B	1185B	1185B	1185B	1185B	1185B	1185B
8R-2	10R-1	11R-1	3R-1	4R-4	5R-5	6R-4	8R-2	11R-1	15R-2	17R-3	19R-3	20R-2	21R-4	22R-3	22R-7	24R-2	28R-1
80–85	24–30	37–42	55–62	114–118	138–142	90–95	87–91	24–30	108–113	34–39	23–30	41–46	38–44	56–61	14–17	100–108	51–56
310.70	322.84	328.07	316.25	323.67	335.01	342.82	360.07	386.74	417.88	437.45	452.23	455.56	467.55	476.52	481.78	494.99	518.11
1	4	5	1	2	2	5	7	9	9	10	10	11	11	11	11	12	12
Kr	Kr	Kr	Kr	Kr	Kr	Kr	Kr	Kr	Kr	Kw	Kw	Kw	Kw	Kw	Kw	Kw	Kw
49.27	49.26	49.32	49.35	49.57	49.02	48.42	48.79	48.97	49.03	50.16	50.19	50.30	50.40	49.71	50.21	50.14	50.11
15.00	15.09	15.31	15.02	15.07	14.65	14.51	14.91	15.19	14.92	14.05	14.00	14.02	13.97	13.99	14.15	14.58	13.97
10.85	10.99	10.51	10.98	10.71	10.92	10.86	10.51	10.65	10.78	12.60	12.52	12.36	12.58	12.13	12.25	11.05	12.51
0.170	0.180	0.180	0.177	0.165	0.173	0.170	0.174	0.196	0.176	0.200	0.180	0.190	0.190	0.190	0.200	0.220	0.200
9.09	9.60	8.90	9.57	8.89	10.44	10.91	8.45	8.93	9.56	7.40	7.78	7.45	7.61	8.23	7.67	7.29	7.55
12.50	12.49	12.42	12.54	12.43	11.98	11.69	12.67	12.72	12.48	12.19	11.58	12.06	12.04	11.58	11.99	12.41	12.00
1.67	1.62	1.71	1.46	1.63	1.57	1.52	1.80	1.79	1.60	2.05	2.02	1.96	1.98	2.00	2.05	2.18	2.03
0.050	0.160	0.060	0.114	0.075	0.073	0.058	0.333	0.081	0.031	0.070	0.080	0.070	0.070	0.080	0.040	0.050	0.070
0.720	0.730	0.740	0.727	0.734	0.713	0.687	0.705	0.732	0.720	1.057	1.079	1.073	1.061	1.030	1.078	1.119	1.060
0.060	0.060	0.060	0.062	0.061	0.057	0.056	0.058	0.061	0.059	0.087	0.086	0.089	0.083	0.079	0.088	0.089	0.085
0.37	−0.33	0.62	−0.45	0.39	0.50	0.97	1.73	0.58	0.45	0.04	0.10	0.00	−0.07	0.48	0.08	0.63	0.07
99.74	99.85	99.83	99.55	99.73	100.10	99.85	100.13	99.90	99.81	99.91	99.61	99.57	99.92	99.49	99.81	99.76	99.65
38.2	44.6	41.4	42.8	40.9	34.8	34.1	37.7	37.6	40.5	39.7	44.6	42.4	42.1	40.8	45.5	47.7	40.2
184.3	188.4	184.6	188.2	188.0	215.5	221.0	226.8	199.3	195.9	98.5	94.6	99.7	101.0	112.0	101.4	101.7	99.1
101.4	104.9	100.5	105.4	103.5	95.0	89.5	85.1	106.3	102.7	147.0	153.8	158.0	150.0	146.2	152.8	165.5	146.3
75.5	80.2	70.6	79.4	76.8	69.8	62.6	81.3	74.8	74.1	81.3	80.8	87.3	82.9	74.1	87.0	93.3	75.1
464.0	507.2	465.1	498.9	478.8	479.6	525.2	482.5	478.6	478.4	163.3	184.1	166.2	167.1	178.3	183.5	185.4	176.0
252.8	260.8	254.0	259.1	258.7	237.5	223.0	243.1	256.9	251.1	317.4	316.5	324.6	310.9	299.9	330.3	345.0	302.3
0.9	2.6	1.1	1.8	1.5	1.5	1.5	7.4	1.6	0.5	0.7	1.0	0.4	0.7	1.3	0.4	0.4	0.8
80.3	79.7	82.9	80.1	81.1	78.4	75.1	83.2	81.9	79.0	106.1	106.1	107.5	106.4	103.5	109.7	113.7	107.3
16.9	17.3	17.2	17.4	17.5	16.7	15.9	16.4	17.2	16.8	23.7	23.6	23.9	23.1	23.0	23.6	23.3	23.0
40.9	41.4	40.8	41.2	41.4	39.6	37.9	39.0	41.3	40.4	59.6	58.5	60.9	59.6	56.6	60.4	62.8	59.2
2.3	2.3	2.3	2.3	2.3	2.3	2.1	2.2	2.3	2.3	3.1	2.9	3.2	3.1	2.9	3.1	3.4	3.1
4.9	6.9	4.0	8.4	4.9	11.2	8.7	1.6	3.9	7.0	20.2	11.1	7.8	14.3	11.0	10.2	15.9	16.3
8.8	6.1	9.0	5.0	7.8	8.4	7.6	44.7	8.2	7.9	5.8	5.3	6.8	6.4	7.8	6.9	10.1	6.0
37.8	37.4	38.0	36.8	37.8	36.5	36.1	35.4	38.2	38.7	43.0	44.8	44.5	43.9	42.5	44.2	45.4	44.0
5100	4900	4900	4800	4900	4700	4615	4540	4900	5100	7150	7500	7350	7120	6900	7350	7450	7200
50.9	49.5	50.8	49.9	50.8	51.6	53.2	52.5	52.2	53.2	49.6	50.0	50.5	50.2	50.9	50.6	50.6	50.7
102.4	98.8	99.9	101.7	101.0	96.9	94.9	76.8	105.3	103.7	151.8	159.3	156.3	152.9	154.1	158.7	158.7	155.3
0.84	2.64	1.03	1.82	1.43	1.47	1.46	7.80	1.61	0.52	0.88	0.91	0.67	0.70	1.11	0.36	0.34	0.78
74.0	76.0	78.2	78.2	77.9	72.8	70.7	78.5	78.2	77.2	108.8	103.0	101.5	99.9	95.1	102.0	106.0	96.1
16.9	17.6	17.6	17.7	17.7	16.7	16.5	16.2	17.5	17.1	24.4	23.1	23.6	22.5	22.8	23.4	23.7	22.6
39.60	40.53	41.27	41.32	41.45	39.05	38.16	37.46	40.29	40.84	65.91	63.77	60.99	60.44	55.73	58.82	61.56	56.15
2.27	2.30	2.36	2.36	2.33	2.15	2.12	2.06	2.24	2.45	3.79	3.69	3.49	3.41	3.19	3.27	3.43	3.08
0.032	0.049	0.021	0.031	0.035	0.040	0.037	0.241	0.028	0.018	0.007	0.008	0.007	0.007	0.015	0.005	0.005	0.007
6.49	13.83	9.08	13.66	9.60	9.46	8.77	5.49	5.24	7.23	18.02	16.55	16.05	16.62	15.20	14.31	14.46	15.84
2.25	2.03	1.96	2.00	2.00	1.96	1.91	1.84	2.02	1.96	3.18	2.97	3.10	2.96	2.95	2.99	3.24	3.00
5.32	5.63	5.48	5.55	5.57	5.33	5.22	5.21	5.60	5.33	8.62	8.21	8.33	8.13	7.97	8.44	8.83	8.26
0.84	0.88	0.85	0.88	0.87	0.84	0.83	0.82	0.88	0.86	1.37	1.32	1.34	1.30	1.26	1.35	1.40	1.31
4.50	4.69	4.67	4.68	4.66	4.45	4.34	4.31	4.65	4.58	7.25	6.90	6.92	6.77	6.66	7.01	7.28	6.89
1.47	1.52	1.51	1.50	1.52	1.46	1.41	1.41	1.51	1.47	2.31	2.19	2.23	2.16	2.14	2.27	2.38	2.28
0.61	0.63	0.62	0.62	0.62	0.60	0.59	0.58	0.63	0.62	0.94	0.91	0.91	0.87	0.88	0.92	0.98	0.91
1.97	2.07	2.05	2.09	2.04	1.98	1.93	1.92	2.08	2.02	3.11	2.98	3.03	2.89	2.86	3.00	3.14	2.96
0.38	0.40	0.40	0.40	0.40	0.39	0.38	0.37	0.40	0.40	0.59	0.56	0.57	0.55	0.54	0.57	0.59	0.56
2.52	2.65	2.61	2.59	2.60	2.53	2.47	2.45	2.62	2.64	3.78	3.64	3.71	3.52	3.48	3.69	3.83	3.67
0.60	0.63	0.63	0.62	0.62	0.60	0.58	0.57	0.62	0.61	0.88	0.83	0.83	0.81	0.81	0.85	0.85	0.84
1.74	1.81	1.80	1.79	1.81	1.75	1.71	1.66	1.80	1.76	2.52	2.39	2.41	2.33	2.29	2.45	2.46	2.35
0.26	0.27	0.27	0.27	0.27	0.26	0.26	0.25	0.27	0.27	0.36	0.35	0.35	0.34	0.34	0.36	0.36	0.35
1.69	1.79	1.74	1.76	1.76	1.71	1.66	1.65	1.78	1.73	2.34	2.27	2.28	2.23	2.19	2.31	2.31	2.23
0.27	0.29	0.29	0.29	0.29	0.28	0.27	0.26	0.28	0.28	0.38	0.37	0.37	0.36	0.36	0.38	0.37	0.37
1.09	1.13	1.14	1.12	1.12	1.07	1.05	1.06	1.12	1.10	1.82	1.72	1.68	1.67	1.61	1.72	1.79	1.70
0.131	0.134	0.134	0.136	0.137	0.129	0.127	0.128	0.134	0.137	0.210	0.204	0.200	0.195	0.190	0.202	0.204	0.197
0.165	0.168	0.168	0.167	0.157	0.151	0.147	0.147	0.155	0.171	0.287	0.280	0.263	0.260	0.246	0.258	0.261	0.241
0.051	0.050	0.051	0.049	0.045	0.042	0.043	0.058	0.043	0.053	0.090	0.086	0.077	0.075	0.071	0.074	0.086	0.066

Table 2. *continued*

Sample	1187-1	1187-2	1187-3	1187-4	1187-5	1187-6	1187-7	1187-8	1187-9	1187-10	1187-11	1187-12	1187-13	1187-14	1187-15
Site, hole	1187A	1187A	1187A	1187A	1187A	1187A	1187A	1187A	1187A	1187A	1187A	1187A	1187A	1187A	1187A
Core-section	2R-2	3R-2	3R-4	5R-4	6R-6	7R-7	9R-4	10R-1	10R-3	13R-2	13R-4	13R-6	14R-3	15R-4	16R-3
Interval (cm)	76–83	45–52	95–98	60–65	98–106	90–94	20–25	114–118	70–75	24–29	11–16	5–9	55–60	35–41	41–46
Depth (mbsf)	367.44	376.41	379.84	398.71	410.51	421.23	436.30	443.04	445.60	472.23	474.78	477.22	483.73	493.82	502.44
Subunit	1	2	3	3	3	4	5	6	6	6	7	8	10	11	12
Magma type	Kr	Kr	Kr	Kr	Kr	Kr	Kr	Kr	Kr	Kr	Kr	Kr	Kr	Kr	Kr
XRF (wt%)															
SiO_2	49.33	48.53	48.39	49.53	49.17	49.34	48.72	49.22	48.79	48.52	48.88	48.90	49.35	48.80	49.21
Al_2O_3	15.46	15.58	15.53	15.66	14.91	15.33	14.78	14.87	14.80	14.71	14.89	15.28	15.06	14.87	15.56
Fe_2O_3T	11.17	10.86	11.22	10.09	10.92	10.66	10.78	10.85	10.80	10.72	10.77	10.98	10.59	10.78	10.67
MnO	0.185	0.166	0.168	0.185	0.168	0.195	0.170	0.172	0.168	0.169	0.170	0.192	0.191	0.170	0.195
MgO	7.44	7.13	8.47	7.89	9.69	8.49	9.97	9.89	9.52	9.96	9.85	8.26	8.69	10.03	8.03
CaO	12.84	13.19	11.69	12.85	12.41	12.74	12.38	12.39	12.52	12.40	12.36	12.48	12.52	12.32	12.91
Na_2O	1.82	1.82	1.69	1.78	1.67	1.72	1.62	1.64	1.66	1.66	1.65	1.80	1.84	1.63	1.79
K_2O	0.298	0.234	0.876	0.045	0.045	0.046	0.047	0.090	0.063	0.065	0.052	0.312	0.061	0.044	0.045
TiO_2	0.750	0.750	0.755	0.766	0.730	0.750	0.716	0.746	0.735	0.718	0.724	0.743	0.742	0.720	0.760
P_2O_5	0.066	0.063	0.085	0.065	0.062	0.065	0.061	0.060	0.061	0.057	0.061	0.062	0.061	0.063	0.063
LOI	0.72	1.17	1.03	0.78	0.25	0.52	0.29	–0.51	0.38	0.55	0.37	0.80	0.99	0.30	0.64
Total	100.08	99.49	99.90	99.64	100.03	99.86	99.53	99.42	99.50	99.53	99.78	99.81	100.10	99.73	99.87
XRF (ppm)															
Sc	39.9	42.4	46.2	40.6	38.4	37.7	36.9	43.0	36.9	36.0	37.0	39.2	38.1	38.0	40.2
Ni	170.8	189.3	174.8	205.5	204.7	219.1	194.0	199.6	193.8	205.2	184.4	195.2	201.2	201.9	213.0
Cu	107.5	130.7	100.4	103.5	96.3	96.8	97.6	100.9	92.8	97.6	93.2	64.9	113.4	95.4	98.2
Zn	80.7	81.6	94.4	81.7	79.0	77.0	70.4	78.5	72.9	69.3	71.0	72.8	74.2	75.7	80.9
Cr	475.8	521.9	540.9	501.0	490.5	499.8	472.3	506.7	480.3	489.4	459.2	489.0	490.0	489.8	494.3
V	264.5	268.7	269.1	255.8	240.1	247.1	238.4	261.2	239.8	244.5	242.1	249.9	255.4	245.0	256.2
Rb	7.4	5.8	10.5	0.7	0.8	0.5	0.8	1.7	1.0	1.5	1.1	7.0	1.0	0.8	0.7
Sr	98.2	96.7	96.6	91.1	82.0	88.0	82.7	81.6	84.2	85.5	81.5	95.0	92.2	81.3	93.7
Y	16.9	17.2	18.0	17.3	16.7	17.2	16.6	16.8	16.7	16.4	16.6	17.1	16.8	16.6	17.3
Zr	41.4	41.7	41.2	41.7	40.0	41.3	39.8	41.3	40.1	39.4	40.0	41.1	40.6	40.0	41.8
Nb	2.4	2.3	2.4	2.4	2.2	2.4	2.1	2.3	2.2	2.3	2.1	2.3	2.3	2.2	2.3
Ba	27.5	1.2	11.3	4.2	2.7	0.0	2.9	11.2	4.5	3.6	10.5	0.2	1.6	10.5	0.0
ICP-MS (ppm)															
Li	21.6	23.0	8.7	9.5	7.6	8.4	7.3	4.0	8.2	9.4	8.1	23.1	11.8	8.0	8.3
Sc	41.3	41.7	40.8	41.8	39.7	40.2	40.1	40.1	39.1	38.3	41.0	41.8	40.8	40.2	42.0
Ti	3900	3750	3690	3760	3570	3650	3510	3620	3520	3410	4000	3930	3800	3760	3930
Co	49.7	49.4	47.9	51.5	50.1	52.3	50.2	50.2	50.3	49.3	49.9	50.6	51.8	50.9	55.3
Cu	104.1	131.5	101.9	105.4	101.2	103.8	102.1	103.9	95.8	97.3	101.4	70.2	120.7	102.9	105.5
Rb	8.56	6.42	11.82	0.62	0.76	0.58	0.89	1.80	0.98	1.38	0.93	7.01	1.01	0.73	0.63
Sr	89.6	89.3	90.0	85.9	74.8	80.5	76.9	75.8	74.36	80.8	83.2	96.0	91.5	80.3	92.2
Y	16.6	16.5	17.6	17.5	17.0	17.6	17.3	17.2	16.1	17.7	18.7	18.8	18.3	17.8	18.6
Zr	38.97	38.94	39.73	41.08	38.28	39.66	39.19	39.04	36.77	39.74	43.16	43.17	41.76	40.65	42.52
Nb	2.17	2.22	2.23	2.36	2.20	2.27	2.22	2.27	2.04	2.29	2.48	2.47	2.34	2.24	2.40
Cs	0.248	0.199	0.206	0.014	0.019	0.012	0.018	0.026	0.014	0.048	0.026	0.203	0.027	0.022	0.015
Ba	25.63	4.72	12.29	5.62	5.18	4.65	4.98	13.46	5.51	7.75	8.83	5.73	4.97	12.10	5.10
La	1.92	1.92	2.11	2.04	1.86	1.91	1.87	1.90	1.90	1.95	1.99	2.07	2.02	2.01	2.07
Ce	5.50	5.51	5.54	5.68	5.20	5.39	5.19	5.41	5.31	5.51	5.40	5.60	5.48	5.42	5.69
Pr	0.84	0.84	0.86	0.87	0.81	0.84	0.80	0.83	0.82	0.85	0.85	0.89	0.85	0.86	0.89
Nd	4.62	4.60	4.81	4.80	4.44	4.63	4.41	4.55	4.32	4.66	4.45	4.62	4.52	4.51	4.68
Sm	1.53	1.56	1.58	1.61	1.51	1.50	1.46	1.49	1.42	1.55	1.46	1.53	1.48	1.49	1.58
Eu	0.64	0.64	0.65	0.66	0.60	0.63	0.61	0.62	0.61	0.63	0.60	0.61	0.61	0.61	0.62
Gd	2.30	2.26	2.33	2.37	2.20	2.25	2.18	2.23	2.20	2.30	2.14	2.23	2.14	2.14	2.22
Tb	0.41	0.40	0.42	0.42	0.39	0.40	0.39	0.40	0.39	0.41	0.39	0.40	0.38	0.39	0.40
Dy	2.98	2.93	3.03	3.00	2.81	2.88	2.78	2.88	2.84	3.00	2.80	2.86	2.77	2.83	2.97
Ho	0.62	0.63	0.65	0.65	0.61	0.63	0.61	0.63	0.61	0.64	0.58	0.62	0.60	0.59	0.63
Er	1.78	1.79	1.86	1.87	1.73	1.80	1.73	1.77	1.70	1.82	1.73	1.78	1.74	1.71	1.80
Tm	0.27	0.27	0.28	0.28	0.26	0.27	0.26	0.27	0.25	0.28	0.27	0.28	0.27	0.27	0.28
Yb	1.73	1.73	1.79	1.80	1.69	1.75	1.66	1.70	1.59	1.73	1.68	1.75	1.67	1.70	1.76
Lu	0.28	0.28	0.29	0.30	0.28	0.28	0.28	0.29	0.27	0.30	0.29	0.29	0.29	0.29	0.30
Hf	1.18	1.16	1.17	1.22	1.12	1.17	1.15	1.18	1.09	1.21	1.07	1.09	1.08	1.07	1.12
Ta	0.142	0.138	0.137	0.143	0.134	0.139	0.138	0.141	0.131	0.139	0.145	0.147	0.140	0.143	0.148
Th	0.156	0.163	0.161	0.168	0.162	0.163	0.159	0.161	0.154	0.174	0.165	0.170	0.161	0.159	0.167
U	0.072	0.048	0.067	0.047	0.047	0.052	0.049	0.045	0.044	0.064	0.051	0.053	0.050	0.049	0.052

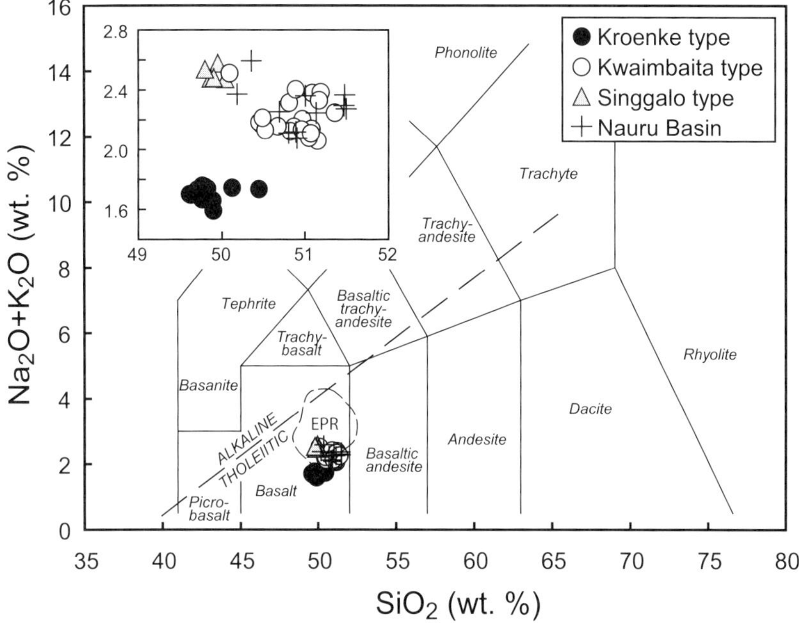

Fig. 3. A plot of total alkalis against SiO_2 (after Le Bas *et al.* 1986), showing the remarkably small range of composition in basalt from the OJP and Nauru Basin. The analyses have been filtered to exclude all rocks with LOI >0.5 wt% and K_2O/P_2O_5 >2, and recalculated to 100% totals, with Fe as FeO. The broken line separates Hawaiian tholeiitic and alkaline basalts (from Macdonald & Katsura 1964). The field labelled EPR encloses *c.* 3100 analyses of basaltic glass from the East Pacific Rise spreading centre (LDEO database: http://petdb.ldeo.columbia.edu/petdb/).

the three compositional types: Kroenke, Kwaimbaita and Singgalo. The compositional plots in this paper differentiate these types with different symbols. Three additional symbols identify volcaniclastic rocks from Sites 1183 and 1184 and basalt from the Nauru Basin. Significant differences between samples of the same type at different sites are indicated by fields drawn around data points.

Most of the OJP and Nauru Basin rock samples are altered through prolonged contact with sea water (Banerjee *et al.* 2004), and most major elements are likely to have been affected to some degree. However, by rejecting all the volcaniclastic rocks (26 samples), 20 lava samples with loss on ignition (LOI) >0.5 weight% (wt%) and a further seven with anomalously high K (K_2O/P_2O_5 >2), we are left with 49 analyses that we consider fresh enough to plot on an alkali–silica diagram (Fig. 3). Figure 3 shows the remarkably small compositional range of the OJP and Nauru Basin rocks, and the similarity of the latter to Kwaimbaita-type basalt from the OJP. Despite some scatter in the data (probably due to alteration), it is clear that the Singgalo-type basalt samples have higher total alkalis and lower silica than does Kwaimbaita-type basalt. For comparison, the compositional range of glass samples from the East Pacific Rise is also shown on Figure 3.

The difference between Kwaimbaita- and Singgalo-type basalt is also seen clearly on a plot of SiO_2 against MgO (Fig. 4). That some scatter in SiO_2 remains after filtering the data is shown by the three samples from the single flow unit drilled at Site 289. However, the consistently low SiO_2 content of the Singgalo-type basalt (unit A at Site 807) suggests that the Singgalo-type magma had lower SiO_2 than Kwaimbaita-type magma at similar MgO. The olivine control line shown in Figure 4 was calculated by olivine addition and subtraction from the composition of fresh glass sample 1187–8 (1187A-10R-1, 114–118). It shows that Kroenke-type magma could have evolved by olivine crystallization toward the composition of Kwaimbaita-type, but not Singgalo-type, evolved magma.

Primitive-mantle-normalized incompatible-element concentrations in all of the rocks analysed are shown in Figure 5. The patterns for Kwaimbaita-type basalt are reproduced, as grey lines, on all the other diagrams for comparison.

Table 3. *Analyses of basalt from the Ontong Java Plateau (Sites 289, 803 and 807) and the Nauru basin (Site 462)*

Sample	803-1	803-2	803-3	807-1	807-2	807-3	807-4	807-5	807-6	807-7	807-8	807-9	807-10	807-11	807-12
Leg	130	130	130	130	130	130	130	130	130	130	130	130	130	130	130
Site, hole	803D	803D	803D	807C	807C	807C	807C	807C	807C	807C	807C	807C	807C	807C	807C
Core-section	69R-3	70R-2	71R-1	75R-2	75R-3	78R-1	78R-3	79R-5	81R-2	83R-1	85R-1	87R-1	88R-2	90R-1	93R-3
Interval (cm)	87–95	69–75	59–65	81–86	107–112	55–61	87–93	4–10	99–104	18–24	83–89	99–105	11–17	29–34	10–16
Depth (mbsf)	635.20	643.22	651.39	1387.47	1389.20	1410.05	1413.20	1420.35	1436.09	1447.38	1466.43	1485.99	1495.92	1509.39	1521.65
Magma type	Kw	Kw	Kw	Sg	Sg	Sg	Sg	Sg	Kw	Kw	Kw	Kw	Kw	Kw	Kw
XRF (wt%)															
SiO_2	48.50	50.01	49.25	48.87	48.97	49.26	49.05	48.94	50.25	50.30	50.28	49.46	50.26	50.42	50.44
Al_2O_3	15.19	16.50	14.84	14.44	14.16	14.22	14.27	14.21	14.60	14.54	13.93	13.68	14.55	13.86	13.89
Fe_2O_3T	11.73	10.21	11.79	12.92	13.27	13.43	13.19	13.12	10.72	11.41	12.53	13.86	11.56	13.12	12.64
MnO	0.159	0.140	0.192	0.228	0.195	0.203	0.207	0.234	0.223	0.230	0.194	0.236	0.192	0.191	0.198
MgO	5.69	5.56	6.73	6.31	6.72	7.04	6.79	6.97	7.68	7.97	7.94	7.15	7.11	7.37	7.09
CaO	12.97	10.48	12.52	12.07	11.75	11.82	11.96	11.90	11.77	11.87	11.62	11.87	12.15	11.92	11.80
Na_2O	2.18	2.33	2.19	2.40	2.28	2.33	2.34	2.32	2.29	2.24	2.01	2.11	2.19	2.06	2.13
K_2O	0.456	1.407	0.357	0.122	0.140	0.117	0.098	0.173	0.427	0.136	0.066	0.280	0.099	0.058	0.077
TiO_2	1.332	1.459	1.308	1.623	1.566	1.573	1.572	1.590	1.187	1.184	1.046	1.114	1.173	1.121	1.124
P_2O_5	0.118	0.193	0.112	0.146	0.138	0.137	0.135	0.135	0.100	0.100	0.087	0.097	0.099	0.092	0.093
LOI	1.09	1.30	0.45	0.40	0.36	0.06	0.50	–0.14	0.33	0.23	0.17	–0.24	0.05	0.05	0.19
Total	99.42	99.59	99.74	99.53	99.55	100.19	100.11	99.45	99.58	100.21	99.87	99.62	99.43	100.26	99.67
XRF (ppm)															
Sc	47.3	57.2	46.6	44.4	42.4	38.7	37.8	41.7	52.4	48.8	41.7	49.0	48.7	42.9	43.8
Ni	114.7	99.2	121.6	102.7	99.4	96.5	99.5	117.4	116.1	106.8	94.8	90.5	99.5	89.0	90.1
Cu	63.7	82.3	65.6	142.2	144.1	134.6	138.7	142.3	46.5	154.3	136.0	147.7	155.7	147.1	148.0
Zn	112.7	79.3	101.3	103.1	100.7	102.8	97.0	103.5	101.9	104.9	106.3	100.2	119.8	90.2	90.8
Cr	259.4	277.8	254.7	174.5	166.0	160.2	157.9	166.9	156.0	148.6	186.7	165.5	154.9	151.0	143.6
V	351.1	315.3	330.7	338.8	320.9	309.7	307.9	342.0	376.4	375.9	311.1	354.6	369.1	328.5	341.9
Rb	9.2	15.7	7.4	1.7	2.5	2.1	1.3	3.4	6.4	1.5	0.7	6.3	1.1	0.5	1.0
Sr	167.6	183.4	151.4	208.7	173.1	163.8	168.8	171.6	120.6	120.9	102.7	103.8	110.9	105.4	107.9
Y	26.6	25.9	26.0	30.6	29.6	29.6	29.7	30.1	24.1	24.6	22.8	24.3	27.7	24.4	24.6
Zr	77.8	83.7	76.2	99.3	95.9	96.3	96.1	98.0	64.7	65.0	56.7	62.1	65.4	61.6	63.4
Nb	4.5	4.9	4.5	6.3	6.1	6.0	5.9	6.1	3.6	3.6	3.0	3.4	3.6	3.3	3.4
Ba	15.6	29.2	8.9	14.5	18.9	21.4	19.7	23.2	18.5	11.4	11.7	8.2	6.2	16.1	20.2
ICP-MS (ppm)															
Li	14.9	10.9	17.3	7.2	10.5	6.5	9.9	7.7	7.7	8.2	9.0	5.8	7.7	4.9	4.4
Sc	47.3	51.3	45.8	44.6	43.0	42.7	42.9	43.2	49.5	49.1	46.0	46.2	45.4	45.9	47.0
Ti	6800	7400	6700	8690	8170	8060	8000	8100	5990	5840	5240	5500	5840	5480	5500
Co	56.9	46.8	56.3	51.5	50.0	49.7	50.2	54.8	56.9	53.6	49.2	47.3	47.9	49.1	50.5
Cu	69.3	88.8	71.3	150.8	149.3	145.5	147.0	147.8	47.4	154.6	151.0	146.3	150.2	145.2	157.9
Rb	9.13	15.84	7.21	1.59	2.31	1.96	1.21	3.20	6.25	1.35	0.65	5.61	1.01	0.42	0.78
Sr	148.4	164.4	137.8	190.8	166.5	151.4	151.8	154.4	110.6	108.6	89.3	92.8	101.7	92.1	96.9
Y	26.9	26.7	26.4	31.2	31.5	30.0	29.2	29.5	23.8	23.9	21.8	23.3	26.2	22.8	23.4
Zr	76.36	83.81	74.45	105.62	106.91	101.15	98.60	98.80	66.15	64.61	55.15	59.84	61.62	58.08	60.74
Nb	4.38	4.76	4.25	6.65	6.73	6.26	6.17	6.09	3.61	3.50	2.91	3.14	3.30	2.98	3.13
Cs	0.294	0.388	0.219	0.036	0.056	0.052	0.043	0.083	0.081	0.012	0.011	0.144	0.006	0.004	0.010
Ba	13.96	36.27	11.70	21.15	22.82	25.77	22.81	25.85	16.83	10.96	13.20	10.92	9.59	15.01	16.43
La	4.10	4.27	4.11	5.55	5.34	5.37	5.38	5.60	3.08	3.11	2.99	3.09	3.34	3.25	3.22
Ce	11.20	11.28	11.22	15.22	14.55	14.74	14.66	15.30	8.96	9.24	8.38	8.72	9.87	9.05	8.78
Pr	1.75	1.86	1.74	2.22	2.13	2.16	2.14	2.21	1.40	1.41	1.26	1.34	1.51	1.36	1.34
Nd	8.94	9.62	8.95	11.02	10.45	10.60	10.66	10.94	7.14	7.16	6.45	6.87	7.78	7.01	7.19
Sm	2.83	2.99	2.79	3.20	3.05	3.09	3.10	3.17	2.31	2.28	2.06	2.17	2.50	2.20	2.31
Eu	1.09	1.16	1.09	1.30	1.24	1.28	1.26	1.31	0.99	0.99	0.89	0.95	1.01	0.96	0.94
Gd	3.84	3.96	3.83	4.52	4.33	4.44	4.42	4.60	3.50	3.51	3.20	3.44	3.69	3.45	3.46
Tb	0.66	0.69	0.66	0.75	0.72	0.72	0.73	0.75	0.61	0.61	0.55	0.60	0.67	0.60	0.59
Dy	4.62	4.77	4.57	5.18	5.04	5.08	5.07	5.22	4.32	4.39	4.02	4.29	4.88	4.23	4.22
Ho	0.94	0.96	0.95	1.05	1.03	1.05	1.02	1.06	0.89	0.92	0.83	0.90	1.03	0.89	0.87
Er	2.65	2.68	2.64	2.81	2.72	2.78	2.76	2.82	2.37	2.47	2.27	2.42	2.84	2.45	2.45
Tm	0.41	0.40	0.40	0.41	0.40	0.41	0.41	0.42	0.36	0.37	0.34	0.37	0.43	0.37	0.35
Yb	2.46	2.42	2.44	2.45	2.32	2.37	2.40	2.41	2.20	2.24	2.07	2.20	2.53	2.18	2.33
Lu	0.42	0.40	0.41	0.41	0.40	0.40	0.40	0.41	0.37	0.38	0.35	0.38	0.41	0.37	0.37
Hf	2.09	2.21	2.04	2.53	2.45	2.48	2.47	2.55	1.75	1.75	1.57	1.69	1.74	1.68	1.71
Ta	0.298	0.323	0.294	0.380	0.365	0.366	0.372	0.381	0.234	0.222	0.197	0.217	0.220	0.212	0.209
Th	0.317	0.340	0.309	0.537	0.535	0.516	0.511	0.507	0.286	0.284	0.236	0.262	0.264	0.245	0.269
U	0.175	0.324	0.151	0.135	0.139	0.128	0.130	0.124	0.196	0.128	0.065	0.064	0.125	0.063	0.074

mbsf is metres below sea floor; Fe_2O_3T is total Fe expressed as Fe_2O_3; LOI is weight loss on ignition at 1100°C.
Magma types: Kw, Kwaimbaita type; Sg, Singgalo type.

Table 3. *continued*

289-1	289-2	289-3	462-1	462-2	462-3	462-4	462-5	462-6	462-7	462-8	462-9	462-10	462-11	462-12
30	30	30	61	61	61	61	61	61	61	61	89	89	89	89
289	289	289	462A	462A	462A	462A	462A	462A	462A	462A	462A	462A	462A	462A
132R-2	132R-3	132R-4	19R-1	23R-4	38R-1	45R-3	52R-3	65R-2	79R-4	88R-2	94R-2	97R-2	102R-2	108R-3
114–118	20–25	108–113	23–25	74–76	96–99	0–3	19–21	94–96	105–107	64–67	133–136	1–6	96–100	82–85
1262.14	1262.70	1265.08	578.23	602.24	691.96	735.82	790.51	878.84	990.50	1034.64	1079.43	1105.51	1145.28	1193.32
Kw	Kw	Kw	Kw	Kw	Kw	Kw	Kw	Kw	Kw	Kw	Kw	Kw	Kw	Kw
49.55	49.34	50.29	49.42	48.93	50.63	50.15	49.95	50.17	49.91	50.55	50.56	50.31	50.84	49.73
14.28	14.22	13.81	13.61	14.42	14.10	14.27	14.05	14.10	14.07	13.56	13.70	13.87	13.69	13.76
12.57	13.25	13.33	13.80	13.57	12.38	11.69	11.96	12.25	12.13	13.52	13.36	13.42	13.24	13.80
0.187	0.245	0.186	0.226	0.199	0.191	0.177	0.184	0.183	0.188	0.204	0.192	0.203	0.198	0.228
6.57	6.96	6.91	7.54	6.73	7.43	7.96	8.04	7.77	7.76	6.88	6.77	7.04	7.25	7.07
11.15	11.65	11.04	11.50	10.88	12.04	12.43	12.24	12.19	12.32	11.29	11.32	11.49	11.37	11.66
2.39	2.35	2.24	2.26	2.65	2.17	2.01	2.04	2.05	2.18	2.19	2.27	2.27	2.18	2.50
0.584	0.124	0.101	0.075	0.087	0.055	0.036	0.036	0.038	0.040	0.065	0.055	0.059	0.065	0.059
1.571	1.552	1.538	1.329	1.701	1.171	0.900	0.943	0.984	0.998	1.223	1.234	1.220	1.140	1.231
0.136	0.134	0.132	0.103	0.138	0.089	0.066	0.072	0.076	0.076	0.095	0.103	0.094	0.088	0.100
0.45	0.11	−0.10	0.19	0.66	−0.06	0.18	0.00	−0.02	0.00	−0.15	−0.09	0.15	0.16	0.04
99.44	99.94	99.48	100.05	99.97	100.19	99.87	99.52	99.79	99.67	99.43	99.47	100.13	100.22	100.18
48.9	45.8	44.8	40.9	48.7	39.0	43.3	39.9	41.0	39.3	42.6	41.8	41.7	40.6	45.4
85.5	111.4	92.7	93.3	102.1	112.2	120.6	123.9	112.4	115.3	91.7	96.2	101.2	91.4	90.1
92.2	201.0	192.9	143.0	222.0	152.5	133.1	131.9	137.1	135.1	166.6	174.4	177.1	151.2	178.4
112.7	114.2	105.3	88.3	110.5	101.9	74.8	82.3	80.9	76.0	92.4	87.4	95.5	94.4	112.5
248.8	226.7	230.4	141.2	171.5	159.9	353.6	340.3	308.4	276.1	162.2	157.4	166.7	153.8	124.2
399.3	391.5	379.3	335.5	432.8	331.7	315.8	317.6	324.2	320.0	368.1	377.1	387.4	344.4	396.5
13.4	2.4	1.0	1.3	0.6	0.6	0.5	0.5	0.6	0.6	0.7	0.5	0.6	0.4	0.4
144.2	141.9	122.5	128.0	130.0	112.1	86.8	90.3	90.3	94.9	97.0	99.0	97.2	96.7	99.4
32.8	33.3	33.1	28.3	39.0	25.5	20.8	21.5	22.2	22.4	28.3	28.4	27.9	26.2	28.7
91.4	90.0	89.4	77.2	97.9	65.9	49.7	51.6	54.6	55.0	68.3	68.8	67.9	62.4	70.4
5.1	5.1	4.9	4.3	5.2	3.3	2.5	2.6	2.8	2.8	3.4	3.5	3.4	3.1	3.7
8.8	7.8	17.2	13.2	18.0	12.0	7.4	6.9	1.4	5.1	14.3	15.1	9.6	8.6	6.6
10.0	9.0	5.4	8.6	6.3	5.4	5.0	5.7	5.2	4.1	5.0	3.5	4.2	5.5	3.8
46.8	45.8	46.2	40.5	43.9	40.5	39.7	39.6	41.2	39.8	42.0	42.0	42.0	40.5	42.0
7530	7250	7300												
46.3	51.2	48.0	49.0	49.0	48.1	45.0	47.0	46.0	45.0	47.2	47.0	46.0	45.0	46.0
104.8	215.3	209.6	150.1	220.7	151.3	126.0	127.9	136.3	130.9	167.8	166.5	166.2	153.4	167.9
13.14	2.19	0.82	1.08	0.64	0.68	0.42	0.32	0.43	0.34	0.51	0.41	0.37	0.28	0.20
127.6	130.1	107.1	120.0	124.0	107.0	86.4	88.3	87.2	91.8	92.2	92.5	92.3	91.4	94.7
31.9	33.3	31.0	27.1	37.1	24.7	20.7	21.2	21.9	21.9	26.4	26.5	26.2	25.3	27.3
91.56	92.76	85.56	70.75	91.12	61.68	46.32	49.12	51.17	51.88	63.63	63.33	63.01	60.70	65.69
4.87	4.81	4.59	4.17	5.25	3.36	2.48	2.66	2.76	2.77	3.43	3.34	3.35	3.22	3.54
0.165	0.030	0.005	0.030	0.012	0.007	0.014	0.010	0.011	0.008	0.006	0.009	0.012	0.006	0.003
11.42	10.19	17.36	12.15	16.40	12.59	6.97	8.13	9.03	8.22	14.05	10.67	11.50	9.12	8.37
4.46	4.69	4.30	4.01	5.19	3.22	2.44	2.47	2.68	2.59	3.05	3.23	3.23	3.05	3.30
12.51	12.99	11.79	10.75	13.69	8.90	6.66	6.92	7.28	7.29	8.71	8.80	9.55	8.47	9.26
1.94	2.01	1.85	1.63	2.08	1.38	1.01	1.07	1.10	1.10	1.33	1.35	1.45	1.28	1.40
10.38	10.81	9.80	8.60	10.98	7.52	5.58	5.86	5.99	6.09	7.38	7.33	7.71	6.97	7.65
3.23	3.41	3.07	2.77	3.54	2.42	1.87	1.91	1.98	2.00	2.40	2.43	2.48	2.28	2.48
1.25	1.34	1.21	1.08	1.31	0.96	0.75	0.79	0.76	0.77	0.94	0.95	0.96	0.90	0.96
4.56	4.90	4.48	3.74	4.90	3.37	2.62	2.83	2.86	2.92	3.53	3.50	3.55	3.28	3.59
0.80	0.84	0.78	0.66	0.85	0.61	0.48	0.50	0.52	0.51	0.63	0.63	0.65	0.59	0.65
5.57	5.92	5.51	4.66	6.09	4.31	3.51	3.56	3.59	3.60	4.42	4.45	4.68	4.22	4.58
1.16	1.24	1.13	0.97	1.29	0.89	0.73	0.75	0.77	0.78	0.95	0.93	0.99	0.88	0.96
3.24	3.41	3.18	2.85	3.74	2.58	2.12	2.20	2.14	2.14	2.60	2.63	2.74	2.48	2.72
0.46	0.50	0.47	0.40	0.53	0.38	0.32	0.33	0.31	0.32	0.40	0.39	0.41	0.36	0.40
3.06	3.18	3.04	2.57	3.41	2.39	2.06	2.10	2.09	2.10	2.58	2.54	2.54	2.39	2.63
0.49	0.51	0.49	0.42	0.55	0.39	0.33	0.33	0.34	0.33	0.42	0.43	0.41	0.40	0.43
2.51	2.65	2.46	2.11	2.70	1.81	1.39	1.47	1.38	1.45	1.78	1.76	1.70	1.65	1.81
0.316	0.330	0.303	0.257	0.327	0.202	0.156	0.157	0.177	0.182	0.224	0.217	0.213	0.205	0.228
0.403	0.421	0.390	0.335	0.420	0.263	0.191	0.205	0.211	0.218	0.251	0.264	0.246	0.246	0.273
0.159	0.116	0.102	0.106	0.117	0.078	0.077	0.064	0.059	0.058	0.071	0.072	0.071	0.066	0.073

Table 4. *Analyses of volcaniclastic rocks from Site 1184 (OPD Leg 192)*

Sample	1184-1	1184-2	1184-3	1184-4	1184-5	1184-6	1184-7	1184-8	1184-9	1184-10	1184-11	1184-12	1184-13	1184-14
Site, hole	1184A	1184A	1184A	1184A	1184A	1184A	1184A	1184A	1184A	1184A	1184A	1184A	1184A	1184A
Core-section	11R-3	12R-5	14R-1	17R-1	19R-1	21R-4	22R-4	25R-5	30R-2	30R-5	31R-1	32R-7	35R-2	36R-7
Interval (cm)	28–34	37–43	120–127	74–81	21–27	22–34	105–111	40–47	128–135	33–43	133–139	11–18	106–123	0–6
Depth (mbsf)	209.34	216.93	231.20	249.94	268.71	292.07	301.62	331.09	377.48	380.72	385.73	402.66	425.61	441.00
Subunit	A	A	A	B	B	B	B	C	C	C	D	D	D	D
Magma type	Kr	Kr	Kr	Kw	Kw	Kw	Kw	hi-Nb	hi-Nb	hi-Nb	Kw	Kw	Kw	Kw
XRF (wt%)														
SiO_2	46.90	47.47	44.68	48.19	48.41	46.92	48.60	45.39	46.37	47.70	48.05	47.84	47.21	47.54
Al_2O_3	14.22	14.20	13.06	13.18	12.82	12.60	13.14	13.95	14.84	14.47	12.80	12.31	12.82	13.05
Fe_2O_3T	10.16	10.39	9.95	12.55	12.36	12.23	12.41	11.84	12.41	11.06	12.83	12.39	13.06	13.33
MnO	0.123	0.111	0.151	0.195	0.183	0.150	0.173	0.207	0.223	0.195	0.208	0.192	0.200	0.210
MgO	13.55	11.99	11.49	9.45	9.54	9.41	10.68	6.58	3.69	5.37	5.90	5.32	6.94	6.66
CaO	0.81	0.64	4.87	3.09	2.28	3.71	1.59	5.74	6.40	6.31	7.25	8.48	8.09	7.28
Na_2O	5.47	6.43	4.39	5.06	4.55	5.68	6.20	5.07	4.33	5.37	4.67	3.35	3.97	4.20
K_2O	0.538	0.401	0.775	0.919	1.544	1.038	0.276	0.673	0.590	0.551	0.551	0.791	0.229	0.129
TiO_2	0.813	0.854	0.759	1.061	1.050	1.072	1.071	1.255	1.551	0.932	1.144	1.120	1.138	1.209
P_2O_5	0.066	0.071	0.070	0.081	0.087	0.097	0.088	0.151	0.224	0.175	0.084	0.113	0.092	0.094
LOI	7.54	7.25	9.74	6.33	7.19	6.86	6.17	8.74	9.33	7.35	6.12	7.79	5.69	6.33
Total	100.19	99.81	99.94	100.11	100.01	99.77	100.40	99.60	99.96	99.48	99.61	99.70	99.44	100.03
XRF (ppm)														
Sc	54.4	55.8	49.3	57.0	58.2	58.9	59.5	45.6	38.1	42.5	51.9	47.7	52.3	51.3
Ni	166.9	167.5	157.4	99.1	95.7	96.4	97.3	65.1	17.5	75.4	88.3	87.7	96.6	108.2
Cu	125.7	133.2	128.7	164.5	167.0	176.4	164.0	108.9	65.5	124.3	266.8	181.1	182.1	234.9
Zn	90.1	79.4	74.0	101.8	90.1	94.3	101.0	109.2	123.7	117.1	106.0	99.7	106.3	127.6
Cr	467.3	479.4	450.8	139.1	129.5	139.8	139.4	153.2	35.6	185.4	157.1	154.1	181.6	168.9
V	292.1	324.8	304.4	354.7	344.4	415.2	403.8	270.2	160.6	318.7	363.5	284.3	419.3	435.0
Rb	8.3	6.1	7.6	9.4	19.4	4.6	2.9	4.8	2.8	4.6	5.4	6.4	2.7	2.0
Sr	101.2	102.0	219.2	646.3	1021.2	36.3	36.0	132.9	254.0	110.5	97.7	157.7	106.5	105.7
Y	17.2	16.7	18.9	23.8	23.3	24.3	24.0	33.7	46.0	40.1	28.9	25.3	26.6	29.3
Zr	41.4	43.9	38.8	56.3	56.0	57.2	56.9	82.0	111.0	104.9	63.7	60.4	62.6	65.8
Nb	2.2	2.4	2.1	3.0	2.9	3.2	3.1	6.7	10.2	7.8	3.4	3.2	3.3	3.6
Ba	1.9	3.3	12.5	23.8	25.9	7.3	5.7	19.3	34.9	22.7	13.2	27.3	17.6	19.3
ICP-MS (ppm)														
Li	19.8	16.1	8.0	8.9	7.8	6.0	9.9	7.0	6.8	6.9	4.2	6.8	5.0	5.2
Sc	38.8	43.3	38.6	42.2	41.4	41.0	41.2	34.6	32.9	43.4	43.4	43.5	42.9	42.3
Ti	5200	5790	5200	6700	6700	6500	6900	7900	11400	7500	7600	8200	7400	7500
Co	44.9	48.4	46.6	44.5	44.2	43.8	43.3	34.0	30.0	42.2	47.6	45.0	45.9	48.9
Cu	120.6	149.7	136.6	162.9	170.2	165.1	160.4	104.6	80.1	156.2	272.0	188.1	179.0	220.9
Rb	7.76	5.53	7.10	8.80	18.83	4.23	2.76	4.43	2.36	4.17	4.83	6.01	2.13	1.39
Sr	94.0	95.6	194.3	578.4	939.3	34.1	33.4	118.9	215.8	102.3	90.1	140.2	100.0	98.5
Y	17.0	16.7	18.8	23.9	23.8	23.8	23.4	32.4	43.8	39.2	28.3	24.9	27.2	29.3
Zr	39.63	41.87	36.78	52.76	53.52	53.67	52.63	74.56	97.05	96.58	58.11	57.10	61.65	63.38
Nb	2.17	2.35	2.14	2.97	3.04	3.05	2.96	5.94	8.88	7.28	3.17	3.08	3.77	3.91
Cs	1.625	0.290	0.030	0.018	0.014	0.024	0.063	0.060	0.021	0.052	0.056	0.016	0.022	0.018
Ba	11.86	5.40	20.62	24.17	29.18	13.19	8.23	26.24	34.94	26.22	13.88	28.09	22.57	20.88
La	2.35	2.43	2.48	2.94	3.01	2.86	2.93	5.26	7.64	6.64	3.64	3.16	3.26	3.59
Ce	5.98	5.97	5.97	7.97	8.02	7.99	7.96	12.64	17.30	16.00	10.30	8.78	9.00	9.80
Pr	0.92	0.94	0.96	1.20	1.23	1.24	1.22	1.81	2.36	2.24	1.56	1.35	1.39	1.46
Nd	5.09	5.23	5.21	6.62	6.67	6.76	6.67	9.25	12.38	11.33	8.48	7.26	7.51	7.84
Sm	1.65	1.69	1.71	2.15	2.14	2.19	2.16	2.88	3.71	3.42	2.74	2.35	2.35	2.46
Eu	0.67	0.69	0.69	0.86	0.85	0.84	0.84	1.10	1.38	1.21	1.04	0.90	0.92	0.94
Gd	2.21	2.25	2.30	2.87	2.87	2.88	2.86	3.86	4.89	4.53	3.63	3.26	3.14	3.37
Tb	0.41	0.42	0.44	0.54	0.54	0.55	0.55	0.73	0.98	0.86	0.68	0.59	0.60	0.64
Dy	2.67	2.68	2.87	3.62	3.57	3.59	3.53	4.86	6.34	5.67	4.44	4.08	3.90	4.19
Ho	0.61	0.60	0.66	0.83	0.82	0.83	0.82	1.15	1.49	1.34	1.01	0.90	0.90	0.99
Er	1.72	1.68	1.94	2.40	2.38	2.38	2.36	3.40	4.50	3.93	2.92	2.63	2.57	2.85
Tm	0.25	0.24	0.29	0.36	0.35	0.36	0.35	0.52	0.70	0.60	0.44	0.39	0.39	0.42
Yb	1.54	1.54	1.84	2.34	2.26	2.31	2.28	3.45	4.54	3.93	2.80	2.54	2.42	2.66
Lu	0.24	0.25	0.31	0.38	0.37	0.37	0.37	0.56	0.75	0.66	0.46	0.41	0.41	0.44
Hf	1.22	1.30	1.17	1.62	1.61	1.62	1.62	2.25	2.93	2.77	1.80	1.77	1.78	1.87
Ta	0.139	0.149	0.132	0.193	0.187	0.196	0.189	0.373	0.517	0.447	0.209	0.200	0.203	0.213
Th	0.204	0.210	0.185	0.266	0.260	0.257	0.261	0.533	0.812	0.694	0.261	0.243	0.282	0.296
U	0.206	0.246	0.176	0.074	0.084	0.092	0.092	0.125	0.223	0.205	0.065	0.052	0.083	0.155

mbsf is metres below sea floor; Fe_2O_3T is total Fe expressed as Fe_2O_3; LOI is weight loss on ignition at 1100°C.
Subunits A–F are defined on lithological and chemical changes (Thordarson, 2004).
Magma types: Kr, Kroenke type; Kw, Kwaimbaita type; hi-Nb, high-Nb type.

Table 4. *continued*

1184-15	1184-16	1184-17	1184-18	1184-19	1184-20	1184-21	1184-22	1184-23	1184-24
1184A	1184A	1184A	1184A	1184A	1184A	1184A	1184A	1184A	1184A
38R-2	39R-4	41R-2	41R-2	42R-5	43R-3	43R-6	44R-3	45R-2	46R-1
35–48	78–85	0–6	70–85	59–69	8–19	0–6	50–65	69–75	54–60
454.17	467.04	482.61	483.31	496.83	502.81	506.78	513.25	521.34	529.74
E	E	E	E	E	E	E	E	E	F
Kw	Kw	Kw	Kw	Kw	Kw	Kw	Kw	Kw	hi-Nb
47.79	47.32	46.78	46.62	46.83	46.59	46.87	47.03	46.65	46.06
13.15	13.35	13.15	13.11	13.09	13.35	13.44	13.19	13.36	17.55
11.60	11.27	11.27	11.12	11.23	11.49	11.66	11.18	11.32	13.74
0.183	0.178	0.185	0.187	0.178	0.170	0.182	0.179	0.178	0.223
7.26	7.48	7.86	7.53	7.43	7.33	7.64	7.42	7.49	4.64
7.52	7.67	10.44	11.26	10.81	8.83	10.25	11.10	11.03	8.78
4.79	4.61	1.77	1.74	1.72	3.71	1.80	1.61	1.58	2.76
0.135	0.076	0.076	0.078	0.079	0.083	0.093	0.080	0.092	0.109
0.979	0.951	0.947	0.938	0.945	0.975	0.987	0.949	0.968	1.608
0.070	0.075	0.078	0.076	0.076	0.077	0.079	0.077	0.077	0.173
6.49	6.48	6.88	6.89	7.07	7.02	6.94	6.96	6.70	3.77
99.97	99.46	99.44	99.55	99.46	99.63	99.94	99.78	99.45	99.41
51.8	50.1	50.1	49.3	50.8	50.2	51.4	49.7	50.5	38.3
122.2	117.8	116.2	115.9	116.3	111.9	116.3	116.7	116.5	39.7
169.5	155.1	150.4	147.4	154.9	154.4	160.3	149.0	150.2	89.8
89.6	92.3	95.4	88.6	87.2	88.9	85.8	91.1	87.3	146.1
358.4	345.7	350.1	356.9	342.1	325.1	342.7	348.0	345.4	48.4
320.3	296.3	341.6	334.3	336.9	340.9	348.1	341.1	335.8	221.1
2.2	1.1	1.4	1.1	1.4	1.9	1.6	1.2	1.2	0.6
93.0	97.5	76.3	99.8	93.1	65.1	97.7	97.5	89.8	100.7
21.8	21.3	20.9	20.9	21.3	21.6	21.5	20.8	21.0	42.7
52.3	51.1	50.1	49.7	50.8	52.4	51.9	50.2	50.3	110.5
2.9	2.8	2.8	2.7	2.8	2.9	2.9	2.8	2.8	10.5
12.4	7.8	15.3	20.3	11.9	7.4	13.2	12.0	11.8	32.3
4.7	4.5	5.0	3.9	4.2	3.7	4.7	3.9	4.0	9.2
40.5	40.0	40.0	39.6	39.0	39.6	40.3	39.1	38.5	29.1
6000	5600	5600	5300	5800	6000	5900	5500	5700	9300
44.3	43.0	42.5	42.2	41.9	42.6	43.4	42.3	41.6	35.5
159.6	148.6	145.9	142.3	146.5	149.8	155.4	142.5	142.0	91.3
1.73	0.77	0.95	0.92	1.07	1.46	1.17	0.99	0.93	0.67
85.4	89.5	70.6	90.4	84.4	60.1	88.7	89.0	81.6	89.7
21.6	20.9	20.3	20.1	20.4	21.4	20.8	20.0	19.9	39.7
49.29	48.19	46.97	45.73	47.05	49.08	48.09	46.22	45.97	97.39
2.94	2.93	2.89	2.77	2.83	2.97	2.85	2.73	2.64	10.05
0.024	0.033	0.008	0.008	0.008	0.065	0.009	0.011	0.009	0.006
15.24	14.06	16.69	25.80	14.22	13.35	15.57	14.50	14.86	32.52
2.58	2.56	2.50	2.47	2.52	2.62	2.62	2.52	2.58	8.83
7.23	7.07	6.90	6.80	6.93	7.34	7.25	6.93	6.94	19.35
1.13	1.10	1.07	1.06	1.08	1.14	1.11	1.08	1.08	2.59
6.09	5.94	5.78	5.71	5.86	6.10	6.02	5.83	5.77	12.77
1.97	1.89	1.89	1.84	1.86	1.94	1.94	1.89	1.88	3.76
0.78	0.76	0.73	0.73	0.74	0.78	0.77	0.73	0.73	1.33
2.62	2.54	2.47	2.48	2.49	2.63	2.58	2.51	2.46	4.86
0.50	0.48	0.47	0.47	0.47	0.50	0.49	0.48	0.47	0.91
3.28	3.15	3.05	3.09	3.05	3.27	3.17	3.06	3.06	5.98
0.75	0.72	0.71	0.71	0.70	0.76	0.73	0.71	0.72	1.38
2.20	2.08	2.04	2.01	2.05	2.15	2.12	2.05	2.06	4.07
0.32	0.31	0.30	0.30	0.30	0.32	0.32	0.31	0.30	0.61
2.05	1.97	1.88	1.91	1.93	2.04	1.98	1.95	1.93	3.91
0.33	0.32	0.31	0.31	0.31	0.33	0.32	0.32	0.31	0.65
1.51	1.43	1.40	1.39	1.42	1.49	1.47	1.43	1.42	2.94
0.173	0.170	0.160	0.159	0.167	0.171	0.177	0.165	0.169	0.570
0.238	0.212	0.211	0.206	0.211	0.222	0.204	0.194	0.199	0.752
0.068	0.070	0.072	0.088	0.070	0.078	0.072	0.070	0.063	0.304

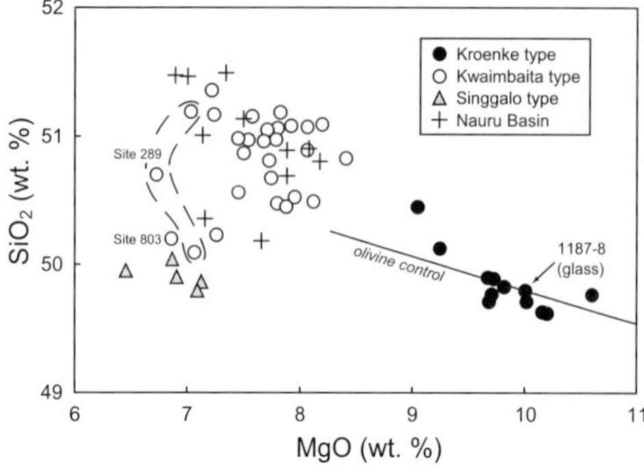

Fig. 4. MgO v. SiO_2 showing the consistently low SiO_2 content of Singgalo-type basalt, and similar MgO content to Kwaimbaita-type basalt. Analyses have been recalculated to 100% totals, with Fe as FeO, after excluding samples with LOI >0.5 wt%. The olivine control line was calculated by olivine addition and subtraction from the composition of fresh glass sample 1187–8 (1187A-10R-1, 114–118). Scatter in the SiO_2 content of the three samples from the single flow unit drilled at Site 289 is probably due to alteration.

Several important features can be seen in these diagrams.

(1) Despite the scatter in Cs, Rb, Ba, U and Sr (due to the mobility of these elements during alteration), the rocks show only a small range in composition, and most have nearly flat primitive-mantle-normalized patterns for all but the most incompatible elements (e.g. Th).
(2) Singgalo-type basalt (Site 807, unit A) has a slight relative enrichment in the more incompatible elements compared with Kwaimbaita-type basalt.
(3) The patterns for Kwaimbaita- and Kroenke-type basalt are very similar, and differ only in that the former has higher concentrations than the latter.
(4) Most of the volcaniclastic rocks from Site 1184 are similar to Kwaimbaita- and Kroenke-type basalt.
(5) Four samples from Site 1184 (identified on the diagram as the high-Nb group) have higher concentrations of Th, U, Nb, Ta and La.
(6) The two samples from the thin volcaniclastic layers just above basement at Site 1183 have patterns similar to those from Singgalo-type basalt.

Points (1) and (2) had been noted in previous studies (e.g. Mahoney *et al.* 1993; Tejada *et al.* 1996, 2002; Neal *et al.* 1997).

The differences and similarities between the three basalt types and the volcaniclastic rocks can be seen on a plot of Nb against Zr (Fig. 6). Of all the elements determined in this study, these are probably the least affected by sea-water alteration. This plot shows that Kwaimbaita-type, Kroenke-type and Nauru Basin basalt, and most of the volcaniclastic rocks from Site 1184, have a near-constant value of Nb/Zr (0.055) suggesting that they could be related by fractional crystallization of a common parental magma type. Three of the Site 1184 rocks plot with Kroenke-type basalt, and most of the rest with Kwaimbaita-type basalt. The four high-Nb samples plot well away from the main trend. Singgalo-type basalt (Site 807, unit A) and one of the two samples of volcaniclastic rock from Site 1183 likewise have higher Nb/Zr. The inset diagram in Figure 6 compares the OJP and Nauru Basin rocks with data fields for normal mid-ocean ridge basalt (N-MORB), ocean island basalt (OIB) and basalt from the active volcanic zones in Iceland. Rocks from the OJP and the Nauru Basin plot between the fields of Iceland and N-MORB but have a much smaller range in Nb, Zr and Nb/Zr.

The volcaniclastic succession drilled at Site 1184 includes a larger range of compositions than at any other drill site on the OJP (Figs 5 and 6). It can be divided into six distinct lithological subunits (A–F; Thordarson 2004) on the basis of facies variation and chemical composition. Fragments of wood, implying nearby emergent land surfaces, were found at the top of subunits C, D, E and F. Compositional variation between the subunits is shown in Figure 7; the high-Nb samples are from subunits C and F, and all the other samples have the uniform Nb/Zr (0.055) of Kroenke- and Kwaimbaita-type basalt. The three samples in Figure 6 that plot with Kroenke-type basalt are from subunit A, which has the highest concentrations of Ni and Cr (Fig. 7), and probably of MgO before alteration.

Figure 8 summarizes the compositional variation of the OJP and Nauru Basin rocks. The Kroenke- and Kwaimbaita-type basalts and the Nauru Basin samples form a cluster with a narrow range of Nb/Zr and Ce/Yb. The small but systematic increase in Ce/Yb from Kroenke- to Kwaimbaita-type basalt correlates well with Ce concentration and can be explained by fractional crystallization of augite, which is a liquidus or near-liquidus phase in Kwaimbaita-type magma (Sano & Yamashita 2004). Basalt from Site 803 plots away from the main cluster, suggesting that it may have evolved from a slightly different parental magma type. Singgalo-type basalt samples from Site 807 (unit A) cluster with the two samples from the thin volcaniclastic layers

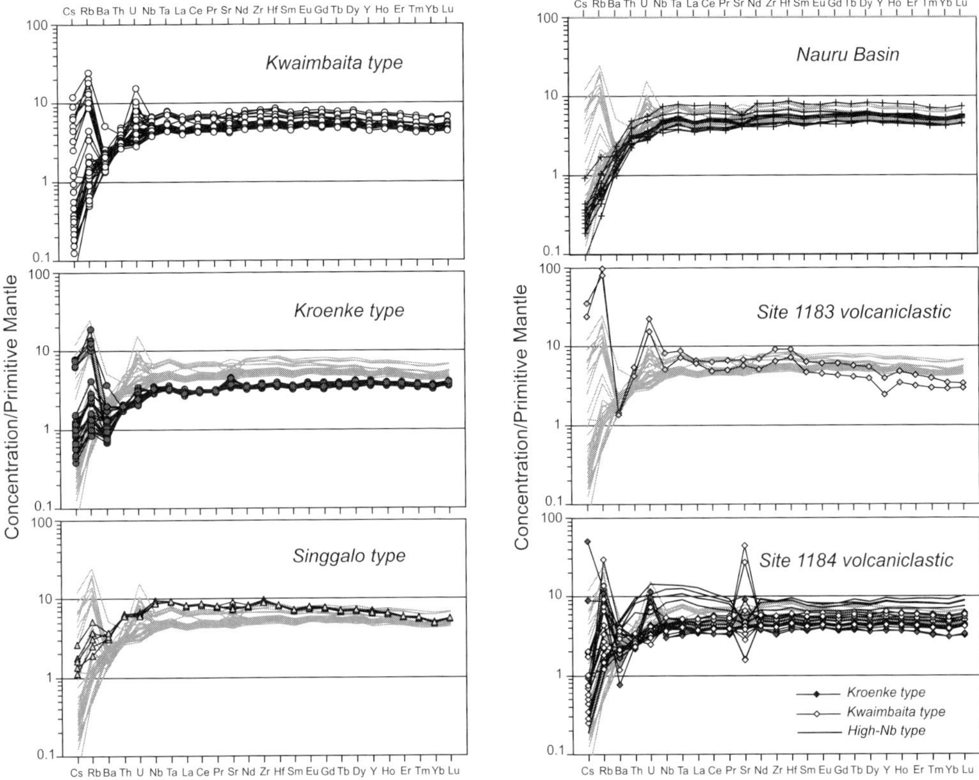

Fig. 5. Primitive-mantle-normalized incompatible-element concentrations (ICP-MS data for all elements) in: Kwaimbaita-, Kroenke- and Singgalo-type basalt samples from the OJP, and basalt samples from the Nauru Basin, and volcaniclastic rocks from ODP Sites 1183 and 1184. Normalizing values are taken from Sun & McDonough (1989). The patterns for Kwaimbaita-type basalt are reproduced, as grey lines, on all the other diagrams for comparison. Note the similarity in the shape of the patterns for Kwaimbaita- and Kroenke-type basalt, and the slight relative enrichment in the more incompatible elements shown by the Singgalo-type basalts (Site 807, unit A). Most of the volcaniclastic rocks from Site 1184 are similar to Kwaimbaita- and Kroenke-type basalt, but four samples (the high-Nb group) have higher concentrations of Th, U, Nb, Ta and La. Two samples from the thin volcaniclastic layers just above basement at Site 1183 have patterns that resemble Singgalo-type basalt. The scatter in Cs, Rb, Ba, U and Sr are due to mobility of these elements during alteration.

above basement at Site 1183. These layers also have the distinctive Singgalo-type isotopic characteristics (Tejada *et al.* 2004). The high-Nb samples from Site 1184 are clearly different from the other OJP and Nauru Basin samples, and represent a magma type not seen elsewhere in the region.

Discussion

Geographical distribution of magma types

With eight drill sites on the plateau, two in adjacent basins and on-land exposures in the Solomon Islands, the OJP is now the best-sampled oceanic plateau on Earth. Although drilling can do no more than scratch the surface, the results of ODP Leg 192 have advanced our knowledge of the stratigraphy and geographical distribution of basalt types on the OJP considerably. The distribution of magma types on and around the OJP is summarized in Figure 9.

Kwaimbaita-type basalt was found at all OJP drill sites except Site 1187, and also at Site 802 in the East Mariana Basin (Castillo *et al.* 1994) and Site 462 in the Nauru Basin. It forms a thick (>2.7 km) succession on Malaita and is also found on Santa Isabel (Tejada *et al.* 1996, 2002). Kwaimbaita-type basalt probably represents the dominant magma type on the plateau and in the adjacent basins.

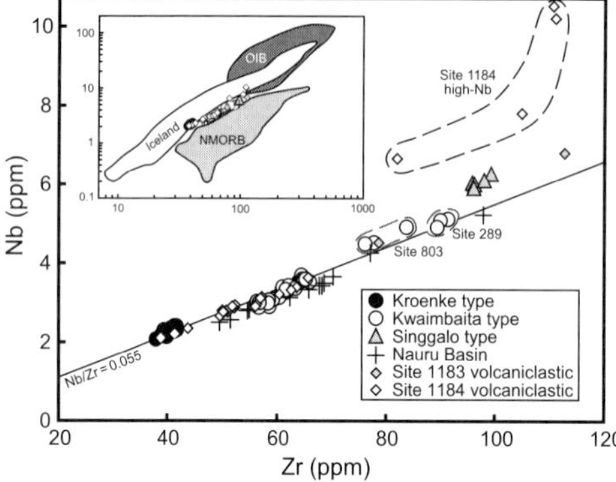

Fig. 6. Nb v. Zr (XRF data) for basalt and volcaniclastic rocks from the OJP and basalt from the Nauru Basin. The inset diagram compares the OJP and Nauru Basin rocks with data fields for normal mid-ocean ridge basalt (N-MORB), ocean island basalt (OIB) and basalt from the active volcanic zones in Iceland. Data sources: OIB, J. G. Fitton & D. James (unpublished); N-MORB, East Pacific Rise (Mahoney *et al.* 1994), North Atlantic (Hanan *et al.* 2000; R. N. Taylor unpublished), SW Indian Ridge (C. J. Robinson & J. G. Fitton unpublished), AAD (Kempton *et al.* 2002), North Chile Ridge (Bach *et al.* 1996); Iceland, Fitton *et al.* (2003) and references therein.

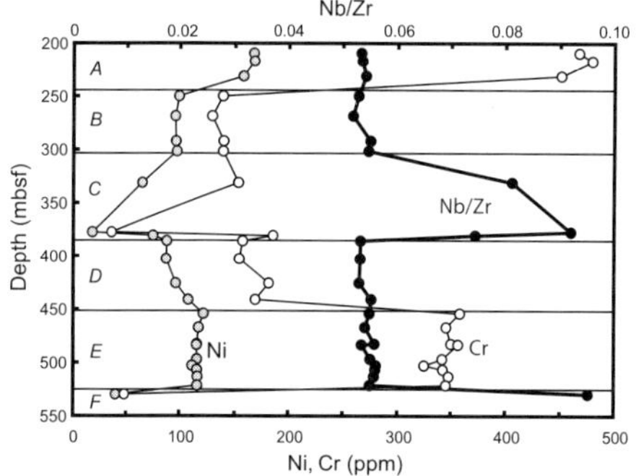

Fig. 7. Stratigraphic variation (depth in metres below sea floor (mbsf)) in Ni, Cr and Nb/Zr (XRF data) through the volcaniclastic succession at Site 1184 on the eastern salient of the OJP. Horizontal lines represent lithological subunit boundaries (A–F; Thordarson 2004). The high-Nb samples identified in Figure 6 are from subunits C and F. Subunit A has the composition of Kroenke-type basalt, and the other subunits have Kwaimbaita-type composition.

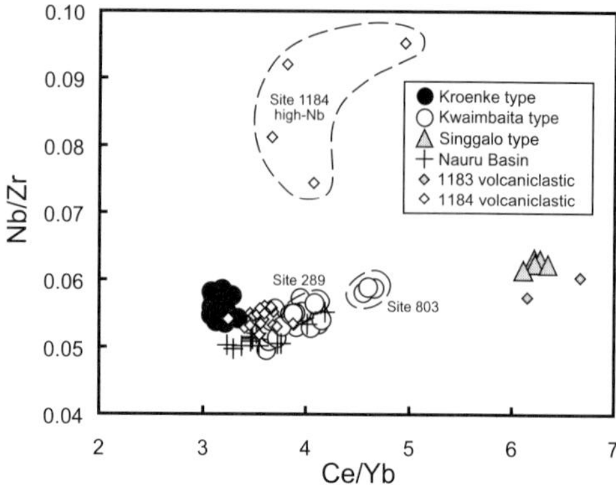

Fig. 8. Nb/Zr (XRF data) v. Ce/Yb for OJP and Nauru Basin basalt and volcaniclastic rocks. The rocks form three distinct clusters: Kroenke- and Kwaimbaita-type basalt; Singgalo-type basalt and Site 1183 volcaniclastic rocks; and the high-Nb type volcaniclastic rocks at Site 1184.

Singgalo-type basalt is found overlying Kwaimbata-type basalt at Site 807 and on the islands of Malaita and Santa Isabel (Tejada *et al.* 2002). The thin volcaniclastic turbidite layers on top of basement at Site 1183 are also of Singgalo type (Fig. 8) (Tejada *et al.* 2004), implying that Singgalo-type magma was erupted after Kwaimbaita-type magma near the crest of the plateau. Singgalo-type basalt, although present over a wide area in the northern, western and SW parts of the plateau, probably represents a minor magma type, erupted at a late stage in the formation of these parts of the plateau. It should be noted, however, that there is no detectable age difference between Singgalo- and Kwaimbaita-type basalt (Mahoney *et al.* 1993; Tejada *et al.* 2002).

Thick successions of the magnesian Kroenke-type basalt were found at Sites 1185 and 1187, 146 km apart on the eastern flank of the plateau. It overlies Kwaimbaita-type basalt at Site 1185, but its base was not reached at Site 1187. Biostratigraphic evidence (Bergen 2004; Sikora & Bergen 2004) suggests a Late Aptian age for the Kroenke-type basalt at these two sites. If correct, this would mean that the Kroenke-type magma was erupted up to 10 Ma later than the main plateau-forming episode (*c.* 122 Ma), an age difference that is consistent with the degree of alteration that affected the immediately underlying Kwaimbaita-type flows at Site 1185 (Mahoney *et al.* 2001; Banerjee *et al.* 2004). However, Os-isotope data from Kroenke-type basalt from Sites 1185 and 1187 plot on the same 122 Ma-isochron as data from Kwaimbaita-type basalt from Sites 1183 and 1186 (Parkinson *et al.* 2002). It is possible that eruption of the primitive magma represented by Kroenke-type basalt was one of the last magmatic events on the eastern side of the plateau, and it may be significant that

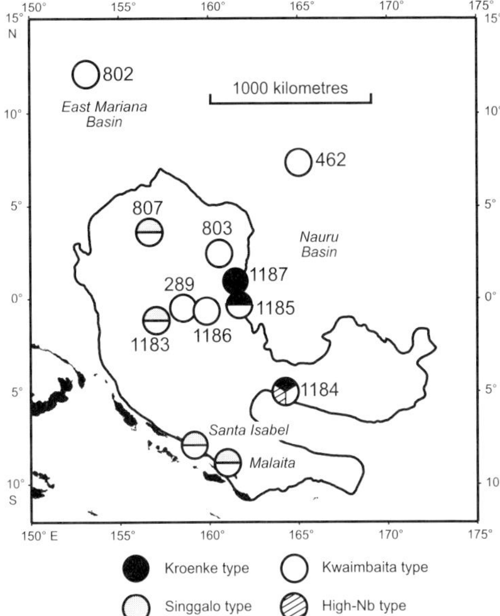

Fig. 9. Map showing distribution of magma types on and around the OJP. Symbols with more than one ornament reflect the stratigraphic sequence. Thus, Singgalo-type basalt overlies Kwaimbaita-type basalt at Site 807 and in the Solomon Islands, and thin Singgalo-type volcaniclastic units overlie Kwaimbaita-type basalt at Site 1183. Interbedded Kwaimbaita- and high-Nb-type volcaniclastic subunits are overlain by a Kroenke-type subunit at Site 1184.

the last volcaniclastic subunit at Site 1184 was also of Kroenke-type basalt composition. As drilling has barely scratched the surface of the OJP, it is possible that Kroenke-type flows also occur deeper within the volcanic succession.

The high-Nb volcaniclastic subunits at Site 1184 (Fig. 7) are unlike any other igneous rocks found on the plateau. They form two out of the six subunits and are interbedded with volcaniclastic layers of Kwaimbaita type.

Magmatic evolution

The discovery of 125 m of high-Mg Kroenke-type basalt at Site 1185 and more than 136 m at Site 1187, 146 km to the north, is important to our understanding of the evolution of OJP magmas. Because Kroenke- and Kwaimbaita-type basalts are isotopically indistinguishable (Tejada *et al.* 2004) and have similar primitive-mantle-normalized incompatible-element patterns (Fig. 5), the former provides a plausible primitive magma composition for the evolution of the latter. As it is unlikely that these two thick successions of Kroenke-type basalt represent small isolated outcrops, the volume of primitive magma erupted on the eastern flank of the plateau must have been immense.

Kroenke-type basalt contains small, sparse phenocrysts of olivine, and experimental studies (Sano & Yamashita 2004) show that olivine alone crystallizes on its liquidus at low pressures (0.1–190 MPa). The primary magma composition can therefore be estimated by incrementally adding equilibrium olivine until the calculated liquid is in equilibrium with mantle olivine. The composition of fresh glass sample 1187–8 from Site 1187 (Table 2) was used, with FeO estimated by taking Fe_2O_3/total Fe (by mass) = 0.1. Equilibrium olivine composition was calculated by using values of $K_D{}^{Ol/L}{}_{FeO/MgO}$ from Herzberg & O'Hara's (2002) parameterization of experimental data, and added iteratively in increments of 0.1 wt% until it reached the composition of residual mantle olivine. The calculated primary magma has 15.6 wt% MgO in equilibrium with Fo_{90} and 20.4 wt% in equilibrium with Fo_{92}, and these require the addition of 17 and 36 wt% of olivine, respectively. The residual mantle olivine composition was constrained by the forward- and inverse-modelling method developed by Herzberg & O'Hara (2002). With a fertile peridotite starting composition (Kettle River; table 1 in Herzberg & O'Hara 2002) and a liquid with the composition of 1187–8, the residual mantle olivine would have a composition of $Fo_{90.5}$ with perfect fractional melting and $Fo_{91.6}$ for equilibrium melting (Herzberg 2004). The corresponding MgO contents of the primary magmas would be 16.8 and 19.3 wt% respectively.

A possible liquid line of descent from the calculated primary magma is shown in Figure 10, in which Nb is plotted against MgO for samples with LOI <0.5 wt%. Niobium is used because it is immobile during alteration, and almost completely incompatible in olivine, plagioclase and augite. The olivine control line in Figure 10 passes through the data points for Kroenke-type basalt and extends to 8.5 wt% MgO and 2.36 ppm Nb, at which point plagioclase and augite join the crystallizing assemblage. The ol+plag+cpx (olivine+plagioclase+clinopyroxene) line in Figure 10 was constructed by joining the average composition of Kwaimbaita-type basalt (excluding samples from Sites 289 and 803) to a point representing the bulk composition of the olivine–plagiclase–augite cotectic. Analyses of phases crystallizing from Kwaimbaita-type basalt in Sano & Yamashita's (2004) experiments, weighted by the low-pressure cotectic phase proportions in tholeiitic

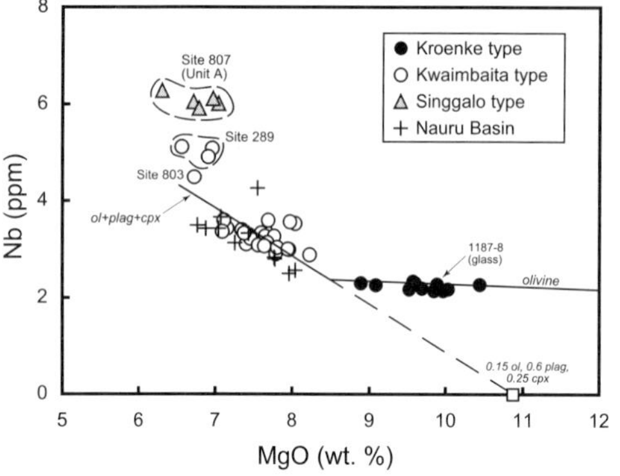

Fig. 10. Nb (XRF data) v. MgO for OJP and Nauru Basin samples (LOI <0.5 wt%) showing a likely liquid line of descent from a calculated primary magma composition (off the diagram at higher MgO).

basalt (0.15 olivine, 0.6 plagioclase, 0.25 augite; Cox & Bell 1972), were used to calculate the MgO content of the cotectic assemblage. Kwaimbaita-type basalt has sparse olivine and plagioclase phenocrysts (±augite), and these phases crystallize on or within 20°C of its liquidus temperature (Sano & Yamashita 2004). Cognate plagioclase-rich gabbroic cumulate xenoliths are common in Kwaimbaita-type basalt (Mahoney *et al.* 2001) and provide further evidence for this fractional crystallization stage.

The liquid line of descent in Figure 10 passes through the main cluster of data points for Kwaimbaita-type and Nauru Basin basalt, and extends to a point close to the composition of evolved basalt from Site 803, but slightly below the composition of basalt from Site 289. Basalt from Sites 289 and 803 are Kwaimbaita-type in their isotopic composition (Tejada *et al.* 2004), and the Site 289 basalt has incompatible-element ratios that are indistinguishable from the other Kwaimbaita-type basalt samples (e.g. Fig. 8). The ol+plag+cpx line in Figure 10 is shown as a straight line but should really be a concave-upwards curve because the MgO content of the cotectic assemblage will fall as fractional crystallization proceeds. The Site 289 basalt may therefore be related by fractional crystallization to the main group of Kwaimbata-type samples. Basalt from Site 803, however, has higher Ce/Yb than the other Kwaimbaita-type samples (Fig. 8), suggesting that it may have evolved from a slightly different parental magma type. Singgalo-type basalt has lower $^{143}Nd/^{144}Nd$ and higher $^{87}Sr/^{86}Sr$ than Kwaimbaita-type basalt (Tejada *et al.* 2004), consistent with its slight relative enrichment in the more incompatible elements (Figs 5, 6, 8 and 10). We have no samples of primitive basalt of Singgalo type but its liquid line of descent probably ran parallel to that in Figure 10, but with Nb about 1 ppm higher.

Although alteration of the rock samples precludes a quantitative treatment of magmatic evolution based on major-element composition, we can use the concentration of Nb to estimate the degree of fractional crystallization. The Nb content of our primary magmas can be calculated by diluting the Nb in 1187–8 (2.27 ppm) with olivine to give values of 1.94 ppm (17% olivine) and 1.67 ppm Nb (36% olivine) for primary magmas in equilibrium with Fo_{90} and Fo_{92}, respectively. Variation of trace-element concentrations during fractional crystallization is described by the Rayleigh fractionation equation

$$C_e/C_p = F^{D-1}$$

where C_p is the concentration of the element in the parental magma, C_e is its concentration in the evolved magma, F is the mass fraction of magma remaining after fractional crystallization and D is the distribution coefficient. For an incompatible element such as Nb (D = 0), the equation reduces to

$$C_e/C_p = 1/F$$

and this can be used to estimate the extent of fractional crystallization represented by the data in Figure 10. Our primary magmas with 1.94 and 1.67 ppm Nb (15.6 and 20.9 wt% MgO) would evolve to a magma with 2.33 ppm Nb (8.5 wt% MgO) through the fractional crystallization of 18 and 29 wt% of olivine, respectively. Plagioclase begins to crystallize at this point, followed by augite (Sano & Yamashita 2004), and the magma would then evolve by fractional crystallization of a cotectic assemblage of the three phases. To produce a residual liquid with 5 ppm Nb (6.8 wt% MgO), equivalent to the basalt at Site 289 (Fig. 10), from magma with 8.5 wt% MgO requires the fractional crystallization of 53% of the cotectic assemblage.

Nature of the mantle source and degree of melting

In the previous section we showed that Kroenke-type basalt represents a plausible parental magma for the bulk of the OJP. Primitive-mantle-normalized incompatible-element patterns (Fig. 5) for samples of Kroenke-type basalt are essentially flat and show depletion in only the most incompatible elements. The implication that the OJP mantle source was slightly depleted is consistent with the radiogenic isotope ratios in Kroenke- and Kwaimbaita-type basalt samples (Tejada *et al.* 2004). These authors show that the data can be modelled by a primitive mantle source from which 1% of melt had been extracted around 3 billion years ago. This mantle source is less depleted than the convecting upper mantle because basalt from the OJP has lower $^{143}Nd/^{144}Nd$ and higher $^{87}Sr/^{86}Sr$ than Pacific N-MORB (Tejada *et al.* 2004).

A mantle source that is less depleted than the N-MORB source is also apparent from incompatible-element concentrations and ratios in OJP basalt. For example, OJP basalt has comparable Zr contents to N-MORB but significantly more Nb (Fig. 6). OJP basalt also has relatively high Nb/Ta (16.3±0.8), a value that is higher than the average value of N-MORB (14.5±1.5, Bodinier *et al.* 2002; 15.5±0.5, Jochum & Hofmann 1988) and in the same range as OIB (16±0.5 for French Polynesian OIB, Bodinier *et al.* 2002).

The concentrations of moderately incompatible elements (Zr–Lu) in Kroenke-type basalt are fairly constant at approximately four times the primitive-mantle values (Fig. 5). As *c.* 30 wt% of equilibrium olivine must be added to the Kroenke basalt to produce magmas in equilibrium with residual mantle olivine with a composition of $Fo_{91.6}$, the primary OJP magma would have had approximately three times the primitive-mantle concentrations of these elements. Because moderately incompatible (unlike highly incompatible) elements are relatively unaffected by mantle depletion processes (e.g. Hofmann 1988) we can use their concentration to give a rough estimate of degree of mantle melting. Assuming $D \approx 0$ in residual harzburgite gives an upper limit of approximately 33% melting. We can refine this estimate through more detailed modelling, but we must first justify our assumption that Kroenke-type basalt can be used to estimate the composition of the primary magma.

Basalt lava flows with concentrations of moderately incompatible elements much lower than those in Kroenke-type basalt are found in the rift zones of Iceland (Fig. 6). For example, the international geochemical reference standard BIR-1 (from the Reykjanes Peninsula in Iceland) has 15.5 ppm Zr (Table 1) compared with 40 ppm in Kroenke-type basalt. Taking these depleted basalt flows to represent primitive Icelandic magma requires either unreasonable degrees (up to 100%) of mantle melting or a very depleted mantle source. This question can be resolved by considering the volumes of magma involved. Lava flows of depleted basalt in Iceland are rare, always small (<0.15 km^3) and confined to the rift axes (Hardarson & Fitton 1997). They are isotopically distinct from the bulk of Icelandic basalt (they have higher $^{143}Nd/^{144}Nd$, for example) and Fitton *et al.* (2003) have argued that they represent small-volume instantaneous melts produced during the advanced stages of dynamic melting of a heterogeneous mantle source. By contrast, Kroenke-type basalt is isotopically identical to the voluminous Kwaimbaita-type basalt (Tejada *et al.* 2004) and was erupted in large-volume flows. Its occurrence in thick (>100 m) units at sites 146 km apart on the eastern flanks of the plateau suggests that huge volumes of Kroenke-type magma were erupted on the OJP. Taking Kroenke-type basalt to represent parental OJP magma therefore seems reasonable.

With increasing incompatibility, the more incompatible elements in Kroenke- and Kwaimbaita-type basalt show a progressive depletion relative to primitive mantle (Fig. 5). This trend is obscured somewhat by mobility of Cs, Rb, Ba and U, but is clear from the abundances of Th and the light rare-earth elements. Figure 11 shows primitive-mantle-normalized abundances of the immobile incompatible trace elements in primary magma calculated from Kroenke-type basalt, plotted on a linear scale for clarity. The constant normalized abundance of the moderately incompatible elements, and the progressive depletion in the more incompatible elements, is clearly seen in this plot. Nb and Ta are not as depleted as the other highly incompatible elements, suggesting that extraction of continental crust (which is deficient in Nb and Ta) could provide a likely mechanism for the depletion of the OJP mantle source.

Figure 11 also shows a best-fit model composition in which primitive mantle (McDonough & Sun 1995) is depleted through the extraction of 1% by mass of average continental crust (Rudnick & Fountain 1995; Nb and Ta from Barth *et al.* 2000), and then batch melted to 30% leaving a harzburgite residue. Olivine and orthopyroxene partition coefficients estimated by Bedini & Bodinier (1999) were used in these calculations. The fit is very good considering the uncertainties in the composition of primitive mantle and continental crust.

The depletion and melting stages in the model are independently constrained, as the former controls the shape of the pattern and the latter the overall abundance level. The degree of melting represented by the calculated primary

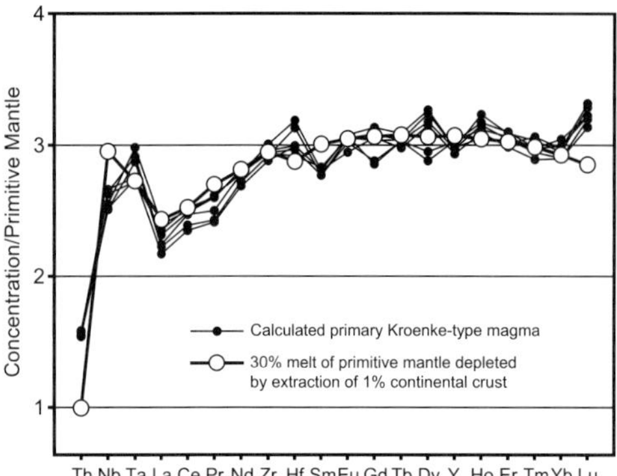

Fig. 11. Primitive-mantle-normalized incompatible-element concentrations (ICP-MS data except for Nb, Zr and Y) in OJP primary magma compared with model concentrations. Primary magma compositions were calculated by incremental addition of equilibrium olivine to analyses of fresh Kroenke-type basalt until they were in equilibrium with $Fo_{91.6}$. Equilibrium melting (Shaw 1970) leaving a harzburgite residue (75% olivine, 25% orthopyroxene) was assumed in calculating the melt composition. Distribution coefficients from Bedini & Bodinier (1999); primitive-mantle values from McDonough & Sun (1995); average continental crust composition from Rudnick & Fountain (1995) and Barth *et al.* (2000).

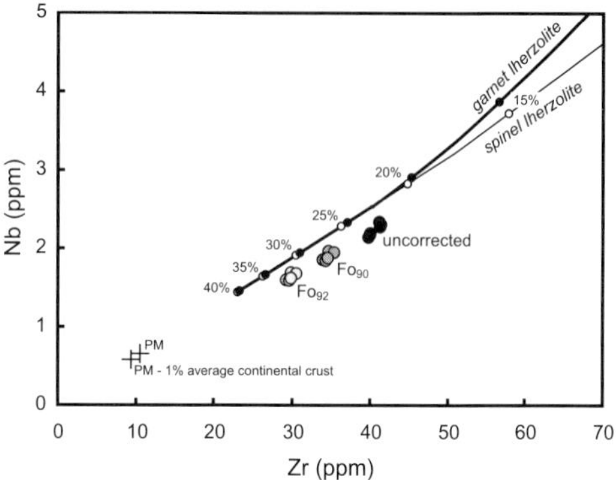

Fig. 12. Nb and Zr (XRF data) contents of primary magma calculated by incremental addition of equilibrium olivine to analyses of the freshest samples of Kroenke-type basalt. 'Uncorrected' data points represent analyses before olivine addition; other points are calculated compositions in equilibrium with residual mantle olivine with Fo_{90} and Fo_{92} composition. Melting curves for depleted garnet- and spinel-lherzolite are calculated for non-modal equilibrium melting (Shaw 1970). The two melting curves converge at 23% melting when garnet, clinopyroxene and spinel are exhausted; the single curve thereafter represents melting of harzburgite residue. Mantle mineral proportions from McKenzie & O'Nions (1991); melting modes from Johnson *et al.* (1990); garnet and clinopyroxene distribution coefficients from Johnson (1998); olivine and orthopyroxene values from Bedini & Bodinier (1999).

magma composition, although insensitive to assumptions about the depletion mechanism, is dependent on assumptions about the composition of olivine in the residual mantle. We used $Fo_{91.6}$ in our calculations because this is the composition calculated for equilibrium (batch) melting by the inverse- and forward-modelling method of Herzberg & O'Hara (2002). The same method gives a less forsteritic residual olivine ($Fo_{90.4}$) under conditions of perfect fractional melting.

The calculated abundances of incompatible elements in primary magma are essentially the same, at large melt fractions, for equilibrium and accumulated fractional melts, but the composition of residual olivine is sensitive to the melting model. Perfect fractional melting seems less likely than equilibrium melting in the present case where enormous volumes of magma are generated in a short period by large degrees of melting, but we must allow for the possibility in our calculations. Figure 12 shows the effect of varying the residual olivine composition between Fo_{90} and Fo_{92} on the calculated Zr and Nb contents of the primary magma. The compositions of primitive mantle and our calculated depleted mantle (primitive mantle minus 1% of continental crust) are shown along with melting curves for the depleted mantle in spinel- and garnet-lherzolite facies. The data points are displaced from the melting curves because our model overestimates the Nb content of the primary magma but gives a good fit for Zr (Fig. 11). Using the calculated Zr contents gives degrees of melting ranging from 27% with Fo_{90} to 31% with Fo_{92} (Fig. 12), values that bracket the range of residual olivine compositions calculated for fractional ($Fo_{90.5}$) and equilibrium ($Fo_{91.6}$) melting of fertile mantle by the Herzberg & O'Hara (2002) method (Herzberg 2004). This procedure also gives an independent estimate of degree of melting, and applying it to the major-element composition of Kroenke-type basalt glass sample 1187–8 gives values of 27% for fractional and 30% for equilibrium melting (Herzberg 2004). These values are in excellent agreement with those derived from incompatible trace elements, and strengthen our conclusion that the OJP primary magma was generated by around 30% melting of a peridotite mantle source. Large degrees of melting are also implied by the low concentrations of total alkalis in the OJP basalts, at the low end of the range for MORB glasses from the East Pacific Rise (Fig. 3).

Our value of 30% is at the upper end of the range of previously published estimates for degree of melting (20–30%). Mahoney *et al.* (1993), for example, used the McKenzie & O'Nions (1991) rare-earth element inversion technique to estimate melt fraction as a function of depth and found a maximum of 30% melting at the top of the melt column. Other authors (e.g. Tejada *et al.* 1996; Neal *et al.* 1997) obtained similar values (20–30%). All these estimates were based on *relative* incompatible-element concentrations in evolved basalt because the composition of the parental OJP magma was unknown at the time. The discovery of primitive, Kroenke-type basalt during ODP Leg 192 allows much more robust estimates of degree of melting to be derived from the *absolute* major and incompatible trace-element composition of primary OJP magma.

Singgalo-type basalt is compositionally and isotopically distinct from Kroenke- and Kwaimbaita types. It has consistently higher concentrations of the more incompatible elements (Fig. 5), and lower $^{143}Nd/^{144}Nd$ and higher $^{87}Sr/^{86}Sr$ (Tejada *et al.* 2004), suggesting a less depleted mantle source. Unfortunately, we have no samples of parental Singgalo-type basalt and so we cannot model the mantle source as we did with the magnesian Kroenke-type basalt. The data show that a less depleted mantle source was tapped during the waning phase of magmatism and that this produced a veneer of Singgalo-type basalt along the western flank of the OJP (Fig. 9). Primitive-mantle-normalized incompatible-element concentrations fall steadily with increasing compatibility from Nb to Lu (Fig. 5), suggesting that the degree of melting was less and the melt zone was probably deeper (leaving garnet as a residual phase) than in the case of the dominant Kroenke- and Kwaimbaita-type magmas. The lower SiO_2 content (at similar MgO) of Singgalo-type compared with Kwaimbaita-type basalt (Fig. 4) is also consistent with derivation of the former at a higher pressure in the mantle (e.g. Falloon & Green 1988).

Singgalo-type magma erupted at a late stage in the evolution of the OJP has some similarity to basalt erupted on the rift flanks in Iceland. This off-axis basalt has higher concentrations of the more incompatible elements and lower $^{143}Nd/^{144}Nd$ and higher $^{87}Sr/^{86}Sr$, than does basalt erupted in the rift axes (Hémond *et al.* 1993; Fitton *et al.* 2003). Smaller degrees of melting beneath thicker lithosphere on the rift flanks seems to be preferentially sampling less depleted and more easily fusible parts of a heterogeneous mantle source (Fitton *et al.* 2003). Progressively smaller-degree melting during the waning phases of OJP magmatism on the western flanks of the plateau may similarly be sampling a less depleted and isotopically distinct component in a heterogeneous mantle source. With our limited data, however, we cannot quantify the composition of the mantle source for the Singgalo-type magmas or the degree of melting involved in their generation.

Quantifying mantle-source composition and degree of melting is likewise impossible for the high-Nb magma type represented by two volcaniclastic subunits at Site 1184 on the eastern salient of the OJP (Figs 7 and 9). Alteration has affected the major-element composition of these rocks, but their low Cr and Ni (Fig. 7) and high incompatible-element (Fig. 5) content suggests that the original magmas were fairly evolved. The rocks have flat primitive-mantle-normalized abundances of moderately incompatible elements (Zr–Lu) and, in this respect, they resemble Kroenke- and Kwaimbaita-type basalt. They are also similar to these basalts in $^{143}Nd/^{144}Nd$ (White *et al.* 2004) and so may have had a similar mantle source. They differ, however, in their Th, Nb, Ta and La contents, which are higher than in any of the other OJP rocks analysed in this study (Fig. 5). Thus, they have comparable Ce/Yb to Kwaimbaita-type basalt, but much higher Nb/Zr (Fig. 8). Enrichment of Kroenke-type mantle by small amounts of a small melt fraction, or mixing between Kwaimbaita-type magma and a small-degree melt, are possible explanations for the high-Nb magmas. With limited data from very altered rocks, however, we can do no more than speculate on their mantle source.

Mantle temperature

Trace- and major-element modelling both require around 30% melting of a fertile peridotitic mantle source to produce the primary magma for the bulk of the observed OJP. This value will be slightly lower if the melt was generated by perfect fractional melting, and will increase if the mantle source had residual olivine more magnesian than Fo_{92}, but is unlikely to be outside the range 25–35% (Fig. 12). We can use our estimate for the degree of melting to estimate the potential temperature of mantle melting during decompression.

Using the parameterization of the melting interval of fertile peridotite given by McKenzie & Bickle (1988) and the entropy change (ΔS)

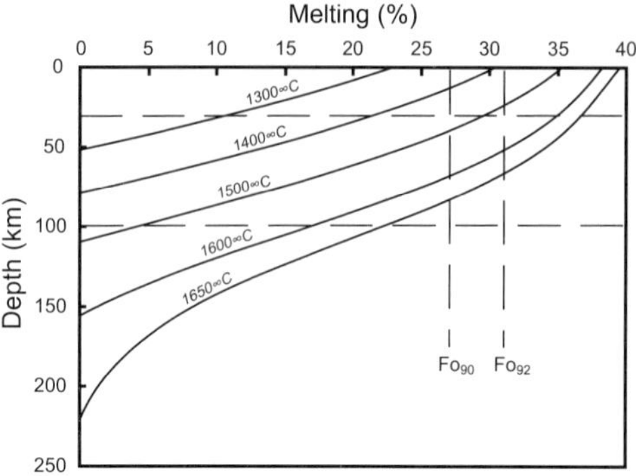

Fig. 13. Variation of degree of equilibrium melting (%) with depth for fertile mantle with potential temperatures ranging from 1300 to 1650°C. The McKenzie & Bickle (1988) parameterization of the melt interval and a value of $\Delta S = 350$ J kg^{-1} K^{-1} (Kojitani & Akaogi 1995) were used in the calculations. Perfect fractional melting would reduce the melt fraction by 5–15% of its value along each curve (Iwamori *et al.* 1995). Vertical broken lines show the calculated range of melting (27–31%) for OJP primary magmas, assuming residual mantle olivine compositions of Fo_{90} and Fo_{92} (equilibrium and fractional melting respectively; Herzberg 2004). The horizontal broken lines mark the depth to the base of the OJP crust (30 km) and oceanic lithosphere (100 km).

associated with melting, it is possible to relate melt fraction to depth and potential temperature (T_p). Figure 13 shows a set of melt fraction–depth profiles for T_p ranging from 1300 to 1650°C, assuming $\Delta S = 350$ J kg^{-1} K^{-1} (Kojitani & Akaogi 1995). These profiles show the extent to which mantle of a given T_p would melt if decompressed instantaneously to the depth shown. Thus, mantle with T_p = 1500°C would begin to melt at a depth of about 110 km, and melting would increase with decreasing depth until it reached 35% at the surface. McKenzie & Bickle (1988) applied similar melting curves to the case of passive decompression at mid-ocean ridges where the melt zone is triangular in cross-section with maximum decompression and melting under the ridge axis, decreasing to zero with distance away from the axis. All the melt is focused into axial magma reservoirs and the residual mantle moves sideways with the accreting plates. McKenzie & Bickle (1988) derived the average melt composition by summing all instantaneous melts in the melt zone, which is equivalent to integrating all melt fractions along one of the curves in Figure 13 (Langmuir *et al.* 1992). The average degree of melting along each curve in Figure 13 will be roughly equal to half the maximum degree in the case of equilibrium (batch) melting and a little less than half in the case of fractional melting (Langmuir *et al.* 1992). In the equilibrium case, the melt stays with the matrix during decompression and is then released at the top of each column in the triangular melt zone. In the fractional case, melt is released continuously during decompression and accumulates near the surface. In both cases it is assumed that melt from all parts of the melt zone is extracted.

It is clear from the profiles in Figure 13 that the melt fractions required to produce the OJP primary magma cannot be achieved by mantle of any reasonable temperature decompressing beneath oceanic lithosphere of normal thickness (100 km). The lithosphere must have been thinner than normal but cannot have been thinner than the OJP crust (30 km) by the time the lava flows analysed in this study were erupted. If the OJP formed from hot mantle decompressing to a depth of 30 km (i.e. at or close to a spreading centre) then the mantle potential temperature must have been at least 1500°C to reach a *maximum* of 27–31% melting. This temperature would need to be about 50°C higher if the melts were produced by fractional melting (Iwamori *et al.* 1995). *Average* melt fractions, produced by integrating the melt profiles in Figure 13, are unlikely to reach 27–31% melting at any potential temperature. One way in which the average melt fraction can approach the maximum melt fraction is for the mantle to be actively decompressed, where residual mantle returns to depth after leaving the melting region rather than being incorporated in horizontal plate flow as in passive upwelling (Langmuir *et al.* 1992). This situation might apply in the case of a start-up plume head (e.g. Richards *et al.* 1989) in which mantle is actively and rapidly fed into the melt zone.

The residual mantle left behind after the extraction of 30% melt should be refractory harzburgite with Fe-depleted olivine. Such compositions are generally thought to result in fast seismic velocity anomalies as, for example, in cratonic mantle (e.g. Jordan 1988). Paradoxically, recent seismic tomography experiments showed a rheologically strong but seismically *slow* upper-mantle root extending to about 300 km depth beneath the OJP (e.g. Richardson *et al.* 2000; Klosko *et al.* 2001). This suggests that the observed seismic mantle root does not represent the residue of the partial melting process that produced the OJP basalts. A residual harzburgite root left behind after 30% mantle melting to produce 30–35 km of basaltic crust should be about 80 km thick rather than the 300 km observed (Neal *et al.* 1997; Klosko *et al.* 2001).

A mantle plume origin for OJP magmatism?

Decompression melting of peridotite mantle with $T_p > 1500$°C in a start-up plume head rising beneath or close to a spreading centre can account for the composition and volume of OJP basalt. In this situation we should expect the ocean floor to be elevated above sea level, as in Iceland, and yet we know that nearly all of the observed products of OJP volcanism were erupted well below sea level (Michael 1999; Roberge *et al.* 2004). There is no evidence for the existence of the clastic apron that would result if large parts of the OJP had ever formed a subaerial volcanic edifice. Potential solutions to this paradox fall into three categories: a more fertile mantle; a hydrous mantle; and extraterrestrial causes for OJP magmatism.

Mantle plumes need not be composed entirely of peridotite but may contain significant amounts of embedded eclogite derived from ancient subducted ocean crust (Hofmann & White 1982). Because eclogite melts at a lower temperature than does peridotite (Yasuda *et al.* 1994), a composite plume could, in theory, generate large volumes of magma at a lower T_p than one composed entirely of peridotite (e.g. Leitch & Davies 2001). Melt production will be enhanced further if the embedded eclogite bodies can melt independently using latent heat supplied by conduction from the surrounding unmelted peridotite (Cordery *et al.* 1997). Takahashi *et al.* (1998) have shown that the Columbia River basalt magmas could have been produced by 30–50% melting of a MORB-like basalt source at 2 GPa and 1300–1350°C (i.e. below the peridotite solidus). They argue that similar melting processes may have operated during the formation of other large igneous provinces (LIPs) and that composite plumes need not be as hot or as large as those composed only of peridotite. Large-scale heterogeneity is not necessary for composite plumes to produce voluminous melt at a lower T_p. Yaxley (2000) has shown that a homogeneous mixture of peridotite + 30% basalt has a lower solidus temperature and a much smaller melting interval than does peridotite alone, and would therefore produce more melt during decompression.

The composite-plume model seems to offer a solution to the OJP paradox of high melt productivity without significant uplift because it allows large degrees of melting in plumes with modest T_p. However, this can only work efficiently if isolated pods of eclogite can melt independently of the more refractory peridotite host (Cordery *et al.* 1997). We can rule out a pure eclogite source for the OJP magmas because eclogite melts to a liquid with higher SiO_2 and lower MgO than the parental Kroenke-type magma (Takahashi *et al.* 1998; Yaxley & Green 1998). Tejada *et al.* (2002) modelled the major- and incompatible trace-element composition of OJP basalt by mixing large-degree melts from eclogite with small-degree peridotite melts, but found that melting of peridotite alone (Mahoney 1993; Tejada *et al.* 1996; Neal *et al.* 1997) fits the data better. Similarly, Figure 11 shows a very good fit between the estimated composition of the primary OJP magma and that of a 30% equilibrium melt of a peridotite source. The remarkable agreement in degrees of melting calculated independently through phase equilibria (Herzberg 2004) and trace-element modelling provides compelling evidence in favour of a peridotite source. Both approaches lead to values of approximately 27% for fractional melting and approximately 30% for equilibrium melting. An enriched mantle source composed of peridotite + basalt (e.g. Yaxley 2000) could produce voluminous magma at lower T_p, but such a source would inevitably have higher concentrations of incompatible elements. It would therefore require correspondingly higher degrees of melting to produce magma with the low concentrations of incompatible elements that are found in Kroenke-type basalt. Any temperature advantage gained through the more efficient melting of an enriched source would be lost by the need for a higher T_p to produce larger-degree melts. Melting needs latent heat, and this can only be supplied by a high-T_p mantle source.

Hydrous mantle provides an equally unsatisfactory solution to the paradox because studies on OJP basaltic glass (Michael 1999; Roberge *et al.* 2004) show that the water content of the magmas was very low. Small amounts of water in the mantle would, in any case, cause the formation of small melt fractions at greater depth than with anhydrous mantle and therefore *decrease* the average melt fraction rather than raise it (Asimow & Langmuir 2003).

Finally, the impact of an asteroid could possibly trigger large-scale mantle melting without causing uplift (e.g. Rogers 1982; Jones *et al.* 2002; Ingle & Coffin 2004), but there is no evidence in the Aptian sedimentary record for an impact of the required magnitude. Furthermore, the impact would normally be expected to produce magma with the chemical and isotopic characteristics of N-MORB as it is the upper mantle that would be melted. OJP basalts, however, are isotopically distinct from modern (Tejada *et al.* 2004) and pre-OJP (Janney & Castillo 1997; Mahoney *et al.* 1998) Pacific N-MORB. Their Os-isotope composition (Parkinson *et al.* 2002) and Nb/Ta are likewise distinct from those in N-MORB, but similar to values inferred for primitive mantle. Tejada *et al.* (2004) discuss the impact hypothesis in some detail and conclude that it is unlikely to be able to resolve the paradox.

In the absence of a viable alternative, we have to conclude that a peridotite mantle plume with T_p >1500°C provides the only plausible explanation for the formation of the OJP. We are therefore left in the unsatisfactory position of not being able to explain why the OJP was emplaced below sea level.

Conclusions

Ocean Drilling Program Leg 192 has extended our knowledge of the composition and origin of OJP magmas considerably. Kwaimbaita-type basalt has now been found at nine of the 10 DSDP and ODP drill sites on and around the plateau, it forms thick successions on-land in Malaita and Santa Isabel in the Solomon Islands, and almost certainly represents the dominant magma type. Subordinate Singgalo-type basalt is found capping Kwaimbaita-type basalt in Malaita, Santa Isabel and at ODP Site 807. Thin volcaniclastic turbidite layers above basement at Site 1183 also have the composition of Singgalo-type basalt. The Singgalo magma type is distinguished from the dominant Kwaimbaita type by its small relative enrichment in the more incompatible elements, lower $^{206}Pb/^{204}Pb$ and $^{143}Nd/^{144}Nd$, and higher $^{87}Sr/^{86}Sr$.

The discovery of magnesian, Kroenke-type basalt at Sites 1185 and 1187 (and as a volcaniclastic subunit at Site 1184) was particularly significant. Kroenke- and Kwaimbaita-type basalt are isotopically identical, and the former represents the likely parental magma for the bulk of the OJP. The primary magmas evolved along a simple liquid line of descent involving the crystallization of olivine followed by olivine, plagioclase and augite.

Identification of a parental magma type allows the primary magma composition to be estimated by incremental addition of equilibrium olivine until the residual mantle olivine composition is reached. Taking residual olivine compositions ranging from Fo_{90} (for fractional melting) to Fo_{92} (equilibrium melting) gives primary magma with MgO ranging from 15.6 to 20.4 wt%, respectively. Incompatible-element contents in the calculated primary magma, coupled with radiogenic isotope ratios, are consistent with a mantle source consisting of primitive mantle depleted through the extraction of 1% by mass of average continental crust. The degree of melting required to produce the primary magma from this source ranges from approximately 27% (fractional melting) to approximately 30% (equilibrium melting). We cannot estimate the primary Singgalo-type magma composition because we have no samples of the corresponding high-Mg basalt. However, the isotopic and chemical composition of Singgalo-type basalt suggest that it represents smaller degrees of melting generated at greater depth from less depleted portions of a heterogeneous mantle source.

Mantle with a potential temperature of approximately 1500°C will melt to a *maximum* extent of about 30% if decompressed to 30 km-depth (i.e. at or close to a spreading centre). To achieve an *average* of 30% melting requires that the mantle be hotter than 1500°C and actively and rapidly fed into the melt zone, and a start-up mantle plume provides the most obvious mechanism. This should have caused uplift well above sea level, but the abundance of essentially non-vesicular pillow lava and the absence of any basalt showing signs of subaerial weathering show that the OJP was emplaced below sea level. Volatile concentrations in basaltic glass (Michael 1999; Roberge *et al.* 2004) support this conclusion.

We have not yet been able to resolve the paradox of apparent high mantle potential temperature coupled with submarine emplacement. An eclogitic source cannot supply the high-MgO parental Kroenke-type magma, and an enriched mantle source would need higher degrees of melting (and therefore higher T_p) to produce magma with concentrations of incompatible elements as low as those in Kroenke-type magma. We can also rule out a hydrous mantle source because the magmas have very low volatile contents. Widespread melting of the mantle following the impact of an asteroid provides an attractive means of avoiding uplift, but the magma would be generated entirely within the upper mantle and ought to have the chemical and isotopic characteristics of N-MORB. OJP basalt is isotopically and chemically distinct from N-MORB and seems to have a lower-mantle source.

This research used samples provided by the Ocean Drilling Program (ODP). ODP is sponsored by the US

National Science Foundation (NSF) and participating countries under management of Joint Oceanographic Institutions (JOI), Inc. We thank the ODP and Transocean/Sedco-Forex staff on board the *JOIDES Resolution* for their considerable contribution to the success of ODP Leg 192. We are also grateful to D. James for help with the XRF analyses, to S. Pourtales and O. Bruguier for technical assistance with the ICP-MS analyses, and to J. Maclennan for calculating the melt profiles used in Figure 13. F. Frey and M. Wilson are thanked for their thoughtful and constructive reviews; and J. Mahoney, P. Wallace, J. Maclennan and C. Herzberg for helpful discussions. Grants from the UK Natural Environment Research Council part-funded the XRF analytical facility at Edinburgh and covered some of the analytical costs involved in this work. Acquisition of ICP-MS data was supported by a grant from the Groupe ad hoc 'OCEANS' INSU/CNRS (AO 2001).

References

ANDREWS, J.E., PACKHAM, G.H. *et al.* 1975. *Proceedings of the Deep Sea Drilling Project, Initial Reports*, **30**.

ASIMOW, P.D. & LANGMUIR, C.H. 2003. The importance of water to oceanic mantle melting regimes. *Nature*, **421**, 815–820.

BACH, W., ERZINGER, J., DOSSO, L., BOLLINGER, C., BOUGAULT, H., ETOUBLEAU, J. & SAUERWEIN, J. 1996. Unusually large Nb–Ta depletions in North Chile ridge basalts at 36°50 to 38°56S: major element, trace element, and isotopic data. *Earth and Planetary Science Letters* **142**, 223–240.

BANERJEE, N.R., HONNOREZ, J. & MUEHLENBACHS, K. 2004. Low-temperature alteration of submarine basalts from the Ontong Java Plateau. *In*: FITTON, J.G., MAHONEY, J.J., WALLACE, P.J. & SAUNDERS, A.D. (eds) *Origin and Evolution of the Ontong Java Plateau*. Geological Society, London, Special Publications, **229**, 259–273.

BARTH, M.G., MCDONOUGH, W.F. & RUDNICK, R.L. 2000. Tracking the budget of Nb and Ta in the continental crust. *Earth and Planetary Science Letters*, **165**, 197–213.

BEDINI, R.-M. & BODINIER, J.-L. 1999. Distribution of incompatible trace elements between the constituents of spinel peridotite xenoliths: ICP-MS data from the East African Rift. *Geochimica et Cosmochimica Acta*, **63**, 3883–3900.

BERGEN, J.A. 2004. Calcareous nannofossils from ODP Leg 192, Ontong Java Plateau. *In*: FITTON, J.G., MAHONEY, J.J., WALLACE, P.J. & SAUNDERS, A.D. (eds) *Origin and Evolution of the Ontong Java Plateau*. Geological Society, London, Special Publications, **229**, 113–132.

BODINIER, J.-L., KALFOUN, F., GODARD, M., BARSCZUS, H.G. & SABATÉ, P. 2002. Nb/Ta geochemical reservoirs. *Geochimica et Cosmochimica Acta*, **66**, Supplement, A85.

CASTILLO, P.R., PRINGLE, M.S. & CARLSON, R.W. 1994. East Mariana Basin tholeiites: Jurassic ocean crust or Cretaceous rift basalts related to the Ontong Java plume? *Earth and Planetary Science Letters*, **123**, 139–154.

CHAMBERS, L.M., PRINGLE, M.S. & FITTON, J.G. 2002. Age and duration of magmatism on the Ontong Java Plateau: ^{40}Ar-^{39}Ar results from ODP Leg 192. Abstract V71B-1271. *Eos, Transactions of the American Geophysical Union*, **83**, F47.

CHAMBERS, L.M., PRINGLE, M.S. & FITTON, J.G. 2004. Phreatomagmatic eruptions on the Ontong Java Plateau: an Aptian $^{40}Ar/^{39}Ar$ age for volcaniclastic rocks at ODP Site 1184. *In*: FITTON, J.G., MAHONEY, J.J., WALLACE, P.J. & SAUNDERS, A.D. (eds) *Origin and Evolution of the Ontong Java Plateau*. Geological Society, London, Special Publications, **229**, 325–331.

COFFIN, M.F. & ELDHOLM, O. 1994. Large igneous provinces: crustal structure, dimensions, and external consequences. *Reviews of Geophysics*, **32**, 1–36.

CORDERY, M.C., DAVIES, G.F. & CAMPBELL, I.H. 1997. Genesis of flood basalts from eclogite-bearing mantle plumes. *Journal of Geophysical Research*, **102**, 20 179–20 198.

COX, K.G. & BELL, J.D. 1972. A crystal fractionation model for the basaltic rocks of the New Georgia Group, British Solomon Islands. *Contributions to Mineralogy and Petrology*, **37**, 1–13.

FALLOON, T.J. & GREEN, D.H. 1988. Anhydrous partial melting of peridotite from 8 to 35 kb and the petrogenesis of MORB. *In*: MENZIES, M.A. & COX, K.G. (eds) *Oceanic and Continental Lithosphere: Similarities and Differences. Journal of Petrology*, Special Volume, 379–414.

FITTON, J.G., SAUNDERS, A.D., KEMPTON, P.D. & HARDARSON, B.S. 2003. Does depleted mantle form an intrinsic part of the Iceland plume? *Geochemistry, Geophysics, Geosystems*, **4**, 1032, 2002GC000424.

FITTON, J.G., SAUNDERS, A.D., LARSEN, L.M., HARDARSON, B.S. & NORRY, M.J. 1998. Volcanic rocks from the southeast Greenland margin at 63°N: composition, petrogenesis and mantle sources. *Proceedings of the Ocean Drilling Program, Scientific Results* **152**, 331–350.

GLADCZENKO, T.P., COFFIN, M.F. & ELDHOLM, O. 1997. Crustal structure of the Ontong Java Plateau: modeling of new gravity and existing seismic data. *Journal of Geophysical Research*, **102**, 22 711–22 729.

GODARD, M., JOUSSELIN, D. & BODINIER, J.-L. 2000. Relationships between geochemistry and structure beneath a palaeo-spreading centre: A study of the mantle section in the Oman Ophiolite. *Earth and Planetary Science Letters*, **180**, 133–148.

GOVINDARAJU, K. 1994. 1994 compilation of working values and sample description for 383 geostandards. *Geostandards Newsletter*, **18**, 1–158.

GRADSTEIN, F.M., AGTERBERG, F.P., OGG, J.G., HARDENBOL, J., VAN VEEN, P., THIERRY, J. & HUANG, Z. 1995. A Triassic, Jurassic and Cretaceous time scale. *In*: BERGGREN, W.A., KENT, D.V., AUBRY, M.P. & HARDENBOL, J. (eds) *Geochronology, Time Scales and Global Stratigraphic Correlation*. SEPM, Special Publication, **54**, 95–126.

HANAN, B.B., BLICHERT-TOFT, J., KINGSLEY, R. & SCHILLING, J.-G. 2000. Depleted Iceland mantle plume geochemical signature: Artifact of multicomponent mixing? *Geochemistry, Geophysics, Geosystems*, **1**, 1999GC000009.

HARDARSON, B.S. & FITTON, J.G. 1997. Mechanisms of crustal accretion in Iceland. *Geology*, **25**, 1043–1046.

HARLAND, W.B., ARMSTRONG, R.L., COX, A.V., CRAIG, L.E., SMITH, A.G. & SMITH, D.G. 1990. *A Geologic Time Scale 1989*. Cambridge University Press, Cambridge.

HÉMOND, C., ARNDT, N., LICHTENSTEIN, U., HOFMANN, A.W., OSKARSSON, N. & STEINTHORSSON, S. 1993. The heterogeneous Iceland plume: Nd–Sr–O isotopes and trace element constraints. *Journal of Geophysical Research*, **98**, 15 833–15 850.

HERZBERG, C. 2004. Partial melting below the Ontong Java Plateau. *In*: FITTON, J.G., MAHONEY, J.J., WALLACE, P.J. & SAUNDERS, A.D. (eds) *Origin and Evolution of the Ontong Java Plateau*. Geological Society, London, Special Publications, **229**, 179–183.

HERZBERG, C. & O'HARA, M.J. 2002. Plume-associated ultramafic magmas of Phanerozoic age. *Journal of Petrology*, **43**, 1857–1883.

HOFMANN, A. 1988. Chemical differentiation of the Earth: the relationship between mantle, continental crust, and oceanic crust. *Earth and Planetary Science Letters*, **90**, 297–314.

HOFMANN, A.W. & WHITE, W.M. 1982. Mantle plumes from ancient oceanic crust. *Earth and Planetary Science Letters*, **57**, 421–436.

INGLE, S.P. & COFFIN, M.F. 2004. Impact origin for the greater Ontong Java Plateau? Geochemical and petrological evidence. *Earth and Planetary Science Letters*, **218**, 123–134.

IONOV, D.A., SAVOYANT, L. & DUPUY, C. 1992. Application of the ICP-MS technique to trace element analysis of peridotites and their minerals. *Geostandards Newsletter*, **16**, 311–315.

IWAMORI, H., MCKENZIE, D. & TAKAHASHI, E. 1995. Melt generation by isentropic mantle upwelling. *Earth and Planetary Science Letters*, **134**, 253–266.

JANNEY, P.E. & CASTILLO, P.R. 1997. Geochemistry of Mesozoic Pacific MORB: constraints on melt generation and the evolution of the Pacific upper mantle. *Journal of Geophysical Research*, **102**, 5207–5229.

JOCHUM, K.P. & HOFMANN, A.W. 1998. Nb/Ta in MORB and continental crust: Implication for a superchondritic Nb/Ta reservoir in the mantle. Abstract V21B-04, *Eos, Transactions of the American Geophysical Union*, **79**, 354.

JOHNSON, K.T., DICK, H.J.B. & SHIMIZU, N. 1990. Melting in the oceanic upper mantle: an ion microprobe study of diopsides in abyssal peridotites. *Journal of Geophysical Research*, **95**, 2661–2678.

JOHNSON, K.T.M. 1998. Experimental determination of partition coefficients for rare earth and high-field-strength elements between clinopyroxene, garnet, and basaltic melt at high pressures. *Contributions to Mineralogy and Petrology*, **133**, 60–68.

JONES, A.P., PRICE, G.D., PRICE, N.J., DECARLI, P.S. & CLEGG, R.A. 2002. Impact induced melting and the development of large igneous provinces. *Earth and Planetary Science Lettes*, **202**, 551–561.

JORDAN, T.H. 1988. Structure and formation of the continental tectosphere. *In*: MENZIES, M.A. & COX, K.G. (eds) *Oceanic and Continental Lithosphere: Similarities and Differences. Journal of Petrology*, Special Volume, 11–37.

KEMPTON, P.D., PEARCE, J.A., BARRY, T.L., FITTON, J.G., LANGMUIR, C. & CHRISTIE, D.M. 2002. Sr–Nd–Pb–Hf isotope results from ODP Leg 187: Evidence for mantle dynamics of the Australian–Antarctic discordance and origin of the Indian MORB source. *Geochemistry, Geophysics, Geosystems*, **3**, 2002GC000320.

KLOSKO, E.R., RUSSO, R.M., OKAL, E.A. & RICHARDSON, W.P. 2001. Evidence for a rheologically strong chemical mantle root beneath the Ontong–Java Plateau. *Earth and Planetary Science Letters*, **186**, 347–361.

KOJITANI, H. & AKAOGI, M. 1995. Measurement of heat of fusion of model basalt in the system diopside–forsterite–anorthite. *Geophysical Research Letters*, **22**, 2329–2332.

KROENKE, L.W., BERGER, W., JANECEK, T.R. *et al*. 1991. *Proceeding of the Ocean Drilling Program, Initial Reports*, **130**.

KROENKE, L.W., WESSEL, P. & STERLING, A. 2004. Motion of the Ontong Java Plateau in the hot-spot frame of reference: 122 Ma–present. *In*: FITTON, J.G., MAHONEY, J.J., WALLACE, P.J. & SAUNDERS, A.D. (eds) *Origin and Evolution of the Ontong Java Plateau*. Geological Society, London, Special Publications, **229**, 9–20.

LANCELOT, Y., LARSON, R.L. *et al*. 1990. *Proceedings of the Ocean Drilling Program, Initial Reports*, **129**.

LANGMUIR, C.H., KLEIN, E.M. & PLANK, T. 1992. Petrological systematics of mid-ocean ridge basalts: constraints of melt generation beneath ocean ridges. *In*: PHIPPS MORGAN, J., BLACKMAN, D.K. & SINTON, J.M. (eds) *Mantle Flow and Melt Generation at Mid-Ocean Ridges*. American Geophysical Union, Geophysical Monograph, **71**, 183–280.

LARSON, R.L., SCHLANGER, S.O. *et al*. 1981. *Initial Reports of the Deep Sea Drilling Project*, **61**.

LE BAS, M.J., LE MAITRE, R.W., STRECKEISEN, A. & ZANETTIN, B. 1986. A chemical classification of volcanic rocks based on the total alkali–silica diagram. *Journal of Petrology*, **27**, 745–750.

LEITCH, A.M. & DAVIES, G.F. 2001. Mantle plumes and flood basalts: Enhanced melting from plume ascent and an eclogite component. *Journal of Geophysical Research*, **106**, 2047–2059.

MACDONALD, G.A. & KATSURA, T. 1964. Chemical composition of Hawaiian lavas. *Journal of Petrology*, **5**, 82–133.

MAHONEY, J.J., STOREY, M., DUNCAN, R.A., SPENCER, K.J. & PRINGLE, M. 1993. Geochemistry and geochronology of the Ontong Java Plateau. *In*: PRINGLE, M., SAGER, W., SLITER, W. & STEIN, S. (eds) *The Mesozoic Pacific. Geology, Tectonics, and Volcanism*. American Geophysical Union, Geophysical Monograph, **77**, 233–261.

MAHONEY, J.J., SINTON, J.M., MACDOUGALL, J.D., SPENCER, K.J. & LUGMAIR, G.W. 1994. Isotope and trace element characteristics of a super-fast spreading ridge: East Pacific rise, 13–23°S. *Earth and Planetary Science Letters*, **121**, 173–193.

MAHONEY, J.J., FREI, R., TEJADA, M.L.G., MO, X.X., LEAT, P.T. & NÄGLER, T.F. 1998. Tracing the Indian Ocean mantle domain through time: isotopic results from old West Indian, East Tethyan, and South Pacific seafloor. *Journal of Petrology*, **39**, 1285–1306.

MAHONEY, J.J., FITTON, J.G., WALLACE, P.J. *et al.* 2001. *Proceedings of the Ocean Drilling Program, Initial Reports*, **192** [Online]. Available from World Wide Web: http://www-odp.tamu.edu/publications/192_IR/192ir.htm

MCDONOUGH, W.F. & SUN, S.-S. 1995. The composition of the Earth. *Chemical Geology* **120**, 223–253.

MCKENZIE, D.P. & BICKLE, M.J. 1988. The volume and composition of melt generated by extension of the lithosphere. *Journal of Petrology*, **29**, 625–679.

MCKENZIE, D & O'NIONS, R.K. 1991. Partial melt distributions from inversion of rare-earth element data. *Journal of Petrology*, **32**, 1021–1091.

MICHAEL, P.J. 1999. Implications for magmatic processes at Ontong Java Plateau from volatile and major element contents of Cretaceous basalt glasses. *Geochemistry, Geophysics, Geosystems*, **1**, 1999GC000025.

MOBERLEY, R., SCHLANGER, S.O. *et al.* 1986. *Initial Reports of the Deep Sea Drilling Project*, **89**.

NEAL, C.R., MAHONEY, J.J., KROENKE, L.W., DUNCAN, R.A. & PETTERSON, M.G. 1997. The Ontong Java Plateau. *In*: MAHONEY, J.J. & COFFIN, M.F. (eds) *Large Igneous Provinces: Continental, Oceanic, and Planetary Flood Volcanism*. American Geophysical Union, Geophysical Monograph, **100**, 183–216.

PARKINSON, I.J., SCHAEFER, B.F. & ARCULUS, R.J. 2002. A lower mantle origin for the world's biggest LIP? A high precision Os isotope isochron from Ontong Java Plateau basalts drilled on ODP Leg 192. *Geochimica et Cosmochimica Acta*, **66**, Supplement, A580.

PETTERSON, M.G. 2004. The geology of north and central Malaita, Solomon Islands: the thickest and most accessible part of the world's largest (Ontong Java) ocean plateau. *In*: FITTON, J.G., MAHONEY, J.J., WALLACE, P.J. & SAUNDERS, A.D. (eds) *Origin and Evolution of the Ontong Java Plateau*. Geological Society, London, Special Publications, **229**, 63–81.

PETTERSON, M.G., BABBS, T., NEAL, C.R., MAHONEY, J.J., SAUNDERS, A.D., DUNCAN, R.A., TOLIA, D., MAGU, R., QOPOTO, C., MAHOA, H. & NATOGGA, D. 1999. Geological–tectonic framework of Solomon Islands, SW Pacific: crustal accretion and growth within an intra-oceanic setting. *Tectonophysics*, **301**, 35–60.

RICHARDS, M.A., DUNCAN, R.A. & COUIRTILLOT, V. 1989. Flood basalts and hot-spot tracks: plume heads and tails. *Science*, **246**, 103–107.

RICHARDSON, W.P., OKAL, E.A. & VAN DER LEE, S. 2000. Rayleigh-wave tomography of the Ontong Java Plateau. *Physics of the Earth and Planetary Interiors*, **118,** 29–61.

RIISAGER, P., HALL, S., ANTRETTER, M. & ZHAO, X. 2004. Early Cretaceous Pacific palaeomagnetic pole from Ontong Java plateau basement rocks. *In*: FITTON, J.G., MAHONEY, J.J., WALLACE, P.J. & SAUNDERS, A.D. (eds) *Origin and Evolution of the Ontong Java Plateau*. Geological Society, London, Special Publications, **229**, 31–44.

ROBERGE, J., WHITE, R. & WALLACE, P. 2004. Volatiles in submarine basaltic glasses from the Ontong Java Plateau (ODP Leg 192): implications for magmatic processes and source region compositions. *In*: FITTON, J.G., MAHONEY, J.J., WALLACE, P.J. & SAUNDERS, A.D. (eds) *Origin and Evolution of the Ontong Java Plateau*. Geological Society, London, Special Publications, **229**, 239–257.

ROGERS, G.C. 1982. Oceanic plateaus as meteorite impact signatures. *Nature*, **299**, 341–342.

RUDNICK, R.L. & FOUNTAIN, D.M. 1995. Nature and composition of the continental crust: a lower crustal perspective. *Reviews of Geophysics*, **33**, 267–309.

SANO, T & YAMASHITA, S. 2004. Experimental petrology of basement lavas from Ocean Drilling Program Leg 192: Implications for differentiation processes in Ontong Java Plateau magmas. *In*: FITTON, J.G., MAHONEY, J.J., WALLACE, P.J. & SAUNDERS, A.D. (eds) *Origin and Evolution of the Ontong Java Plateau*. Geological Society, London, Special Publications, **229**, 185–218.

SAUNDERS, A.D. 1986. Geochemistry of basalts from the Nauru Basin, Deep Sea Drilling Project Legs 61 and 89: implications for the origin of oceanic flood basalts. *In*: MOBERLEY, R., SCHLANGER, S.O. *et al. Initial Reports of the Deep Sea Drilling Project*, **89**, 499–517.

SHAW, D.M. 1970. Trace element fractionation during anatexis. *Geochimica et Cosmochimica Acta*, **34**, 237–243.

SIKORA, P.J. & BERGEN, J.A. 2004. Lower Cretaceous planktonic foraminiferal and nonnofossil biostratigraphy of Ontong Java sites from DSDP Leg 30 and ODP Leg 192. *In*: FITTON, J.G., MAHONEY, J.J., WALLACE, P.J. & SAUNDERS, A.D. (eds) *Origin and Evolution of the Ontong Java Plateau*. Geological Society, London, Special Publications, **229**, 83–111.

SMITH, W.H.F. & SANDWELL, D.T. 1997. Global seafloor topography from satellite altimetry and ship depth soundings. *Science*, **277P** 1956–1962.

SUN, S.-S. & MCDONOUGH, W.F. 1989. Chemical and isotopic systematics of oceanic basalts: implications for mantle composition and processes. *In*: SAUNDERS, A.D. & NORRY, M.J. (eds) *Magmatism in the Ocean Basins*. Geological Society, London, Special Publications, **42**, 313–345.

TAKAHASHI, E., NAKAJIMA, K. & WRIGHT, T.L. 1998. Origin of the Columbia River basalts: melting of a heterogeneous plume head. *Earth and Planetary Science Letters*, **162**, 63–80.

TEJADA, M.L.G., MAHONEY, J.J., DUNCAN, R.A. & HAWKINS, M.P. 1996. Age and geochemistry of basement and alkalic rocks of Malaita and Santa Isabel, Solomon Islands, southern margin of

Ontong Java Plateau. *Journal of Petrology*, **37,** 361–394.

TEJADA, M.L.G., MAHONEY, J.J., NEAL, C.R., DUNCAN, R.A. & PETTERSON, M.G. 2002. Basement geochemistry and geochronology of Central Malaita, Solomon Islands, with implications for the origin and evolution of the Ontong Java Plateau. *Journal of Petrology*, **43**, 449–484.

TEJADA, M.L.G., MAHONEY, J.J., CASTILLO, P.R., INGLE, S.P., SHETH, H.C. & WEIS, D. 2004. Pin-pricking the elephant: evidence on the origin of the Ontong Java Plateau from Pb–Sr–Hf–Nd isotopic characteristics of ODP Leg 192 basalts. *In*: FITTON, J.G., MAHONEY, J.J., WALLACE, P.J. & SAUNDERS, A.D. (eds) *Origin and Evolution of the Ontong Java Plateau*. Geological Society, London, Special Publications, **229**, 133–150.

THORDARSON, T. 2004. Accretionary-lapilli-bearing pyroclastic rocks at ODP Leg 192 Site 1184: a record of subaerial phreatomagmatic eruptions on the Ontong Java Plateau. *In*: FITTON, J.G., MAHONEY, J.J., WALLACE, P.J. & SAUNDERS, A.D. (eds) *Origin and Evolution of the Ontong Java Plateau*. Geological Society, London, Special Publications, **229**, 275–306.

WHITE, R.V., CASTILLO, P.R., NEAL, C.R., FITTON, J.G. & GODARD, M.M. 2004. Phreatomagmatic eruptions on the Ontong Java Plateau: chemical and isotopic relationship to Ontong Java Plateau basalts. *In*: FITTON, J.G., MAHONEY, J.J., WALLACE, P.J. & SAUNDERS, A.D. (eds) *Origin and Evolution of the Ontong Java Plateau*. Geological Society, London, Special Publications, **229**, 307–323.

YASUDA, A., FUJII, T, & KURITA, K. 1994. Melting phase relations of anhydrous mid-ocean ridge basalt from 3 to 20 GPa: Implications for the behaviour of subducted ocean crust in the mantle. *Journal of Geophysical Research*, **99**, 9401–9414.

YAXLEY, G.M. 2000. Experimental study of the phase and melting relations of homogeneous basalt + peridotite mixtures and implications for the petrogenesis of flood basalts. *Contributions to Mineralogy and Petrology*, **139**, 326–338.

YAXLEY, G.M. & GREEN, D.H. 1998. Reactions between eclogite and peridotite: Mantle refertilisation by subducted oceanic crust. *Schweizerische Mineralogische und Petrographische Mitteilungen*, **78**, 243–255.

Partial melting below the Ontong Java Plateau

CLAUDE HERZBERG

Department of Geological Sciences, Rutgers University, New Brunswick, NJ 08903, USA
(e-mail: Herzberg@rci.rutgers.edu)

Abstract: Primary magma compositions for Kroenke-type basalts from the Ontong Java Plateau (OJP) have been estimated using a hybrid forward and universe model. For accumulated fractional melting of a fertile peridotite source, the primary magma had 16.8% MgO and lost 18% olivine by fractional crystallization to produce Kroenke-type basalts; the melt fraction was 0.27 and the potential temperature was 1500°C. For equilibrium melting of a fertile peridotite source, the primary magma had 19.3% MgO and lost 25% olivine by fractional crystallization to produce Kroenke-type basalts; the melt fraction was 0.30 and the potential temperature was 1560°C. The model peridotite source composition, melt fraction and potential temperature required to produce the primary OJP magmas are in excellent agreement with those that have been independently estimated from incompatible trace-element concentrations.

Basic and ultrabasic primary magmas are formed within the partial melting regime of the mantle during decompression, and under unusual circumstances they can erupt as lava flows on the surface. But the primary magma composition can lose its identity by olivine crystallization. When this occurs, reconstruction of the primary magma composition can be made with substantial confidence by incrementally adding olivine back into a derivative liquid. The calculation is stopped when there is identity between the compositions of the computed olivine and observed olivines with the maximum mg-number (e.g. Albarède 1992; Nisbet *et al.* 1993). In this situation the computed primary magma composition is the same as the parental magma composition, which is the most primitive eruptive from which all others are derived by partial crystallization. Komatiites from Gorgona are an example where the primary and parental magma compositions are similar in composition (Nisbet *et al.* 1993; Kerr *et al.* 1996; Révillon *et al.* 2000; Herzberg & O'Hara 2002).

More typically, primary magmas begin to lose their identity below the surface by partial crystallization of olivine, troctolite, olivine gabbro and wehrlite assemblages in either crustal magma chambers or mantle melt conduits (e.g. Larsen & Pedersen 2000). Liquids erupted at the surface will already be derivative, with MgO contents that are lower than that of the primary magma, and the parental magma will precipitate olivine with an mg-number that is lower than that of the primary magma. In this situation, the reconstructed parental magma composition will have an MgO content that is lower than that of the primary magma. As discussed below, the Ontong Java Plateau (OJP) provides a good example of very different primary and parental magma compositions. OJP lavas are mostly basalts with a petrological record of partial crystallization of olivine followed by olivine gabbro in shallow crustal magma chambers (Sano & Yamashita 2004).

It is clear from the above example that primary magma compositions cannot necessarily be constrained from the compositions of observed olivine phenocrysts in surface lava flows. In addition, there is some evidence that observed olivines may be either xenocrysts from the melting regime or they may be phenocrysts from advanced instantaneous melts produced by dynamic fractional melting (Herzberg & O'Hara 2002). For these reasons, it is desirable to have a method of computing primary magma composition that is independent of observed olivine composition. This is the basis of the hybrid forward and inverse model offered by Herzberg & O'Hara (2002). The purpose of the present paper is to apply the model to lavas from the OJP. Model solutions provide primary magma composition, melt fraction, residuum mineralogy, mantle potential temperature and temperature of eruption.

A method for computing a model primary magma composition for the Ontong Java Plateau

Primary magmas are estimated with a hybrid forward and inverse model discussed in detail in Herzberg & O'Hara (2002). The forward model component identifies all potential primary

From: FITTON, J. G., MAHONEY, J. J., WALLACE, P. J. & SAUNDERS, A. D. (eds) 2004. *Origin and Evolution of the Ontong Java Plateau*. Geological Society, London, Special Publications, **229**, 179–183. 0305-8719/$15.00

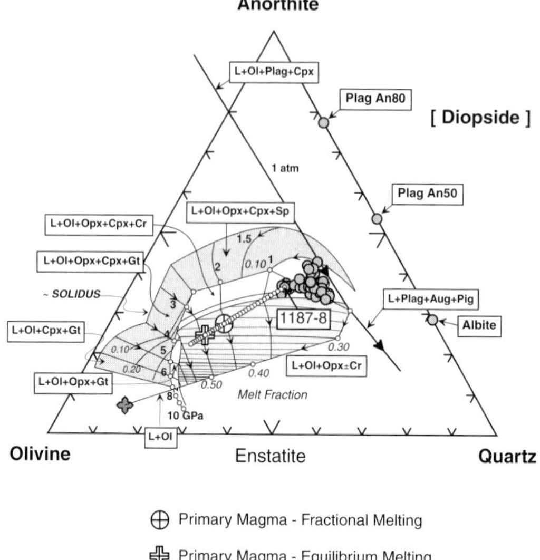

Fig. 1. A projection of partial melt compositions of fertile peridotite KR-4003 from Herzberg & O'Hara (2002), and OJP lava compositions (filled circles; Fitton & Godard 2004) superposed. Small white circles, incremental additions of olivine to OJP lava 1187–8 (Table 1). These provide unique solutions to primary magma compositions when the melt fractions are the same as those displayed in the FeO–MgO plots shown in Figures 2 and 3. Cross-in-circle and cross-in-cross, primary magma compositions for accumulated fractional and equilibrium melting, respectively (Table 1). Pressures of the various cotectic equilibria are indicated in GPa, and apply only for the case of equilibrium melting. Details of the projection are found in Herzberg & O'Hara (2002).

magma compositions as functions of melt fraction for an assumed peridotite source and melting mechanism, based on a parameterization of experimental data. These are displayed in diagrams of CMAS-type projections (Fig. 1) and FeO–MgO plots (Figs 2 and 3). The inverse component assumes that olivine is the only crystallizing phase, and selects a derivative liquid in an erupted lava suite into which olivine is incrementally added; this provides an array of possible primary magma compositions shown as dotted arrays in Figures 1–3. In the OJP case, the derivative liquid is represented by a sparsely olivine-phyric Kroenke-type basalt glass, sample 1187–8 from Fitton & Godard (2004). The potential primary magmas in both forward and inverse models are compared, and common melt fractions in both projection and FeO–MgO plots are used to seek a unique primary magma solution.

The peridotite used in the forward model is Kettle River peridotite KR-4003 (Table 1), a composition that is slightly less fertile than the model pyrolite composition of McDonough & Sun (1995). There are several advantages to considering this particular composition. Experimental and model partial melt compositions are available for KR-4003 (Walter 1998; Herzberg & O'Hara 2002); variations in FeO–MgO and projected major-element composition are shown in Figures 1–3 as functions of melt fraction (Herzberg & O'Hara 2002). The major element composition of KR-4003 can be modelled by 1.2% MORB (mid-ocean ridge basalt) extraction from the primitive McDonough & Sun (1995) composition. Primitive-mantle-normalized incompatible-element concentrations in Kroenke-type basalts from the Ontong Java Plateau can also be reasonably reproduced by 30% equilibrium melting of a primitive mantle source that had been modified by extraction of 1% continental crust (Fitton & Godard 2004). In addition, Tejada *et al.* (2004) modelled radiogenic isotope ratios of the OJP mantle source by 1% melt extraction from primitive mantle around 3 billion years ago. Therefore, peridotite KR-4003 is an ideal model initial source composition for major and trace elements.

For the inverse model, it is necessary to compute the MgO and FeO components in the lavas that are directly exchangeable with olivine. This requires that a distinction be made between the true amount of FeO in the lava, and the total iron that is often measured as Fe_2O_3. As FeO for OJP basalts has not been directly measured by wet chemistry, it is assumed that $Fe^{3+}/\Sigma Fe = 0.10$ for the parental magma 1187-8, as shown in

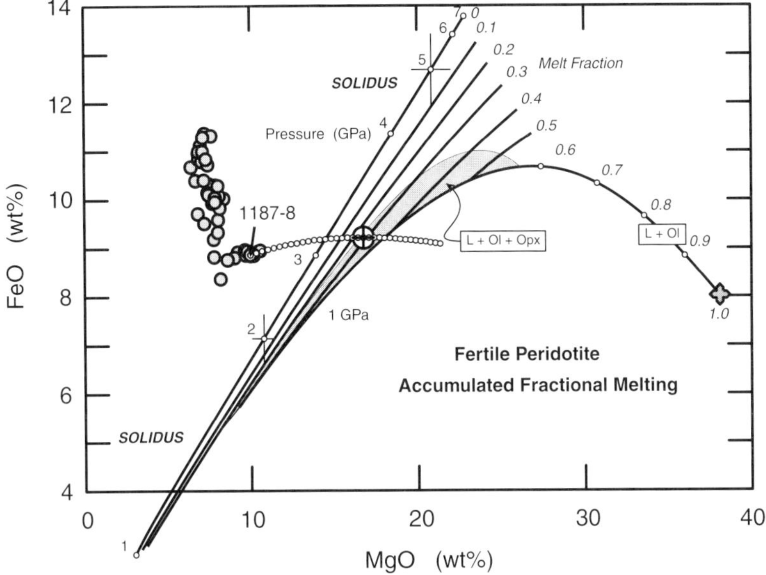

Fig. 2. FeO–MgO contents of accumulated fractional partial melt compositions of fertile peridotite KR-4003 from Herzberg & O'Hara (2002), and OJP lava compositions (filled circles) superposed. FeO and Fe_2O_3 contents of lavas are calculated using $Fe^{3+}/\Sigma Fe = 0.10$, with FeO_T from Fitton & Godard (2004). Small white circles, incremental additions of olivine to OJP lava 1187–8 (Table 1). These provide a unique solution to the primary magma composition when the melt fraction is the same as that displayed in the projection in Figure 1. Cross-in-circle, primary magma composition for accumulated fractional melting at 0.27 mass fractions of melting (Table 1). Uncertainty bars arise from $\pm 1\sigma$ in experimental determination of FeO–MgO parititioning between olivine and liquid (Herzberg & O'Hara 2002).

Table 1. This value has been directly measured by wet chemistry for Kilauea lavas (Wright *et al.* 1975), and is slightly higher than $Fe^{3+}/\Sigma Fe = 0.07$ for MORB (Christie *et al.* 1986).

Reference to Figure 1 illustrates that many of the OJP basalts project close to the olivine gabbro cotectic at 1 atm, in agreement with the experimental observations and conclusions of Sano & Yamashita (2004). Other lavas, such as the olivine phyric Kroenke-type basalt 1187–8, are displaced to the olivine-rich side of the cotectic. When viewed strictly in projection, this lava composition is similar to those of liquids that might be produced by the partial melting of mantle peridotite. However, reference to Figures 2 and 3 demonstrate clearly that all OJP lavas are too low in MgO to be the partial melt products of mantle peridotite at any pressure. The effects of incremental additions of olivine to 1187–8 are shown in both projection and FeO–MgO plots. For accumulated fractional melting, addition of 18% olivine yields a liquid that could be formed by 0.27 mass fractions of melting of fertile mantle peridotite in both projection (Fig. 1) and in FeO–MgO space (Fig. 2). For equilibrium melting, addition of 25% olivine yields a liquid that could be formed by 0.30 mass fractions of melting in both projection (Fig. 1) and in FeO–MgO space (Fig. 3). The amount of olivine that is added in each case is constrained by internal consistency of partial melt compositions in both FeO–MgO and *n*-component projection space. Here *n* refers to SiO_2, TiO_2, Al_2O_3, Fe_2O_3, FeO, MnO, MgO, CaO, Na_2O, K_2O and NiO components of peridotite partial melts that are represented in three-dimensional projection space. Internal consistency is defined by liquid compositions with common mass fractions of melting in both projection and FeO–MgO plots. These are model primary magmas whose compositions are listed in Table 1.

Results

Accumulated fractional melting

The primary OJP magma would have formed by 0.27 mass fractions of accumulated fractional melting, it would have contained 16.8% MgO and it would have precipitated olivine with an mg-number = 90.5 (Table 1). Had such a magma erupted to the surface rather than enter a crustal magma chamber, it would have had a liquidus

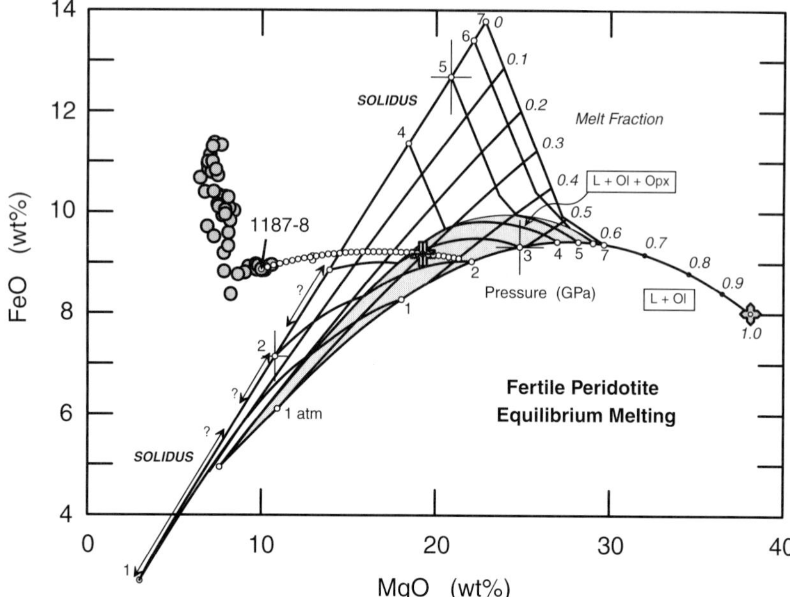

Fig. 3. FeO–MgO contents of equilibrium partial melt compositions of fertile peridotite KR-4003 from Herzberg & O'Hara (2002), and OJP lava compositions (filled circles) superposed. FeO and Fe_2O_3 contents of lavas are calculated using $Fe^{3+}/\Sigma Fe = 0.10$, with FeO_T from Fitton & Godard (2004). Small white circles, incremental additions of olivine to OJP lava 1187–8 (Table 1). These provide a unique solution to the primary magma composition when the melt fraction is the same as that displayed in the projection in Figure 1. Cross-in-cross, primary magma composition for equilibrium melting at 0.30 mass fractions of melting (Table 1). Uncertainty bars arise from $\pm 1\sigma$ in experimental determination of FeO–MgO paritioning between olivine and liquid (Herzberg & O'Hara 2002).

Table 1. *Model peridotite and OJP primary magma compositions*

	Pyrolite model[1]	Kettle R[2]	Kroenke 1187–8[3]	Primary FR[4]	Primary EQ[5]
SiO_2	44.87	44.90	49.69	48.0	47.5
TiO_2	0.20	0.16	0.75	0.62	0.58
Al_2O_3	4.43	4.26	15.01	12.3	11.4
Cr_2O_3	0.38	0.41	0.08	0.07	0.06
Fe_2O_3	0.00	0	1.10	0.90	0.83
FeO	8.04	8.02	8.87	9.2	9.2
MnO	0.13	0.13	0.17	0.17	0.16
MgO	37.75	38.12	9.98	16.8	19.3
CaO	3.54	3.45	12.51	10.3	9.5
Na_2O	0.36	0.22	1.66	1.36	1.26
K_2O	0.03	0.09	0.091	0.075	0.069
NiO	0.25	0.25	0.025	0.098	0.136
Olivine mg-number	89.3	89.5	86.5	90.5	91.6
Melt fraction	0	0	–	0.27	0.30
Eruption temperature (°C)[a]	–	–	1236	1382	1426
Potential temperature (°C)[b]	–	–	–	1500	1560

[1] Pyrolite model (McDonough & Sun 1995).
[2] KR-4003 Kettle River Peridotite (Walter 1998; Herzberg & O'Hara 2002).
[3] Kroenke 1187–8 composition from Fitton & Godard (2004); $Fe^{3+}/\sum Fe = 0.1$.
[4] Primary FR, primary magma composition for accumulated fractional melting of Kettle R peridotite.
[5] Primary EQ, primary magma composition for equilibrium melting of Kettle R peridotite.
[a] Eruption temperature, liquidus temperature at 1 atm from Beattie (1993) and Herzberg & O'Hara (2002).
[b] Potential temperature from Herzberg & O'Hara (2002, p. 1879, fig. 13).

temperature of 1382°C. Mantle potential temperature would have been 1500°C and melting would have initiated at about 3.6 GPa (Herzberg & O'Hara 2002, p. 1879, fig. 13). Harzburgite is the residuum mineral assemblage (Figs 1 and 2). Fractional crystallization of 18% olivine from the primary magma would have yielded OJP Kroenke-type basalt 1187-8. Computed olivine crystallization temperature for 1187-8 at 1 atm is 1236°C, in good agreement with experiment (1240–1250°C; Sano & Yamashita 2004).

Equilibrium melting

In the hypothetical case of equilibrium melting, the OJP primary magma would have formed by about 0.30 mass fractions of melting, similar to that for accumulated fractional melting. It would have contained 19.3% MgO. Harzburgite is the residuum mineral assemblage (Figs 1 and 3). Mantle potential temperature would have been 1560°C and melting would have initiated at 4.4 GPa (Herzberg & O'Hara 2002, p. 1879, fig. 13). Fractional crystallization of 25% olivine would have yielded OJP Kroenke-type basalt 1187-8.

Conclusions

Using the method of Herzberg & O'Hara (2002) Kroenke-type basalts from the Ontong Java Plateau can be modelled as derivative liquids that lost between 18 and 25% olivine from picritic primary magmas. Model primary magma compositions depend on the assumed melting mechanism, with MgO ranging from 16.8% for accumulated fractional melting to 19.3% for equilibrium melting. They also depend on the peridotite source composition, which is assumed to be similar to fertile Kettle River peridotite KR-4003. This source composition is similar to a residue produced by about 1% melt extraction from the primitive mantle composition of McDonough & Sun (1995). Melt fractions of KR-4003 range from 0.27 for accumulated fractional melting to 0.30 for equilibrium melting. Potential temperatures ranged from 1500°C for accumulated fractional melting to 1560°C for equilibrium melting. The model peridotite source composition, melt fraction and potential temperature required to produce the primary OJP magmas are in excellent agreement with those that have been independently estimated from incompatible trace-element concentrations (Fitton & Godard 2004).

I thank G. Fitton for discussions and the National Science Foundation (EAR 0228592) for partial support. Thanks are also extended to L. Larsen and E. Nisbet for critical reviews.

References

ALBARÈDE, F. 1992. How deep do common basaltic magmas form and differentiate? *Journal of Geophysical Research*, **97**, 10 997–11 009.

BEATTIE, P. 1993. Olivine-melt and orthopyroxene-melt equilibria. *Contributions to Mineralogy and Petrology*, **115**, 103–111.

CHRISTIE, D.M., CARMICHAEL, I.S.E. & LANGMUIR, C.H. 1986. Oxidation states of mid-ocean ridge basalt glasses. *Earth and Planetary Science Letters*, **79**, 397–411.

FITTON, J.G. & GODARD, M. 2004. Origin and evolution of magmas on the Ontong Java Plateau. *In*: FITTON, J.G., MAHONEY, J.J., WALLACE, P.J. & SAUNDERS, A.D. (eds) *Origin and Evolution of the Ontong Java Plateau*. Geological Society, London, Special Publications, **229**, 151–178.

HERZBERG, C. & O'HARA, M.J. 2002. Plume-associated ultramafic magmas of Phanerozoic age. *Journal of Petrology*, **43**, 1857–1883.

KERR, A.C., MARRINER, G.F., ARNDT, N.T., TARNEY, J., NIVIA, A., SAUNDERS, A.D. & DUNCAN, R.A. 1996. The petrogenesis of Gorgona komatiites, picrites and basalts: new field, petrographic and geochemical constraints. *Lithos*, **37**, 245–260.

LARSEN, L.M. & PEDERSEN, A.K. 2000. Processes in high-Mg, high-T magmas: evidence from olivine, chromite and glass in Palaeogene picrites from West Greenland. *Journal of Petrology*, **41**, 1071–1098.

McDONOUGH, W.F. & SUN, S.-S. 1995. The composition of the Earth. *Chemical Geology*, **120**, 223–253.

NISBET, E.G., CHEADLE, M.J., ARNDT, N.T. & BICKLE, M.J. 1993. Constraining the potential temperature of the Archaean mantle: a review of the evidence from komatiites. *Lithos*, **30**, 291–307.

RÉVILLON, S., ARNDT, N.T., CHAUVEL, C. & HALLOT, E. 2000. Geochemical study of ultramafic volcanic and plutonic rocks from Gorgona Island, Colombia: the plumbing system of an oceanic plateau. *Journal of Petrology*, **41**, 1127–1153.

SANO, S. & YAMASHITA, S. 2004. Experimental petrology of basement lavas from Ocean Drilling Program Leg 192: implications for differentiation processes in Ontong Java Plateau magmas. *In*: FITTON, J.G., MAHONEY, J.J., WALLACE, P.J. & SAUNDERS, A.D. (eds) *Origin and Evolution of the Ontong Java Plateau*. Geological Society, London, Special Publications, **229**, 185–218.

TEJADA, M.L.G., MAHONEY, J.J., CASTILLO, P.R., INGLE, S.P., SHETH, H.C. & WEISS, D. 2003. Pin-pricking the elephant: evidence on the origin of the Ontong Java Plateau from Pb–Sr–Hf–Nd isotopic characteristics of the ODP Leg 192 basalt. *In*: FITTON, J.G., MAHONEY, J.J., WALLACE, P.J. & SAUNDERS, A.D. (eds) *Origin and Evolution of the Ontong Java Plateau*. Geological Society, London, Special Publications, **229**, 133–150.

WALTER, M.J. 1998. Melting of garnet peridotite and the origin of komatiite and depleted lithosphere. *Journal of Petrology*, **39**, 29–60.

WRIGHT, T., SWANSON, D.A. & DUFFIELD, W.A. 1975. Chemical compositions of Kilauea east-rift lava, 1968–1971. *Journal of Petrology*, **16**, 110–133.

Experimental petrology of basement lavas from Ocean Drilling Program Leg 192: implications for differentiation processes in Ontong Java Plateau magmas

TAKASHI SANO[1] & SHIGERU YAMASHITA[2]

[1]*College of Environment and Disaster Research, Fuji Tokoha University, 325 Ohbuchi, Fuji 417–0801, Japan (e-mail: sano@fuji-tokoha-u.ac.jp)*

[2]*Institute for Study of the Earth's Interior, Okayama University, 827 Yamada, Misasa, Tottori 682–0193, Japan (e-mail: shigeru@misasa.okayama-u.ac.jp)*

Abstract: Melting relations of the basement lavas drilled from the Ontong Java Plateau during ODP Leg 192 were experimentally determined at 1150–1250°C and 0.1–190 MPa under the oxygen fugacity along the fayalite–magnetite–quartz (FMQ) and cobalt–cobalt oxide (CCO) buffers. The basement lavas were classified into two types according to phenocryst assemblage and whole-rock composition: one type is low in MgO (<8 wt%) and olivine + plagioclase + augite-phyric (Kwaimbaita type); and the other is rich in MgO (>8 wt%) and olivine-phyric (Kroenke type). One sample was chosen from each type as a starting material of the melting experiments. The experimental results demonstrate that the variations in phenocryst assemblage and whole-rock composition in the basement lavas can be modelled adequately by fractional crystallization processes in a shallow magma chamber (<6 km in depth). The experimentally determined mineral–melt equilibria, in combination with detailed petrographical investigation, revealed that the vast majority of phenocrysts are in equilibrium with their host magma composition, but some are not. The latter include unusually An-rich parts of plagioclase phenocrysts in the Kwaimbaita-type lavas. These An-rich parts probably crystallized in a mushy boundary layer along the wall of the magma chamber where the melt was relatively rich in H_2O. Some olivine phenocrysts in the Kroenke-type lavas show reverse zoning, with core compositions that can be in equilibrium with the Kwaimbaita-type magmas. The cores of these olivine phenocrysts were most probably assimilated from a solidified pile of the Kwaimbaita-type lavas when the Kroenke-type magmas ascended through it.

The Ontong Java Plateau (OJP), the world's most voluminous igneous province, was formed by eruption and intrusion of a large amount of magma (Kroenke 1972; Mahoney 1987; Coffin & Eldholm 1993). Geochemical studies of the OJP samples have suggested that much of the upper part of the OJP seems to consist of rocks with differentiated (MgO <8wt%) and homogeneous compositions (Mahoney & Spencer 1991; Mahoney *et al.* 1993; Tejada *et al.* 1996, 2002; Neal *et al.* 1997; Petterson *et al.* 1997, 1999). The compositions were probably controlled by differentiation processes (fractional crystallization, magma mixing, assimilation and reaction with cumulates) in a shallow chamber (e.g. Michael 1999), but further investigation has been hampered by lack of samples having relatively primitive compositions.

Recently, a series of basement lavas were successfully cored from the OJP during Ocean Drilling Program (ODP) Leg 192 (Mahoney *et al.* 2001). Some of those have relatively primitive compositions (MgO >10 wt%; Cr >450 ppm) not previously found. In this study, we performed a series of experimental determinations of melting relations in the OJP magmas of both relatively primitive and differentiated compositions. The experimental results, in combination with detailed petrographical investigation, tightly constrain the physicochemical properties of the magma differentiation processes beneath the OJP.

Geological setting and samples

The OJP, defined by the 4000 m-depth contours, is located in the west-central Pacific Ocean and bounded to the SW by the >3000 m-deep North Solomon Trench and Solomon Islands (Coffin & Eldholm 1993). This large igneous province, covering an area of 2.0×10^6 km^2, consists of two parts: the main plateau and the eastern lobe (Fig. 1a). Seismic and gravity data yield an average depth to the Moho of approximately 30 km and a depth to the top of Layer 3A of 10–16 km (Furumoto *et al.* 1976; Gladczenko *et*

From: FITTON, J. G., MAHONEY, J. J., WALLACE, P. J. & SAUNDERS, A. D. (eds) 2004. *Origin and Evolution of the Ontong Java Plateau*. Geological Society, London, Special Publications, **229**, 185–218. 0305-8719/$15.00

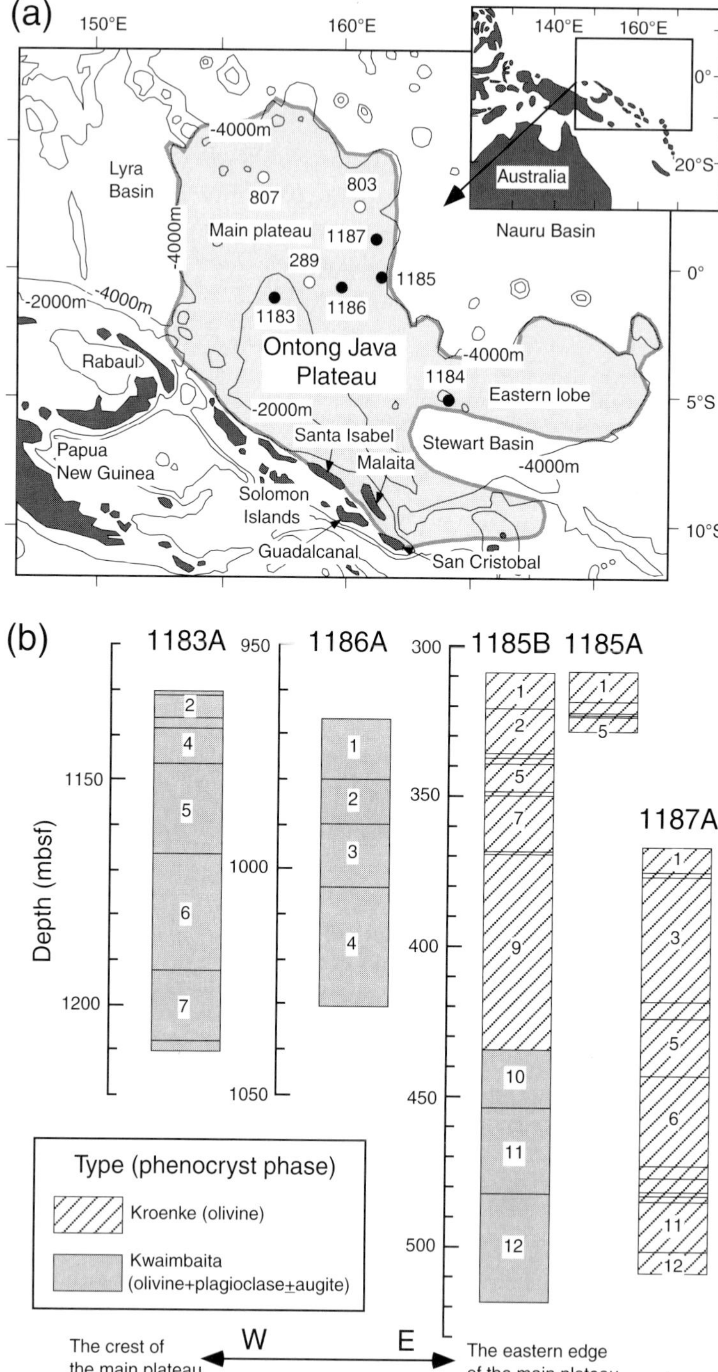

Fig. 1. (**a**) Map of the Ontong Java Plateau showing ODP drill sites and major bathymetric features in the western Pacific Ocean. Solid circles, locations of sites drilled during Leg 192. Open circles, locations of previous ODP and DSDP drill sites that reached basement. The inset shows the location of the main map (box). (**b**) Basement holes drilled during Leg 192. The depth (mbsf, metres below seafloor), unit number and dominant mineralogy are shown.

al. 1997; Neal *et al.* 1997; Phinney *et al.* 1999; Richardson *et al.* 2000).

The islands of Malaita and the northern half of Santa Isabel are considered to be exposed sections of the plateau resulting from accretion of the Malaita accretionary prism during SW-directed subduction of the Pacific plate beneath the island arc (Hughes & Turner 1977; Neal *et al.* 1997; Petterson *et al.* 1997, 1999). Lava piles in central Malaita are divided into two stratigraphic formations, the lower thick (>2.7 km) Kwaimbaita Formation and the upper thin (750 m) Singgalo Formation, based on both field observations and geochemical characteristics (Tejada *et al.* 2002). Geochemical studies on basement lavas from ODP Leg 130, Santa Isabel, and south Malaita also suggested that the upper part of the OJP consists of Kwaimbaita-type lavas that are capped by the thin Singgalo-type lavas (Mahoney & Spencer 1991; Mahoney *et al.* 1993; Tejada *et al.* 1996; Neal *et al.* 1997; Petterson *et al.* 1997, 1999).

ODP Leg 192 obtained drill cores of basaltic basement from two sites (Sites 1183 and 1186) on the central part of the main plateau and two sites (Sites 1185 and 1187) at the edge of the eastern flank of the main plateau (Fig. 1a). Volcaniclastic rocks, containing fresh basaltic glasses (Roberge *et al.* 2004; Thordarson 2004; White *et al.* 2004), were also cored from one site (Site 1184) on the northern ridge of the eastern lobe. The basaltic basement at Sites 1183, 1185, 1186 and 1187 are subdivided into eight, 12, five and 12 units, respectively (Fig. 1b). The basement units are composed of small (10–150 cm) pillows and massive lava flows of 10–50 m in thickness (Mahoney *et al.* 2001).

^{40}Ar–^{39}Ar dating results on the OJP lavas revealed bimodal ages of 122 ± 3 and 90 ± 4 Ma (Mahoney *et al.* 1993; Tejada *et al.* 1996), and the 122 Ma-event seems to be significantly larger than the 90 Ma-event (Tejada *et al.* 1996). ^{40}Ar–^{39}Ar ages of the Leg 192 basement lavas and volcaniclastic rocks fit into the 122 Ma-event (Chambers *et al.* 2002).

We have carried out petrographical studies on a total of six, five, six and seven lava samples from Sites 1183, 1185, 1186 and 1187, respectively, and included five volcaniclastic rocks from Site 1184. Seventeen basement lavas from the above were selected for detailed petrological study. Modal compositions of the 17 basement lavas were determined by point counting on thin sections (2000 points per section) and are listed in Table 1.

The Leg 192 basement lavas are divided into two types based on phenocryst assemblages and geochemical characteristics (Mahoney *et al.* 2001). The first type has phenocrysts of plagioclase, augite and altered olivine (Table 1). As the phenocryst assemblage and geochemical characteristics of the first type are very similar to those forming the Kwaimbaita Formation on Malaita, this is called the Kwaimbaita-type (Tejada *et al.* 2002). The second, Kroenke-type (Fitton & Godard 2004), has phenocrysts of olivine and microphenocrysts of spinel, and is the most basic basalt type yet found on the plateau. Singgalo-type basalt (Tejada *et al.* 2002) was not obtained during Leg 192. The Kwaimbaita-type was found at Site 1183, Site 1186 and the lower part of Site 1185 (Fig. 1b). The Kroenke-type lavas overlie the Kwaimbaita-type lavas at the eastern edge of the main plateau (Fig. 1b), implying that the Kroenke-type erupted on the edge of the main plateau during the last stage of the 122 Ma-event.

Petrography

Analytical methods

Whole-rock major element composition was determined by X-ray fluorescence (XRF) analysis (Philips PW-2510) at the Earthquake Research Institute of the University of Tokyo (ERI), following the method of Sano (2002). Altered parts were carefully removed and the samples were then ground to powder in a tungsten carbide mill and dried for 24 h in an oven at 110°C. Glass beads were prepared from these powders for XRF analyses, using anhydrous lithium tetraborate ($Li_2B_4O_7$) as a flux. Each sample powder was mixed with the $Li_2B_4O_7$ flux in the proportion of 1 to 5, and then fused at 1100°C in a $Pt_{95}Au_5$ crucible and cast into a glass bead.

Compositional zoning of phenocryst minerals in the basement lavas was first examined with backscattered electron images (BEIs) at Fuji Tokoha University. Quantitative determinations of compositions of the phenocryst minerals and basaltic glasses were carried out at the ERI, using a JEOL JXA-8800R electron microprobe analyser (EPMA) with a beam current of 12 nA, focused beam of approximately 1 μm (for minerals) or defocused beam of 10–20 μm (for glass), and accelerating voltage of 20 kV (for olivine) or 15 kV (for plagioclase, clinopyroxene and glass). The counting time was 20 s for major elements (SiO_2, FeO* and MgO in olivine; SiO_2, Al_2O_3, CaO and Na_2O in plagioclase; SiO_2, FeO*, MgO and CaO in clinopyroxene; SiO_2, Al_2O_3, FeO*, MgO, CaO and Na_2O in glass) and 60 s for minor elements. ZAF (atomic number, absorption, fluorescence) correction was applied to all analyses.

Table 1. *Whole-rock compositions and modal compositions of Leg 192 basement lavas*

Hole	1183A	1183A	1183A	1183A	1183A	1183A	1185A	1185B	1185B	1185B	1185B	1186A	1186A
Core-Section	55R-2	55R-3	57R-3	60R-1	65R-3	67R-3	9R-2	3R-1	17R-3	20R-1	22R-6	32R-2	34R-5
Interval	57–60	91–96	58–65	142–147	21–26	133–139	92–94	32–35	107–110	85–89	19–23	95–99	134–140
Piece	3	5A	8	19	2	18	5	7	10	11B	2	6	11
Unit	3B	4B	5B	6	7	8	2	1	10	11	11	1	3
Whole-rock composition (wt%)													
SiO_2	50.03	50.76	50.25	49.52	49.92	50.09	49.24	49.31	50.50	49.79	50.15	50.39	50.86
TiO_2	1.13	1.18	1.13	1.11	1.09	1.12	0.74	0.75	1.13	1.19	1.09	1.14	1.17
Al_2O_3	14.39	14.92	14.28	14.16	14.07	14.09	15.00	15.03	14.63	15.17	13.96	14.76	15.18
Cr_2O_3	0.031	0.031	0.029	0.030	0.029	0.029	0.066	0.065	0.028	0.028	0.026	0.029	0.030
FeO*	10.79	9.08	11.00	11.36	11.13	11.10	9.98	9.98	10.28	9.84	11.38	9.96	9.29
MnO	0.19	0.17	0.19	0.21	0.20	0.21	0.18	0.18	0.20	0.18	0.18	0.20	0.24
MgO	7.31	7.79	7.58	7.94	7.62	7.72	9.76	9.56	7.54	7.81	7.83	7.72	7.33
CaO	11.86	11.94	12.15	12.35	12.58	12.31	12.49	12.51	12.26	12.25	11.72	12.10	11.94
Na_2O	2.11	2.21	2.07	2.00	2.02	2.03	1.61	1.48	2.11	2.24	2.06	2.22	2.31
K_2O	0.59	0.18	0.04	0.05	0.04	0.07	0.09	0.10	0.33	0.11	0.07	0.25	0.23
P_2O_5	0.09	0.10	0.09	0.09	0.09	0.09	0.06	0.06	0.10	0.10	0.09	0.09	0.10
Total	98.52	98.36	98.81	98.82	98.79	98.86	99.22	99.03	99.11	98.71	98.56	98.86	98.68
Modal composition (vol. %)[1]													
ol	1	1	1			1	3	3	1		tr	tr	1
sp							tr	tr					
pl	tr[4]	1	tr			tr			1		1	1	1
aug	tr								tr				1
gm	99	98	99			99	97	97	98		99	99	97

Hole	1186A	1186A	1186A	1186A	1187A	1187A	1187A	1187A	1187A	1187A	1187A	JB-1[2]	XRF[3]
Core-Section	37R-1	37R-3	38R-3	39R-4	4R-3	6R-6	9R-4	10R-7	13R-3	15R-3	16R-3		(1σ)
Interval	13–19	50–55	61–66	40–45	116–120	109–114	133–139	31–38	91–95	67–72	31–34		
Piece	2	1E	3D	3	12	2C	3B	1B	6	3D	3A		
Unit	4	4	4	4	3	3	5	6	7	11	12		
Whole-rock composition (wt%)													
SiO_2	50.37	50.03	50.04	50.08	48.48	48.93	49.06	48.97	49.18	49.17	49.20	52.55	0.24
TiO_2	1.10	1.02	1.03	1.07	0.77	0.72	0.75	0.74	0.74	0.77	0.78	1.32	0.023
Al_2O_3	14.20	14.17	14.28	14.19	15.16	14.57	14.85	14.94	14.88	15.38	15.33	14.48	0.12
Cr_2O_3	0.028	0.034	0.035	0.029	0.069	0.076	0.070	0.068	0.067	0.069	0.069	0.067	0.003
FeO*	11.00	10.93	11.04	11.00	10.10	9.91	9.97	9.82	9.90	9.81	9.63	8.13	0.14
MnO	0.20	0.20	0.19	0.20	0.19	0.18	0.17	0.18	0.18	0.19	0.20	0.16	0.004
MgO	7.77	7.82	7.43	7.84	9.71	10.17	9.39	9.22	9.32	8.13	8.15	7.78	0.140
CaO	11.87	12.55	12.34	12.07	12.25	12.15	12.53	12.51	12.48	12.78	12.93	9.40	0.05
Na_2O	2.07	1.95	1.99	2.10	1.67	1.57	1.70	1.69	1.67	1.78	1.80	2.74	0.05
K_2O	0.04	0.05	0.07	0.07	0.34	0.05	0.05	0.05	0.06	0.04	0.05	1.42	0.02
P_2O_5	0.09	0.08	0.08	0.09	0.06	0.06	0.06	0.06	0.06	0.06	0.07	0.26	0.007
Total	98.74	98.83	98.53	98.74	98.80	98.39	98.60	98.25	98.54	98.18	98.21		
Modal composition (vol. %)[1]													
ol	1				4	6	2	2	2		2		
sp					tr	tr	tr	tr	tr		tr		
pl	1												
aug													
gm	98				96	94	98	98	98		98		

* Total iron as FeO.
[1] Abbreviations of phenocrysts are ol, olivine; sp, spinel; pl, plagioclase; aug, clinopyroxene; gm, groundmass.
[2] An average value of six measurements of standard JB-1 (a standard rock from the Geological Survey of Japan).
[3] One standard deviation (1σ) of a calibration line for each element.
[4] ≪1.

Whole rock and basaltic glass compositions

The whole-rock compositions of 24 basement lava samples cored at Sites 1183, 1185, 1186 and 1187 are given in Table 1. Compositions of fresh glasses of five volcaniclastic rocks cored at Site 1184 are listed in Table 2. We assume that whole-rock compositions represent the melt compositions because all samples in this study are aphyric basalts (phenocryst ≤6 vol.%).

The major-element compositions of the basement lavas change in accordance with their phenocryst assemblage; the Kroenke-type basalt (MgO >8 wt%, Cr_2O_3 >450 ppm) containing phenocrysts only of olivine has less differentiated compositions compared with the Kwaimbaita-type basalt (MgO <8 wt%, Cr_2O_3 <300 ppm) containing olivine, plagioclase and clinopyroxene (Table 1, Fig. 2). A clear compositional gap between the Kroenke-type basalt and the Kwaimbaita-type basalt was observed for some elements (e.g. TiO_2, Cr_2O_3) plotted in Figure 2.

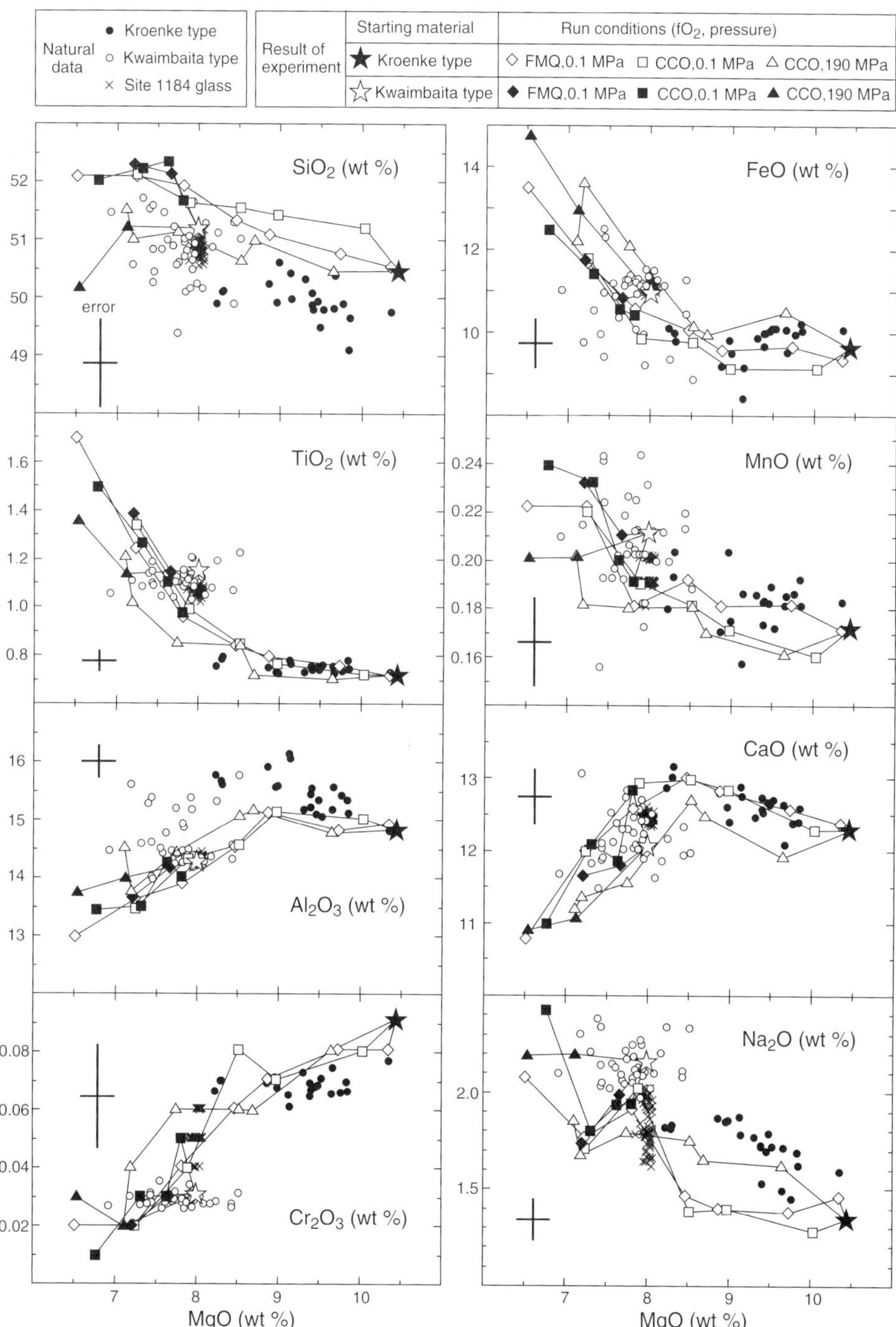

Fig. 2. Plots of major elements v. wt% MgO for basement lavas from Sites 1183, 1185, 1186 and 1187; and basaltic glasses from Site 1184. To compare the natural data with the experimental results, melt compositions obtained in the experiments on the two starting materials (Kroenke-type basalt: 1187A-6R-6, 109–114; and Kwaimbaita-type basalt: 1186A-37R-1, 13–19) are also shown. Data are from Mahoney *et al.* (2001) and this study. Error bars are $\pm 1\sigma$ errors of analyses (more than six points) of experimental glasses that have the most heterogeneous compositions (Table 8). We assume the errors are experimental errors.

Table 2. *Electron microprobe analyses of Site 1184 basaltic glasses*

Hole-Core-Section (interval: cm)	Glass name	n^1	SiO_2 wt%	TiO_2 wt%	Al_2O_3 wt%	Cr_2O_3 wt%	FeO* wt%	MnO wt%	MgO wt%	CaO wt%	Na_2O wt%	K_2O wt%	Total wt%
1184A-39R-7(92–95)	gl01	6	50.36(10)[2]	1.06(2)	14.17(16)	0.06(2)	10.92(25)	0.19(1)	7.94(15)	12.44(9)	1.86(13)	0.09(0)	99.09
1184A-39R-7(92–95)	gl02	6	50.49(14)	1.06(3)	14.15(15)	0.05(2)	11.16(17)	0.19(2)	7.92(6)	12.31(17)	1.65(4)	0.08(1)	99.06
1184A-39R-7(92–95)	gl03	8	50.33(16)	1.06(2)	14.25(8)	0.06(1)	11.06(11)	0.19(1)	7.98(8)	12.41(13)	1.60(6)	0.09(1)	99.03
1184A-39R-7(92–95)	gl04	9	50.41(23)	1.07(2)	14.28(14)	0.05(2)	11.08(16)	0.19(1)	7.94(28)	12.37(11)	1.91(17)	0.09(1)	99.39
1184A-39R-7(92–95)	gl05	9	50.43(17)	1.07(3)	14.22(11)	0.05(1)	11.11(13)	0.19(1)	7.99(7)	12.34(15)	1.99(5)	0.09(1)	99.48
1184A-39R-7(92–95)	gl06	8	50.44(22)	1.07(2)	14.19(21)	0.05(1)	11.07(21)	0.19(1)	7.95(14)	12.46(7)	1.90(5)	0.09(1)	99.41
1184A-39R-7(92–95)	gl07	9	50.37(27)	1.05(3)	14.17(12)	0.05(1)	11.18(13)	0.19(1)	8.01(10)	12.34(15)	1.89(11)	0.09(1)	99.34
1184A-39R-7(92–95)	gl08	10	50.27(19)	1.08(1)	14.21(10)	0.05(2)	11.14(10)	0.19(1)	7.95(12)	12.49(8)	1.97(6)	0.09(1)	99.44
1184A-39R-7(92–95)	gl09	10	50.44(17)	1.07(3)	14.15(13)	0.05(1)	11.14(15)	0.19(1)	7.99(10)	12.39(14)	1.90(4)	0.09(1)	99.41
1184A-39R-7(92–95)	gl10	10	50.32(13)	1.07(4)	14.16(19)	0.06(1)	11.13(9)	0.19(1)	8.02(8)	12.39(13)	1.95(4)	0.09(1)	99.38
1184A-39R-7(92–95)	gl11	11	50.16(37)	1.06(3)	14.33(8)	0.05(1)	11.06(27)	0.19(1)	7.99(8)	12.36(11)	1.91(6)	0.09(1)	99.20
1184A-39R-7(92–95)	gl12	11	50.31(18)	1.08(3)	14.25(12)	0.05(1)	11.09(20)	0.19(1)	7.97(10)	12.34(23)	1.94(5)	0.08(1)	99.30
1184A-39R-7(92–95)	gl13	10	50.33(24)	1.06(2)	14.23(19)	0.05(1)	11.02(16)	0.19(1)	7.89(13)	12.42(10)	2.00(5)	0.09(1)	99.28
1184A-42R-2(60–65)	gl01	12	50.33(15)	1.07(3)	14.29(15)	0.05(2)	11.12(16)	0.20(1)	7.83(11)	12.45(24)	1.97(6)	0.09(1)	99.40
1184A-42R-2(60–65)	gl02	12	50.35(17)	1.05(2)	14.23(18)	0.05(1)	11.02(25)	0.19(1)	7.95(17)	12.43(11)	1.95(6)	0.09(1)	99.31
1184A-42R-2(60–65)	gl03	10	50.17(40)	1.05(3)	14.24(14)	0.05(2)	10.93(32)	0.20(1)	8.00(12)	12.43(11)	1.91(4)	0.09(0)	99.07
1184A-42R-2(60–65)	gl04	11	50.42(40)	1.05(2)	14.28(10)	0.04(1)	11.02(25)	0.19(1)	7.91(9)	12.40(11)	1.93(5)	0.09(1)	99.33
1184A-42R-2(60–65)	gl05	12	50.31(33)	1.05(4)	14.23(12)	0.04(1)	11.06(13)	0.20(1)	7.93(13)	12.39(15)	1.95(5)	0.09(1)	99.25
1184A-42R-2(60–65)	gl06	12	50.41(18)	1.06(3)	14.24(14)	0.05(1)	11.02(16)	0.19(1)	7.85(17)	12.35(15)	1.82(12)	0.09(1)	99.08
1184A-42R-2(60–65)	gl07	12	50.55(17)	1.05(2)	14.16(17)	0.05(1)	11.14(18)	0.19(1)	7.94(8)	12.41(19)	1.76(4)	0.09(1)	99.34
1184A-42R-2(60–65)	gl08	12	50.49(28)	1.05(3)	14.25(21)	0.05(1)	11.08(19)	0.20(1)	7.91(19)	12.47(12)	1.77(5)	0.09(1)	99.36
1184A-42R-2(60–65)	gl09	12	50.55(21)	1.06(3)	14.21(19)	0.05(1)	10.97(24)	0.19(1)	7.79(18)	12.40(18)	1.77(4)	0.09(1)	99.08
1184A-42R-2(60–65)	gl10	11	50.58(28)	1.06(4)	14.25(15)	0.05(2)	11.08(30)	0.19(1)	7.89(14)	12.39(22)	1.79(4)	0.09(1)	99.37
1184A-42R-2(60–65)	gl11	12	50.54(23)	1.03(2)	14.25(23)	0.06(1)	11.17(9)	0.20(1)	8.01(8)	12.42(18)	1.76(3)	0.09(1)	99.53
1184A-42R-2(60–65)	gl12	12	50.51(27)	1.05(3)	14.28(12)	0.05(1)	11.02(16)	0.20(1)	7.92(22)	12.40(11)	1.76(3)	0.09(1)	99.28
1184A-42R-5(83–87)	gl01	12	50.69(21)	1.06(3)	14.26(12)	0.05(2)	11.12(21)	0.19(1)	8.03(7)	12.37(21)	1.75(5)	0.08(1)	99.60
1184A-42R-5(83–87)	gl02	12	50.89(20)	1.04(2)	14.30(15)	0.05(1)	11.20(12)	0.19(1)	8.02(16)	12.36(31)	1.72(5)	0.08(0)	99.85
1184A-42R-5(83–87)	gl03	12	50.84(17)	1.05(2)	14.25(22)	0.04(1)	11.16(14)	0.19(1)	7.97(12)	12.42(17)	1.78(5)	0.09(1)	99.79
1184A-42R-5(83–87)	gl04	12	50.83(19)	1.02(2)	14.30(19)	0.05(1)	11.05(23)	0.20(1)	8.01(9)	12.40(11)	1.74(6)	0.09(1)	99.69
1184A-42R-5(83–87)	gl05	12	50.87(28)	1.04(2)	14.24(17)	0.05(1)	11.10(16)	0.20(1)	7.99(17)	12.48(12)	1.73(5)	0.09(1)	99.79
1184A-42R-5(83–87)	gl06	12	50.91(17)	1.05(2)	14.29(14)	0.05(1)	11.07(18)	0.20(2)	8.03(9)	12.42(11)	1.74(4)	0.09(1)	99.85
1184A-42R-5(83–87)	gl08	12	50.98(17)	1.05(4)	14.30(14)	0.05(1)	11.09(29)	0.20(1)	8.04(10)	12.37(16)	1.73(12)	0.09(1)	99.90
1184A-42R-5(83–87)	gl09	12	50.87(25)	1.05(2)	14.30(13)	0.05(1)	11.14(19)	0.19(1)	8.02(7)	12.33(27)	1.66(4)	0.09(1)	99.70
1184A-42R-5(83–87)	gl10	22	50.88(25)	1.05(2)	14.31(18)	0.05(1)	11.12(22)	0.19(1)	7.94(16)	12.44(14)	1.64(3)	0.09(1)	99.71
1184A-42R-5(83–87)	gl11	12	50.86(25)	1.05(2)	14.31(20)	0.05(1)	10.91(33)	0.19(1)	8.02(7)	12.28(36)	1.65(6)	0.09(1)	99.41
1184A-42R-5(83–87)	gl12	12	50.69(22)	1.06(3)	14.20(31)	0.05(1)	11.11(17)	0.19(2)	7.94(16)	12.42(11)	1.63(6)	0.09(1)	99.38
1184A-42R-6(17–22)	gl01	12	50.67(26)	1.05(1)	14.24(18)	0.05(1)	11.12(16)	0.20(1)	8.00(13)	12.42(7)	1.79(6)	0.08(0)	99.62

1184A-42R-6(17–22)	gl02	12	50.39(27)	1.05(2)	14.24(21)	0.05(1)	11.03(12)	0.20(2)	8.01(8)	12.34(25)	1.79(5)	0.09(1)	99.19
1184A-42R-6(17–22)	gl03	12	50.55(21)	1.07(3)	14.26(19)	0.05(1)	11.03(25)	0.19(1)	7.98(13)	12.38(19)	1.78(4)	0.09(1)	99.38
1184A-42R-6(17–22)	gl04	12	50.65(12)	1.06(2)	14.26(17)	0.05(1)	10.96(17)	0.19(2)	7.98(9)	12.35(25)	1.77(5)	0.09(1)	99.36
1184A-42R-6(17–22)	gl05	12	50.54(26)	1.05(2)	14.21(27)	0.05(1)	11.04(12)	0.19(1)	8.00(12)	12.39(18)	1.75(3)	0.09(1)	99.31
1184A-42R-6(17–22)	gl06	10	50.67(14)	1.06(2)	14.34(12)	0.05(1)	11.12(18)	0.20(1)	7.97(8)	12.35(25)	1.79(4)	0.09(1)	99.64
1184A-42R-6(17–22)	gl07	12	50.73(13)	1.07(3)	14.28(12)	0.06(1)	11.09(18)	0.19(1)	8.00(8)	12.30(21)	1.80(5)	0.09(0)	99.61
1184A-42R-6(17–22)	gl08	12	50.63(22)	1.08(3)	14.28(12)	0.05(1)	11.08(20)	0.20(1)	8.04(11)	12.40(17)	1.81(4)	0.09(1)	99.66
1184A-42R-6(17–22)	gl09	12	50.43(48)	1.06(2)	14.21(11)	0.05(1)	11.07(27)	0.19(1)	7.98(11)	12.37(30)	1.80(6)	0.09(0)	99.25
1184A-42R-6(17–22)	gl10	12	50.65(22)	1.08(3)	14.24(17)	0.05(1)	11.08(19)	0.20(1)	7.95(15)	12.37(13)	1.79(4)	0.09(1)	99.50
1184A-42R-6(17–22)	gl11	12	50.72(13)	1.07(2)	14.25(23)	0.05(1)	11.07(22)	0.19(1)	7.99(9)	12.40(16)	1.82(4)	0.09(1)	99.65
1184A-42R-6(17–22)	gl12	12	50.77(20)	1.07(2)	14.27(11)	0.05(2)	11.14(13)	0.19(1)	7.99(15)	12.45(26)	1.79(3)	0.08(1)	99.80
1184A-44R-8(66–69)	gl01	20	50.49(20)	1.05(2)	14.20(12)	0.05(1)	10.79(35)	0.19(1)	7.87(19)	12.45(26)	1.98(9)	0.09(1)	99.16
1184A-44R-8(66–69)	gl02	9	50.53(18)	1.05(3)	14.17(15)	0.05(1)	11.05(10)	0.19(1)	7.94(17)	12.35(5)	1.97(6)	0.09(1)	99.39
1184A-44R-8(66–69)	gl03	10	50.63(21)	1.06(3)	14.18(12)	0.05(1)	11.05(16)	0.19(1)	7.98(7)	12.36(16)	1.97(5)	0.09(1)	99.56
1184A-44R-8(66–69)	gl04	10	50.55(27)	1.08(3)	14.22(10)	0.06(1)	10.80(29)	0.19(1)	7.94(8)	12.26(20)	2.00(4)	0.09(1)	99.19
1184A-44R-8(66–69)	gl05	10	50.42(28)	1.04(3)	14.16(16)	0.05(2)	10.89(26)	0.19(1)	7.87(19)	12.33(29)	1.94(4)	0.09(1)	98.98
1184A-44R-8(66–69)	gl06	10	50.58(13)	1.05(3)	14.24(9)	0.05(1)	11.01(9)	0.20(1)	7.91(20)	12.31(18)	1.97(5)	0.09(0)	99.41
1184A-44R-8(66–69)	gl07	10	50.41(29)	1.05(3)	14.22(10)	0.05(1)	11.07(11)	0.18(1)	7.90(13)	12.31(12)	2.01(4)	0.09(1)	99.29
1184A-44R-8(66–69)	gl08	10	50.72(25)	1.05(2)	14.21(21)	0.06(1)	10.92(19)	0.19(1)	7.99(12)	12.26(21)	1.85(13)	0.09(1)	99.34
1184A-44R-8(66–69)	gl09	10	50.39(15)	1.04(2)	14.16(16)	0.05(1)	10.87(23)	0.20(2)	7.90(17)	12.46(12)	1.67(5)	0.09(1)	98.83
1184A-44R-8(66–69)	gl10	10	50.41(37)	1.06(2)	14.15(18)	0.04(1)	10.92(20)	0.19(1)	7.95(22)	12.37(13)	1.70(6)	0.09(1)	98.88
1184A-44R-8(66–69)	gl11	10	50.62(36)	1.06(3)	14.06(27)	0.05(1)	10.87(17)	0.20(1)	7.95(12)	12.38(13)	1.64(6)	0.09(1)	98.92
1184A-44R-8(66–69)	gl12	10	50.35(27)	1.06(4)	14.18(6)	0.05(1)	11.00(13)	0.20(1)	7.91(19)	12.34(10)	1.68(5)	0.09(1)	98.86

* Total iron as FeO.

[1] Number of analyses.

[2] Numbers in parentheses adjacent to the analyses are 1 SD on the basis of replicate analyses; e.g. 50.36(10) represents 50.36 ± 0.10 wt%.

While the Al_2O_3 and CaO contents slightly increase with a decrease of MgO content in the Kroenke-type basalt, they decrease in the Kwaimbaita-type basalt. The FeO* content of the Kroenke-type basalt remains unchanged over the range of MgO contents between 8 and 10 wt%, whereas the FeO* content of the Kwaimbaita-type basalt increases with decreasing MgO.

The basaltic glasses in Site 1184 volcaniclastic rocks have homogeneous compositions (Table 2, Fig. 2) within the precision of the EPMA technique. Analyses of Site 1184 basaltic glasses plot in the field of the Kwaimbaita-type basalt, indicating that the Kwaimbaita-type basalt is also present in the eastern lobe.

Mineralogy of phenocrysts

Olivine zoning pattern. Olivine phenocrysts (100–600 μm) are euhedral–subhedral in the Kroenke-type basalt (Fig. 3a–c). Based on the BEIs for the phenocrysts in eight lavas, their zoning pattern was examined. Most olivine crystals (approximately 300 grains) exhibit normal zoning (Fig. 3a), but rare olivine crystals (10 grains) show reverse zoning (Fig. 3b, c).

The normal-zoned crystals are composed of cores and surrounding rims (Fig. 4a). The cores have uniform mg-number [$100 \times Mg/(Mg+Fe^{2+})$] (86.8–88.1); and MnO (0.17–0.21 wt%), CaO (0.27–0.37 wt%) and NiO (0.22–0.34 wt%) contents (Table 3, Fig. 5).

The reverse-zoned crystals are large (>240 μm in diameter) and usually rounded (Fig. 3c). Generally M-shaped zoning profiles in mg-number and NiO are observed in the reverse-zoned olivine (e.g. Fig. 4c). The M-shaped zoning is composed of core, uniform margin and rim. The compositions of the uniform margin are akin to those of the uniform core of the normal-zoned olivine (Fig. 5). The largest olivine (1185A-9R-2, 92–94, ol1 in Fig. 3b) among the reverse-zoned olivine is skeletal with sieve texture in the core. The largest olivine has a uniform core whose mg-number is 81.8–82.5 (Table 3, Fig. 4b), but uniform cores are not seen in the other reverse-zoned olivine phenocrysts (e.g. 1187A-9R-4, 133–139, ol10 in Fig. 4c). The core mg-number of the other reverse-zoned olivines is 84.0–86.4 (Fig. 5).

Plagioclase zoning pattern. Plagioclase phenocrysts (50–400 μm wide) form euhedral–subhedral tabular crystals in the Kwaimbaita-type basalt (Fig. 3d-f). Based on observation of the BEIs of more than 300 plagioclase phenocrysts in nine lavas, the plagioclase phenocrysts were classified into two types, type I and type II (Fig. 3d–f). Electron microprobe profiles representative of the two types are shown in Figure 4d–f.

The type I plagioclase phenocrysts are mostly euhedral (Fig. 3d) and exhibit normal zoning (Fig. 4d). The EPMA analyses indicate that cores of type I have uniform An [$100 \times Ca/(Ca+Na)$] content, most of which range from 73 to 78 (Table 4, Fig. 6).

The type II plagioclase phenocrysts, showing complex zoning patterns, are subdivided into two types, the reverse-zoned type and the oscillatory zoned type (Fig. 3e, f). The reverse-zoned type is characterized by a low-An core (An_{73-78}) surrounded by a high-An mantle (An_{80-84}) (Figs 3e and 4e), with a marked compositional change at the boundary between core and mantle (Fig. 4e). The oscillatory zoned type is composed of a low-An core, high-An mantles and low-An mantles (Figs 3f and 4f). The low-An parts (low-An core and/or mantles) of the oscillatory zoned type and the low-An parts of the reverse-zoned type have nearly identical compositions (An_{73-78}, MgO 0.25–0.45 wt%). The high An parts (high-An mantles) of the oscillatory zoned crystals are similar in compositions (An_{80-84}, MgO 0.20–0.30 wt%) to the high-An parts of the reverse-zoned type.

The low-An parts of the type II plagioclase have a similar range in An (An_{73-78}) and MgO (0.25–0.45 wt%) contents to the type I cores (Table 4, Fig. 6), implying similar origin to the type I plagioclase. The type II plagioclase is present in all the Kwaimbaita-type lavas and shows systematic spatial distribution; they are more abundant at Site 1183 than at Site 1186,

Fig. 3. Backscattered electron (BSE) microprobe images of olivine, plagioclase and clinopyroxene phenocrysts. (**a**) 1187A-6R-6 (109–114) ol2, normal-zoned olivine; (**b**) 1185A-9R-2 (92–94) ol1, reverse-zoned olivine with skeletal sieve texture in the core; (**c**) 1187A-9R-4 (133–139) ol10, reverse-zoned olivine; (**d**) 1186A-34R-5 (134–140) pl1, type I plagioclase with normal zoning; (**e**) 1185B-17R-3 (107–110) pl13, type II plagioclase with reverse zoning showing a high-An mantle that surrounds a low-An core; (**f**) 1183A-67R-3 (133–139) pl17, type II plagioclase with oscillatory zoning that consists of a low-An core, high-An mantles and low-An mantles; (**g**) 1185B-17R-3 (107–110) cp5, clinopyroxene phenocryst with normal zoning; (**h**) 1186A-34R-5 (134–140) pl13 and cp5, type I plagioclase and clinopyroxene phenocrysts. Continuous lines indicate locations of line profiles shown in Figure 4.

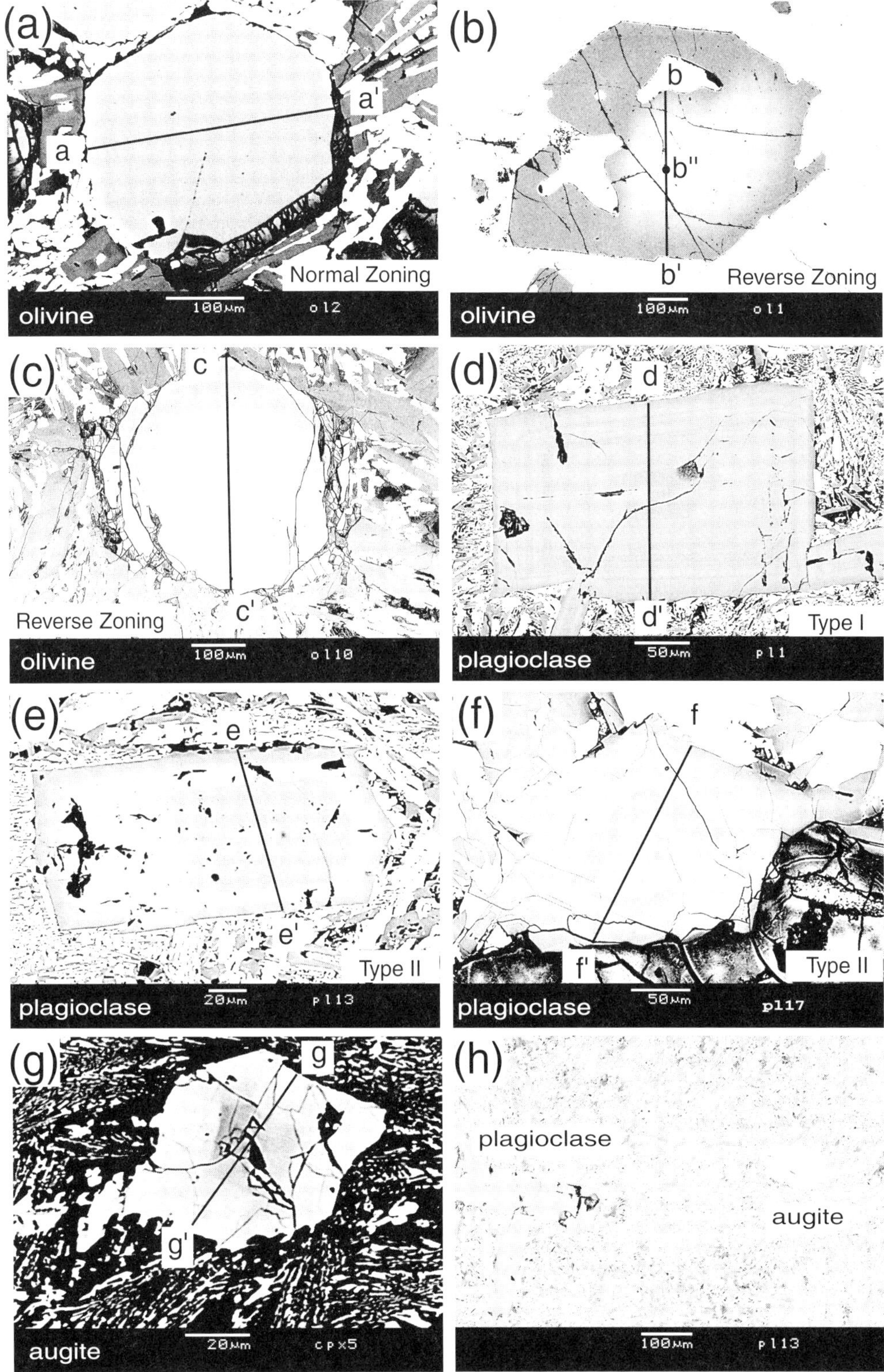
(a)
a
a'
Normal Zoning
olivine
100µm
ol2
(b)
b
b''
b'
Reverse Zoning
olivine
100µm
ol1
(c)
c
c'
Reverse Zoning
olivine
100µm
ol10
(d)
d
d'
Type I
plagioclase
50µm
pl1
(e)
e
e'
Type II
plagioclase
20µm
pl13
(f)
f
f'
Type II
plagioclase
50µm
pl17
(g)
g
g'
augite
20µm
cpx5
(h)
plagioclase
augite
100µm
pl13

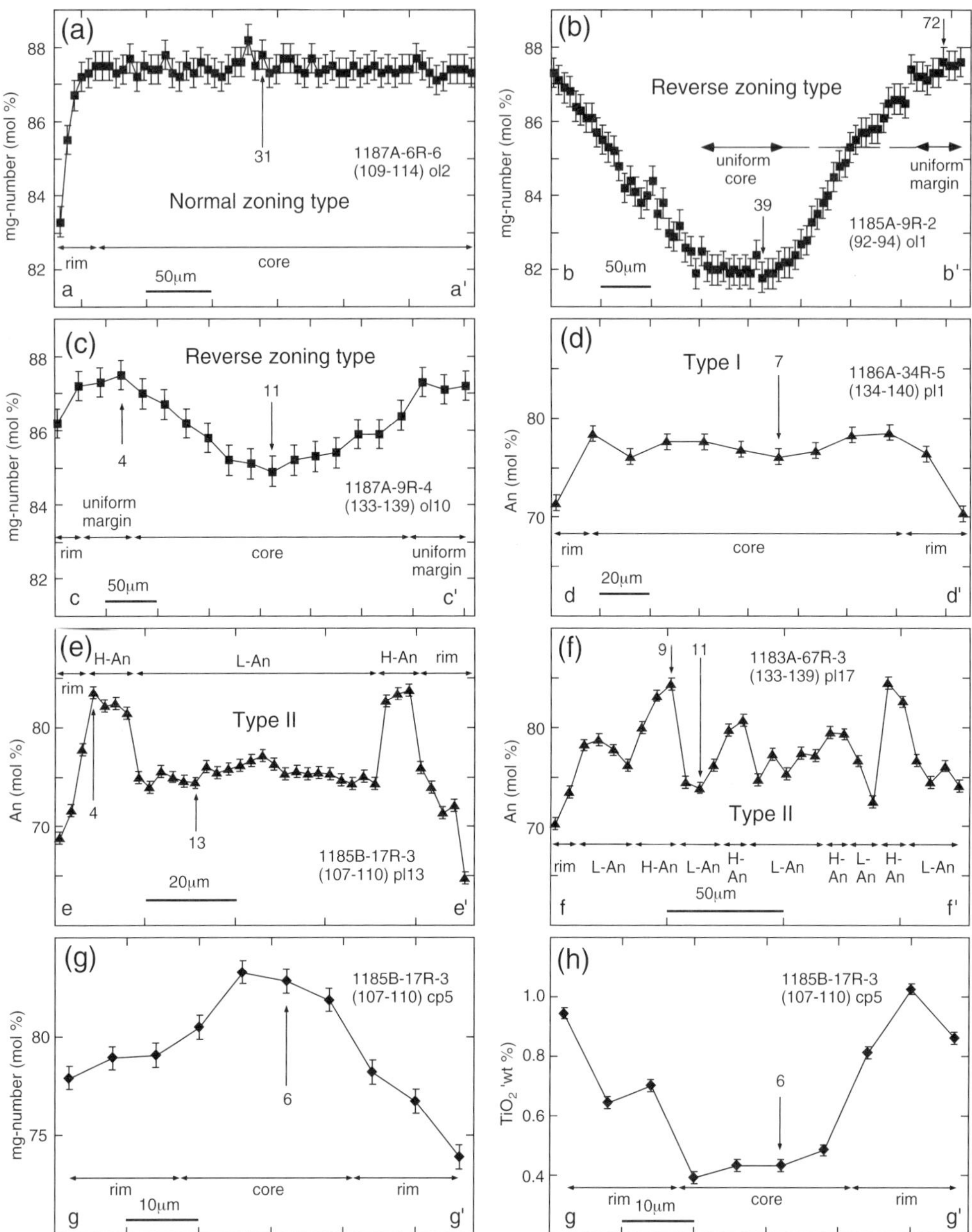

Fig. 4. Line profiles of mg-number of olivine and clinopyroxene, and An content of plagioclase. Positions of each profile are shown in Figure 3. Numbers correspond to the analytical data cited in Tables 3–5. H-An, high-An mantle; L-An, low-An core and/or mantle. Error bar shows standard deviation uncertainty ($\pm 1\sigma$) due to counting statistics.

and their abundance increases with increasing depth at Site 1186 (Fig. 6).

Clinopyroxene zoning pattern. Some clinopyroxene crystals (50–100 μm) are strained and partially resorbed, implying that they are probably xenocrysts. However, clinopyroxene phenocrysts with euhedral and subhedral habit are also present in three Kwaimbaita-type samples (e.g. Fig. 3g, h). Selected electron microprobe analyses of the clinopyroxene phenocrysts are listed in Table 5.

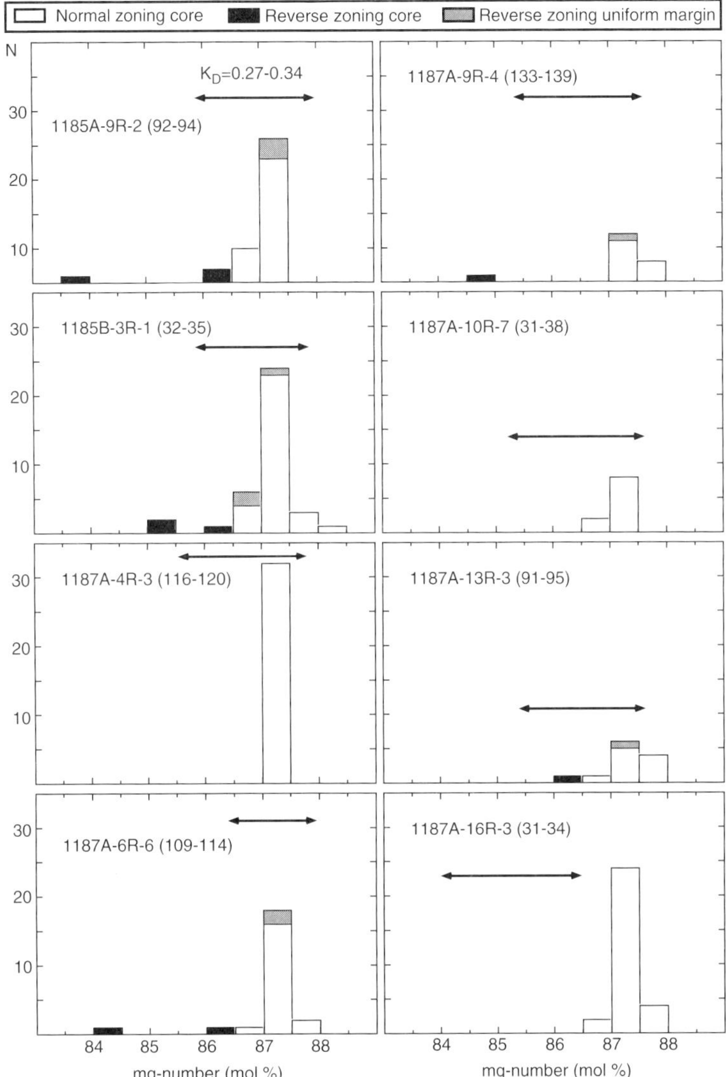

Fig. 5. Histogram of mg-number for olivine phenocrysts in the eight basalt samples used in this study. Arrows correspond to the mg-number ranges of olivine in equilibrium with the host magma calculated using our experimental results.

All clinopyroxene phenocrysts exhibit normal zoning (Fig. 4g, h). The phenocrysts have cores of high mg-number (82–83) and rims of low mg-number (76–82) that show the same range as groundmass clinopyroxene (Fig. 7). The core compositions (TiO_2 0.2–0.5 wt%, Al_2O_3 2–4 wt%) are obviously distinct from the surrounding rim compositions (TiO_2 0.5–1.2 wt%; Al_2O_3 4–8 wt%). The cores are typical augite composition, having Ca/(Ca+Mg+Fe) molar ratios of 0.3–0.4. Therefore, the clinopyroxene is called augite in the following paragraphs.

Melting experiments

Starting materials

The starting materials of melting experiments were two basalt samples; one is 1187A-6R-6, 109–114 of the Kroenke-type, and the other is 1186A-37-R-1, 13–19 of the Kwaimbaita-type. The Kroenke-type basalt is sparsely olivine–spinel-phyric (6 vol.%, Table 1) and has the highest MgO content (10.2 wt%) of our samples (Table 1). The Kwaimbaita-type sample

Table 3. *Electron microprobe analyses of representative olivine phenocrysts*

Hole	1185A	1185A	1185A	1185B	1185B	1185B	1187A	1187A	1187A	1187A	1187A	1187A	1187A	1187A	1187A	1187A
Core-Section	9R-2	9R-2	9R-2	3R-1	3R-1	3R-1	4R-3	6R-6	6R-6	6R-6	9R-4	9R-4	9R-4	10R-7	13R-3	16R-3
Interval	92–94	92–94	92–94	32–35	32–35	32–35	116–120	109–114	109–114	109–114	133–139	133–139	133–139	31–38	91–95	31–34
Mineral	ol14	ol1	ol1	ol31	ol55	ol55	ol16	ol2	ol6	ol6	ol1	ol10	ol10	ol1	ol9	ol30
Size (μm)	102	411	411	156	355	355	151	318	430	430	309	403	403	196	154	220
Point[1]	2	39	72	1	11	2	6	31	1	2	7	11	4	5	1	2
Type[2]	N	R	R	N	R	R	N	N	R	R	N	R	R	N	N	N
Condition[3]	C	C	UM	C	C	UM	C	C	C	UM	C	C	UM	C	C	C
SiO_2 (wt%)	40.36	39.20	40.11	40.50	39.93	40.26	40.26	40.02	39.77	40.38	40.35	40.00	40.47	40.60	40.20	40.09
FeO* (wt%)	12.15	17.24	11.96	12.29	14.21	12.29	12.17	11.66	39.77	40.38	40.35	40.00	40.47	40.60	40.20	40.09
MnO (wt%)	0.19	0.27	0.19	0.18	0.21	0.18	0.19	0.19	0.23	0.18	0.19	0.21	0.19	0.20	0.19	0.19
MgO (wt%)	46.99	43.37	47.59	47.29	45.05	46.67	46.81	47.28	44.71	46.89	47.05	45.14	47.07	46.95	47.62	47.10
CaO (wt%)	0.31	0.27	0.30	0.28	0.31	0.30	0.27	0.30	0.28	0.30	0.28	0.32	0.30	0.29	0.29	0.29
NiO (wt%)	0.29	0.18	0.26	0.27	0.24	0.29	0.26	0.28	0.20	0.25	0.26	0.20	0.26	0.26	0.27	0.25
Total (wt%)	100.29	100.53	100.41	100.81	99.95	99.99	99.96	99.73	100.36	100.46	100.25	100.19	100.29	100.54	100.74	100.38
mg-number	87.3	81.8	87.6	87.3	85.0	87.1	87.3	87.8	84.0	87.0	87.4	84.9	87.5	87.2	87.5	87.1

* total iron as FeO.

[1] Analysed points of 1185A-9R-2 (92–94) ol1, 1187A-6R-6 (109–114) ol2 and 1187A-9R-4 (133–139) ol10 are shown in the line profiles of Figure 4.

[2] Type of olivine phenocryst: N, normal zoning; R, reverse zoning.

[3] Condition of analysed point: C, core; UM, uniform margin of the reverse-zoning olivine.

Table 4. *Electron microprobe analyses of representative plagioclase phenocrysts*

Sample	1183A-55R-2 (57–60)			1183A-55R-3 (91–96)			1183A-57R-3 (58–65)		
Mineral	pl8	pl3	pl3	pl24	pl13	pl13	pl9	pl3	pl3
Size (μm)	97	156	156	82	84	84	111	96	96
Point[1]	2	4	9	2	10	11	7	2	4
Type[2]	I	II	II	I	II	II	I	II	II
Condition[3]	C	H-An	L-An	C	H-An	L-An	C	H-An	L-An
SiO_2 (wt%)	50.11	48.42	50.17	49.61	48.22	48.68	50.77	46.97	49.66
Al_2O_3 (wt%)	30.39	32.01	30.79	30.72	31.80	31.24	30.31	32.85	30.64
FeO* (wt%)	0.53	0.53	0.46	0.60	0.55	0.56	0.80	0.65	0.62
MgO (wt%)	0.29	0.25	0.27	0.32	0.27	0.29	0.36	0.23	0.29
CaO (wt%)	14.96	16.45	14.68	15.34	16.30	15.69	14.52	17.04	15.26
Na_2O (wt%)	2.85	2.07	2.71	2.73	2.21	2.62	3.00	1.70	2.63
K_2O (wt%)	0.03	0.02	0.03	0.03	0.01	0.03	0.03	0.01	0.01
Total (wt%)	99.16	99.75	99.11	99.35	99.36	99.11	99.79	99.45	99.11
An (mol%)	74.4	81.5	75.0	75.6	80.3	76.8	72.8	84.7	76.2

Sample	1183A-67R-3 (133–139)			1185B-17R-3 (107–110)			1185B-22R-6 (19–23)		
Mineral	pl11	pl17	pl17	pl17	pl13	pl13	pl1	pl5	pl5
Size (μm)	54	177	177	61	92	92	87	73	73
Point[1]	3	9	11	3	4	13	5	3	5
Type[2]	I	II	II	I	II	II	I	II	II
Condition[3]	C	H-An	L-An	C	H-An	L-An	C	H-An	L-An
SiO_2 (wt%)	50.14	47.73	50.38	49.18	47.34	49.96	49.26	47.71	50.09
Al_2O_3 (wt%)	30.27	32.62	30.82	31.29	32.60	30.51	31.06	32.42	30.71
FeO* (wt%)	0.59	0.51	0.55	0.57	0.55	0.54	0.60	0.62	0.60
MgO (wt%)	0.29	0.22	0.29	0.29	0.24	0.30	0.29	0.23	0.31
CaO (wt%)	15.18	16.60	14.81	15.78	16.82	15.13	15.13	16.56	14.93
Na_2O (wt%)	2.68	1.69	2.90	2.52	1.84	2.87	2.69	1.97	2.94
K_2O (wt%)	0.03	0.01	0.02	0.03		0.03	0.02	0.01	0.02
Total (wt%)	99.18	99.38	99.77	99.66	99.39	99.34	99.05	99.52	99.60
An (mol%)	75.8	84.4	73.8	77.6	83.5	74.4	75.7	82.3	73.7

Sample	1186A-32R-2 (95–99)			1186A-34R-5 (134–140)			1186A-37R-1 (13–19)		
Mineral	pl17	pl2	pl2	pl1	pl23	pl23	pl5	pl23	pl23
Size (μm)	119	100	100	170	80	80	112	227	227
Point[1]	1	4	2	7	2	1	4	3	7
Type[2]	I	II	II	I	II	II	I	II	II
Condition[3]	C	H-An	L-An	C	H-An	L-An	C	H-An	L-An
SiO_2 (wt%)	49.76	48.35	49.77	50.12	48.97	51.38	48.98	46.96	50.09
Al_2O_3 (wt%)	30.82	31.81	30.94	30.81	31.24	29.98	31.39	32.80	30.53
FeO* (wt%)	0.59	0.49	0.50	0.54	0.56	0.65	0.65	0.51	0.60
MgO (wt%)	0.30	0.25	0.30	0.31	0.29	0.34	0.29	0.24	0.32
CaO (wt%)	15.02	16.52	15.42	15.41	16.07	14.53	15.94	17.03	15.05
Na_2O (wt%)	2.64	2.16	2.73	2.68	2.36	3.03	2.53	1.72	2.84
K_2O (wt%)									
Total (wt%)	99.13	99.58	99.66	99.87	99.49	99.91	99.78	99.26	99.43
An (mol%)	75.9	80.9	75.7	76.1	79.0	72.6	77.7	84.6	74.5

* Total iron as FeO.
[1] Analysed points of 1183A-67R-3 (133–139) pl17, 1185B-17R-3 (107–110) pl13 and 1186A-34R-5 (134–140) pl1 are shown in the line profiles of Figure 4.
[2] Type of plagioclase phenocryst: I, type I plagioclase; II, type II plagioclase.
[3] Condition of analysed point: C, core of type I plagioclase; H-An, high-An mantle of type II plagioclase; L-An, low-An core and/or mantle of type II plagioclase.

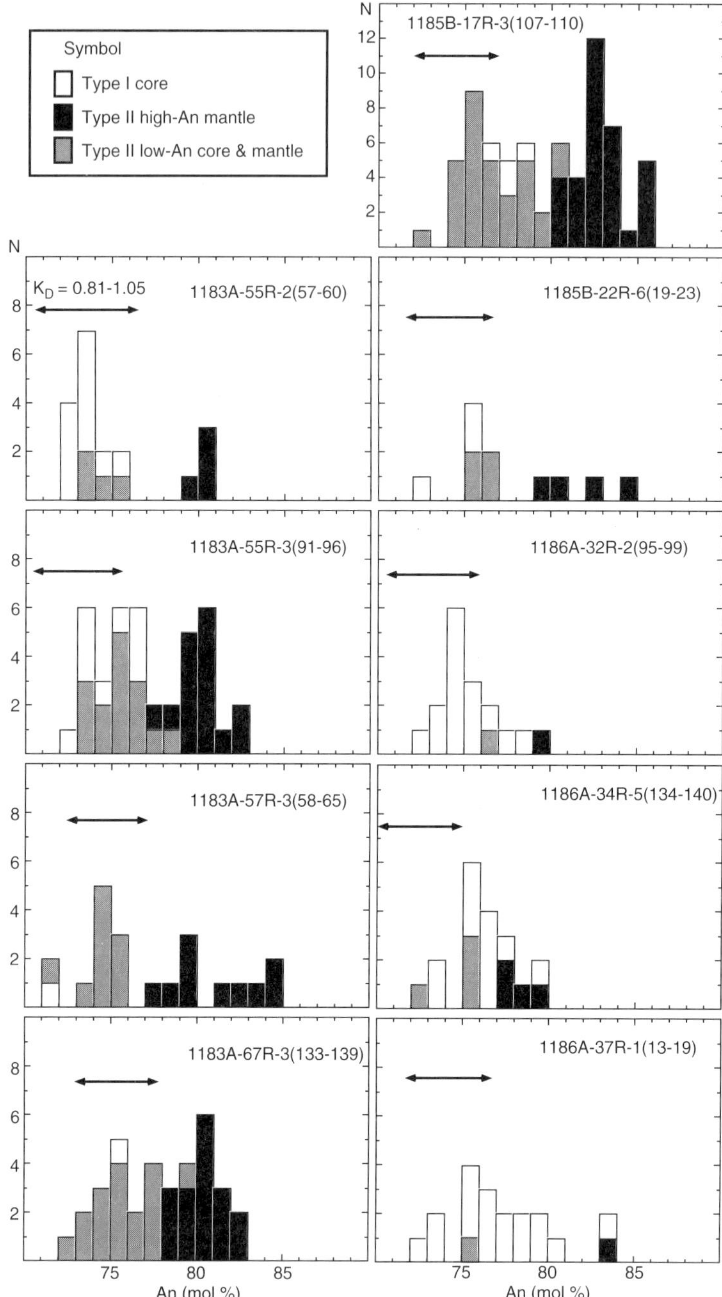

Fig. 6. Histogram of An content for plagioclase phenocrysts in the nine basalt samples used in this study. Arrows correspond to the An ranges of plagioclase in equilibrium with the host magma calculated using our experimental results.

is sparsely olivine–plagioclase-phyric (2 vol.%, Table 1). This sample was chosen because it contains small amounts of the type II plagioclase (Fig. 6) that is not in equilibrium with liquid composition (described later). The sample 1186A-37-R-1, 13–19 has major-element compositions nearly identical with the average compositions of the Kwaimbaita-type lavas (Fig. 2).

All melting experiments were performed under anhydrous conditions because the OJP

Table 5. *Electron microprobe analyses of representative core compositions of clinopyroxene phenocrysts*

Hole	1183A	1183A	1185B	1185B	1186A	1186A
Core-Section	55R-2	55R-2	17R-3	17R-3	34R-5	34R-5
Interval	57–60	57–60	107–110	107–110	134–140	134–140
Mineral	cp1	cp2	cp5	cp6	cp2	cp5
Size (μm)	52	43	57	46	106	126
Point	1	2	6	5	4	4
SiO_2 (wt%)	51.54	51.40	51.08	52.80	53.26	52.99
TiO_2 (wt%)	0.49	0.41	0.43	0.30	0.23	0.24
Al_2O_3 (wt%)	3.74	3.92	3.70	2.16	1.70	1.82
Cr_2O_3 (wt%)	0.46	0.50	0.84	0.58	0.40	0.42
FeO* (wt%)	6.62	7.23	6.29	7.96	6.90	7.69
MnO (wt%)	0.20	0.23	0.14	0.23	0.22	0.24
MgO (wt%)	18.03	18.19	16.98	20.54	19.82	20.16
CaO (wt%)	18.15	17.34	19.39	14.61	17.32	16.33
Na_2O (wt%)	0.21	0.17	0.19	0.16	0.10	0.09
Total (wt%)	99.44	99.39	99.04	99.34	99.95	99.98
mg-number (mol%)	82.9	81.8	82.8	82.1	83.7	82.4

Table 6. *Run conditions and run products in the 0.1 MPa experiments on Ontong Java Plateau basalt*

Starting material	Run No.	*T* (°C)	fO_2[1]	Duration (h)	Run products[2]	Phase proportion[3]	SSR[4]	% Fe loss[5]	% Na loss[5]
Kroenke type	1B22	1250	FMQ	13	gl				
(1187A-6R-6, 109–114)	1B02-1	1240	FMQ	24	gl,ol,sp	100:tr:tr	0.09	3	
	1B19	1220	FMQ	22	gl,ol,sp	98:2:tr	0.08		
	1B01	1200	FMQ	48	gl,ol,sp,pl	95:3:tr:2	0.08		
	1B21	1190	FMQ	44	gl,ol,sp,pl	85:7:tr:8	0.02		
	1B10	1180	FMQ	49	gl,ol,pl,cpx	73:9:14:4	0.18		
	1B06	1165	FMQ	46	gl,ol,pl,cpx	56:10:23:11	0.06		
	1B07	1150	FMQ	48	gl,ol,pl,cpx	39:9:29:23	0.14		
	1B16	1220	CCO	44	gl,ol,sp	98:2:tr	0.57	4	5
	1B11	1200	CCO	45	gl,ol,sp,pl	94:4:1:1	0.18	2	
	1B18	1190	CCO	43	gl,ol,sp,pl	86:7:tr:7	0.02	1	1
	1B28	1180	CCO	48	gl,ol,pl,cpx	80:9:10:1	0.39	1	
	1B26	1165	CCO	48	gl,ol,pl,cpx	56:11:22:11	0.08		
Kwaimbaita type	1B09	1180	FMQ	50	gl				
(1186A-37R-1, 13–19)	1B15	1170	FMQ	48	gl,ol,pl	95:2:3	0.70	2	8
	1B05	1160	FMQ	50	gl,ol,pl,cpx	84:4:9:3	0.68	1	19
	1B13	1180	CCO	32	gl				
	1B12	1175	CCO	49	gl,ol,pl	97:2:1	1.03	5	11
	1B14	1170	CCO	41	gl,ol,pl	96:2:2	1.19	4	11
	1B20	1160	CCO	48	gl,ol,pl,cpx	88:4:8:tr	0.34	1	16
	1B25	1150	CCO	72	gl,ol,pl,cpx	74:3:14:9	0.27	1	

[1] Abbreviations: FMQ, fayalite–magnetite–quartz; CCO, cobalt–cobalt oxide.
[2] Abbreviations: gl, glass; ol, olivine; sp, spinel; pl, plagioclase; aug, augite; pig, pigeonite.
[3] Experimental phase proportions estimated in wt% by materials balance calculation.
[4] Residual sum of squares (SSR) in the materials balance calculation.
[5] Per cent relative Fe and Na loss estimated with the material balance calculation. Blank indicates that the Fe or Na loss was $\ll 1$.

magmas are believed to be very poor in H_2O (0.1–0.2 wt%, Roberge *et al.* 2004), and the effect of such a small amount of H_2O on melting relations in basaltic systems is negligibly small (Danyushevsky 2001).

Experimental and analytical methods

The melting experiments at atmospheric pressure (0.1 MPa) were carried out at the Institute for Study of the Earth's Interior (ISEI),

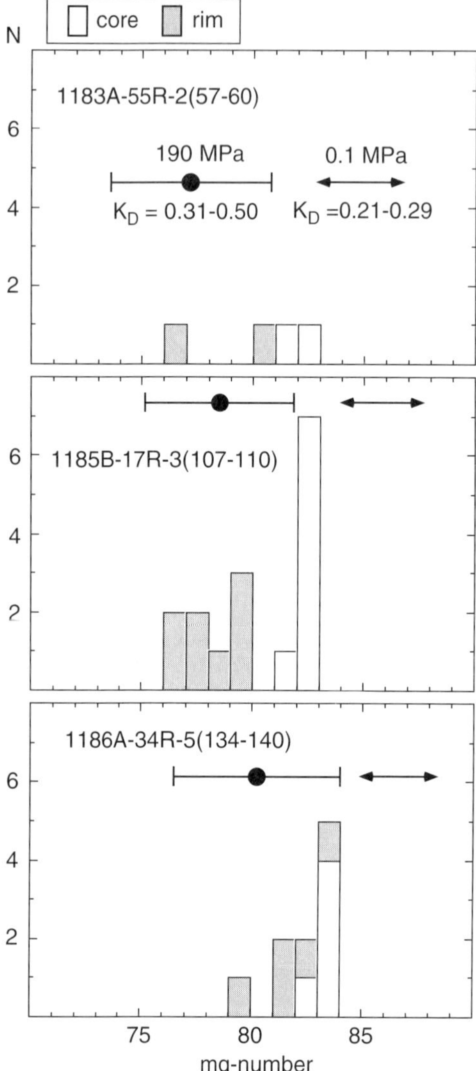

Fig. 7. Histogram of mg-number for clinopyroxene phenocrysts in the three basalt samples used in this study. Horizontal lines around solid circles correspond to the mg-number ranges of clinopyroxene in equilibrium with host magma at 190 MPa, and arrows correspond to the mg-number ranges of clinopyroxene in equilibrium with the host magma at 0.1 MPa.

Okayama University, using a H_2–CO_2 gas mixing furnace. The starting materials were finely ground to powder and dried at 110°C for more than 24 h. A pellet of about 4 g was formed from each sample powder with a hydraulic press and then crushed to small pieces (20–30 mg). Each piece was bound with a thin Pt wire-loop (0.050 mm in diameter) and fused at a temperature above the liquidus (1300°C for the Kroenke-type sample and 1250°C for the Kwaimbaita-type sample).

The quenched glass product (still suspended in the Pt wire-loop) was used as a starting sample of the 0.1 MPa experiments. The glass was hung in the hot spot of a H_2–CO_2 gas mixing furnace, and held for 13–72 h at the desired temperatures and oxygen fugacities. Temperature was measured with a Pt–$Pt_{87}Rh_{13}$ thermocouple with accuracy of better than ±2°C. All experiments were performed under the oxygen fugacity (fO_2) along the fayalite–magnetite–quartz (FMQ) and cobalt–cobalt oxide (CCO) buffers. The oxygen fugacity was monitored with a ZrO_2–CaO electrolyte cell. At the end of the run, the sample was quenched in water. Conditions of the 0.1 MPa experiments are reported in Table 6.

The melting experiments at 190 MPa were carried out in a Kobelco internally heated pressure vessel at the ISEI, following the technique described by Yamashita (1999). Pure Ar gas was used as the pressure medium, and the double capsule method (Sisson & Grove 1993; Yamashita 1999) was employed to control the oxygen fugacity in the experiments (Fig. 8a). The inner capsule was a Fe–Pt alloy-lined Pt capsule containing the sample powder that was dried at 900°C for 24 h under the oxygen fugacity along the CCO buffer. The outer capsule was a Pt tube (4.7 mm inner diameter, 5.0 mm outer diameter, 13–14 mm length) containing Co and CoO powders as a buffering assemblage. The inner capsule was crimped and twisted at its end, but not welded.

The inner capsule was prepared by a method modified after Chen & Lindsley (1983) and van der Laan & van Groos (1991). A Pt tube (1.4 mm inner diameter, 1.6 mm outer diameter, approximately 7 mm length) was welded and flattened at one end. The Grove (1981) empirical expression of the Fe partitioning between basaltic melts and Fe–Pt alloy was employed to calculate the composition of Fe–Pt alloy that is in equilibrium with the starting materials under CCO-buffered conditions. A suitable weight (6–10 mg) of Fe was electroplated onto the surface of the Pt tube. The Fe-plated Pt tube was annealed for 120 h at 1300°C in a Ar-H_2 gas mixing furnace to fabricate homogeneous Fe–Pt alloy. The Fe–Pt alloy tube was then inserted into a Pt tube (2.0 mm inner diameter, 2.3 mm outer diameter, approximately 8 mm length) to insulate the Fe–Pt alloy from the buffering assemblage.

The capsule was hung by Mo wire in the hot spot of a Mo furnace within the pressure vessel, and held for 10–20 h at temperatures between

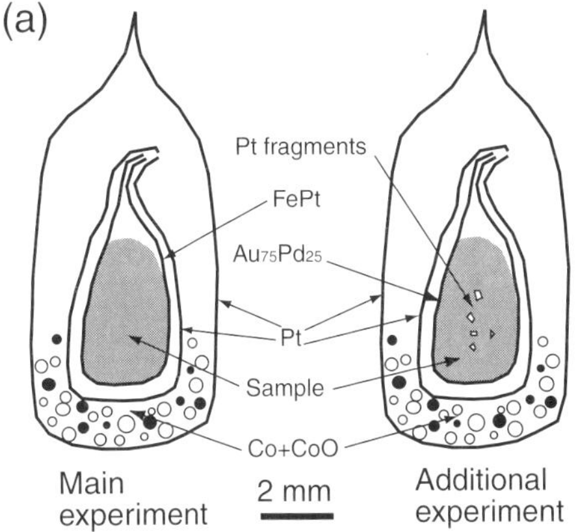

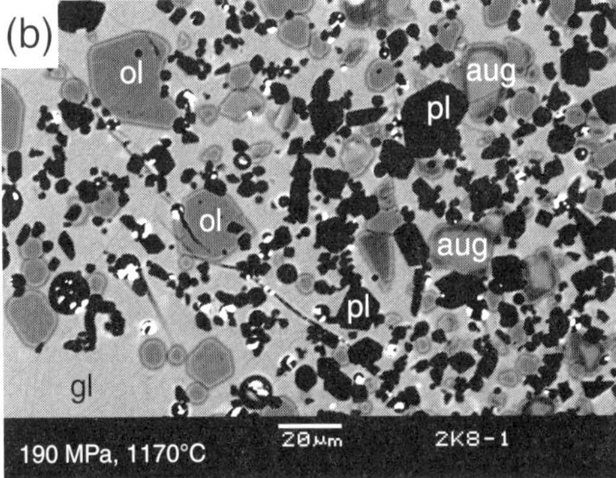

Fig. 8. (**a**) Schematic cross-sections of capsules used in our main experiments (left-hand side) and additional experiments (right-hand side) at 190 MPa. (**b**) Backscattered electron (BSE) microprobe image of Run 2K8–1 (190 MPa, 1170°C). Note that glass, olivine and plagioclase crystals have homogeneous compositions except for quench rims, but augite crystals have relatively heterogeneous compositions (see text for details). Abbreviations used for phases are the same as in Table 6.

1160 and 1220°C. Pressure was measured with a strain-gauge pressure transducer. Temperature was monitored with a $W_5Re–W_{26}Re$ thermocouple, which was calibrated to the melting point of Au. Temperature gradient was less than 10°C over the length of the capsule (Yamashita 1999). At the end of the run, the hanging wire was broken with a surging current, thereby letting the capsule fall into the cold (<300°C) bottom of the vessel. After quenching, the capsules were punctured under a microscope, and the presence of both Co and CoO was confirmed. Conditions of the 190 MPa experiments are reported in Table 7.

To test for achievement of successful control of the oxygen fugacity, two additional experiments were carried out with the same technique. In these experiments, however, the inner capsule was a $Au_{25}Pd_{75}$-lined Pt tube and its contents were tiny (< 10 μm) Pt fragments, the sample powder was dried at 900°C under either the CCO-buffered condition or in air (Fig. 8a). The capsule was held for 20 h at 1200°C and then quenched. The Fe partitioning between the quenched glasses and the Pt fragments remains unchanged regardless of the oxidation–reduction state of the starting materials. The oxygen fugacity calculated from the Fe partitioning was −8.3 to −8.9 in log units, a value close to the oxygen fugacity along the CCO buffer (−9.1 at 1200°C; Myers & Gunter 1979). This observation indicates that the oxygen fugacity was successfully controlled in our experiments. In the experiments with the $Au_{25}Pd_{75}$-lined Pt capsule, a significant amount of FeO (10–20%) was lost from the charge, however. Therefore, we do not use the results of these experiments in the later discussion.

Compositions of minerals and glasses in the run products were determined by the JEOL JXA-8800R EPMA at ISEI with the same procedure as that employed for the basement samples.

Experimental results

The results of the melting experiments are summarized in Tables 6 and 7. Average composition of each phase in an experimental product is given in Table 8. A least-squares approach was employed to calculate mass balance among all phases in an experimental product, which thereby yields the proportion of each phase (Tables 6 and 7).

Preliminary considerations. The duration of the 0.1 MPa experiments was as long as possible, but was limited by possible volatilization of Na from the charge into $CO_2–H_2$ gas, and possible loss of Fe into the Pt wire-loop. The mass-balance calculation ensures that both of these effects are negligible during most of the 0.1 MPa experiments (Table 6). The rationale for this observation is that the duration of each experiment was less than 72 h, and the proportion of the Pt wire-loop to the sample was one tenth or less in weight. Some experimental products are believed to have greater than 10% of Na loss (1B05, 1B12, 1B14 and 1B20 in Table 6) and are thus discarded from the later discussion.

In the 190 MPa experiments, the duration was limited by possible Fe diffusion from the Fe–Pt

Table 7. *Run conditions and run products in the 190 MPa experiments on Ontong Java Plateau basalt*

Starting material	Run No.	T (°C)	Duration (h)	Run products[1]	Phase proportion	SSR	% Fe in Fe–Pt alloy		
							Start[2]	End[3]	n^4
Kroenke type	2K5–1	1220	20	gl,ol,sp	98:2:tr	1.25	6.8	6.2(1.5)	10
(1187A-6R-6, 109–114)	2K5–2	1220	20	gl,ol,sp			4.9	6.0(0.9)	18
	2K10–2	1200	20	gl,ol,sp,pl	91:5: tr:4	0.29	8.1	8.6(1.2)	29
	2K10–1	1200	20	gl,ol,sp,pl			7.3	8.4(1.0)	21
	2K12	1190	10	gl,ol,sp,pl,aug	84:6:tr:6:4	0.40	7.4	7.3(0.5)	44
	2K6–1	1180	20	gl,ol,pl,aug	56:8:20:16	0.40	7.1	8.3(1.2)	21
	2K8–1	1170	20	gl,ol,pl,aug	41:11:28:20	1.31	8.7	9.2(0.6)	36
	2K11–1	1160	20	gl,ol,pl,aug	52:9: 21:18	0.66	7.8	8.6(1.0)	22
	2K7–2	1160	20	gl,ol,pl,aug			9.9	8.7(0.7)	14
Kwaimbaita type	2K6–2	1180	20	gl			7.1	6.9(0.8)	12
(1186A-37R-1, 13–19)	2K8–2	1170	20	gl,ol,pl,aug	73: 1: 12: 14	0.11	6.9	8.2(1.6)	18
	2K11–2	1160	20	gl,pl,aug,pig	55: 21: 15: 9	0.44	7.6	8.7(1.2)	17

[1] Abbreviations of phases are the same as Table 6.
[2] Fe contents (wt%) in Fe–Pt alloy before the experiments.
[3] Fe contents (wt%) in Fe–Pt alloy after the experiments determined by electron probe analysis.
[4] Number of the electron probe analyses.

alloy liner to the Pt insulator (e.g. van der Laan & van Groos 1991). This effect was negligible in our experiments because little change in Fe contents in the Fe–Pt alloy liner capsules was observed during the experiments (Table 7). This observation also suggests that the Fe exchange between the charge and the Fe–Pt alloy was minor in most cases, which in turn provides further evidence for successful control of the oxygen fugacity in our experiments. The Fe exchange partitioning in the run product 2K5–1 indeed yields the oxygen fugacity of –8.6 in log units, a value very close to the oxygen fugacity along the CCO buffer (–8.9 at 1220°C; Myers & Gunter 1979). When the Fe–Pt alloy showed either a significant gain (2K5–2, 2K10–1; Table 7) or loss (2K7–2; Table 7) of Fe during the experiments, we have not used the results of those experiments in the later discussion.

It was also found that in the 190 MPa melting experiments a large amount of NiO (> 60%) was lost from the charges and added to the Fe–Pt alloy-lined Pt capsule (see NiO contents of olivine crystals in Table 8). Therefore, we did not use the NiO contents in the run products of 190 MPa experiments in the following discussion. The NiO loss would not affect the phase appearance and the mineral–melt partitioning of major elements because NiO contents in the basaltic melts are generally negligible.

Attainment of equilibrium. Longer duration experiments than those carried out here (>72 h at 0.1 MPa and >20 h at 190 MPa) are required to achieve total homogeneity of the phases (e.g. Falloon *et al.* 1999; Sugawara 2000*b*). However, for the reasons outlined below, we conclude that our experiments approached equilibrium sufficiently to allow the understanding of the petrological aspect of the OJP basalts.

First, in all experimental products, each phase has the same composition within the analytical precision, except for augite crystals in the 190 MPa experiments (Table 8, Fig. 8b). The heterogeneity probably arises because the augite crystals are not in full equilibrium with the melt because of the relatively short (10–20 h) duration of the 190 MPa experiments, compared with the duration of the 0.1 MPa experiments (more than 46 h in runs where augite crystallized). The heterogeneity is reflected in the large 1σ errors of analyses (more than six points) of the augite crystals (Table 8). The 1σ errors are considered in the later discussion.

Secondly, values of exchange equilibria for Fe–Mg in olivine and for Ca–Na in plagioclase are similar for all experiments (Table 8). Under both the FMQ and CCO buffers, the olivine–melt Fe–Mg partitioning coefficient [K_D: $(Fe/Mg)^{crystal}/(Fe^{2+}/Mg)^{melt}$)] is 0.31–0.34 in the 0.1 MPa experiments and 0.27–0.30 in the 190 MPa experiments. The K_D values were calculated by using an Fe^{3+} content appropriate for the FMQ and CCO buffers (Sack *et al.* 1980), and are in good agreement with those reported from previous equilibrium experiments (e.g. Tormey *et al.* 1987; Yang *et al.* 1996). The plagioclase–melt Ca–Na partitioning coefficient [K_D: $(Ca/Na)^{crystal}/(Ca/Na)^{melt}$)] is 0.84–1.05 in the 0.1 MPa experiments and 0.81–0.98 in the

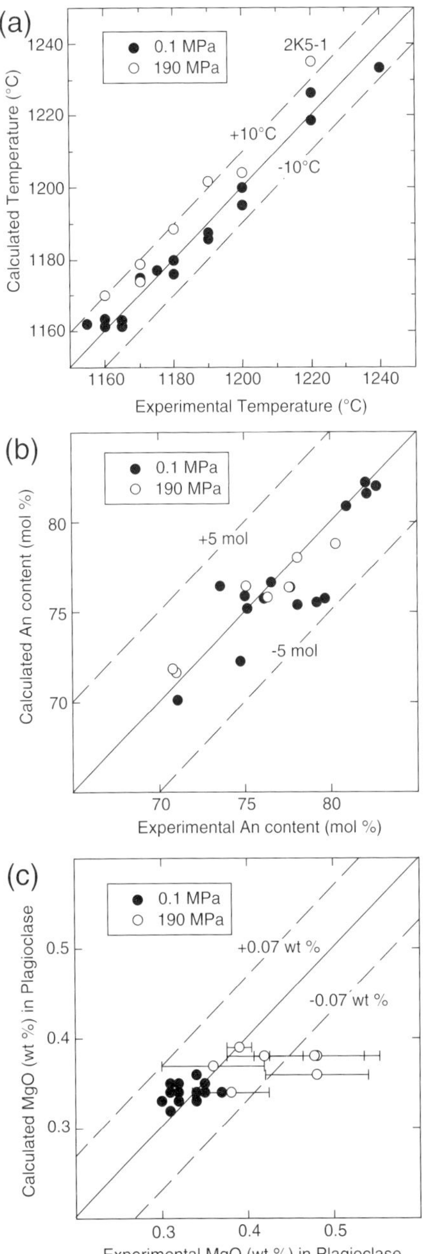

Fig. 9. (**a**) Observed temperatures v. predicted temperatures calculated using the geothermometer of Ford *et al.* (1983). (**b**) Observed An contents of plagioclase v. predicted An contents of plagioclase calculated using the method of Grove *et al.* (1992). (**c**) Observed MgO contents of plagioclase v. predicted MgO contents calculated by using the method of Sugawara (2000*b*). Only MgO content errors at 190 MPa are shown as error bars (± 1 SD of the mean on six or more analyses of plagioclase crystals in each charge).

190 MPa experiments. The plagioclase K_D values also agree well with those determined by previous equilibrium experiments (e.g. Tormey *et al.* 1987; Sano *et al.* 2001). In the 0.1 MPa experiments the K_D for augite–melt $[(Fe/Mg)^{crystal}/(Fe^{2+}/Mg)^{melt})]$ is nearly constant (0.21–0.29), and identical to that reported by previous equilibrium experiments (e.g. Yang *et al.* 1996; Sano *et al.* 2001). On the other hand, the K_D for augite–melt in the 190 MPa experiments has a slightly greater range (0.31–0.50) than that in the 0.1 MPa experiments because of the augite heterogeneity.

Thirdly, the experimental temperatures coincide well with predicted temperatures calculated using the geothermometer of Ford *et al.* (1983). They reported that temperatures would be obtained within ±10°C by using major-element compositions of magmas equilibrated with olivine crystals. When we consider the accuracy of temperature measurements in our experiments (±2°C for the 0.1 MPa experiments and <10°C for the 190 MPa experiments), calculated temperatures are all within 10°C of the experimental temperatures (Fig. 9a).

Fourthly, compositions of plagioclase in the experiments are in agreement with those calculated using the methods of previous studies (Grove *et al.* 1992; Sugawara 2000*b*). Grove *et al.* (1992) have proposed that the An content of plagioclase would be calculated within an error of ±5 mol% as a function of pressure and melt compositions. Figure 9b shows that the experimental An contents and calculated An contents are the same within the ±5 mol% error. Sugawara (2000*b*) also suggested that the MgO content of plagioclase was calculated within an error of ±0.07 wt% when pressure and melt compositions are known. Considering the calculated error and the analytical precision, our calculated MgO contents coincide with the experimental MgO contents (Table 8, Fig. 9c).

Mineral appearance sequences and melt compositions. At 0.1 MPa, changes in the oxygen fugacity (fO_2) did not change mineral appearance sequences, mineral compositions and melt compositions (Tables 6 and 8, Fig. 10), implying that effects of the fO_2 were negligible within the fO_2 range (the QFM–CCO buffer).

Olivine is the liquidus phase for the sample 1187A-6R-6, 109–114 (Kroenke-type basalt) in both the 0.1 and 190 MPa experiments (Tables 6 and 7, Fig. 10), which is consistent with the petrographical observations (Table 1). The olivine liquidus temperature is approximately 1240°C in the 0.1 MPa experiments, and the liquidus temperature at 190 MPa is not known in detail because of the small data set. Spinel is also a

Table 8. *Electron microprobe analyses of run products of the melting experiments (oxides reported as weight per cent)*

Phase[1]	Run	*n*	SiO_2	TiO_2	Al_2O_3	Cr_2O_3	FeO*	MnO	MgO	CaO	Na_2O	K_2O	NiO	Total	K_D[2]
Kroenke type: 1187A-6R-6,109–114 (0.1 MPa, FMQ)															
gl	1B0–1[3]	10	49.98(7)	0.71(2)	14.67(10)	0.09(1)	9.53(12)	0.17(1)	10.35(8)	12.17(5)	1.33(7)	0.04(1)		99.04	
	1B02–1	12	50.13(84)	0.71(2)	14.80(27)	0.08(1)	9.27(36)	0.17(1)	10.27(22)	12.27(39)	1.45(10)	0.06(1)		99.21	
	1B19	12	50.35(27)	0.75(2)	14.69(13)	0.08(1)	9.57(38)	0.18(1)	9.66(11)	12.47(13)	1.37(5)	0.07(1)		99.19	
	1B01	12	50.71(19)	0.79(3)	15.02(11)	0.07(2)	9.50(11)	0.18(2)	8.81(10)	12.74(15)	1.39(6)	0.07(1)		99.28	
	1B21	10	50.84(21)	0.83(2)	14.39(13)	0.06(1)	9.95(14)	0.19(2)	8.39(9)	12.88(15)	1.45(2)	0.08(1)		99.06	
	1B10	10	51.55(15)	0.95(2)	13.81(17)	0.04(1)	10.51(16)	0.18(2)	7.77(7)	12.49(12)	1.90(5)	0.09(1)		99.29	
	1B06	10	51.58(20)	1.23(5)	13.50(16)	0.02(1)	11.56(16)	0.22(2)	7.17(11)	11.87(13)	1.75(5)	0.12(1)		99.02	
	1B07	9	51.58(36)	1.68(3)	12.86(9)	0.02(1)	13.35(37)	0.22(3)	6.45(13)	10.69(29)	2.06(7)	0.14(1)		99.05	
ol	1B02–1	7	40.56(30)			12.17(34)	0.19(1)	47.11(51)	0.35(3)				0.21(2)	100.59	0.32(2)
	1B19	7	40.42(39)			12.63(21)	0.19(1)	46.10(77)	0.37(3)				0.24(1)	99.95	0.31(2)
	1B01	6	39.56(27)			14.06(8)	0.21(1)	45.80(39)	0.40(4)				0.22(1)	100.25	0.32(1)
	1B21	12	39.74(26)			14.32(38)	0.22(1)	44.32(14)	0.34(3)				0.19(1)	99.13	0.31(1)
	1B10	12	39.54(23)			16.57(16)	0.24(1)	43.84(36)	0.41(3)				0.18(1)	100.78	0.32(1)
	1B06	10	38.38(27)			19.47(32)	0.29(1)	41.01(41)	0.42(4)				0.15(2)	99.72	0.34(1)
	1B07	11	38.21(37)			23.31(35)	0.33(1)	38.03(30)	0.44(2)				0.14(1)	100.46	0.34(1)
sp	1B2–1	4	0.16(4)	0.45(3)	27.73(35)	37.40(65)	12.93(1.75)	0.24(2)	19.02(1.06)	0.27(2)				98.20	
	1B19	4	0.31(9)	0.45(1)	24.31(27)	39.72(10)	17.92(26)	0.24(1)	14.08(33)	0.37(2)				97.40	
	1B01	6	0.42(18)	0.50(2)	26.86(58)	36.05(77)	20.11(22)	0.24(2)	14.00(24)	0.41(3)				98.59	
	1B21	10	0.29(5)	0.51(1)	26.68(39)	35.91(61)	20.44(46)	0.25(1)	13.44(12)	0.36(2)				97.88	
pl	1B01	3	49.09(38)		31.71(20)		0.69(2)		0.32(3)	16.3(13)	2.02(5)	0.02(1)		100.15	0.88(4)
	1B21	10	48.53(33)		31.29(24)		0.65(6)		0.35(1)	16.15(21)	2.11(10)	0.04(1)		99.12	0.86(5)
	1B10	9	48.67(16)		31.20(17)		0.67(4)		0.34(2)	16.00(18)	2.33(5)			99.21	1.05(4)
	1B06	11	50.55(24)		30.58(19)		0.83(6)		0.35(3)	15.31(20)	2.44(18)	0.02(1)		100.08	0.92(7)
	1B07	11	50.04(38)		30.29(28)		0.82(8)		0.32(4)	15.07(20)	2.82(18)			99.36	1.03(8)
aug	1B10	7	51.71(36)	0.22(2)	3.37(26)	1.10(7)	5.87(28)	0.17(2)	18.01(38)	19.00(60)	0.16(2)			99.61	0.22(1)
	1B06	7	53.43(38)	0.32(4)	2.31(27)	0.57(10)	7.10(24)	0.20(2)	17.80(31)	17.99(50)				99.72	0.29(1)
	1B07	6	51.63(92)	0.32(6)	3.69(2.19)	0.34(5)	8.38(1.27)	0.23(5)	19.02(83)	15.60(1.03)	0.14(2)			99.35	0.21(4)
Kwaimbaita type: 1186A-37R-1,13–19 (0.1 MPa, FMQ)															
gl	1B15	11	51.97(42)	1.14(3)	14.13(17)	0.03(1)	10.80(19)	0.21(2)	7.64(22)	11.77(25)	1.98(4)	0.05(1)		99.72	
	1B05	10	51.77(21)	1.37(2)	13.51(8)	0.02(1)	11.62(17)	0.23(2)	7.14(12)	11.53(23)	1.72(11)	0.09(1)		99.00	
ol	1B15	5	39.26(36)				17.05(18)	0.27(1)	43.60(43)	0.34(1)			0.12(0)	100.64	0.32(1)
	1B05	8	38.77(32)				19.22(20)	0.30(1)	41.64(62)	0.35(1)			0.10(1)	100.38	0.33(1)
pl	1B15	9	50.73(25)		30.54(28)		0.72(4)		0.31(1)	14.78(13)	2.70(6)			99.78	0.92(4)
	1B05	10	51.69(36)		29.77(38)		0.80(9)		0.31(2)	14.56(24)	2.90(13)	0.02(1)		100.05	0.75(6)
aug	1B05	7	53.31(49)	0.42(8)	2.35(17)	0.28(3)	7.41(50)	0.22(3)	18.27(1.09)	17.16(1.57)	0.16(2)			99.58	0.29(3)

Kroenke type: 1187A-6R-6,109–114 (0.1 MPa, CCO)															
gl	1B16	12	51.03(28)	0.72(2)	14.98(23)	0.08(1)	9.11(18)	0.16(2)	10.00(9)	12.25(35)	1.28(5)	0.05(1)		99.66	
	1B11	10	51.04(22)	0.76(4)	15.03(18)	0.07(1)	9.07(13)	0.17(2)	8.91(9)	12.75(30)	1.39(4)	0.07(1)		99.26	
	1B18	10	51.05(24)	0.84(1)	14.44(9)	0.08(1)	9.69(10)	0.18(1)	8.43(8)	12.87(24)	1.37(4)	0.09(0)		99.04	
	1B28	12	51.55(17)	0.99(2)	14.30(19)	0.04(1)	9.83(43)	0.19(1)	7.88(8)	12.92(17)	2.02(10)	0.12(1)		99.84	
	1B26	12	52.04(33)	1.34(3)	13.45(20)	0.02(1)	11.77(36)	0.22(1)	7.22(11)	11.98(13)	1.71(4)	0.09(1)		99.84	
ol	1B16	8	40.06(30)				12.47(20)	0.19(1)	47.14(11)	0.35(3)			0.18(1)	100.39	0.31(1)
	1B11	2	40.86(38)				13.34(8)	0.20(0)	45.58(20)	0.36(0)			0.13(1)	100.47	0.31(1)
	1B18	10	39.54(24)				14.64(21)	0.21(1)	44.12(19)	0.34(2)			0.17(1)	99.02	0.31(1)
	1B28	9	39.65(13)				15.63(24)	0.24(1)	44.14(14)	0.41(3)			0.14(2)	100.21	0.31(2)
	1B26	9	39.04(19)				19.52(25)	0.29(1)	40.84(26)	0.43(3)			0.13(1)	100.25	0.32(1)
sp	1B11	1	0.17	0.48	27.47	36.90	18.19	0.22	14.35	0.41				98.19	
	1B18	8	0.25(4)	0.52(3)	27.63(61)	36.85(93)	18.72(36)	0.25(1)	13.62(10)	0.35(3)				98.19	
pl	1B11	6	48.34(43)		32.31(59)		0.48(3)		0.34(2)	16.43(16)	1.93(10)			99.83	0.93(6)
	1B18	11	48.33(26)		31.44(38)		0.57(3)		0.34(1)	16.42(23)	1.98(11)	0.05(0)		99.13	0.87(6)
	1B28	13	49.53(32)		31.20(26)		0.59(9)		0.34(2)	15.58(40)	2.42(18)	0.01(1)		99.67	1.01(9)
	1B26	12	50.10(29)		30.75(32)		0.74(6)		0.37(3)	15.36(22)	2.61(14)	0.02(1)		99.95	0.84(5)
aug	1B28	9	52.48(63)	0.29(5)	3.17(70)	1.18(21)	5.27(9)	0.16(1)	17.60(51)	19.70(46)	0.16(1)			100.01	0.26(2)
	1B26	6	52.34(50)	0.40(6)	3.35(55)	0.64(15)	6.61(34)	0.18(1)	17.53(95)	18.70(1.26)	0.16(2)			99.91	0.25(2)
Kwaimbaita type: 1186A-37R-1,13–19 (0.1 MPa, CCO)															
gl	1B0–2[3]	12	50.95(37)	1.14(2)	14.20(10)	0.03(1)	10.90(36)	0.21(1)	7.96(11)	11.98(36)	2.14(6)	0.04(1)		99.55	
	1B12	8	51.33(20)	0.97(4)	13.93(8)	0.05(1)	10.35(11)	0.19(1)	7.75(9)	12.75(13)	1.93(4)	0.09(1)		99.34	
	1B14	10	52.23(17)	1.10(4)	14.24(15)	0.03(1)	10.54(14)	0.20(1)	7.61(15)	11.83(13)	1.93(5)	0.05(0)		99.76	
	1B20	10	51.64(19)	1.25(4)	13.36(11)	0.03(1)	11.31(29)	0.23(2)	7.22(13)	11.96(5)	1.78(5)	0.08(0)		98.86	
	1B25	16	52.11(45)	1.50(5)	13.48(16)	0.01(1)	12.51(36)	0.24(1)	6.77(14)	11.02(12)	2.43(10)	0.10(1)		100.17	
ol	1B12	12	39.25(13)				16.14(35)	0.23(1)	43.58(44)	0.38(3)			0.17(2)	99.75	0.31(1)
	1B14	5	39.01(25)				17.22(13)	0.27(1)	43.28(18)	0.34(2)			0.13(2)	100.25	0.32(1)
	1B20	7	39.05(21)				18.19(21)	0.29(1)	41.11(36)	0.36(3)			0.11(1)	99.11	0.31(1)
	1B25	6	38.44(19)				21.69(32)	0.35(2)	39.55(27)	0.36(2)			0.00(0)	100.39	0.33(1)
pl	1B12	10	48.41(28)		31.66(18)		0.60(5)		0.35(1)	16.10(20)	2.28(8)			99.40	
	1B14	6	50.41(24)		30.50(36)		0.65(8)		0.30(2)	14.86(15)	2.56(9)			99.28	
	1B20	10	50.12(64)		30.21(41)		0.71(5)		0.32(1)	14.87(32)	2.78(9)	0.04(1)		99.05	0.79(4)
	1B25	12	51.08(28)		30.25(22)		0.72(7)		0.31(1)	14.35(13)	3.20(8)	0.02(1)		99.93	0.99(5)
aug	1B20	7	52.27(47)	0.53(8)	2.77(34)	0.23(5)	7.58(73)	0.22(3)	18.14(1.00)	17.54(1.67)	0.15(1)			99.43	0.28(3)
	1B25	14	52.48(37)	0.49(6)	2.64(36)	0.20(7)	7.79(88)	0.23(2)	18.25(84)	17.34(1.56)	0.16(2)			99.58	0.24(3)

Table continued overleaf

Table 8. *continued*

Phase[1]	Run	n	SiO_2	TiO_2	Al_2O_3	Cr_2O_3	FeO*	MnO	MgO	CaO	Na_2O	K_2O	NiO	Total	K_D[2]
Kroenke type: 1187A-6R-6,109–114 (190 MPa, CCO)															
gl	2K05–1	12	49.99(24)	0.70(2)	14.66(13)	0.08(1)	10.42(60)	0.16(1)	9.56(17)	11.81(29)	1.61(6)	0.04(1)		99.03	
	2K10–2	11	50.83(20)	0.72(3)	15.14(12)	0.06(1)	9.95(20)	0.17(1)	8.66(13)	12.44(19)	1.65(4)	0.05(1)		99.67	
	2K12	13	50.30(24)	0.84(3)	14.97(23)	0.06(1)	10.09(63)	0.18(1)	8.47(9)	12.62(32)	1.74(6)	0.05(1)		99.32	
	2K06–1	10	50.90(19)	0.85(3)	14.39(27)	0.06(2)	12.05(56)	0.18(1)	7.72(5)	11.52(28)	1.79(7)	0.06(1)		99.52	
	2K08–1	10	50.49(36)	1.01(2)	13.63(15)	0.04(1)	13.49(31)	0.18(2)	7.12(11)	11.25(14)	1.66(4)	0.07(1)		98.94	
	2K11–1	5	50.97(13)	1.20(2)	14.38(10)	0.02(2)	12.10(17)	0.20(2)	7.03(20)	11.09(12)	1.84(5)	0.07(1)		98.90	
ol	2K05–1	6	40.04(23)				12.42(30)	0.18(1)	46.49(29)	0.38(1)			0.03(1)	99.54	0.27(2)
	2K10–2	8	39.83(24)				14.43(28)	0.20(1)	45.74(31)	0.41(3)			0.02(1)	100.63	0.30(1)
	2K12	7	39.80(22)				14.92(55)	0.21(1)	45.13(23)	0.39(3)			0.06(3)	100.51	0.30(2)
	2K06–1	9	40.01(22)				17.12(79)	0.24(1)	42.70(46)	0.39(3)			0.00(0)	100.46	0.28(2)
	2K08–1	6	38.74(11)				20.61(31)	0.24(1)	39.66(23)	0.40(1)			0.04(1)	99.69	0.30(1)
	2K11–1	5	39.24(54)				19.27(18)	0.27(1)	41.12(88)	0.47(12)			0.07(1)	100.44	0.30(1)
sp	2K5–1	3	0.21(4)	0.43(3)	24.77(27)	37.07(36)	22.43(43)	0.20(0)	11.17(47)	0.28(3)				96.56	
	2K10–2	3	0.19(2)	0.43(0)	25.42(21)	38.06(40)	20.46(75)	0.23(1)	13.82(2)	0.28(5)				98.89	
	2K12	2	0.13(0)	0.48(1)	25.30(21)	36.19(2)	22.23(32)	0.25(1)	13.57(16)	0.18(1)				98.33	
pl	2K10–2	2	48.66(23)		31.44(81)		0.75(5)		0.39(0)	16.11(30)	2.19(27)	0.01(0)		99.55	0.98(13)
	2K12	7	49.88(56)		30.40(55)		0.99(11)		0.48(8)	15.25(41)	2.37(43)	0.02(1)		99.39	0.88(17)
	2K6–1	10	50.79(51)		30.44(41)		0.99(10)		0.48(5)	15.08(18)	2.41(14)	0.02(1)		100.21	0.98(7)
	2K8–1	11	50.80(66)		29.80(51)		1.06(11)		0.42(4)	14.80(45)	2.71(33)	0.02(1)		99.61	0.81(10)
	2K11–1	9	49.88(45)		30.28(34)		1.09(14)		0.48(6)	15.23(26)	2.64(7)	0.02(1)		99.62	0.96(4)
aug	2K12	10	50.06(85)	0.87(14)	4.60(92)	0.22(11)	9.12(92)	0.23(3)	14.60(73)	19.88(84)	0.20(3)			99.78	0.50(7)
	2K6–1	6	50.31(41)	0.90(7)	4.99(86)	0.30(13)	9.39(97)	0.22(3)	14.59(32)	18.89(73)	0.20(2)			99.79	0.45(5)
	2K8–1	8	50.35(65)	0.86(17)	4.83(61)	0.28(15)	9.29(84)	0.22(4)	14.14(32)	19.60(60)	0.21(2)			99.78	0.38(4)
	2K11–1	6	49.75(61)	0.93(16)	5.07(71)	0.33(12)	8.56(65)	0.21(1)	15.12(81)	19.29(1.17)	0.25(7)			99.51	0.31(3)
Kwaimbaita type: 1186A-37R-1,13–19 (190 MPa, CCO)															
gl	2K8–2	12	50.81(46)	1.13(3)	13.88(21)	0.02(1)	12.86(40)	0.20(1)	7.07(22)	10.98(15)	2.18(5)	0.05(1)		99.18	
	2K11–2	12	49.86(22)	1.35(3)	13.67(16)	0.03(1)	14.68(33)	0.20(1)	6.50(15)	10.84(16)	2.18(5)	0.05(1)		99.36	
ol	2K08–2	6	38.98(26)				20.02(62)	0.27(1)	40.67(51)	0.47(27)			0.02(1)	100.43	0.30(2)
pl	2K8–2	8	51.64(85)		29.58(79)		1.02(11)		0.36(6)	13.89(43)	3.14(30)	0.02(1)		99.65	0.88(9)
	2K11–2	6	51.28(57)		29.61(34)		1.09(6)		0.38(4)	14.19(36)	3.24(20)	0.02(2)		99.81	0.88(6)
aug	2K8–2	12	51.22(60)	0.81(16)	4.22(98)	0.14(8)	9.05(89)	0.23(3)	16.01(76)	17.62(1.21)	0.20(3)			99.50	0.34(4)
	2K11–2	8	50.19(62)	0.89(15)	4.20(84)	0.10(6)	10.13(1.98)	0.23(5)	15.46(1.57)	18.25(2.24)	0.20(3)			99.65	0.27(7)
pig	2K11–2	7	54.08(21)	0.16(2)	0.98(8)	0.09(2)	14.16(54)	0.32(2)	25.07(26)	4.72(17)	0.06(1)			99.63	

* Total iron as FeO.
[1] Abbreviations used for phases are the same as in Table 6.
[2] $K_D = [(Fe/Mg)^{ol}/(Fe^{2+}/Mg)^{gl}]$ for olivine, $K_D = [(Ca/Na)^{pl}/(Ca/Na)^{gl}]$ for plagioclase, $K_D = [(Fe^{2+}/Mg)^{aug}/(Fe^{2+}/Mg)^{gl}]$ for augite.
[3] Compositions of starting samples in the experiments.

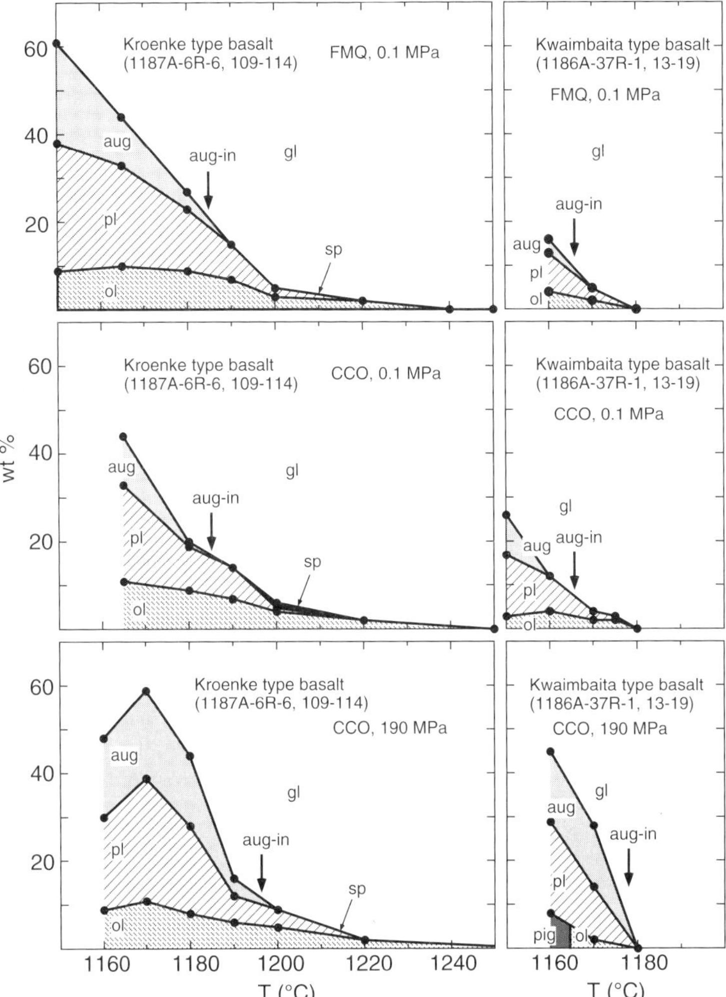

Fig. 10. Results of melting experiments at atmospheric pressure (0.1 MPa) and 190 MPa performed on two Ontong Java Plateau basalts. Solid circles, experimental temperature and estimated phase proportions for each experiment. Abbreviations used for phases and fO_2 are given in Table 6.

near-liquidus phase in both the 0.1 and 190 MPa experiments. The crystallization sequence for the Kroenke-type basalt sample is olivine, plagioclase and augite in both the 0.1 and 190 MPa experiments. Plagioclase crystallizes at ≤1200°C in both the 0.1 and 190 MPa experiments (Tables 6 and 7, Fig. 10), indicating that we cannot detect the effect of pressure on the temperature of plagioclase saturation within the pressure range. On the other hand, as the pressure increased (from 0.1 to 190 MPa) the temperature of augite saturation also increased (from 1180 to 1190°C). These facts are consistent with studies of basalt melting as a function of pressure (e.g. Bender *et al.* 1978; Presnall *et al.* 1979); augite appearance temperature has a steeper dT/dP slope (where T is temperature and P is pressure) than either olivine or plagioclase.

For the Kwaimbaita-type basalt sample, olivine and plagioclase are liquidus phases at approximately 1170°C, and augite appears at approximately 1160°C in the 0.1 MPa experiments (Table 6, Fig. 10). On the other hand, olivine, plagioclase and augite crystallize simultaneously at about 1170°C in the 190 MPa experiments (Table 7, Fig. 10). These results also indicate that the temperature at which augite appears increases with increasing pressure. Olivine, plagioclase and augite crystallize even at temperatures as low as 1150°C in the 0.1 MPa experiments, but olivine is unstable and pigeonite appears at 1160°C in the 190 MPa experiments (Fig. 10).

Discussion

Depth and temperature of magma chamber

Using the K_D value (0.27–0.34) for olivine–melt in the experiments, we can judge whether or not natural olivine crystals are in equilibrium with their host magma compositions (i.e. the whole-rock compositions). The Kroenke-type basalt has normal-zoned olivine crystals whose mg-numbers imply equilibrium with whole rocks, except for sample 1187A-16R-3, 31–34 (Fig. 5). As MgO content of the samples may have been affected by alteration processes (Mahoney *et al.* 2001), the mg-number of the whole rock could have changed from an original value. The cores of the reverse-zoned olivine phenocrysts are not in equilibrium with the whole-rock compositions (Fig. 5). They must be in equilibrium with more-differentiated basalt (e.g. the Kwaimbaita-type basalt) in terms of the mg-number, as we discuss below.

The plagioclase K_D values (0.81–1.05) also show that the plagioclase An content is in equilibrium with the whole-rock An content. The cores of the type I plagioclase and the low-An parts of the type II have An content in equilibrium with the whole-rock (the Kwaimbaita-type basalt) An content. On the other hand, the high-An parts of the type II plagioclase are not in equilibrium with the Kwaimbaita-type basalt in terms of the An content (Fig. 6).

While Leg 192 basement lavas contain some phenocrysts whose compositions are not in equilibrium with whole-rock compositions, we assume that the whole-rock compositions are the same as the melt compositions in a magma chamber. This is because contents of the disequilibrium phenocrysts are negligible ($\ll$1 vol%).

The chemical trends observed in Leg 192 basement lavas are approximately reproduced by liquid lines of descent (LLD) obtained from the experiments (Fig. 2), indicating that basement lavas experienced fractional crystallization at low pressure (0.1–190 MPa) under relatively reducing conditions (the FMQ–CCO buffer). Al_2O_3 content of the LLD is just a little lower than that of Leg 192 basement lavas. This observation is simply due to Al_2O_3 depletion of the starting composition (Fig. 2). The experimental results (Tables 6–8) show that one of the most differentiated Kwaimbaita-type magmas, with 7.0 wt% MgO, can be produced by subtraction of approximately 10 wt% of olivine, 20 wt% of plagioclase and 10 wt% of augite from the most Mg-rich basalt (1187A-6R-6, 109–114: the Kroenke-type basalt starting material) among the Leg 192 basement lavas. The result is similar to that Fitton & Godard (2004) who discuss the behaviour of incompatible trace elements during the fractional crystallization process. The fractional crystallization model is also supported by isotopic data; the Kroenke-type lavas are similar in isotopic composition to the Kwaimbaita-type lavas (Tejada *et al.* 2004).

The depth of the magma chamber was estimated by using Kwaimbaita-type basalt melt compositions projected into an olivine–clinopyroxene–quartz (ol–cpx–qtz) pseudoternary diagram (Grove *et al.* 1992) as a geobarometer. Melting experiments for mid-ocean ridge basalt (MORB) lavas have shown that the cotectic line of olivine–plagioclase–augite (ol–pl–cpx) saturated melt is a function of pressure (e.g. Grove *et al.* 1992; Kinzler & Grove 1992; Yang *et al.* 1996). Experimentally determined compositions of the ol–pl–cpx saturated melts projected into the ol–cpx–qtz pseudoternary (Tormey *et al.* 1987) are shown in Figure 11, together with the compositions of the Kwaimbaita-type lavas that have phenocrysts of olivine, plagioclase and augite. For the experiments on the OJP lavas, the effect of increasing pressure is to move the ol–pl–cpx saturation line toward the ol apex (Fig. 11a). This move is similar to that in MORB. In the ol–cpx–qtz pseudoternary system, the Kwaimbaita-type compositions project to a cluster near the ol–pl–cpx saturation lines predicted for the pressure range from 0.1 to 190 MPa (Fig. 11b). This observation implies that the depth of the magma chamber is shallower than 6 km.

The shallow-chamber model is supported by augite–melt partition coefficients, which can be used to estimate pressure (Blundy *et al.* 1995; Putirka *et al.* 1996). Our experimental results also found effects of pressure on augite–melt partition coefficients of Ti and Al. Figure 12 shows histograms of the Ti and Al partition coefficients between the augite core and whole-rock Kwaimbaita-type basalt host, with the experimentally determined Ti and Al partition coefficients. We can confirm that the partition coefficients of the Kwaimbaita-type basalt have the partition coefficients predicted for the pressure range from 0.1 to 190 MPa (Fig. 12). The augite K_D value is also a function of

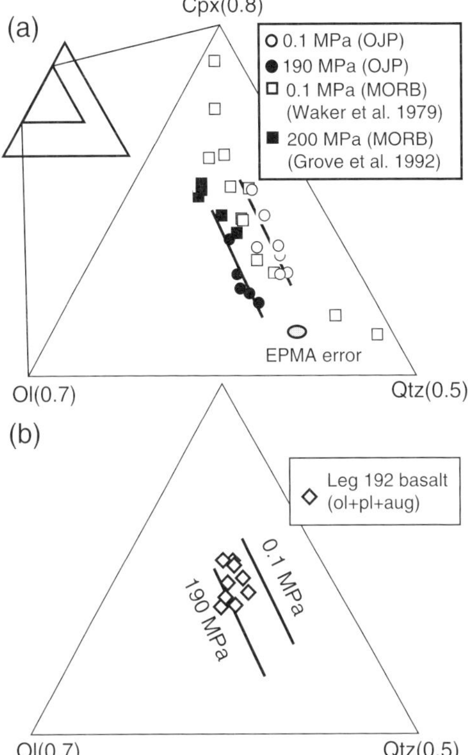

Fig. 11. (**a**) Compositions of experimentally produced melts that coexist with olivine, plagioclase and augite at 0.1 and 190 MPa (or 200 MPa) projected into the pseudoternary olivine–clinopyroxene–quartz (ol–cpx–qtz) system. Experiments are from Walker *et al.* (1979) and Grove *et al.* (1992) on mid-ocean ridge basalts (MORBs), and from this study on the Ontong Java Plateau (OJP) basalts. Normative components are calculated following expressions given by Tormey *et al.* (1987), with the proportions of FeO and Fe_2O_3 recalculated for the experimental fO_2 conditions (Sack *et al.* 1980). The 2σ error ellipses (EPMA error) are calculated for standard deviation of the mean on replicate electron microprobe analysis of the 1186A-37R-1 (12–19) starting composition (Table 8). Note that experimental results of the OJP basalts are nearly identical with those of MORBs. (**b**) Compositions of the OJP basalts that have phenocrysts of olivine, plagioclase and augite, projected into the pseudoternary ol–cpx–qtz system. The 0.1 and 190 MPa cotectics for olivine–plagioclase–augite in the OJP melts are also shown. Data are from Mahoney *et al.* (2001) and this study.

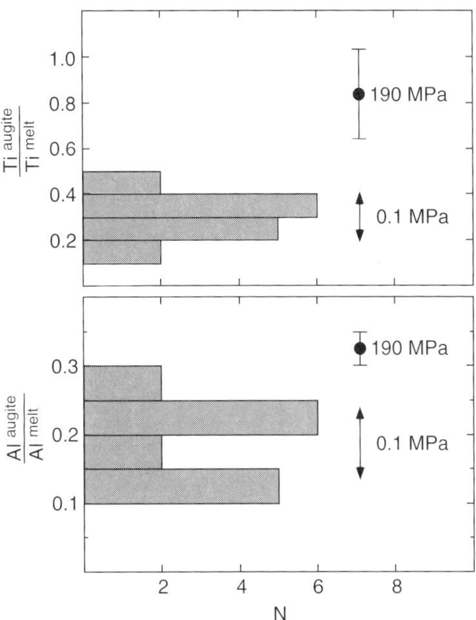

Fig. 12. Histogram of Ti and Al partition coefficients between augite and whole rock for the three basalt samples analysed in this study. The Ti and Al partition coefficients determined by experiments at 190 MPa (lines with solid circles) and 0.1 MPa (arrows) are also shown.

pressure; the K_D values in the 190 MPa experiments are higher than those in the 0.1 MPa experiments (Fig. 7). The mg-numbers of cores of augite phenocrysts are in equilibrium with those of whole rock at pressures between 0.1 and 190 MPa (Fig. 7).

As the MgO content of the majority of the Kwaimbaita type basalt is 7.0–8.0 wt% (Fig. 2), the temperature range during crystallization is estimated to be 1160–1180°C based on the MgO geothermometer (see the Appendix). For the Kroenke-type basalt, the temperature range during crystallization is estimated to be 1210–1230°C, using the MgO content of fresh basalts (MgO 9.5–10.0 wt%). The estimated temperatures are in the same ranges as the temperatures calculated using the geothermometer of Ford *et al.* (1983): 1220–1240°C for the fresh Kroenke-type basalt; 1170–1190°C for the Kwaimbaita-type basalt. The crystallization temperature of type I plagioclase and low-An parts of type II plagioclase in the Kwaimbaita-type basalt were also estimated to be 1160–1190°C based on the anorthite–albite (An–Ab) geothermometer (see the Appendix).

The above geothermometers predict that the temperature range during crystallization of phenocryst phases in the magma chambers was 1160–1240°C. As the main body of the chambers would be filled with Kwaimbaita-type magma, the temperature of the main body was probably 1160–1190°C.

Origin of high-An parts of plagioclase

Extremely calcic plagioclase (>An_{80}) is common in island arc lavas (Arculus & Wills 1980; Crawford *et al.* 1987; Kuritani 1998) and is also found in MORB lavas (e.g. Meyer & Shibata 1990). To clarify the magmatic differentiation for the OJP, it is crucial to explain the origin of the high-An parts of type II plagioclase phenocrysts and we therefore examine the problem based on the experimentally determined compositional relationship between melt and plagioclase.

Figure 13a illustrates the schematic An–Ab binary loops projected from a natural multi-component system. The loop A (under dry conditions) corresponds to the main body of a magma chamber (Fig. 14). The low-An parts of the type II plagioclase (P1 in Fig. 13a) are in equilibrium with the Kwaimbaita-type magma (M1 in Fig. 13a). The high-An parts of the type II plagioclase (P2 or P3 in Fig. 13a) cannot be crystallized from the Kwaimbaita-type magma.

One possible origin of the high-An parts is plagioclase crystallization from a high-temperature magma (Fig. 13a, Case 1). The high-An parts (P2 in Fig. 13a) are in equilibrium with the high-temperature ($T_{H\text{-}An1}$ in Fig. 13a) magma (M2 in Fig. 13a). When a plagioclase with low-An content (P1 in Fig. 13a) in the main body (M1 in Fig. 13a) is brought into the high-temperature part (M2 at $T_{H\text{-}An1}$ in Fig. 13a) of the chamber, the high-An parts (P2 in Fig. 13a) would replace the low-An parts. Our melting experiments also show that high-An (>An_{80}) plagioclase crystals are present in melts whose temperature is higher than 1190°C. However, the Case 1 model cannot explain MgO contents in the high-An parts (Fig. 13b). As shown in Figure 13b, a negative correlation between the MgO and An contents is found in the OJP plagioclase phenocrysts. On the other hand, the experiments at dry conditions show that the MgO content is nearly constant with increasing An content (Fig. 13b). The low MgO content in the high-An parts does not occur even if we assume high-pressure (>200 MPa) crystallization, because MgO content of plagioclase at high-pressure (190 MPa) is higher than that of plagioclase at atmospheric pressure (Fig. 13b).

The high-An with low MgO can be explained by plagioclase crystallization under H_2O-rich conditions (Kuritani 1998). Figure 13c shows experimental results of a high-alumina basalt (35–79 g) under dry (Grove *et al.* 1982; Bartels *et al.* 1991) and H_2O-saturated (Sisson & Grove

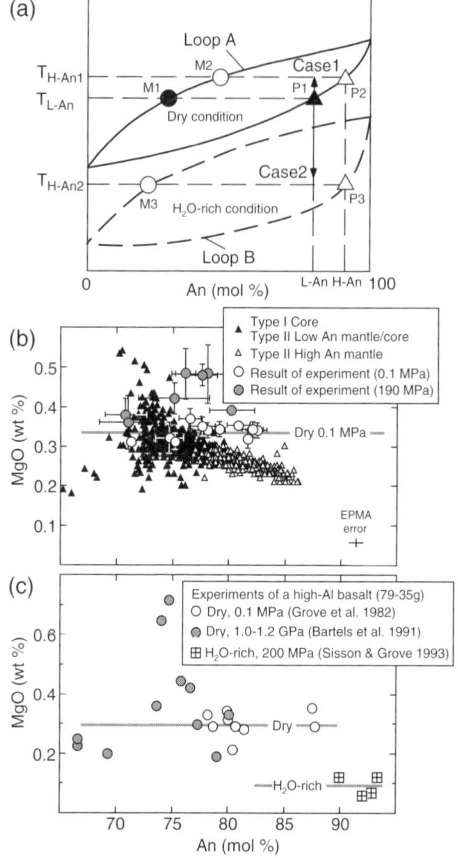

Fig. 13. (**a**) Schematic albite–anorthite diagram projected from a multi-component system. Loop A (solid lines) is for plagioclase and surrounding melt in the main body of the magma chamber. Loop B (broken lines) is for plagioclase and surrounding H_2O-rich melt in a mushy boundary layer. M1, M2 and M3 are melt compositions, and P1, P2 and P3 are corresponding plagioclase compositions at each temperature, $T_{L\text{-}An}$, $T_{H\text{-}An1}$ and $T_{H\text{-}An2}$. Previous melting experiments (Yoder *et al.* 1957; Housh & Luhr 1991; Sisson & Grove 1993; Panjasawatwong *et al.* 1995) have demonstrated that the loop under H_2O-rich conditions (Loop B) is shifted downward in temperature compared to the loop under dry conditions (Loop A). (**b**) An content v. MgO for each type of plagioclase in the nine basalt samples analysed in this study. To compare the natural data with the experimental results, plagioclase compositions obtained in the experiments on the Ontong Java Plateau basalts are also shown. Error bars on the experimental results show ± 1SD of the mean on six or more analyses of plagioclase crystals in each charge. EPMA error: standard deviation uncertainty (± 1σ) due to counting statistics. (**c**) An content v. MgO for plagioclase crystals that are determined by melting experiments on an island arc basalt (a high-Al basalt; 35–79 g). Data are from Grove *et al.* (1982), Bartels *et al.* (1991) and Sisson & Grove (1993).

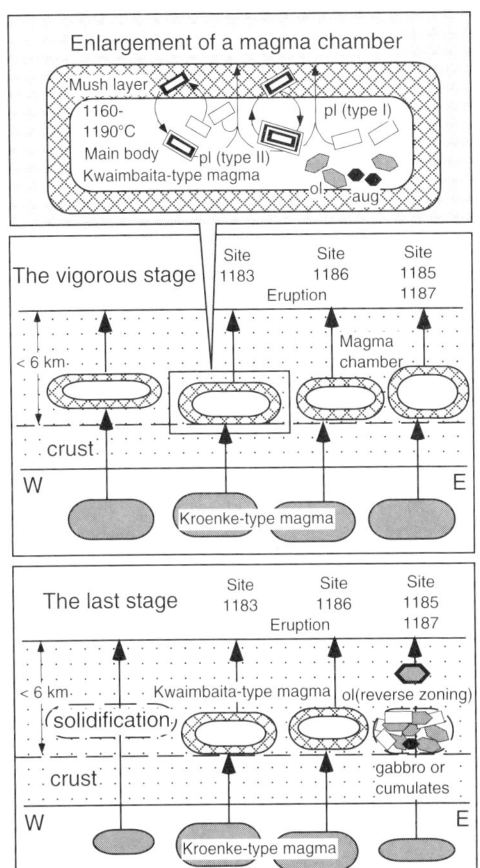

Fig. 14. Schematic illustration of the differentiation processes in a magma chamber (see text for explanation).

1993) conditions. The results suggest that plagioclase formed under H_2O-rich conditions has higher An and lower MgO contents than the plagioclase formed under dry conditions.

The H_2O-rich condition is achieved most plausibly in a mushy layer (Fig. 14) that is formed between the wall and a well-convecting main magma body (e.g. Brandeis & Jaupart 1986; Turner *et al.* 1986). The mushy layer composed of crystals and fractionated interstitial melt is likely to be enriched in H_2O because of the addition of H_2O expelled from the chilled margin and because of significant fractionation of crystal phases (McBirney *et al.* 1985).

The An–Ab binary system of the mushy layer corresponds to loop B in Figure 13a (Case 2). In the mushy layer, the interstitial melt has a more differentiated composition (M3 in Fig. 13a) than the melt of the main body (M1 in Fig. 13a). The plagioclase of high-An content (P3 in Fig. 13a) is in equilibrium with the interstitial melt (M3 in Fig. 13a) at low temperature ($T_{H\text{-}An2}$ in Fig. 13a). When plagioclase with a low-An composition (P1 in Fig. 13a) in the main magma body (M1 at $T_{L\text{-}An}$ in Fig. 13a) is transported to the mushy layer (M3 in Fig. 13a), high-An parts (P3 in Fig. 13a) are formed around the low-An plagioclase.

Although crystallization in the mushy layer would increase H_2O content, the mushy layer was probably under H_2O-undersaturated conditions ($\ll$6 wt%) because H_2O content of the main magma body was small (0.1–0.2 wt%; Roberge *et al.* 2004). Sisson & Grove (1993) have reported that the plagioclase K_D values $[(Ca/Na)^{crystal}/(Ca/Na)^{melt}]$ show a progressive increase with melt H_2O from approximately 1.7 for melts with 2 wt% H_2O, to 3–4 for melts with 4 wt% H_2O, and 5.5 for melts with 6 wt% H_2O. As plagioclase K_D values for the high-An parts of type II plagioclase phenocrysts are >1.3, H_2O content of the mushy layer would be ≤2 wt%. However, the estimation should be checked by melting experiments for the OJP compositions, because melting experiments of Sisson & Grove (1993) use high-Al basalts as starting materials, and compositional (e.g. SiO_2 and Al_2O_3) effects on the plagioclase K_D values has been proposed (Panjasawatwong *et al.* 1995).

Origin of reverse-zoned olivine

Reverse-zoned olivine is also found in MORB (Nabelek & Langmuir 1986; Humler & Whitechurch 1988; Pan & Batiza 2002) and is most probably formed by incorporation of olivine xenocrysts during ascent of the host magma (Nabelek & Langmuir 1986). Assuming that reaction of the xenocrysts with the host magma leads to the reverse-zoning profiles, residence times between incorporation of the xenocrysts and quenching on the seafloor can be estimated by a geospeedometry technique (Crank 1975). The geospeedometry studies of MORB olivines have yielded residence time estimates of a few days to a few years (Nabelek & Langmuir 1986; Humler & Whitechurch 1988; Pan & Batiza 2002).

The reverse-zoned olivine (Figs 3b, c and 4b, c) in the Kroenke-type basalt could also be xenocrysts assimilated from solidified Kwaimbaita-type magma for the following reasons. The first is that the Kroenke-type basalt at Site 1185 erupted after the eruption of Kwaimbaita-type basalt (Fig. 1b). The Kroenke-type magma must have ascended through the solidified or partially solidified Kwaimbaita-type magma (gabbro or olivine cumulates shown in Fig. 14), and thus the Kroenke-type magma would have incorporated

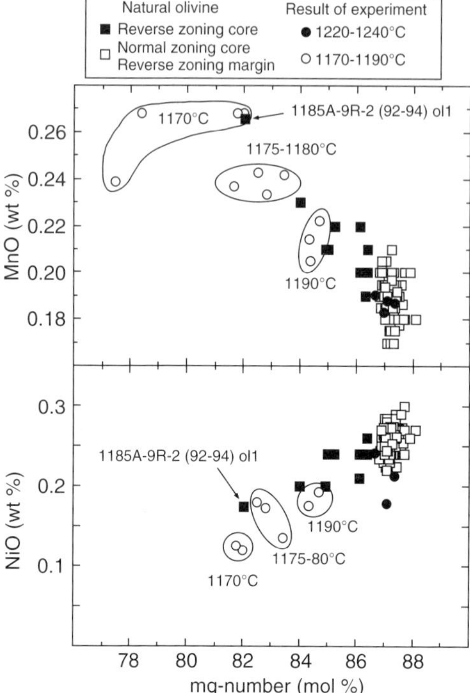

Fig. 15. Plots of wt% MnO and wt% NiO v. mg-number for each type of olivine. To compare the natural data with the experimental results, olivine compositions obtained in the experiments on the Ontong Java Plateau basalts are also shown.

olivine that had crystallized from the Kwaimbaita-type magma. The second reason is that a uniform core of the largest of the reverse-zoned olivines (1185A-9R-2, 92–94, ol1) is in equilibrium with Kwaimbaita-type basalt. The uniform core composition (mg-number 82, MnO 0.27, CaO 0.27, NiO 0.18) of the largest olivine is similar to the composition of olivine crystallized during the melting experiments on Kwaimbaita-type starting material (Fig. 15). As the geospeedometer that was used in the MORB studies (e.g. Nabelek & Langmuir 1986; Humler & Whitechurch 1988; Pan & Batiza 2002) can be applied to the OJP olivine, we have estimated the residence time by using the mg-number, and MnO and NiO zoning patterns.

The mg-number, and MnO and NiO profiles of the largest olivine are shown in Figure 16 as data points, with solid curves to show calculated compositional profiles resulting from diffusion. To calculate the diffusion profiles, we used a three-dimensional spherical diffusion equation (Crank 1975). The concentrations at different radii and periods of time can be calculated from:

$$\frac{C - C_c}{C_0 - C_C} = 1 + \frac{2a}{\pi \cdot r} \sum_{n=1}^{\infty} \frac{(-1)^n}{n} \sin\left(\frac{n\pi \cdot r}{a}\right) \exp\left(\frac{-Dn^2\pi^2 t}{a^2}\right)$$

where C is the concentration at some distance r from the centre of the crystal, C_0 is the surface concentration, C_C is the initial concentration of the crystal, a is the radius of the reversed diffusion profile, D is the diffusion coefficient (the chemical diffusion coefficient of Mg–Fe, Mg–Mn and Mg–Ni exchanges in olivine), t is the diffusion time and n is the summation integer. In the diffusion calculation we assume that there has been no overgrowth or dissolution of the olivine crystal, and that the observed compositional zoning (Fig. 16) is due entirely to diffusion. Further, we assume that the measured rim composition is a fixed boundary condition for the diffusion calculation. The calculated diffusion time could be somewhat in error because of the uncertainty of the diffusion coefficient; the diffusion coefficient varies with conditions such as temperature, composition, crystallographic direction and oxygen fugacity (Buening & Buseck 1973; Misener 1974; Morioka 1981; Chakraborty 1997). While we fix the conditions, proposed values of the diffusion coefficients still have wide ranges. For example, at 1200°C, Fo_{87}, along the c crystallographic direction and with an oxygen fugacity of –12 in log units, published values for D range from 1.9×10^{-12} cm^2 s^{-1} (Misener 1974) to 5.2×10^{-13} cm^2 s^{-1} (Chakraborty 1997). When we use the values in the diffusion calculation, the mg-number profile of the largest olivine is reproduced by the calculated diffusion profiles at times between a few months and a few years (Fig. 16a, b).

The diffusion calculations from the MnO and NiO profiles yield the same residence times as the estimation from the mg-number profile (Fig. 16c, d). In the MnO and NiO diffusion calculations, we used the value of 5.6×10^{-12} cm^2 s^{-1} for the D of Mn and 4.4×10^{-13} cm^2 s^{-1} for the D of Ni at 1200°C (Morioka 1981). These D values might have errors, however, because the experiments of Morioka (1981) were conducted on Fe-free olivine.

The core compositions (mg-number, MnO and NiO) of all reverse-zoned olivines are plotted against radii as data points in Figure 17. A reverse correlation between the mg-number and radius, a slight normal correlation between

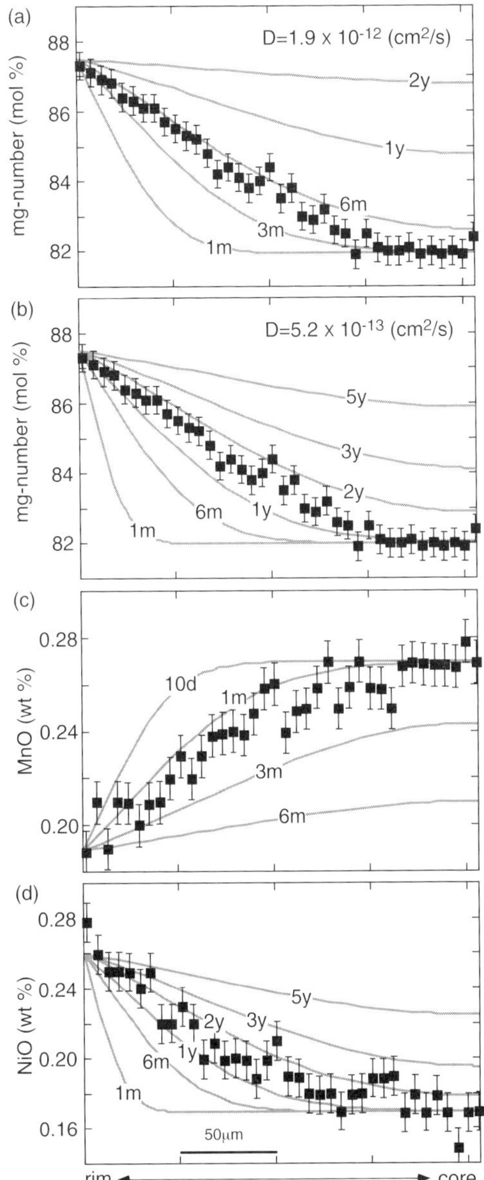

Fig. 16. Line profiles from rim (left) to core (right) for reverse-zoned olivine (1185A-9R-2, 92–94, ol1) compositions. The diffusion-controlled zoning of olivine calculated for periods of time between 1 month and 5 years are also shown. Error bars show standard deviation uncertainty ($\pm 1\sigma$) due to counting statistics. (**a**) mg-number profile. The diffusion controlled zoning is calculated using the diffusion coefficient (D) value of 1.9×10^{-12} cm^2 s^{-1} (Misenar 1974). (**b**) mg-number profiles. The diffusion controlled zoning is calculated using the diffusion coefficient value of 5.2×10^{-13} cm^2 s^{-1} (Chakraborty 1997). (**c**) MnO profile. (**d**) NiO profile.

the MnO and radius, and a reverse correlation between the NiO and radius are observed. The correlation suggests that the chemical diffusion would change the original core composition to the current core composition, except for the largest olivine. The relationships between core compositions and radius are also used to estimate the residence time. The calculated compositional profiles resulting from diffusion are shown as solid curves in Figure 17. The calculation also estimates the residence time of a few months to a few years.

In summary, the estimated residence times are not significantly different from those estimated for MORB (Nabelek & Langmuir 1986; Humler & Whitechurch 1988; Pan & Batiza 2002), implying that the Kroenke-type magma did not stay for long in the crust during its ascent.

Magma differentiation model of the Ontong Java Plateau basalt

On the basis of our experimental study, in combination with petrographical investigations, we propose the following differentiation history of magmas beneath the OJP, as illustrated in Figure 14. To simplify our explanation, we divided the 122 Ma-event into two stages: the vigorous stage; and the last stage. A boundary between the vigorous stage and the last stage corresponds to a boundary between the Kwaimbaita-type and Kroenke-type basalt units at Site 1185. We assume that individual batches of parental magma evolved in several magma chambers (Fig. 14) because the isotopic ratios vary somewhat from place to place (Tejada *et al.* 2004).

During the vigorous stage, the Kwaimbaita-type magma with homogeneous composition was present in each magma chamber beneath the whole OJP, including the main plateau and eastern lobe. The Kwaimbaita-type magma was formed by fractional crystallization from Kroenke-type magma that was injected into each magma chamber (Fig. 14). The depth of the magma chambers was probably less than 6 km. In the main bodies of the magma chambers, olivine crystals with an mg-number of 80–84 and plagioclase crystals with a composition of An_{73-78} were crystallized over the temperature range of 1160–1190°C. When the temperature of the main bodies was lower than 1180°C, augite crystals with an mg-number of 81–83 were also formed. Some plagioclase crystals in the main bodies were transported into the mushy boundary layers along the chamber walls, and high-An (An_{80-84}) mantles were developed around the

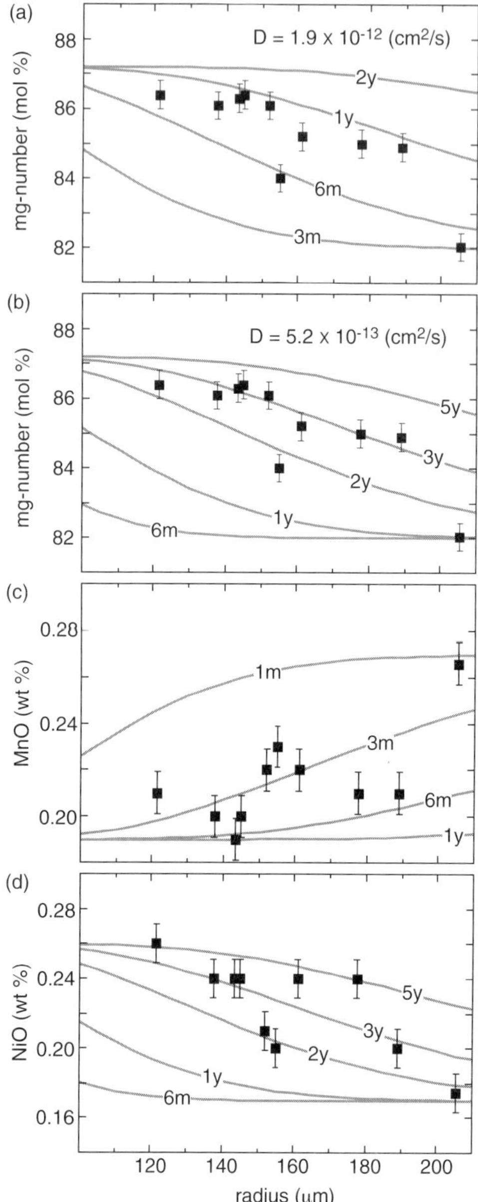

Fig. 17. Plots of (**a**, **b**) mg-numbers, (**c**) wt% MnO and (**d**) wt% NiO v. crystal radius for cores of the reverse-zoned olivine crystals. Diffusion-controlled zoning profiles of olivine core calculated for periods of time between 1 month and 5 years are also shown. The diffusion-controlled zoning is calculated using the diffusion coefficient (D) values of (**a**) 1.9×10^{-12} cm^2 s^{-1} for and (**b**) 5.2×10^{-13} cm^2 s^{-1}. Error bars show standard deviation uncertainty ($\pm 1\sigma$) due to counting statistics.

plagioclase crystals because of a relatively H_2O-rich environment. Cycling in and out of the mushy boundary layer could have caused the oscillatory zoning of the type II plagioclase phenocrysts (Fig. 14).

At the last stage, the Kroenke-type magma erupted on the eastern edge of the main plateau (Fig. 1). The Kroenke-type magma ascending beneath the eastern edge would not be trapped by the magma chambers, however, because the magma chamber at the edge would have contracted or disappeared by this stage (Fig. 14). Instead of the magma chambers, the solidified pile of Kwaimbaita-type lavas should have existed beneath the eastern edge. The reverse-zoned olivine in the Kroenke-type basalt was presumably formed by incorporation of the olivine crystal from solidified Kwaimbaita-type magma into the Kroenke-type magma. Beneath the crest of the main plateau, on the other hand, the Kroenke-type magmas were trapped within the magma chambers, and the Kwaimbaita-type magma erupted at the surface (Fig. 14).

Conclusions

Basement lavas recovered from the Ontong Java Plateau at Ocean Drilling Program Leg 192 were separated into two types: relatively primitive basalt lavas (MgO 8–10 wt%; Kroenke type) and differentiated basalt lavas (MgO 7–8 wt%; Kwaimbaita type). While the Kwaimbaita-type basalt forms much of the upper part of the OJP (Tejada *et al.* 2002), the Kroenke-type basalt has only been found on the eastern edge of the plateau (Mahoney *et al.* 2001).

Low-pressure (0.1–190 MPa) phase equilibrium experiments on both a Mg-rich (MgO 10.2 wt%) Kroenke-type basalt and a typical Kwaimbaita-type basalt (MgO 7.8 wt%) have suggested that the Kwaimbaita-type magmas were probably crystallized in shallow magma chambers (< 6 km in depth) under relatively reducing (FMQ–CCO buffer) conditions. The experimental results also show that the temperature range in the main bodies of the chambers was 1160–1190°C.

Comparison of the experimentally determined mineral–melt equilibria with detailed petrographical investigation made it clear that almost all phenocrysts were in equilibrium with their host magma. However, the comparison found two cases where compositions are not in equilibrium with their host magma compositions: high An mantles of plagioclase in the Kwaimbaita-type basalt and cores of reverse-zoned olivine in the Kroenke-type basalts. On the basis of comparison with experimental phase equilibria, the high-An parts appear to have been formed in the mushy boundary layer, which had a relatively high H_2O content. The reverse-zoned olivine was interpreted to have been xenocrysts assimilated from the solidified Kwaimbaita-type magma. Assuming that zoning in the reverse-zoned olivine was formed due to diffusion after assimilation, residence times of the reverse-zoned olivine in the Kroenke-type magma were calculated to be a few months to a few years.

The authors are grateful to the officers and crew of the R.V. *JOIDES Resolution* for the help in obtaining drill cores during ODP Leg 192. The authors would also like to thank S. Nakada for his help with the XRF analysis, and T. Fujii, A. Yasuda, and M. Kitamura for discussions. C. Herzberg and one anonymous reviewer are thanked for their constructive comments. This paper presents a result of a joint research programme carried out at the Institute for Study of the Earth's Interior, Okayama University. This research used samples and data provided by the Ocean Drilling Program (ODP). The ODP is sponsored by the US National Science Foundation (NSF) and participating countries under management of Joint Oceanographic Institutions (JOI), Inc. Funding for this research was provided by a Grant-in Aid from the Ministry of Education, Science and Culture of Japan (14740307) to T. Sano.

Appendix

Geothermometer for the OJP lavas

Previous melting experiments show that variation of the MgO content of glasses produced experimentally from basaltic starting compositions varied linearly with temperature (Helz & Thornber 1987; Montierth *et al.* 1995; Sugawara 2000*a*). The MgO geothermometer is well calibrated when the glasses are in equilibrium with olivine crystals. We therefore investigated the experimentally calibrated MgO geothermometer for basaltic lavas from the OJP. The MgO contents of the experimental glasses are plotted against temperature in Figure A1a, with the least-squares linear regression shown as a thick line. The weight per cent of MgO in the glass varies nearly linearly with temperature over the entire range investigated (1150–1250°C; Fig. A1a) and thus provides a useful empirical geothermometer for the OJP lavas. This linear relationship is best fit by the equation (residual sum of squares = 0.967):

$$T\,(^\circ\mathrm{C}) = 21.61 \times \mathrm{MgO} + 1009$$

where MgO is the weight per cent of MgO in the glass. The difference between the calculated temperature and true temperature for the experiments is 0.3–9.6°C, with a standard deviation of 2.5°C. The uncertainties in temperature estimates generated by the electron probe analysis error (1 SD uncertainties due to counting statistics) are 1.1–4.8°C. We therefore assume that the uncertainty in the calculated temperature is within ±10°C.

A simple binary anorthite–albite (An–Ab) system shows that the composition of a melt or a plagioclase in equilibrium will be sensitive to temperature at a given pressure (e.g. Yoder *et al.* 1957). The An–Ab geothermometer has been applied to natural systems (Drake 1976; Smith 1983; Housh & Luhr 1991; Panjasawatwong *et al.* 1995). We therefore used our experimental results to examine the An–Ab geothermometer for basaltic lavas from the OJP. There is a good correlation between the experimental temperature and the An content of the experimental

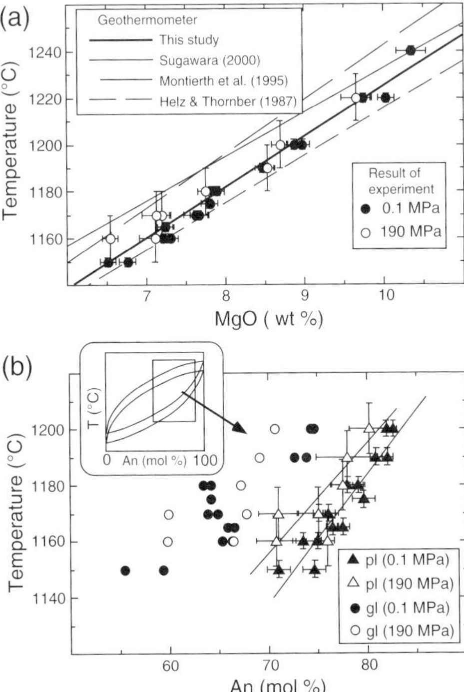

Fig. A1. (**a**) Weight per cent MgO in experimental glasses plotted as a function of temperature. The lines representing the calibration of the previous studies (Helz & Thornber 1987; Montierth *et al.* 1995; Sugawara 2000*a*) are shown for comparison. (**b**) Experimental glasses and coexisting plagioclase compositions projected onto the anorthite–albite join. Multi-component glass compositions are plotted as 100 × [normative an/(an + ab)].

plagioclase, whereas the An content of the glasses is scattered. This fact indicates that the simple binary An–Ab system can be applied to plagioclase composition from the OJP lavas. As temperature effect is small (*c*. 10°C) within the pressure range (0.1–190 MPa), we assign a conservative error of ±20°C to the geothermometer.

References

ARCULUS, R.J. & WILLS, K.J.A. 1980. The petrology of plutonic blocks and inclusions from the Lesser Antilles island arc. *Journal of Petrology*, **21**, 743–799.

BARTELS, K.S., KINZLER, R.J. & GROVE, T.L. 1991. High pressure phase relations of primitive high-alumina basalts from Medicine Lake volcano, northern California. *Contributions to Mineralogy and Petrology*, **108**, 253–270.

BENDER, J.F., HODGES, F.N. & BENCE, A.E. 1978. Petrogenesis of basalts from the project FAMOUS area: experimental study from 0 to 15 kbars. *Earth and Planetary Science Letters*, **41**, 277–302.

BLUNDY, J.D., FALLOON, T.J., WOOD, B.J. & DALTON, J.A. 1995. Sodium partitioning between clinopyroxene and silicate melts. *Journal of Geophysical Research*, **100**, 15 501–15 515.

BRANDEIS, G. & JAUPART, C. 1986. On the interaction between convection and crystallization in cooling magma chambers. *Earth and Planetary Science Letters*, **82**, 345–361.

BUENING, D.K. & BUSECK, P.R. 1973. Mg–Fe lattice diffusion in olivine. *Journal of Geophysical Research*, **78**, 6852–6862.

CHAKRABORTY, S. 1997. Rates and mechanisms of Mg–Fe interdiffusion in olivine at 980°C–1300°C. *Journal of Geophysical Research*, **102**, 12 317–12 331.

CHAMBERS, L.M., PRINGLE, M.S. & FITTON, J.G. 2002. Age and duration of magmatism on the Ontong Java Plateau: ^{40}Ar–^{39}Ar results from ODP Leg 192. Abstract V71B-1271. *Eos, Transactions of the American Geophysical Union*, **83**, F47.

CHEN, H.-K. & LINDSLEY, D.H. 1983. Apollo 14 very low titanium glasses: melting experiments in iron-platinum alloy capsules. *Journal of Geophysical Research*, **88**, B335–B342.

COFFIN, M.E. & ELDHOLM, O. 1993. Scratching the surface: estimating dimension of large igneous provinces. *Geology*, **21**, 515–518.

CRANK, J. 1975. *The Mathematics of Diffusion,* 2nd Edition. Oxford University Press, Oxford.

CRAWFORD, A.J., FALLOON, T.J. & EGGINS, S. 1987. The origin of island arc high-alumina basalts. *Contributions to Mineralogy and Petrology*, **97**, 417–430.

DANYUSHEVSKY, L.V. 2001. The effect of small amounts of H_2O on crystallisation of mid-ocean ridge and backarc basin magmas. *Journal of Volcanology and Geothermal Research*, **110**, 265–280.

DRAKE, M.J. 1976. Plagioclase–melt equilibria. *Geochimica et Cosmochimica Acta*, **40**, 457–465.

FALLOON, T.J., GREEN, D.H., DANYUSHEVSKY, L.V. & FAUL, U.H. 1999. Peridotite melting at 1.0 and 1.5 GPa: an experimental evaluation of techniques using diamond aggregates and mineral mixes for determination of near-solidus melts. *Journal of Petrology*, **40**, 1343–1375.

FITTON, J.G. & GODARD, M. 2004. Origin and evolution of magmas on the Ontong Java Plateau. *In*: FITTON, J.G., MAHONEY, J.J., WALLACE, P.J. & SAUNDERS, A.D. (eds) *Origin and Evolution of the Ontong Java Plateau*. Geological Society, London, Special Publications, **229**, 151–178.

FURUMOTO, A.S., WEBB, J.P., ODEGARD, M.E. & HUSSONG, D.M. 1976. Seismic studies on the Ontong Java Plateau, 1970. *Tectonophysics*, **34**, 71–79.

FORD, C.E., RUSSELL, D.G., CRAVEN, J.A. & FISK, M.R. 1983. Olivine–liquid equilibria: Temperature, pressure and composition dependence of the crystal/liquid cation partition coefficients for Mg, Fe^{2+}, Ca and Mn. *Journal of Petrology*, **24**, 256–265.

GLADCZENKO, T.P., COFFIN, M.E. & ELDHOLM, O. 1997. Crustal structure of the Ontong Java Plateau: modeling of new gravity and existing seismic data. *Journal of Geophysical Research*, **102**, 22 711–22 729.

GROVE, T.L. 1981. Use of FePt alloys to eliminate the iron loss problem in 1 atmosphere gas mixing experiments: theoretical and practical considerations. *Contributions to Mineralogy and Petrology*, **78**, 298–304.

GROVE, T.L., GERLACH, D.C. & SANDO, T.W. 1982. Origin of calc-alkaline series lavas at Medicine Lake volcano by fractionation, assimilation and mixing. *Contributions to Mineralogy and Petrology*, **80**, 160–182.

GROVE, T.L., KINZLER, R.J. & BRYAN, W.B. 1992. Fractionation of mid-ocean ridge basalt (MORB). *In*: MORGAN, J.P., BLACKMAN, D.K. & SINTON, J.M. (eds) *Mantle Flow and Melt Generation at Mid-ocean Ridges*. American Geophysical Union, Geophysical Monograph, **71**, 281–310.

HELZ, R.T. & THORNBER, C.R. 1987. Geothermometry of Kilauea Iki lava lake, Hawaii. *Bulletin of Volcanology*, **49**, 651–668.

HOUSH, T.B. & LUHR, J.F. 1991. Plagioclase–melt equilibria in hydrous systems. *American Mineralogist*, **76**, 477–492.

HUMLER, E. & WHITECHURCH, H. 1988. Petrology of basalts from the Central Indian Ridge (lat. 25°23′S, long. 70°04′E): estimations of frequencies and fractional volumes of magma injections in a two-layered reservoir. *Earth and Planetary Science Letters*, **88**, 169–181.

HUGHES, G.W. & TURNER, C.C. 1977. Upraised Pacific Ocean floor, southern Malaita, Solomon Islands. *Geological Society of American Bulletin*, **88**, 412–424.

KINZLER, R.J. & GROVE, T.L. 1992. Primary magmas of mid-ocean ridge basalts 1. experiments and methods. *Journal of Geophysical Research*, **97**, 6885–6906.

KROENKE, L.W. 1972. *Geology of the Ontong Java Plateau*. PhD Thesis, University of Hawaii.

KURITANI, T. 1998. Boundary layer crystallization in a basaltic magma chamber: evidence from Rishiri volcano, Northern Japan. *Journal of Petrology*, **39**, 1619–1640.

MAHONEY, J.J. 1987. An isotopic survey of Pacific ocean plateaus: implications for their nature and origin. *In*: KEATING, B.H., FRYER, P., BATIZA, R. & BOEHLERT, G.W. (eds) *Seamounts, Islands, and Atolls*. American Geophysical Union, Geophysical Monograph, **43**, 207–220.

MAHONEY, J.J. & SPENCER, K.J. 1991. Isotopic evidence for the origin of the Manihiki and Ontong-Java oceanic plateaus. *Earth and Planetary Science Letters*, **104**, 196–210.

MAHONEY, J.J., STOREY, M., DUNCAN, R.A., SPENCER, K.J. & PRINGLE, M. 1993. Geochemistry and geochronology of Leg 130 basement lavas: nature and origin of the Ontong Java Plateau. *Proceedings of the Ocean Drilling Program, Scientific Results*, **130**, 3–22.

MAHONEY, J.J., FITTON, J.G., WALLACE, P.J. *et al.* 2001. *Proceedings of the Ocean Drilling Program, Initial Reports*, **192** [Online]. Available from World Wide Web: http: www-odp.tamu.edu/publications/192ir.htm

MCBIRNEY, A.R., BAKER, B.H. & NILSON, R.H. 1985. Liquid fractionation. Part 1: basic principles and experimental simulations. *Journal of Volcanology and Geothermal Research*, **24**, 1–24.

MEYER, P.S. & SHIBATA, T. 1990. Complex zoning in plagioclase feldspars from ODP Site 648. *Proceedings of the Ocean Drilling Program, Scientific Results*, **106/109**, 123–142.

MICHAEL, P.J. 1999. Implications for magmatic processes at Ontong Java Plateau from volatile and major elements contents of Cretaceous basalt glasses. *Geochemistry, Geophysics, Geosystems*, **1**, 1999GC000025.

MISENER, D.J. 1974. Cationic diffusion in olivine to 1400°C and 25 kbar. *In*: HOFMANN, A.W. (ed.) *Geochemical Transport and Kinetics*. Carnegie Institute of Washington Publication, **634**, 117–129.

MONTIERTH, C., JOHNSTON, A.D. & CASHMAN, K.V. 1995. An empirical glass-composition-based geothermometer for Mauna Loa lavas. *In*: RHODES, J.M. & LOCKWOOD, J.P. (eds) *Mauna Loa Revealed: Structure, Composition, History, and Hazards*. American Geophysical Union, Geophysical Monograph, **92**, 207–217.

MORIOKA, M. 1981. Cation diffusion in olivine – II. Ni–Mg, Mn–Mg, and Mg and Ca. *Geochimica et Cosmochimica Acta*, **45**, 1573–1580.

MYERS, J. & GUNTER, W.D. 1979. Measurement of the oxygen fugacity of the cobalt–cobalt oxide buffer assemblage. *American Mineralogist*, **64**, 224–228.

NABELEK, P.I. & LANGMUIR, C.H. 1986. The significance of unusual zoning on olivines from FAMOUS area basalt 527-1-1. *Contributions to Mineralogy and Petrology*, **93**, 1–8.

NEAL, C.R., MAHONEY, J.J., KROENKE, L.W. & DUNCAN, R.A. 1997. The Ontong Java Plateau. *In*: MAHONEY, J.J. & COFFIN, M.F. (eds) *Large Igneous Provinces: Continental, Oceanic, and Planetary Flood Volcanism*. American Geophysical Union, Geophysical Monograph, **100**, 183–216.

PAN, Y. & BATIZA, R. 2002. Mid-ocean ridge magma chamber processes: constrains from olivine zonation in lavas from the East Pacific Rise at 9°30′N and 10°30′N. *Journal of Geophysical Research*, **107**, B1, ECV9.1–9.13.

PANJASAWATWONG, Y., DANYUSHEVSKY, L.V., CRAWFORD, A.J. & HARRIS, K.L. 1995. An experimental study of the effects of melt composition on plagioclase-melt equilibria at 5 and 10 kbar: implications for the origin of magmatic high-An plagioclase. *Contributions to Mineralogy and Petrology*, **118**, 420–432.

PETTERSON, M.G., NEAL, C.R., MAHONEY, J.J., KROENKE, L.W., SAUNDERS, A.D., BABBS, T., DUNCAN, R.A., TOLIA, D. & MCGRAIL, B. 1997. Stracture and deformation of north and central Malaita, Solomon Islands: tectonic implications for the Ontong Java Plateau–Solomon arc collision, and for the fate of oceanic plateaus. *Tectonophysics*, **283**, 1–33.

PETTERSON, M.G., BABBS, T., NEAL, C.R., MAHONEY, J.J., SAUNDERS, A.D., DUNCAN, R.A., TOLIA, D., MAGU, R., OOPOTO, C., MAHOA, H. & NATOGGA, D. 1999. Geological–tectonic framework of Solomon Islands, SW Pacific: crustal accretion and growth within an intra-oceanic setting. *Tectonophysics*, **301**, 35–60.

PHINNEY, E.J., MANN, P., COFFIN, M.F. & SHIPLEY, T.H. 1999. Sequence stratigraphy, structure, and tectonic history of the southwestern Ontong Java Plateau adjacent to the North Solomon Trench and Solomon Islands arc. *Journal of Geophysical Research*, **104**, 20 449–20 466.

PRESNALL, D.C., DIXON, J.R., O'DONNELL, T.H. & DIXON, S.A. 1979. Genesis of mid-ocean ridge tholeiites. *Journal of Petrology*, **20**, 3–35.

PUTIRKA, K., JOHNSON, M., KINZLER, R., LONGHI, J. & WAKER, D. 1996. Thermobarometry of mafic igneous rocks based on clinopyroxene-liquid equilibria, 0–30 kbar. *Contributions to Mineralogy and Petrology*, **123**, 92–108.

RICHARDSON, W.P., OKAL, E. & VAN DER LEE, S. 2000. Rayleigh-wave tomography of the Ontong-Java Plateau. *Physics of the Earth and Planetary Interiors*, **118**, 29–51.

ROBERGE, J., WHITE, R.V. & WALLACE, J.P. 2004. Volatiles in submarine basaltic glasses from the Ontong Java Plateau (ODP Leg 192): implications for magmatic processes and source regions compositions. *In*: FITTON, J.G., MAHONEY, J.J., WALLACE, P.J. & SAUNDERS, A.D. (eds) *Origin and Evolution of the Ontong Java Plateau*. Geological Society, London, Special Publications, **229**, 239–257.

SACK, R.O., CARMICHAEL, I.S.E., RIVERS, M. & GHIORSO, M.S. 1980. Ferric-ferrous equilibria in natural silicate liquids at 1 bar. *Contributions to Mineralogy and Petrology*, **75**, 369–376.

SANO, T., FUJII, T., DESHMUKH, S.S., FUKUOKA, T. & ARAMAKI, S. 2001. Differentiation processes of Deccan Trap basalts: contribution from geochemistry and experimental petrology. *Journal of Petrology*, **42**, 2175–2195.

SANO, T. 2002. Determination of major and trace element contents in igneous rocks by X-ray fluorescence spectrometer analysis. (In Japanese with English abstract.) *Bulletin of Fuji Tokoha University*, **2**, 43–59.

SISSON, T.W. & GROVE, T.L. 1993. Experimental investigations of the role of H_2O in calc-alkaline differentiation and subduction zone magmatism. *Contributions to Mineralogy and Petrology*, **113**, 143–166.

SMITH, M.P. 1983. A feldspar-liquid geothermometer. *Geophysical Research Letters*, **10**, 193–195.

SUGAWARA, T. 2000*a*. Empirical relationships between temperature, pressure, and MgO content in olivine and pyroxene saturated liquid. *Journal of Geophysical Research*, **105**, 8457–8472.

SUGAWARA, T. 2000*b*. Thermodynamic analysis of Fe and Mg partitioning between plagioclase and silicate liquid. *Contributions to Mineralogy and Petrology*, **138**, 101–113.

TEJADA, M.L.G., MAHONEY, J.J., DUNCAN, R.A. & HAWKINS, M. 1996. Age and geochemistry of basement and alkalic rocks of Malaita and Santa Isabel, Solomon Islands, southern margin of Ontong Java Plateau. *Journal of Petrology*, **37**, 361–394.

TEJADA, M.L.G., MAHONEY, J.J., NEAL, C.R., DUNCAN, R.A. & PETTERSON, M.G. 2002. Basement geochemistry and geochronology of central Malaita, Solomon Islands, with implications for the origin and evolution of the Ontong Java Plateau. *Journal of Petrology*, **43**, 449–484.

TEJADA, M.L.G., MAHONEY, J.J., CASTILLO, P., INGLE, S., SHETH, H.C. & WEIS, D. 2004. Pin-pricking the elephant: evidence on the origin of the Ontong Java Plateau from Pb–Sr–Hf–Nd isotopic characteristics of ODP Leg 192 basalts. *In*: FITTON, J.G., MAHONEY, J.J., WALLACE, P.J. & SAUNDERS, A.D. (eds) *Origin and Evolution of the Ontong Java Plateau*. Geological Society, London, Special Publications, **229**, 133–150.

THORDARSON, T. 2004. Accretionary-lapilli-bearing pyroclastic rocks at ODP Leg 192 Site 1184: a record of subaerial phreatomagmatic eruptions on the Ontong Java Plateau. *In*: FITTON, J.G., MAHONEY, J.J., WALLACE, P.J. & SAUNDERS, A.D. (eds) *Origin and Evolution of the Ontong Java Plateau*. Geological Society, London, Special Publications, **229**, 275–306.

TORMEY, D.R., GROVE, T.L. & BRYAN, W.B. 1987. Experimental petrology of normal MORB near the Kane Fracture Zone: 22°–25°N, mid-Atlantic ridge. *Contributions to Mineralogy and Petrology*, **96**, 121–139.

TURNER, J.S., HUPPERT, H.E. & SPARKS, R.S.J. 1986. Komatiites II: experimental and theoretical investigations of post-emplacement cooling and crystallization. *Journal of Petrology*, **27**, 397–437.

VAN DER LAAN, S.R. & VAN GROOS, A.F.K. 1991. PtFe alloys in experimental petrology applied to high-pressure research on Fe-bearing systems. *American Mineralogist*, **76**, 1940–1949.

WALKER, D., SHIBATA, T. & DELONG, E. 1979. Abyssal tholeiites from the oceanographic fracture zone II, Phase equilibria and mixing. *Contributions to Mineralogy and Petrology*, **70**, 111–125.

WHITE, R.V., CASTILLO, P.R., NEAL, C.R., FITTON, J.G. & GODARD, M. 2004. Phreatomagmatic eruptions on the Ontong Java Plateau: chemical and isotopic relationship to Ontong Java Plateau basalts. *In*: FITTON, J.G., MAHONEY, J.J., WALLACE, P.J. & SAUNDERS, A.D. (eds) *Origin and Evolution of the Ontong Java Plateau*. Geological Society, London, Special Publications, **229**, 307–323.

YAMASHITA, S. 1999. Experimental study of the effect of temperature on water solubility in natural rhyolite melt to 100 MPa. *Journal of Petrology*, **40**, 1497–1507.

YANG, H.-J., KINZLER, R.J. & GROVE, T.L. 1996. Experiments and models of anhydrous, basaltic olivine–plagioclase–augite saturated melts from 0.001 to 10 kbar. *Contributions to Mineralogy and Petrology*, **124**, 1–18.

YODER, H.S., STEWART, D.B. & SMITH, J.R. 1957. Ternary feldspars. *Carnegie Institution of Washington Year Book*, **56**, 206–214.

Large igneous province magma petrogenesis from source to surface: platinum-group element evidence from Ontong Java Plateau basalts recovered during ODP Legs 130 and 192

WILLIAM J. CHAZEY, III & CLIVE R. NEAL

Department of Civil Engineering and Geological Sciences, University of Notre Dame, Notre Dame, IN 46556, USA (e-mail: chazey.1@nd.edu)

Abstract: A total of 16 Ontong Java Plateau (OJP) basalt samples from Ocean Drilling Program Legs 192 and 130 were analysed for major, trace and platinum-group elements (PGEs; Ir, Ru, Rh, Pt and Pd). Major- and trace-element compositions determined by our study confirm Leg 192 shipboard analyses that indicated a new group of more primitive or 'Kroenke-type' basalts, with higher MgO, Ni and Cr, and lower incompatible-element, abundances than the more common Kwaimbaita-type basalts. The PGE abundances quantified here extend the range of the continuum of compositions found in previously analysed OJP basalts and are similar to those present in some komatiites. The PGEs, therefore, cannot be used to differentiate definitively between OJP basalts groups. The two samples analysed from Leg 130 (one from Site 803 and one from Site 807) are akin to the Kwaimbaita-type basalts.

Low-temperature alteration has not affected Pd abundances in the Leg 192 basalts as it has in the Solomon Island and the Leg 130 samples. Elemental abundances and ratios along with petrography reveal that the OJP basalts have not experienced sulphide saturation. Positive correlations of Ir and Ru with Cr and Ni attest to the lithophile behaviour of the PGEs and lend more credence to studies suggesting compatibility of these elements in oxide and silicate phases, such as Cr-spinel and olivine. Estimates of sulphur abundance in the mantle, degree of partial melting and pressure of melt initiation were used in conjunction with the model of Mavrogenes & O'Neill to calculate a minimum initial excess temperature of +185–+235°C (1465–1515°C at 3.5–4.0 GPa) above ambient mantle for the OJP source. This is in broad agreement with a fossil geotherm preserved in megacrysts and peridotite xenoliths found in pipe-like intrusives of alnöite that outcrop on the island of Malaita, Solomon Islands. Using the PGEs as a guide, the OJP basalts were modelled using a three-source component melt mix: a 10% garnet peridotite melt of primitive mantle composition, which then passed through the garnet–spinel transition and melted a further 20%, a 30% partial melt of fertile upper mantle and 0–1% of outer core material. The core component was included only in the plume source, and the ratio of plume source to upper mantle source was 19: 1. It is evident from this study that the PGE contents of at least some of the OJP basalts are too high to be generated by primitive mantle sources alone. A PGE-enriched component is required and we suggest that this is outer core material. While a sulphide-rich mantle component could also increase the PGE abundances (assuming that the sulphide is exhausted during partial melting), the sulphur-undersaturated nature of these basalts argues against this. Variations in the amount of outer core in the source (from 0 to 1 wt%) and degree of fractional crystallization can account for the entire range in PGE abundances of OJP basalts.

Large igneous provinces (LIPs) are massive crustal emplacements of predominantly mafic igneous rock (extrusive and intrusive), which are generated at rates far quicker than can be accounted for by plate margin volcanism (e.g. Coffin & Eldholm 1993, 1994). They include such manifestations as continental flood basalts, volcanic passive margins, oceanic plateaus, submarine ridges and seamount groups. The origin of LIPs has generally been attributable to surfacing mantle plumes, which originate via heat transfer across a mechanical boundary layer, causing thermal instability and rising of superheated mantle material forming a plume head (e.g. Campbell & Griffiths 1990), the melting of which produces a flood basalt province. This is followed by a less voluminous and longer-lived 'plume tail' stage (e.g. Richards *et al.* 1989). Alternative models have been proposed, such as the rift-induced melting of fixed, anomalously hot upper-mantle 'hot cells' (e.g. Anderson *et al.* 1992) and magmatism induced by meteorite impact (Rogers 1982; Jones *et al.* 2002).

In terms of eruptive mechanism and duration,

From: FITTON, J. G., MAHONEY, J. J., WALLACE, P. J. & SAUNDERS, A. D. (eds) 2004. *Origin and Evolution of the Ontong Java Plateau*. Geological Society, London, Special Publications, **229**, 219–238. 0305-8719/$15.00

there are two types of oceanic LIPs: those that occur on or near mid-ocean ridge axes (e.g. Iceland) and those that occur mid-plate (e.g. Hawaii). Both the Hawaiian and Icelandic hot spots have been almost continuously active for millions of years (e.g. Clague and Dalrymple 1987, 1989; Chalmers *et al.* 1995). The Kerguelen Plateau hot spot has displayed both periods of pulsed and continuous volcanism; the plateau formed both off-ridge and near the SE Indian Ridge during periods of its evolution (Royer *et al.* 1991; Frey & Weis 1995; Pringle & Duncan 2000; Duncan 2002). The Ontong Java Plateau, however, appears to have erupted in at least two and perhaps four discrete pulses at approximately 122, 90, 62 and 34 Ma (Mahoney *et al.* 1993; Tejada *et al.* 1996, 2002; Birkhold 2000), although the initial pulse at *c.* 122 Ma seems to be by far the most voluminous. This causes difficulties in using a plume model for OJP petrogenesis (see Tejada *et al.* 2004 for further details), although Bercovici & Mahoney (1994) developed a model that could accommodate two magmatic events approximately 30 Ma apart within the framework of a rising plume head. Birkhold (2000) suggested further magmatic events that tap the OJP source could occur through plate reorganization producing pressure-release partial melting in the fossil plume head material that sits beneath the OJP (cf. Richardson *et al.* 2000; Klosko *et al.* 2001). These post-90 Ma events would be progressively smaller in magnitude and in degree of partial melting as the source is progressively depleted.

The depth of origin and chemical composition of plumes are intrinsically linked. The predominance of hot upper mantle in the Icelandic plume has been used to suggest a partial origin at the 670 km discontinuity (Fitton *et al.* 1997). Osmium, O, Sr, Nd, Pb and Hf isotopes have been used to suggest that ancient recycled crustal material has been incorporated into the Hawaiian plume (Hauri *et al.* 1996; Lassiter *et al.* 1996; Lassiter & Hauri 1998; Blichert-Toft *et al.* 1999). Others have suggested that outer core material is responsible for suprachondritic $^{187}Os/^{188}Os$ (Widom & Shirey 1996), and coupled suprachondritic $^{186}Os/^{188}Os$ and $^{187}Os/^{188}Os$, in both Hawaii and other LIPs (Walker *et al.* 1995, 1997; Brandon *et al.* 1999). High $^{3}He/^{4}He$ values are also interpreted to suggest that the Hawaiian plume originated in the undegassed lower mantle (e.g. Kurz *et al.* 1996; Brandon *et al.* 1999). Geophysical and experimental evidence suggests that core–mantle interaction is not only feasible, but occurs in the D″ layer (Shannon & Agee 1998; Vidale & Hedlin 1998; Vinnik *et al.* 1998). Ely & Neal (2003) have modelled the platinum-group elements (PGEs) of OJP basalts cropping out on the Solomon Islands to infer a core–mantle boundary origin for the OJP, assuming an initial origin as a rising plume head. If the OJP were a product of either hot-cell activity or meteorite impact, only upper-mantle material would be involved and the erupted lavas would be expected to be MORB-like in character; Sr, Nd and Pb isotopes demonstrate that this is not the case (Mahoney & Spencer 1991, Mahoney *et al.* 1993; Tejada *et al.* 1996, 2002; Neal *et al.* 1997). While there are still problems with a simple plume origin for the OJP (see Tejada *et al.* 2004), it appears that these are less severe than with the alternatives. Therefore, for this chapter we assume the origin of the OJP via a surfacing plume head.

Owing to their siderophile nature, the PGEs would be partitioned strongly into the metallic core during planetary differentiation. Incorporation of less than 1% of such material into the plume source would dominate the PGE budget. However, it has been suggested that the lower mantle may equilibrate with the core with respect to Os isotopes, but that admixtures resulting in high overall PGE abundances are unlikely (Puchtel & Humayun 2000). In this paper, we attempt to test the idea that PGE abundances can be used to indicate a core–mantle boundary origin for plumes, and expand upon the work of Ely & Neal (2003) using OJP basalts recovered from Ocean Drilling Program (ODP) Legs 130 and 192.

Geological background

Located in the SW Pacific (Fig. 1), the OJP is the world's most expansive oceanic large igneous province at approximately 2.0×10^6 km^2, with a total volume of *c.* 5×10^7 km^3 (e.g. M. Coffin pers. comm. 2003). The OJP is almost entirely submarine, although subaerial outcrops occur on the islands of Malaita, Makira, Ulawa and Santa Isabel in the Solomon Islands. This exposure is a result of uplift after collision between the OJP and the Australian plate (Coleman & Kroenke 1981; Petterson *et al.* 1997, 1999).

Remarkably, the chemical variation of the OJP basalts is extremely limited given the size of this LIP. Prior to ODP Leg 192, only two significant types of basalt were known for the OJP. Tejada *et al.* (2002) used the terms Singgalo Formation and Kwaimbaita Formation (from type localities exposed on Malaita) for the two isotopically distinct basalt types. Singgalo-type basalts are characterized by $(^{87}Sr/^{86}Sr)_t$ = 0.7040–0.7042, $(\varepsilon_{Nd})_t$ = +3.8–+5.4 and $^{206}Pb/^{204}Pb$ = 18.245–18.521, and include the Unit A basalt at Site 807, whereas Kwaimbaita-type basalt

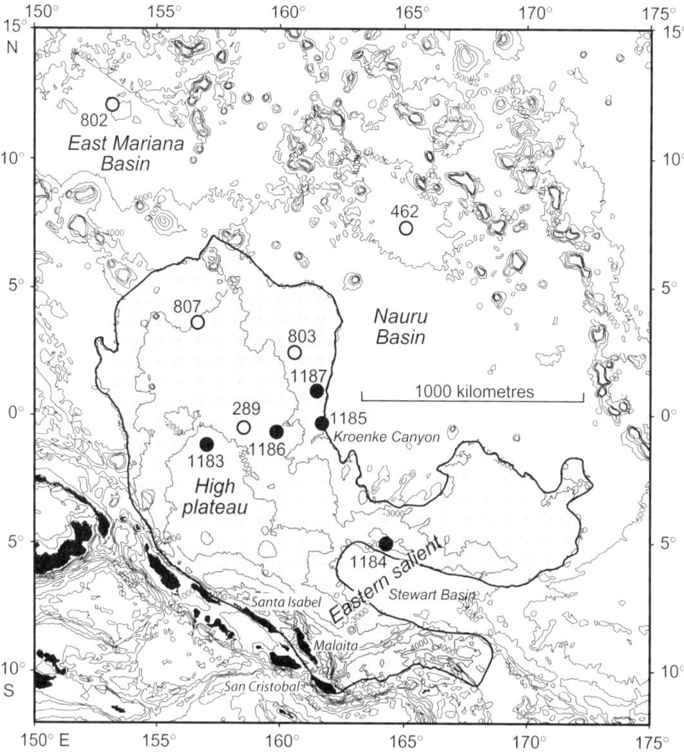

Fig. 1. Map showing the locations of all Deep Sea Drilling Project (DSDP) and Ocean Drilling Program (ODP) drill sites on the Ontong Java Plateau (adapted from Mahoney *et al.* 2001). Leg 192 drill sites are marked by black circles; open circles represent pre-Leg 192 drill sites.

incompatible trace-element abundances are slightly depleted relative to those of the Singgalo-type basalts, and have $(^{87}Sr/^{86}Sr)_t$ = 0.7034–0.7039, $(\varepsilon_{Nd})_t$ = +5.4–+6.5 and $^{206}Pb/^{204}Pb$ = 18.626–18.708, and include basalts of Units C–G at Site 807 and the *c.* 90 Ma basalts at Site 803 (Mahoney 1987; Mahoney & Spencer 1991; Mahoney *et al.* 1993; Tejada *et al.* 1996, 2002; Neal *et al.* 1997). The term $(\varepsilon_{Nd})_t$ represents $^{143}Nd/^{144}Nd$ normalized to the chondritic uniform reservoir, both being corrected for the age of the sample. Abundances of MgO are slightly lower in Singgalo-type basalts, typically showing a range of 6–7.3 wt%, while the range in the Kwaimbaita-type basalts is usually from 7 to 8 wt%. Notably, Singgalo-type basalts have been found only on the northern and southern margins of the plateau, at ODP Site 807 and as subaerial outcrops in the Solomon Islands, respectively. Kwaimbaita-type basalts appear to be present across the plateau, being found in cores from Deep Sea Drilling Program Site 289, and ODP Sites 803, 807, 1183, 1185 and 1186, as well as subaerial outcrops on the Solomon Islands (Fig. 1). Birkhold (2000) described basalts on Makira from the Wairahito Formation, which contain higher abundances of the incompatible trace elements than other OJP basalts, but were isotopically similar to Kwaimbaita basalts. These are interpreted as a more evolved lava type (MgO = 4.5–6.6 wt%) derived from the Kwaimbaita source and, to date, basalts of this type have only been described from Makira. Several basalt flows recovered during ODP Leg 192 at Sites 1185 and 1187 have higher MgO abundances (8–11 wt %), and lower contents of incompatible trace elements than those of the Kwaimbaita-type basalts (Nb = *c.* 2.1–2.4 v. *c.* 2.9–3.5 ppm, respectively; Fitton & Godard 2004). However, these primitive basalts are isotopically similar to the more evolved Kwaimbaita-type basalts (Tejada *et al.* 2004).

Analytical techniques

Sixteen basalt samples were analysed for PGEs, and major and trace elements. Sample collection began onboard the *JOIDES Resolution* during Leg 192, where precautions against contamination were taken from the moment the cores were brought onboard. Staff and Leg 192 scientists removed all jewellery and, where possible, contact with metal was avoided. Samples were selected from each volcanic unit as determined on board the ship, making sure to select only the least altered samples. Samples were selected adjacent to those used in Fitton & Godard's (2004) study. Each sample was then cut from the working half of the core with a diamond-impregnated saw and sealed in a plastic bag for transport. Once at Notre Dame, samples were first cut into approximately 0.5 cm-thick wafers with a diamond-impregnated saw. Care was taken to avoid veins, especially those with black or brown haloes, in which secondary sulphide (pyrite) or Fe-oxyhydroxides, respectively, were observed in thin section. All surfaces were subsequently ground away with a diamond grinding disk to remove any contamination imparted by the drill bit or the saw during cutting. The samples were then crushed using alumina jaw crushers, washed in dilute (0.6 M) HCl in an ultrasonic bath for 20 min to remove secondary minerals and then rinsed several times with 18 MΩ H_2O, until all flocculants were removed.

Heterogeneous distribution of trace primary sulphides in basaltic samples can produce the 'nugget' effect, so care was taken to obtain a sample that truly represents the whole-rock composition. Reducing this effect is accomplished by choosing fine-grained samples and/or powdering a large amount of sample (10–20 g) in an alumina ball mill, followed by remixing each powdered sample (if it was left unagitated for more than a week) before analysing a given aliquot. This study used the PGE standard UMT-1, which was re-mixed on a roller for 24 h before each split was taken. Two-hundred-and-fifty milligrams of each sample powder was then dissolved in sequential stages of 16 M HNO_3 + 29 M HF, aqua regia (made from 16 M HNO_3 and 12 M HCl), concentrated HCl and finally 0.6 M HCl (see Ely *et al.* 1999; Ely & Neal 2002 for full details of the PGE analytical techniques and data reduction methods employed). All acids used were double distilled in-house from reagent-grade stocks. For details regarding the analytical procedures used in determining the major and trace elements of the samples discussed in this paper, see Jenner *et al.* (1990), Neal (2001) and Shafer *et al.* (2004).

Where possible, multiple isotopes of each of the PGEs were analysed and the interference-free isotope of highest abundance used to quantify PGE contents (Table 1). For example, while 193 is the most abundant Ir isotope (62.7%), it is interfered with by $^{177}Hf^{16}O$ if Hf is present in the whole rock at >1 ppm, as Hf is not completely removed by the cation-exchange procedure (Ely *et al.* 1999). The PGE abundances are based on results obtained for ^{102}Ru, ^{103}Rh, ^{105}Pd and ^{191}Ir. The Pt isotope used depends on the abundance of Hf in the sample (and the magnitude of the HfO interference), but is either ^{195}Pt or ^{198}Pt. The low abundance of ^{198}Pt generates larger errors than those with ^{195}Pt, but the oxide interferences are typically present on ^{195}Pt ($^{179}Hf^{16}O$) and absent on ^{198}Pt ($^{182}W^{16}O$), as very little W is usually present in the sample, especially as the sample was prepared using alumina crushing and powdering equipment. Mercury was also monitored as ^{198}Hg interferes with ^{198}Pt, but was always below limits of detection. The ^{195}Pt isotope was used when Pt abundances derived for that isotope were within statistical error of those for ^{198}Pt. Limits of detection and quantification (Table 1) were calculated from a procedural blank analysed immediately before the sample. In this way, memory effects can be monitored (although with the wash-out procedure described in Ely *et al.* 1999, they were non-existent) and, if needed, applied to the calculation of PGE abundances in the sample. Oxide formation was monitored pre-run using a 1 ppb tuning solution containing Ce. The instrument was tuned such that CeO was no greater than 1–2% of the Ce counts before a run proceeded; typically CeO was around 0.5% of the Ce counts.

Results

Trace-element and selected major-element data for the samples analysed are presented in Table 1. Major- and trace-element data presented are from this study (all data for Leg 130 samples; V, Cr and Zn for all samples), M. Godard (all remaining Leg 192 trace-element data; pers. comm.) and Fitton & Godard (all Leg 192 major-element data; 2004); MgO for some samples are presented from within the same unit, within a few centimetres of the sample selected for PGE analysis. The primitive or 'Kroenke-type' basalts from Sites 1185 and 1187 are characterized by MgO and TiO_2 abundances of 8.0–10.9 wt% and 0.69–0.76 wt%, while those of Kwaimbaita-type basalts range from 7.3 to 8.0 wt% and 1.04–1.10 wt%, respectively. The Kroenke-type basalts are named after the location of Site 1185 adjacent to Kroenke Canyon, a large submarine canyon just south of this site. Kroenke-type basalts contain Ni and Cr abundances (177–250 and 371–494 ppm, respectively) that indicate that these lavas have experienced a lower degree of fractional crystallization

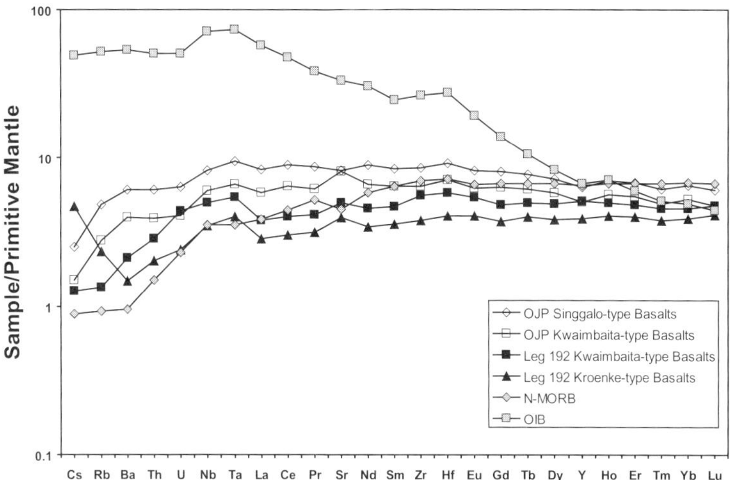

Fig. 2. Incompatible trace-element abundances normalized to primitive mantle for the Ontong Java Plateau basalts. Data are from Mahoney *et al.* (1993), Tejada *et al.* (1996, 2002), Neal *et al.* (1997) and Fitton & Godard 2004. Normalization values are from Sun & McDonough (1989).

than the Kwaimbaita-type basalts (122–135 and 142–294 ppm, respectively). Incompatible trace-element abundances for both the Kroenke-type and Kwaimbaita-type basalts are also consistent with these observations (Fig. 2). In addition to these Leg 192 samples, two samples from ODP Leg 130 (one each from Sites 803 and 807) were also analysed, both of which have major and trace elements in the range of Kwaimbaita-type basalts. The Site 803 basalt, however, is apparently approximately 32 Ma younger (*c.* 90 Ma) than the other Kwaimbaita-type basalts analysed (*c.* 122 Ma; e.g. Mahoney *et al.* 1993).

Platinum-group element abundance data are presented in Table 1. All values are reported with $\pm 2\sigma$ errors (after Ely & Neal 2002). Values falling below the limit of detection (LOD: background + 3σ) are reported as 'BDL' (below detection limit). Values presented in italics fall between the LOD and the limit of quantification (LOQ: background + 10σ). The LOD and LOQ are calculated using a blank analysed immediately before the sample or reference material. Average values for UMT-1 are also reported. As the PGEs were quantified using standard addition, reference materials were analysed at least once for every two unknown samples. For this study, UMT-1 data are reported as averages of 13 separate analyses and the errors here represent $\pm 2\sigma$ of the mean value.

In general, the Leg 192 Kroenke- and Kwaimbaita-type basalts contain overlapping PGE abundances (Fig. 3a–d), which are roughly equivalent to those found in komatiites (e.g. Brügmann *et al.* 1987; Rehkämper *et al.* 1999; Puchtel & Humayun 2001). While Pt and Pd values for these basalt types are essentially identical, a slight distinction can be made between Kwaimbaita- and Kroenke-type basalts in terms of Ir, Ru and Rh in Hole 1185B. Positive correlations exist between Ni and Cr v. Ir and Ru (Fig. 4a–d), with the primitive Kroenke-type basalts containing higher abundances. The pattern breaks down, however, if basalts external to this site are also included. Primitive-mantle-normalized plots of the PGEs (Fig. 3a–d) exhibit a generally positive slope. Slight negative Pd anomalies also occur in some samples, but most have a flat–slightly positive transition from Pt to Pd to Y (Fig. 3a–d). The Leg 130 Kwaimbaita-type basalts have slightly lower Ru and Rh abundances than the Leg 192 basalts, but have similar Pt and Pd, although both Leg 130 basalts also exhibit Pd depletions that produce negative anomalies in normalized profiles (Fig. 3b).

The PGE compositions of Leg 130 and Leg 192 OJP basalts are similar to those from the Solomon Islands (Fig. 5) (Ely & Neal 2003), regardless of age. The Leg 192 Kwaimbaita-type basalts generally have lower PGE abundances relative to the Kroenke-type basalts, but they overlap with PGE abundances of Kwaimbaita- and Singgalo-type basalts from the Solomon Islands (Ely & Neal 2003). The clinopyroxene–plagioclase–titanomagnetite cumulates ML-475 and ML-476 (Malaita; Ely & Neal 2003) contain PGE abundances that are similar to the Leg 192 basalts, but contain twice the Pd content. While

Table 1. *Platinum-group (ppb), major- (wt%), and trace-element (ppm) abundances for the ODP Leg 130 and Leg 192 basalts analysed in this study**

Leg	Site	Hole	Core	Section	Piece	Interval (cm)	Unit	Type	Ir	2σ	Ru	2σ	Rh	2σ	Pt	2σ	Pd	2σ
192	1185	A	10R	3	1A	0–20	5B	Kroenke	*0.5*	*0.1*	*0.9*	*0.4*	0.9	0.4	8.5	0.5	13.4	2.3
192	1185	B	5R	5	2C	143–149	2	Kroenke	2.4	0.7	1.5	0.4	1.3	0.1	13.6	2.0	18.2	4.5
192	1185	B	6R	4	2	101–120	5	Kroenke	*1.5*	*0.1*	*1.3*	*0.5*	1.2	0.3	11.7	1.3	13.9	3.6
192	1185	B	11R	1	3A	13–17	9	Kroenke	1.4	0.3	1.3	0.4	1.1	0.1	12.8	4.1	13.7	2.1
192	1187	A	6R	6	2B	92–97	3B	Kroenke	*0.6*	*0.1*	1.0	0.2	0.9	0.1	11.3	1.9	11.5	0.8
192	1187	A	13R	2	1B	50–55	6	Kroenke	*1.1*	*0.8*	*1.1*	*0.5*	1.2	0.3	9.4	1.0	*11.7*	*5.2*
192	1187	A	13R	2	1B	50–55	6	Kroenke	1.7	0.4	1.1	0.2	0.9	0.2	13.8	2.9	14.3	1.6
192	1187	A	16R	3	3A	34–39	12	Kroenke	BDL		1.4	0.2	1.0	0.1	11.2	5.2	24.8	3.0
192	1183	A	59R	1	3B	40–45	5B	Kwaimbaita	BDL		1.7	0.7	1.0	0.3	28.8	3.6	18.9	2.1
192	1183	A	65R	1	5D	112–128	7	Kwaimbaita	1.1	0.5	1.9	0.7	1.3	0.8	19.1	3.4	16.7	6.6
192	1185	B	17R	3	3	39–45	10	Kwaimbaita	1.4	0.8	0.9	0.4	0.9	0.1	18.7	4.3	13.3	4.3
192	1185	B	21R	4	1B	45–50	11	Kwaimbaita	1.0	0.1	*1.2*	*0.4*	0.8	0.2	10.1	0.7	10.9	1.2
192	1185	B	24R	2	6	87–93	12	Kwaimbaita	*0.9*	*0.1*	1.1	0.2	BDL		12.5	0.4	9.1	3.1
192	1186	A	32R	2	6	101–106	1	Kwaimbaita	BDL		1.2	0.1	0.8	0.3	16.3	6.9	8.5	1.6
192	1186	A	39R	1	10	42–58	4	Kwaimbaita	1.4	0.2	1.5	0.2	1.1	0.1	13.6	0.7	11.4	0.8
130	803	D	70R	3	7	93–98	9B	Kwaimbaita	BDL		*0.7*	*0.1*	0.3	0.1	17.3	2.0	7.3	0.8
130	807	C	93R	3	1	16–19	4G	Kwaimbaita	BDL		*0.6*	*0.1*	0.5	0.04	18.4	0.4	8.8	0.9
								UMT-1 (*n* = 13)	9.4	0.8	9.7	1.3	9.8	1.4	113	12	95	11
								UMT-1 (cert.)	8.8	0.6	10.9	1.5	9.5	1.5	129	5	106	3

Leg	Site	Hole	Core	Section	Piece	Interval (cm)	Unit	Type	MgO	TiO_2	Fe_2O_3	Sc	V	Cr	Co	Ni	Cu	Zn	Rb	Sr
192	1185	A	10R	3	1A	0–20	5B	Kroenke	8.90	0.74	10.99	40	248	404	51	189	101	81	1.34	73
192	1185	B	5R	5	2C	143–149	2	Kroenke	10.44	0.71	10.92	41	543	397	56	238	103	101	1.34	72
192	1185	B	6R	4	2	101–120	5	Kroenke	10.91	0.69	10.86	42	260	443	63	250	102	96	1.21	85
192	1185	B	11R	1	3A	13–17	9	Kroenke	8.93	0.73	10.65	42	294	403	51	177	100	88	4.13	77
192	1187	A	6R	6	2B	92–97	3B	Kroenke	9.69	0.73	10.92	40	236	371	49	199	96	71	0.61	82
192	1187	A	13R	2	1B	50–55	6	Kroenke	9.96	0.72	10.72	40	307	469	51	207	100	95	0.73	80
192	1187	A	16R	3	3A	34–39	12	Kroenke	8.03	0.76	10.67	42	n.a.	494	52	212	100	n.a.	0.49	82
192	1183	A	59R	1	3B	40–45	5B	Kwaimbaita	7.59	1.10	12.07	48		212	51	135	149		0.55	110
192	1183	A	65R	1	5D	112–128	7	Kwaimbaita	7.64	1.08	12.26	46	459	282	48	131	150	128	0.46	102
192	1185	B	17R	3	3	39–45	10	Kwaimbaita	7.40	1.06	12.60	45	265	142	49	127	149	83	0.76	95
192	1185	B	21R	4	1B	45–50	11	Kwaimbaita	7.61	1.06	12.58	47	295	240	49	129	141	89	0.62	96
192	1185	B	24R	2	6	87–93	12	Kwaimbaita	7.29	1.12	11.05	49	n.a.	230	51	122	136	n.a.	0.16	96
192	1186	A	32R	2	6	101–106	1	Kwaimbaita	7.34	1.09	12.36	47	n.a.	218	51	131	243	n.a.	2.25	105
192	1186	A	39R	1	10	42–58	4	Kwaimbaita	7.96	1.04	12.23	49	425	294	50	137	126	112	0.85	87
130	803	D	70R	3	7	93–98	9B	Kwaimbaita	6.53	1.29	10.25	59	297	275	n.a.	142	n.a.	n.a.	3.69	169
130	807	C	93R	3	1	16–19	4G	Kwaimbaita	7.22	1.11	12.62	58	348	178	n.a.	117	n.a.	n.a.	0.43	112

									Y	Zr	Nb	Cs	Ba	La	Ce	Pr	Nd	Sm	Eu	Gd
192	1185	A	10R	3	1A	0–20	5B	Kroenke	16.8	38	2.22	0.032	9.2	1.73	4.72	0.76	4.01	1.40	0.61	1.90
192	1185	B	5R	5	2C	143–149	2	Kroenke	16.0	37	2.16	0.040	9.4	1.64	4.47	0.72	3.88	1.30	0.58	1.85
192	1185	B	6R	4	2	101–120	5	Kroenke	14.8	41	2.15	0.043	19.7	1.86	5.20	0.85	4.42	1.60	0.65	2.21
192	1185	B	11R	1	3A	13–17	9	Kroenke	17.3	42	2.30	0.091	5.8	1.78	5.02	0.79	4.31	1.48	0.64	1.98
192	1187	A	6R	6	2B	92–97	3B	Kroenke	17.7	40	2.45	0.018	5.9	2.03	5.11	0.80	4.35	1.39	0.64	2.03
192	1187	A	13R	2	1B	50–55	6	Kroenke	17.5	42	2.51	0.021	11.9	2.00	5.42	0.85	4.59	1.51	0.66	2.14
192	1187	A	16R	3	3A	34–39	12	Kroenke	17.3	40	2.31	0.015	6.2	1.94	5.38	0.86	4.61	1.52	0.65	2.09
192	1183	A	59R	1	3B	40–45	5B	Kwaimbaita		65	3.51	0.006	17.2	2.54	7.09	1.10	6.03	1.97	0.87	2.64
192	1183	A	65R	1	5D	112–128	7	Kwaimbaita	21.9	61	3.47	0.006	15.0	2.49	6.92	1.07	5.87	1.93	0.84	2.70
192	1185	B	17R	3	3	39–45	10	Kwaimbaita	21.6	56	3.15	0.005	16.5	2.66	7.01	1.11	5.84	1.93	0.84	2.63
192	1185	B	21R	4	1B	45–50	11	Kwaimbaita	27.3	60	3.37	0.007	15.9	2.26	5.98	0.94	5.17	1.76	0.82	2.43
192	1185	B	24R	2	6	87–93	12	Kwaimbaita	20.4	56	3.12	0.003	9.1	2.16	6.08	0.98	5.33	1.82	0.82	2.52
192	1186	A	32R	2	6	101–106	1	Kwaimbaita	22.5	62	3.57	0.032	10.3	2.64	7.40	1.15	6.36	2.09	0.88	2.99
192	1186	A	39R	1	10	42–58	4	Kwaimbaita	19.7	49	2.83	0.011	13.6	2.47	6.76	1.05	5.67	1.86	0.76	2.60
130	803	D	70R	3	7	93–98	9B	Kwaimbaita	25.4	72	4.27	0.059	13.0	4.04	11.72	1.92	8.95	2.84	1.13	3.89
130	807	C	93R	3	1	16–19	4G	Kwaimbaita	21.7	55	3.16	0.014	15.4	3.52	9.55	1.59	7.48	2.38	1.01	3.43

									Tb	Dy	Ho	Er	Tm	Yb	Lu	Hf	Ta	Pb	Th	U
192	1185	A	10R	3	1A	0–20	5B	Kroenke	0.37	2.47	0.58	1.65	0.24	1.62	0.26	1.16	0.13	0.31	0.14	0.04
192	1185	B	5R	5	2C	143–149	2	Kroenke	0.36	2.39	0.57	1.60	0.24	1.58	0.26	1.13	0.13	0.48	0.14	0.05
192	1185	B	6R	4	2	101–120	5	Kroenke	0.45	2.84	0.64	1.85	0.29	2.05	0.32	1.13	0.21	0.37	0.17	0.06
192	1185	B	11R	1	3A	13–17	9	Kroenke	0.39	2.56	0.60	1.79	0.26	1.67	0.27	1.17	0.15	0.29	0.17	0.04
192	1187	A	6R	6	2B	92–97	3B	Kroenke	0.40	2.57	0.62	1.78	0.26	1.75	0.28	1.11	0.14	0.30	0.17	0.05
192	1187	A	13R	2	1B	50–55	6	Kroenke	0.40	2.75	0.62	1.80	0.26	1.74	0.29	1.25	0.14	0.41	0.17	0.05
192	1187	A	16R	3	3A	34–39	12	Kroenke	0.41	2.62	0.63	1.81	0.27	1.74	0.29	1.18	0.14	0.24	0.17	0.05
192	1183	A	59R	1	3B	40–45	5B	Kwaimbaita	0.51	3.33	0.77	2.23	0.33	2.06	0.34	1.71	0.21	0.31	0.25	0.07
192	1183	A	65R	1	5D	112–128	7	Kwaimbaita	0.50	3.41	0.75	2.20	0.33	2.08	0.33	1.67	0.20	0.45	0.24	0.07
192	1185	B	17R	3	3	39–45	10	Kwaimbaita	0.50	3.22	0.75	2.07	0.30	1.99	0.31	1.66	0.20	0.44	0.22	0.06
192	1185	B	21R	4	1B	45–50	11	Kwaimbaita	0.46	3.12	0.69	2.02	0.30	1.91	0.31	1.63	0.20	0.30	0.24	0.07
192	1185	B	24R	2	6	87–93	12	Kwaimbaita	0.48	3.11	0.73	2.01	0.29	1.94	0.31	1.70	0.20	0.34	0.20	0.08
192	1186	A	32R	2	6	101–106	1	Kwaimbaita	0.53	3.78	0.80	2.35	0.34	2.24	0.36	1.79	0.23	0.49	0.25	0.22
192	1186	A	39R	1	10	42–58	4	Kwaimbaita	0.47	3.25	0.71	2.02	0.30	1.95	0.31	1.46	0.18	0.432	0.20	0.06
130	803	D	70R	3	7	93–98	9B	Kwaimbaita	0.70	4.32	0.95	2.80	0.40	2.49	0.37	2.10	n.a.	0.95	0.35	0.19
130	807	C	93R	3	1	16–19	4G	Kwaimbaita	0.62	4.01	0.86	2.57	0.39	2.40	0.37	1.78	n.a.	0.88	0.30	0.14

* Concentrations between the limit of detection (LOD; blank + 3σ) and limit of quantification (LOQ; blank plus 10σ) are reported in italics; those below detection limits are listed as BDL. All samples are also reported with 2σ uncertainty. An average (n = 13) of the standard UMT-1 is reported with certified values for comparison. Values listed as 'n.a.' were not analysed. Data are from this study (all Leg 130 data; V, Cr and Zn for all samples), M. Godard (all remaining trace-element data; pers. comm.).

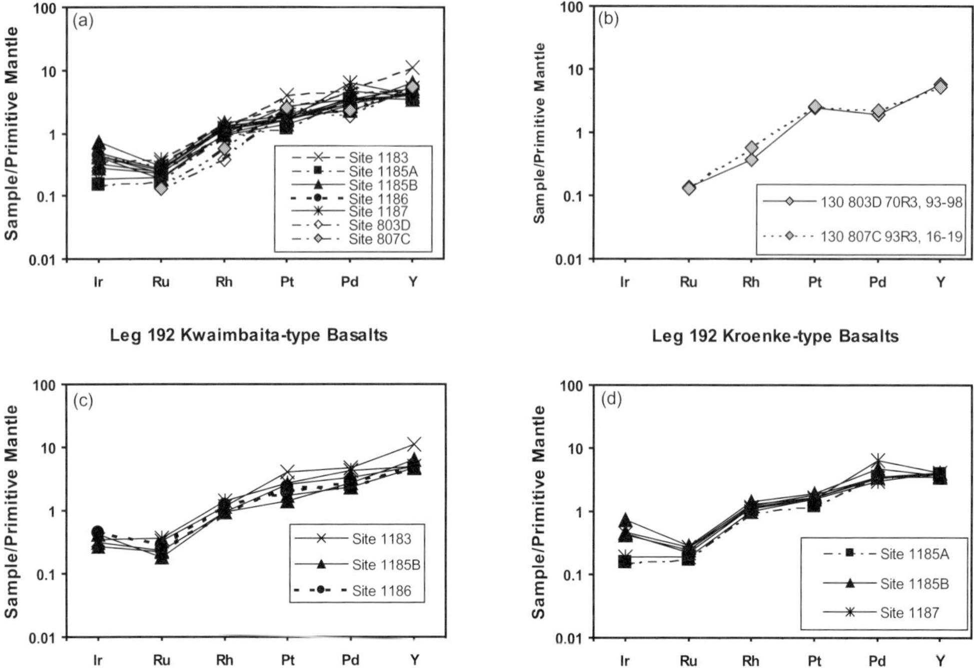

Fig. 3. Primitive-mantle-normalized plots of PGE abundances for (**a**) all of the OJP basalts analysed in this study, (**b**) the Leg 130 basalts, (**c**) the Leg 192 Kwaimbaita-type basalts and (**d**) the Leg 192 Kroenke-type basalts. Only the Leg 130 samples indicate any loss of Pd via low-temperature alteration ($[Pd/Pt]_{pm} < 1$). Normalization values are from McDonough & Sun (1995).

the Leg 192 basalts generally contain more Ir and Rh than those from the Solomon Islands, it is evident that a continuum of compositions exists.

Discussion

Low-temperature alteration and the effects of sulphide immiscibility

Thin-section examination revealed that all samples were subject to at least small degrees of secondary (low-temperature) alteration. Most alteration took place near veins and miarolitic cavities, but mild, incomplete replacement of olivine and the mesostasis by clay minerals was also present. Ely & Neal (2003) reported negative Pd anomalies in several samples from the Solomon Islands, and suggested it may result from the preferential mobilization of Pd during secondary alteration processes. This negative Pd anomaly is also found more commonly in Singgalo-type basalts, which are stratigraphically above the more ubiquitous Kwaimbaita-type basalts and have been exposed to secondary alteration–weathering processes for a longer period. However, while this feature is present in both samples analysed from Leg 130 (Fig. 3b), it is generally absent from the Leg 192 basalts (Fig. 3c, d). The presence of a limestone unit (Unit B) between the Singgalo- and Kwaimbaita-type basalts at Site 807 and the lack of any Singgalo-type basalt flows at Site 803 suggests that the Kwaimbaita-type basalts at these two sites were similarly exposed to secondary alteration–weathering processes. Plots of MgO v. Pt and Pd (Fig. 6a, b) demonstrate that several OJP samples from the Solomon Islands have similar Pt abundances to those from Leg 192, but Pd data for the former are scattered and are sometimes as low as mid-ocean ridge basalt (MORB) values. While the separation of immiscible sulphide could have this effect on Pd for the now subaerially exposed basalts from the Solomon Islands, this would affect all of the PGEs, not just Pd (Ely & Neal 2003). Basalts, such as MORBs, that have experienced the separation of immiscible sulphide are lower in overall PGE abundance, and have significant vertical scatter for Pt and Pd at a given MgO abundance (Fig. 6a, b). The lack of negative Pd

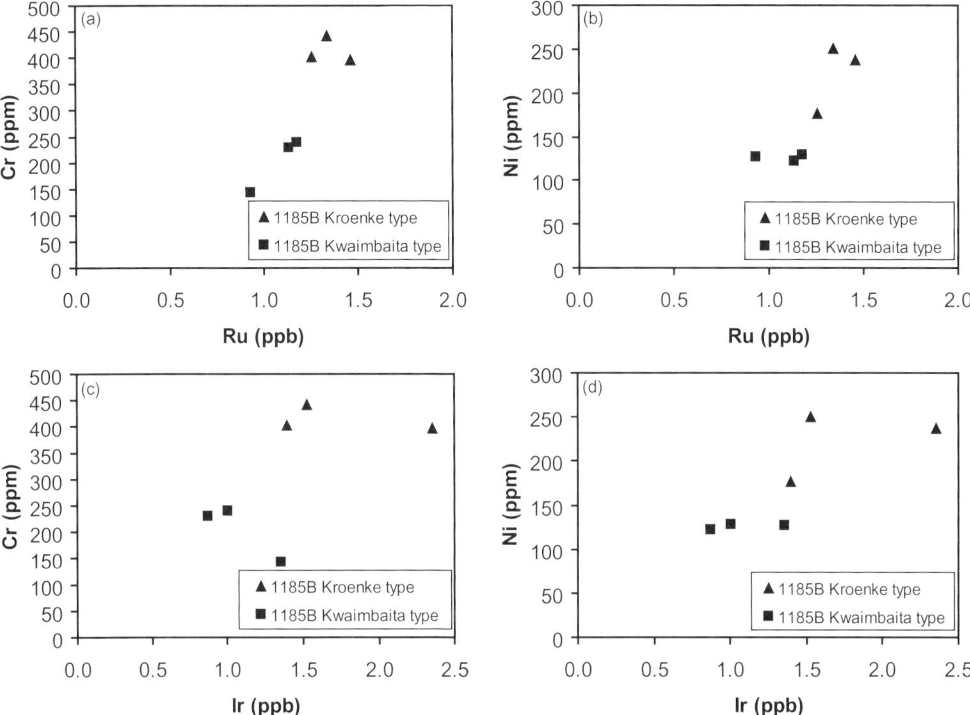

Fig. 4. (**a–d**) Plots of Ir and Ru v. Cr and Ni. Positive correlation between these elements demonstrates that these PGEs are exhibiting lithophile behaviour and are being controlled by olivine and/or Cr-spinel crystallization. See text for discussion.

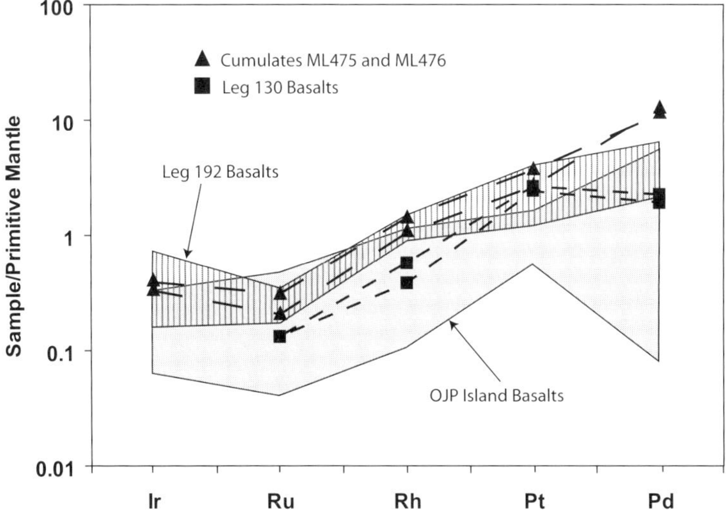

Fig. 5. Platinum-group element profiles, normalized to primitive mantle, for all OJP basalts analysed to date. The Leg 192 Kroenke- and Kwaimbaita-type basalts make up the high end of the range of PGEs found in the OJP, but are not completely distinct from Kwaimbaita-type basalts from elsewhere on the plateau or the Singgalo-type basalts. A continuum of compositions is evident, with Singgalo-type basalts tending towards lower PGE abundances, and Kroenke-type basalts tending towards higher. Data for the Solomon Island basalts are from Ely & Neal (2003). Primitive-mantle-normalization values are from McDonough & Sun (1995).

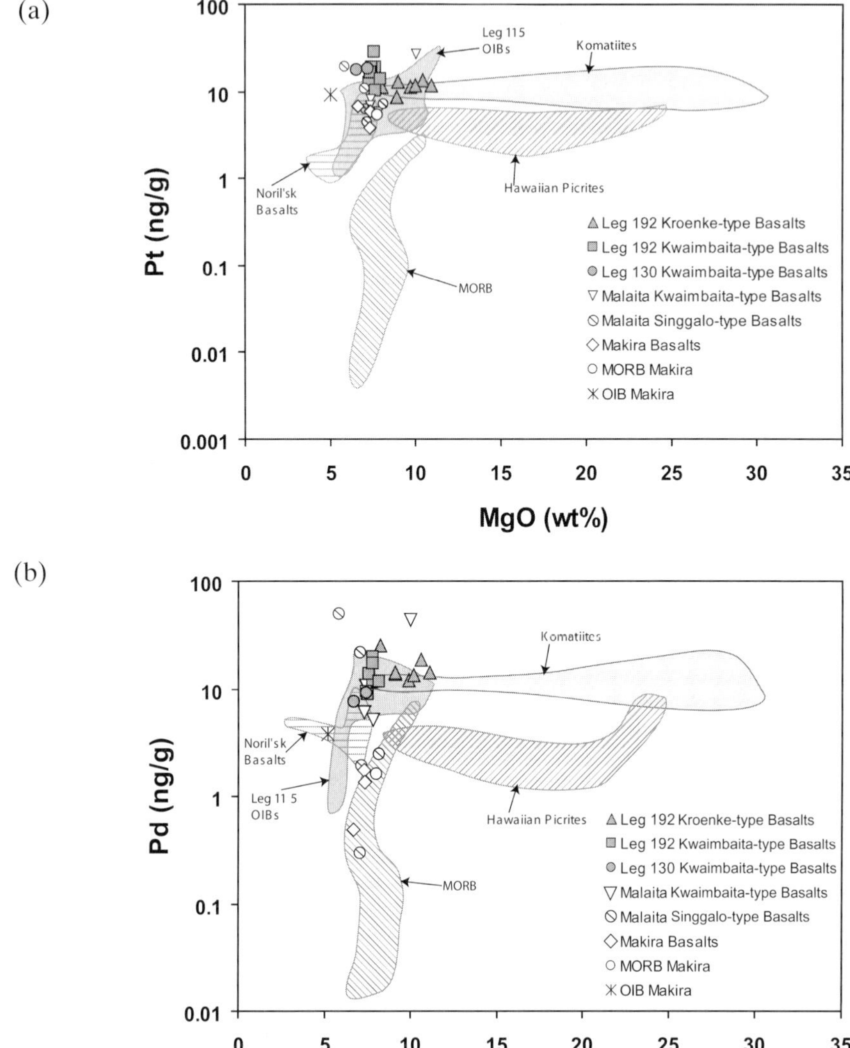

Fig. 6. (**a**) MgO (wt%) v. Pt (ppb) and (**b**) MgO (wt%) v. Pd (ppb) for various types of basalts and komatiites. Platinum abundances for all OJP samples plot in roughly the same region, and are similar to values for komatiites. Palladium abundances for the OJP basalts have a significantly larger range than those of Pt, with many Solomon Islands basalts having experienced Pd loss via low-temperature alteration; sulphide removal would have affected all of the PGEs, as it has in MORB (see text for discussion). The Pd values for the Leg 192 basalts appear unaffected by low-temperature alteration. Data are from Arndt & Nesbitt (1984), Arndt (1986), Lightfoot *et al.* (1990), Fryer & Greenough (1992), Brügmann *et al.* (1993), Devey *et al.* (1994), Puchtel *et al.* (1996), Norman & Garcia (1999), Rehkämper *et al.* (1999), Tatsumi *et al.* (1999), Bennett *et al.* (2000), Puchtel & Humayun (2001), Tejada *et al.* (1996, 2002), Ely & Neal (2003) and Fitton & Godard (2004).

anomalies in the Leg 192 basalt PGE profiles indicates that the data reflect primary igneous processes and are not the result of low-temperature alteration.

The PGEs partition strongly into sulphides, with partition coefficients between sulphide and silicate melts of greater than 1×10^4 (e.g. Barnes *et al.* 1985; Peach *et al.* 1990, 1994; Fleet *et al.*

1991, 1996; Bezmen *et al.* 1994). If a silicate magma were to reach saturation with respect to sulphur under typical mantle oxygen fugacities (e.g. Lee *et al.* 2003 and references therein), an immiscible sulphide liquid would form, removing chalcophile elements such as the PGEs, Fe, Ni and Cu. However, compositional data indicate that the OJP basalts were sulphur-undersaturated at the time of eruption (cf. Michael & Cornell 1996; Roberge *et al.* 2004). Indeed, thin-section examination reveals only traces of primary sulphide, which is present as micron-size interstitial blebs. This could indicate that immiscibility has occurred and practically all sulphur has been removed or that immiscibility was only reached once the bulk of the basalt flow had crystallized. Ratios such as Cu/Pd are similar to those in primitive mantle (7300) for the Kroenke-type basalts (*c.* 4100–8350) and are generally higher for all Kwaimbaita-type basalts (8000–28 500). Singgalo-type basalts from the Solomon Islands vary dramatically, but both Cu and Pd may have been affected by alteration in these samples (Ely & Neal 2003). Primitive-mantle-normalized PGE profiles (Fig. 3a–d) demonstrate the difference between the OJP basalts and MORB, where the latter has experienced sulphur saturation, followed by immiscibility and separation of a sulphide-rich liquid (e.g. Keays 1995; Roy-Barman *et al.* 1998; Rehkämper *et al.* 1999). During crystal fractionation, Pd and Y should have similar incompatibilities in a sulphur-undersaturated system, resulting in $[Pd/Y]_{pm}$ ratios of approximately 1 (Brügmann *et al.* 1993; where pm stands for 'normalized to primitive mantle'). This is the case for the OJP basalts, whereas in MORB $[Pd/Y]_{pm} \ll 1$. Experimental observations of the sulphur concentration at sulphide saturation (SCSS) by Mavrogenes & O'Neill (1999) and Holzheid & Grove (2002) indicate that many magmas may initially be close to sulphur saturation, but a strong negative correlation between SCSS and pressure suggests that most should arrive at the surface sulphur-undersaturated. Only large degrees of crystal fractionation, rapid temperature decrease due to assimilation, or assimilation of a relatively sulphur-rich component (such as continental crust) would push the system towards sulphur-saturation. Although Kwaimbaita- and Singgalo-type basalts have experienced relatively large degrees of crystal fractionation, the $[Pt/Y]_{pm}$ and $[Pd/Y]_{pm}$ ratios have not been overly fractionated. This demonstrates that the PGEs are behaving as lithophile elements, and indicates a lack sulphur saturation and subsequent removal of a sulphide-rich immiscible melt.

On the basis of the above discussion, temperature estimates of the OJP source (our assumption is that this is a plume head) can now be made. Using major- and trace-element models, pressure estimates for the initial melt of the OJP are *c.* 3.5–4.0 GPa (Mahoney *et al.*, 1993; Farnetani & Richards, 1994; Neal *et al.* 1997; Tejada *et al.* 2002). Estimates for sulphur in primitive mantle are about 250 ppm (Keays 1995), and with the average degree of partial melting estimated to be approximately 30% (Fitton & Godard 2004), the models of Mavrogenes & O'Neill (2002) and Holzheid & Grove (2002) can be used to estimate the minimum temperature of the plume head required to generate the OJP basalts. One assumption in this temperature calculation is that the sulphur content of the initial melt can be projected from that of the source and the degree of partial melting. Sulphide in the source is assumed to be exhausted at no more than approximately 20–23% partial melting (e.g. Rehkämper *et al.* 1999). Therefore, at 30% partial melting, sulphur is exhausted, and the maximum concentration of sulphur in the melt is approximately 830 ppm. Using equation (10) from Mavrogenes & O'Neill (1999):

$$\ln[S(ppm)] = \frac{A}{T} + B + \frac{CP}{T} + \ln a_{FeS}^{sulphide}$$

and assuming $a_{FeS}^{sulphide} = 1$ (where $a_{FeS}^{sulphide}$ is the activity of FeS in the sulphide liquid), and the basaltic values for the parameters A (enthalpy; −6684K), B (entropy; 11.52), and C (volume; −0.047K/bar) along with the pressure (P) estimates outlined above, the minimum temperature (T) required for sulphur to remain undersaturated at a given pressure can be calculated. Pressure estimates for the Ontong Java Plateau based on phase equilibria suggest melting in a range of 1.0–6.0 GPa, but with the majority of data indicating pressures between 1.0 and 4.0 GPa (Tejada *et al.* 2002). Other methods used to estimate pressure of melting (e.g. Farnetani & Richards 1994) are consistent with these values (i.e. *c.* 0.6–4.0 GPa). For melts originating at 1.0 and 6.0 GPa, the minimum calculated temperatures are approximately 1170°C and 1650°C, respectively. Trace-element modelling seems to constrain much of the melting to between 3.5 and 4.0 GPa (Mahoney *et al.* 1993; Neal *et al.* 1997; Tejada *et al.* 2002), and calculated temperatures for melts originating at these pressures are approximately 1465–1515°C. However, variation in the sulphur content of the source could cause these temperature estimates to change significantly. At 200 ppm sulphur the calculated temperatures are *c.* 1385–1435°C, while at 300 ppm sulphur they would be *c.* 1535–1585°C to prevent saturation. Assuming a potential temperature for ambient mantle of 1280°C (McKenzie & Bickle 1988), the model temperatures for 250 ppm sulphur at 3.5–4.0 GPa are lower than the ΔT of +350°C excess temperature estimates (without lithospheric extension) of Farnetani & Richards (1994) that were required to produce magma volumes of 5×10^7 km^3 at 35% partial melting, but overlap with those of Griffiths & Campbell (1990) who estimated a 100–200°C initial excess temperature for a plume head that is beginning to impinge on the rigid lithosphere. Farnetani & Richards (1994) also use an ambient mantle temperature of 1280°C, making their temperature estimates for large volume melts ($\geq 1 \times 10^7$ km^3) in excess of 1600°C. The model of Farnetani & Richards (1994) takes account of the presence of lithosphere, which necessitates large temperature differentials to produce high-volume melts. While the constants A, B and C (and thus the SCSS) can be affected by variations in magma composition, especially Fe and Ni, and would therefore need to be experimentally determined for OJP basalts to ensure more accurate temperature estimates, the +185–+235°C initial excess temperatures of this study are comparable to the those for the OJP (>1500°C; Fitton & Godard 2004) and other plumes (e.g. Iceland; 1480–1520°C or a ΔT of 180–240°C; Maclennen *et al.* 2001). The temperatures may thus be underestimated considering the high degree of partial melting for the OJP (*c.* 30%; Fitton & Godard 2004) relative to other plumes. However, it must be remembered that these estimates are *minimum* temperatures because they are the minimum required for sulphur to remain undersaturated.

One possible problem with the model is that melting is polybaric. In a triangular melting regime, where the degree of partial melting decreases to zero at the base, it is possible that sulphur saturation may be reached. Assuming that the degree of partial melting varies linearly with depth, that the initial sulphur content of the mantle is constant through the melt column and the maximum degree of partial melting is 35%, then at 4.0 GPa the degree of partial melting would be 14%, and the resulting magma concentration for a source containing 250 ppm sulphur would be 1786 ppm (assuming sulphide exhaustion), requiring unrealistically high temperatures in excess of 1800°C to remain undersaturated. However, there are several problems with this analysis. First, at 14% partial melting, the sulphides in the source may or may not have been exhausted. As noted above, it has been estimated that the *maximum* degree of partial melting at which sulphides are exhausted in a mantle source is approximately 20–23%, although the exact point at which this occurs will be model dependent (i.e. source composition, modal mineralogy, etc.). If we assume that sulphide is exhausted at 20% and all of the sulphur is in the sulphides, then sulphur concentrations could be much lower. Another factor to take into account is that, due to the relative homogeneity of the OJP basalts, the magma(s) that produced them must have been well mixed. If the resulting basalts indicate that the *average* degree of partial melting was 30% (Fitton & Godard 2004), then the maximum degree of melting in a triangular melting regime would have to be much higher (*c.* 40–50%?). Thus, at 4.0 GPa, assuming partial melting is linear over the melting depths, the degree of partial melting would be 16–20%, the sulphur concentration would be 1250–1563 ppm for a 250 ppm source (assuming source exhaustion of sulphides) and would require temperatures of approximately 1680–1780°C for sulphur to remain undersaturated. These temperatures exceed even mantle potential temperature of Farnetani & Richards (1994). In addition, the maximum degree of partial melting would be unreasonably high. Finally, we envision the source of the OJP to have risen while it melted, such that a given part of the source actually melts over a range of depths and pressures. Thus, at melt initiation, the degree of partial melting may be smaller than the average, but during ascent it is envisaged that the melt produced would have been well mixed, thus producing basalts with compositions consistent with derivation by an average of 30% partial melting.

Supporting evidence for the calculated temperature excess comes from the fossil geotherm preserved by megacrysts and peridotite xenoliths found in pipe-like intrusives of alnöite cropping out on the island of Malaita, Solomon Islands (Allen & Deans 1965; Nixon & Coleman 1978). The geotherm thus recorded gives a temperature at 4.0 GPa of 1320–1350°C (Nixon & Coleman 1978; Nixon and Boyd 1979). Phase relations in the peridotite xenoliths are consistent with a cooling event in the mantle prior to inclusion in the erupting alnöite. For example, spinel in some of the peridotite xenoliths has a corona of garnet and such textures have been interpreted as indicating a cooling of the mantle (Neal & Nixon 1985; Nixon & Neal 1987), which occurred after the dissipation of heat that produced the large degrees of partial melting needed to form the OJP. Hence, while the temperatures recorded in the fossil

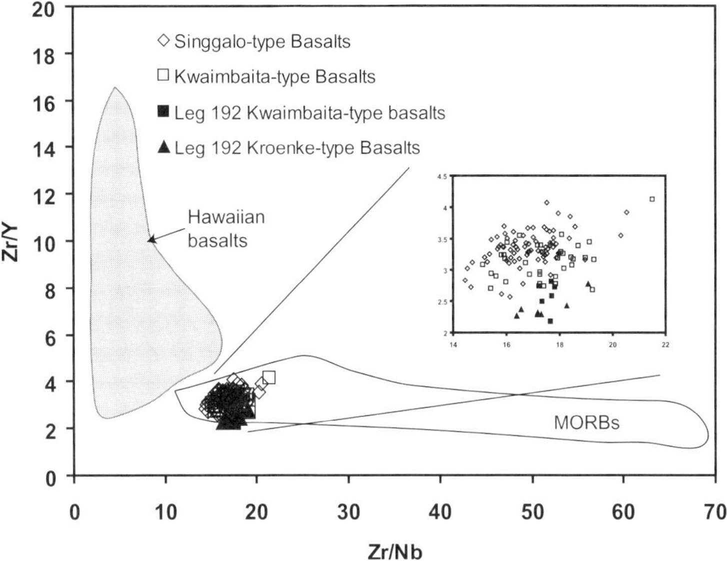

Fig. 7. Zr/Y v. Zr/Nb for the OJP basalts. The inset is an expanded view of the OJP sample set. New data from Leg 192 are similar to those of previous studies. Leg 192 basalts, like other OJP basalts, exhibit relatively limited variability in Zr/Y and Zr/Nb, which can be generated by initial melting in the garnet stability field followed by continued melting in the spinel stability field as the plume head continued to rise. See text for discussion. Data are from Tejada *et al.* (1996, 2002) and Fitton & Godard (2004). The MORB field represents East Pacific Rise MORB (J. Sinton unpublished data). The Hawaiian basalt field is generated by data from references too numerous for them all to be included here.

geotherm are about 150°C cooler than the estimates presented here, they are consistent with a large temperature perturbation associated with OJP formation.

Source modelling

Previous source modelling used a hybrid source composed of lower- and upper-mantle components to generate the OJP basalts (Mahoney *et al.* 1993; Farnetani & Richards 1994; Farnetani *et al.* 1996; Neal *et al.* 1997; Tejada *et al.* 2002; Ely & Neal 2003). New estimates indicate that the maximum degree of partial melting that formed OJP basalts may have been as high as 35%, although the average is still about 30% (Fitton & Godard 2004). The new samples from Leg 192 can now be used to refine the overall mineralogy of the proposed hybrid source. For example, Leg 192 basalts exhibit relatively limited variability in Zr/Y and Zr/Nb (Fig. 7). Neal *et al.* (1997) showed that such limited ranges in these ratios could be generated by initial melting in the garnet stability field of a composite source (primitive (lower) mantle ± MORB source) followed by melting of this source in the spinel stability field as the plume head surfaced. Following this, Ely & Neal (2003) successfully modelled the PGE abundances and patterns in OJP basalts exposed on the Solomon Islands using this composite source but with the addition of 0–0.5% of outer core material. Recently, it has been argued that there is no isotopic evidence for the presence of a depleted MORB source component in the OJP basalts (Tejada *et al.* 2002). However, if the plume model is to be used to explain OJP origin, entrainment of some upper-mantle material into the plume head would have occurred (cf. Campbell & Griffiths 1990), even though it may not be depleted MORB source. Therefore, the model presented here assumes an average degree of partial melting of 30% (cf. Fitton & Godard 2004), which is achieved through 10% melting of a garnet peridotite of primitive mantle composition, which then rises through the garnet–spinel transition and melts a further 20%. In addition to this plume component, an average fertile upper-mantle spinel lherzolite (Rekhämper *et al.* 1999) representing entrained upper mantle is melted to 30% and mixed into the plume melt at a ratio of 1: 19. The primitive mantle composition of McDonough & Sun (1995) is used. Source mineralogy for the garnet lherzolite is

Table 2. *Partition coefficients and concentrations (ppb) for the OJP basalt modelling*

	Ir	Ru	Rh	Pt	Pd	Reference*
Olivine	0.77	1.7	1.8	0.08	0.03	1
OPX	1.8	1.9	0.8	2.2	0.3	2
CPX	1.8	1.9	0.8	0.8	0.3	3
Cr-spinel	100	151	63	3.3	1.6	4
Sulphide	4400	2400	3000	6900	6300	5
Plagioclase	0.3	0.3	0.4	0.3	0.2	6
Ti-mag	500	300	130	3	1.1	7
Spinel lherzolite	2.46	4.29	0.8	4.34	2.48	8
Primitive mantle	3.2	5.0	0.9	7.1	3.9	9
Outer core	188	669	134	776	572	10

* References for data are as follows: 1, Capobianco *et al.* (1994), Puchtel & Humayun (2001); 2, Ely & Neal (2003); 3, Capobianco *et al.* (1990, 1991), Capobianco & Drake (1994); 4, Capobianco *et al.* (1990), Puchtel & Humayun (2001); Capobianco et al. (1990); 5, Bezmen *et al.* (1994), Peach *et al.* (1994), Tredoux *et al.* (1995); 6, Capobianco *et al.* (1990, 1991), Capobianco & Drake (1994); 7, Capobianco *et al.* (1994), Neal *et al.* (2002); 8, Rehkämper *et al.* (1997); 9, McDonough & Sun (1995); 10, Snow & Schmidt (1998).
OPX, orthopyroxene; CPX, clinopyroxene; Ti-mag, titanomagnetite.

taken to be olivine, orthopyroxene, clinopyroxene, sulphide and garnet (60: 14.94: 20: 0.06: 5). For both compositions of spinel lherzolite, the mineralogy was taken to be olivine, orthopyroxene, clinopyroxene, sulphide and Cr-spinel (60: 14.94: 20: 0.06: 5). The modal mineralogies are similar to the peridotite xenoliths brought to the surface in the alnöite pipes on Malaita, Solomon Islands (Neal 1986; Neal & Davidson 1988) The model employs non-modal batch melting for all sources, and initial melting ratios for garnet peridotite (36.9: 20: 27.5: 0.6: 15) and spinel lherzolite (34.4: 20: 25: 0.6: 20) both reflect sulphide completely entering the melt at 10% melting. It should be noted that partition coefficients are not available for all of the PGEs in orthopyroxene or garnet. For orthopyroxene, it is assumed that partition coefficients are comparable to clinopyroxene (cf. Ely & Neal 2003). For modelling purposes, the PGEs are assumed to be perfectly incompatible in garnet (K_D = 0; where K_D is the bulk distribution coefficient). While this may not be strictly true, Mitchell & Keays (1981) suggested that the PGEs partition very poorly into garnet, and therefore that any contribution to the bulk distribution coefficient would be swamped by those of other phases. A complete list of partition coefficients used in the modeling can be found in Table 2.

PGE partitioning and fractionation

The three different groups of basalts from the OJP have different phenocryst assemblages. The Kroenke-type basalts from Leg 192 are olivine-phyric, with olivine phenocryst abundances of up to 10 modal% (Sano & Yamashita 2004). These basalts also contain small Cr-spinels that are commonly included in the olivine phenocrysts. Kwaimbaita-type basalts from Leg 192 have significantly less olivine, ranging from 1–3 modal%, but also have occurrences of clinopyroxene and plagioclase phenocrysts, usually less than 2 modal% each. Cr-spinel is typically absent. Singgalo-type basalts, although not recovered by Leg 192, contain only rare olivine phenocrysts, but more commonly are sparsely clinopyroxene–plagioclase-phyric (Petterson 1995; Ely & Neal 2003). Published partition coefficients (Table 2) indicate that Ru and Rh are slightly–moderately compatible in both olivine and clinopyroxene (Table 2). Positive correlations of Ru with Cr and Ni (Fig. 4a, b) support such partition coefficients. Similar positive correlations are also seen with Ir (Fig. 4c, d), and we suspect this is due to the influence of Cr-spinel (cf. Puchtel & Humayun 2001; Righter *et al.* 2004).

The ODP Leg 192 Kroenke-type basalts, with 8–11 wt% MgO and approximately 10 modal% olivine, are obviously less evolved than other OJP basalt types and confirm that spinel was an early fractionating phase. Experimental results indicate that the major-phase fractionating sequence for the leg 192 basalts was olivine, olivine + plagioclase, olivine + clinopyroxene + plagioclase (Sano & Yamashita 2004). Clinopyroxene–plagioclase–titanomagnetite cumulates from the Solomon Islands indicate that additional stages of crystallization may have occurred for some parts of the OJP (Neal *et al.*

1997; Ely & Neal 2003). Thus, a combination of these models was employed in this study. Differentiation was assumed to have been accomplished in five stages of fractional crystallization, each crystallizing 10% of the magma volume (cf. Neal *et al.* 1997). The stages proceed sequentially in the following manner: olivine (100%); olivine (95%) + Cr-spinel (5%); olivine (50%) + plagioclase (50%); olivine (20%) + clinopyroxene (60%) + plagioclase (20%); clinopyroxene (60%) + plagioclase (39%) + titanomagnetite (1%).

Model results

Figure 8a shows the results for the mixed source parental magma and the effect of the fractionation sequence on the PGEs. Variation in the proportions of silicate phases during crystal fractionation has only a minor effect on the PGE model results; the opposite is true for the oxide minerals (Table 2). The primary melt generated by our model falls within the field of the OJP basalts, but cannot account for the range of abundances or Pt/Ru and Pt/Ir ratios, even if the degree of partial melting is varied by 10–15%. Given that the Kroenke-type basalts were fractionating olivine and spinel immediately before eruption, the model clearly underestimates Ir, Ru, Rh and Pt for these lavas. The model Pd abundances are within error of the abundances determined in the Leg 192 samples.

When 1% of outer core material (cf. Snow & Schmidt 1998; see Table 2) is included (Fig. 8b), the primary magma thus generated has PGE abundances of similar magnitude to the most enriched Leg 192 samples. The slight overestimate for Ru is inconsequential, as at least some crystal fractionation must have occurred before eruption of the Kroenke-type samples in order to reduce the MgO abundance to 11 wt% (the highest MgO value; sample 192-1185B-6R-4, 101–120 cm, Unit 5). It is likely that Cr-spinel fractionation will decrease $[Ru/Ir]_{pm}$, based on published partition coefficients (Puchtel & Humayun 2001; see Table 2), which results in a characteristic 'kick-up' in Ir on the normalized PGE profiles. It is possible, however, that the 'kick-up' on Ir is a result of the sample concentration being close to the LOD for the analysis, resulting in larger uncertainties than for the other PGEs. The fractionated magmas properly estimate the highest Pt abundances for the OJP basalts. Late-stage fractionation of titanomagnetite can also help increase $[Ru/Ir]_{pm}$, resulting in the relatively flat transitions from Ir to Ru seen in some OJP samples (i.e. compare the Kroenke-type basalts to the Kwaimbaita-type

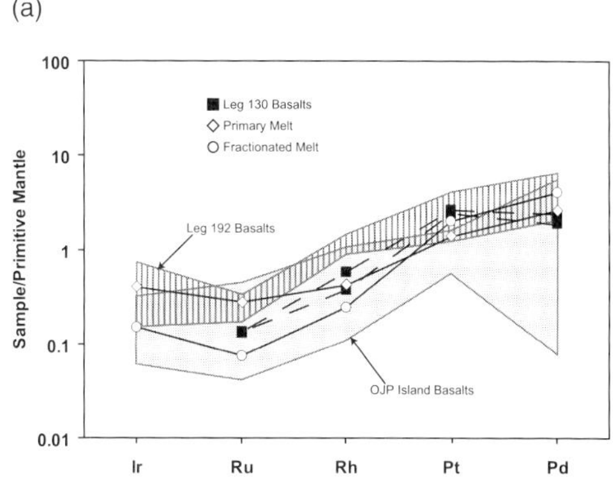

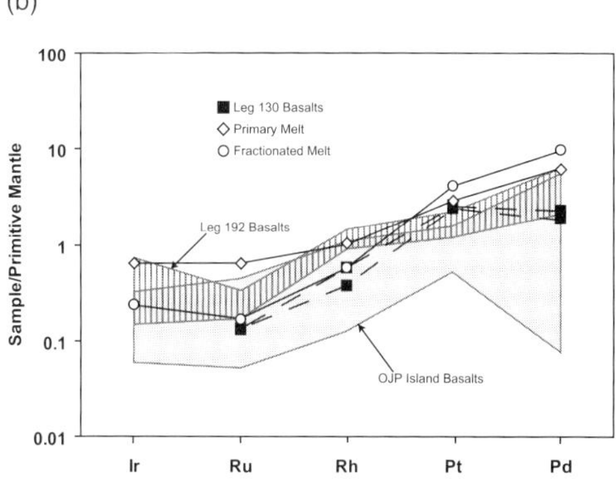

Fig. 8: (**a**) Modelling results for the PGE using a composite source. The parental melt contains a mixture of melts from a plume component (95%) and a fertile upper-mantle component (5%). The plume component is a 10% melt of a garnet peridotite, which then passes into the spinel stability field and is melted a further 20%. Added to this is a 30% melt of spinel lherzolite of fertile upper-mantle composition. This models a surfacing plume head. Crystal fractionation is accomplished in five stages of 10% of each of the following: olivine; olivine + Cr-spinel; olivine + plagioclase; olivine + clinopyroxene + plagioclase; clinopyroxene + plagioclase + titanomagnetite. For full details, see text. This model approximates the compositions of basalts with relatively low PGE abundances, but fails to generate the basalts with higher abundances. (**b**) Model using the same parameters as in (a), except for an addition of 1% outer core material to the sources with otherwise primitive mantle composition. All of the PGE abundances in the OJP basalts are successfully modelled, assuming that the outer core material varies from 0 to 1%, and that the degree of fractionation may also vary; the Leg 192 basalts, for instance, only involve three of the five stages of fractional crystallization.

basalts at Site 1185, Hole B; Fig. 3c, d). While we recognize the model-dependence of our results, our modelling is consistent with the conclusion of Ely & Neal (2003): a small amount of an outer core component in the OJP source is necessary to account for the observed PGE abundances in some, but not all, of the erupted basalts. The maximum amount of outer core (1%) is double that proposed in the model of Ely & Neal (2003), as a result of the generally greater PGE abundances of the Leg 192 basalts, especially the Kroenke-type. As only some basalts require this component to account for PGE abundances, it is likely to have been heterogeneously distributed in the otherwise relatively homogeneous OJP source. Inclusion of 1% outer core material (assuming it to be approximately 100% Fe metal) is unlikely to have a definitive effect on Fe abundances in the OJP basalts because, at its maximum, it would contribute about 1 wt% FeO to the magma; OJP basalts analysed as part of this study have Fe (as Fe_2O_3) ranging from 10.3 to 12.6 wt% (Table 1). The model developed above is also consistent with the Re–Os isotopic analyses of Leg 192 basalts. Parkinson *et al.* (2001) concluded that in the limited number of samples they analysed a core component was not evident, at least from Re–Os isotopes. Our modelling demonstrates that *some* of the basalts do not require this outer core component to account for the PGE abundances. While we recognize the dependence of our model on the parameters chosen, it is evident that, overall, that this model successfully accounts for all measured PGE abundances in the OJP basalts from ODP Leg 192, it is consistent with other models that replicate major- and trace-element abundances for the OJP (e.g. Neal *et al.* 1997; Tejada *et al.* 2002), and it does not violate isotopic or petrological considerations. Basically, the PGE concentrations from ODP Legs 130 and 192 basalts are consistent with a heterogeneously distributed outer core component in the OJP source region. The sample from Site 803, which is the only sample analysed from the 90 Ma-event, is in the middle of the range for the OJP and can be adequately modelled without the use of a core component (Fig 8a). While it is difficult to draw the conclusion that the source for the 90 Ma-event contained no core component based on one sample, it may be a worthwhile avenue for future investigation. For example, is the 90 Ma-event the result of a separate plume or is it the result of remelting of the 'fossil' plume head beneath the OJP (cf. Richardson *et al.* 2000; Klosko *et al.* 2001)? Within the constraints of the modelling presented above, the core component within the original plume material would have been exhausted during the *c.* 122 Ma melting event.

Summary and conclusions

- The OJP basalts form a continuum of PGE compositions and, as such, no discrete groups can be defined on the basis of PGE abundances or ratios as is possible with radiogenic isotope ratios and incompatible trace elements.
- Although the PGE compositions of the OJP basalts are similar even if they are of different ages, the 90 Ma sample from ODP Site 803 can be modelled without the incorporation of a core component.
- Unlike the OJP basalts from ODP Leg 130 and the Solomon Islands, low-temperature alteration does not appear to have affected the Leg 192 basalt Pd abundances.
- The melt that produced the OJP basalts was not saturated with respect to sulphur; therefore, the PGEs exhibit lithophile behaviour with oxide phases exerting the greatest influence.
- At the time of melting, the minimum initial excess temperature of the OJP plume head was +185–+235°C (above ambient), with a minimum temperature range of 1465–1515°C at 3.5–4.0 GPa. These results are consistent with other estimates for OJP plume-head temperatures, including that of Fitton & Godard (>1500°C, 2004). The results of this study are also broadly consistent with the fossil geotherm defined by megacrysts and peridotite xenoliths from alnöite pipes on Malaita, Solomon Islands.
- In order to generate the observed range of PGE abundances in ODP Leg 192 basalts, a small outer core component is required in some but not all samples, suggesting that this component is heterogeneously distributed in the OJP basalt source.

We are deeply indebted to the officers and crew of the R.V. *JOIDES Resolution* for the help and professionalism in obtaining drill cores during ODP Leg 192. We would also like to thank J. Seidler for assistance in sample preparation, and J. Shafer for assistance with ICP-AES analyses. Thoughtful and thorough reviews by M. Rehkämper, G. Fitton and G. Pearson greatly improved this contribution. The ODP is sponsored by the US National Science Foundation (NSF) and participating countries under the management of Joint Oceanographic Institutions (JOI), Inc. Funding for this research was provided by grants from the United States Science Support Program to C.R. Neal.

References

ALLEN, J.B. & DEANS, T. 1965. Ultrabasic eruptives with alnöitic–kimberlitic affinities from Malaita, Solomon Islands. *Mineralogical Magazine*, **34**, 16–34.

ANDERSON, D.L., ZHANG, Y.-S. & TANIMOTO, T. 1992. Plume heads, continental lithosphere, flood basalts, and tomography. *In*: STOREY, B.C., ALABASTER, T. & PANKHURST, R.J. (eds) *Magmatism and Causes of Continental Break-up*. Geological Society, London, Special Publications, **68**, 99–124.

ARNDT, N.T. 1986. Differentiation of komatiite flows. *Journal of Petrology*, **27**, 279–301.

ARNDT, N.T. & NESBITT, R.W. 1984. Magma mixing in komatiitic lavas from Munro Township, Ontario. *In*: KRÖNER, A., HANSON G.N. & GOODWIN, A.M. (eds) *Archaen Geochemistry*, Springer-Verlag, Berlin, 99–114.

BARNES, S.-J., NALDRETT, A.J. & GORTON, M.P. 1985. The origin of fractionation of platinum group elements in terrestrial magmas. *Chemical Geology*, **53**, 303–323.

BENNETT, V.C., NORMAN, M.D. & GARCIA, M.O. 2000. Rhenium and platinum group element abundances correlated with mantle source components in Hawaiian picrites: sulphides in the plume. *Earth and Planetary Science Letters*, **183**, 513–526.

BERCOVICI D. & MAHONEY J.J. 1994. Double flood-basalt events and the separation of mantle-plume heads at the 660 km discontinuity. *Science*, **266**, 1367–1369.

BEZMEN, N.I., ASIF, M., BRÜGMANN, G.E., ROMANENKO, I.M. & NALDRETT, A.J. 1994. Distribution of Pd, Rh, Ru, Ir, Os, and Au between sulphide and silicate melts. *Geochimica et Cosmochimica Acta*, **58**, 1251–1260.

BIRKHOLD, A. 2000. *A geochemical investigation of San Cristobal (Makira) basement: evidence for the presence of Ontong Java Plateau crust. PhD Thesis*, University of Notre Dame.

BLICHERT-TOFT, J., FREY, F.A. & ALBARÈDE, F. 1999. Hf isotope evidence for pelagic sediments in the source of Hawaiian basalts. *Science*, **285**, 879–882.

BRANDON, A.D., NORMAN, M.D., WALKER, R.J. & MORGAN, J.W. 1999. ^{186}Os–^{187}Os systematics of Hawaiian picrites. *Earth and Planetary Science Letters*, **174**, 25–42.

BRÜGMANN, G.E., ARNDT, N.T., HOFMANN, A.W. & TOBSCHALL, H.J. 1987. Noble metal abundances in komatiite suites from Alexo, Ontario, and Gorgona Island, Colombia. *Geochimica et Cosmochimica Acta*, **51**, 2159–2169.

BRÜGMANN, G.E., NALDRETT, A.J., ASIF, M., LIGHTFOOT, P.C., GORBACHEV, N.S. & FERORENKO, V.A. 1993. Siderophile and chalcophile metals as tracers of the evolution of the Siberian Trap in the Noril'sk region, Russia. *Geochimica et Cosmochimica Acta*, **57**, 2001–2018.

CAMPBELL, I.H. & GRIFFITHS, R.W. 1990. Implications of mantle plume structure for the evolution of flood basalts. *Earth and Planetary Science Letters*, **99**, 79–93.

CAPOBIANCO, C.J. & DRAKE, M.J. 1994. Partitioning and solubility of PGEs in oxides and silicates. *Mineralogical Magazine*, **58A**, 144–145.

CAPOBIANCO, C.J., DRAKE, M.J. & ROGERS, P.S.Z. 1990. Experimental solubilities and partitioning behavior of noble metals among lithophile magmatic phases. *Lunar and Planetary Science*, **21**, 166–167.

CAPOBIANCO, C.J., DRAKE, M.J. & ROGERS, P.S.Z. 1991. Crystal/melt partitioning of Ru, Rh, and Pd for silicate and oxide basaltic liquidus phases. *Lunar and Planetary Science*, **22**, 179–180.

CAPOBIANCO, C.J., HERVIG, R.L. & DRAKE, M.J. 1994. Experiments on crystal/liquid partitioning of Ru, Rh, and Pd for magnetite and hematite solid solutions crystallized from silicate melt. *Chemical Geology*, **113**, 23–43.

CHALMERS, J.A., LARSEN, L.M. & PEDERSEN, A.K. 1995. Widespread Paleocene volcanism around the northern North Atlantic and Labrador Sea: evidence for a large, hot early, plume head. *Journal of the Geological Society, London*, **152**, 965–969.

CLAGUE, D.A. & DALRYMPLE, G.B. 1987. The Hawaiian–Emperor volcanic chain; Part I, Geologic evolution. *In*: DECKER, R.W., WRIGHT, T.L. & STAUFFER, P.H. (eds) *Volcanism in Hawaii*. US Geological Society, Professional Paper, **5–73**.

CLAGUE, D.A. & DALRYMPLE, G.B. 1989. Tectonics, geochronology, and origin of the Hawaiian–Emperor volcanic chain. *In*: WINTERER, E.L., HUSSONG, D.M. & BECKER, R.W. (eds) *The Eastern Pacific Ocean and Hawaii*, Geological Society of America, Denver, 188–217.

COFFIN, M.F. & ELDHOLM, O. 1993. Scratching the surface: Estimating dimensions of large igneous provinces. *Geology*, **21**, 515–518.

COFFIN, M.F. AND ELDHOLM, O. 1994. Large igneous provinces: Crustal structure, dimensions, and external consequences. *Reviews in Geophysics*, **32**, 1–36.

COLEMAN, P.J. AND KROENKE, L.W. 1981. Subduction without volcanism in the Solomon Islands arc. *Geo-Marine Letters*, **1**, 129–134.

DEVEY, C.W., GARBE-SCHÖNBERG, C.-D., STOFFERS, P., CHAUVEL, C. & MERTZ, D.F. 1994. Geochemical effects of dynamic melting beneath ridges: Reconciling major and trace element variations in Kolbeinsey (and global) mid-ocean ridge basalt. *Journal of Geophysical Research*, **99**, 9077–9095.

DUNCAN, R.A. 2002. A time frame for the construction of the Kerguelen Plateau and Broken Ridge. *Journal of Petrology*, **43**, 1109–1119.

ELY, J.C. & NEAL, C.R. 2002. Method of data reduction and uncertainty estimation for platinum-group element (PGE) data using inductively coupled plasma-mass spectrometry (ICP-MS). *Geostandards Newsletter*, **26**, 31–39.

ELY, J.C. AND NEAL, C.R. 2003. Using platinum-group elements to investigate the origin of the Ontong Java Plateau, SW Pacific. *Chemical Geology*, **196**, 235–257.

ELY, J.C., NEAL, C.R., O'NEILL, J.A. & JAIN, J.C. 1999. Quantifying the platinum group elements (PGEs) and gold in geological samples using cation exchange pretreatment and ultrasonic nebulization inductively coupled plasma-mass spectrometry (USN-ICP-MS). *Chemical Geology*, **157**, 219–234.

FARNETANI, C.G. & RICHARDS, M.A. 1994. Numerical investigations of the mantle plume initiation model for flood basalts events. *Journal of Geophysical Research*, **99**, 13 813–13 833.

FARNETANI, C.G., RICHARDS, M.A. & GHIORSO, M.S. 1996. Petrological models of magma evolution and deep crustal structure beneath hotspots and flood basalt provinces. *Earth and Planetary Science Letters*, **143**, 81–94.

FITTON, J.G., SAUNDERS, A.D., NORRY, M.J., HARDASON, B.S. & TAYLOR, R.N. 1997. Thermal and chemical structure of the Iceland plume. *Earth and Planetary Science Letters*, **153**, 197–208.

FITTON, J.G. & GODARD, M. 2004. Origin and evolution of magmas on the Ontong Java Plateau. *In*: FITTON, J.G., MAHONEY, J.J., WALLACE, P.J. & SAUNDERS, A.D. (eds) *Origin and Evolution of the Ontong Java Plateau*. Geological Society, London, Special Publications, **229**, 151–178.

FLEET, M.E., CROCKET, J.H. & STONE, W.E. 1996. Partitioning of platinum-group elements (Os, Ir, Ru, Pt, Pd) and gold between sulphide liquid and basalt melt. *Geochimica et Cosmochimica Acta*, **60**, 2397–2412.

FLEET, M.E., STONE, W.E. & CROCKET, J.H. 1991. Partitioning of palladium, iridium, and platinum between sulphide liquid and basalt melt: effects of melt composition, concentration, and oxygen fugacity. *Geochimica et Cosmochimica Acta*, **55**, 2545–2554.

FREY, F.A. & WEIS, D. 1995. Temporal evolution of the Kerguelen plume: geochemical evidence from ~38 to 82 Ma lavas forming the Ninetyeast Ridge. *Contributions to Mineralogy and Petrology*, **121**, 12–28.

FRYER, B.J. & GREENOUGH, J.D. 1992. Evidence for mantle heterogeneity from platinum-group element abundances in Indian Ocean basalts. *Canadian Journal of Earth Science*, **29**, 2329–2340.

GRIFFITHS, R.W. & CAMPBELL, I.H. 1990. Stirring and structure in mantle starting plumes. *Earth and Planetary Science Letters*, **99**, 66–78.

HAURI, E.H., LASSITER, J.C. & DEPAOLO, D.J. 1996. Osmium isotope systematics of drilled lavas from Mauna Loa, Hawaii. *Journal of Geophysical Research*, **101**, 11 793–11 806.

HOLZHEID, A. & GROVE, T.L. 2002. Sulfur saturation limits and their implications for core formation scenarios for terrestrial planets. *American Mineralogist*, **87**, 227–237.

JENNER, G.A., LONGERICH, H.P., JACKSON, S.E. & FRYER, B.J. 1990. ICP-MS – A powerful tool for high-precision trace-element analysis in Earth sciences: Evidence from analysis of selected U.S.G.S. reference samples. *Chemical Geology*. **83**, 133–148.

JONES, A.P., PRICE, G.D., PRICE. N.J., DECARLI, P.S. & CLEGG, R.A. 2002. Impact induced melting and the development of large igneous provinces. *Earth and Planetary Science Letters*, **202**, 551–561.

KEAYS, R.R. 1995. The role of komatiitic and picritic magmatism and S-saturation in the formation of ore deposits. *Lithos*, **34**, 1–18.

KLOSKO, E.R., RUSSO, R.M., OKAL, E.A. & RICHARDSON, W.P. 2001. Evidence for a rheologically strong chemical mantle root beneath the Ontong-Java Plateau. *Earth and Planetary Science Letters*, **186**, 347–361.

KURZ, M.D., KENNA, T.C., LASSITER, J.C. & DEPAOLO, D.J. 1996. Helium isotopic evolution of Mauna Kea volcano: First results from the 1-km drill core. *Journal of Geophysical Research*, **101**, 11 781–11 791.

LASSITER, J.C. & HAURI, E.H. 1998. Osmium-isotope variations in Hawaiian lavas: evidence for recycled oceanic lithosphere in the Hawaiian plume. *Earth and Planetary Science Letters*, **164**, 483–496.

LASSITER, J.C., DEPAOLO, D.J. & TATSUMOTO, M. 1996. Isotopic evolution of Mauna Kea volcano: results from the initial phase of the Hawaii Scientific Drilling Project. *Journal of Geophysical Research*, **101**, 11 769–11 780.

LEE, C.-T.A., BRANDON, A.D. & NORMAN, M. 2003. Vanadium in peridotites as a proxy for paleo-fO_2 during partial melting: Prospects, limitations, and implications. *Geochimica et Cosmochimica Acta*, **67**, 3045–3064.

LIGHTFOOT, P.C., NALDRETT, A.J., GORBACHEV, N.S., DOHERTY, W. & FEDORENKO, V.A. 1990. Geo chemistry of the Siberian Trap of the Noril'sk area, USSR, with implications for the relative contributions of crust and mantle to flood basalt magmatism. *Contributions to Mineralogy and Petrology*, **104**, 631–644.

MACLENNAN, J., MCKENZIE, D. & GRONVÖLD, K. 2001. Plume-driven upwelling under central Iceland. *Earth and Planetary Science Letters*, **194**, 67–82.

MAHONEY, J.J. 1987. An isotopic survey of Pacific oceanic plateaus: implications for their nature and origin. *In*: KEATING, B., FRYER, P., BATIZA, R. & BOELHERT, G. (eds) *Seamounts, Islands, and Atolls*. American Geophysical Union, Geophysical Monograph, **43**, 207–220.

MAHONEY, J.J. & SPENCER, K.J. 1991. Isotopic evidence for the origin of the Manihiki and Ontong Java oceanic plateaus. *Earth and Planetary Science Letters*, **104**, 196–210.

MAHONEY, J.J., FITTON, J.G., WALLACE, P.J. *et al.* 2001. Basement drilling of the Ontong Java Plateau, Sites 1183–1187, 8 September–7 November, 2000. *Proceedings of the Ocean Drilling Program, Initial Reports*, **192**.

MAHONEY, J.J., STOREY, M., DUNCAN, R.A., SPENCER, K.J. & PRINGLE, M. 1993. Geochemistry and age of the Ontong Java Plateau. *In*: *The Mesozoic Pacific: Geology, Tectonics, and Volcanism*, American Geophysical Union, Geophysical Monograph, **77**, 233–261.

MAVROGENES, J.A. & O'NEILL, H.S.C. 1999. The relative effects of pressure, temperature and oxygen fugacity on the solubility of sulphide in mafic magmas. *Geochimica et Cosmochimica Acta*, **63**. 1173–1180.

MCDONOUGH, W.F. & SUN, S.-S. 1995. The composition of the Earth. *Chemical Geology*, **120**, 223–253.

MCKENZIE, D.P. AND BICKLE, M.J. 1988. The volume

and composition of melt generated by extension of the lithosphere. *Journal of Petrology*, **29**, 625–679.

MICHAEL, P.J. & CORNELL, W.C. 1996. H_2O, CO_2, Cl, and S contents in 122 Ma glasses from Ontong Java Plateau, ODP 807C: implications for mantle and crustal processes. *Eos, Transactions of the American Geophysical Union*, **77**, Fall Meeting Suppl., F714.

MITCHELL, R.H. & KEAYS, R.R. 1981. Abundance and distribution of gold, palladium, and iridium in some spinel and garnet lherzolites: implications for the nature and origin of precious metal-rich intergranular components in the upper mantle. *Geochimica et Cosmochimica Acta*, **45**, 2425–2442.

NEAL, C.R. 1986. *Mantle studies in the western Pacific and kimberlite-type intrusives*. PhD Thesis, University of Leeds.

NEAL, C.R. 2001. The interior of the moon: The presence of garnet in the primitive, deep lunar mantle. *Journal of Geophysical Research*, **106**, 27 865–27 885.

NEAL, C.R. & DAVIDSON, J.P. 1989. An unmetasomatized source for the Malaitan alnöite (Solomon Islands): Petrogenesis involving zone refining, megacryst fractionation, and assimilation of oceanic lithosphere. *Geochimica et Cosmochimica Acta*, **53**, 1975–1990.

NEAL C.R. & NIXON P.H. 1985. Spinel–garnet relationships in mantle xenoliths from the Malaitan alnöite. *Transactions of the Geological Society of South Africa*, **88**, 347–354.

NEAL, C.R., MAHONEY, J.J. & CHAZEY, W.J., III. 2002. Mantle sources and the highly variable role of continental lithosphere in basalt petrogenesis of the Kerguelen Plateau and Broken Ridge LIP: Results from ODP Leg 183. *Journal of Petrology*, **43**, 1177–1205.

NEAL, C.R., MAHONEY, J.J., KROENKE, L.W., DUNCAN, R.A. & PETTERSON, M.G. 1997. The Ontong Java Plateau. *In*: *Large Igneous Provinces: Continental, Oceanic, and Planetary Flood Volcanism*. American Geophysical Union, Geophysical Monograph, **100**, 183–216.

NIXON, P.H. & BOYD, F.R. 1979. Garnet-bearing lherzolites and discrete nodule suites from the Malaita alnöite, Solomon Islands, SW Pacific, and their bearing on oceanic mantle composition and geotherm. *In:* BOYD, F.R. & MEYER, H.O.A. (eds) *The Mantle Sample: Inclusions in Kimberlites and Other Volcanics*. American Geophysical Union, Washington, DC, 400–423.

NIXON, P.H. & COLEMAN, P.J. 1978. Garnet-bearing lherzolites and discrete nodule suites from the Malaita alnöite, Solomon Islands, and their bearing on the nature and origin of the Ontong Java Plateau. *Australian Society of Exploration Geophysical Bulletin*, **9**, 103–106.

NIXON, P.H. & NEAL, C.R. 1987. Ontong Java Plateau: Deep seated xenoliths from thick oceanic lithosphere. *In*: NIXON, P.H. (ed.) *Mantle Xenoliths*. Wiley, Chichester, 335–345.

NORMAN, M.D. & GARCIA, M.O. 1999. Primitive magmas and source characteristics of the Hawaiian plume: petrology and geochemistry of shield picrites. *Earth and Planetary Science Letters*, **168**, 27–44.

PARKINSON, I.J., SCHAEFER, B.F. & OJP LEG 192 SHIPBOARD SCIENTIFIC PARTY. 2001. A lower mantle origin for the world's biggest LIP? A high precision OS isotope isochron from Ontong Java Plateau basalts drilled on OJP Leg 192. *Transactions of the American Geophysical Union*, **82**, Fall Meeting Supplement, F1398.

PEACH, C.L., MATHEZ, E.A. & KEAYS, R.R. 1990. Sulphide melt-silicate melt distribution coefficients for noble metals and other chalcophile elements as deduced from MORB: Implications for partial melting. *Geochimica et Cosmochimica Acta*, **54**, 3379–3389.

PEACH C.L., MATHEZ E.A., KEAYS, R.R. & REEVES, S.J. 1994. Experimentally determined sulphide melt–silicate melt partition coefficients for iridium and palladium. *Chemical Geology*, **117**, 361–377.

PETTERSON, M.G. 1995. *The Geology of North and Central Malaita, Solomon Islands. (Including Implications of Geological Research on Makira, Savo, Isabel, Guadalcanal, and Choiseul Between 1992 and 1995)*. **1/95**.

PETTERSON, M.G., BABBS, T., NEAL, C.R., MAHONEY, J.J., SAUNDERS, A.D., DUNCAN, R.A., TOLIA, D., MAGU, R., QOPOTO, C., MAHOA, H. & NATOGGA, D. 1999. Geological-tectonic framework of Solomon Islands, SW Pacific: crustal accretion and growth within an intra-oceanic setting. *Tectonophysics*, **301**, 35–60.

PETTERSON, M.G., NEAL, C.R., MAHONEY, J.J., KROENKE, L.W., SAUNDERS, A.D., BABBS, T.L., DUNCAN, R.A., TOLIA, D. & MCGRAIL, B. 1997. Structure and deformation of north central Malaita, Solomon Islands: tectonic implications for the Ontong Java Plateau–Solomon arc collision, and for the fate of oceanic plateaus. *Tectonophysics*, **283**, 1–33.

PRINGLE, M.S. & DUNCAN, R.A. 2000. Basement ages from the southern and central Kerguelen plateau: Initial products of the Kerguelen Large Igneous Province. *Eos, Transactions American Geophysical Union 2000 Spring Meeting*, Supplement, **81**, S424.

PUCHTEL, I.S., HOFMANN, A.W., MEZGER, K., SHCHIPANSKY, A.A., KULIKOV, V.S. & KULIKOVA, V.V. 1996. Petrology of a 2.41 Ga remarkably fresh komatiitic basalt lava lake in Lion Hills, central Vetreny Belt, Baltic Shield. *Contributions to Mineralogy and Petrology*, **124**, 273–290.

PUCHTEL, I. & HUMAYUN, M. 2000. Platinum group elements in Kostomuksha komatiites and basalts: Implications for oceanic crust recycling and core–mantle interaction. *Geochimica et Cosmochimica Acta*, **64**, 4227–4242.

PUCHTEL, I. & HUMAYUN, M. 2001. Platinum group element fractionation in a komatiitic basalt lava lake. *Geochimica et Cosmochimica Acta*, **65**, 2979–2993.

REHKÄMPER, M., HALLIDAY, A.N., BARFOD, D., FITTON, J.G. & DAWSON, J.B. 1997. Platinum-group element abundance patterns in different mantle environments. *Science*, **278**, 1595–1598.

REHKÄMPER, M., HALLIDAY, A.N., FITTON, J.G., LEE, D.-C., WIENEKE, M. & ARNDT, N.T. 1999. Ir, Ru, Pt, and Pd in basalts and komatiites: New constraints for the geochemical behavior of the platinum-group elements in the mantle. *Geochimica et Cosmochimica Acta*, **63**, 3915–3934.

RICHARDS, M.A., DUNCAN, R.A. & COURTILLOT, V.E. 1989. Flood basalts and hot-spot tracks: plume heads and tails. *Science*, **246**, 103–107.

RICHARDSON, W.P., OKAL, E.A. & VAN DER LEE, S. 2000. Rayleigh-wave tomography of the Ontong Java Plateau. *Physics of Earth Planetary Interiors*, **118**, 29–51.

RIGHTER, K., CAMPBELL, A.J., HUMAYUN, M. & HERVIG, R.L. 2004. Partitioning of Ru, Rh, Pd, Re, Ir, and Au between Cr-spinel, olivine, pyroxene and silicate melts. *Geochimica et Cosmochimica Acta*, **68**, 867–880.

ROBERGE, J., WHITE, R.V. & WALLACE, P.J. 2004. Volatiles in submarine basaltic glasses from the Ontong Java Plateau (ODP Leg 192): implications for magmatic processes and source region compositions. *In*: FITTON, J.G., MAHONEY, J.J., WALLACE, P.J. & SAUNDERS, A.D. (eds) *Origin and Evolution of the Ontong Java Plateau*. Geological Society, London, Special Publications, **229**, 239–257.

ROGERS, G.C. 1982. Oceanic plateaus as meteorite impact signatures. *Nature*, **299**, 341–342.

ROY-BARMAN, M., WASSERBURG, G.J., PAPANASTASSIOU, D.A. & CHAUSSIDON, M. 1998. Osmium isotopic compositions and Re-Os concentrations in sulphide glodules from basaltic glasses. *Earth and Planetary Science Letters*, **154**, 331–347.

ROYER, J.-Y., PIERCE, J.W. & WEISSEL, J.K. 1991. Tectonic constraints on hotspot formation of the Ninetyeast Ridge. *Proceedings of the Ocean Drilling Program, Scientific Results*, **121**, 763–776.

SANO, T. & YAMASHITA, S. 2004. Experimental petrology of basement lavas from Ocean Drilling Program Leg 192: implications for differentiation processes in Ontong Java Plateau magmas. *In*: FITTON, J.G., MAHONEY, J.J., WALLACE, P.J. & SAUNDERS, A.D. (eds) *Origin and Evolution of the Ontong Java Plateau*. Geological Society, London, Special Publications, **229**, 185–218.

SHAFER J.S., NEAL C.R. & CASTILLO P.R. 2004. Compositional variability in lavas from the Ontong Java Plateau: results from basalt clasts within the volcaniclastic sequence of Ocean Drilling Program Site 1184. *In*: FITTON, J.G., MAHONEY, J.J., WALLACE, P.J. & SAUNDERS, A.D. (eds) *Origin and Evolution of the Ontong Java Plateau*. Geological Society, London, Special Publications, **229**, 333–351.

SHANNON, M.C. & AGEE, C.B. 1998. Percolation of core melts at lower mantle conditions. *Science*, **280**, 1059–1061.

SNOW, J.E. & SCHMIDT, G. 1998. Constraints on Earth accretion deduced from noble metals in the ocean mantle. *Nature*, **391**, 166–169.

SUN, S.-S. & MCDONOUGH, W.F. 1989. Chemical and isotopic systematics of oceanic basalts: Implications for mantle composition and processes. *In:* SAUNDERS, A.D. & NORRY, M.J. (eds) *Magmatism in the Ocean Basins*. Geological Society, London, Special Publications, **42**, 313–345.

TATSUMI, Y., OGURI, K. & SHIMODA, G. 1999. The behaviour of platinum-group elements during magmatic differentiation in Hawaiian tholeiites. *Geochemistry Journal*, **33**, 237–247.

TEJADA, M.L.G., MAHONEY, J.J., DUNCAN, R.A. & HAWKINS, M.P. 1996. Age and geochemistry of basement alkalic rocks of Malaita and Santa Isabel, Solomon Islands, Southern Margin of Ontong Java Plateau. *Journal of Petrology*, **37**, 361–394.

TEJADA, M.L.G., MAHONEY, J.J., NEAL. C.R., DUNCAN, R.A. & PETTERSON, M.G. 2002. Basement geochemistry and geochronology of central Malaita, Solomon Islands, with implications for the origin and evolution of the Ontong Java Plateau. *Journal of Petrology*, **43**, 449–484.

TEJADA, M.L.G., MAHONEY, J.J., CASTILLO, P.R., INGLE, S., SHETH, H.C. & WEIS, D. 2004. Pin-pricking the elephant: evidence on the origin of the Ontong Java Plateau from Pb-Sr-Hf-Nd isotopic characteristics of ODP Leg 192 basalts. *In*: FITTON, J.G., MAHONEY, J.J., WALLACE, P.J. & SAUNDERS, A.D. (eds) *Origin and Evolution of the Ontong Java Plateau*. Geological Society, London, Special Publications, **229**, 133–150.

TREDOUX, M., LINDSAY, N.M., DAVIES, G. & MCDONALD, I. 1995. The fractionation of platinum-group elements in magmatic systems, with the suggestion of a novel causal mechanism. *South African Journal of Geology*, **98**, 157–167.

VIDALE, J.E. & HEDLIN, M.A.H. 1998. Evidence for partial melt at the core–mantle boundary north of Tonga from the strong scattering of seismic waves. *Nature*, **391**, 682–685.

VINNIK, L., BREGER, L. & ROMANOWICZ, B. 1998. Anisotropic structures at the base of the Earth's mantle. *Nature*, **393**, 564–567.

WALKER, R.J., MORGAN, J.W. & HORAN, M.F. 1995. Osmium-187 enrichment in some plumes: evidence for core–mantle interaction? *Science*, **269**, 819–821.

WALKER, R.J., MORGAN, J.W., HANSKI, E.J. & SMOLKIN, V.F. 1997. Re–Os systematics of Earyl Proterozoic ferropicrites, Pechenga Complex, northwestern Russia: Evidence for ancient ^{187}Os-enriched plumes. *Geochimica et Cosmochimica Acta*, **61**, 3145–3160.

WIDOM, E. & SHIREY, S.B. 1996. Os isotope systematics in the Azores: implications for mantle plume sources. *Earth and Planetary Science Letters*, **142**, 451–465.

Volatiles in submarine basaltic glasses from the Ontong Java Plateau (ODP Leg 192): implications for magmatic processes and source region compositions

JULIE ROBERGE[1], ROSALIND V. WHITE[2] & PAUL J. WALLACE[1]

[1]*Department of Geological Sciences, 1272 University of Oregon, Eugene, OR 97403-1272, USA (e-mail: pwallace@darkwing.uoregon.edu)*

[2]*Department of Geology, University of Leicester, University Road, Leicester LE1 7RH, UK*

Abstract: Submarine basaltic glasses from five widely separated sites on the Ontong Java Plateau (OJP) were analysed for major and volatile elements (H_2O, CO_2, S, Cl). At four of the sites (1183, 1185, 1186, 1187) the glass is from pillow basalt rims, whereas at Site 1184 the glass occurs as non-vesicular glass shards in volcaniclastic rocks. Glassy pillow rims from Site 1187 and the upper group of flows at Site 1185 have 8.3–9.3 wt% MgO compared with values of 7.2–8.0 wt% MgO for glasses from Sites 1183, 1184 1186, and the lower group of flows at Site 1185. Low-MgO glasses have slightly higher H_2O contents (average 0.22 wt% H_2O) than high-MgO glasses (average 0.19 wt%), with the exception of Site 1184, where low-MgO glasses have lower H_2O (average 0.16 wt%). Average S concentrations are 910 ± 60 ppm for the high-MgO glasses v. 1030 ± 60 ppm for the low-MgO glasses. When compared with mid-ocean ridge basalt (MORB), the OJP glasses have lower S at comparable FeO^T. This suggests that OJP basaltic magmas were not saturated with immiscible sulphide liquid during crystallization, but small decreases in S/K_2O and S/TiO_2 with decreasing MgO require some sulphide fractionation. Measurements of the wavelength of the S $K\alpha$ peak in the glasses indicate low oxygen fugacities comparable to MORB values. Chlorine contents of the glasses are very high compared with MORB, and Cl/K ratios for all glasses are relatively high (>0.7). This ratio is sensitive to assimilation of hydrothermally altered material, so the high values indicate assimilation during shallow-level crystallization of OJP magmas. Ratios of H_2O to Ce, which have similar incompatibility to each other, are higher than most depleted and enriched MORB. However, these high H_2O/Ce values are probably also caused by the same assimilation process that results in high Cl. The water content of the high MgO-magmas before contamination is estimated to be approximately 0.07 wt% H_2O, corresponding to H_2O/Ce of 135 for OJP basalts, a value at the low end of the range for Pacific MORB. There is no evidence for high H_2O contents that would have significantly increased extents of mantle melting beneath the OJP, and the estimated H_2O content of the OJP mantle source region (170 ± 30 ppm H_2O) is similar to that of the depleted MORB source (140 ± 40 ppm H_2O). Instead, large extents of melting beneath the OJP must have been caused by a relatively high mantle potential temperature, consistent with upwelling of a hot mantle plume.

The Ontong Java Plateau (OJP) is the largest volcanic oceanic plateau and may represent the largest magmatic event on Earth in the last 200 Ma. The OJP is located in the SW Pacific and is believed to have formed in response to the emplacement of a mantle plume head (e.g. Mahoney & Spencer 1991; Richards *et al.* 1991). Large igneous provinces such as the OJP are important because they provide information on mantle processes and compositions, and because their formation may have global environmental consequences (Larson & Erba 1999; Courtillot & Renne 2003). Magmatic volatiles are especially important in many aspects of large igneous province formation. Basaltic magmas related to mantle plumes commonly have higher H_2O than depleted mid-ocean ridge basalt (MORB), suggesting that the excess magmatism associated with mantle plumes could be caused, at least in part, by the effect of higher H_2O on mantle melting (Schilling *et al.* 1980; Bonatti 1990; Nichols *et al.* 2002). A greater H_2O content for the lower mantle, from which mantle plumes are probably derived (Hofmann 1997), could also indicate the involvement of undegassed primitive mantle (based on high $^3He/^4He$), or could result from recycling of subducted oceanic crust and sediments into the lower mantle. At the Earth's surface release of volatiles such as CO_2, S, Cl and F during eruption of enormous

From: Fitton, J. G., Mahoney, J. J., Wallace, P. J. & Saunders, A. D. (eds) 2004. *Origin and Evolution of the Ontong Java Plateau*. Geological Society, London, Special Publications, **229**, 239–257. 0305-8719/$15.00

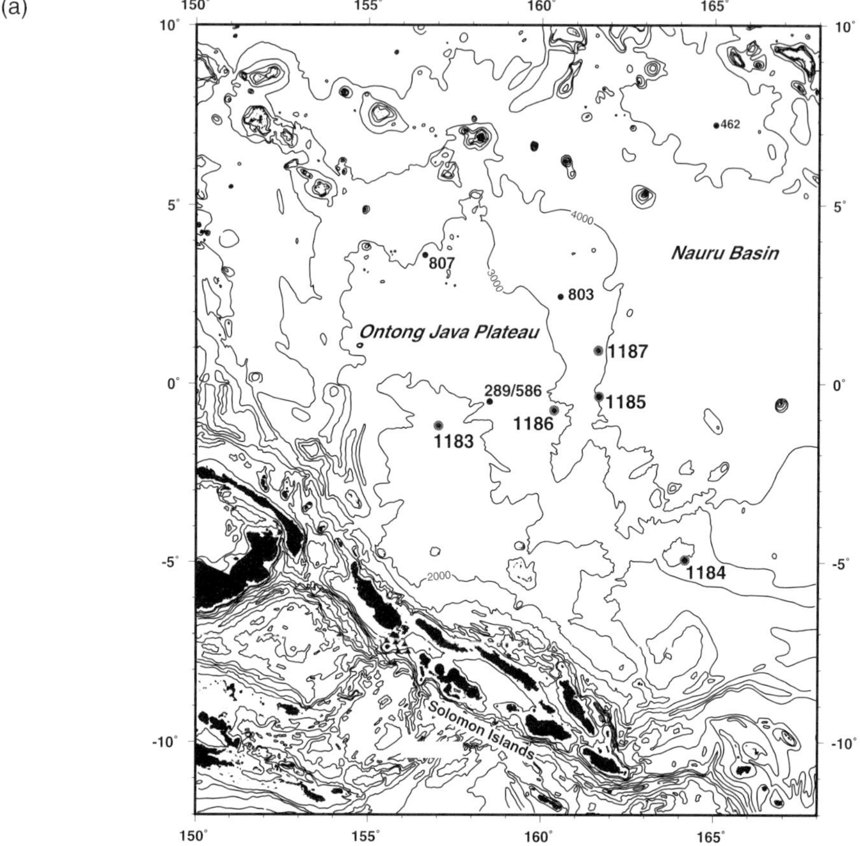

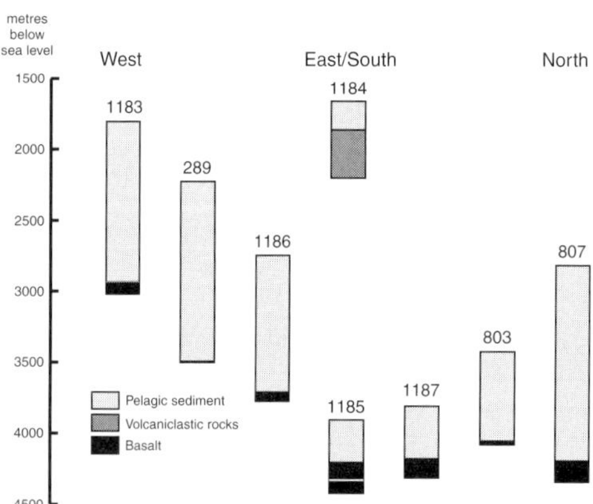

Fig. 1. (**a**) ETOPO5 bathymetric map of the Ontong Java Plateau showing locations of Leg 192 drill sites (large solid circles). Locations of previous ODP and DSDP drill sites that reached basement (small solid circles) are also shown. Depth contours are in metres below sea level. (**b**) Stratigraphic sections drilled during Leg 192 and at previous DSDP/ODP Ontong Java Plateau basement sites (modified from Mahoney *et al.* 2001).

volumes of basaltic magma in large igneous provinces may have significant environmental consequences.

During Ocean Drilling Program (ODP) Leg 192, igneous rock and sediment cores were obtained from five widely separated sites in previously unsampled regions across the OJP (Fig. 1). An exciting result of Leg 192 was the discovery that the basement at Site 1187 and the upper group of flows at Site 1185 are composed of high-MgO, incompatible-element-poor basalt that is unlike the more evolved basalts found elsewhere on the OJP. Because they are less differentiated, these high-MgO basalts are especially valuable in providing information on parental magma compositions. Basaltic lavas recovered during Leg 192, as well as lavas sampled previously in DSDP/ODP drill sites and on land in the Solomon Islands, were all erupted well below sea level, except for a basaltic volcaniclastic sequence erupted in shallow water at Site 1184 (Mahoney *et al.* 2001). Submarine-erupted lavas are particularly valuable for studying magmatic volatiles because quenched glassy pillow rims preserve information on pre-eruption volatile contents.

In this chapter we present major- and volatile- (H_2O, CO_2, S, Cl) element concentrations of fresh glass derived from pillow basalt rims (Sites 1183, 1185, 1186 and 1187) and non-vesicular glass shards in volcaniclastic rocks (Site 1184). Our results complement previously published data for glasses recovered from Sites 803 and 807 on the OJP (Michael 1999). We use the new and published data to infer the H_2O content of the source region for OJP basaltic magmas, to assess magmatic processes such as pressures of fractionation and assimilation in crustal magma chambers, and to discuss possible environmental effects of released volatiles during formation of the OJP. Volatile solubilities in magma are pressure dependent, and thus knowing the volatile content of the quenched glass also allows us to estimate the original eruption depth. A discussion of eruption depths and their implications for plateau subsidence will be presented elsewhere (Roberge *et al.* 2004).

Geological setting and sample characteristics

The Ontong Java Plateau (OJP) covers an area of approximately 2.0×10^6 km^2 and is delimited by the Lyra Basin to the NW, the East Mariana Basin to the north, the Nauru Basin to the NE, and the Ellice Basin to the SE (Fig. 1). ^{40}Ar–^{39}Ar geochronology suggests that the bulk volume of the plateau was formed in a single episode around 122 Ma (Mahoney *et al.* 2001; Chambers *et al.* 2002). A second, volumetrically minor episode happened at about 90 Ma, but none of the lavas recovered during Leg 192 were of this age (Mahoney *et al.* 2001). After its formation, the OJP collided with the Solomon Islands arc and now lies between the Pacific and Australian plate, resisting subduction (Neal *et al.* 1997). The igneous rocks that form the OJP are now covered with thick accumulations of pelagic sediment. At its highest point, the upper surface of sediment on the plateau is approximately 1700 m below sea level, but elsewhere lies between a depth of 2 and 3 km.

During ODP Leg 192, five widely spaced sites were drilled (Fig. 1) (Mahoney *et al.* 2001). Site 1183 is located on the northern part of the high plateau. The sediment sequence overlying the basement is approximately 1130 m thick, and about 80 m of basement rocks consisting of pillow basalt were penetrated. Basaltic glasses analysed from this site come from throughout the recovered basement sequence. Site 1184 is on the northern ridge of the eastern lobe of the OJP. The recovered section at this site contains 337 m of volcaniclastic sequences formed by hydroclastic eruptions in shallow water, but the hole did not penetrate into the underlying igneous basement, nor is the depth to basement known (Mahoney *et al.* 2001). The entire sequence recovered is altered to varying degrees. Unaltered, non-vesicular glass shards from Subunits IIA, IID and IIE were used for analysis; Subunits IIB and IIC do not contain any unaltered glass shards. At Site 1185, which lies on the eastern edge of the OJP, 216 m of basement rock was recovered beneath a 309 m-thick sediment sequence. The basement rocks at this site consist of pillow basalt and massive basalt; because abundant pillow basalt is present only in the first *c.* 150 m of the hole, most glasses sampled for analysis come from this interval. However, we did sample glass from two quenched margins associated with the underlying sheet flows. Site 1186 is located on the eastern slope of the OJP between Sites 1183 and 1185. The sediment sequence at this site is 968 m thick, and about 65 m of basement rocks were penetrated. The basement rocks consist of pillow basalt alternating with massive basalt. Again, because pillow basalt is present only in the upper part of the drill hole, the samples analysed from Site 1186 represent only the uppermost 50 m of the recovered sequence. Finally, Site 1187 is located about 100 km north of Site 1185 on the eastern edge of the OJP. At this site the sediment thickness is 372 m and the basement penetration was 135 m, consisting almost entirely of pillow basalt. Therefore, the samples analysed from this site represent the entire sequence of the recovered basement.

Based on whole-rock analyses (Mahoney *et al.* 2001; Fitton & Godard 2004), basalts at Sites 1183, 1186 and the lower part of 1185 are homogeneous, moderately evolved, low-K tholeiites, whereas the low-K basalts found in the upper part of Site 1185 and at Site 1187 have higher MgO (between 8 and 10 wt%).

Analytical methods

Sample preparation

Fragments of visually unaltered glass from pillow margins or volcaniclastic sediment (Site 1184) were selected for infrared (IR) spectroscopy. Pieces of glass containing hair-like tubules that are the result of microbial alteration (Fisk *et al.* 1998; Banerjee & Muehlenbachs 2003) were avoided. The pieces of glass chosen were mounted on a glass slide using acetone-soluble cement. The samples were then doubly ground and polished into wafers with two parallel sides. The thickness of each glass wafer was measured using a micrometer with a precision of ±2 μm.

Infrared spectroscopy

A Nicolet Magna 560 Fourier transform IR spectrometer interfaced with a Spectra-Tech Nic-Plan microscope was used at Texas A&M University to obtain transmission IR spectra. Two individual spectra, taken on different areas of each glass sample, were acquired using a circular aperture 100 μm in diameter.

Band assignments for dissolved water and carbonate in basaltic glass are based on Dixon *et al.* (1995). Quantitative measurements of dissolved total H_2O, molecular H_2O and carbonate (CO_3^{2-}) were obtained using Beer's law:

$$c = \frac{M\,A}{\rho\,d\,\varepsilon}$$

where c is the concentration (weight fraction) of the absorbing species, M is the molecular weight of H_2O (18.02) or CO_2 (44.00), A is the absorbance intensity of the band of interest, ρ is room temperature density of the basaltic glass (2800 kg m^{-3} was used for all glasses), d is the thickness of the glass wafer and ε is the molar absorption coefficient.

Total dissolved H_2O was measured using the intensity of the band centred at 3550 cm^{-1}, which corresponds to the fundamental O–H stretching vibration (Ihinger *et al.* 1994). On a printed copy of the spectra the background was drawn as a smooth curve and graphically subtracted from the peak height to measure the absorbance intensity of the 3550 cm^{-1} band. The total dissolved water contents (Table 1) were calculated using a molar absorption coefficient of 63 ± 3 l mol^{-1} cm^{-1} from Ihinger *et al.* (1994).

To examine the speciation of water in the glasses as a means of screening for low-temperature hydration, concentrations of dissolved molecular H_2O were measured using the intensity of the 1630 cm^{-1} absorption band. Unlike the molar absorptivity for the 3550 cm^{-1} band, which is relatively independent of composition for basaltic glasses, the molar absorptivity for molecular water is compositionally dependent (Dixon *et al.* 1995). Using the method described in Dixon *et al.* (1995), the molar absorptivity of the 1630 cm^{-1} band for OJP basalt glasses is 25 ± 1 l mol^{-1} cm^{-1}.

Dissolved carbonate was measured from the absorbance of the 1515 and 1430 cm^{-1} bands, which correspond to distorted asymmetric stretching of carbonate groups (Ihinger *et al.* 1994). Because the shape of the background in the region of the carbonate doublet is complex, it is necessary to subtract a carbonate-free reference spectrum to obtain a flat background (Dixon *et al.* 1995). We measured absorbance intensities of the 1515 and 1430 cm^{-1} bands using a peak-fitting program that fits the sample spectrum with a straight line, a devolatilized basaltic glass spectrum, a pure 1630 cm^{-1} band for molecular H_2O and a pure carbonate doublet (unpublished program by S. Newman). The molar absorption coefficient of carbonate in basaltic glass is compositionally dependent and was derived using the average composition of the glasses and the linear equation reported in Dixon & Pan (1995). The dissolved carbonate content was calculated using a molar absorption coefficient of 384 l mol^{-1} cm^{-1} for the high-MgO glasses and 370 l mol^{-1} cm^{-1} for the low-MgO glasses.

Based on replicate analyses, precision (2σ) for total H_2O is <16% (relative) and <11% for CO_2. Accuracy for these techniques is estimated to be ±10% for total H_2O and ±20% for CO_2 (Dixon & Clague 2001).

Electron microprobe

Major elements, S and Cl in most glass samples were analysed using a Cameca SX-50 electron microprobe at the University of Oregon (Table 1). The major elements were acquired using an electron beam diameter of 10 μm with an accelerating voltage of 15 kV and a beam current of 10 nA. Both glass and mineral standards were used, and US National Museum glass standard VG-2 was used to assess analytical accuracy (Table 1). Sulphur was analysed using an anhydrite standard and an S *Kα* wavelength position measured on pyrite, which corresponds approximately to the $S^{6+}/\Sigma S$ ratio expected for basaltic glass equilibrated at the fayalite–magnetite–quartz (FMQ) oxygen buffer (Wallace & Carmichael 1994). For all elements, five spots on each glass sample were analysed, and the average value is reported in Table 1. Chlorine was analysed using an electron beam diameter of 20 μm with an accelerating voltage of 15 kV and a beam current of 100 nA. The counting time for chlorine was 200 s on peak and 200 s on background for each spot analysed. Sulphur speciation was also determined for selected samples by measuring the S *Kα* X-ray wavelength position as described by Wallace & Carmichael (1994), using sphalerite for S^{2-} and anhydrite for S^{6+}. The operating conditions for S speciation measurements were 10 μm electron beam diameter with an accelerating voltage of 15 kV and a beam current of 30 nA. Wavelength scans were performed using counting times of 40 s per

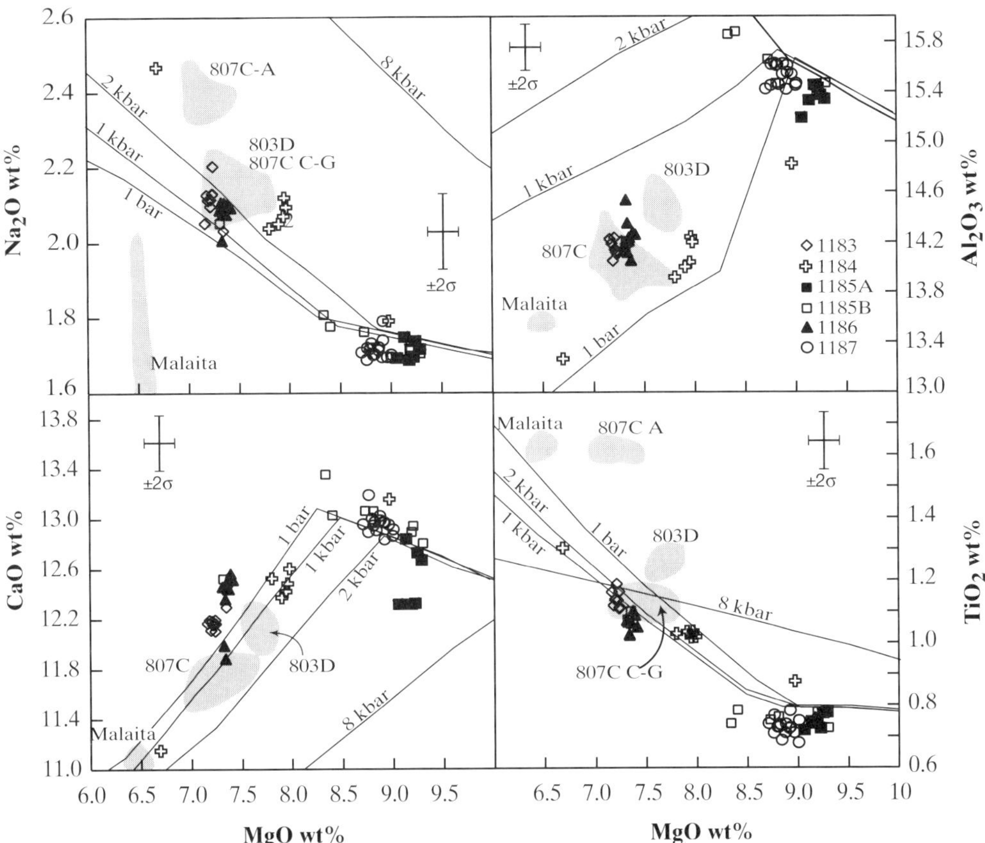

Fig. 2. Major-element compositions of Ontong Java Plateau basaltic glasses. Data from Sites 803 and 807 and the island of Malaita are from Michael (1999). Lines show fractional crystallization paths for a parental magma with 17.6 wt% MgO calculated as described in the text. Crystallization of this parental composition at pressures of 1 bar–2 kbar can largely reproduce the observed range of major-element compositions. Under these conditions, the crystallization sequence is olivine, followed by olivine + plagioclase, followed by olivine + plagioclase + clinopyroxene.

wavelength step on unknowns and 5 s per step on standards. The sample was moved slightly relative to the beam after each step in the wavelength scan to avoid problems associated with beam damage to the glass (Wallace & Carmichael 1994). Major elements and S in the Site 1184 glasses were analysed at the University of Leicester using procedures described in White *et al.* (2004). Chlorine in Site 1184 glasses was measured at the University of Oregon as described above.

Two-sigma (2σ) precision based on counting statistics is <2.5% (relative) for major elements and <30% for minor elements (K, Mn, Na, Ti, P). For Cl and S the 2σ precision is <9% and <22%, respectively.

Results

All 55 samples of unaltered basaltic glass, representing all sites drilled during ODP Leg 192, are tholeiitic basalts based on a total alkalies v. silica diagram (not shown). Two types of basalt could be observed based on MgO contents. Basement at Site 1187 and the upper group of flows at Site 1185 is composed of high-MgO basalt (average of 9 wt% MgO), compared to 7.2–8.0 wt% MgO for basalt found elsewhere on the OJP (Fig. 2). The low-MgO glasses from Sites 1183, 1186 and the lower part of 1185 are similar in composition to those found in the northern part of the OJP (Hole 803D and units C–G of ODP Hole 807C). All glasses have relatively low Na_2O, K_2O and

Table 1. *Compositions of basaltic glasses from the Ontong Java Plateau*

Sample	SiO_2	TiO_2	Al_2O_3	FeO^T	MnO	MgO	CaO	Na_2O	K_2O	P_2O_5	S (ppm)	Cl (ppm)	H_2O	CO_2 (ppm)	Total	Sat. *P** (bars)
1183A-55R-1 32	50.58	1.12	14.19	11.25	0.19	7.18	12.19	2.13	0.11	0.09	948	673	0.24	45	99.42	
1183A-55R-2 134	50.46	1.14	14.23	11.33	0.21	7.21	12.18	2.12	0.11	0.08	1029	664	0.23	44	99.47	
1183A-56R-1 10	51.06	1.16	14.22	11.35	0.21	7.16	12.17	2.05	0.12	0.07	1156	601	0.20	43	99.97	
1183A-57R-2 97	50.54	1.19	14.14	11.26	0.21	7.21	12.16	2.10	0.10	0.09	1013	638	0.24	46	99.41	
1183A-58R-1 97	50.89	1.14	14.05	11.29	0.18	7.19	12.11	2.11	0.11	0.11	1131	667	0.23	47	99.59	
1183A-60R-1 36	51.00	1.11	14.11	11.25	0.22	7.23	12.19	2.13	0.11	0.09	957	666	0.22	46	99.82	
1183A-62R-2 1	50.86	1.14	14.18	11.12	0.21	7.34	12.30	2.03	0.11	0.08	1083	713	0.22	57	99.77	
1183A-64R-1 112	50.90	1.14	14.12	11.25	0.19	7.22	12.16	2.12	0.11	0.10	1139	691	0.23	51	99.72	
1183A-65R-2 19	51.15	1.16	14.10	11.13	0.22	7.23	12.10	2.13	0.11	0.10	982	678	0.23	43	99.83	
1183A-67R-3 60	50.97	1.11	14.13	11.27	0.22	7.24	12.16	2.20	0.11	0.08	1021	737	0.23	41	99.91	
Average 1183A	**50.84**	**1.14**	**14.15**	**11.25**	**0.21**	**7.22**	**12.17**	**2.11**	**0.11**	**0.09**	**1046**	**673**	**0.23**	**46**	**99.69**	107±9
1184A-13R-3 142	49.86	0.88	14.82	9.50	0.17	8.97	13.16	1.79	0.08	n.a.	641	495	0.17	19	99.52	
1184A-31R-7 53	50.63	1.30	13.27	12.32	0.23	6.70	11.15	2.47	0.14	n.a.	692	3171	0.14	47	98.73	
1184A-39R-7 95	49.82	1.03	13.92	10.24	0.18	7.81	12.53	2.04	0.10	n.a.	725	1708	0.16	31	98.07	
1184A-41R-6 67	50.81	1.03	14.20	10.59	0.18	7.98	12.61	2.09	0.11	n.a.	717	1678	0.16	27	99.99	
1184A-42R-5 69	50.21	1.04	14.01	10.44	0.20	7.91	12.37	2.05	0.10	n.a.	720	1685	0.16	24	98.73	
1184A-44R-3 78	50.80	1.04	14.23	10.59	0.17	7.96	12.49	2.12	0.10	n.a.	708	1690	0.16	30	99.91	
1184A-45R-2 75	50.35	1.02	14.03	10.46	0.19	7.95	12.42	2.07	0.10	n.a.	705	1699	0.17	29	99.01	
Average 1184A	**50.36**	**1.05**	**14.07**	**10.59**	**0.19**	**7.90**	**12.39**	**2.09**	**0.10**		**700**	**1700**	**0.16**	**30**	**99.13**	54±21
1185A-9R-1 130	49.43	0.75	15.33	9.64	0.18	9.13	12.84	1.75	0.07	0.06	895	812	0.18	94	99.56	
1185A-9R-3 88	49.74	0.78	15.34	9.69	0.17	9.28	12.67	1.72	0.07	0.05	922	778	0.19	106	99.90	
1185A-9R-4 100	49.94	0.77	15.38	9.73	0.17	9.24	12.73	1.74	0.07	0.06	876	777	0.19	100	100.19	
1185A-10R-1 10	50.41	0.73	15.43	9.64	0.18	9.22	12.33	1.70	0.08	0.07	902	756	0.18	92	100.13	
1185A-10R-1 40	48.87	0.72	15.20	9.91	0.17	9.06	12.32	1.69	0.08	0.07	814	798	0.18	98	98.45	
1185A-10R-2 1	50.33	0.74	15.46	9.67	0.20	9.18	12.32	1.69	0.07	0.06	867	759	0.20	102	100.10	
Average 1185A	**49.79**	**0.75**	**15.36**	**9.71**	**0.18**	**9.18**	**12.54**	**1.71**	**0.08**	**0.06**	**879**	**780**	**0.18**	**99**	**99.72**	219±11
1185B-3R-1 54	49.13	0.78	15.45	9.87	0.18	9.19	12.90	1.72	0.07	0.06	964	885	0.19	101	99.73	
1185B-5R-7 19	49.59	0.73	15.47	9.82	0.16	9.30	12.80	1.71	0.07	0.08	818	936	0.19	104	100.09	
1185B-6R-1 95	49.83	0.76	15.46	9.92	0.19	9.20	12.93	1.71	0.07	0.07	865	919	0.18	97	100.52	
1185B-9R-1 55	49.19	0.76	15.47	9.87	0.16	8.81	13.06	1.72	0.08	0.07	957	929	0.21	104	99.61	
1185B-9R-3 pc3	49.78	0.78	15.88	9.64	0.20	8.41	13.03	1.78	0.08	0.07	929	980	0.19	95	100.04	
1185B-15R-1 149	49.86	0.75	15.66	9.71	0.18	8.73	13.07	1.76	0.08	0.06	832	836	0.20	105	100.24	
1185B-16R-1 64	49.47	0.74	15.86	9.56	0.20	8.34	13.35	1.81	0.07	0.07	875	987	0.18	105	99.85	
Average upper 1185B	**49.55**	**0.76**	**15.61**	**9.77**	**0.18**	**8.85**	**13.02**	**1.74**	**0.08**	**0.07**	**891**	**924**	**0.19**	**102**	**100.01**	221±7

1185B-23R-1 4	50.58	1.07	14.21	11.48	0.17	7.31	12.52	2.05	0.11	0.10	1050	915	0.24	104	100.05	
1185B-23R-1 78	50.71	1.10	14.20	11.35	0.19	7.35	12.44	2.10	0.11	0.09	1072	779	0.21	100	100.04	
Average lower 1185B	**50.65**	**1.09**	**14.21**	**11.41**	**0.18**	**7.33**	**12.48**	**2.08**	**0.11**	**0.09**	**1061**	**847**	**0.22**	**102**	**100.05**	223±8
1186A-31R-2 47	50.87	1.07	14.18	11.15	0.21	7.31	12.46	2.09	0.10	0.08	1034	735	0.19	87	99.90	
1186A-31R-3 34	51.01	1.06	14.24	11.34	0.22	7.35	12.50	2.11	0.10	0.11	1097	717	0.18	92	100.41	
1186A-31R-3 47	51.06	1.09	14.28	11.38	0.18	7.39	12.57	2.10	0.10	0.07	972	717	0.23	107	100.62	
1186A-32R-3 68	50.23	1.10	14.05	11.39	0.20	7.37	12.44	2.08	0.11	0.09	1041	740	0.23	101	99.49	
1186A-32R-3 76	50.62	1.02	14.12	11.33	0.25	7.34	12.36	2.08	0.09	0.10	953	743	0.22	99	99.72	
1186A-33R-1 27	52.07	1.07	14.54	11.29	0.21	7.32	11.99	2.11	0.10	0.10	962	687	0.22	97	101.19	
1186A-34R-4 98	51.39	1.03	14.35	11.44	0.19	7.33	11.88	2.01	0.11	0.11	1031	693	0.23	102	100.25	
1186A-34R-5 57	50.92	1.05	14.26	11.43	0.20	7.42	12.51	2.09	0.11	0.10	1023	907	0.19	89	100.49	
Average 1186A	**51.05**	**1.06**	**14.27**	**11.38**	**0.21**	**7.36**	**12.29**	**2.08**	**0.10**	**0.10**	**997**	**748**	**0.22**	**99**	**100.29**	215±14
1187A-3R-2 104	49.70	0.73	15.42	9.54	0.17	8.91	12.98	1.70	0.07	0.06	881	946	0.17	118	99.64	
1187A-4R-2 70	49.67	0.68	15.45	9.57	0.17	9.01	12.87	1.70	0.07	0.08	967	954	0.19	109	99.66	
1187A-5R-7 62	49.65	0.71	15.54	9.54	0.17	8.97	12.96	1.69	0.07	0.05	962	955	0.21	117	99.76	
1187A-6R-7 102	49.94	0.78	15.55	9.66	0.16	8.92	12.97	1.79	0.07	0.05	964	963	0.21	110	100.32	
1187A-7R-2 53	49.63	0.74	15.42	9.56	0.16	8.71	12.96	1.71	0.07	0.09	769	998	0.18	87	99.42	
1187A-7R-5 106	50.12	0.71	15.45	9.47	0.15	8.76	12.90	1.69	0.07	0.04	881	960	0.20	112	99.76	
1187A-8R-2 133	50.04	0.69	15.45	9.52	0.14	8.84	12.91	1.70	0.08	0.06	799	974	0.21	116	99.83	
1187A-9R-6 103	50.20	0.75	15.47	9.50	0.18	9.01	12.92	1.69	0.08	0.07	930	960	0.20	109	100.26	
1187A-10R-4 63	49.82	0.74	15.62	9.73	0.18	8.88	12.99	1.72	0.07	0.05	972	980	0.20	109	100.22	
1187A-11R-2 1	50.05	0.74	15.61	9.69	0.17	8.82	12.98	1.70	0.07	0.05	961	1016	0.20	115	100.29	
1187A-12R-1 145	49.98	0.73	15.61	9.71	0.17	8.81	12.95	1.73	0.08	0.04	945	1005	0.21	115	100.22	
1187A-13R-6 39	50.16	0.73	15.61	9.68	0.18	8.92	12.84	1.74	0.07	0.07	951	993	0.19	109	100.39	
1187A-14R-4 35	49.75	0.71	15.54	9.63	0.15	8.87	13.03	1.71	0.07	0.06	969	1021	0.20	111	99.94	
1187A-15R-1 83	50.11	0.77	15.62	9.54	0.17	8.76	13.19	1.72	0.07	0.07	980	1012	0.22	111	100.46	
1187A-16R-5 96	49.94	0.73	15.62	9.68	0.17	8.80	13.00	1.72	0.07	0.08	935	999	0.23	111	100.25	
Average 1187A	**49.92**	**0.73**	**15.53**	**9.60**	**0.17**	**8.87**	**12.96**	**1.71**	**0.07**	**0.06**	**924**	**982**	**0.20**	**111**	**100.03**	245±7
Average 803D	50.09	1.26	14.44	11.60	0.22	7.72	12.11	2.13	0.13	0.09	n.a.	370	0.27	201	100.12	
Average 807CA	49.83	1.69	14.13	12.07	0.24	7.14	11.71	2.38	0.20	0.12	n.a.	580	0.41	54	99.99	
Average 807CC–G	50.56	1.14	13.93	11.93	0.23	7.35	11.76	2.15	0.11	0.09	n.a.	850	0.24	158	99.59	
Average Malaita	50.37	1.63	13.57	12.90	0.22	6.49	11.07	1.72	0.18	0.11		600	0.91		99.23	
Analysed VG-2	50.44	1.89	14.04	11.82	0.21	6.77	11.23	2.68	0.20	0.20	1549	270			99.66	
Standard deviation	0.54	0.10	0.26	0.24	0.04	0.11	0.26	0.07	0.02	0.03	166	57				
VG-2 USNM†	50.81	1.85	14.06	11.84	0.22	6.71	11.12	2.62	0.19	0.20					99.62	

All data are in wt% except where noted. Data for glasses from Sites 803 and 807 and from the Island of Malaita are from Michael (1999).
n.a., not analysed.
* Saturation pressures were calculated using VolatileCalc 1.1 (Newman & Lowenstern 2002).
† Data from Jarosewich et al. (1979)

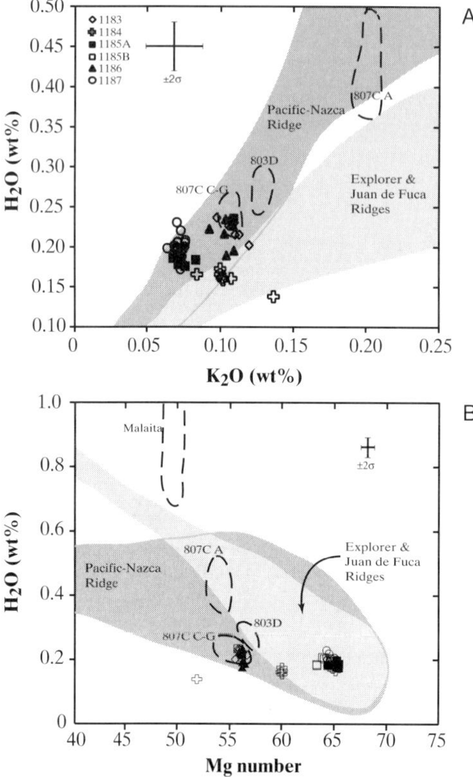

Fig. 3. (**A**) H_2O v. K_2O and (**B**) H_2O versus Mg number for OJP basaltic glasses. For comparison, the shaded fields show data for MORB from the Pacific–Nazca, Explorer and Juan de Fuca ridges (Michael 1995), and the dashed lines represent data from Sites 803 and 807, and the island of Malaita (Michael 1999).

P_2O_5, and high FeO^T, consistent with derivation of OJP basaltic magmas by large degrees of partial melting (Mahoney *et al.* 1993; Michael 1999; Fitton & Godard 2004).

Pillow rim glasses from a given site have a very restricted range of composition, except at Site 1185, which has both high- and low-MgO glasses (Fig. 2). Within each of these two chemical types, however, the glasses are homogeneous. A similar homogeneity of composition was found at Sites 803 and 807 by Michael (1999), who noted that the OJP shows considerably less geochemical variability than is found in drilled sections of MORB. The homogeneity suggests that the lavas recovered from a given site (except Site 1185) may represent single large eruptions with multiple flows or lobes, resulting in multiple glass-bounded cooling units and interbedded sheet flows within a given hole. In contrast to the homogeneity observed for the pillow rim glasses from most sites, glass shards from the volcaniclastic sequence found at Site 1184 fall into three distinct compositional groups spanning a range of MgO contents (Fig. 2; see also fig. 4 of White *et al.* 2004).

Water concentrations are relatively low, similar to normal MORB (N-MORB) values, in all glass samples (Fig. 3A). Low-MgO glasses have slightly higher H_2O contents (average 0.22 wt% H_2O) than high-MgO glasses (average 0.19 wt%), with the exception of Site 1184, where low-MgO glasses have lower H_2O (average 0.16 wt%; Fig. 3B). All glasses in Table 1 have very low amounts of molecular H_2O (mostly below detection), consistent with expectations for high-temperature equilibrium speciation in basaltic melts with low total H_2O (Dixon *et al.* 1995). Some of the glasses analysed had much higher molecular H_2O and anomalously high total H_2O, which we attribute to low-temperature hydration of the glass. These glasses were excluded from the data set in Table 1.

Carbon dioxide concentrations of OJP glasses average 47 ± 5 ppm for Site 1183, 30 ± 9 ppm for Site 1184, 101 ± 4 ppm for Site 1185, 99 ± 7 for Site 1186 and 110 ± 7 ppm for Site 1187 (Fig. 4). Using the method of Dixon *et al.* (1995), vapour saturation pressures were calculated for all sites (Table 1) and then converted into eruption depths (1 bar = 10 m of water depth) assuming that the volatiles were not supersaturated at the time of eruption. As expected, glass shards from the volcaniclastic deposits at Site 1184 have low saturation pressures, indicating an average quenching depth of approximately 540 ± 210 m. Site 1183 glasses, which come from the shallowest water site on the crest of the plateau, also have relatively low saturation pressures of 107 bars (1070 ± 90 m), whereas Sites 1185, 1186 and 1187 have saturation pressures of 215–245 bars, yielding estimated eruption depths of 2150–2450 (±100) m (Fig. 4). These estimated depths must be viewed with caution, however, because submarine basaltic pillow rims, particularly MORB samples, are typically supersaturated with CO_2 (Dixon & Stolper 1995).

Compared with MORB, the OJP glasses have lower S (910 ± 60 ppm for high-MgO glasses, 1030 ± 60 ppm for low-MgO glasses) at comparable FeO^T (Fig. 5). Sulphur contents of MORB magmas are usually controlled by saturation with immiscible sulphide (Fe–S–O) liquid (Wallace & Carmichael 1992), so the lower S contents of OJP glasses suggest that OJP basaltic magmas may not have been saturated with immiscible sulphide liquid during crystallization. However, if OJP basaltic magmas had

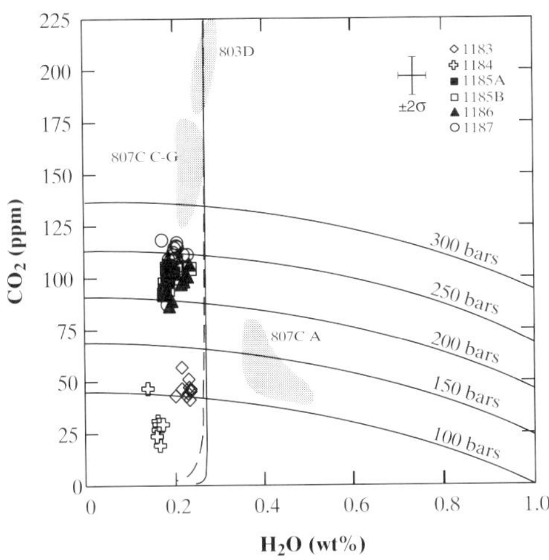

Fig. 4. CO_2 v. H_2O for OJP basaltic glasses. Shaded fields as in Figure 2. Vertical lines represent degassing paths for basaltic melts with initial CO_2 contents of 225 ppm (solid line; value was chosen based on the maximum concentration in OJP glasses) and 2000 ppm (dashed line). Also shown are vapour-saturation curves for basaltic melts at pressures from 100 to 300 bars. All calculations were made using VolatileCalc 1.1 (Newman & Lowenstern 2002).

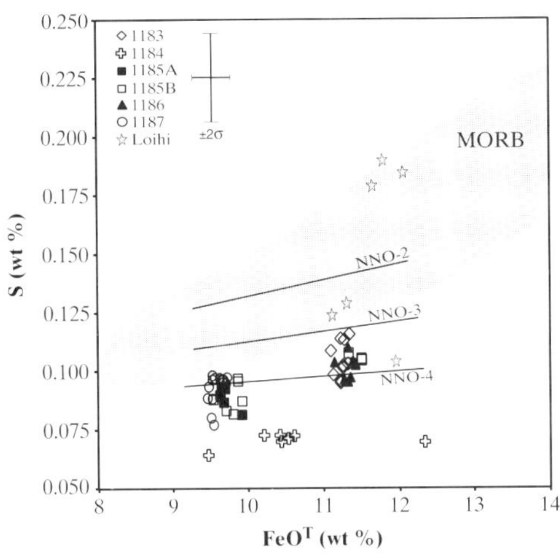

Fig. 5. S v. FeO^T for OJP basaltic glasses. The shaded field shows data for Pacific MORB glasses (MORB database, http://petdb.ldeo.columbia.edu/petdb), and small stars show Loihi glasses (Wallace & Carmichael 1992). Lines show 1 bar sulphide saturation limits for OJP basaltic melts at relative oxygen fugacities ranging from 2 (NNO-2) to 4 (NNO-4) log units more reduced than the Ni–NiO buffer. Saturation limits were calculated using the thermodynamic model of Wallace & Carmichael (1992) updated to incorporate the temperature dependence from Mavrogenes & O'Neill (1999).

lower oxygen fugacities than MORB, they could have been sulphide-liquid saturated because lower oxygen fugacity reduces the sulphur solubility in basaltic melts (Fig. 5). Measurements of the wavelength of the S $K\alpha$ peak in the glasses indicate low oxygen fugacities (fO_2), broadly comparable to MORB values, but this technique lacks the precision at low fO_2 necessary to resolve whether the OJP glasses are more reduced than MORB (Wallace and Carmichael 1994). Ratios of S to other incompatible elements (S/K_2O, S/TiO_2) in the glass samples decrease slightly with decreasing MgO, consistent with some fractionation of sulphide liquid, but sulphides have not been observed in quenched glass samples and platinum-group element systematics show no evidence for sulphide-liquid fractionation (Chazey & Neal 2004). Given this contradictory evidence, the issue of whether or not OJP basaltic magmas were sulphide-liquid saturated remains unresolved at this time. The low S contents of OJP basalts are probably caused by the high degree of melting, which is estimated from major- and trace-element modelling to be approximately 30% (Fitton & Godard 2004). Such large degrees of melting would probably exhaust residual sulphide in the mantle source during melting, in contrast to MORB magmas, which are generated by lower degrees of melting and probably form in equilibrium with residual mantle sulphide.

All OJP basaltic glasses have high Cl, and glasses from a given site are relatively uniform, except for Site 1184 (Table 1). Average values for each site are 670 ± 40 ppm Cl (Site 1183), 1470 ± 600 ppm (Site 1184), 860 ± 80 ppm (Site 1185), 750 ± 70 ppm (Site 1186) and 980 ± 25 ppm (Site 1187). Melt inclusions in olivine from a high-MgO sample from Hole 1185A mostly have Cl contents (*c.* 900 ppm) similar to the glass, but some inclusions have lower Cl (*c.* 300 ppm; J. Roberge unpublished data). There is no correlation between Cl and K_2O, despite their similar incompatibility. OJP basaltic glasses are highly enriched in Cl relative to MORB, as was found previously for glasses from Sites 803 and 807 (Michael 1995). The cause of these very high Cl contents and their implications for OJP magmatic processes are discussed later.

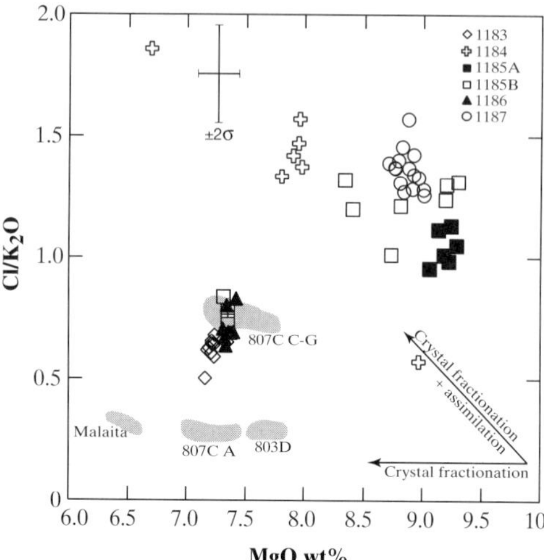

Fig. 6. Cl/K_2O v. MgO for OJP glasses showing the effects of crystal fractionation and assimilation of material with high Cl derived from a sea-water or brine component. Shaded fields as in Figure 2.

Discussion

Depths of crystallization

Basaltic glasses from the OJP preserve a record of quenched liquid compositions spanning a range of MgO contents, making them useful for constraining differentiation processes on the plateau. We estimated the composition of a primary magma by taking the most MgO-rich glass (9.3 wt% MgO; equilibrium olivine Fo_{86}) and adding equilibrium olivine in 0.1 wt% increments until reaching a composition in equilibrium with mantle olivine of Fo_{91}, a value that is constrained by modelling presented in Fitton & Godard (2004). This calculated primary magma contains 17.6 wt% MgO and requires addition of 27% olivine to the most MgO-rich glass. We modelled the liquid line of descent from this starting composition at a range of pressures (Fig. 2) using the MELTS program (Ghiorso & Sack 1995; Asimow & Ghiorso 1998). Previous modelling of crystallization used an assumed picritic composition with approximately 16 wt% MgO because of lack of information about primary magma compositions (Farnetani *et al.* 1996; Neal *et al.* 1997; Michael 1999).

Equilibrium crystallization of the estimated primary magma (17.6 wt% MgO) at relatively low pressure (1 bar–2 kbar) yields residual liquids that match the compositions of OJP basalts (Fig. 2), in agreement with results based on experimental phase equilibria (Sano & Yamashita 2004). These results indicate that derivation of OJP magmas from likely primary compositions could have occurred at low pressures within the upper crust. A similar conclusion was reached by Michael (1999) for glasses from Sites 803 and 807 using the crystallization program of Weaver & Langmuir (1990). These pressures are at the low end of the range found for MORB worldwide; specifically, they are similar to MORB from fast-spreading ridges and from robust slow-spreading ridges like the Kolbeinsey Ridge north of Iceland (Michael & Cornell 1998; Michael 1999). An alternative possibility is that crystallization took place in two stages, with crystallization of picritic liquids occurring first at deeper (e.g. Moho) levels followed by a second stage of crystallization in the upper crust (Farnetani *et al.* 1996; Michael 1999). However, as shown by Michael (1999), and reinforced by our results, no more than about 20–30% of crystallization of primary liquids could have happened at high pressures because the increase in clinopyroxene stability results in residual liquid compositions that do not match the observed values for OJP basalts. Thus, much of the crystallization of magmas parental to the erupted lavas must have occurred in shallow-level magma chambers, resulting in the formation of extensive volumes of cumulates within the OJP crust.

Cause of high chlorine in Ontong Java Plateau basalts

Chlorine contents of normal MORB magmas are generally low (<50 ppm) reflecting low Cl in the upper-mantle source region (Michael & Cornell 1998). Like H_2O, Cl is incompatible during partial melting in the mantle and crystallization of basaltic magmas (Schilling *et al.* 1980), so variations in Cl can be best understood by comparing them with other incompatible elements such as K (Fig. 6). The Cl/K ratio varies from below detection limits (*c.* 0.01) in normal MORB to approximately 0.07 in enriched MORB magmas (Michael & Cornell 1998). Both Cl contents and Cl/K ratios are highly sensitive to assimilation involving sea water or sea-water-derived brine because of their high Cl contents (Figs 6 and 7). Such assimilation is common in regions of the mid-ocean ridge system where magma chambers are relatively shallow, and in submarine oceanic islands (Michael & Cornell 1998; Kent *et al.* 1999).

The very high Cl contents of OJP basalts were first discovered by Michael (1999) from analyses

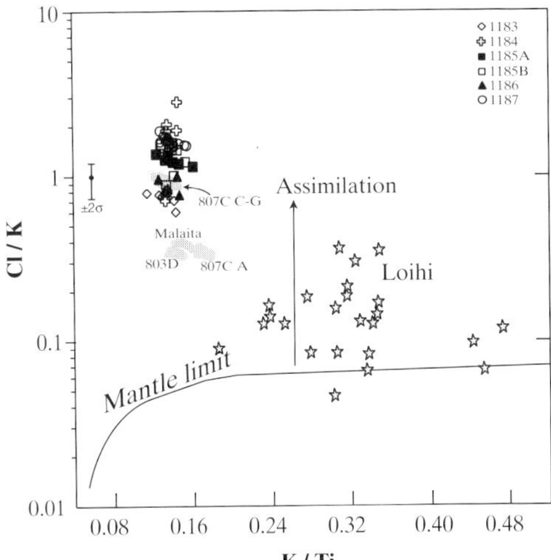

Fig. 7. Cl/K v. K/Ti showing the effect of contamination with hydrothermally altered materials containing sea water or sea-water-derived brines. Loihi data are from Kent *et al.* (1999). Uncontaminated MORB lavas generally plot below the line labelled 'mantle limit' (Michael & Cornell 1998; Michael 1999). K/Ti in MORB glasses generally correlates with La/Sm, and thus is an indicator of the relative enrichment or depletion of the mantle source (Michael & Cornell 1998). Shaded fields as in Figure 2.

of glasses from Sites 803 and 807. He attributed the high Cl to shallow-level assimilation, a hypothesis that we discuss further below. An alternative possibility is that the mantle source region for OJP basaltic magmas is enriched in Cl, perhaps by addition of subducted materials. It is well documented that mafic magmas associated with convergent plate margins have elevated Cl compared to MORB (Anderson 1974; Schilling *et al.* 1978; Ito *et al.* 1983). Elevated Cl contents are also found in back-arc basin basalts (e.g. Stolper & Newman 1994). However, basaltic magmas in oceanic islands with mantle source regions believed to contain a deeply recycled crustal component do not generally show elevated Cl/K compared to MORB, suggesting that this process does not strongly increase mantle Cl (Kent *et al.* 1999; Lassiter *et al.* 2002). Given this, and the observation that Sr–Nd–Pb–Hf isotope systematics of OJP basalts show little or no evidence of a recycled (enriched mantle (EM) or high μ, where $\mu = {}^{238}U/{}^{204}Pb$(HIMU)) component (Tejada *et al.* 2004; next section), we think it is unlikely that the OJP source region contains elevated Cl relative to MORB and other ocean island basalt (OIB) mantle sources.

All OJP basaltic glasses have high Cl, so if this is caused by assimilation of material with a sea-water or brine component then such a process was a widespread phenomenon and relatively uniform in OJP magma chambers. It is possible to infer the composition of the contaminant by comparing Cl/K_2O and H_2O/K_2O for basaltic glasses (Fig. 8) (Kent *et al.* 1999). The result suggests that the OJP basaltic magmas are typically contaminated with material that contains relatively concentrated brine with 50–60 wt% NaCl. Data from a few sites (803, 807 Unit C and 1184) indicate more or less concentrated brine. We find it remarkable that data from such geographically widespread sites all show evidence of interaction with brine of a similar composition. Interestingly, many of the contaminated basalts at the Loihi seamount also show evidence of interaction with similarly concentrated brine components, although the overall range of values at Loihi includes less concentrated brines and sea water as contaminants (Kent *et al.* 1999). It is important to note, however, that only very small amounts of this brine phase (<0.4 wt%) need to be incorporated to explain the range of variation in Cl/K_2O and H_2O/K_2O for OJP basalts. Such small amounts are consistent with the hypothesis that the brine phase is present within inclusions and along grain boundaries in altered basalts that become assimilated (e.g. Michael & Schilling 1989).

Brines in subseafloor hydrothermal systems form by high-temperature phase separation of sea water into gas plus a small amount of saline liquid (Bischoff & Rosenbauer 1987; Fournier 1987). One possibility is that OJP magmas became contaminated during eruption and transport on the seafloor, as magma moved through tube and/or sheet-flow systems. However, this is unlikely because the seafloor eruption pressures (<250 bars) for most sites (estimated from glass CO_2 data) are too low for brines with 50–60 wt% NaCl to form by phase separation (Fournier 1987). Moreover, contamination during eruption would be expected to result in a much wider variation of Cl contents. A more likely possibility is that the contamination occurred in well-mixed subseafloor magma chambers. Fluid inclusions from ophiolites and altered mid-ocean ridge gabbros contain as much as 52 wt% NaCl (Kelley & Delaney 1987; Nehlig 1991), but the highest NaCl contents found for vent fluids are only about twice the sea water value of 3.5 wt% (Von Damm & Bischoff 1987). Owing to the difference in density of the saline fluid and gas that form during phase separation of sea water, hydrothermal systems commonly form layered systems in which brine underlies fluid. However, mixing of these layers and further mixing with sea water probably occurs as fluids move to the seafloor, which might explain why high-salinity fluids have not been discovered at seafloor vents.

Formation of brine with >50 wt% NaCl by phase separation of sea water requires pressures greater than 250 bars and temperatures higher than 450°C (Fournier 1987; Nehlig 1991), which is consistent with contamination occurring either in the conduit system or in a shallow magma reservoir. The highest temperatures observed for hydrothermal fluids (*c.* 420°C; Berndt & Seyfried 1997) is believed to be limited by rock properties because at temperatures higher than about 450°C quasi-plastic behaviour closes off permeability (Fournier 1987). Higher temperatures for hydrothermal fluids are only possible if repeated magmatic intrusions preheat the surrounding wallrocks and repeated fracturing of the hot rock allows water to circulate (Fournier 1987). Given the large volume of OJP lava flows and the presumed large volumes and frequent magmatic recharge of subseafloor reservoirs, these conditions may be quite common, creating rocks surrounding the magma reservoirs with concentrated brine in fractures and along grain boundaries. Only small amounts of brine are needed to explain the Cl and H_2O data. The very homogeneous nature of large volume flows on the OJP suggests convective homogenization in magma chambers during crystal fractionation, so this might explain the observation that glasses from a given site are rather uniformly contaminated.

A final possibility is that the saline brines have a magmatic origin rather than being formed by phase separation of sea water (Nehlig 1991). Fractionation of olivine, plagioclase and clinopyroxene are required to derive the uniformly fractionated composition of OJP basalts from likely parent compositions, so enormous volumes of crystal mush must have been present within the OJP crust. Continued crystallization of crystal mush zones underlying magma chambers would result in exsolving H_2O–CO_2–Cl–S fluids that are relatively H_2O-rich. A brine with 50–60 wt% NaCl could form from this fluid phase under a wide range of conditions (from 450 to 550°C and *c.* 400 bars to temperatures higher than 1000°C at pressures greater than 1.6 kbar; Fournier 1987). However, given that Cl isotopes in MORB glasses indicate that the brine contaminant in mid-ocean ridge environments is derived from sea water (Magenheim *et al.* 1997), and the likelihood of large hydrothermal systems associated with OJP magma chambers, we think it is more plausible that the brine contaminants in the OJP were derived from sea water.

Glass shards from the shallow-water hydroclastic volcanic deposits at Site 1184 are anomalous in terms of Cl/K_2O and H_2O/K_2O (Fig. 8). Their composition could be explained either by contamination involving a very concentrated brine or partial degassing of H_2O during shallow-water eruption and quenching. The latter seems unlikely because the shards still contain some CO_2, which has much lower solubility than H_2O and should therefore be totally degassed before any significant H_2O is lost. Disequilibrium degassing caused by the slower diffusion of CO_2 in the melt relative to H_2O might allow H_2O to be exsolved while CO_2 is retained, but such partial degassing of H_2O is difficult to reconcile with the constant H_2O/K_2O and H_2O/TiO_2 of most of the shards. Brines with >60 wt% NaCl would be difficult to form in hydrothermal systems for reasons of temperature and quasi-plastic rock behaviour, as discussed above. However, at the low hydrostatic pressures at the seafloor and upper part of the conduit system for this shallow-water eruption, contact of hot magma with sea water could cause flash vaporization of sea water to form either gas plus solid salt or gas plus highly saline brine (>80 wt% NaCl; Fournier 1987). Thus, contamination of the magma erupted at Site 1184 could possibly have happened in the very shallow conduit system. If so, one would expect this to be a common process in shallow-water hydrovolcanic eruptions, but we are not aware of Cl-rich glass shards having been reported from other such deposits.

Water in the mantle source region for Ontong Java Plateau basaltic magmas

Ocean island basaltic magmas typically have higher H_2O than depleted MORB, suggesting that the excess magmatism associated with mantle plumes could be caused in part by the effect of H_2O on mantle melting (Schilling *et al.* 1980; Bonatti 1990; Nichols *et al.* 2002). The higher H_2O content of mantle-plume-derived magmas could result from either the involvement of undegassed primitive mantle (with high $^3He/^4He$) or deep recycling of subducted oceanic crust and sediments back into the mantle. The presence of such recycled lithospheric components in the source regions of mantle-plume-related basalts has been recognized on the basis of radiogenic Pb, Nd and Sr isotopes in some OIBs (Hofmann 1997). However, based on a detailed comparison of H_2O in plume-influenced

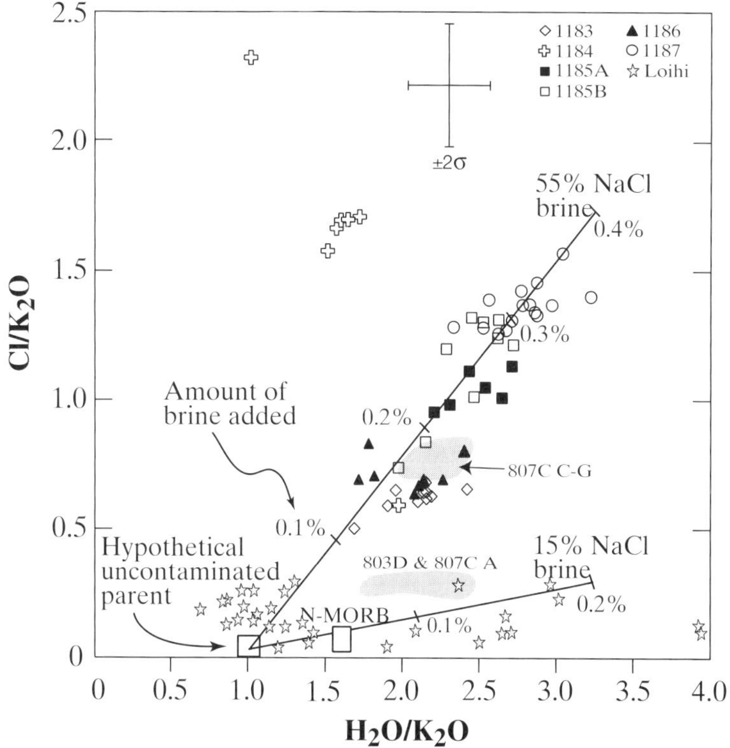

Fig. 8. Cl/K_2O v. H_2O/K_2O for basaltic glass from OJP. Lines show expected trends of composition for mixing between basaltic melt and small amounts of brine containing 15 and 55 wt% NaCl. The proportion of brine added (also in wt%) is noted by small tick marks along the lines. Loihi data are from Kent *et al.* (1999). Shaded fields as in Figure 2.

MORB glasses, it appears that basalts associated with mantle plume components containing recycled subducted lithosphere contain less H_2O than those without recycled components (Dixon *et al.* 2002). The likely explanation for this is that lithospheric components are efficiently dehydrated during subduction so that they are relatively poor in H_2O.

Before using the OJP glasses to constrain mantle volatile contents for the Ontong Java plume, it is important to ascertain whether significant H_2O was lost by degassing prior to eruption and quenching of the glass. Given the estimated pressures of eruption, the relatively low H_2O contents and the presence of CO_2, which has much lower solubility, it is unlikely that any significant loss of H_2O occurred before eruption. Closed-system degassing calculations (Fig. 4) suggest that less than 10% of the original primary H_2O would be lost by degassing, even if the primary magmas contained relatively high initial CO_2. However, the Cl and H_2O data (Fig. 8) show that assimilation involving brine has increased the H_2O content of all OJP basaltic glasses, including those previously analysed from Sites 803 and 807. The effect of this enrichment in H_2O must be considered before making an assessment of mantle source concentrations.

To estimate the H_2O content of the uncontaminated parent magma, we assume that it had Cl/K of 0.03 (Cl/K_2O = 0.025), intermediate between the values for uncontaminated normal (N-MORB) and enriched MORB. The basis for this assumption is that OJP basalts have incompatible-element abundances intermediate between those of N-MORB and many oceanic island tholeiites (Tejada *et al.* 1996; see also fig. 6 in Fitton & Godard 2004). Given the relations between Cl/K_2O and H_2O/K_2O for OJP basalts (Fig. 8), this requires that the parental magmas have H_2O/K_2O of approximately 1. Thus, for the high-MgO glasses, uncontaminated magmas would contain about 0.07 wt% H_2O. This is at the very low end of the range for N-MORB glasses that show no evidence of brine

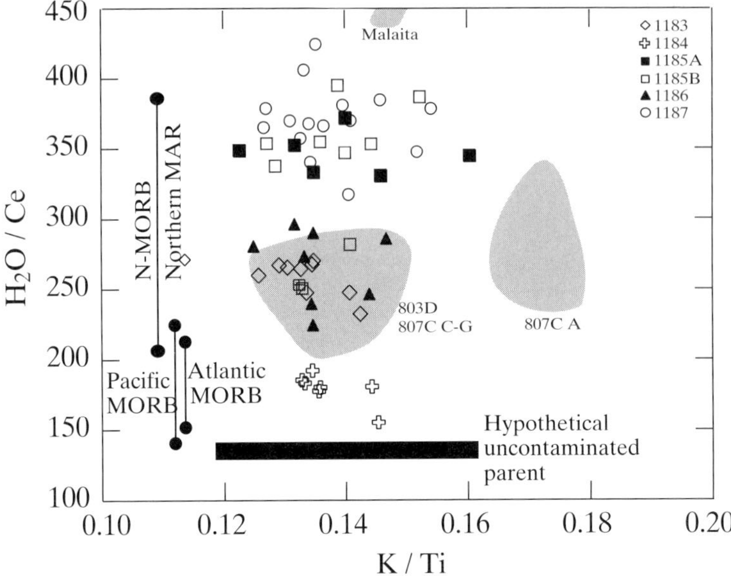

Fig. 9. H_2O/Ce v. K/Ti for OJP basaltic glasses. Shown for comparison are H_2O/Ce ranges for MORB glasses from various regions (MAR, Mid-Atlantic Ridge). The horizontal black bar shows the H_2O/Ce ratio estimated as described in the text for uncontaminated OJP magmas. Shaded fields as in Figure 2.

contamination (Michael 1995; Michael & Cornell 1998). Using the large degrees of mantle melting estimated for OJP basalts (*c.* 30%; Fitton & Godard 2004), an estimate of 27% crystallization to get from a primary composition in equilibrium with mantle olivine to the high-MgO glasses, and a partition coefficient of 0.01 for H_2O (Michael 1995), we calculate a source region H_2O concentration of 170 ± 30 ppm. This value is similar to the estimate for the depleted upper-mantle source for N-MORB (140 ± 40 ppm H_2O) and much lower than estimates for enriched MORB sources (350 ± 100 ppm H_2O; Michael 1988). Given the very low H_2O content inferred for the OJP mantle source, the large extents of melting (*c.* 30%) inferred from major- and trace-element data must have been caused by high mantle temperatures (>1500°C; Fitton & Godard 2004). Water does not appear to play a role in promoting large degrees of melting for the OJP.

To understand H_2O in mantle source regions it is useful to compare H_2O/Ce ratios because these elements have a similar incompatibility to one another during mantle melting and fractional crystallization (Michael 1995). Using our estimate of the H_2O content of high-MgO magmas at Sites 1185A and 1187 before contamination, and whole-rock Ce values (Fitton & Godard 2004; whole rock and glass from these sites have identical K_2O and TiO_2, consistent with a very low phenocryst content), we calculate H_2O/Ce to be 135 for uncontaminated OJP basalts (Fig. 9). However, it is difficult to know the uncertainty in this value because it is essentially based on our assumption that uncontaminated Cl/K is 0.03 for these samples. Mean values of H_2O/Ce for Pacific MORB glasses with ≤20 ppm Ce (to exclude highly enriched samples) are 145 ± 12 (Galapagos Spreading Centre), 150 ± 10 (Easter microplate), 180 ± 37 (Explorer and Juan de Fuca ridges), 189 ± 19 (Pacific–Nazca Ridge) and 194 ± 44 (East Pacific Rise; Michael 1995; Simons *et al.* 2002). Southern Mid-Atlantic Ridge MORB have comparable values (183 ± 30; Michael 1995), but MORB from the northernmost Mid-Atlantic Ridge and from around Iceland have distinctly higher values of 220–380 (Michael 1995). Our results suggest that H_2O/Ce for uncontaminated OJP basalts is at the low end of the range for Pacific MORB glasses.

Incompatible trace elements and Sr–Nd–Pb–Hf isotopic characteristics of the volumetrically dominant OJP basalt type (Kwaimbaita-type, named from occurrences in the Solomon Islands) show that they come from a mantle source region that is distinct in composition from the upper-mantle source for Pacific MORB, and there is no evidence for mixing involving a MORB end member (Tejada *et al.* 1996, 2004). Mixing involving an EM-1-like recycled lithospheric component can explain OJP isotopic ratios, but the homogeneity of major- and trace-element data for the OJP basalts shows no evidence of such mixing between several end members (Tejada *et al.* 2004). If an EM-1-like component was involved, the proportion of this recycled component would be very small (J. Mahoney pers. comm.). The OJP basalts are also isotopically distinct from the common mantle component (focus zone (FOZO) or common mantle component (C)) for OIBs (Hauri *et al.* 1994; Hanan & Graham 1996), which Simons *et al.* (2002) suggest has H_2O/Ce of approximately 210. The simplest explanation that accounts for both trace-element and isotopic data is that the OJP basalts were derived from a primitive mantle source that underwent a minor fractionation event (removal of *c.* 1% partial melt) at approximately 3 Ga (Tejada *et al.* 2004). A partial melting event would not significantly fractionate H_2O from Ce because of their similar incompatibility. Our results suggest that there is a major Pacific mantle reservoir, probably in the lower mantle, that has H_2O/Ce lower than most Pacific and all Atlantic MORB, and lower than the common mantle plume component for OIBs. Alternatively, if the OJP source region does contain an EM-1-like recycled component, then the lower H_2O/Ce relative to most Pacific MORB could be caused by the near-total dehydration of this component during subduction, as has been inferred for other EM-plumes like Hawaii (Dixon *et al.* 2002).

Volatile release to the environment during formation of the Ontong Java Plateau

OJP basaltic lavas sampled at DSDP/ODP drill sites and on land in the Solomon Islands were all erupted well below sea level, except for the volcaniclastic deposits erupted in shallow water at Site 1184 (Mahoney *et al.* 2001). Even at Site 1183, which lies on the crest of the main plateau and should have originally had the greatest elevation, pillow lava rims are poorly vesicular and CO_2 contents of the glass indicate eruption at about 1100 m water depth. Because nearly all of the plateau formed at substantial water depths, there would have been limited release of environmentally important volatiles such as S, Cl and F compared to subaerial flood basalt provinces or oceanic plateaus that formed subaerially, like the Kerguelen Plateau. High hydrostatic pressure during deep submarine eruptions and low magmatic H_2O contents would result in very little vesiculation, and these volatiles would have remained quenched in glassy pillow rims or would have entered accessory phases like sulphide during crystallization of pillow and sheet-flow interiors. In contrast, the very low solubility of CO_2 in silicate melts would result in most of the primary magmatic CO_2 being degassed even during deep-water eruptions or intrusive solidification. CO_2 released into the ocean would also affect atmospheric CO_2 given that the ocean–atmosphere system equilibrates on a geologically short timescale.

If OJP basaltic magmas contained as much CO_2 as other tholeiitic basalts derived from mantle plumes (Kilauea, Reunion), then the magmas may have contained as much as 7000 ppm CO_2 (Bureau *et al.* 1999; Gerlach *et al.* 2002). Given a crustal volume of approximately 5×10^7 km^3 (Coffin & Eldholm 1993), formation of the OJP could have released a maximum of about 1×10^{21} g CO_2. However, the time period over which the plateau was formed is poorly constrained. Isotopic data (^{40}Ar–^{39}Ar and Re–Os) suggest that the uppermost part of the plateau formed over about 1 Ma or less (Chambers *et al.* 2002; Parkinson *et al.* 2002), but whether more deeply buried extrusive and intrusive rocks formed over a longer period of time is unknown. If most of the volume of the plateau formed during a 1 Ma time period, then the annual CO_2 flux would have been approximately 1×10^{15} g CO_2 year^{-1}, equivalent to about 10 times the annual CO_2 flux from the global mid-ocean ridge system. Note that we have made assumptions about magma CO_2 content and eruption duration for the plateau that make our estimate for CO_2 flux during formation of the OJP a maximum value.

Another major uncertainty in assessing the environmental effects of plateau formation and volatile release is lack of information on the magnitude and extent of hydrothermal systems associated with shallow-crustal magma chambers. Basalts recovered from the OJP show only relatively minor, low-temperature alteration, and there is no evidence from recovered materials or on-land sections for extensive hydrothermal alteration like that associated with mid-ocean ridge volcanism (Mahoney *et al.* 2001). This is likely to be due to the limited sampling of the plateau rather than an absence of such hydrothermal activity because the large volume and fractionated nature of OJP basalts both suggest the existence of large, relatively shallow-crustal magma reservoirs. These large hydrothermal systems are important for leaching metals from basalt and releasing them to the ocean. Most importantly for assessing volatile release, hydrothermal systems precipitate carbonate and sulphate minerals in veins in altered basalt, resulting in significant transfer of C and S from sea water to altered crust. The magnitude of this transfer can be large, as there is good evidence from the oceanic crust that hydrothermal systems associated with mid-ocean ridge volcanism consume as much or more C and S from sea water than they release from magmatism (Staudigel *et al.* 1989; Alt & Teagle 1999).

Conclusions

Submarine basaltic glasses from five widely separated sites on the Ontong Java Plateau all have relatively low H_2O contents (0.16–0.24 wt%) that are similar to N-MORB values. In contrast, Cl contents and Cl/K ratios of the glasses are very high compared with MORB, probably as a result of assmilation involving small amounts of relatively concentrated brine. Such brines could have formed by high-temperature phase separation of sea water in large hydrothermal systems associated with shallow-crustal magma chambers on the OJP, and were stored in fractures and along grain boundaries of rocks that were assimilated by crystallizing magmas. The existence of large, upper-crustal magma chambers on the OJP is supported by crystal fractionation modelling, which shows that relatively low pressures (1 bar–2 kbar) are required to derive OJP basalts from likely primary magmas.

Ratios of H_2O to Ce, which have similar incompatibility to each other, are higher than most depleted and enriched MORB. However, positive correlation between H_2O/K_2O and Cl/K_2O in the glasses suggests that magmatic H_2O contents were increased by the same assimilation process that results in high Cl. Therefore, the high H_2O/Ce values of the glasses are probably an artifact of contamination. The water content of the high MgO-magmas before contamination is estimated to be approximately 0.07 wt% H_2O, corresponding to H_2O/Ce of 135 for OJP basalts, a value at the low end of the range for Pacific MORB. There is no evidence for high H_2O contents that would have significantly increased extents of mantle melting beneath the OJP, and the estimated H_2O content of the OJP mantle source region (170 ± 30 ppm H_2O) is similar to that of the depleted MORB source (140 ± 40 ppm H_2O). Large extents of melting beneath the OJP must have been caused by a relatively high mantle potential temperature, consistent with upwelling of a hot mantle plume.

OJP glasses have lower S (910–1030 ppm) at comparable FeO^T than MORB glasses. This indicates that OJP basaltic magmas could not have been saturated with immiscible sulphide liquid during crystallization unless they had significantly lower oxygen fugacities than MORB. Available data from measurements of S $K\alpha$ peak positions in the glasses show that oxygen fugacities of OJP magmas are relatively low, but this technique does not offer the precision necessary to resolve whether OJP magmas are more reduced than MORB magmas. Small decreases in S/K_2O and S/TiO_2 with decreasing MgO require some sulphide fractionation, but sulphides have not been observed in quenched glass samples, and platinum-group element systematics show no evidence for sulphide liquid fractionation. The low S contents of OJP basalts are probably caused by the high degree of melting (*c.* 30%) of the OJP source. Such large degrees of melting would probably exhaust residual sulphide in the mantle source during melting, in contrast to MORB magmas, which are generated by lower degrees of melting and probably form in equilibrium with residual mantle sulphide.

Because nearly all of the OJP formed at substantial water depths, there would have been limited release of environmentally important volatiles such as S, Cl and F compared to subaerial flood basalt provinces or oceanic plateaus that formed subaerially. However, the very low solubility of CO_2 in silicate melts would result in most CO_2 being degassed even during deep-water eruptions or intrusive solidification. The magnitude of CO_2 released during formation of the OJP is difficult to assess due to lack of information on primary magmatic CO_2 contents and magma output rates, but we estimate a maximum possible value that is about 10 times the annual CO_2 flux from the global mid-ocean ridge system.

We would like to thank J. Donovan for assistance with electron probe analyses, M. Ghiorso for his patient help with numerous questions and discussion about the MELTS program, and M. Coombs, P. Michael and M. Reed for suggestions and helpful discussions about Cl in hydrothermal fluids and the behaviour of S in OJP basalts. We are especially grateful to S. Newman for providing the program for peak fitting of IR spectra. We would also like to thank C. Macpherson and P. Michael for their thorough and constructive reviews. Funding was provided by the US Science Support Program (USSSP). The Ocean Drilling Program is sponsored by the National Science Foundation and participating countries under management of Joint Oceanographic Institutions, Inc. RV White is supported by a Royal Society Dorothy Hodgkin Research Fellowship and a NERC/ODP Rapid Response Grant.

References

ALT, J.C. & TEAGLE, D.A. 1999. The uptake of carbon during alteration of oceanic crust. *Geochimica et Cosmochimica Acta*, **63**, 1527–1535.

ANDERSON, A.T. 1974. Chlorine, sulfur, and water in magmas and oceans. *Geological Society of America Bulletin*, **85**, 1485–1492.

ASIMOW, P.D. & GHIORSO, M.S. 1998. Algorithmic modifications extending MELTS to calculate subsolidus phase relations. *American Mineralogist*, **83**, 1127–1132.

BANERJEE, N.R & MUEHLENBACHS, K. 2003. Tuff life: Bioalteration in volcaniclastic rocks from the Ontong Java Plateau. *Geochemistry Geophysics Geosystems*, **4**, doi 10.1029/2002GC000470.

BERNDT, M.E. & SEYFREID, W.E.J. 1997. Calibration of Br/Cl fractionation during subcritical phase-separation of seawater: Possible halite at 9 to 10°N East Pacific Rise. *Geochimica et Cosmochimica Acta*, **61**, 2849–2854.

BISCHOFF, J.L. & ROSENBAUER, R.J. 1987. Phase separation in seafloor geothermal systems: an experimental study of the effects on metal transport. *American Journal of Science*, **287**, 953–978.

BONATTI, E. 1990. Not so hot 'hotspots' in the oceanic mantle. *Science*, **250**, 107–111.

BUREAU, H., METRICH, N., SEMET, M.P. & STAUDACHER, T. 1999. Fluid-magma decoupling in a hot-spot volcano. *Geophysical Research Letters*, **26**, 3501–3504.

CHAMBERS, L.M., PRINGLE, M. S. & FITTON J.G. 2002. Age and duration of magmatism on the Ontong Java Plateau: ^{40}Ar–^{39}Ar results from ODP Leg 192. Abstract V71B-1271. *Eos, Transactions American Geophysical Union*, **83**, F47.

CHAZEY, W.J., III, NEAL, C.R. 2004. Large igneous province magma petrogenesis from source to surface: platinum-group element evidence from Ontong Java Plateau basalts recovered during Legs 130 and 192. *In*: FITTON, J.G., MAHONEY, J.J., WALLACE, P.J. & SAUNDERS, A.D. (eds) *Origin and Evolution of the Ontong Java Plateau*. Geological Society, London, Special Publications, **229**, 219–238.

COFFIN, M.F. & ELDHOLM, O. 1993. Scratching the surface: estimating dimensions of large igneous provinces. *Geology*, **21**, 515–518.

COURTILLOT, V.E. & RENNE, P.R. 2003. On the ages of flood basalt events. *Comptes Rendus Geoscience*, **335**, 113–140.

DIXON, J.E. & CLAGUE, D.A. 2001. Volatiles in basaltic glasses from Loihi Seamount: evidence for a relatively dry plume source. *Journal of Petrology*, **42**, 627–654.

DIXON, J.E. & PAN, V. 1995. Determination of the molar absortivity of dissolved carbonate in basanitic glass. *American Mineralogist*, **80**, 1339–1342.

DIXON, J.E. & STOLPER, E.M. 1995. An experimental study of water and carbon dioxode solubilities in Mid-Ocean Ridge basaltic liquids. Part II: applications to degassing. *Journal of Petrology*, **36**, 1633–1646.

DIXON, J.E., LEIST, L., LANGMUIR, C. & SCHILLING, J.-G. 2002. Recycled dehydrated lithosphere observed in plume-influenced mid-ocean-ridge basalt. *Nature*, **420**, 385–389.

DIXON, J.E., STOLPER, E.M. & HOLLOWAY, J.R. 1995. An experimental study of water and carbon dioxode solubilities in Mid-Ocean Ridge basaltic liquids. Part I: calibration and solubility models. *Journal of Petrology*, **36**, 1607–1631.

FARNETANI, C. G., RICHARDS, M.A. & GHIORSO, M.S. 1996. Petrological models of magma evolution and deep crustal structure beneath hotspots and flood basalt provinces. *Earth and Planetary Science Letters*, **143**, 81–96.

FITTON, J.G. & GODARD, M. 2004. Origin and evolution of magmas on the Ontong Java Plateau. *In*: FITTON, J.G., MAHONEY, J.J., WALLACE, P.J. & SAUNDERS, A.D. (eds) *Origin and Evolution of the Ontong Java Plateau*. Geological Society, London, Special Publications, **229**, 151–178.

FISK, M.R., GIOVANNONI, S.J. & THORSETH, I.H. 1998. Alteration of oceanic volcanic glass: Textural evidence of microbial activity. *Science*, **281**, 978–980.

FOURNIER, R.O. 1987. *Conceptual models of Brine Evolution in Magmatic–Hydrothermal Systems*. US Geological Survey, Professional Paper, **1350**, 1487–1506.

GERLACH, T.M., MCGEE, K.A., ELIAS, T., SUTTON, A.J. & DOUKAS, M.P. 2002. Carbon dioxide emission rate of Kilauea Volcano: Implications for primary magma and the summit reservoir. *Journal of Geophysical Research*, **107**, doi: 10.1029/2001JB000407.

GHIORSO, M.S. & SACK, R.O. 1995. Chemical mass transfer in magmatic processes; IV. A revised and internally consistent thermodynamic model for the interpolation and extrapolation of liquid–solid equilibria in magmatic systems at elevated temperatures and pressures. *Contribution to Mineralogy and Petrology*, **119**, 197–212.

HANAN, B.B. & GRAHAM, D.W. 1996. Lead and helium isotope evidence from oceanic basalts for a common deep source of mantle plumes. *Science*, **272**, 991–995.

HAURI, E.H., WHITEHEAD, J.A. & HART, S.R. 1994. Fluid dynamic and geochemical aspects of entrainment in mantle plumes. *Journal of Geophysical Research*, **99**, 24 275–24 300.

HOFMANN, A.W. 1997. Mantle geochemistry: the message from oceanic volcanism. *Nature*, **385**, 219–229.

IHINGER, P.D., HERVIG, R.V. & MCMILLAN, P.F. 1994. Analytical methods for volatiles in glasses. *In*: CAROLL, M. (ed.) *Volatiles in Magmas*. Reviews in Mineralogy. Mineralogical Society of America, Washington, DC, **30**, 67–121.

ITO, E., HARRIS, D.M. & ANDERSON, A.T. 1983. Alteration of oceanic crust and geologic cycling of chlorine and water. *Geochimica et Cosmochimica Acta*, **47**, 1613–1624.

JAROSEWICH, E., NELEN, J.A. & NORBERG, J.A. 1979. Electron microprobe reference samples for mineral analyses. *Contributions to the Earth Sciences*, **22**, 68–69.

KELLEY, D.S. & DELANEY, J.R. 1987. 2-Phase separation and fracturing in mid-ocean ridge gabbros at temperatures greater than 700°C. *Earth and Planetary Science Letters*, **83**, 53–66.

KENT, A.J.R., NORMAN, M.D., HUTCHEON, I.D. & STOLPER, E.M. 1999. Assimilation of seawater-derived components in an oceanic volcano: Evidence from matrix glasses and glass inclusions from Loihi seamount, Hawaii. *Chemical Geology*, **156**, 299–319.

LARSON, R.L. & ERBA, E. 1999. Onset of the mid-Cretaceous greenhouse in the Barremian–Aptian: Igneous events and the biological, sedimentary, and geochemical responses. *Paleoceanography*, **14**, 663–678.

LASSITER, J.C., HAURI, E.H., NIKOGOSIAN, I.K. & BARSCZUS, H.G. 2002. Chlorine–potassium variations in melt inclusions from Raivavae and Rapa, Austral Islands: Constraints on chlorine recycling in the mantle and evidence for brine-induced melting of oceanic crust. *Earth and Planetary Science Letters*, **202**, 525–540.

MAGENHEIM, A.J., HANAN, B.B. & PERFIT, M.R. 1997. Chlorine contamination of mid-ocean ridge lavas: New insight from chlorine stable isotope ratios. *In*: *Seventh Annual V.M. Goldschmidt Conference*. Lunar and Planetary Institute, Houston, TX, 130.

MAHONEY, J.J. & SPENCER, K.J. 1991. Isotopic evidence for the origin of the Manihiki and Ontong Java oceanic plateaus. *Earth and Planetary Science Letters*, **104**, 196–210.

MAHONEY, J.J., FITTON, J.G., WALLACE, P.J. *et al.*, 2001. *Proceedings of the Ocean Drilling Program, Initial Report*, **192**.

MAHONEY, J.J., STOREY, M., DUNCAN, R.A., SPENCER, K.J. & PRINGLE, M. 1993. Geochemistry and geochronology of the Ontong Java Plateau. *In*: PRINGLE, M., SAGER, W., SLITER, W. & STEIN, S. (eds) *The Mesozoic Pacific: Geology, Tectonics, and Volcanism*. American Geophysical Union, Geophysical Monograph, **77**, 233–261.

MAVROGENES, J.A. & O'NEILL, H.ST.C. 1999. The relative effects of pressure, temperature and oxygen fugacity on the solubility of sulfide in mafic magmas. *Geochimica et Cosmochimica Acta*, **63**, 1173–1180.

MICHAEL, P.J. 1988. The concentration, behavior and storage of H_2O in the suboceanic upper mantle. Implications for mantle metasomatism. *Geochimica et Cosmochimica Acta*, **52**, 555–566.

MICHAEL, P.J. 1995. Regionally distinctive sources of depleted MORB: evidence from trace elements and H_2O. *Earth and Planetary Science Letters*, **131**, 301–320.

MICHAEL, P.J. 1999. Implications for magmatic processes at Ontong Java Plateau from volatile and major element contents of Cretaceous basalt glasses. *Geochemistry Geophysics Geosystems*, **1**, doi: 10.1029/1999GC000025.

MICHAEL, P.J. & CORNELL, W.C. 1998. Influence of spreading rate and magma supply on crystallization and assimilation beneath mid-ocean ridges: Evidence from chlorine and major element chemistry of mid-ocean ridge basalts. *Journal of Geophysical Research*, **103**, 18 325–18 356.

MICHAEL, P.J. & SCHILLING, J.-G. 1989. Chlorine in mid-ocean ridge magmas: Evidence for assimilation of seawater-influenced components. *Geochimica et Cosmochimica Acta*, **53**, 3131–3143.

NEAL, C.R., MAHONEY, J.J., KROENKE, L.W., DUNCAN, R.A. & PETTERSON, M.G. 1997. The Ontong Java Plateau. *In*: MAHONEY, J.J. & COFFIN, M.F. (eds) *Large Igneous Provinces: Continental, Oceanic, and Planetary Flood Volcanism*. Geophysical Monograph. American Geophysical Union, Geophysical Monograph, **100**, 183–216.

NEHLIG, P. 1991. Salinity of oceanic hydrothermal fluids: a fluid inclusion study. *Earth and Planetary Science Letters*, **102**, 310–325.

NEWMAN, S. & LOWENSTERN, J.B. 2002. VOLATILECALC: a silicate melt–H_2O–CO_2 solution model written in Visual Basic for excel. *Computers and Geosciences*, **28**, 597–604.

NICHOLS, A.R.L., CARROLL, M.R. & HOSKULDSSON, A. 2002. Is the Iceland hot spot also wet? Evidence from the water contents of undegassed submarine and subglacial pillow basalts. *Earth and Planetary Science Letters*, **202**, 77–87.

PARKINSON, I.J., SCHAEFER, B.F. & ARCULUS, R.J. 2002. A lower mantle origin for the world's biggest LIP? A high precision Os isotope isochron from Ontong Java Plateau basalts drilled on ODP Leg 192. *Geochimica et Cosmochimica Acta*, **66**, Supplement, A580.

RICHARDS, M.A., JONES, D.L., DUNCAN, R.A. & DEPAOLO, D.J. 1991. A mantle plume initiation model for the Wrangellia flood basalt and other oceanic plateaus. *Science*, **254**, 263–267.

ROBERGE, J., WALLACE, P.J., WHITE, R.V. & COFFIN, M.F. (2004) Anomalous uplift and subsidence of the Ontong Java Plateau inferred from CO_2 contents of submarine basaltic glasses. Submitted to Geology.

SANO, T. & YAMASHITA, S. 2004. Experimental petrology of basement lavas from Ocean Drilling Program Leg 192: implications for differentiation processes of Ontong Java Plateau magmas. *In*: FITTON, J.G., MAHONEY, J.J., WALLACE, P.J. & SAUNDERS, A.D. (eds) *Origin and Evolution of the Ontong Java Plateau*. Geological Society, London, Special Publications, **229**, 185–218.

SCHILLING, J.G., UNNI, C.K. & BENDER, M.L. 1978. Origin of chlorine and bromine in the oceans. *Nature*, **273**, 631–636.

SCHILLING, J.G., BERGERON, M.B. & EVANS, R. 1980. Halogens in the mantle beneath the North Atlantic. *Philosophical Transactions of the Royal Society of London, Series A*, **297**, 147–178.

SIMONS, K., DIXON, J., SCHILLING, J.-G., KINGSLEY, R. & POREDA, R. 2002. Volatiles in basaltic glasses from the Easter-Salas y Gomez seamount chain and Easter Microplate: Implications for geochemical cycling of volatile elements. *Geochemistry Geophysics Geosystems*, **3**, doi 10.1029/2001GC000173

STAUDIGEL, H., HART, S.R., SCHMINCKE, H.-U. & SMITH, B.M. 1989. Cretaceous ocean crust at DSDP Sites 417 and 418: Carbon uptake from weathering versus loss by magmatic outgassing. *Geochimica et Cosmochimica Acta*, **53**, 3091–3094.

STOLPER, E.M. & NEWMAN, S. 1994. The role of water in the petrogenesis of Mariana trough magmas. *Earth and Planetary Science Letters*, **121**, 293–325.

TEJADA, M.L.G., MAHONEY, J.J., DUNCAN, R.A. & HAWKINS, M. 1996. Age and geochemistry of basement rocks and alkalic lavas of Malaita and Santa Isabel, Solomon Islands, southern margin of Ontong Java Plateau. *Journal of Petrology*, **37**, 361–394.

TEJADA, M.L.G., MAHONEY, J.J., CASTILLO, P.R., INGLE, S.P., SHETH, H.C. & WEIS, D. 2004. Pin-pricking the elephant: evidence on the origin of the Ontong Java Plateau from Pb–Sr–Hf–Nd isotopic characteristics of ODP Leg 192 basalts. *In*: FITTON, J.G., MAHONEY, J.J., WALLACE, P.J. & SAUNDERS, A.D. (eds) *Origin and Evolution of the Ontong Java Plateau*. Geological Society, London, Special Publications, **229**, 133–150.

VON DAMM, K.L.V. & BISCHOFF, J.L. 1987. Chemistry of hydrothermal solutions from the southern Juan de Fuca Ridge. *Journal of Geophysical Research*, **92**, 11 334–11 346.

WALLACE, A. & CARMICHAEL, I.S.E. 1992. Sulfur in basaltic magmas. *Geochimica et Cosmochimica Acta*, **56**, 1863–1874.

WALLACE, P.J. & CARMICHAEL, I.S.E. 1994. S speciation in submarine basaltic glasses as determined by measurements of SKa X-Ray wavelength shifts. *American Mineralogist*, **79**, 161–167.

WEAVER, J.S. & LANGMUIR C. 1990. Calculation of phase equilibrium in mineral–melt systems. *Computers and Geosciences*, **16**, 1–19.

WHITE, R.V., CASTILLO, P.R., NEAL, C.R., FITTON, J.G. & GODARD, M. 2004. Phreatomagmatic eruptions on the Ontong Java Plateau: chemical and isotopic relationship to Ontong Java Plateau basalts. *In*: FITTON, J.G., MAHONEY, J.J., WALLACE, P.J. & SAUNDERS, A.D. (eds) *Origin and Evolution of the Ontong Java Plateau*. Geological Society, London, Special Publications, **229**, 307–323.

Low-temperature alteration of submarine basalts from the Ontong Java Plateau

NEIL R. BANERJEE[1,2], JOSÉ HONNOREZ[2] & KARLIS MUEHLENBACHS[1]

[1]*Department of Earth and Atmospheric Sciences, University of Alberta, Edmonton, Alberta, Canada T6G 2E3 (e-mail: banerjee@ualberta.ca; karlis.muehlenbachs@ualberta.ca)*

[2]*Centre de Géochimie de la Surface, Ecole et Observatoire des Sciences de la Terre, Université Louis Pasteur, Strasbourg, 67084 France (e-mail: honnorez@illite.u-strasbg.fr)*

Abstract: We present a detailed mineralogical and petrological description of the low-temperature alteration patterns in basalts from four new sites drilled during ODP Leg 192 on the Early Cretaceous Ontong Java Plateau. Three main alteration types have been identified: pervasively altered dark grey basalt; black or dusky green halos; and brown halos. Dark grey basalts are the most common and represent the least intensive, but most pervasive, alteration phase. Early interaction of the basalts with low-temperature sea-water-derived hydrothermal fluids lead to the development of black and dusky green halos characterized by the replacement of groundmass and olivine phenocrysts by celadonitic phyllosilicates and smectite. Later interaction of basalts with cold oxidizing sea water produced brown halos characterized by replacement of primary phases and mesostasis by smectite and iron oxyhydroxides. Secondary minerals in order of decreasing abundance include phyllosilicates, calcite, iron oxyhydroxides, pyrite, chalcedony, quartz and zeolites. Veins, resulting from symmetrical infilling of open cracks, commonly contain phyllosilicates, iron oxyhydroxide or pyrite, and late calcite. Carbonate veins cross-cut all other alteration features and stable isotope analyses of vein carbonates indicate formation from marine bicarbonate below about 40°C. A positive correlation between vein density and overall degree of alteration is observed resulting in pervasive development of brown alteration halos in highly fractured rocks. Overall, alteration of basalts from the Ontong Java Plateau is similar to that observed from other DSDP/ODP sites throughout the oceans.

Low-temperature alteration of basalts by sea water has a significant influence on chemical fluxes between the oceanic crust and sea water. Most of our knowledge of low-temperature basalt alteration comes from studies of oceanic crust formed at mid-ocean ridges (MOR) (e.g. Böhlke *et al.* 1980; Honnorez 1981; Alt & Honnorez 1984; Alt *et al.* 1986*a*). However, relatively little is known about sea water–rock interactions during emplacement of oceanic large igneous provinces (LIPs) such as the Ontong Java Plateau (OJP). In this paper we describe alteration data collected from four new sites drilled on the OJP during Ocean Drilling Program (ODP) Leg 192. The main focus of Leg 192 was to core 100 m or so of basaltic basement from a number of new, geographically separated sites. This is significant because more than 1 km of pelagic sediments blanket the OJP in most areas making it difficult to sample the underlying basaltic basement (Mahoney *et al.* 2001). Two previous cruises (Deep Sea Drilling Project Leg 30 and ODP Leg 130) recovered basement at only three holes, two of which had very limited penetration. Despite basement penetration at Site 807 of 149 m, no detailed alteration study has been undertaken. As a result, the cores recovered during Leg 192 provide an excellent sample set to determine the effects of low-temperature sea-water–rock exchange during eruption of the OJP and submarine LIPs in general. Here we present a detailed mineralogical and petrological description of the alteration patterns observed in basalts erupted on the OJP.

Geological setting and samples

The OJP is the world's largest volcanic oceanic plateau covering approximately 2.0×10^6 km^2 and with an estimated crustal volume of 40×10^6–50×10^6 km^3 (Coffin & Eldholm 1993). Preliminary data from Leg 192 and previous sampling of the OJP suggest the majority of the plateau formed during a single, relatively rapid volcanic episode around 120 Ma, making it possibly the largest magmatic event on the Earth in the last 200 Ma (Mahoney *et al.* 2001;

From: FITTON, J. G., MAHONEY, J. J., WALLACE, P. J. & SAUNDERS, A. D. (eds) 2004. *Origin and Evolution of the Ontong Java Plateau*. Geological Society, London, Special Publications, **229**, 259–273. 0305-8719/$15.00

Chambers *et al.* 2002). The samples studied were recovered from Sites 1183, 1185, 1186 and 1187. Basaltic basement at these sites is made up of mainly pillow basalts and thin flows (Mahoney *et al.* 2001).

Methods

In order to quantify the alteration types and provide a consistent alteration characterization of the samples, the distribution of alteration types and secondary minerals filling veins were measured on a piece-by-piece scale by observation of cut wet surfaces of core hand specimens (Banerjee & Honnorez 2004). Estimates of vein density were calculated on the basis of recovery and are regarded as minima, as recovery is commonly poor in highly fractured/veined rock. All descriptions were performed on the archive half of the core, although observations were aided by viewing the working halves (which were later sampled) where specific features of interest may have been better preserved. Initial observations were conducted on hand-specimen scale by naked eye, ×10 magnifying lens and binocular microscope. Bulk rock and halo colours described below are from the Munsell Soil Color Charts (1975).

Primary mineralogical identification was conducted on polished thin sections by transmitted and reflected light microscopy. For this reason, specific mineral identifications are tentative (particularly for phyllosilicate minerals) except where unequivocal identification was possible based on optical properties alone. Clay minerals identified as smectites or celadonitic phyllosilicates are probably mixed layered and/or mixtures of saponite, nontronite, celadonite and, possibly, other phyllosilicates. It should be noted that the presence of celadonite is likely to be underestimated by hand-specimen identification because both smectite and celadonite appear black on wet, cut core surfaces. Mineral identification by shipboard X-ray diffraction (XRD) analysis was used to help confirm hand-specimen and thin-section identifications on a limited number of samples.

Carbonate minerals were separated from veins for isotopic analysis by hand picking. Stable C- and O-isotope analyses of carbonates were performed by pouring 100% phosphoric acid on powdered material under vacuum (McCrea 1950) and analysing the exsolved CO_2 on a Finnegan MAT 252 mass spectrometer. The data are reported in the usual delta-notation with respect to VPDB for carbon and VSMOW for oxygen. Reproducibility of replicate analyses of samples and standards is better than 0.1‰ for both carbon- and oxygen-isotope compositions.

Basalt alteration characteristics

The entire sequence of basaltic basement rocks recovered during Leg 192 has undergone low-temperature water–rock interactions, resulting in complete replacement of olivine and mesostasis. Clay minerals are the most abundant secondary minerals tentatively identified as saponite, nontronite and celadonite. We compared the phyllosilicates observed with well-studied clay minerals identified in other sections of the oceanic crust (e.g. Böhlke *et al.* 1980; Honnorez 1981; Alt & Honnorez 1984; Alt *et al.* 1986*a*). Based on core descriptions and thin-section observations, we have identified three major types of low-temperature alteration processes that correspond to the variously coloured halos observed in hand specimen: (1) dark grey basalt; (2) black halos or dusky green halos; and (3) brown halos.

Dark grey basalt refers to the normal grey colour (commonly dark grey but also varying to light grey) of the least-altered basalts from the inner portions of cooling units, commonly adjacent to the variously coloured halos described below. Pervasively altered dark grey basalt is the most abundant alteration type. Clay minerals (smectite ± celadonitic phyllosilicates) commonly replace olivine microphenocrysts, rare interstitial glass and mesostasis in the groundmass (Figs 1 and 2). Less commonly iron oxyhydroxides and calcite or (more rarely) pyrite are found in pseudomorphs of olivine and, rarely, of plagioclase phenocrysts. Similar secondary mineral assemblages fill miarolitic cavities and rare vesicles. The overall alteration of the grey basalts ranges from 5 to 30% and averages approximately 20%.

Centimetre-scale black halos and dusky green halos are observed along surfaces previously exposed to sea water and, less commonly, along the margins of veins. Within these halos, olivine phenocrysts are commonly altered to clay minerals and iron oxyhydroxides (Fig. 1). Bands of pyrite (and/or marcasite) are commonly observed at the margins of black halos, indicating the terminus of a reduction front. The dusky green halos are very similar to black halos but are interpreted to represent either an extreme case of replacement of primary basaltic phases by higher proportions of celadonite, which impart the dusky green colour, or intensely green colorations developed in more coarsely granular basalt with larger and more closely spaced intergranular pore spaces filled by green

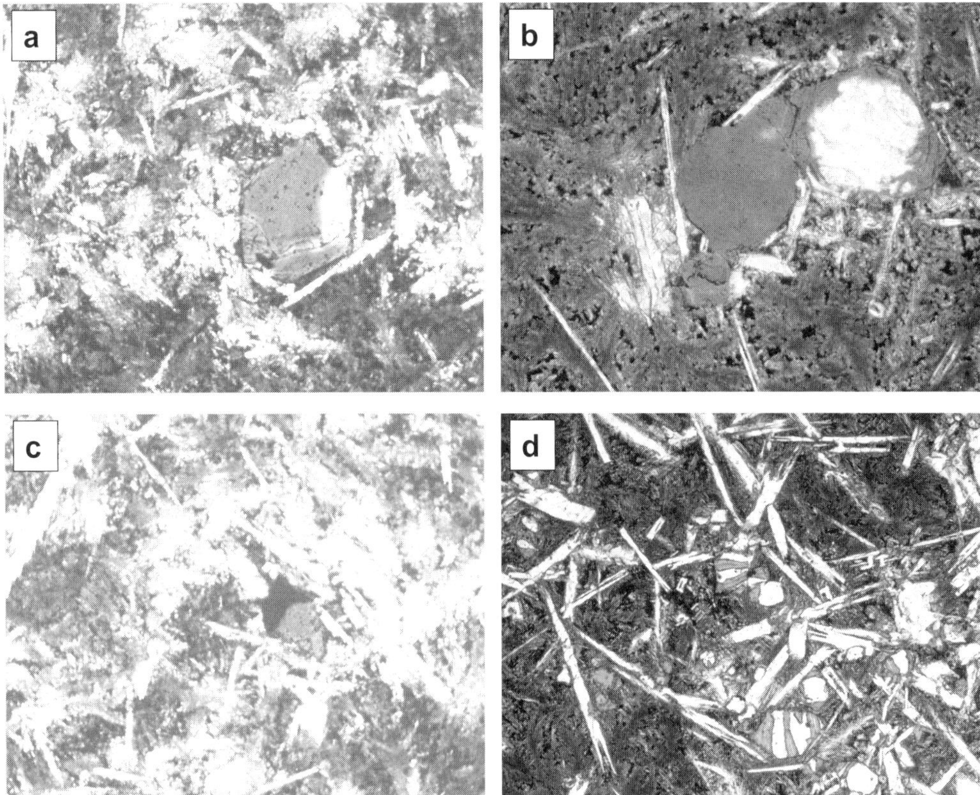

Fig. 1. Photomicrographs illustrating common secondary minerals replacing olivine (from Mahoney *et al.* 2001). (**a**) Euhedral olivine pseudomorph (replaced by celadonite) in a black halo (field of view is 1.4 mm). Sample 192-1183A-64R-2 (15–17 cm). (**b**) Euhedral olivine pseudomorph (right – containing celadonite and calcite) next to a vesicle (centre – containing celadonite) in a dusky green halo (field of view is 1.4 mm). Sample 192-1186A-31R-1 (44–48 cm). (**c**) Euhedral olivine pseudomorph (containing pyrite and celadonite) in a black halo (field of view is 1.4 m). Sample 192-1183A-64R-2 (15–17 cm). (**d**) Olivine phenocrysts partially altered to smectite in dark grey basalt (field of view is 2.8 mm). Sample 192-1187A-15R-4 (42–44 cm).

secondary phyllosilicates. The overall alteration rarely exceeds 30% in either dusky green or black halos. The boundary beyond the edges of these halos and the adjacent grey rock is commonly marked by fine-grained disseminated pyrite and/or marcasite in the groundmass. Calcite is less common in these halos than in the normal grey basalt. The contact between black halos and the grey interior is very sharp both in hand specimen and in thin section. This is a result of strong changes in chemical conditions across the alteration front during the formation of the dusky green and black halos.

Brown halos (yellowish-brown to brown to olive) are generally formed parallel to, or concentric with, complex smectite ± celadonite ± calcite veins, iron oxyhydroxide-bearing veins or glassy pillow margins, and they surround the least-altered dark grey basalt. These halos result from complete replacement of olivine and the pervasive alteration of groundmass. Their colour generally ranges, from the (originally) exposed surface inward, from light yellow–brown to dark yellow–brown (and more rarely dark brown) at the exterior, to olive, and finally grading to grey or dark grey in the inner parts of the cooling unit. In contrast to the sharp contact observed between the black or dusky green halos and the normal grey basalt, the transition between brown halos and both black halos and grey interiors is not sharp but generally gradational. The various brown colours result from variable proportions of iron oxyhydroxide and tan–brown smectite that stain and partly to totally replace olivine phenocrysts, the primary minerals of the basalt groundmass, and fill the

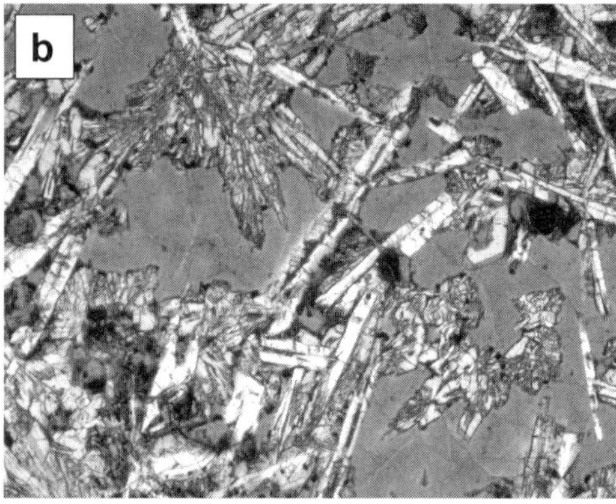

Fig. 2. Photomicrographs displaying common groundmass alteration patterns of typical dark grey basalt in which olivine phenocrysts and interstitial glass are completely replaced by smectite. Plagioclase and clinopyroxene are unaltered. Field of view for (**a**) is 5.5 mm and for (**b**) is 2.8 mm. Samples (a) 192-1185B-17R-3 (1–3 cm) and (b) 192-1187A-12R-5 (90–93 cm) (from Mahoney *et al.* 2001).

interstitial voids in the groundmass. Within these halos, olivine phenocrysts are totally replaced by iron oxyhydroxide, smectite and minor calcite. Miarolitic cavities and rare vesicles in the brown halos are filled with the same secondary minerals. Smectite and/or iron oxyhydroxide replace as much as 90% of the basalt groundmass in the lighter coloured halos. The overall alteration in the brown halos ranges from 30 to almost 100%.

Basaltic glass is present either in pillow rims or as shards in hyaloclastites, possibly associated with inter-pillow cavities. Fresh glassy mesostasis was not observed in pillow interiors. Alteration of basaltic glass to phyllosilicates ranges from 20 to 100%. Pillow-rim glass is generally the least altered because of its low permeability. Glass shards in the hyaloclastites are almost always completely replaced by phyllosilicates, except where cemented by micritic calcite. The association of unaltered glass clasts cemented by calcite is commonly observed in oceanic basalts, regardless of the age of the hyaloclastite or the environment in which it formed (Honnorez 1967, 1972).

Carbonate and smectite are the most commonly observed secondary minerals filling veins. Most veins contain the following succession of secondary minerals, from vein walls to centre: smectite ± celadonite, iron oxyhydroxide or pyrite, calcite, with minor zeolites and silica minerals (quartz and chalcedony). Disseminated pyrite grains commonly line the walls of smectite and/or celadonite veins. Evidence for the successive reopening and filling of veins is often clear, particularly in the case of carbonate deposition, because of the contrast in colour between the carbonate and the other vein-filling secondary minerals. Rare calcareous-sediment filled veins were observed throughout the basement sections. In the following sections we provide detailed descriptions of alteration patterns on a site-by-site basis.

Site 1183

Alteration of basalt from Hole 1183A ranges from <5 to 30 vol%, estimated visually by colour distribution. Low-temperature alteration in these samples has generally resulted in complete replacement of olivine and mesostasis (Fig. 1). Clinopyroxene and plagioclase only show alteration in some halos close to veins and miarolitic cavities. Smectite and celadonitic phyllosilicates are the most abundant secondary minerals. Other secondary minerals include iron oxyhydroxides, calcite, pyrite, chalcedony and quartz. These minerals are less abundant and have more restricted distributions.

The most common alteration type results in basalt colours ranging from dark to light grey (Fig. 3). Secondary smectite and celadonitic phyllosilicates in grey basalts replace interstitial glass and mesostasis in the groundmass. Smectite and celadonitic phyllosilicate minerals, and less commonly iron oxyhydroxides and calcite or (more rarely) pyrite, replace euhedral olivine microphenocrysts and, rarely, plagioclase phenocrysts (Fig. 1).

Very dark grey to very dark green to black halos are observed along exposed surfaces and, less commonly, along the walls of veins. Within black halos, olivine phenocrysts are commonly altered to smectite, celadonitic phyllosilicate and iron oxyhydroxides with rare pyrite (Fig. 1). Pyrite also occurs as grains scattered in the grey basalt groundmass beyond the black halo alteration front. Calcite is less common in these halos

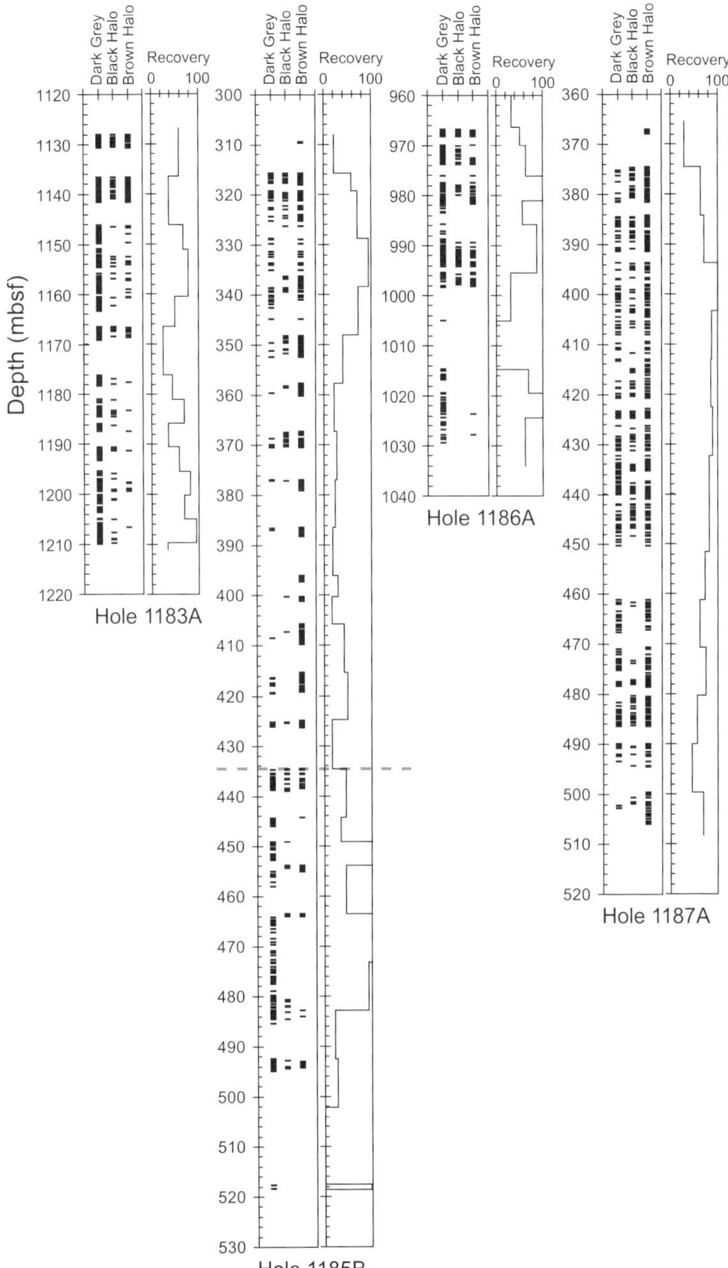

Fig. 3. Down-hole distribution (mbsf) of common alteration types in Holes 1183A, 1185B, 1186A and 1187A. Each occurrence of an alteration type is plotted (i.e. samples with multiple alteration types are plotted more than once). A dashed line marks the division between the upper and the lower alteration zones in Hole 1185B. Core recovery (%) is plotted at right.

than in dark grey altered basalt. Black halos decrease in abundance down-hole but less so than the brown halos described below (Fig. 3).

Brown–olive halos commonly occur in zones adjacent to iron oxyhydroxide-bearing veins. These halos overprint both dark grey altered basalt and black halos. Brown halos decrease in abundance down-hole (Fig. 3).

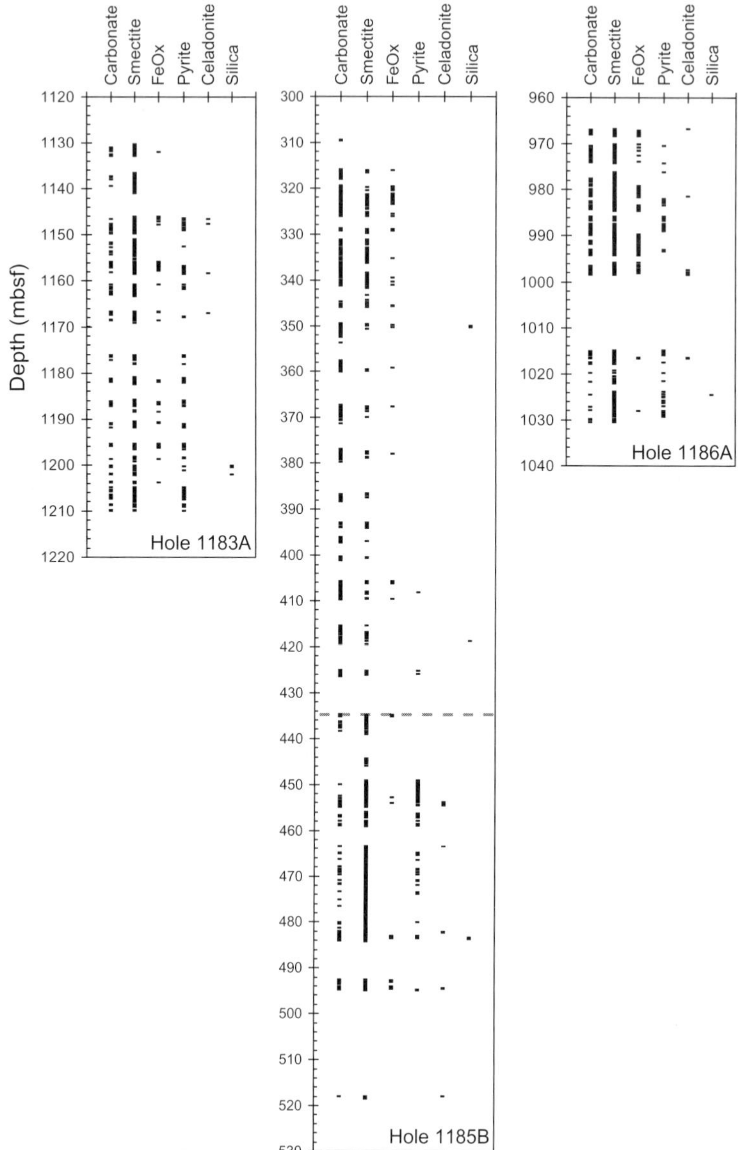

Fig. 4. Down-hole distribution (mbsf) of vein minerals in Holes 1183A, 1185B, 1186A and 1187A. Each occurrence of a vein mineral is plotted (i.e. veins containing more than one mineral are plotted more than once). The dashed line marks the division between the upper and the lower alteration zones in Hole 1185B. FeOx, iron oxyhydroxides; Silica, chalcedony ± quartz.

Alteration of basaltic glass is generally pervasive. One exception is a hyaloclastite found in Section 192–1183A-60R-1 (Piece 1) that contains pillow-rim material composed mostly of unaltered glass cemented by calcite (Mahoney *et al.* 2001).

Veins and miarolitic voids. We measured 849 veins in Hole 1183A at an average frequency of 19 veins m^{-1} (Banerjee & Honnorez 2004). Most of the veins result from symmetrical infilling of open cracks with minor or no replacement of the wall rock. Smectite is the most common vein mineral (Fig. 4) and most of the veins contain the following succession of secondary minerals, from vein walls to centres: smectite ± celadonitic phyllosilicate, iron oxyhydroxide or pyrite and calcite (Fig. 5). Evidence for the successive

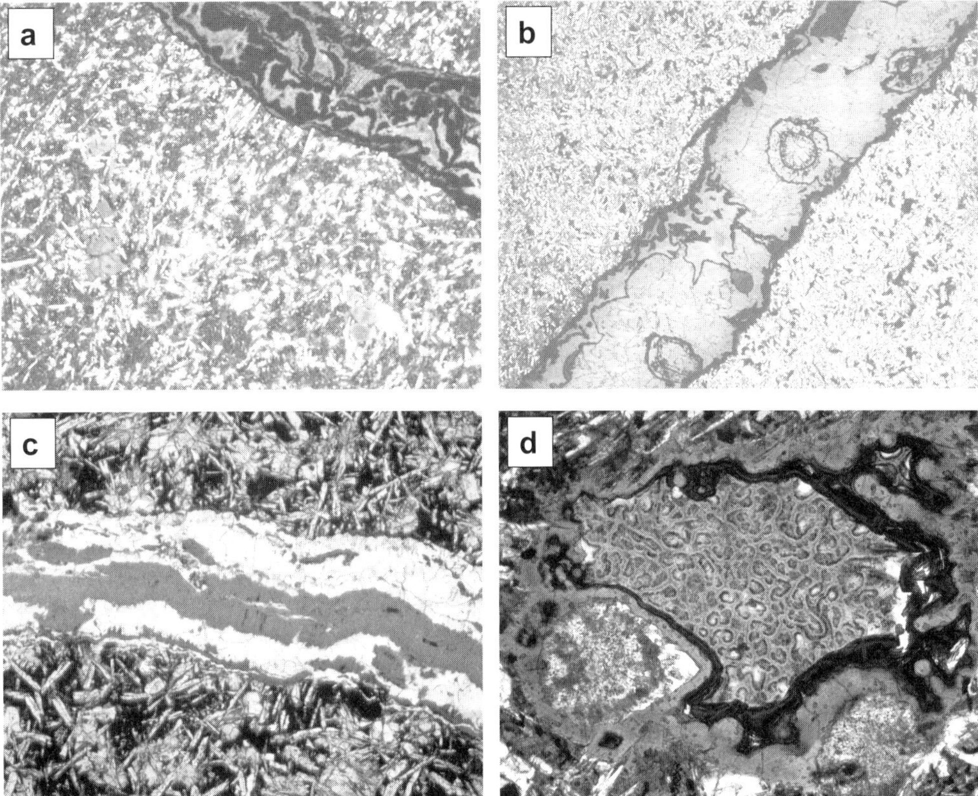

Fig. 5. Photomicrographs illustrating common secondary minerals in veins and miarolitic cavities (from Mahoney *et al.* 2001). (**a**) Vein containing nontronite and minor celadonite cutting across dark grey pillow basalt interior. Euhedral olivine crystals are replaced with the same secondary minerals (field of view is 5.5 mm). Sample 192-1183A-57R-3 (15–17 cm). (**b**) Vein containing celadonite and goethite cutting across dark grey basalt (field of view is 5.5 mm). Sample 192-1183A-65R-2 (84–88 cm). (**c**) Vein showing evidence of multiple openings/fillings with celadonite and calcite. Celadonite was deposited in both stages, first at the margins then in the centre. Brown halos have been developed in the wall rock on both sides of the vein. Within the brown halo the groundmass is altered to smectite and iron oxyhydroxides (field of view is 5.5 mm). Sample 192-1186A-35R-2 (77–80 cm). (**d**) Miarolitic cavity filled with a complex sequence of secondary minerals (from walls to centre): smectite, iddingsite, goethite, celadonite and calcite (field of view is 1.4 mm). Sample 192-1185B-20R-1 (27–29 cm).

reopening and filling of veins is often clear. In hand specimen, disseminated pyrite grains commonly line the walls of smectite and/or celadonitic phyllosilicate veins cracked open during drilling. However, observation of these veins in thin section by optical microscopy shows that pyrite commonly completely fills the width of these veins.

Carbonate and smectite does not show any systematic variation with depth (Fig. 4). Pyrite first appears below 1145 metres below seafloor (mbsf) and shows a slight increase in abundance with depth. Like calcite and smectite, iron oxyhydroxides are unevenly distributed with no systematic variation with depth. Rare occurrences of native copper occur in the uppermost basement cores. Silica minerals (chalcedony and quartz) were only observed in the lower part of the cored basement below 1200 mbsf (Fig. 6). Miarolitic voids are common throughout the cores, whereas vesicles are present only rarely. The largest dimension of the miarolitic cavities ranges from a few tenths of a millimetre to 4 mm. Vesicles are commonly < 1 mm in diameter. Vesicles and miarolitic cavities are generally completely filled by the same secondary minerals as those observed in the veins.

Site 1185

The basement section at Site 1185 was cored in two holes: A and B. Alteration of basalts in

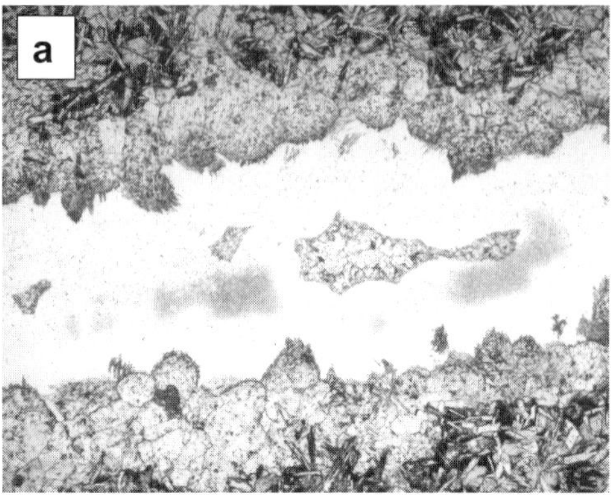

Fig. 6. Photomicrographs of vein filled with calcite (near walls) with chalcedony and quartz (centre). Sample 192-1183A-66R-2 (60–64 cm). Quartz is visible with crossed polars (**b**). Field of view is 5.5 mm for both photomicrographs (from Mahoney *et al.* 2001).

these holes can be divided into two superposed zones that display the effects of low-temperature submarine alteration under different conditions that probably corresponds to a difference in primary permeability of the basalt. The upper alteration zone consists of all basalts in Hole 1185A and basalts above about 435 mbsf in Hole 1185B. These basalts are dominated by pillow-lava flows. The lower alteration zone corresponds to basalts below *c.* 435 mbsf in Hole 1185B, which are mainly massive lava flows. Alteration characteristics common to both zones include broad (cm-scale), dusky green halos developed near veins. Reduction fronts that extend a few millimetres to centimetres beyond these halos that consist of scattered pyrite grains are common in the groundmass of the least altered dark grey rock interiors. The dusky green halos may be equivalent to the black halos described from Site 1183 with the greenish shade resulting from a higher abundance of celadonite or possibly a change in granularity of the basalts leading to larger pore spaces and a stronger visibility of secondary mineral colour. Smaller (mm-scale) dark yellowish brown–olive halos are developed close to veins and occur both as discrete features and in association with dusky green halos. Where these two halo types are associated, brown halos are commonly located proximal to the vein and the dusky green halos extend toward the least altered dark grey basalt interiors. Dusky green halos commonly preserve sharp contacts with the dark grey interiors, whereas the brown halo contacts are generally more diffuse.

Alteration of all basalts from Hole 1185A and above about 435 mbsf in Hole 1185B are characterized by the development of light and dark yellow–brown alteration halos (Fig. 3). The aphanitic basalt adjacent to glassy pillow rims is generally almost black (corresponding to least altered), with the exception of 3–6 mm brown spherulites. Within spherulites olivine is replaced by smectite and iron oxyhydroxide, whereas acicular plagioclase and clinopyroxene crystals radiating from the olivine nuclei are stained by iron oxyhydroxide. Where abundant, spherulites give the basalt a striking mottled appearance. Unaltered olivine is only preserved in the least altered, aphanitic, very dark grey–black areas close to glassy pillow rims. Rare replacement of the wall rock proximal to veins by an unidentified zeolite is observed in thin section. Alteration of all basalts from Hole 1185A and the upper part of Hole 1185B (above *c.* 435 mbsf) occurred under highly oxidizing conditions and with high water–rock ratios. These characteristics differ significantly from those in Hole 1183A.

The top of Core 1185B-17R contains breccia consisting of angular basalt fragments cemented by micritic sediment. These fragments represent the most pervasively altered basaltic material observed from all sites during Leg 192 (Mahoney *et al.* 2001). Below this interval a dramatic change in basalt alteration character occurs to a type similar to that observed in Hole 1183A. The least altered basalt in pillow interiors below Core 1185B-16R (*c.* 435 mbsf) is dark grey and halos are commonly dark yellowish brown–olive.

In the upper alteration zone, basalts commonly have partially altered glassy pillow margins that grade into black aphanitic pillow interiors. Rare hyaloclastites cemented by calcite are also present in this zone (Fig. 7). Secondary minerals replacing glass are predominantly smectite and iron oxyhydroxide. Unaltered glass is scarce below about 435 mbsf due to

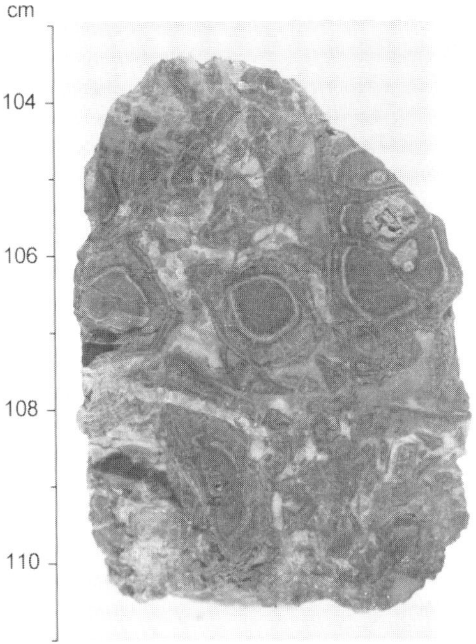

Fig. 7. Photograph of hyaloclastite showing complete replacement of glass fragments by smectite, and cemented by carbonate. Core interval 192-1185B-4R-1 (103–111 cm) (from Mahoney *et al.* 2001).

the abundance of thicker massive igneous units without glassy margins. Spherulites are absent from the massive lava flows in the lower part of Hole 1185B.

Veins and miarolitic voids. We measured 308 veins in Hole 1185A at an average frequency of 28 veins m^{-1} and 2282 veins in Hole 1185B at an average frequency of 25 veins m^{-1} (Banerjee & Honnorez 2004). However, within Hole 1185B vein density is much higher in the upper alteration zone (average frequency 30 veins m^{-1}) than the lower alteration zone (average frequency 19 veins m^{-1}). We noticed a clear relationship between vein density and host-rock alteration colour (Fig. 4). The lighter yellow–brown colours in the basalts are generally associated with portions of cores displaying more horizontal and subhorizontal veins; that is, from three to six veins per 10 cm of core length. The highest degree of alteration, as evidenced by the lightest colour, occurs in rocks with the highest permeability, such as highly veined and fractured pillow lavas, hyaloclastites and breccias.

Veins throughout Holes 1185A and 1185B predominantly contain carbonate, smectite, iron oxyhydroxide (and possibly manganese oxyhydroxide), pyrite, unidentified zeolite (Fig. 8), and rare celadonitic phyllosilicates and silica. Iron oxyhydroxide veins are most common in the upper alteration zone, Hole 1185A and above about 435 mbsf in Hole 1185B, corresponding to more oxidizing conditions. Conversely, pyrite is abundant in the lower alteration zone, below approximately 435 mbsf in Hole 1185B, probably due to locally more reducing conditions. Secondary mineral fillings in miarolitic cavities generally follow a similar pattern as vein minerals (Fig. 5). Silica minerals in veins are rare and show no systematic variation with depth. Some veins are filled with micritic pink carbonate that contains iron oxyhydroxide pellets and foraminifer ghosts, indicating that the veins are sediment-filled open fissures. These veins range in width from a few centimetres to <1 mm, and are located both near the upper boundaries of units and within units.

Site 1186

Alteration of the basaltic basement in Hole 1186A is similar to that in Hole 1183A. Clinopyroxene and plagioclase generally remain unaltered. Least-altered basalts, from the inner portions of cooling units, are characteristically dark grey with an overall alteration ranging from 5 to 30% (average *c.* 20%).

Like the lower alteration zone of Hole 1185B, dusky green halos are commonly present in the basalts from Hole 1186A instead of the black halos observed at Site 1183. As with Site 1185, we interpret the dusky green halos as equivalent to the black halos observed in Site 1183. In a few cases, pyrite or calcite is associated with celadonitic phyllosilicates in olivine pseudomorphs (Fig. 1). More commonly, fine-grained pyrite, or more rarely marcasite, is disseminated in the groundmass of the adjacent grey basalt beyond the front of the dusky green halos. Calcite is less common in these halos than in the brown halos (see below) and the dark grey basalts. Like Hole 1183A, we observe a decrease of these halos with depth (Fig. 3).

Brown halos are common in the upper *c.* 40 m of the basaltic sequence in Hole 1186A and occur in only two samples below this point (Fig. 3). The brown halos are generally parallel to, or concentric with, smectite ± celadonite ± carbonate veins (Fig. 5) or with glassy pillow margins, and they surround the least-altered dark grey inner basalt. Miarolitic cavities and rare vesicles in the brown halos are filled with the same secondary minerals present in the brown halos. Smectite and/or iron oxyhydroxide replace as much as 90% of the basalt groundmass in the lighter coloured halos.

Veins and miarolitic voids. We measured 715 veins in Hole 1186A, representing an average of 18 veins m^{-1} (Banerjee & Honnorez 2004). Most veins contain smectite and/or celadonitic phyllosilicates, iron oxyhydroxide or pyrite, and calcite in sequential order from vein walls to centre. Carbonate and iron oxyhydroxide tend to decrease in abundance with depth, whereas pyrite tends to increase (Fig. 4). The abundance of smectite and celadonite shows no down-hole trend. Chalcedony associated with calcite spherules was observed in one sample from the bottom of the section. Disseminated pyrite and/or marcasite grains are commonly found scattered in the walls of smectite veins cracked open during drilling. Where the core is fractured perpendicularly to veins surrounded by dusky green halos, bright bands of pyrite (and/or marcasite) are commonly observed at the margins of the halos. Veins filled with pinkish micritic carbonate, interpreted as sediment at Site 1185, are also observed in Hole 1186A. The walls of these veins are commonly lined with crystalline calcite, suggesting successive filling and reopening.

Site 1187

Basalt in Hole 1187 appears to have undergone the greatest overall alteration of any Leg 192 site. This is characterized by the common development of light–dark yellow–brown alteration halos (Fig. 3). The high degree of alteration is consistent with the dominantly pillowed nature of the basalts recovered in this hole, with massive basalt forming only a minor component. The light–dark yellow–brown colours in the outer portions of the pillow lavas result from the complete replacement of olivine and partial–complete alteration of groundmass by brown–tan smectite and iron oxyhydroxide (Fig. 2). Smectite, calcite and minor goethite fill miarolitic cavities, which are commonly surrounded by 2–10 mm-wide brown halos. Black or dusky green halos are rare and commonly thin or poorly preserved. Away from pillow margins, the colour grades into dark grey within several coarser-grained pillow interiors. Although olivine phenocrysts are generally

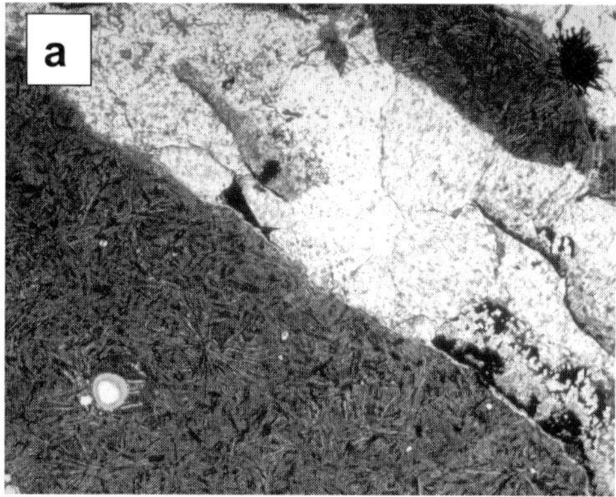

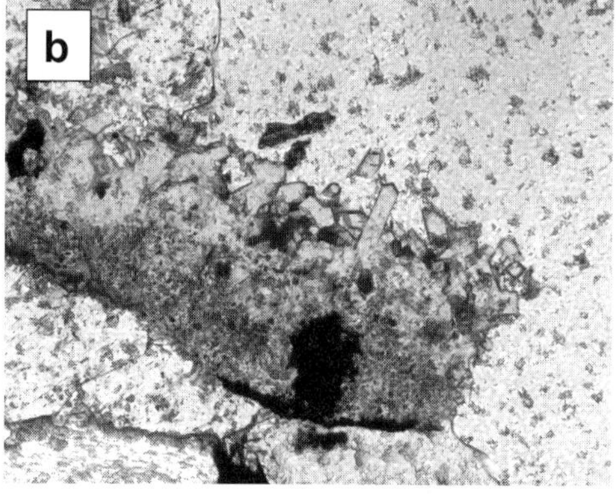

Fig. 8. (**a**) Vein containing (from walls to centre): thin tan smectite edge, botryoidal concretions of opaque minerals (possibly manganese oxyhydroxides), discontinuous screens of blocky zeolites with opaque margins and sparry calcite. Sample 192-1185B-5R-8 (31–33 cm). Approximately 85% of the vein contains calcite. The zeolite screens represent fragments of the wall lining plucked away during successive reopening. The vesicle in the lower left is rimmed by smectite. Field of view is 5.5 mm. (**b**) Close-up of a zeolite screen from (a) showing prismatic, longitudinal and square cross-sections (probably phillipsite). The zeolite screen is rimmed at the bottom by opaque manganese oxyhydroxides (?) and is surrounded by calcite. Field of view is 1.4 mm. (**c**) Close-up of botryoidal aggregate of opaque material (possibly manganese oxyhydroxides) from the upper-right corner of (a). Basalt fragments are lined by smectite and surrounded by calcite. Field of view is 2.8 mm (from Mahoney *et al.* 2001).

completely replaced, unaltered or only incipiently altered olivine is present in several dark grey basalt intervals interpreted as either the interiors of large pillows or as massive flows (Fig. 1). In this respect, Site 1187 is unique among the Leg 192 sites.

Unaltered glass is more abundant here than at any other Leg 192 site, despite the pervasive alteration of pillow margins. For example, vertical sections through complete pillows including both upper and lower glassy margins and the aphanitic interior (altered to iron oxyhydroxide and parallel to the pillow margin) are preserved (Mahoney *et al.* 2001). Secondary minerals replacing glass are predominantly smectite and iron oxyhydroxide.

Veins and miarolitic voids. A vein log was not compiled for Hole 1187A so comparisons of vein density with the other sites cannot be made (Banerjee & Honnorez 2004). Veins throughout Hole 1187A are mostly filled with calcite, unidentified zeolite, smectite, iron oxyhydroxide, and rare celadonitic phyllosilicate, pyrite and marcasite. Abundant subhorizontal calcite veins are present in light–dark yellow–brown pillow margins, and calcite veins radiate from pillow interiors toward the margins in well-preserved pillows (Fig. 9), possibly filling radial thermal contraction fractures. As with the upper basement alteration zone at Site 1185, we again noticed a clear relationship between vein abundance and the colour of host-rock alteration. The lighter yellow–brown colours in the basalts at both sites are generally associated with the parts of cores that display the most abundant horizontal and subhorizontal veins. Because the degree of alteration is a function of permeability, the abundance of veins, pervasiveness of the alteration, and development of light and dark yellow–brown alteration colours indicate that basement in Hole 1187 may have been initially the most permeable of any site drilled during Leg 192.

Stable isotopes

Stable isotopic data for vein carbonates are illustrated in Figure 10. Vein carbonates typically have $\delta^{13}C$ ratios between –2 and 4‰, typical of previous studies of vein carbonates in oceanic basalts (Muehlenbachs 1977, 1979; Muehlenbachs & Hodges 1978; Alt *et al.* 1986*b*; Teagle *et al.* 1996). Rare values below –2 ‰ occur in Hole 1187A and the upper portion of Hole 1185B. These rare depleted values correspond to intervals of sediments intercalated with the basalts and probably represent oxidized organic carbon

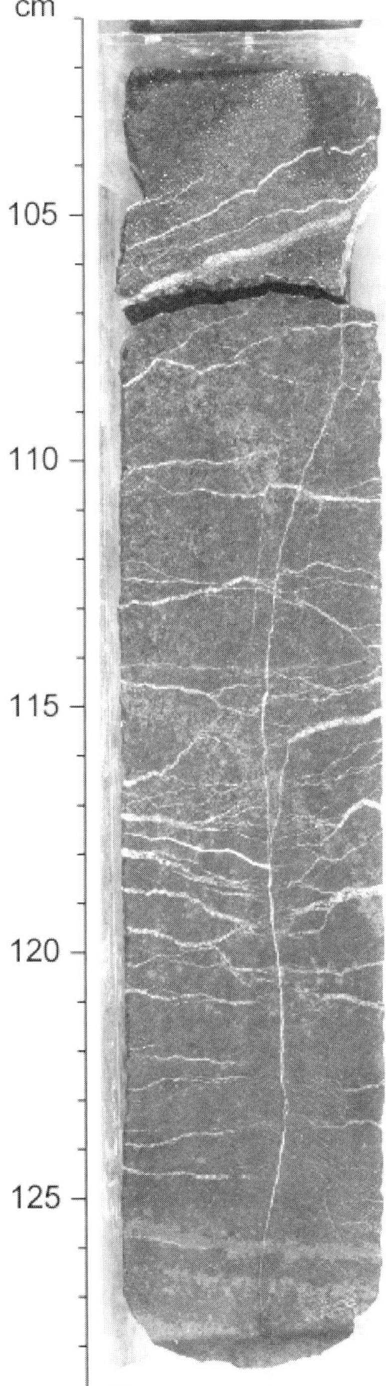

Fig. 9. Photograph of pervasively altered basalt core cut by numerous carbonate-bearing veins. Core interval 192-1187A-5R-6 (102–128 cm) (from Mahoney *et al.* 2001).

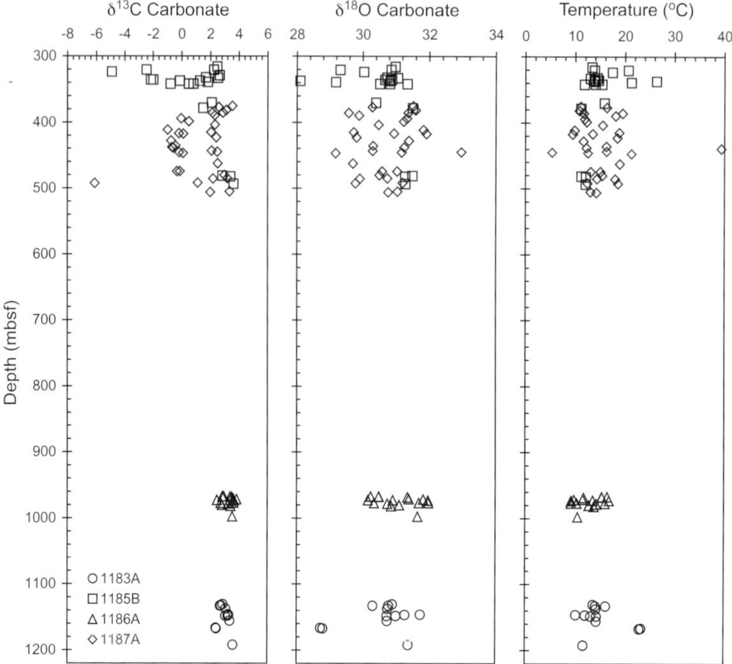

Fig. 10. Isotopic data and calculated temperatures of formation for vein carbonates from Holes 1183A, 1185B, 1186A and 1187A. Temperatures are calculated assuming equilibrium with normal sea water (0‰ VSMOW; O'Neil *et al.* 1969).

incorporated from sedimentary material. Oxygen isotope ratios in vein carbonates vary from 28 to 34‰, also typical of previous studies (Muehlenbachs 1977, 1979; Muehlenbachs & Hodges 1978; Alt *et al.* 1986*b*; Teagle *et al.* 1996). Both $\delta^{13}C$ and $\delta^{18}O$ ratios show considerable scatter and neither show systematic trends with depth within individual holes, although $\delta^{13}C$ ratios at shallower depths below sea floor (Sites 1185 and 1187) extend to slightly more depleted values than at the deeper sites (Fig. 10). Other studies of vein carbonate stable isotopic compositions commonly exhibit a decrease in $\delta^{18}O$ with depth (see 'Temperatures of alteration' later), corresponding to increasing temperature (e.g. Muehlenbachs 1977, 1979; Muehlenbachs & Hodges 1978; Alt *et al.* 1986*b*; Teagle *et al.* 1996). This trend is not observed in Leg 192 samples probably due to the relatively shallow depths of penetration.

Discussion

Basalts from Sites 1183, 1185, 1186 and 1187 on the OJP all show similar effects of low-temperature alteration resulting in the development of variously coloured halos along exposed surfaces and veins. The general sequence of halos, from cores to exterior, is dark grey basalt, followed by black or dusky green halos, and finally brown halos. This distribution is related to the early development of the black or dusky green halos that are later overprinted by brown halos. The following sections provide a brief explanation of the mechanisms and conditions of formation of the halos and veins.

Pervasively altered dark grey basalt

The dark grey colour results from extended and pervasive interaction between basalt and sea-water-derived fluid (evolved sea water) under anoxic–suboxic conditions at low temperature (probably 10–50°C). These fluids probably reacted previously with basaltic crust (e.g. with the rock of the halo adjacent to the grey basalt; Alt *et al.* 1986*a*). The water–rock ratio during such interaction is probably low. This alteration stage ceases once secondary minerals fill fluid pathways, sealing the formation and permeability.

Black and dusky green halos

This stage of alteration results from early interaction between basalt and warm (<60°C) sea-water-derived fluids (e.g. Böhlke *et al.* 1980; Honnorez 1981; Laverne 1987; Buatier 1989). Such fluids are supplied by diffuse warm springs of shimmering water that have been conductively cooled by mixing with bottom sea water (James & Elderfield 1996). The formation of black and probably dusky green halos are characteristic of an early alteration process initiated during cooling of the lava within much less than 1 Ma of basalt emplacement (Honnorez unpulished data). Further effusions of lava and/or injection of magma during dyking events can reactivate this alteration process.

As noted earlier, dusky green and black halos that surround veins at the other basement sites are rare or poorly developed in basalt from Hole 1187A, in which the alteration is characterized by extensive brown halos. Reduction fronts containing pyrite and/or marcasite are similarly absent. Two possible explanations are: (1) that neither feature ever formed; or (2) both may have formed but were overprinted by the later pervasive oxidative alteration that produced the various brown-coloured halos. We prefer the second explanation because black and dusky green halos are ubiquitous and form quickly after basalt emplacement.

Brown halos

Brown halos result from basalt–sea water reaction referred to as halmyrolysis or submarine weathering. This form of alteration takes place at bottom sea water temperature (*c.* 2°C), under oxidizing conditions, and generally large water/rock ratios (Honnorez 1981 and references therein). This corresponds to passive alteration of basalt by bottom sea water circulating through the crust to depths of several hundred metres. In the most permeable basaltic formations, such as pillow lavas, hyaloclastites and breccias, halmyrolysis may lead to intense alteration as seen in Hole 1187A. Brown halos represent the last, oxidizing, low-temperature alteration stage, which ceases when the oceanic crust is sealed off from overlying sea water by a sufficiently thick and comparatively impermeable sediment cover.

Veins and pore space fillings

Most of the veins result from symmetrical infilling of open cracks. Vein widths vary from <1 mm (hairline veins) to tens of millimetres. We observed a bimodal distribution of vein densities. Holes 1183A, 1186A and the lower alteration zone of Hole 1185B have an average frequency of about 19 veins m^{-1}. Hole 1185A and the upper alteration zone on Hole 1185B have higher vein densities (*c.* 25 veins m^{-1}), and correspondingly higher degrees of alteration. Although a vein log was not performed for Hole 1187A, we infer that vein density would be similar to the upper alteration zone observed at Site 1185 based on similarity of alteration patterns.

Temperatures of alteration

Stable isotopic mineral–water fractionation factors are a useful tool for determining the temperatures of formation of secondary minerals during low-temperature alteration of basalts. Although systematic stable isotopic analyses of secondary mineral separates were not performed in this study, comparison of mineral assemblages with previous studies of basalts altered at low temperature can provide some constraints. All of the secondary minerals (e.g. smectite, celadonitic phyllosilicates, iron oxyhydroxides, zeolites and calcite) observed in basement during Leg 192 are known to form at low temperatures (less than about 100°C) during basalt alteration (e.g. Honnorez 1967, 1972; Muehlenbachs 1977, 1979; Muehlenbachs & Hodges 1978; Böhlke *et al.* 1980; Honnorez 1981; Alt & Honnorez 1984; Alt *et al.* 1986*a*). Our only direct evidence for alteration temperatures in the basement sequence of the OJP comes from calculations based on carbonate–water isotope fractionation factors assuming the carbonates formed in equilibrium with normal sea water (O'Neil *et al.* 1969). Temperatures calculated from carbonate oxygen isotope data yield values between 0 and 40°C, but most values cluster between approximately 10 and 20°C (Fig. 10) (O'Neil *et al.* 1969). Owing to the scatter in the data discussed above, there is no clear evidence of a decrease in $\delta^{18}O$ of vein carbonates with depth that would correspond to an increase in temperature with depth. This is probably due to the relatively shallow depths of penetration and the probable high original permeabilities, particularly in Hole 1187 and the upper portion of Hole 1185B. The $\delta^{18}O$ ratios of vein carbonates from OJP are well within the range observed for the upper few hundred metres of Hole 504B and other studies of upper oceanic crust (e.g. Muehlenbachs 1977, 1979; Muehlenbachs & Hodges 1978; Alt *et al.* 1986*b*; Teagle *et al.* 1996). Oxygen isotope ratios of carbonates in Holes 504B and 896A only tend to significantly decrease below about 28‰ at depths of greater than about 300 m sub-basement, which is considerably deeper than the holes from this study (Alt *et al.* 1986*b*; Teagle *et al.* 1996). The low temperatures recorded by vein carbonates are not surprising given that they cross-cut all other features and represent the latest sealing of the crust from overlying sea water. Furthermore, calcite precipitation is not necessarily tied to the processes described above leading to the development of alteration halos but can occur long after the crust has formed (e.g. Staudigel & Hart 1985; Alt & Teagle 1999).

Conclusions

Alteration of basalts from the upper oceanic crust formed at the OJP can be classified into three main types: pervasively altered dark grey basalt; black or dusky green halos; and brown halos. These alteration types all result from low-temperature fluid–rock interactions. Dark grey basalts represent the least intensive, but most pervasive, alteration phase. Early interaction of low-temperature sea-water-derived hydrothermal fluids under anoxic conditions (at possibly low water/rock) lead to the replacement of groundmass and olivine phenocrysts by celadonitic phyllosilicates and smectite, resulting in the formation of black or dusky green halos. Pyrite bands are commonly observed beyond the alteration front marked by the black and dusky green halos. Later interaction of basalts with cold oxidizing sea water (at possibly higher water/rock) produced brown halos characterized by replacement of primary phases by smectite and iron oxyhydroxides. Phyllosilicates are the most abundant secondary minerals with lesser amounts of calcite, iron oxyhydroxides, pyrite, chalcedony, quartz and zeolites. Veins, resulting from symmetrical infilling of open cracks, commonly contain smectite and/or celadonite, iron oxyhydroxide or pyrite and calcite (as the final phase). The higher vein densities observed in Hole 1185A, Hole 1187A and the upper alteration zone of Hole 1185B correspond to a higher proportion of brown alteration halos resulting in higher overall degrees of alteration and replacement of the groundmass. Isotopic analyses of vein carbonates indicate that they predominantly precipitated from marine bicarbonate, and calculated temperatures indicate that they formed below about 40°C. Calcite veins clearly cross-cut all other alteration features and represent the final alteration process that continues long after basalt eruption on the seafloor (Staudigel & Hart 1985; Alt & Teagle 1999).

Our results suggest that nothing with respect to mineralogical, petrographical or chemical characteristics can differentiate alteration processes operating within the uppermost portion of the OJP from oceanic crust formed at MOR spreading centres. This is to be expected as all of the effects observed in MOR basalts, from the development of black halos to halmyrolysis, result from low-temperature interaction between sea water or sea-water-derived fluids and tholeiitic basalts, which is exactly the same for the OJP. This similarity could be limited only to the uppermost sequence of basaltic basement of the OJP or it might also be characteristic of basalt alteration at depth. The question still remains whether or not alteration of the huge volume of basalts that make up the OJP is similar to MOR-generated basalts as a whole or if the relatively rapid rate of eruption resulted in special alteration conditions. We believe that low-temperature alteration is probably very similar to MOR basalts, even at greater depths (< 500 m), but that higher-temperature hydrothermal alteration may have been more affected by the special tectonic conditions and rapid emplacement of this oceanic plateau. Answers to questions regarding alteration of the OJP as a whole will have an important impact on calculations of chemical fluxes and budgets of sea-water–crust exchange but can only be addressed once deeper drilling is achieved.

We thank O. Levner and T. Huynh for help with stable isotope analyses; S. Laurence for help with data entry; and the captain, crew, technicians and science party of ODP Leg 192 aboard the drill ship *JOIDES Resolution* who made Leg 192 a success. We also thank J. Alt and D. Teagle for helpful reviews. This research used samples provided by the Ocean Drilling Program (ODP). ODP is sponsored by the US National Science Foundation (NSF) and participating countries under management of Joint Oceanographic Institutions (JOI), Inc. Participation in Leg 192 for NRB was funded by Canada ODP and by ODP France for J. Honnorez. Funding for this research was provided by the National Sciences and Engineering Research Council of Canada and the Institut National des Sciences de l'Univers of France.

References

ALT, J.C. & HONNOREZ, J. 1984. Alteration of the upper oceanic crust, DSDP Site 417: mineralogy and chemistry. *Contributions to Mineralogy and Petrology*, **87**, 149–169.

ALT, J.C. & TEAGLE, D.A.H. 1999. Uptake of carbon during alteration of oceanic crust. *Geochimica et Cosmochimica Acta*, **63**, 1527–1535.

ALT, J.C., HONNOREZ, J., LAVERNE, C. & EMMERMANN, R. 1986*a*. Hydrothermal alteration of a 1 km section through the upper oceanic crust, Deep Sea Drilling Project Hole 504B: mineralogy, chemistry, and evolution of seawater–basalt interactions. *Journal of Geophysical Research*, **91**, 10 309–10 335.

ALT, J.C., MUEHLENBACHS, K. & HONNOREZ, J. 1986*b*.

An oxygen isotope profile through the upper kilometer of the oceanic crust, DSDP Hole 504B. *Earth and Planetary Science Letters*, **80**, 217–229.

BANERJEE, N.R. & HONNOREZ, J. 2004. Low-temperature alteration of upper oceanic crust from the Ontong Java Plateau, Leg 192: Alteration and vein logs. Data report. *In*: FITTON, J.G., MAHONEY, J.J., WALLACE, P.J. & SAUNDERS, A.P. (eds). *Proceedings of the Ocean Drilling Program, Scientific Results*, **192**. [Online]. Available from World Wide Web: http://www-odp.tamu.edu/publications/192_SR/VOLUME/CHAPTERS/103.PDF.

BÖHLKE, J.K., HONNOREZ, J. & HONNOREZ-GUERSTEIN, B.M. 1980. Alteration of basalts from site 396B, DSDP: petrographic and mineralogical studies. *Contributions to Mineralogy and Petrology*, **73**, 341–364.

BUATIER, M. 1989. *Genèse et évolution des argiles vertes hydrothermales océaniques: les monts du rift des galapagos (Pacifique équatorial)*. PhD Thesis, Université Louis Pasteur.

COFFIN, M.F. & ELDHOLM, O. 1993. Scratching the surface: estimating dimensions of large igneous provinces. *Geology*, **21**, 515–518.

CHAMBERS, L.M., PRINGLE, M.S. & FITTON, J.G. 2002. Age and duration of magmatism on the Ontong Java Plateau: $^{40}Ar/^{39}Ar$ results from ODP Leg 192, *Eos Transactions of the American Geophysical Union*, **83**, Fall Meeting Supplement, Abstract V71B, 1271.

HONNOREZ, J. 1967. *La palagonitisation: l'alteration sous-marine du verre volcanique basique de Palagonia (Sicile)*. PhD Thesis, Université Libre de Bruxelles.

HONNOREZ, J. 1972. *La Palagonitisation: l'Alteration Sous-marine du Verre Volcanique Basique de Palagonia (Sicile)* [*Palagonitization: the Submarine Alteration of Basic Volcanic Glass in Palagonia, Sicily*]. Vulkaninst. Immanuel Friedlaender, 9.

HONNOREZ, J. 1981. The aging of the oceanic crust at low temperature. *In*: EMILIANI, C. (ed.) *The Sea* (Vol. 7): *The Oceanic Lithosphere*. Wiley, New York, 525–587.

JAMES R.H. & ELDERFIELD H. 1996. Chemistry of ore-forming fluids and mineral formation rates in an active hydrothermal sulfide deposit on the Mid-Atlantic Ridge. *Geology*, **24**, 1147–1150.

LAVERNE, C. 1987. *Les altérations des basaltes en domaine océanique: minéralogie, pétrologie et géochimie d'un système hydrothermal: le puits 504B, Pacifique oriental.* PhD Thesis, Université Aix-Marseille III.

MAHONEY, J.J., FITTON, J.G., WALLACE, P.J. *et al.* 2001. *Proceedings of the Ocean Drilling Program, Initial Report*, **192** [Online]. Available from World Wide Web: http://www-odp.tamu.edu/publications/192_IR/192ir.htm.

MCCREA, J.M. 1950. On the isotope chemistry of carbonates and a paleotemperature scale. *Journal of Chemical Physics*, **18**, 849–857.

MUEHLENBACHS, K. 1977. Oxygen isotope geochemistry of DSDP Leg 37 rocks. *Initial Reports of the Deep Sea Drilling Project*, **37**, 617–620.

MUEHLENBACHS, K. 1979. The alteration and aging of the basaltic layer of the seafloor: Oxygen isotope evidence from DSDP/IPOD Legs 51, 52, and 53. *Initial Reports of the Deep Sea Drilling Project*, **51–53**, 1159–1167.

MUEHLENBACHS, K. & HODGES, F.N. 1978. Oxygen isotope geochemistry of rocks from DSDP Leg 46. *Initial Reports of the Deep Sea Drilling Project*, **46**, 253–256.

MUNSELL COLOR COMPANY, INC. 1975. *Munsell Soil Color Charts*. Munsell, Baltimore, MD.

O'NEIL, J.R., CLAYTON, R.N. & MAYEDA, T.K. 1969. Oxygen isotope fractionation in divalent metal carbonates. *Journal of Chemical Physics*, **51**, 5547–5558.

STAUDIGEL, H. & HART, S.R. 1985. Dating of ocean crust hydrothermal alteration: Strontium isotope ratios from Hole 504B carbonates and the re-interpretation of Sr isotope data from Deep Sea Drilling Project Sites 105, 332, 417, and 418. *Initial Reports of the Deep Sea Drilling Project*, **83**, 297–303.

TEAGLE, D.A.H., ALT, J.C., BACH, W., HALLIDAY, A.N. & ERZINGER, J. 1996. Alteration of upper ocean crust in a ridge-flank hydrothermal upflow zone: Mineral, chemical and isotopic constraints from ODP Hole 896A. *Proceedings of the Ocean Drilling Program, Scientific Results*, **148**, 119–150.

Accretionary-lapilli-bearing pyroclastic rocks at ODP Leg 192 Site 1184: a record of subaerial phreatomagmatic eruptions on the Ontong Java Plateau

THOR THORDARSON[1, 2]

[1]*Department of Geology and Geophysics, School of Ocean and Earth Sciences and Technology, University of Hawaii at Manoa, Manoa, HI 96822, USA (e-mail: moinui@soest.hawaii.edu)*

[2]*Science Institute, University of Iceland, Dunhagi 6, Reykjavík, IS101, Iceland (e-mail: torvth@hi.is)*

Abstract: Detailed analysis of lithologies and lithofacies associations within the 337.7 m thick basement volcaniclastic succession, recovered during Ocean Drilling Program (ODP) Leg 192 at Site 1184 on the Ontong Java Plateau, shows that in bulk it is made up of pyroclastic deposits of phreatomagmatic origin. The succession is essentially made up of two lithologies: lapilli tuff (59% of the total recovered core length) and tuff (34%), consisting almost entirely of juvenile clasts (>97%) and containing significant amounts of matrix-supported accretionary and/or armoured lapilli clasts. The evidence indicates that the succession was formed by at least six (and perhaps as many as 10) major phreatomagmatic eruptions that were subaerial and associated with the main phase of volcanism on the Ontong Java Plateau.

Recent studies indicate that phreatomagmatic basalt volcanism has contributed significantly to the construction of flood basalt provinces (e.g. Hanson & Elliot 1996; Pedersen *et al.* 1997; Fedorenko & Czamanske 2000; White & McClintock 2001; McClintock *et al.* 2002). With this perspective in mind, it is interesting that ODP Leg 192 drilled approximately 338 m into the volcanic basement at Site 1184 on the Ontong Java Plateau (Fig. 1), yielding the first record of emergent phreatomagmatic volcanism within this submarine lava plateau (Mahoney *et al.* 2001). The volcaniclastic succession at Site 1184 was interpreted by the Shipboard Scientific Party to consist of submarine density-current deposits formed by erosion and reworking of nearby Surtseyan edifices over a period of about 3 Ma. Very rare and poorly preserved nannofossils are present throughout the volcaniclastics, consisting of an assemblage of early Miocene–late Cretaceous species. Nevertheless, a mid-Eocene age (NP-zone 16) is inferred for the volcaniclastic succession (Shipboard Scientific Party 2001; Bergen 2004).

However, the origin of these deposits remains ambiguous because of conflicting evidence and uncertainties on the relationship of the Site 1184 volcanic basement to other parts of the Ontong Java Plateau. It may have been constructed by contemporaneous or subsequent plume-tail Ontong Java volcanism or by volcanic activity related to the opening of the Stewart Basin (Yan & Kroenke 1993; Kroenke & Mahoney 1996; Mahoney *et al.* 2001). Furthermore, the fossil and radiometric ages obtained from the succession give results that differ by as much as 75 Ma, because the latter suggests a lower Cretaceous age of *c.* 123 Ma (Chambers *et al.* 2004). Chemical and isotopic compositions of the Site 1184 volcaniclastics closely match other Ontong Java basalts, indicating a common magmatic source and possibly coeval eruptions (Fitton & Godard 2004; Roberge *et al.* 2004; White *et al.* 2004). Also, steep palaeomagnetic inclinations are more compatible with a Cretaceous age (Riisager *et al.* 2004).

In view of these uncertainties and conflicting evidence it is important to critically assess the nature of the volcaniclastic succession at Site 1184. Is it essentially epiclastic in origin or is it effectively composed of pyroclastic deposits? An epiclastic origin for the volcaniclastic succession implies deposition resulting from erosion and reworking of pre-existing Ontong Java volcanics, and therefore could reconcile the discrepancies between the obtained fossil and radiometric ages. Pyroclastic origin, on the other hand, indicates that the succession was either formed directly by eruptions or by syn-eruption resedimentation or both. Consequently, the succession would have to have been formed either during lower Cretaceous or mid-Eocene times.

From: FITTON, J. G., MAHONEY, J. J., WALLACE, P. J. & SAUNDERS, A. D. (eds) 2004. *Origin and Evolution of the Ontong Java Plateau*. Geological Society, London, Special Publications, **229**, 275–306. 0305-8719/$15.00

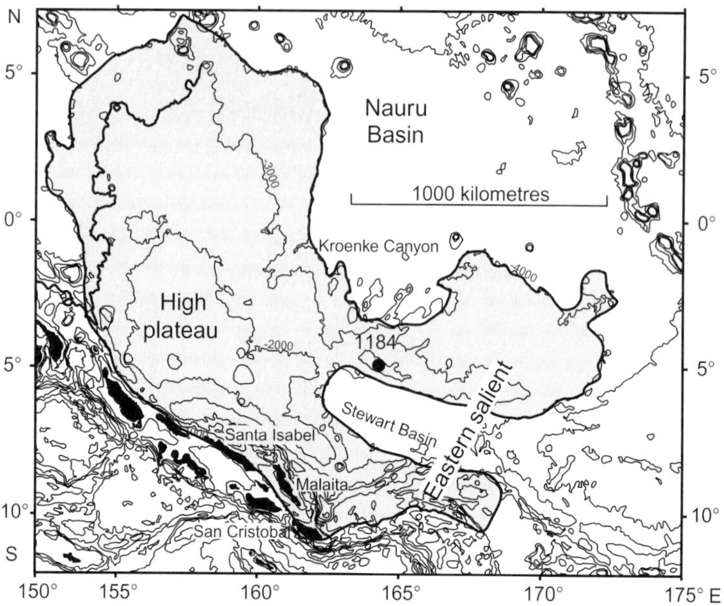

Fig. 1. Bathymetric map of the Ontong Java Plateau showing the location of Site 1184 of Leg 192 (filled circle) on the crest of the unnamed northern ridge of the eastern salient. The plateau outlines are also shown. Predicted bathymetry after Smith & Sandwell (1997).

This information is important for evaluating the depositional and volcanic environments, as well as assessing the timing and nature of the volcanism that constructed this part of the Ontong Java Plateau.

This paper aims to answer the question by volcanological assessment of the lithologies at Site 1184 by systematic documentation of the deposit textures, structures and components across the whole drill-core sequence, as well as by interpretation of principal lithofacies and their associations. This is a challenging task because the one-dimensional core view is deprived of the often critical geometrical constraints provided by two- and three-dimensional surface outcrops. Therefore the assessment relies heavily on component and textural analysis, in addition to interpreting vertical lithofacies associations.

Geological setting

Site 1184 is located at water depths of approximately 1662 m on the crest of an unnamed submarine ridge within the eastern lobe of the Ontong Java Plateau and paralleling the northern margins of the down-faulted Stewart Basin (Fig. 1). Seismic reflection records show that a *c.* 250 m-thick Miocene sedimentary sequence laps onto a volcanic basement fault block with a surface dip of 4° to the NNE (N18°E). They also show that the volcanic basement is characterized by subparallel internal reflectors with a regional dip of 9° to the north, but contain no indications of the presence of volcanic edifices in the vicinity of Site 1184 (Mahoney *et al.* 2001). However, several distinct, elliptical bathymetric and free-air gravity highs are situated within 15–35 km of Site 1184 (Fig. 1) (see also Mahoney *et al.* 2001). These gravity anomalies may represent major volcanic edifices and, if they are such structures, they could be the source for the Site 1184 volcaniclastics.

ODP Leg 192 to the Ontong Java Plateau drilled a 538.8 m-deep hole at Site 1184 and the recovered sequence consisted of two lithological units. Unit I, 134.4–201.1 mbsf (metres below seafloor) consists of lower Miocene pelagic calcareous ooze that rests unconformably on Unit II (201.1–538.8 mbsf), which is a basement sequence of volcaniclastic deposits that may be ≥1600 m thick (Mahoney *et al.* 2001). Total penetration into Unit II was 337.7 m and the core recovery was excellent, being 82.5% on average. The Unit II succession consists of tuffs, lapilli tuffs, and lapillistone containing accretionary lapilli, armoured lapilli and charred wood. These deposits also contain very rare nannofossils, but are barren of foraminifers and other marine

Table 1. *Terminology and class intervals of the pyroclastic grade scale*

φ	Size (mm)	Class
<3	<0.12	very fine ash
3–1	0.12–0.5	fine ash
1–0	0.5–1	medium ash
0 –1	1–2	coarse ash
–1 to –3	2–8	fine lapilli
–3 to –5	8–32	medium lapilli
–5 to –6	32–64	coarse lapilli
> –6	>64	blocks and bombs

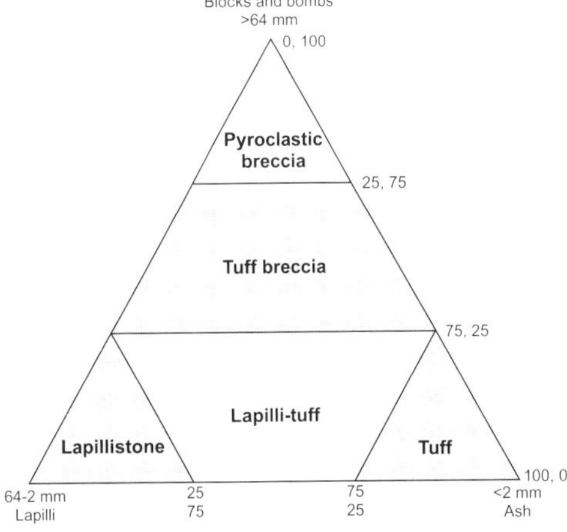

Fig. 2. Mixture and end-member rock names for pyroclastic deposits. After Fisher (1966).

fossils (Shipboard Scientific Party 2001). Onboard measurements of inclined layers within Unit II indicate NW-trending dips in the range of 2–30°. Dips in the top 75 m of the succession cluster around a mean of 5° (range 3–9°), but in the lowest 150 m they are more variable and significantly steeper (mean *c.* 12°, range 0–29°) (Shipboard Scientific Party 2001). These dips and dip directions are compatible with independent estimates of the stratigraphic dips of 4–9°N within the Site 1184 fault block, implying that regional tectonic movements can explain the bulk of the tilting within the Site 1184 volcaniclastic sequence (Mahoney *et al.* 2001).

Methods and terminology

The Leg 192 Site 1184 volcaniclastic succession was examined at the ODP core repository in College Station, Texas. Detailed dm-scale core measurements and descriptions were transcribed to a graphic log, documenting the deposit structures, textures and components, in addition to the nature of unit contacts. In addition, digital photographs were taken of the whole succession at 20 cm-intervals, and 122 samples were collected for future reference and supplementary analysis of deposit properties. In total 77 thin sections, representing all lithologies, were examined for details of deposit textures, framework and components, and clast morphologies. The relative abundance of components was estimated with the aid of standard visual reference charts and the reported per cent proportions are calculated on a cement-free basis. The clast vesicularity index is descriptive and uses the terms non-vesicular (<1% vesicles), sparsely (1–10%), poorly (10–30%), moderately (30–50%) and highly (>50%) vesicular to indicate vesicle abundances. Dips of bedding planes and contacts were measured whenever possible. All dips are measured relative to the vertical axis of the core. Thus, errors (assumed to be on the order of a few degrees) caused by erratic deviation of the drill-string from perfect vertical are included in these measurements.

Pyroclastic terminology is used here to describe the Site 1184 volcaniclastic succession because in bulk the deposits consist of pyroclasts that exhibit well-preserved igneous textures in spite of pervasive alteration (Shipboard Scientific Party 2001). The pyroclast size terms are given in Table 1 and the basic nomenclature scheme, which uses the relative proportions of ash, lapilli and bombs in classifying the deposits, is shown in Figure 2. A distinction is also made between ash-grade deposits containing <25 and >25 modal% of the very fine to fine ash (i.e. the fraction <0.5 mm). This distinction is important because the lithofacies association of fines-poor tuffs is vastly different to that of fines-rich tuffs.

Common pyroclastic terminology is used to describe and classify the volcanic grains (e.g. Fisher & Schmincke 1984; Cas & Wright 1987), and includes the five principal phreatomagmatic clast types of Wohletz (1983). Note that the terms sideromelane and tachylite refer to quenched pyroclasts. The former describes grains with groundmass of glass (or pseudomorphs thereof) and the latter refers to glassy-looking, opaque or near-opaque cryptocrystalline grains with groundmass consisting mainly of feathery pyroxene and plagioclase microlites (e.g. Kawachi *et al.* 1983). The terms scoria and basaltic-pumice are used to indicate sparsely to moderately vesicular and highly vesicular juvenile clasts, respectively. Also in this paper, the term 'accretionary lapilli' is used to describe subspherical ash aggregates that

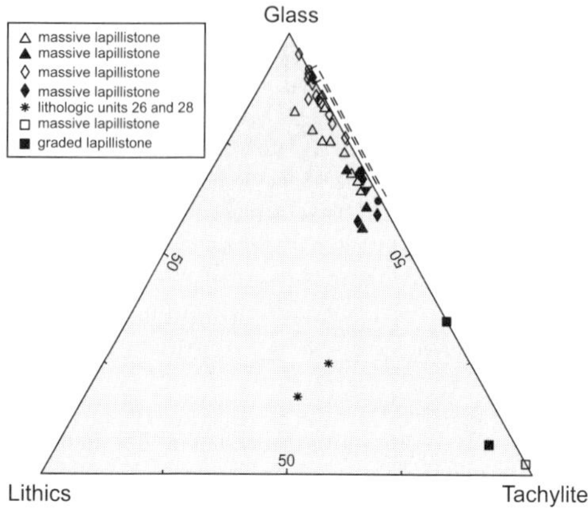

Fig. 3. Ternary plot showing the relative abundance of juvenile clasts (crystal fragments are excluded) and lithic fragments. The broken lines connect lithofacies pairs of: (1) accretionary-lapilli tuff (open circle) and fines-poor tuff (filled circle); and (2) massive lapilli tuff (open inverse triangle) and thin beds of normally graded or massive lapillistone (closed inverse triangle). Stars indicate lithostratigraphic units 26 and 28. See key for other symbols.

exhibit concentric internal layering, whereas the term ash-pellets is used to describe subspherical and structureless ash aggregates that lack concentric fine-ash rims. Similarly, ash-cluster is used for structureless loosely-packed aggregate with irregular outlines.

On the basis of a three-tier system of deposit grain size, structure and components, the Site 1184 volcaniclastic succession is divided into cm- to m-thick depositional units, each represented by an event of continuous deposition, and defined by upper and lower contacts. As such, the depositional units are the basic building blocks of the succession and form the basis for the lithofacies classification presented here. Depositional units of the same lithofacies are commonly stacked into packages within the succession. Such packages are grouped into informal lithostratigraphic units, which in accordance with ODP convention are labelled in numerical order from the top down.

Volcaniclastic components

Site 1184 volcaniclastics are composed of basaltic ash- to lapilli-size sideromelane, tachylite and crystal fragments, along with variable amounts of ash aggregates, armoured lapilli and lithic fragments, all resting in a cement of zeolites, clay and calcite. The original igneous textures of the clasts are well preserved, despite fairly strong post-emplacement alteration of the succession. Thin sections reveal a clast population that is dominated by solitary grains of angular sideromelane and tachylite that exhibit vitroclastic textures indicative of quenching. These clasts typically exceed 95 modal% (Fig. 3, Table 2). Most of these grains are blocky and irregular, with angular or cuspate shapes and curviplanar fracture surfaces (Fig. 4a). These characteristics indicate explosive phreatomagmatic fragmentation (e.g. Morrissey *et al.* 2000). Furthermore, mineralogical changes in sideromelane and tachylite grains, such as the appearance or disappearance of particular phenocryst phases, follow each other throughout the succession. This suggests that at any level within the core the sideromelane and tachylite clasts are co-magmatic juvenile clasts. The sideromelane and tachylite grains, together with subordinate amount of plagioclase and clinopyroxene (and rarely olivine pseudomorph) crystal fragments, constitute the juvenile clast population in the Site 1184 volcaniclastic deposits.

Aphyric to sparsely phyric sideromelane clasts, or pseudomorphs thereof, are on average the most common clast type in the Site 1184 volcaniclastic deposits, with modal abundances in excess of 60% in three-quarters of the lithostratigraphic units analysed (Fig. 3, Table 2). Blocky sparsely to non-vesicular sideromelane grains are most common, but scoria and basalt pumice grains are also present in substantial amounts (Fig. 4b). The scoria and basalt pumice clasts typically feature spherical to oval vesicles and often have fluidal outlines (some have scalloped edges). The least-altered sideromelane clasts feature a core of fresh glass and a thin (<400 μm thick) rim of brown smectite, which is a pseudomorph after fibrous palagonite and indicative of an early plagonitization phase (e.g. Jakobsson 1972; Jakobsson & Moore 1986). Similar smectite rims are found on strongly altered sideromelane clasts that also often feature an inner rim of green celadonite. The core of these completely altered clasts is either replaced by zeolites, analcime, or variably yellow, brown or green clay minerals.

Tachylite is the second most abundant clast type found in the Site 1184 basement rocks and typically accounts for 10–35% of the clast population (Fig. 3, Table 2). These clasts are characterized by either angular margins and clean fracture surfaces or highly irregular and convolute outlines (i.e. Type 3 clast of Wohletz 1983). They are typically poorly to moderately vesicular, with numerous highly irregular micro-vesicles. They also often incorporate smaller grains of sideromelane or tachylite, or, more

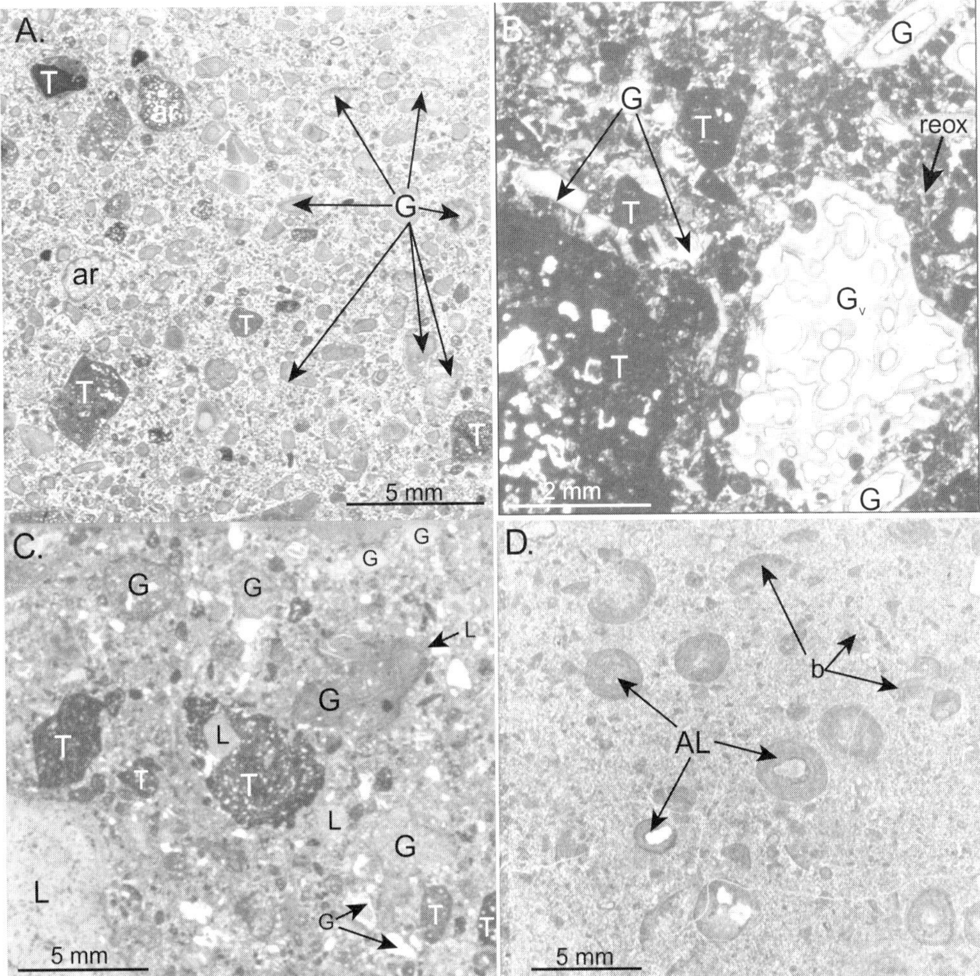

Fig. 4. Photomicrographs showing the type of clasts that make up the Site 1184 pyroclastic rocks: (**a**) blocky grains of sideromelane glass (G), lesser tachylite (T) and armoured lapilli (ar) from depositional unit of massive lapilli tuff with normally graded top within lithostratigraphic unit 22; (**b**) lapilli-size grains of non-vesicular (G) and highly vesicular (G_v) sideromelane glass and vesicular tachylite (T) from a depositional unit of massive lapilli tuff within lithostratigraphic unit 32; (**c**) lapilli-size grains of glass (G), tachylite (T), and lithics (L) from a depositional unit of massive lithic-rich lapilli tuff within lithostratigraphic unit 28. Note the lithic inclusion in the tachylite grain in the centre; (**d**) whole (AL) and broken (b) accretionary lapilli from lithostratigraphic unit 8.

rarely, accidental lithic fragments (Fig. 4c). Together, these features clearly indicate that the tachylite grains were produced by brittle and viscous disintegration of rapidly cooled magma, confirming their juvenile status.

Submillimetre angular crystal fragments of plagioclase, clinopyroxene and, more rarely, olivine pseudomorphs are present in trace or minor amounts (5 modal%) in almost all lithostratigraphic units (Table 2).

The other pyroclastic components are armoured lapilli and ash aggregates. Armoured lapilli, which are essentially confined to the lapilli tuffs, are coarse-ash- to lapilli-size grains with a complete submillimetre to millimetres-thick coating of very-fine- to fine-grained ash made up of juvenile shards. Ash aggregates of very- fine- to fine-grained ash made up of sideromelane and, to a lesser degree, tachylite shards are present in abundance within the

Table 2. *Component abundances (modal%) in individual lithostratigraphic units based on estimates in thin section*

Member	Lithostr. unit	Lithofacies	Components Glass (%)	Tachylite (%)	Crystals (%)	Lithics (%)	Juvenile (%)	Components Matrix dominant type	Characteristic components Coarse fraction vesicular	blocky	accr. lapilli	ash clusters	arm. lapilli
	1A-C	Tm, Ttb,Tal	93.4	6.0	0.6	0.0	99.4	b,(sp,v)	xx	xxxx	T [xxx]		T
	1D-E	Tngr, Tm-fr	68.2	30.2	0.1	1.5	98.4	b,(sp,v)	xx	xxx			T
	2	Tngr	46.2	53.8	0.0	0.0	100.0	b	xx	xxx			
	3	Tm-fr	89.0	10.2	0.7	0.1	99.2	b,sp	T	xxxx		[xxx]	
	4	Tal, Tm-pf	95.2	4.5	0.0	0.3	99.7	b,(sp)	x	xxxx	xxx	xx	[x]
IIA	5	Ttb-p/x	52.3	46.2	1.5	0.0	98.5	b,sp	x	xxxx			
	6	Tm-fr	58.0	38.1	1.6	2.3	96.1	b	x	xxxx	[xx]	[x]	
	7	Ttb-p/x	66.2	31.3	1.3	1.2	97.5	b	x	xxxx			
	8A-B	Tal	68.0	30.4	0.3	1.3	98.4	b,sp,(v)	xx	xxxx	xxx	xxx	
	8H-L	Tal	90.0	9.0	0.6	0.4	99.0	b,sp,(v)	xx	xxxx	xxx	xxx	x
	8M	Tal	90.6	7.7	1.5	0.2	98.3	b,sp,(v)	xx	xxx	xxx	xxx	x
	8M	Tm-fp	60.9	36.4	1.8	0.9	97.3	b	xx	xxxx			x
	10	LTm	74.9	20.8	0.1	4.2	95.7	b	xx	xxxx		x	x
	11	Tal	88.7	9.3	1.0	1.0	98.0	b,(sp,v)	xx	xxxx	xxx	xx	x
	13	LTm	70.1	24.1	3.3	2.5	94.2	b,(v)	xx	xxxx	[x]		[T]
	14	cT/LTm	66.7	26.6	2.7	4.0	93.3	b, (v)	x	xxxx	T		
	15	Tal	77.5	18.8	2.4	1.3	96.3	b,sp,(v)	T	xxxx	xxx	xx	x
	16	Tal	77.5	16.7	5.0	0.8	94.2	b,sp	xx	xxxx	xxx	xx	x
IIB	17	Tal	82.7	13.3	2.7	1.3	96.0	b,sp,(v)	x	xxxx	xxx		x
	18	LTm	75.6	15.5	2.4	6.5	91.1	b,(sp,v)	xx	xxxx			
	22	LTm	70.9	18.2	5.5	5.4	89.1	b,(v)	xx	xxxx			x
	24	LTm	76.5	9.6	6.2	7.7	86.1	b,(sp,v)	xx	xxxx	xxx		xx
	25	LSTm	2.5	97.3	0.1	0.1	100.0	b		xxxx			
	26	LTm	17.4	42.6	1.4	38.6	60.0	b	xx	xxxx	x		x
IIC	28	LTm	24.8	45.0	1.6	28.6	69.8	b	xx	xxxx	xx		x
	29	LSTngr	34.8	65.0	0.1	0.1	99.8	b	x	xxxx			
	30	Tal	82.2	11.1	3.4	3.3	93.3	b,(sp,v)		xxx	xxx	x	

	43	cT/LTm	56.0	33.3	6.7	4.0	89.3	b	x	xxxx	x-xx		x
	45	LTm	65.9	28.1	2.6	3.4	94.0	b	x	xxxx	[x]		x
IID	46	cT/LTm	63.3	33.3	1.7	1.7	96.6	b,(v)	xx	xxx			
	47	LTngr	88.0	10.6	0.8	0.6	98.6	b,(sp,v)	x	xxxx	xxxx	[x]	x
	47	LTngr (base)	64.0	33.4	1.3	1.3	97.4	b,(v)	xx	xxx	T	xxx	x
	55	Tal	85.0	12.6	1.1	1.3	97.6	b,(sp,v)	xx	xxxx	xxx	xx	d
	58	Tm	55.9	34.4	2.8	6.9	90.3	b,sp	xx	xxxx	x		x
	60	cT/LTm	52.9	35.7	4.3	7.1	88.6	b,sp,(v)	xx	xxxx	xxx		T
IIE	67	LTm	66.4	30.9	0.9	1.8	97.3	b,sp	T	xxxx			
	68	Tal	82.3	14.1	2.4	1.2	96.4	b,sp	xx	xxxx	xxx	x	
	70	Tal	75.2	23.1	1.5	0.2	98.3	b,sp,(v)	xx	xxxx	xxx	x	xxx
	71	LTm	80.0	15.4	3.1	1.5	95.4	b,sp	xx	xxxx	x		x
	73	LTm	82.0	13.3	4.0	0.7	95.3	b,sp	xx	xxx	x		x
IIF	76	LST-sgr	6.3	81.2	7.5	5.0	87.5	b	x	xx x			

Key to abbreviations: abundance of characteristic grains in the coarse fraction: T, trace; x, minor; xx, present in significant amount; xxx, common; xxxx, major component. Shard types: b, blocky and curvi-planar shards; sp, splinter-like or cuspate shards; v, vesicular shards. Coarse fraction: vesicular, vesicular juvenile clasts; blocky, non-vesicular, blocky clasts; accr. lapilli, accretionary lapilli and ash-pellets; arm lapilli, armoured lapilli. () in column 9 indicate shards present as minor component and [ ] in columns 12–14 indicates that component is present in some depositional units.

succession, especially in the tuffs. They are represented by three aggregate types: ash-clusters, ash-pellets and accretionary lapilli (Fig. 4d). Ash-clusters are distinct submillimetre irregular-shape domains of loosely packed 200–300 μm shards. Ash-pellets are present as featureless 2–16 mm spherical and loosely packed aggregates consisting entirely of very fine to fine ash (≤400 μm). Fragments of pellets are also common. Whole and broken accretionary lapilli always occur together and are the most common type of ash aggregates in the Site 1184 succession. Intact accretionary lapilli are 1–12 mm spherical to ovoid pea-like grains consisting of a submillimetre thick shell of very-fine-grained ash (grains <50 μm) completely enclosing a massive or concentrically laminated core of very-fine to fine ash that coarsens toward the centre to shard size up to 350 μm. Commonly, the accretionary lapilli contain a nucleus of a single, moderately to highly vesicular sideromelane or tachylite fragment. More rarely, the nucleus consists of an open-framework cluster of coarse ash grains or a lithic fragment. Often the accretionary lapilli are broken, i.e. occur as dish-like fragments featuring the outermost shell of the original accretionary lapilli and parts of the interior. The contact between the original accretionary-lapilli interiors and the surrounding matrix is most commonly diffuse where shards from the accretionary-lapilli core are mixed with the tuff matrix. This relationship indicates that upon deposition these fragments were loosely bound and unlithified aggregates.

Lithic fragments are present in trace to minor amounts (0.1–5 modal%) in the majority of the lithostratigraphic units (Fig. 3, Table 2). They are usually most abundant in lapilli tuffs where they are present as angular to subangular, <1–20 mm fragments of crystalline basalt and lithified tuff, along with very rare clastic sedimentary fragments. The exceptions are lithostratigraphic units 26 and 28 (Fig. 3). Here the lithic fragments are a major component (35 modal%, range 19–39%) and range in size from <1 to 65 mm. They are most commonly angular fragments of holocrystalline basalt and, to a lesser extent, lithified tuff and lapilli tuff. Some fragments are essentially unaltered; some are stained red by oxidation, while in others their components are completely replaced by clay minerals and zeolites. Also, lithic fragments from the tuff and lapilli tuff are held together by cement that differs from that of the host volcaniclastic deposit (i.e. analcime–clay cement v. zeolite–calcite cement).

The component analysis shows that the deposits of the Site 1184 volcanic basement primarily consist of juvenile sideromelane and tachylite clasts, along with a variety of ash aggregates. In other words, phreatomagmatic pyroclasts comprise >95 modal% of the clast population in deposits accounting for more than 82% of the stratigraphic thickness. These results strongly favour pyroclastic origin for the succession, because most epiclastic deposits (*sensu stricto*), and certainly depositional sequences hundreds of metres thick, would be characterized by a clast population consisting of much more diverse volcanic and non-volcanic lithologies (e.g. Fisher & Schmincke 1984). Juvenile grains are also the major component (>60 modal%) in the remaining 18% (Table 2, see lithostratigraphic units 26 and 28), but these lower juvenile clast proportions do not rule out pyroclastic origin. Abundant non-juvenile lithic fragments are present in many pyroclastic deposits and especially in phreatomagmatic tephra deposits (e.g. White & Houghton 2000). Therefore, from now on the Site 1184 succession will be referred to as pyroclastic.

Lithologies and stratigraphy

Lithologies

The Site 1184 succession is entirely pyroclastic consisting of tuff, lapilli tuff and lapillistone, which in bulk consists of angular to subangular clasts ≤32 mm in diameter. Coarse lapilli (32–64 mm) are present as a minor component within the lapilli tuff and the lapillistone, whereas block- and bomb-size clasts (>64 mm) are virtually absent from the succession. The largest clast found in this approximately 338 m-thick succession is 65 mm. These lithologies alternate throughout the succession, although in variable proportions (Fig. 5, Table 3).

Lapilli tuffs are the most abundant lithology in the Site 1184 pyroclastic succession and account for about 59% of the total thickness of the sequence (Table 4). Excluding the top 22 m, the lapilli tuffs are present throughout the succession as discrete 0.1–41.6 m-thick lithostratigraphic units. Generally, >95 modal% of the clast population is juvenile, with sideromelane fragments at *c.* 71% (range 56–80%), tachylite at 22% (range 7–35%) and crystals at 3% (range 0–7.7%). The remaining 4% (range 0.7–10.8%) are lithic fragments of crystalline basalt and lithified tuff (Fig. 3, Table 2). The only exceptions are lithostratigraphic units 26 and 28, where the juvenile to lithic clast proportions are about 2: 1. These units, along with the top 2 m of lithostratigraphic unit 13, also contain red ash- to lapilli-size juvenile and lithic fragments.

Grains like this, which clearly have been subjected to subaerial oxidation, are often present as minor components in phreatomagmatic deposits. In these cases it is inferred that the red-oxidized grains have been incorporated into the erupted mixture by recycling when earlier-formed cone material slumped into the vents. Consequently, such oxidized grains are a good indicator of a subaerial eruption.

Tuffs are the second-most abundant lithology in the Site 1184 pyroclastic succession, with a collective thickness of approximately 115 m, accounting for about 34% of the thickness of the drilled succession (Table 4). The tuffs are essentially monomictic with juvenile basaltic grains comprising >97 modal% of the clast population. The juvenile population is made up of angular non- to highly vesicular sideromelane (72 ± 8%) and tachylite (26 ± 8%) fragments, with trace amounts of crystal fragments. Lithic fragments are usually present as a trace or minor component (Table 2). The tuffs almost exclusively consist of ash-size fragments, although with variable grain-size distributions and sorting. More often than not, they include a dispersed coarse fraction of fine lapilli (2–8 mm clasts), but coarser clasts such as bombs and blocks are entirely absent.

Lapillistone is present as three, discrete 1.5–12.5 m-thick lithostratigraphic units at the base and near the middle of the Site 1184 pyroclastic succession (Table 3). They also occur as rare cm- to dm-thick beds within lithostratigraphic units dominated by either tuffs or lapilli tuffs. The collective thickness of the lapillistone lithofacies is approximately 24 m, accounting for 7% of the cored thickness (Table 4). All of the lapillistone deposits are essentially monomict, and are primarily made up of juvenile tachylite and lesser sideromelane fragments (Table 2).

The complete absence of block-size clasts in the succession along with the dominant pyroclastic clast population and the overall fine-grained nature of the Site 1184 pyroclastic lithologies, appears to exclude deposition in a near-vent environment (i.e. within the ballistic zone). This, especially the non-existence of block-size clasts, also makes transport and deposition by cohesive density currents (i.e. debris flows) an unlikely explanation (e.g. Cas & Wright 1897; Moulder & Alexander 2001).

Dip of strata

Dips of contacts and inclined layers were measured whenever possible across the entire core and the results are shown in Figure 6. The average of 189 dip measurements is 7.4 ± 5.5° (range −4° to 25°). In the top 100 m of the succession the dip is significantly lower and clusters around a mean of 5.7 ± 4.3°, whereas in the lowest 120 m the dip is steeper and somewhat more variable with a mean of 10.5±5.8°. These results are comparable with those obtained by the Shipboard Scientific Party (Fig. 6). These dips are also similar to the calculated 4°NNE dip of the fault block surface and 9°N for deeper reflectors within the basement (Mahoney *et al.* 2001). These data not only suggest that regional tectonic tilting can explain the bulk of the observed dip in the core, but implies that for the most part the succession was accumulated onto a flat or at best a gently sloping (<5°S) surface.

Lithostratigraphic members

The Site 1184 pyroclastic succession is divided into five subunits, labelled IIA–IIE (Mahoney *et al.* 2001). New major- and trace-element whole-rock analyses of a sample suite representing the entire succession (Fitton & Godard 2004) show that individual subunits are chemically distinct and that four (IIA, IIB, IID and IIE) are essentially homogeneous in chemical composition (Fig. 5a). In addition, microprobe and laser ablation inductively coupled plasma-mass spectrometry (ICP-MS) analysis of fresh glass from several levels within subunits IIA, IID and IIE give the same results and demonstrate that the chemical composition of the sideromelane clast population within individual subunits is uniform (Roberge *et al.* 2004; White *et al.* 2004). In light of these new chemical data, underpinned by re-evaluation of lithofacies associations and contact relationships, minor revisions are made on the original subunit boundaries. The key changes are (see also Fig. 5a):

- the contact between subunits IIC and IID is now placed at a sharp erosional contact at 385.34 mbsf because the chemical composition of the sample at 380.72 mbsf differs significantly from other unit IID samples. It is more in line with the low Zr/Nb (i.e. high Nb) samples of IIC (Fig. 5a), and the top 4.8 m of IID, above the erosional contact, consists of tuff that accumulated as a continuous succession;
- the boundary of IID and IIE is now placed at 451.58 mbsf. This change is made because the original IID–IIE boundary, which was placed at 437.39 mbsf, is a diffuse contact between two related tuff lithofacies and the chemical composition of sample 1184A-36R-7, 0–6 cm (441.0 mbsf), originally thought to be from the top of IIE, is identical to that of IID (Fig. 5a);

(a)

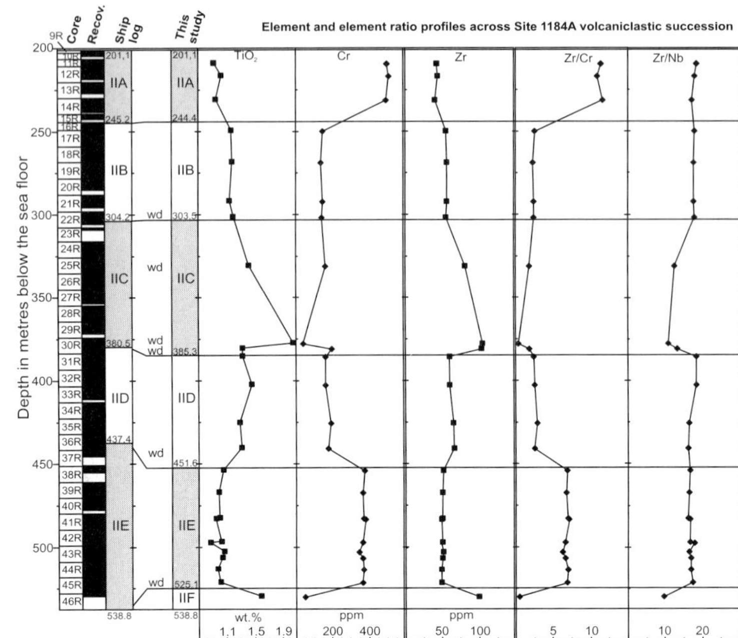

Fig. 5. (**a**) Lithostratigraphic members in Site 1184 pyroclastic succession as defined by concentrations and ratios of highly immobile elements plotted against depth (data are from Fitton & Godard 2004). The label wd, indicates position of charred wood within the succession. (**b**) Graphic logs of the Site 1184 pyroclastic succession depicting stratigraphic relationships of lithostratigraphic members and units. See key for further explanation. See tables for lithofacies codes and names.

- the original subunit IIE is split into two (i.e. IIE and IIF) with the boundary across an erosional contact at 525.13 mbsf (Fig. 5a, Table 3). However, this division should be viewed with caution, because it is based on an analysis of a single sample. A distinctive erosional contact in IIF at 528.28 mbsf, and the fact that it includes an assortment of unrelated lithologies (Table 3), suggests that this unit may be mixed.

The geochemical coherency of subunits IIA, IIB, IID and IIE, in conjunction with the uniform component composition, provides compelling evidence that the deposits of these subunits, especially those of IIA, IIB, IID and IIE, correspond to the products of individual eruptions. It is difficult to imagine how erosion and reworking of pre-existing volcanic strata by epiclastic processes could be so selective as to erode, rework and transport one compositionally uniform stratum at a time and do that recurrently in a systematic fashion. Therefore, subunits IIA–IIF are categorized here as informal lithostratigraphic members (Fig. 5, Table 3).

Lithofacies analysis

The lithofacies classification is based on analyses of grain size, structure and components of individual depositional units, which are the basic building blocks of the succession, each representing an event of continuous deposition. In total 13 lithofacies are identified within the succession; four lapilli tuff, six tuff and three lapillistone lithofacies (Table 5). A lithofacies name is obtained by adding a component or textural modifier to the basic pyroclastic nomenclature. Depositional units of the same lithofacies are commonly stacked, to form intervals decimetres to tens of metres thick that divide the succession into 77 lithostratigraphic units (Fig. 5b, Table 3). Descriptions of clast types and morphologies are generally not included in the lithofacies descriptions because the components of all lithologies are essentially pyroclastic and dominated by

(b)
Key 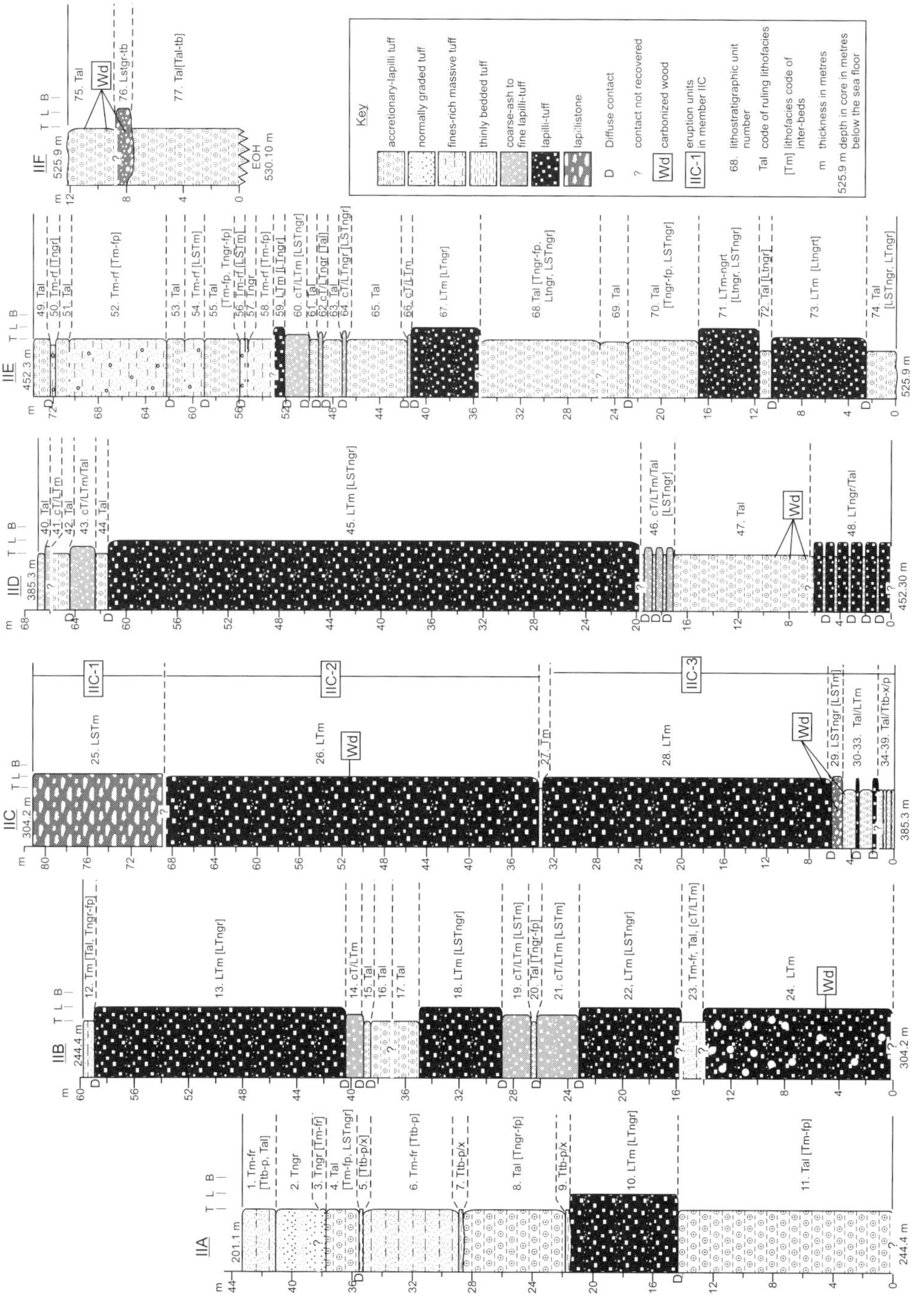
accretionary-lapilli tuff
normally graded tuff
fines-rich massive tuff
thinly bedded tuff
coarse-ash to fine lapilli-tuff
lapilli-tuff
lapillistone
D Diffuse contact
? contact not recovered
Wd carbonized wood
IIC-1 eruption units in member IIC
68. lithostratigraphic unit number
Tal code of ruling lithofacies
[Tm] lithofacies code of inter-beds
m thickness in metres
525.9 m depth in core in metres below the sea floor
IIA
IIB
IIC
IID
IIE
IIF

Table 3. *Stratigraphic log showing distribution of lithostratigraphic members, lithostratigraphic units and lithofacies in the Site 1184 pyroclastic succession*

Depth (mbsl)	Member	Lithostrat. unit	Thickness (cm)	Key lithofacies	Other lithofacies	Description	Unit contacts
201.10	2 cm	M	1.5			manganese crust	contact with unit 1 is sharp
201.12		1	222.5	Tm-fr	Ttb-p, Tal	multiple depositional units of fines-rich massive tuff with interbeds of thinly planar bedded and accretionary-lapilli tuff	contact with unit 2 is sharp
203.34		2	306	Tngr		multiple depositional units of normally graded, near clast-supported, coarse-grained tuff	contact with unit 3 not recovered
206.40		3	47	Tngr	Tm-fr	multiple depositional units of alternating normally graded and fines-rich massive tuff featuring evenly dispersed coarse fraction and rare accretionary lapilli	contact with unit 4 is sharp
206.87		4	228	Tal	Tm-fp, LSTngr	multiple depositional units of massive accretionary-lapilli tuff with ash-cluster horizons at the top and including interbeds of normally graded lapillistone and massive fines-poor tuff	contact with unit 5 is diffuse
209.15		5	22	Ttb-p/x		thinly planar- and cross-bedded tuff featuring alternating beds of fine and coarse ash with rare accretionary lapilli	contact with unit 6 is sharp
209.37	**IIA** 43.3 m	6	633	Tm-fr	Ttb-p	multiple depositional units of massive coarse-grained, fines-rich tuff with dispersed coarse fraction; rare interbeds of accretionary-lapilli and thinly planar-bedded tuff	contact with unit 7 is sharp
215.70		7	23	Ttb-p/x		thinly bedded tuff with discordant bedding (cross-bedded?) consisting of alternating fine and coarse ash beds	contact with unit 8 is scoured
215.93		8	686	Tal	Tngr-fp, Ttb-x	multiple depositional units of massive and normally graded accretionary-lapilli tuff with ash-cluster horizons at top alternating with cm-thick interbeds of lapillistone and massive fines-poor tuff; contains a single 9 cm-thick cross-bedded tuff interval	contact with unit 9 is sharp
222.79		9	19	Ttb-p/x		thinly planar- and cross-bedded tuff featuring inverse-graded ash beds	contact with unit 10 is scoured
222.98		10	705	LTm	LTngr	multiple depositional units of massive lapilli tuff with normal size-graded tops containing interbeds of multiple normally graded lapilli tuff	contact with unit 11 is diffuse
230.03		11	1437	Tal	Tm-fp, Tngr-fp, Ttb-p	multiple depositional units of massive accretionary-lapilli tuff containing dispersed coarse fraction, accretionary lapilli horizons at the base and ash-cluster horizons at the top alternate with cm-thick interbeds of fines-poor tuff	contact with unit 12 not recovered
244.40		12	72	Tm-fr	Tngr-fp, Tal	fines-rich massive tuff with thin interbeds of fines-poor normally graded tuff and accretionary-lapilli tuff	contact with unit 13 is diffuse
245.12		13	1847	LTm	LTngr	multiple depositional units of massive lapilli tuff, where individual units commonly feature cm-thick normally graded tops	contact with unit 13 is diffuse
263.59		14	144	cT/LTm		several depositional units of dm-thick massive very-fine-grained lapilli tuff with normally graded tops	contact with unit 15 is diffuse

265.03		15	53	Tal		accretionary-lapilli tuff with dispersed normally graded coarse fraction	contact with unit 16 is diffuse
265.56		16	166	Tal		several depositional units of massive accretionary-lapilli tuff	contact with unit 17 not recovered
267.22		17	203	Tal		multiple depositional units of accretionary-lapilli tuff with dispersed normally graded coarse fraction	contact with unit 18 is sharp
269.25		18	602	LTm	LSTngr	multiple depositional units of massive lapilli tuff featuring normally graded tops and interbeds of cm-thick normally graded lapillistone	contact with unit 19 is diffuse
275.27	**IIB** 59.8 m	19	208	cT/LTm	LSTm	several depositional units of massive very fine-grained lapilli tuff featuring normal size-graded tops; includes rare cm-thick lapillistone interbeds	contact with unit 20 not recovered
277.35		20	40	Tal	Tngr-fp	several depositional units of accretionary-lapilli tuff featuring normally graded accretionary lapilli horizon the base and ash-cluster horizon at the top	contact with unit 21 is diffuse
277.75		21	316	cT/LTm	LSTm, LSTngr	several depositional units of massive very fine-grained lapilli tuff featuring normally graded tops; includes rare cm-thick lapillistone interbeds	contact with unit 22 is diffuse
280.91		22	759	LTm	LSTngr	multiple depositional units of massive lapilli tuff, some with normally graded tops and a single interbed of normally graded lapillistone	contact with unit 23 not recovered
288.50		23	150	Tm-fr	Tal, cT/LTm	several depositional units of massive coarse-grained, fines-rich tuff with normally graded coarse fraction; includes interbeds of accretionary-lapilli tuff and very-fine-grained lapilli tuff	contact with unit 24 not recovered
290.00		24	1418	LTm		multiple depositional units of massive lapilli tuff, some with normally graded tops; contain scattered 2–16 mm accretionary lapilli	contact with unit 25 not recovered
304.18	**IIC-1** 47.7 m		25	1259	LSTm	massive monomictic lapillistone, almost entirely made up of sub-trachytic textured clasts	contact with unit 26 not recovered
316.77		26	3510	LTm		multiple depositional units of massive, lithic-rich lapilli tuff featuring normally graded tops and more rarely normally graded coarse fraction; contains red-oxidized clasts	contact with unit 27 is sharp
351.87	**IIC-2** 28.7 m	27	5	Tm-fr		fine-grained, massive tuff bound by sharp (erosional ?) contacts	contact with unit 28 is sharp
351.92		28	2706	LTm		multiple depositional units of massive lithic-rich lapilli tuff featuring normally graded tops and more rarely normally graded coarse fraction contains red-oxidized clasts	contact with unit 29 is diffuse
378.98		29	156	LSTngr	LSTm	monomictic lapillistone, consisting of multiple cm- to dm-thick normally graded and massive depositional units	contact with unit 30 is sharp
380.54		30	124	Tal	Tm-fp	several depositional units of massive accretionary-lapilli tuff with cm-thick interbeds of fines-poor massive tuff	contact with unit 31 is diffuse
381.78		31	10	LTngr		normal-graded lapilli tuff	contact with unit 32 is sharp

Table continued overleaf

Table 3. *continued*

Depth (mbsl)	Member	Lithostrat. unit	Thickness (cm)	Key lithofacies	Other lithofacies	Description	Unit contacts
381.88		32	175	Tal		several depositional units of massive accretionary-lapilli tuff containing with up to 10 cm-thick accretionary lapilli horizon at their bases	contact with unit 33 is diffuse
383.63		33	64	cT/LTm		several depositional units of massive very coarse tuff to lapilli tuff featuring normally graded tops	contact with unit 34 not recovered
384.27	**IIC-3**	34	17	Tal		massive accretionary-lapilli tuff	contact with unit 35 is sharp
384.44	4.8 m	35	5	Ttb-p/x		thinly planar-bedded tuff with cross-laminations in top centimetres	contact with unit 36 is sharp
384.49		36	16	Tal		massive accretionary-lapilli tuff with cm-thick cross-bedded top	contact with unit 37 is sharp
384.65		37	8	Ttb-p/x		thinly planar-bedded tuff	contact with unit 38 is sharp
384.73		38	60	Tal		massive accretionary-lapilli tuff	contact with unit 39 is sharp
385.335		39	0.5	Ttb-xl		laminae of cross-laminated tuff	contact with unit 40 is scoured
385.34		40	58	Tal		massive accretionary-lapilli tuff	contact is with unit 41 is sharp
385.92		41	50	ct/LTm		massive very coarse tuff to lapilli tuff	contact with unit 42 not recovered
386.42		42	152	Tal		several depositional units of massive accretionary-lapilli tuff with dispersed coarse fraction, accretionary lapilli horizons at their base and ash-cluster horizons at the top	contact with unit 43 is diffuse
387.94	**IID** 67.0 m	43	201	cT/LTm	Tal	several depositional units of coarse tuff to lapilli tuff, consisting of cm- to dm-thick normally graded units, each capped by accretionary-lapilli tuff	contact with unit 44 is sharp
389.95		44	102	Tal		massive accretionary-lapilli tuff	contact with unit 45 is diffuse
390.97		45	4158	LTm	LSTngr	multiple depositional units of massive lapilli tuff featuring normally graded tops; contains rare accretionary and armoured lapilli; cm-thick sets of normally graded lapillistone	contact with unit 46 not recovered
432.55		46	249	cT/LTngr	Tal, LSTngr	multiple depositional units of coarse tuff to lapilli tuff, consisting of cm- to dm-thick normally graded units, each capped by accretionary-lapilli tuff; includes cm-thick interbeds of normally graded lapillistone	contact with unit 47 is sharp
435.04		47	1073	Tal		multiple depositional units of accretionary-lapilli tuff consisting of massive units commonly with ash-cluster horizon at the top and accretionary lapilli horizon at the base	contact with unit 48 not recovered
445.77		48	653	LTngr	Tal	multiple depositional units of normally graded lapilli tuff where individual units are capped by cm- to dm-thick horizon of accretionary-lapilli tuff	contact with unit 49 not recovered
452.30		49	128	Tal		multiple depositional units of accretionary-lapilli tuff with ash-cluster horizon at the top and accretionary lapilli horizon at the base	contact with unit 50 is diffuse

453.58		50	37	Tm-fr	Tngr	fines-rich massive tuff, with dispersed normally graded coarse fraction and rare accretionary lapilli; includes an interbed of normally graded tuff	contact is with unit 51 is sharp
453.95		51	80	Tal		massive accretionary-lapilli tuff	contact with unit 52 is diffuse
454.75		52	794	Tm-fr	Tm-fp	multiple depositional units of fines-rich massive tuff with evenly dispersed or normally graded coarse fraction; includes cm-thick interbeds of fines-poor, coarse-grained tuff	contact with unit 53 is diffuse
462.69		53	156	Tal		massive accretionary-lapilli tuff with dispersed coarse fraction and several horizons of accretionary lapilli	contact with unit 54 is diffuse
464.25		54	120	Tm-fr	LSTm	fines-rich massive tuff with inversely graded coarse fraction and accretionary lapilli; includes cm-thick interbeds of massive lapillistone	contact with unit 55 is diffuse
465.45		55	334	Tal	Tm-fp, Tngr-fp	multiple depositional units of massive accretionary-lapilli tuff with dispersed coarse fraction, accretionary lapilli horizons at their base and ash-cluster horizons at the top; includes several cm-thick interbeds of fines-poor tuff	contact with unit 56 is diffuse
468.79		56	45	Tm-fr	LSTm	massive tuff with evenly dispersed coarse fraction and scattered accretionary lapilli	contact with unit 57 is diffuse
469.24		57	16	Tngr		normally graded, thinly planar-bedded tuff comprised of cm-thick layers where fines-poor coarse ash grades upwards into fine-ash	contact with unit 58 is diffuse
469.408		58	219	Tm-fr	Tm-fp	multiple depositional units of massive tuff with evenly distributed coarse fraction; includes a few cm-thick interbeds of fines-poor, coarse tuff	contact with unit 59 not recovered
471.59		59	99	LTm	LTngr	several depositional units of massive lapilli tuff with normally graded tops; includes interbeds of normally graded lapilli tuff	contact with unit 60 is diffuse
472.58		60	201	cT/LTm	LSTngr	massive coarse tuff to lapilli tuff with evenly distributed or normally graded coarse fraction and dispersed accretionary lapilli	contact with unit 61 is diffuse
474.59		61	70	Tal		several depositional units of accretionary-lapilli tuff with evenly distributed or normally graded coarse fraction	contact with unit 62 is diffuse
475.29	**IIE** 73.6 m	62	26	cT/LTm	Tal	thinly to medium bedded coarse tuff to lapilli tuff consisting of cm- to dm-thick normally graded depositional units, each capped by accretionary-lapilli tuff	contact with unit 63 is diffuse
475.55		63	155	Tal		several depositional units of accretionary-lapilli tuff with ash-cluster horizon at the top and accretionary-lapilli horizon at the base	contact with unit 64 is diffuse
477.10		64	33	cT/LTngr	LSTngr	thinly to medium bedded coarse tuff to lapilli tuff consisting of dm-thick normally graded depositional units; includes cm-thick interbeds of normally graded lapillistone beds	contact is with unit 65 is sharp
477.43		65	528	Tal	Tngr-fp	multiple depositional units of massive accretionary-lapilli tuff with evenly dispersed or normally graded coarse fraction and accretionary lapilli horizon at the base; includes cm-thick interbeds of fines-poor, coarse tuff	contact with unit 66 is diffuse

Table continued overleaf

Table 3. *Lithostratigraphic members, units and lithofacies*

Depth (mbsl)	Member	Lithostrat. unit	Thickness (cm)	Key lithofacies	Other lithofacies	Description	Unit contacts
482.71		66	36	cT/LTm		massive coarse tuff to lapilli tuff with evenly dispersed coarse fraction	contact with unit 67 is diffuse
483.07		67	763	LTm	LTngr	multiple depositional units of massive lapilli tuff with multiple graded coarse fraction and normally graded tops and rare interbeds of normally graded lapilli-tuff units	contact with unit 68 not recovered
490.70		68	982	Tal	Tngr-fp, LTngr, r LSTng	multiple depositional units of accretionary-lapilli tuff with normally graded coarse fraction and several cm- to dm-thick interbeds normally graded fines-poor tuff, lapilli tuff and lapillistone	contact with unit 69 not recovered
500.22		69	276	Tal		a depositional unit of massive accretionary-lapilli tuff capped by several cm-thick accretionary-lapilli tuff units featuring ash-cluster horizons at their tops	contact with unit 70 is diffuse
503.286		70	579	Tal	Tngr-fp, LSTngr	multiple depositional units of massive accretionary-lapilli tuff with normally graded coarse fraction and ash-cluster horizon at the top. Also contains cm-thick interbeds of normally graded fines-poor tuff and lapillistone	contact is with unit 71 is sharp
509.07		71	506	LTm	LTngr, LSTngr	several depositional units of massive lapilli tuff with normally graded tops, containing scattered accretionary lapilli and rare cm-thick interbeds of normally graded lapilli tuff and lapillistone	contact is with unit 72 is sharp
513.14		72	107	Tal	LTngr	several depositional units of massive accretionary-lapilli tuff with normally graded coarse fraction and includes an interbed of normally graded, thinly bedded lapilli tuff where each bed is capped by cm-thick horizon of accretionary-lapilli tuff	contact with unit 73 is diffuse
515.20		73	716	LTm	LTngr	multiple depositional units of massive lapilli tuff with normally graded tops; contains fair abundance of accretionary lapilli and armoured lapilli	contact with unit 74 is diffuse
522.36		74	349	Tal	LTngr, LSTngr	multiple depositional units of massive accretionary-lapilli tuff with normally graded coarse fraction and cm-thick accretionary lapilli horizons. Includes several thin interbeds of normally graded lapilli tuff and lapillistone	contact with unit 75 is sharp, convoluted and scoured
525.85		75	335	Tal	cT/LTm	multiple dm-thick depositional units of massive accretionary-lapilli tuff with evenly distributed or normally graded coarse fraction and scattered armoured lapilli. Includes a few thin interbeds of massive-very-fine grained	contact with unit 76 not recovered (27 cm gap)
529.20	**IIF** 12.1 m	76	90	LSTsgr		thinly bedded, monomictic, symmetrically graded lapillistone where each bed is bound by sharp contacts and has an inverse graded base and normally graded top	contact with unit 77 is sharp, convoluted and scoured
530.10		77	870	Tal, Ttb		10 cm of massive accretionary-lapilli tuff; the remaining 30 cm of the core are broken, with pieces consisting of thinly bedded tuff	basal contact not recovered

Table 4. *Proportions of individual lithologies and lithofacies in Site 1184 pyroclastic succession*

Lithology	Lithofacies	Code	Cumulative thickness (m)	Per cent of succession
Tuff	accretionary-lapilli tuff	Tal	83.1	24.6
	massive tuff	Tm-fr	25.1	7.4
	graded tuff	Tngr-fr	3.7	1.1
	thinly planar bedded tuff	Ttb-p	2.7	0.8
	cross-bedded tuffs	Ttb-x	0.9	0.3
	Tuff, total	T	115.6	34.2
Lapilli tuff	massive lapilli tuff	LTm	177.9	52.7
	graded lapilli tuff	LTngr	6.6	2.0
	coarse tuff/lapilli tuff	ct/LTm	13.9	4.1
	Lapilli tuff, total	LT	198.4	58.8
Lapillistone	massive lapillistone	LSTm	12.6	3.7
	graded lapillistone	LSTgr	11.2	3.3
	Lapillistone, total	LST	23.8	7.0
	total thickness of succession		337.7	

Table 5. *Codes and names of lithofacies*

Lithology	Lithofacies code	Name
Tuff (T)	Tal	accretionary-lapilli tuff
	Tm-fr	fines-rich massive tuff
	Tm-fp	fines-poor massive tuff
	Tngr-fr	fines-rich normally graded tuff
	Tgr-fp	fines-poor normally graded tuff
	Ttb	very thinly bedded tuff
Lapilli tuff (LT)	LTm	massive lapilli tuff
	LTngr	normally graded lapilli tuff
	cT/LTm	massive coarse tuff to fine lapilli tuff
	cT/LTngr	normally graded coarse tuff to fine lapilli tuff
Lapillistone (LST)	LSTm	massive lapillistone
	LSTngr	normally graded lapillistone
	LSTsgr	symmetrically graded lapillistone

angular–subangular phreatomagmatic clasts (see section on Volcaniclastic components for details).

Lithofacies Tal; accretionary-lapilli tuff

Description. This lithofacies is composed of massive, non-graded or normally graded, poorly sorted, fine-grained tuff containing ash-clusters and accretionary lapilli in excess of 10 modal%. The matrix grain size of the accretionary-lapilli tuff typically spans the ash-grade (from very fine ash to coarse ash) but is skewed towards the finer (<0.5 mm) fractions, which always make up >70 modal% of the total grain population. Almost the entire population is juvenile sideromelane and tachylite fragments, or 98 modal% on average (Fig. 3, Table 2). Lithic and crystal fragments of plagioclase and clinopyroxene are only present in trace amounts.

The accretionary-lapilli tuff lithofacies is present as several cm to dm-thick depositional units (range 1–200 cm) with sharp or diffuse depositional contacts. A simple depositional unit consists of a massive body of very-fine-grained tuff with dispersed whole and broken accretionary lapilli and lesser ash-pellets in abundances that range from fairly low (2–5 modal%) to high (*c.* 30 modal%) on the dm scale. The accretionary lapilli are present exclusively as discrete clasts resting in the ash-grade matrix and typically contain a single grain

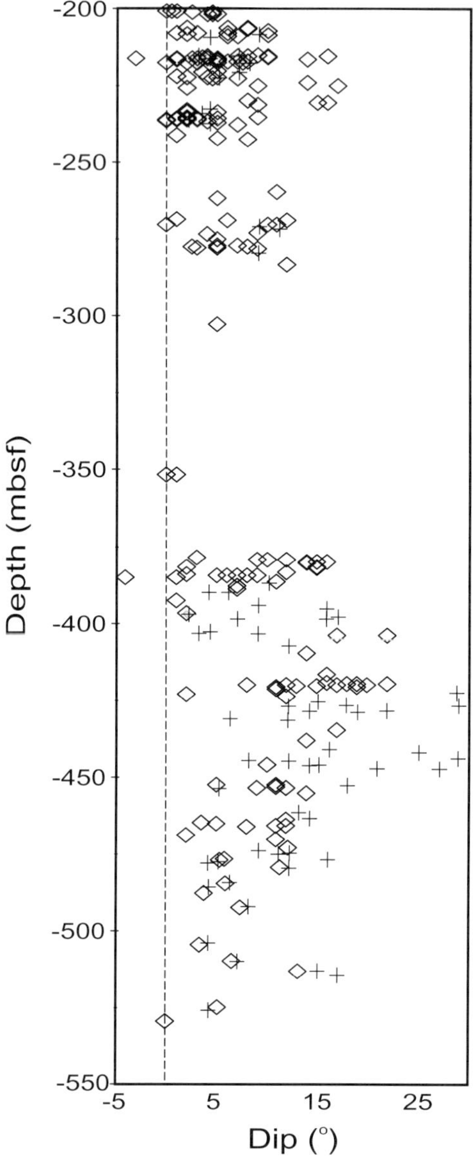

Fig. 6. Dip of inclined layers and bedding planes within the Site 1184 pyroclastic succession; this study (diamonds), Shipboard measurements (pluses).

nucleus of basaltic pumice. The broken accretionary-lapilli population includes slightly curved, 5–10 mm dish-like fragments that usually display bedding-parallel orientation, with either the convex or concave side up, and are never imbricated. The massive body of each set commonly grades upwards into cm-thick open framework horizons of ash-clusters, which usually are draped by mm-thick laminae of very-fine-grained ash (Fig. 7a). The bases of these units usually feature a horizon containing very high concentrations (40–60 modal%) of relatively large accretionary lapilli, characterized by a grain-supported framework with large cement-filled voids (Fig. 7a). These horizons commonly overlie or are mingled with a thin layer of fines-poor, massive (lithofacies Tm-fp) or normally graded (lithofacies Tngr-fp), coarse-grained tuff (Fig 7b, Table 3). The basal accretionary lapilli horizons incorporate a fairly closely packed mixture of perfectly spherical–ovoid accretionary lapilli that often are indented or buckled by those resting on them (Fig. 7a, inset). Spherical and ovoid accretionary lapilli with identical internal structures are commonly found as nearest neighbours in these horizons, making post-emplacement deformation by compaction due to burial and diagenesis an unlikely explanation. The observed relationships of selective compaction and buckling by the nearest neighbour are consistent with impact-induced deformation during deposition. Many of the thicker and more complex depositional units feature multiple graded coarse fractions and several distinctive horizons of whole and broken accretionary lapilli, indicating episodic deposition of the larger ash aggregate clasts (Fig. 7c).

The accretionary-lapilli tuff lithofacies is present throughout the succession as discrete, medium-bedded, 0.37–23 m-thick lithostratigraphic units and is the second most common lithofacies after the massive lapilli-tuff lithofacies. Collectively, it makes up about 24% of the total thickness of the succession and approximately 72% of the cumulative thickness of the tuffs. However, this lithofacies has the highest frequency of occurrence below 435 mbsf, where it constitutes about 65% of the succession (Table 3).

Interpretation. The accretionary-lapilli lithofacies is interpreted to represent phreatomagmatic fall units. In addition to essentially phreatomagmatic pyroclast population and subhorizontal depositional surfaces, the internal structures, such as those described above for the simple depositional unit (Fig. 7a), are entirely consistent with origin by fallout from an ash-cloud where deposition is initiated by flocculation and formation of ash aggregates. Subsequent disintegration of loosely bound aggregates upon deposition offers a logical explanation for the poor sorting and fine-grained nature of these tuffs. Furthermore, the lithostratigraphic units of accretionary-lapilli tuff are characterized by dm-scale plane-parallel stratification and their overall architecture strongly resembles that of shower-bedded phreatomagmatic tuff (e.g. Branney 1991). Frequent alternations of depositional units characterized by sharp grain-size changes to units with more diffuse changes are best explained by variable repose times between successive ash-showers. The thicker multiple-graded depositional units of this lithofacies exhibit structures similar to that of the massive body in the simple depositional unit. Therefore, its mode of deposition is assumed to be the same, but involving semi-continuous fallout from more closely spaced ash-showers.

Ash aggregation takes place in moist subaerial eruption plumes or ash-clouds rising from pyroclastic density currents (e.g. Gilbert & Lane 1994). In such instances the ash aggregation involves firm contact and bonding between liquid-coated ash particles by short-range surface tension (capillary) forces in a moist environment. Experimental results indicate that water content in the range of 10–30% is required for ash aggregation to take place (Schumacher & Schmincke 1995). The clustering process produces a spectrum of aggregate types, ranging from dry high-porosity (i.e. ash-clusters and ash-pellets) through moist low-porosity (i.e. accretionary lapilli) to wet aggregates (i.e. mud lumps), which all have substantially higher fall velocity than their component particles. These aggregates are fragile and they commonly collapse on impact with the surface, especially the 'dry' and 'wet' types. Thus, their preservation potential in the geological record is very low. However, the durability of the 'moist' low-porosity (i.e. accretionary lapilli) aggregates is sometimes enhanced by pre-deposition cementation by secondary minerals and salts (Gilbert & Lane 1994). Ash aggregates and armoured lapilli may form in shallow subaqueous phreatomagmatic eruptions (i.e. the subaqueous stage of Surtseyan events). It is inferred that these eruptions produce a large-scale steam-dominated cupola above the vent that provides the environment for ash aggregation (Kokelaar 1983; White 1996, 2000). Even if the notion of ash aggregation within the steam-cupola is accepted, it is highly unlikely that the newly formed ash aggregates, especially the ash-clusters and ash-pellets, would survive the inevitable transport and deposition through the water column, at least not in the quantities in which they are found in the accretionary-lapilli tuff lithofacies at Site 1184. Therefore, it is unlikely that the Site 1184 accretionary-lapilli tuffs were formed by turbulent or non-cohesive density currents associated with shallow subaqueous or emergent Surtseyan eruptions. For the same reason it is not likely that the succession was formed by immediate erosion and reworking of unconsolidated tephra followed by subaqueous deposition. For example, the subaqueous resedimented facies of the accretionary-lapilli-bearing Hatepe ash in New Zealand are barren of accretionary lapilli (B. Houghton pers. comm 2003). Although other syn-eruption resedimented deposits have been found to contain accretionary lapilli, they are present in only minor or trace amounts (e.g. Branney 1991). Consequently, subaerial eruption and deposition offers the best explanation for the common preservation and abundance of delicate ash-clusters and other ash aggregates within the Site 1184 pyroclastic succession. A subaerial origin is further corroborated by the presence of carbonized fragments of wood (i.e. charred tree-branches?) in the accretionary-lapilli tuff of lithostratigraphic members IIC, IID and IIF (Fig. 5a). In members IIC and IID, the wood horizons are only included in accretionary-lapilli tuff within 7–9 m of basal contact, suggesting that they may be remnants of trees that were buried by heavy tephra fall during the early stages of the eruptions (see also the Discussion).

Lithofacies Tm-fp and Tngr-fp; fines-poor, massive and normally graded tuff

Description. These fines-poor tuff lithofacies represent cm-thick, massive or normally graded, well-sorted beds of medium to very-coarse ash with grain-supported framework. They are present in association with the accretionary-lapilli tuff lithofacies and more rarely massive fines-rich tuff (Table 3). The composition of their clast population is always similar to that of the adjacent tuffs, although the fines-poor tuffs typically contain somewhat higher proportions of non-vesicular tachylite fragments (Fig. 3, Table 2). They commonly include minor amounts of accretionary lapilli and ash-pellets. Individual beds are characterized by both sharp and diffuse bedding contacts. Where the basal contact is diffuse so is the upper contact, and where the basal contact is sharp the upper contact is either sharp or diffuse (Fig. 7b).

Interpretation. The fines-poor, massive or normally graded tuff lithofacies are present as interbeds within lithostratigraphic units dominated by the accretionary-lapilli tuff lithofacies and less commonly the fines-rich massive tuff lithofacies. Contact relationships indicate that these fines-poor tuff beds are always associated with overlying depositional units of the host

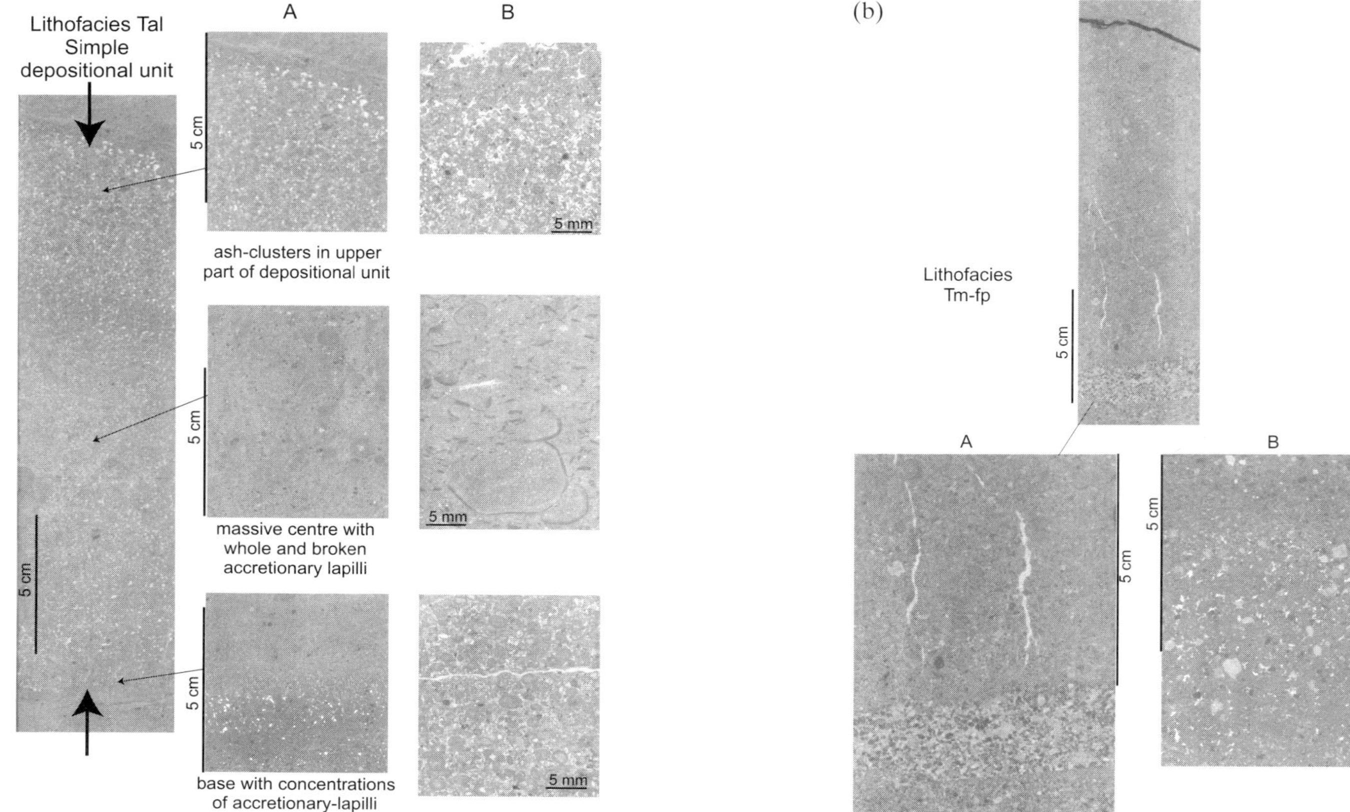

Fig. 7. Examples showing the key features of the accretionary-lapilli tuff lithofacies (Tal). (**a**) A simple depositional set in core (arrows point to bottom and top), enlargments to the right show the structures of the base, centre and top in core and thin section. See text for further details. (**b**) A thin fines-poor tuff bed (Tm-fp) at the base of an accretionary-lapilli-tuff unit, enlargement to left (A) shows the details of bed, whereas the one to right B shows an accretionary-lapilli-rich base of a depositional unit mingled with lapilli-size glass and tachylite grains. (**c**) Part of a complex, multiple-graded depositional unit with several concentrations of whole and broken accretionary lapilli. Note that on these core photographs and the ones that follow, the dip of the beds is apparent because of regional tectonic tilt.

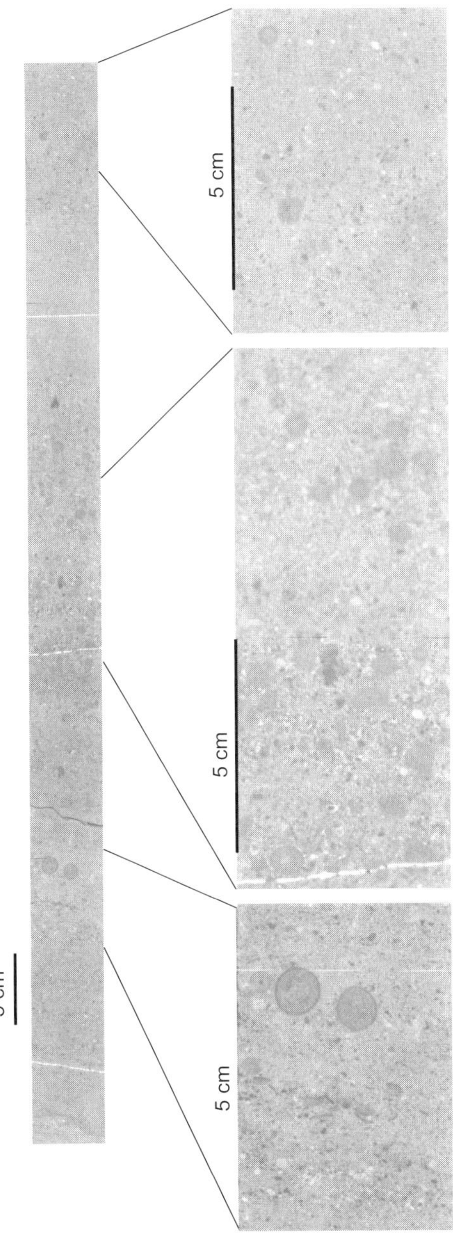

Fig. 7. (**c**)

lithofacies. On the basis of their association these lithofacies are inferred to have been deposited from turbulent suspension and most likely represent fallout from a relatively dry (fines-poor) explosion immediately preceding a wetter one. Alternatively, they could have been produced by segregation of larger and denser grains as a poorly sorted tephra mass from a subaerial fallout or surge event settled through a column of water. The evidence is not conclusive in this regard. Although settling through a water column appears to be a straightforward explanation for the fines-poor nature of these tuff beds, it is difficult to envisage how delicate ash-clusters of the associated accretionary-lapilli tuff units could survive such deposition in abundance (Fig. 7a). Therefore, the subaerial fall origin is adopted.

Lithofacies Tm-fr and Tngr-fr; fines-rich, massive tuff and normally graded tuff

Description. The fines-rich massive and normally graded tuff lithofacies consists of poorly sorted and matrix supported, fine to coarse tuff, where very fine to fine ash makes up 40–70 modal% of the total clast population. It also typically features an evenly dispersed or a normally graded coarse fraction of 0.5–3 mm clasts. These lithofacies are essentially monomictic because juvenile basaltic grains comprise >98 modal% of the clast population, which consist of angular non-vesicular to highly vesicular sideromelane (69%) and tachylite (29%), along with minor amounts of crystal and lithic fragments (Table 5). Accretionary lapilli and, more rarely, ash-clusters are typically present in minor amounts as solitary grains. These lithofacies have a collective thickness of approximately 29 m, representing about 8% of the Site 1184 pyroclastic succession. These lithofacies are the major constituents of 10 lithostratigraphic units spread throughout the succession and ranging in thickness from 0.01 to 7.5 m (Fig. 5b, Table 3). Each of these lithostratigraphic units is a compound package of multiple cm- to dm-thick depositional units bounded by sharp or diffuse subhorizontal contacts. Bedding-plane structures indicative of erosion (e.g. scouring) are absent.

Interpretation. These lithofacies were probably produced by rapid suspension deposition, although the depositional structures provide inconclusive evidence in regards to the mode of transport. However, pyroclastic fall origin is inferred because of the common presence of accretionary lapilli and ash-clusters, and subhorizontal bedding along with complete absence of scoured bedding planes and structures indicative of lateral transport.

Lithofacies Ttb; thinly bedded tuff

Description. The thinly bedded tuff lithofacies is a minor constituent in the Site 1184 pyroclastic succession and collectively it accounts for about 1.1% of its thickness. This lithofacies is characterized by planar-bedded and/or cross-bedded deposits.

The planar-bedded variety is present as 0.07–1.4 m-thick lithostratigraphic units bounded by diffuse or sharp depositional contacts. They are characterized by 0.5–3 cm-thick, crudely normally graded layers and are marked by plane-parallel contacts that in the core exhibit a dip (6–14°) compatible with the regional dip of 4–9° of the succession (Mahoney *et al.* 2001). The tuff consists almost entirely of juvenile sideromelane and tachylite grains, and more often than not contains scattered accretionary lapilli. Elongate, rod or plate-like clasts, when present, are always oriented parallel to bedding planes, but clast imbrication is absent.

The cross-bedded (or discordant-bedded) tuff lithofacies is exceptionally rare in the succession and is present only as cm- to dm-thick lithostratigraphic units (Table 3). They are thinly stratified with discordant and opposite-dipping bedding planes (range, –5° to 16°). The dip of the internal bedding planes generally decreases up-section. The lithofacies is typified by cm-thick layers, where beds of medium-sorted, very-fine- to fine-grained tuff alternate with beds of moderately to well-sorted coarse-grained tuff. Individual layers are massive or inversely graded with scoured basal contacts. More rarely, this cross-bedded tuff lithofacies is present as cm-thick horizons of cross-laminated, moderately sorted fine-grained tuff capping other lithologies.

Interpretation. These are the only lithofacies of the succession that exhibit clear evidence of deposition by tractional transport. They are always present as very thinly to thinly bedded, cm- to dm-thick lithostratigraphic units capping a thick sequence of other lithologies. The decreasing up-section dip within these units suggests that these lithofacies are filling small depressions. These lithofacies were, therefore, most probably formed by localized reworking of unconsolidated tephra, although a pyroclastic-surge origin cannot be ruled out for some units.

Lithofacies LTm and cT/LTm; massive lapilli tuffs

Description. The massive lapilli-tuff lithofacies is by far the most common in the succession. It is the key building block in all of the thicker lapilli-tuff lithostratigraphic units, accounting for approximately 53% of the thickness of the drilled sequence (Table 4). They consist of poorly sorted, fine- to medium-grained lapilli tuff, with grain sizes of the lapilli fraction ranging from 2 to 25 mm (Fig. 8). Armoured lapilli, accretionary lapilli and ash-pellets are common, although present in highly variable proportions that range from trace amounts to 10 modal%. In this regard, lithostratigraphic unit 24 stands out because it contains significant amounts of large (6–18 mm diameter) ash-pellets with abundances as high as 25 modal%. This lithofacies also contains rare out-sized juvenile and lithic fragments with diameters of 35–65 mm. The matrix typically consists of medium to coarse ash, with lesser and highly variable amounts of fine ash. The massive lapilli-tuff lithofacies consists of disorganized deposits often with diffuse internal stratification. This stratification is defined by gradual changes from fine- to medium-grained lapilli on a dm- to m-scale, as well as by changes in the relative abundance of the matrix resulting in frequent alternations from a matrix to a clast-supported framework. When individual depositional units are discernible they are 1 dm- to metres thick (range 0.1–4 m) and bounded by diffuse or sharp contacts. Each depositional unit features a body of disorganized lapilli tuff that is often capped by a 5–40 cm-thick normally graded (lapilli to ash) horizon. Evidence of traction-carpet transport, such as discordant bedding and scoured basal contacts, or of high shear rates within suspension, such as an inversely graded base, are absent. However, these lapilli-tuff units are commonly intercalated with a few cm-thick lapillistone beds (Fig. 8).

The overall characteristics of the coarse-ash/lapilli-tuff lithofacies (cT/LT) are essentially the same as for those of the massive lapilli tuff, except that the grain size of the lapilli fraction is smaller, ranging from 1 to 6 mm.

Interpretation. Pyroclastic origin for these lithofacies is indicated by juvenile-dominated clast population, characterized by clast types indicative of explosive water to magma interaction, and common presence of armoured lapilli and accretionary lapilli. Furthermore, diffuse stratification and poor sorting, along with frequent but non-systematic changes from matrix- to grain-supported framework or graded to non-graded coarse fraction, is indicative of rapid deposition from high-concentration suspension (e.g. Chough & Sohn 1990), which may have resulted from a pyroclastic fall or density currents.

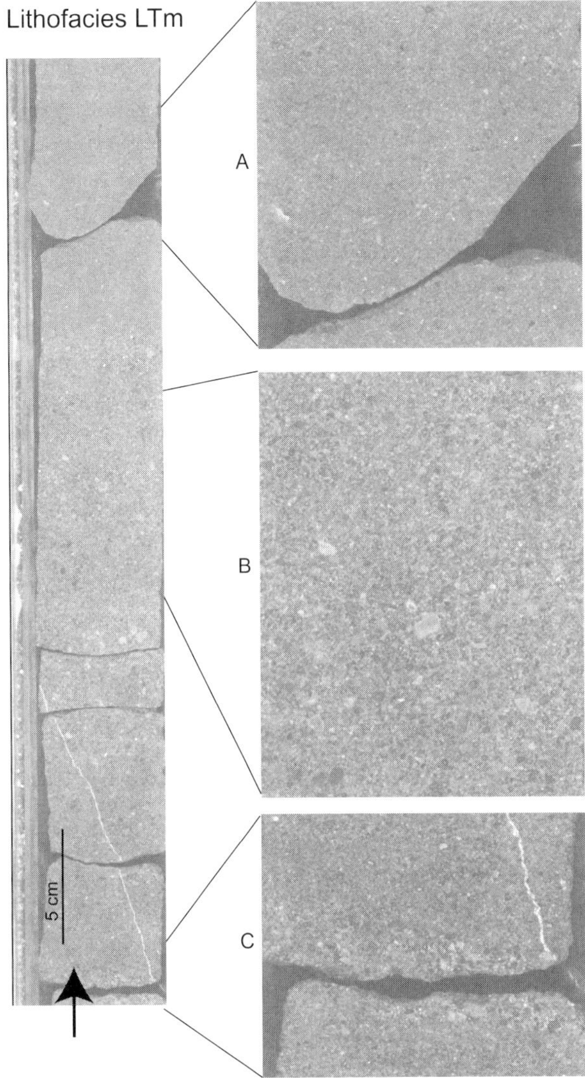

Fig. 8. Example showing the characteristic features of the lapilli-tuff lithofacies; a core section featuring a depositional set of massive lapilli tuff with normal-graded top (LTm). The enlargements to right show normal-graded top (A), massive interior (B) and a thin lapillistone bed (C) at the base.

These lithofacies are the main component in the approximately 5–41 m-thick lapilli-tuff lithostratigraphic units that dominate the succession (Fig. 5b, Table 3). Each of these lithostratigraphic units consists of multiple dm- to m-thick depositional units, yet their grain size is more or less constant throughout. It is effectively truncated at the fine or medium lapilli size fractions. Coarse lapilli clasts are only present in minor amounts, whereas blocks and bombs are absent altogether. Structures indicative of high dispersive shear rates or current transport such as inversely graded bases, scoured bedding planes and cross-bedding are also absent. Phreatomagmatic cones and their near-vent deposits typically consist of a very diverse clast population, where numerous dm- to m-size blocks are dispersed throughout ash and lapilli beds (e.g. White 1996, Houghton *et al.* 1999; Cole *et al.* 2001). Consequently, if the lapilli tuffs at Site 1184 were emplaced by eruption-fed mass flows or syn-eruption sedimentary density by reworking of unconsolidated near-vent deposits, their grain size would be more variable and include some block-size clasts. On the other hand, the uniform grain size and abrupt coarse-tail truncation is a characteristic feature of pyroclastic fall deposits, because of effective size segregation of the largest clasts during atmospheric transport (e.g. Cas & Wright 1987). Therefore, the massive lapilli-tuff lithofacies are inferred to originate as phreatomagmatic fall deposits. This interpretation is strengthened by the observation that consecutive lapilli-tuff and accretionary-lapilli-tuff units are commonly genetically linked in their accumulation, both in terms of contact relationships and chemical composition.

Lithofacies LTngr and cT/LTngr; normally graded lapilli tuffs

Description. The graded lapilli-tuff lithofacies resembles the massive lapilli-tuff lithofacies in almost all respects, except that it is characterized by a normally graded or, more rarely, an inversely graded lapilli fraction in a matrix-supported framework. It is typically present as dm- to m-thick interbeds within lithostratigraphic units characterized by the massive lapilli-tuff lithofacies (Table 3). Therefore, on the basis of association, the normally graded lapilli-tuff lithofacies are inferred to have formed by similar processes to the massive lapilli tuff.

Lithofacies LSTngr, LSTsgr and LSTm; normally graded, symmetrically graded and massive lapillistone

Description of lithofacies LSTngr. This lithofacies consists of normally graded depositional units in which the grain sizes range from fine to medium lapilli at the base to coarse ash at the top (Fig. 9). It is the main constituent of lithostratigraphic unit 29, which is made up of multiple 2–20 cm-thick, well-sorted depositional units of normally graded lapillistone. The grading in individual units can be gradual from the base to the top or may be characterized by an abrupt grain size change in the upper half of the sets. The contacts between depositional units are diffuse or sharp, but show no evidence of scouring. The normally graded lapillistone is almost entirely made up of juvenile angular, moderately to non-vesicular tachylite and sideromelane fragments, with a 2:1 ratio of tachylite to sideromelane. Both clast types include swarms of needle-shaped plagioclase microlites. Lithic fragments of lithified tuff and altered basalt are present in trace amounts.

Each depositional unit features a loosely compacted coarse ash-grade top, where void space is filled by fine ash, and a densely packed lower part of framework-supported lapilli, where cement-filled voids occupy 10–25 modal%. Some, but not all, of the lapilli in the lower part are characterized by large grain-to-grain contacts, with 50–100% of the facing grain surfaces touching (Fig. 9). These clasts often drape underlying clasts as if they were flattened upon impact or by subsequent loading of new material, suggesting that these lapilli grains were deposited in a hot (i.e. viscoplastic) state.

In addition, this lithofacies is present as a minor component within lithostratigraphic units dominated by the lapilli-tuff and accretionary-lapilli-tuff lithofacies, where the cm-thick beds are normally bounded by sharp lower contacts and diffuse upper contacts.

Description of lithofacies LSTsgr, symmetrically graded lapillistone. This lithofacies consists of 3 15 cm-thick depositional units that exhibit an inversely graded base of coarse ash to fine lapilli, a non-graded middle part of fine lapilli and a normally graded top of fine lapilli to medium ash. Juvenile tachylite grains, 0.5–6 mm in diameter, dominate this lithofacies, but it also contains minor amounts of sideromelane and submillimetre plagioclase fragments (Table 2). Lithic fragments are present in trace amounts. The tachylite and the sideromelane grains contain variable amounts of seriate-textured plagioclase laths (≤0.4 mm, 0–50 modal%) and plagioclase phenocrysts up to 1 mm in diameter. The lapilli-dominated middle part of each set is characterized by a compact grain-supported framework with flattened and draping grains identical to that described above for other lapillistone facies. In contrast, the adjacent inversely and normally graded intervals, which are on average finer grained and typified by an open framework fabric, consist of entirely undeformed clasts. These relationships strongly favours hot emplacement for the larger lapilli.

Description of lithofacies LSTm, massive lapillistone. This facies consists of monomict brown–greenish-grey lapillistone that has a tightly packed framework-supported fabric with ≤15 modal% of zeolite-filled intergranular voids (Fig. 9). It consists almost entirely of 3–16 mm moderately to non-vesicular, subtrachytic textured tachylite grains (*c.* 97 modal%) containing up to 60 modal% of submillimetre-long, lath-shaped plagioclase crystals. Plagioclase-bearing,

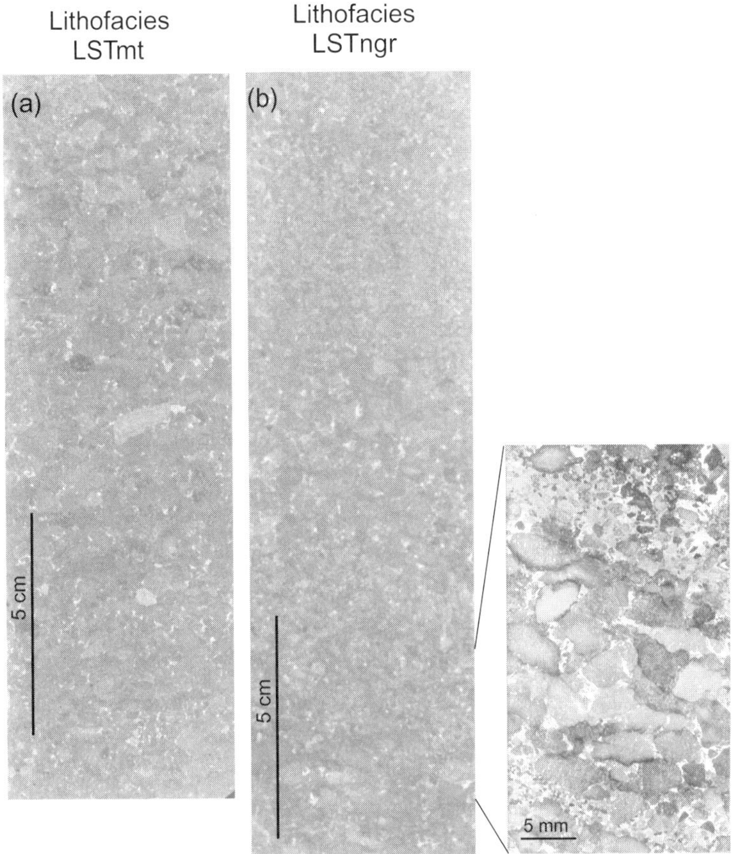

Fig. 9. Example showing the characteristic features of the lapillistone lithofacies: (**a**) massive lapillistone (LSTm) and (**b**) normal-graded lapillistone (LSTngr). Enlargement shows the details of bedding and deposit framework. Note the dense packing of grains, large grain-to-grain contacts and grain deformation indicative of hot emplacement.

sideromelane grains with hyalopilitic texture are present in minor amounts (*c.* 2 modal%), both as lapilli and as ash-size fragments. Crystal (plagioclase) and accidental lithic fragments are present in trace amounts. As in the graded lapillistone lithofacies, the juvenile population recurrently features fragments with large grain-to-grain contacts and flattened grains draping underlying ones, suggesting hot emplacement.

Interpretation of the lapillistone lithofacies. A monomictic population of partly crystalline tachylite and sideromelane clasts, and evidence of hot emplacement suggests that these lithofacies were derived from partly solidified lava disintegrated by explosive activity and subsequently deposited by fallout from relatively 'dry' eruption plumes. These deposits may have been produced by a rootless eruption as lava interacted explosively with surface water or it may be derived from explosive fragmentation of a crust on a lava pond (e.g. Houghton & Schmincke 1986).

Lithostratigraphic members and lithofacies associations

The outstanding and, perhaps, the most astonishing feature of the Site 1184 basement succession is that it is exclusively pyroclastic, no other volcanic or sedimentary rock types, not even one, interrupt this depositional record. It is also dominated by lithologies of ash and lapilli that are spread throughout the succession as discrete

<1–>40 m-thick intervals without any obvious pattern or frequency. Another important aspect is that discrete core intervals consist of an association of depositional units of the same lithofacies and, as such, define the lithostratigraphic units that make up the succession. Individual lithostratigraphic units are essentially uniform in terms of the component proportions (Table 2) and within them the contacts between successive depositional units are either diffuse or sharp but are always non-erosional. Similar contact relationships are most common between successive lithostratigraphic units within individual members, indicating a more or less continuous and rapid deposit accumulation over large stratigraphic intervals corresponding to the lithostratigraphic members (Table 3, Fig. 5b). The lithostratigraphic members IIA, IIB, IID and IIE, which are all chemically uniform and distinct (Fitton & Godard 2004; White *et al.* 2004), consist of alternating tuff and lapilli-tuff units that were emplaced, as far as can be determined, in continuous succession. This evidence strongly suggests that each of these four lithostratigraphic members corresponds to a single eruption. However, this appears not to be the case for lithostratigraphic members IIC and IIF.

Lithostratigraphic member IIA

This member consists of 11 lithostratigraphic units (labelled 1–11) and features an association of five main lithofacies (LTm, cT/LTm, Tal, Tm-fr and Ttb-p/x). It is dominated by thinly stratified accretionary-lapilli tuff in three 7–14 m-thick lithostratigraphic units consisting of multiple cm- to dm-thick fall units (i.e. depositional units). Collectively, they account for 54% of the member thickness (Fig. 5b, Table 3). Other lithologies are a *c.* 7 m of crudely stratified lapilli-tuff unit, three *c.* 0.5–6.5 m-thick stratified fines-rich tuff units, and three dm-thick thinly planar to cross-bedded tuff units. All of the thicker lithostratigraphic units are marked by non-erosional contacts, indicating that the deposition was largely characterized by periods of sustained accumulation (Fig. 5b, Table 3). However, the three dm-thick planar- to cross-bedded tuff units clearly indicate a swift change from one mode of deposit transport and accumulation to another and back. In this type of succession, such abrupt alternations in deposit structures do not necessarily indicate a change in the depositional environment. They are common in phreatomagmatic tephra sequences and a direct consequence of the two characteristic modes by which the tephra is transported from the vents (i.e. fall v. surge; Cas & Wright 1987). Alternatively, they may represent a lull in the eruption marked by a brief period of erosion and reworking of the loose tephra fall deposits. The fact that the composition of these thinly bedded tuffs is identical to other deposits of IIA supports this conclusion. With this in mind, along with the compositional uniformity and observed facies associations, IIA is considered to comprise the products of a single eruption with at least four eruption episodes separated by short breaks.

Lithostratigraphic member IIB

This member consists of 13 lithostratigraphic units (units 12–24) and an association of four main lithofacies (LTm, cT/LTm, Tal and Tm-fr). The lapilli-tuff lithofacies dominates and is present as seven crudely stratified, *c.* 1.5–18.5 m-thick lithostratigraphic units, accounting for 88% of the thickness of IIB (Fig. 5b, Table 3). Other lithologies are units of accretionary lapilli tuff and fines-rich massive tuff, which are several metres thick. Fragments of carbonized wood were found in a lapilli-tuff unit about 5 m above the base of IIB. Diffuse and sharp non-erosional contact relationships indicate more or less uninterrupted accumulation by tephra fall. The recorded lithofacies associations may indicate changes in eruption style or intensity or wind-induced changes in eruption plume dispersal or combinations thereof.

Lithostratigraphic member IIC

This member consists of 15 lithostratigraphic units (units 25–39) and an association of six main lithofacies (LTm, LSTm, LSTngr, Tal, Tm-fr and Ttb-x/p). As in subunit IIB, lapilli tuff is the dominant lithofacies (Fig. 5b, Table 3). From the base upward, IIC consists of a 4.8 m-thick interval comprising dm- to m-thick accretionary-lapilli tuff units alternating with thin cross-bedded tuff and lapilli-tuff units. This sequence is separated from the overlying normally graded lapillistone unit by a sharp contact. Upward the lapillistone unit grades into 27 m of lithic-rich lapilli-tuff unit (unit 28), which in turn is overlain by 5 cm of massive tuff marked by sharp lower and upper contacts (Fig. 5b, Table 3). The tuff layer is overlain by another thick (35 m) lithic-rich lapilli-tuff unit (unit 26), which in turn is capped by an approximately 12 m-thick interval of massive lapillistone. The stratigraphic association of the last two lithostratigraphic units is uncertain because the contact between them was not recovered (Table 3). The lithofacies association in lithostratigraphic member IIC indicates that lithostratigraphic units 39–30 were deposited in a rapid succession by pyroclastic fall and surges. The relationship of this sequence to the overlying lapillistone is uncertain, but both unit 29 (LSTngr) and 30 (Tal) contain fragments of carbonized wood (Fig 6b). Lithostratigraphic units 29 and 28 were deposited as a continuous succession, before the massive tuff (unit 27) was laid down. The small thickness and sharp lower and upper boundaries of this massive tuff suggest that it was formed by a single abrupt event. The strong similarities between the lapilli-tuffs in units 28 and 26 indicate a return to deposition modes similar to those operating before the tuff-forming event. However, the capping by massive lapillistone (unit 25) appears to indicate deposition by an entirely separate event. Three samples were analysed from lithostratigraphic member IIC and reveal significant differences in chemical composition, although all are characterized by uniquely high Nb concentrations (Fig. 5a). The lapilli-tuff units, however, contain high, but variable, proportions of <1–65 mm-large accidental lithics, so these analyses are of questionable value as source discriminators. Nevertheless, the lithofacies associations in IIC suggest that it may have been produced by at least three separate eruptions (Fig 5b, Table 3).

Lithostratigraphic member IID

This member consists of nine lithostratigraphic units (units 40–48) and features an association of four lithofacies (LTm, cT/LTm, LTngr and Tal). The lapilli-tuff lithofacies dominates IID, but the accretionary-lapilli lithofacies has also a strong presence. The lapilli tuff is present as a single, *c.* 41.5 m-thick lithostratigraphic unit characterized by diffuse stratification and as a few, metres-thick units interbedded with 0.5–11 m-thick units of accretionary-lapilli tuff (Fig. 5b, Table 3). The lowest accretionary-lapilli-tuff unit contains fragments of carbonized wood between 7 and 9 m above the base of IID. Diffuse or sharp non-erosional contacts throughout the member indicate an uninterrupted accumulation by tephra fall. The frequent alternations from lapilli tuff to accretionary-lapilli tuff throughout IID is more likely to represent changes in eruption style and intensity rather than changes in the dispersal direction of the eruption plume.

Lithostratigraphic member IIE

This member consists of 26 lithostratigraphic units (units 49–74) and includes an association of five main lithofacies (LTm, cT/LTm, Tm-fr, Tngr-fr and Tal). The accretionary-lapilli tuff is the dominant lithofacies in IIE, accounting for 12, 0.8–10 m-thick lithostratigraphic units each featuring multiple cm- to dm-thick fall units (Fig. 5b, Table 3). Collectively, they make up 51% of the member thickness. Other lithologies include seven, *c.* 0.3–7 m-thick, crudely stratified lapilli-tuff units and five *c.* 0.4–8 m thick stratified fines-rich tuff units and three dm-thick thinly planar to cross-bedded tuff units. Uninterrupted accumulation by tephra fall is indicated by diffuse or sharp non-erosional contacts throughout, and recurrent alternations between lithofacies most probably reflect changes in style or intensity or both.

Lithostratigraphic member IIF

This member consists of three lithostratigraphic units (units 75–77) and includes an association of two lithofacies (LSTsgr and Tal). The top 3.3 m of IIF consist of thinly bedded accretionary-lapilli tuff containing dispersed carbonized wood fragments. The symmetrically graded lapillistone lithofacies (Lithostratigraphic unit 76) is only present in the top 90 cm of core section-46R-1 (the last core section recovered at Site 1184; total length 130 cm). It is composed of framework-supported beds marked by sharp gently dipping contacts (dip 1°–3°), except in the interval 46R-1, 60–85 cm, where the bedding is disturbed. Here, the core cuts through part of a subvertical structure that may be a ballistic-impact sag, because it is filled with massive fine-grained lapilli tuff of the same composition as the surrounding lapillistone beds. The basal contact is a scoured surface, eroded into >10 cm of accretionary-lapilli tuff (unit 77). The remaining 30 cm is broken core, featuring pieces of thinly bedded tuff.

The contact of the lapillistone with an upper accretionary-lapilli-tuff unit was not recovered, but the strongly scoured basal contact indicates a period of erosion before the lapillistone was deposited. Therefore, it is possible that IIF may contain deposits from at least two, and possibly three, separate eruptions.

Discussion and conclusions

Epiclastic or pyroclastic origin?

Recovering a continuous *c.* 340 m-thick section of purely pyroclastic rocks from the Ontong Java Plateau, let alone one containing evidence that can be linked to explosive emergent or subaerial eruptions, was unexpected (Mahoney *et al.* 2001). As stated at the beginning, the first critical issue is to determine whether the origin of the Site 1184 succession is epiclastic or pyroclastic.

Pyroclastic origin is inferred for the succession because it has been shown that the *c.* 43–74 m-thick lithostratigraphic members are essentially composed of pyroclastic grains, and are characterized by uniform chemical and component compositions. Also, the members IIA, IIB, IID and IIE are made up of an association of lithostratigraphic units that were deposited in succession without any obvious long-lasting breaks, suggesting that each of these members corresponds to a single eruption.

Also, it is unlikely that the Site 1184 pyroclastic succession was produced by post-eruption erosion of phreatomagmatic (including Surtseyan) edifices constructed by recurring eruptions. Lithification of wet or moist phreatomagmatic tephra deposits in such structures is, geologically speaking, incredibly fast (i.e. tens to thousands of years). For example, the tuff and lapilli tuffs at Surtsey were almost completely consolidated within 10 years of the eruption (e.g. Jakobsson & Moore 1982). Similarly, an approximately 200 year-old tuff cone largely made up of rather 'wet' phreatomagmatic deposits on the 1783–1784 Laki fissure is completely lithified today (Thordarson & Self 1993). Consequently, a sequence produced by erosion of such structures is expected to include gradually increasing amounts of debris derived from erosion of consolidated tuff and lapilli-tuff fragments, as well as crystalline basalt fragments from feeder dykes and conduits. No such pattern is observed in the lithologies that make up the Site 1184 succession. Therefore, it is concluded that the drilled basement deposits at Site 1184 are syn-eruption accumulations and these deposits provide an important insight into the eruption history of this part of the Ontong Java Plateau.

Significance of accretionary lapilli and other ash aggregates

Ash aggregates are present as solitary grains, together with a dominant phreatomagmatic clast population, in high abundances throughout the Site 1184 pyroclastic succession. The significance of accretionary lapilli and other ash aggregates for assessing the origin of these pyroclastic rocks is that their only proven origin is by flocculation of ash in moist subaerial eruption plumes or in clouds rising from pyroclastic density currents. Ash aggregates are formed as loosely bound and delicate clusters of very fine to fine ash and commonly disintegrate during transport and upon deposition, although some are strengthened by incipient cementation of water-soluble alteration minerals and salts (e.g. Gilbert & Lane 1994). Yet, perfectly preserved ash aggregates of all types are present as individual clasts throughout the Site 1184 pyroclastic succession and were always deposited as solitary grains. Most remarkable is the common occurrence of ash-clusters, which have a very low preservation potential in the rock record because of their fragile nature (e.g. Fig 7a). Ovoid shapes of many accretionary lapilli indicate impact-induced deformation upon deposition. Also, in the case of broken accretionary lapilli, the exposed centres were disintegrating as the fragments were being deposited. This in particular implies that in bulk the accretionary lapilli were still soft and largely unconsolidated at the time of deposition. The inevitable conclusion drawn from these observations, irrespective of the exact processes by which these deposits were accumulated, is that the formation of the ash-aggregate-bearing lithologies in the succession was directly linked to ongoing eruptions. Therefore it provides, for the first time, compelling evidence of subaerial or emergent phreatomagmatic activity within the Ontong Java Plateau (Mahoney *et al.* 2001).

Primary v. secondary eruption processes

The logical question that follows the conclusion above is how much of the Site 1184 pyroclastic succession was produced by primary v. syn-eruption resedimentary processes. Some important conclusions can be drawn from the current observations, but an exact determination of their abundances requires a finer-scale analysis than that presented here.

Lithostratigraphic units of accretionary-lapilli tuff are characterized by subhorizontal plane-parallel stratification on the decimetre scale and are inferred to consist of numerous fall units. Their overall architecture strongly resembles that of shower-bedded phreatomagmatic tuff, where frequent alternations between intervals characterized by sharp grain-size changes to intervals with more diffuse changes are best explained by variable repose times between successive ash-showers (e.g. Branney 1991). The overall fine-grained nature of these deposits, as well as the absence of outsized clasts, suggest that units were deposited outside of the vent structures, most probably at proximal to medial distances (see below).

The Site 1184 lapilli-tuff units are characterized by rather uniform grain size, confined to fine to medium lapilli sizes, poor sorting and a juvenile-dominated clast population. Also, most beds include trace to substantial amounts of armoured lapilli and accretionary lapilli. Individual lithostratigraphic units feature subhorizontal diffuse stratification and frequent, but non-systematic, changes from matrix to framework support or from graded to non-graded coarse fraction. Furthermore, the overall disorganized structure of the lapilli-tuff units indicates high flux as well as semi-steady accumulation, where sharp contacts suggest short breaks in accumulation between successive depositional units and diffuse contacts indicate continuous deposition. All of these features are characteristic of lapilli tephra deposits produced by phreatomagmatic eruptions. By analogy, the Site 1184 lapilli tuffs are inferred to be phreatomagmatic lapilli fall deposits, although concurrent emplacement by non-cohesive density currents cannot be ruled out categorically. This interpretation is somewhat strengthened by the observation that consecutive lapilli-tuff and accretionary-lapilli-tuff units are commonly genetically linked in their accumulation, both in terms of contact relationships and chemical composition (Fig. 5). The majority of the lapilli-tuff lithostratigraphic units are most probably produced by proximal to medial fallout. However, the overall coarseness of lithostratigraphic units 26, 28 and 45 (mean, medium lapilli; range, ≤0.1–65 mm), along with indications of relatively steeply dipping bedding planes (i.e. –4° to 16° after correction for the regional dip), may indicate accumulation on the fringes of a cone.

The evidence of hot emplacement for the lapillistone units provides the strongest argument for interpreting them as pyroclastic deposits produced by fallout from relatively dry eruption plumes. A uniform juvenile clast population and framework-supported fabric are consistent with this interpretation. Furthermore, the diffuse contact relationship observed between lithostratigraphic units 29 and 28 shows that the deposition of the lapillistone (*c*. 1.5 m-thick) was immediately followed by the accumulation of 27 m of lapilli tuff, indicating a genetic association between the two. Shifts from 'dry' to 'wet' eruption style were frequent during the continuous uprush events at Surtsey, producing intercalated beds of agglutinates and poorly sorted lapilli tuff (e.g. Thorarinsson *et al.* 1964; Houghton & Schmincke 1986; Houghton *et al.* 1999).

The examples simply emphasize that wherever the Site 1184 pyroclastic succession is examined the evidence for primary pyroclastic origin is overwhelming and at this stage in the study less than 2% of the succession is categorized as syn-eruptive resedimented deposits. In conclusion, the bulk of the Site 1184 pyroclastic succession is interpreted to consist of phreatomagmatic fall deposits.

Number and size of eruptions

The next logical question is how many eruptions contributed to the construction of the Site 1184 pyroclastic succession and what was their magnitude? An answer to the first part of the question is straightforward. If it is assumed that each lithostratigraphic member corresponds to an eruption, then the total number of eruptions is at least six (Fig. 5). However, as mentioned previously, lithostratigraphic evidence suggests that up to three eruptions were involved in producing each of the deposits of IIC and IIF. Thus, it is possible that the tally of eruptions is as high as 10.

Assessing the size of individual eruptions is very difficult, because the core only provides a one-dimensional view of the products from each eruption and the location of the source vents is essentially unknown. If the free-air gravity highs mentioned earlier are volcanic edifices, then the closest possible source vents that we know of are 15–35 km from Site 1184. If these structures indeed contained the source vents for the Site 1184 succession, then the eruptions responsible for its deposition must have been capable of transporting large quantities of tephra some tens of kilometres. The relatively fine-grained nature of the sequence (i.e. mostly medium lapilli to fine ash), along with the complete absence of ballistics in all lithologies, is consistent with this. However, the great thicknesses of individual eruption packages, as much as 73.5 m in the case of lithostratigraphic member IIE (Fig 5, Table 3), appear to challenge this interpretation. The accumulated thickness of modern phreatomagmatic tephra fall rarely exceeds 1 m outside of the main cone structure, and the range of notable tephra accumulation (i.e. fall deposit thickness >1 cm) rarely extends beyond 30 km from the source (Thorarinsson *et al.* 1964; Sæmundsson 1991; Thordarson & Self 1993; Cole *et al.* 2001). The exceptions are large (volume 1–4 km^3) phreatomagmatic basaltic fissure eruptions in Iceland, which produced bedded tephra fall that completely blanketed the surrounding countryside within 100–200 km radius of the source vents. These tephra layers are up to 14 m thick within 5 km of the source vents, and 1–3 m thick at distances of 10–20 km from the source (Larsen 1984, 2000).

Is it possible that the Site 1184 lithostratigraphic members were produced by flood-basalt-size phreatomagmatic eruptions during the main phase of the Ontong Java volcanism? This is perhaps not such a far-fetched notion because finding 10–14 m thicknesses of shower-bedded accretionary-lapilli tuff in any volcanic environment is rare, even within Surtseyan tuff cones. Needless to say, such sequences are, as far as we know, very rare in the geological record and when present are thought to be a clear indication of a major event. One example comparable with the accretionary-lapilli units in IIE and IID, in terms of their thickness and deposit structures, is the 30 m-thick andesitic Whorneyside Tuff Formation (Ordovician) in NW England. This formation essentially consists of accretionary-lapilli tuff and was produced by a phreatoplinian eruption that was associated with a major caldera collapse event (e.g. Branney 1991). The accretionary-lapilli tuffs in the Prebble Formation of the mid-Jurassic Ferrar igneous complex in the Transantarctic Mountains (e.g. Hanson & Elliot 1996) provide another example. Therefore, the Site 1184 pyroclastic succession may represent the proximal to medial fallout from phreatomagmatic flood basalt fissure eruptions, similar to those of the 1477 AD Veidivotn and 870 AD Vatnaoldur fissure eruptions in Iceland (Larsen 1984), but one–two orders of magnitude larger.

Depositional environment and age

The profusion of ash aggregates and armoured lapilli within all of the lithostratigraphic members at Site 1184 implies that the eruptions that produced these deposits involved subaerial transport within turbulent phreatomagmatic eruption plumes or ash-clouds rising from pyroclastic density currents or both. This indicates an emergent or subaerial eruption environment. This conclusion is consistent with the relatively low volatile abundances in the pyroclasts from the Site 1184 succession (Roberge *et al.* 2004), suggesting formation by phreatomagmatic fragmentation and subsequent arresting of magma degassing at conduit depths of ≤150 m (Thordarson *et al.* 1996).

The presence of nannofossils in the Site 1184 pyroclastics, although in extremely low abundances, led the Shipboard Scientific Party to assume submarine deposition for the succession. However, the succession is completely barren of other marine fossils (e.g. foraminifera and burrows) as well as marine sediments, which is strange because deposition of the succession is inferred to span 3 Ma (Shipboard Scientific Party 2001).

There are two lines of evidence that suggest subaerial deposition for the Site 1184 succession: (a) the abundance of ash-clusters in the accretionary-lapilli tuff, as well as the indication of hot emplacement for the lapillistone units, are more suggestive of on-land deposition; and (b) the presence of tree remains (branches) in the deposits of lithostratigraphic members IIB, IIC, IID and IIF (Fig. 5b). These wood fragments are in deposits near the base of the inferred eruption units but not elsewhere. This is best accounted for if these are remains of standing trees buried by the tephra fallout. Tree moulds after standing trees are not uncommon in fall deposits. For example they are found in the phreatomagmatic fallout tephra from the *c.* 2500 years BP Hverfjall eruption in North Iceland (e.g. Sæmundsson 1991) and in the phreatomagmatic phase of the 934 AD Eldgjá flood lava eruption in South Iceland (unpublished data by the author).

The interpretation of the depositional environment is linked to the discrepancy between the age obtained from nannofossil evidence (middle Eocene; Bergen 2004) and that from radiometric dating (Early Cretaceous; Chambers *et al.* 2004) because the interpretation of submarine deposition hinges on the assumption that the nannofossils were incorporated at the time of deposition. If they were, then the inescapable conclusion is that the whole succession was produced by mid-Eocene eruptions and unrelated to the bulk of the Ontong Java volcanism. However, other results of Leg 192 post-cruise studies overwhelmingly support an Ontong Java origin and age for the succession. The evidence includes steep palaeomagnetic inclination for the Site 1184 basement rocks consistent with an Early Cretaceous age (Riisager *et al.* 2004), and major- and trace-element as well as isotopic compositions identical to those of the basalts elsewhere within the Ontong Java Plateau (e.g. Fitton & Godard 2004; Roberge *et al.* 2004; Shafer *et al.* 2004; White *et al.* 2004).

Is it possible that the nannofossils are a later addition to the succession? In the basement section of Site 1184, the nannofossils are poorly preserved and present in extremely small amounts within sediment pods and stringers (Bergen 2004). Detailed core and thin-section inspection revealed no trace of detrital or carbonate sediments within the succession. Examination of the core at the nannofossil sample intervals indicates that the sediment pods correspond to the cement-filled voids between lapilli or accretionary-lapilli grains, whereas the stringers appear to be small cement-filled veins. Thus, the nannofossils are in the cement. Large calcite domains in otherwise zeolite-dominated cement, as well as irregular patches of calcite or clay in zeolite-filled veins, indicate that the lithification of the succession involved at least two stages of cementation, an earlier zeolite stage and a later carbonate alteration stage. Is it possible that the nannofossils were introduced into the succession during the later alteration stage? Although this may seem unlikely, it is not impossible, especially if it was related to an event that coincided with basement faulting and subsidence associated with the opening of the Stewart Basin. So, then the nannofossils may record the faulting and sinking of the succession as well as providing data on the rate of subsidence.

I am indebted to P. Wallace, J. Mahoney and G. Fitton for suggesting that I take on this study. I also thank J. Miller for invaluable help in photographing the cores and the US Science Support Program (USSSP) for financial support. I also want to thank J. D. L. White, M. J. Branney and G. Fitton for constructive review and useful suggestions for improving the paper. The Ocean Drilling Program is sponsored by the National Science foundation and participating countries under the management of Joint Oceanographic Institutions, Inc.

References

BERGEN, J.A. 2004. Calcareous nannofossils from ODP Leg 192, Ontong Java Plateau. *In*: FITTON, J.G., MAHONEY, J.J., WALLACE, P.J. & SAUNDERS, A.D. (eds) *Origin and Evolution of the Ontong Java Plateau*. Geological Society, London, Special Publications, **229**, 113–132.

BRANNEY, M.J. 1991. Eruption and depositional facies of the Whorneyside Tuff Formation, English lake District: An exceptionally large-magnitude phreatoplinian eruption. *Geological Society of America Bulletin*, **103**, 886–897.

CAS, R.A.F. & WRIGHT, J.V. 1987. *Volcanic Successions: Modern and Ancient*. Allen and Unwin, London.

CHAMBERS, L.M., PRINGLE, M.S. & FITTON, J.G. 2004. Phreatomagmatic eruptions on the Ontong Java Plateau: an Aptian $^{40}Ar/^{39}Ar$ age for the volcaniclastic rocks at ODP Site 1184. *In*: FITTON, J.G., MAHONEY, J.J., WALLACE, P.J. & SAUNDERS, A.D. (eds) *Origin and Evolution of the Ontong Java Plateau*. Geological Society, London, Special Publications, **229**, 325–331.

CHOUGH, S.K. & SOHN, Y.K. 1990. Depositional mechanics and sequences of base surges, Songaksan tuff ring, Cheju island, Korea. *Sedimentology*, **37**, 1115–1135.

COLE, P.D., GUEST, J.E., DUNCAN, A.M. & PACHECO, J.M. 2001. Capelinhos 1957–1958, Faial, Azores: deposits formed by an emergent surtseyan eruption. *Bulletin of Volcanology*, **63**, 204–220.

FEDORENKO, V. & CZAMANSKE, G.K. 2000. Results of new field and geochemical studies of the volcanic and intrusive rocks of the Maymecha-Kotuy area, Siberian flood-basalt province, Russia. *In*: ERNST, G. & COLEMAN, R. (eds) *Tectonic studies of Asia and the Pacific Rim; A Tribute to Benjamin M. Page*. Geological Society of America, International Book Series, **3**, 54–106.

FISHER, R.V. 1966. Rocks composed of volcanic fragments. *Earth Science Review*, **1**, 287–298.

FISHER, R.V. & SCHMINCKE, H.-U. 1984. *Pyroclastic Rocks*. Springer, Heidelberg.

FITTON, J.G. & GODARD, M. 2004. Origin and evolution of magmas on the Ontong Java Plateau. *In*: FITTON, J.G., MAHONEY, J.J., WALLACE, P.J. & SAUNDERS, A.D. (eds) *Origin and Evolution of the Ontong Java Plateau*. Geological Society, London, Special Publications, **229**, 151–178.

GILBERT, J.S. & LANE, S.J. 1994. The origin of accretionary lapilli. *Bulletin of Volcanology*, **56**, 398–411.

HANSON, R.E. & ELLIOT, D.H. 1996. Rift-related Jurrasic basalt phreatomagmatic volcanism in the central Tarnsantarctic Mountains: precursory stage to flood-basalt effusion. *Bulletin of Volcanology*, **58**, 327–347.

HOUGHTON, B.F. & SCHMINCKE, H.-U. 1986. Mixed deposits of simultaneous Strombolian and phreatomagmatic volcanism: Rotenberg volcano, East Eifel volcanic field. *Journal of Volcanology and Geothermal Research*, **30**, 117–130.

HOUGHTON, BF., WILSON, C.J.N. & SMITH, I.E.M. 1999. Shallow-seated controls on styles of explosive basaltic volcanism: a case study from New Zealand. *Journal of Volcanology and Geothermal Research*, **91**, 97–120.

JAKOBSSON, S.P. 1972. On the consolidation and palagonitization of the tephra of the Surtsey volcanic island, Iceland. *Surtsey Research Progress Report*, **6**, 121–128.

JAKOBSSON, S.P. & MOORE, J.G. 1982. The Surtsey research drilling project of 1979. *Surtsey Research Progress Report*, **9**, 76–93.

JAKOBSSON, S.P. & MOORE, J.G. 1986. Hydrothermal minerals and alteration rates at Surtsey volcano, Iceland. *Geological Society of America Bulletin*, **97**, 648–659.

KAWACHI, Y., PRINGLE, I.J. & COOMBS, D.S. 1983. *Pillow Lavas of the Eocene Oamaru Volcano*, North Otago. Pacific Science Congress 1983, Tour Bh 3. Dunedin, New Zealand.

KOKELAAR, B.P. 1983. The mechanism of Surtseyan volcanism. *Journal of the Geological Society, London*, **140**, 939–944.

KROENKE, L.W., & MAHONEY, J.J. 1996. Rifting of the Ontong Java Plateau's eastern salient and seafloor spreading in the Ellice Basin: relation to the 90 Myr eruptive episode on the plateau. *Eos, Transactions of the American Geophysical Union*, **77F**, 713.

LARSEN, G. 1984. Recent volcanic history of the Veidivotn fissure swarm, southern Iceland – An approach to volcanic risk assessment. *Journal of Volcanolology and Geothermal Research*, **22**, 33–58.

LARSEN, G. 2000. Holocene eruptions within the Katla volcanic system, south Iceland: Characteristics and environmental impact. *Jokull*, **49**, 1–28.

MAHONEY, J.J., FITTON, J.G., WALLACE, P.J. *et al.* 2001. *Proceedings of the Ocean Drilling Program, Initial Reports*, **192** (CD-ROM). Available from: Ocean Drilling Program, Texas A&M University, College Station TX 77845–9547, USA.

MCCLINTOCK, M.K., HOUGHTON, B.F., SKILLING, I.P. & WHITE, J.D.L. 2002. The volcaniclastic opening phase of Karoo flood basalt volcanism: Drakensberg Formation, South Africa. American Geophysical Union 2002 Fall Meeting, San Francisco, USA. CD-ROM Abstract V71B-1277. Academic Press, San Diego.

MORRISSEY, M.M., ZIMMANOWSKI, B., WOHLETZ, K. & BUTTNER, R. 2000. Phreatomagmatic fragmentation. *In*: SIGURDSSON, H., HOUGHTON, B.F., MCNUTT, S.R., RYMER, H. & STIX, J. (eds) *Encyclopedia of Volcanoes*, 431–446.

MOULDER, T. & ALEXANDER, J. 2001. The physical character of subaqueous sedimentary density flows and their deposits. *Sedimentology*, **48**, 269–299.

PEDERSEN, A.K., WATT, M., WATT, W.S. & LARSEN, L.M. 1997. Structure and stratigraphy of the early Tertiary basalts of the Blosseville Kyst, east Greenland. *In*: ROBERTS, A.M. & KUSZNIR, N. (eds) *Tectonic, Magmatic and Depositional Processes at Passive Continental Margins*. Geological Society, London, Special Publications, **154**, 565–570.

RIISAGER, P., HALL, S., ANTRETTER, M. & ZHAO, X. 2004. Early Cretaceous Pacific palaeomagnetic pole from Ontong Java basement rocks. *In*: FITTON, J.G., MAHONEY, J.J., WALLACE, P.J. & SAUNDERS, A.D. (eds) *Origin and Evolution of the Ontong Java Plateau*. Geological Society, London, Special Publications, **229**, 31–44.

ROBERGE, J., WHITE, R.V. & WALLACE, P.J. 2004. Volatiles in submarine basaltic glasses from the Ontong Java Plateau (ODP Leg 192): implications for magmatic processes and source region compositions. *In*: FITTON, J.G., MAHONEY, J.J., WALLACE, P.J. & SAUNDERS, A.D. (eds) *Origin and Evolution of the Ontong Java Plateau*. Geological Society, London, Special Publications, **229**, 239–257.

SCHUMACHER, R. & SCHMINCKE, H.-U. 1995. Models for the origin of accretionary lapilli. *Bulletin of Volcanology*, **56**, 626–639.

SHAFER, J.T., NEAL, C.R. & CASTILLO, P.R. 2004. Compositional variability in lavas from the Ontong Java Plateau: results from basalt clasts within the volcaniclastic succession at Ocean Drilling Program Site 1184. *In*: FITTON, J.G., MAHONEY, J.J., WALLACE, P.J. & SAUNDERS, A.D. (eds) *Origin and Evolution of the Ontong Java Plateau*. Geological Society, London, Special Publications, **229**, 331–351.

SHIPBOARD SCIENTIFIC PARTY. 2001. Site 1184. *In*: MAHONEY, J.J., FITTON, J.G., WALLACE, P.J. *et al.* *Proceedings of the Ocean Drilling Program, Initial Reports*, **192**, 1–169 (CD-ROM).

SMITH, W.H.F. & SANDWELL, D.T. 1997. Global seafloor topography from satellite altimetry and ship depth soundings. *Science*, **277**, 1956–1962.

SÆMUNDSSON, K. 1991. The geology of the Katla volcanic system. *In*: GARDARSON, A. & EINARSSON, A. *Náttúra Mývatns*. Hid Islenska Nattutfrædafelag, Reykjavík, 24–95.

THORARINSSON, S., EINARSSON, T., SIGVALDASON, G.E. & ELISSON, G. 1964. The submarine eruption off the Vestmann Islands, 1963–64. *Bulletin of Volcanology*, **27**, 434–445.

THORDARSON, T. & SELF, S. 1993. The Laki (Skaftár Fires) and Grímsvötn eruptions in 1783–1785. *Bulletin of Volcanology*, **55**, 233–263.

THORDARSON, T., SELF, S., OSKARSSON, N. & HULSEBOSCH, T. 1996. Sulfur, chlorine, and fluorine degassing and atmospheric loading by the 1783–1784 AD Laki (Skaftar fires) eruption in Iceland. *Bulletin of Volcanology*, **58**, 205–225.

WHITE, J.D.L. 1996. Pre-emergent construction of a lacustrine basaltic volcano, Pahvant Butte, Utah (USA). *Bulletin of Volcanology*, **58**, 249–262.

WHITE, J.D.L. 2000. Subaqueous eruption-fed density currents and their deposits. *Precambrian Research*, **101**, 87–109.

WHITE, J.D.L. & HOUGHTON, B.F. 2000. Surtseyan and related phreatomagmatic eruptions. *In*: SIGURDSSON, H., HOUGHTON, B.F., MCNUTT, S.R., RYMER, H. & STIX, J. (eds) *Encyclopedia of Volcanoes*. Academic Press, San Diego, 495–512.

WHITE J.D.L. & MCCLINTOCK, M.K. 2001. Immense vent complex marks flood-basalt eruption in a wet, failed rift: Coombs Hills, Antarctica. *Geology*, **29**, 935–938.

WHITE, R.V., CASTILLO, P.R., NEAL, C.R., FITTON, J.G. & GODARD, M. 2004. Phreatomagmatic eruptions on the Ontong Java Plateau: chemical and isotopic relationship to Ontong Java Plateau basalts. *In*: FITTON, J.G., MAHONEY, J.J., WALLACE, P.J. & SAUNDERS, A.D. (eds) *Origin and Evolution of the Ontong Java Plateau*. Geological Society, London, Special Publications, **229**, 307–323.

WOHLETZ, K.H. 1983. Mechanisms of hydrovolcanic pyroclast formation: grain-size, scanning electron microscopy, and experimental studies. *Journal of Volcanology and Geothermal Research*, **17**, 31–63.

YAN, C.Y. & KROENKE, L.W. 1993. A plate tectonic reconstruction of the southwest Pacific, 0–100 Ma. *In*: BERGER, W.H., KROENKE, L.W., MAYER, L.A. *et al.* *Proceedings of the Ocean Drilling Project, Scientific Results*, **130**, 697–709.

Phreatomagmatic eruptions on the Ontong Java Plateau: chemical and isotopic relationship to Ontong Java Plateau basalts

ROSALIND V. WHITE[1], PATERNO R. CASTILLO[2], CLIVE R. NEAL[3], J. GODFREY FITTON[4] & MARGUERITE GODARD[5]

[1]*Department of Geology, University of Leicester, University Road, Leicester LE1 7RH, UK (e-mail: rvw1@le.ac.uk)*

[2]*Geosciences Research Division, Scripps Institution of Oceanography, University of California, San Diego, La Jolla, CA 92093-0212, USA*

[3]*Department of Civil Engineering and Geological Sciences, University of Notre Dame, 156 Fitzpatrick Hall, Notre Dame, IN 46556, USA*

[4]*School of GeoSciences, University of Edinburgh, Grant Institute, West Mains Road, Edinburgh EH9 3JW, UK*

[5]*Laboratoire de Tectonophysique – CNRS UMR 5568, Case 49, Institut des Sciences de la Terre, de l'Eau et de l'Espace de Montpellier, Université de Montpellier II, Place Eugéne Bataillon, F-34095 Montpellier Cedex 5, France*

Abstract: The compositions of glass clasts in volcaniclastic rocks recovered from drilling at Site 1184 on the eastern salient of the Ontong Java Plateau (OJP) are investigated using microbeam analytical methods for major, minor and trace elements. These data are compared with whole-rock elemental and isotopic data for bulk tuff samples, and with data from basalts on the high plateau of the OJP. Three subunits of Hole 1184A contain blocky glass clasts, thought to represent the juvenile magmatic component of the phreatomagmatic eruptions that generated the volcaniclastic rocks. The glass clasts have unaltered centres, and are all basaltic low-K tholeiites, with flat chondrite-normalized rare earth element (REE) patterns. Their elemental compositions are very similar to the Kwaimbaita-type and Kroenke-type basalts sampled on the high plateau. Each subunit has a distinct glass composition and there is no intermixing of glass compositions between subunits, indicating that each subunit is the result of one eruptive phase, and that the volcaniclastic sequence has not experienced reworking. The relative heterogeneity preserved at Site 1184 contrasts with the uniformity of compositions recovered from individual sites on the high plateau, and suggests that the eastern salient of the OJP had a different type of magma plumbing system. Our data support the hypothesis that the voluminous subaerially erupted volcaniclastic rocks at Site 1184 belong to the same magmatic event as the construction of the main Ontong Java Plateau. Thus, the OJP would have been responsible for volatile fluxes into the atmosphere in addition to chemical fluxes into the oceans, and these factors may have influenced the contemporaneous oceanic anoxic event.

The Ontong Java Plateau (OJP), located in the equatorial western Pacific Ocean, is the world's largest oceanic plateau. Covering an area of 2.0×10^6 km^2, or eight times that of the UK, the OJP has an average crustal thickness of 30–35 km (Gladczenko *et al.* 1997; Richardson *et al.* 2000), and, in terms of volume, represents the largest documented igneous event on Earth. An enigmatic characteristic of the OJP is the fact that, despite its great thickness, it does not appear to have experienced sufficient uplift during magmatism to yield widespread subaerially erupted lavas. This is a paradox, as the size and extent of the OJP has led to widespread support for a large and vigorous mantle plume head being responsible for the magmatism (e.g. Richards *et al.* 1991; Saunders *et al.* 1992), in which case considerable uplift might be expected during volcanism. The observation that significant uplift seemingly did not occur suggests either that our models of mantle plume dynamics and/or oceanic plateau subsidence require modification, or even that alternative models for OJP formation should be re-examined. Neal *et al.* (1997) concluded that the impingement of the OJP plume head was insufficient to raise the high plateau above sea level, and the density of a hidden cumulate layer

From: Fitton, J. G., Mahoney, J. J., Wallace, P. J. & Saunders, A. D. (eds) 2004. *Origin and Evolution of the Ontong Java Plateau*. Geological Society, London, Special Publications, **229**, 307–323. 0305-8719/$15.00

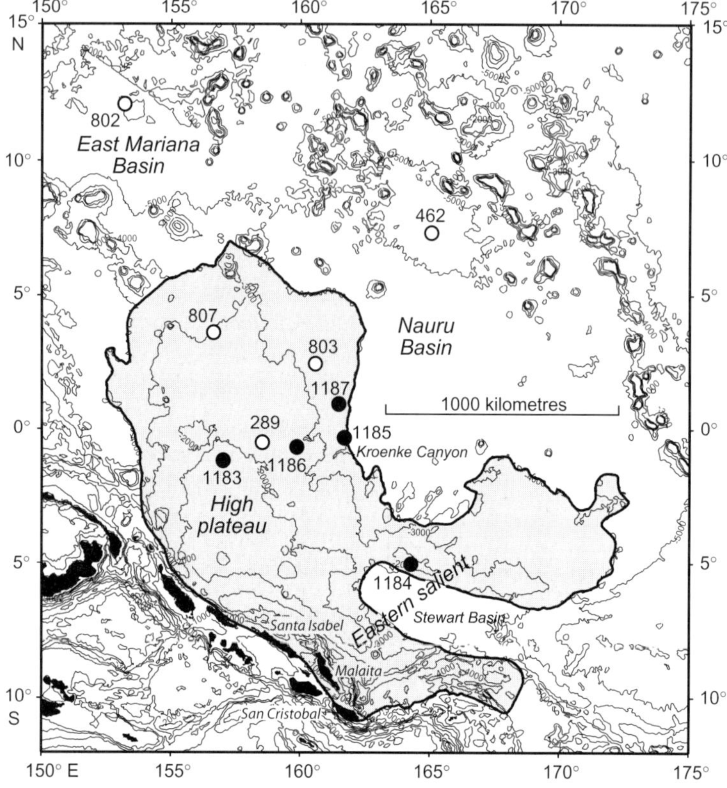

Fig. 1. Predicted bathymetry (after Smith & Sandwell 1997) of the OJP and surrounding areas showing the location of DSDP and ODP basement drill sites. The Ontong Java Plateau is shaded.

(complementary to the fractionated basalts) was sufficient to temper subsidence of the OJP such that it still stands above the surrounding ocean floor. Indeed, it is apparent that, at least along the northern and eastern margins of the OJP, the adjoining abyssal seafloor has subsided relative to the plateau (Kroenke 1972; Kroenke *et al.* 1986; Hagen *et al.* 1993).

Prior to Autumn 2000 and drilling of the OJP during Leg 192 of the Ocean Drilling Program, the volcanic basement of the plateau had been sampled at only three drill sites (DSDP Site 289 and ODP Sites 803 and 807) with limited penetration at each site (Andrews *et al.* 1975; Kroenke *et al.* 1991). This information was supplemented by studies of uplifted portions of the plateau in the Solomon Islands (e.g. Babbs 1997; Petterson *et al.* 1999; Birkhold 2000). One of the primary objectives of Leg 192 was to determine the environment and style of eruption at five widely separated sites in previously unsampled areas across the plateau (Fig. 1). In particular, a site at the current topographic high point of the plateau (Site 1183) was considered the location most likely to yield subaerial volcanic rocks. However, all of the basement sites sampled on the 'high plateau' are characterized by the presence of pillow basalts, with some massive flows and minor sedimentary interbeds. In fact, the basalt flows recovered from Site 1183 were very similar to the deep-water eruptives sampled on Malaita, Solomon Islands (Babbs 1997; Petterson *et al.* 1999; Tejada *et al.* 2002). This suggests that Ontong Java high plateau volcanism was confined largely to the submarine realm. In contrast, drilling at Site 1184, located on the previously unsampled 'eastern salient' of the OJP, cored a sequence of subaerially erupted volcaniclastic rocks at least 338 m thick (Mahoney *et al.* 2001).

Ocean Drilling Program Leg 192, Site 1184

The volcaniclastic sequence at Site 1184 was divided into six subunits (Subunits IIA–IIF;

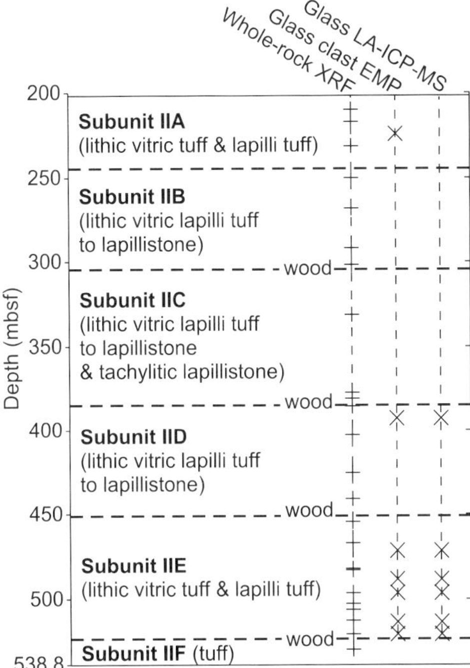

Fig. 2. Simplified section through the volcaniclastic succession at 1184A, showing the thicknesses and *dominant* lithological characteristics (see Thordarson 2004 for more details) of Subunits IIA–IIF, and stratigraphic positions of samples analysed for this study. Stratigraphic positions of wood fragments are also indicated.

Fig. 2) based on sedimentological and chemical characteristics (Fitton & Godard 2004; Thordarson 2004). Lithological features of the Site 1184 rocks are described by Mahoney *et al.* (2001) and Thordarson (2004), and the rocks are interpreted as primary hydromagmatic deposits from Surtseyan eruptions. The sequence consists predominantly of massive coarse lithic vitric tuffs, lapilli tuffs and lapillistones. The existence of blocky, subangular, non-vesicular glass shards implies that the eruptions were phreatomagmatic, and abundant accretionary lapilli indicate that the eruptions formed subaerial eruption columns. Many intervals contain basaltic lithic clasts (see Shafer *et al.* 2004). The presence of wood fragments at the top of Subunits IIC, IID, IIE and IIF shows that parts of the eastern salient formed emergent islands at the time of eruption. Discrete mineral grains (plagioclase and pyroxene) are a volumetrically minor component, and the tuffs are cemented with zeolites or calcite. Glass shards, abundant in Subunits IIA, IIB, IID and IIE, are thought to represent the juvenile magmas that were quenched during these hydroclastic eruptions. Although in general the sequence is heavily altered, some intervals contain glass clasts with unaltered centres.

Site 1184 was chosen to be near the summit of the northern ridge of the eastern salient, a location that may have originally been at relatively shallow water depths. Prior to drilling at Site 1184, the relationship between the eastern salient and the high plateau was unknown. The eastern salient has been proposed as: (1) part of the main plateau (Mahoney *et al.* 2001); (2) belonging to a postulated plume tail (Mahoney *et al.* 2001); (3) a younger section of the plateau formed in response to a change in the stress field of the plateau associated with a major change in Pacific plate motion at approximately 43 Ma (Duncan & Clague 1985); (4) the main locus of approximately 90 Ma volcanism (Tejada *et al.* 1996); or (5) a younger volcanic constructional feature related to passage of the plateau over the Samoan hot spot at *c.* 35–40 Ma (Yan & Kroenke 1993), and which may have also been recorded on Makira (Birkhold 2000). Thus, the presence of a thick volcaniclastic sequence in the eastern salient was surprising.

Biostratigraphical evidence collected during Leg 192 suggested that the basement at Sites 1183, 1185, 1186 and 1187 is Aptian in age, with a 125 m-thick carapace of overlying 'Kroenke-type' basalts at Site 1185 having a poorly constrained age of Cenomanian–Albian (Mahoney *et al.* 2001). However, $^{40}Ar/^{39}Ar$ dating of these basalts (Chambers *et al.* 2002) shows no evidence that the basaltic carapace at Site 1185 is significantly younger than the underlying lavas: all of the basalts sampled during Leg 192 were emplaced during one magmatic event at about 120 Ma (range: 122.0 ± 1.9–105.5 ± 3.9 Ma (1σ)), with no evidence for a volumetrically significant younger phase at *c.* 90 Ma, as had been suggested by previous work (e.g. Mahoney *et al.* 1993; Tejada *et al.* 1996). At Site 1184, shipboard nannofossil data suggested that the volcaniclastic sequence was emplaced during the middle Eocene. However, the weighted mean of six total fusion $^{40}Ar/^{39}Ar$ ages on discrete feldspar crystals is 123.5 ± 1.8 Ma (1σ: Chambers *et al.* 2004), and a study of the volcaniclastic lithologies by Thordarson (2004) suggests that the volcaniclastic rocks are syn-eruption accumulations. Thus, in contrast to the biostratigraphic data, the $^{40}Ar/^{39}Ar$ and lithological evidence point to eruption and emplacement of the volcaniclastic rocks during the main Aptian plateau-building magmatic event. This is supported by a steep palaeomagnetic inclination, consistent with a Cretaceous age (Riisager *et al.* 2004).

It is important to verify that the Site 1184 volcaniclastic rocks are compositionally similar to the main plateau-building basalts because, if so, this would be the first confirmed instance of substantial subaerial eruptions occurring as part of the OJP-forming magmatic event. This has implications for our understanding of the origin of the OJP and its environmental effects. In this paper, we present major- and trace-element data for samples of unaltered glass from clasts in volcaniclastic rocks at Site 1184, and Sr-, Nd- and Pb-isotopic data for bulk tuff samples. The chemical data are compared with whole-rock elemental data for basalts from the Ontong Java Plateau (Fitton & Godard 2004) to investigate any degree of compositional similarity. In addition, a comparison of the major-element data for unaltered glass and altered bulk tuff samples, as well as isotopic data for altered volcaniclastic rocks and basalts from the OJP (cf. Mahoney *et al.* 1993; Tejada *et al.* 1996, 2002, 2004; Neal *et al.* 1997), provide some constraints on the chemical and isotopic changes that occur during sedimentation and alteration, and may assist future studies of similarly altered volcaniclastic rocks. This work is complemented by a study of lithic clasts from the volcaniclastic rocks at Site 1184 (Shafer *et al.* 2004).

Analytical methods

For analysis of glass clasts, shipboard samples were taken at intervals where hand-specimen examination indicated that unaltered glass was present: criteria for determining this included conchoidal fracture, a vitreous lustre and hardness exceeding that of a steel probe. Subunits IIB and IIC of the volcaniclastic succession at Site 1184 did not contain unaltered glass, and clasts from Subunit IIA were heavily fractured. Thus, the samples are biased towards Subunit IIE, reflecting the increasing glass preservation with depth in the core.

To separate glass clasts for laser ablation inductively coupled plasma-mass spectrometry (LA-ICP-MS), samples were soaked in water, subjected to repeated freeze–thaw action over several weeks and then gently crushed using a fly press. Hand-picked, cleaned, intact glass fragments large enough for LA-ICP-MS analysis were obtained from Subunits IID and IIE. Polished epoxy grain mounts were prepared; the smooth surface helps prevent the build-up of ablated particles. LA-ICP-MS analyses were conducted at the University of Notre Dame using a PlasmaQuad II STE quadrupole ICP-MS and a Nd-YAG laser quadrupled to 266 nm. Power used for these analyses was *c.* 0.9 mJ with a spot size of *c.* 25 μm and a repetition rate of 4 Hz. Tuning and standardization was conducted using NIST612 standard glass (cf. Pearce *et al.* 1997). Data reduction was achieved using the LAMTRACE program of Simon Jackson (MacQuarie University). Each sample was analysed twice. The standard NIST612 glass was analysed twice at the beginning and twice at the end of each analytical run, and was also analysed as an unknown during the run as a check of analytical precision and machine stability. Electron microprobe data from the University of Leicester (see below) were used as internal standards.

Polished thin sections were prepared for electron microprobe analysis. This approach was preferred as it minimized any bias towards particular morphologies during picking of glass fragments, and eliminated any risk of mixing grains between samples. Major- and minor-element compositions were determined on a JEOL 8600 electron microprobe at the University of Leicester using a 15 kV accelerating voltage, a 30 nA beam current and a 20 μm-diameter beam. This beam diameter was chosen to be small enough to avoid inclusions and/or heterogeneities in the glass clasts, whilst being large enough to minimize loss of the more volatile elements, notably Na, during analysis. Count times were 20 s on the peak and background for Si, Ti, Al, Fe, Mg, Ca, Na and K. The raw data were corrected using a ZAF (atomic number, absorption, fluorescence) correction procedure.

Whole-rock major- and trace-element data for the bulk tuff samples were analysed by X-ray fluorescence (XRF) at the University of Edinburgh and by ICP-MS using a VG-PQ2 Turbo+ spectrometer at ISTEEM, Montpellier; data and analytical details are reported in Fitton & Godard (2004). Four bulk tuff samples, representing the range of trace-element characteristics of the volcaniclastic rocks, were analysed for Sr-, Nd- and Pb-isotopic composition at the Scripps Institution of Oceanography using the procedure described in Tejada *et al.* (2004; see also Janney & Castillo 1996, 1997). For detailed descriptions of the lithologies from which these samples were taken, see Mahoney *et al.* (2001) and Thordarson (2004). Powders of samples analysed for Sr- and Nd-isotopic ratios were first leached in 4.5 N HCl to mitigate the effects of sea-water alteration (e.g. Mahoney 1987; Janney & Castillo 1996, 1997; Tejada *et al.* 2004); those for Pb-isotopic ratios were not leached because it was suspected that the majority of the Pb was in a leachable phase.

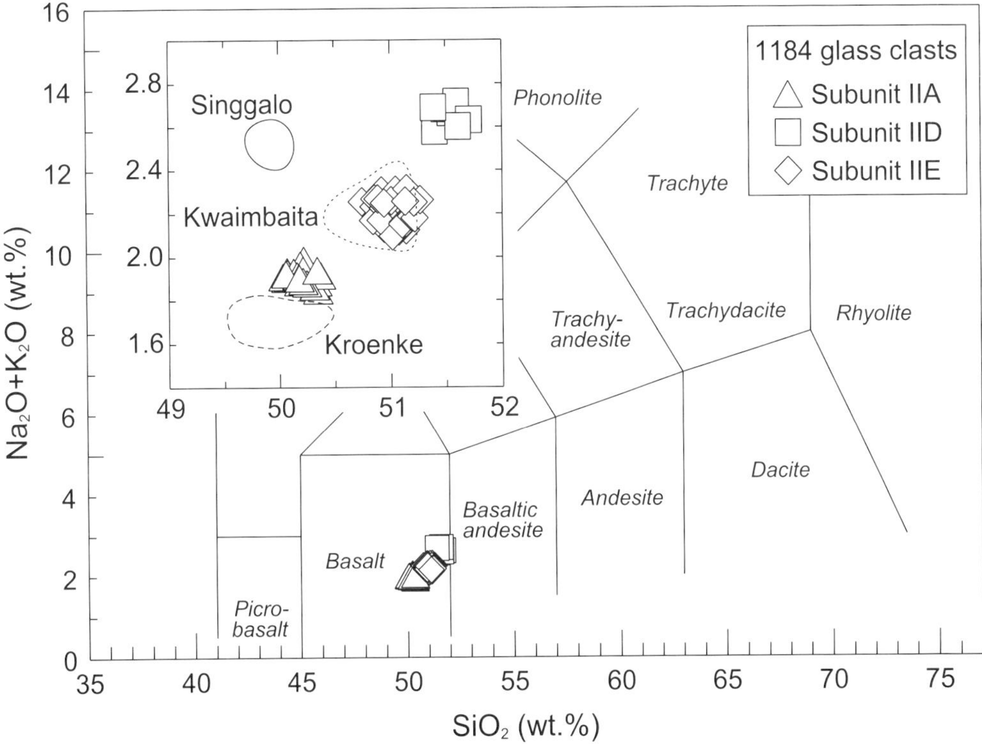

Fig. 3. Total alkalis v. silica plot for glass clasts analysed by electron microprobe (EMP), compared with fields for Kwaimbaita-, Kroenke- and Singgalo-type basalts from the OJP (data for basalts with LOI<0.5 wt% from Fitton & Godard 2004).

Results

Major-element compositions of glass clasts

All of the glass clasts analysed are basaltic low-K tholeiites (Fig. 3, Table 1), with approximately 49.5–51 wt% SiO_2, 6.5–9 wt% MgO and K_2O < 0.2 wt%. Data from all seven samples define three distinct compositional sets, corresponding perfectly to Subunits IIA, IID and IIE. This suggests that volcanic and sedimentary processes did not rework and mix glass populations, and supports the hypothesis that the glass clasts represent the juvenile magmatic component of the volcanic eruptions driving the emplacement of the volcaniclastic sequence (Thordarson 2004).

The uppermost subunit, Subunit IIA, has the most primitive magma composition, with an average MgO content of 9.0 wt% and average TiO_2 of 0.9 wt%. The least primitive compositions are present in Subunit IID, with average MgO of 6.7 wt% and average TiO_2 of 1.3 wt%. Subunit IIE has intermediate compositions, with average MgO of 7.9 wt% and average TiO_2 of 1.0 wt%.

Comparison with OJP basalt compositions. Three distinct basalt types are now recognized on the OJP (for details see Fitton & Godard 2004): (1) abundant Kwaimbaita-type basalts, typically with 6–8 wt% MgO; (2) Kroenke-type basalts, first described during ODP Leg 192, with MgO contents up to 11 wt%; and (3) Singgalo-type basalts, with similar MgO contents to Kwaimbaita-type basalts, but distinct Sr–Nd–Pb isotope ratios from both Kwaimbaita- and Kroenke-type basalts. Figure 4 compares glass clast compositions (open symbols) with fields for OJP Kwaimbaita-, Kroenke- and Singgalo-type basalts that have loss on ignition (LOI) of less than 1% (Fitton & Godard 2004). No Singgalo-type basalts were recovered during Leg 192; those reported by Fitton and Goddard (2004) are from ODP Leg 130 Site 807. Data for Site 1184 bulk tuff samples with LOI ranging

Table 1. *Electron microprobe data (wt%) for glass clasts from Site 1184, Subunits IIA, IID and IIE*

Analysis No.	SiO_2	TiO_2	Al_2O_3	FeO	MnO	MgO	CaO	Na_2O	K_2O	Total
1184A 13R-3, 142–148 cm (224.49 mbsf) Subunit IIA										
RVW022: 31	49.76	0.84	14.92	9.63	0.17	8.84	13.33	1.79	0.08	99.36
RVW022: 32	49.94	0.89	14.71	9.64	0.14	8.99	13.05	1.77	0.06	99.19
RVW022: 33	49.93	0.85	14.85	9.44	0.20	8.95	13.15	1.80	0.09	99.26
RVW022: 34	49.74	0.84	14.74	9.46	0.15	9.02	13.12	1.77	0.07	98.91
RVW022: 36	49.91	0.87	14.81	9.57	0.18	8.98	13.07	1.79	0.08	99.26
RVW022: 37	49.93	0.88	14.83	9.52	0.16	9.01	13.12	1.79	0.08	99.32
RVW022: 38	49.66	0.87	14.74	9.21	0.18	9.01	13.10	1.71	0.09	98.57
RVW022: 39	50.03	0.89	14.72	9.58	0.19	8.83	13.14	1.77	0.08	99.23
RVW022: 41	49.82	0.89	14.79	9.47	0.14	8.76	13.30	1.87	0.08	99.12
RVW022: 42	49.70	0.86	14.81	9.44	0.16	9.04	12.90	1.74	0.08	98.73
RVW022: 43	49.84	0.88	14.85	9.52	0.15	8.98	13.11	1.85	0.08	99.26
RVW022: 44	49.82	0.86	14.75	9.64	0.14	9.01	13.16	1.77	0.08	99.23
RVW022: 45	49.84	0.90	14.79	9.63	0.17	8.78	13.28	1.82	0.08	99.29
RVW022: 46	49.73	0.91	14.95	9.53	0.18	8.95	13.11	1.78	0.09	99.23
RVW022: 47	49.98	0.90	14.77	9.59	0.20	9.03	13.09	1.78	0.09	99.43
RVW022: 48	49.79	0.85	14.86	9.30	0.16	9.04	13.26	1.78	0.10	99.14
RVW022: 49	49.74	0.88	14.77	9.27	0.17	8.92	13.10	1.77	0.10	98.72
RVW022: 50	49.95	0.89	14.91	9.33	0.21	9.02	13.15	1.81	0.10	99.37
RVW022: 51	50.04	0.93	14.88	9.43	0.19	9.05	13.16	1.77	0.08	99.53
RVW022: 52	49.71	0.89	14.78	9.53	0.18	9.03	13.16	1.83	0.07	99.18
RVW022: 53	49.95	0.88	14.89	9.62	0.15	8.99	13.28	1.83	0.07	99.66
RVW022: 55	49.81	0.84	14.84	9.66	0.17	9.18	13.08	1.81	0.07	99.46
RVW022: 56	49.99	0.92	14.86	9.51	0.19	8.78	13.34	1.78	0.08	99.45
RVW022: 57	49.97	0.87	14.82	9.49	0.12	9.11	13.14	1.78	0.10	99.40
RVW022: 58	50.01	0.86	14.89	9.65	0.23	9.04	13.21	1.82	0.08	99.79
RVW022: 59	49.85	0.91	14.80	9.55	0.18	8.97	13.16	1.77	0.10	99.29
RVW022: 60	49.88	0.83	14.79	9.34	0.18	8.93	13.17	1.81	0.09	99.02
Average	**49.86**	**0.88**	**14.82**	**9.50**	**0.17**	**8.97**	**13.16**	**1.79**	**0.08**	**99.24**
1184A 31R-7 53–58 cm (392.85 mbsf) Subunit IID										
RVW020: 05	50.67	1.30	13.26	12.29	0.22	6.78	11.12	2.47	0.15	98.26
RVW020: 06	50.69	1.22	13.29	12.18	0.24	6.61	11.13	2.46	0.12	97.94
RVW020: 07	50.66	1.31	13.27	12.27	0.21	6.68	11.29	2.49	0.15	98.33
RVW020: 08	50.60	1.39	13.30	12.52	0.24	6.71	11.12	2.41	0.13	98.42
RVW020: 09	50.78	1.28	13.23	12.49	0.19	6.69	11.09	2.43	0.13	98.31
RVW020: 10	50.57	1.24	13.24	12.27	0.23	6.68	11.09	2.52	0.14	97.98
RVW020: 11	50.46	1.38	13.29	12.20	0.28	6.72	11.19	2.50	0.14	98.16
Average	**50.63**	**1.30**	**13.27**	**12.32**	**0.23**	**6.70**	**11.15**	**2.47**	**0.14**	**98.20**
1184A 39R-7, 95–99 cm (471.62 mbsf) Subunit IIE										
RVW021: 01	50.09	1.02	14.02	10.35	0.16	7.83	12.74	2.10	0.12	98.43
RVW021: 06	49.97	1.04	13.90	10.23	0.17	7.75	12.43	2.00	0.12	97.62
RVW021: 07	49.77	1.06	13.94	10.14	0.18	7.75	12.61	2.04	0.10	97.60
RVW021: 15	49.73	1.03	13.90	10.23	0.25	7.88	12.36	2.01	0.08	97.46
RVW021: 16	49.52	0.99	13.85	10.27	0.14	7.82	12.50	2.04	0.09	97.22
Average	**49.82**	**1.03**	**13.92**	**10.24**	**0.18**	**7.81**	**12.53**	**2.04**	**0.10**	**97.67**
1184A 41R-6, 67–72 cm (488.72 mbsf) Subunit IIE										
RVW020: 21	50.68	1.06	14.14	10.49	0.21	7.91	12.63	2.11	0.10	99.33
RVW020: 22	50.91	1.04	14.29	10.68	0.13	8.09	12.67	2.09	0.12	100.02
RVW020: 23	50.41	0.97	14.06	10.57	0.19	7.95	12.43	2.06	0.12	98.76
RVW020: 25	51.09	0.99	14.29	10.54	0.17	8.04	12.58	2.07	0.10	99.87
RVW020: 27	51.16	1.10	14.20	10.60	0.23	8.00	12.68	2.10	0.11	100.18
RVW020: 28	50.81	1.01	14.19	10.47	0.13	7.87	12.59	2.08	0.10	99.25
RVW020: 29	50.64	1.02	14.22	10.74	0.18	8.00	12.67	2.15	0.10	99.72
Average	**50.81**	**1.03**	**14.20**	**10.58**	**0.18**	**7.98**	**12.61**	**2.09**	**0.11**	**99.59**
1184A 42R-5, 69–74 cm (496.93 mbsf) Subunit IIE										
RVW021: 23	50.20	1.07	14.01	10.45	0.18	7.96	12.48	2.10	0.10	98.56
RVW021: 24	50.09	0.92	14.04	10.48	0.17	7.88	12.47	2.08	0.10	98.23
RVW021: 26	50.16	0.99	13.84	10.49	0.24	7.86	12.37	1.99	0.10	98.04
RVW021: 27	50.36	1.16	14.09	10.33	0.17	7.85	12.27	2.03	0.11	98.36
RVW021: 30	50.23	1.05	14.07	10.47	0.23	7.97	12.28	2.06	0.11	98.47
Average	**50.21**	**1.04**	**14.01**	**10.44**	**0.20**	**7.91**	**12.37**	**2.05**	**0.10**	**98.33**

Table 1. *continued*

Analysis No.	SiO_2	TiO_2	Al_2O_3	FeO	MnO	MgO	CaO	Na_2O	K_2O	Total
1184A 44R-3, 78–83 cm (513.53 mbsf) Subunit IIE										
RVW022: 02	50.83	1.04	14.27	10.69	0.16	8.04	12.57	2.10	0.08	99.78
RVW022: 03	50.81	1.02	14.34	10.58	0.14	8.10	12.46	2.12	0.11	99.68
RVW022: 06	50.62	1.04	14.25	10.34	0.16	8.33	12.30	2.17	0.12	99.33
RVW022: 07	50.74	1.01	14.21	10.48	0.17	8.12	12.44	2.14	0.09	99.40
RVW022: 09	50.84	1.02	14.27	10.49	0.17	8.04	12.40	2.12	0.09	99.44
RVW022: 10	50.89	1.01	14.21	10.73	0.16	8.10	12.38	2.04	0.09	99.61
RVW022: 11	50.75	1.01	14.27	10.68	0.22	8.08	12.41	2.08	0.10	99.60
RVW022: 13	50.59	1.06	14.18	10.47	0.16	7.99	12.62	2.06	0.11	99.24
RVW022: 14	50.83	0.98	14.20	10.40	0.13	8.03	12.53	2.11	0.10	99.31
RVW022: 15	50.87	1.07	14.18	10.75	0.15	7.58	12.43	2.14	0.09	99.26
RVW022: 16	50.96	1.11	14.23	10.75	0.20	7.62	12.36	2.16	0.09	99.48
RVW022: 17	50.60	1.03	14.20	10.68	0.22	8.02	12.44	2.15	0.10	99.44
RVW022: 18	50.98	1.07	14.12	10.64	0.22	7.72	12.40	2.13	0.12	99.40
RVW022: 20	50.72	1.06	14.17	10.61	0.19	7.74	12.58	2.08	0.09	99.24
RVW022: 21	50.66	1.00	14.26	10.49	0.13	7.96	12.56	2.10	0.11	99.27
RVW022: 22	50.82	1.09	14.12	10.71	0.18	8.07	12.67	2.14	0.09	99.89
RVW022: 23	50.84	1.00	14.27	10.57	0.22	8.02	12.42	2.10	0.11	99.55
RVW022: 24	50.85	1.02	14.12	10.66	0.19	8.05	12.60	2.13	0.09	99.71
RVW022: 25	50.64	1.04	14.19	10.56	0.19	8.12	12.60	2.05	0.11	99.50
RVW022: 26	50.64	1.08	14.24	10.36	0.16	8.08	12.51	2.13	0.10	99.30
RVW022: 27	50.87	1.05	14.42	10.59	0.16	7.44	12.56	2.18	0.12	99.39
RVW022: 28	50.96	1.03	14.37	10.76	0.20	7.82	12.67	2.18	0.10	100.09
RVW022: 29	50.94	0.99	14.18	10.54	0.18	7.98	12.36	2.14	0.10	99.41
RVW022: 30	50.95	1.04	14.26	10.59	0.15	8.06	12.48	2.16	0.09	99.78
Average	**50.80**	**1.04**	**14.23**	**10.59**	**0.18**	**7.96**	**12.49**	**2.12**	**0.10**	**99.50**
1184A 45R-2, 75–81 cm (521.40 mbsf) Subunit IIE										
RVW020: 32	50.86	0.91	14.33	10.60	0.21	8.13	12.52	2.13	0.11	99.80
RVW020: 33	51.05	1.08	14.27	10.77	0.20	8.02	12.51	2.10	0.10	100.10
RVW020: 37	50.47	0.95	14.02	10.45	0.20	8.00	12.44	2.11	0.09	98.73
RVW020: 40	51.13	0.94	14.34	10.64	0.18	8.05	12.57	2.19	0.12	100.16
RVW022: 61	50.18	1.03	14.00	10.60	0.17	7.91	12.42	2.02	0.08	98.41
RVW022: 62	50.27	1.00	14.06	10.51	0.19	7.89	12.38	2.06	0.10	98.46
RVW022: 66	50.12	1.01	14.04	10.33	0.23	7.88	12.51	1.98	0.08	98.18
RVW022: 67	50.31	1.00	13.97	10.42	0.19	7.96	12.36	2.05	0.11	98.37
RVW022: 68	50.29	1.04	13.99	10.28	0.20	7.94	12.45	2.05	0.11	98.35
RVW022: 69	50.37	1.05	13.97	10.28	0.18	7.92	12.66	2.01	0.08	98.52
RVW022: 70	50.37	1.00	14.01	10.33	0.18	8.04	12.40	2.07	0.10	98.50
RVW022: 71	50.44	1.04	14.10	10.49	0.19	7.96	12.34	2.09	0.12	98.77
RVW022: 73	50.35	1.07	13.96	10.53	0.20	7.99	12.27	2.12	0.11	98.60
RVW022: 74	50.32	1.03	14.06	10.36	0.20	7.89	12.29	2.10	0.12	98.37
RVW022: 75	49.80	1.04	13.95	10.35	0.18	7.87	12.44	2.13	0.09	97.85
RVW022: 76	50.34	1.07	13.97	10.53	0.22	7.96	12.36	2.05	0.11	98.61
RVW022: 77	50.20	1.05	13.97	10.49	0.17	7.87	12.42	2.07	0.10	98.34
RVW022: 78	50.40	1.08	14.01	10.61	0.20	7.90	12.32	2.04	0.10	98.66
RVW022: 79	50.14	1.01	14.00	10.42	0.20	7.94	12.49	2.10	0.09	98.39
RVW022: 80	50.11	1.05	13.98	10.44	0.16	7.95	12.50	2.04	0.10	98.33
RVW022: 81	50.30	1.01	13.97	10.55	0.16	7.94	12.28	2.04	0.11	98.36
RVW022: 82	50.27	1.04	14.13	10.41	0.23	7.96	12.47	2.13	0.11	98.75
RVW022: 83	50.37	1.03	14.01	10.48	0.14	7.96	12.43	2.06	0.08	98.56
RVW022: 84	50.22	1.04	14.00	10.44	0.22	7.99	12.49	2.06	0.08	98.54
RVW022: 85	49.93	1.04	13.89	10.12	0.21	7.90	12.49	1.99	0.12	97.69
RVW022: 86	50.25	1.02	14.04	10.28	0.21	7.86	12.43	2.03	0.09	98.21
RVW022: 87	50.53	1.04	14.01	10.73	0.24	7.93	12.39	2.13	0.10	99.10
RVW022: 88	50.54	1.00	13.92	10.56	0.14	7.90	12.41	2.06	0.10	98.63
RVW022: 89	50.30	0.99	14.01	10.23	0.18	7.96	12.34	2.05	0.09	98.15
RVW022: 90	50.35	1.07	14.01	10.46	0.20	7.97	12.32	2.03	0.09	98.50
Average	**50.35**	**1.02**	**14.03**	**10.46**	**0.19**	**7.95**	**12.42**	**2.07**	**0.10**	**98.60**

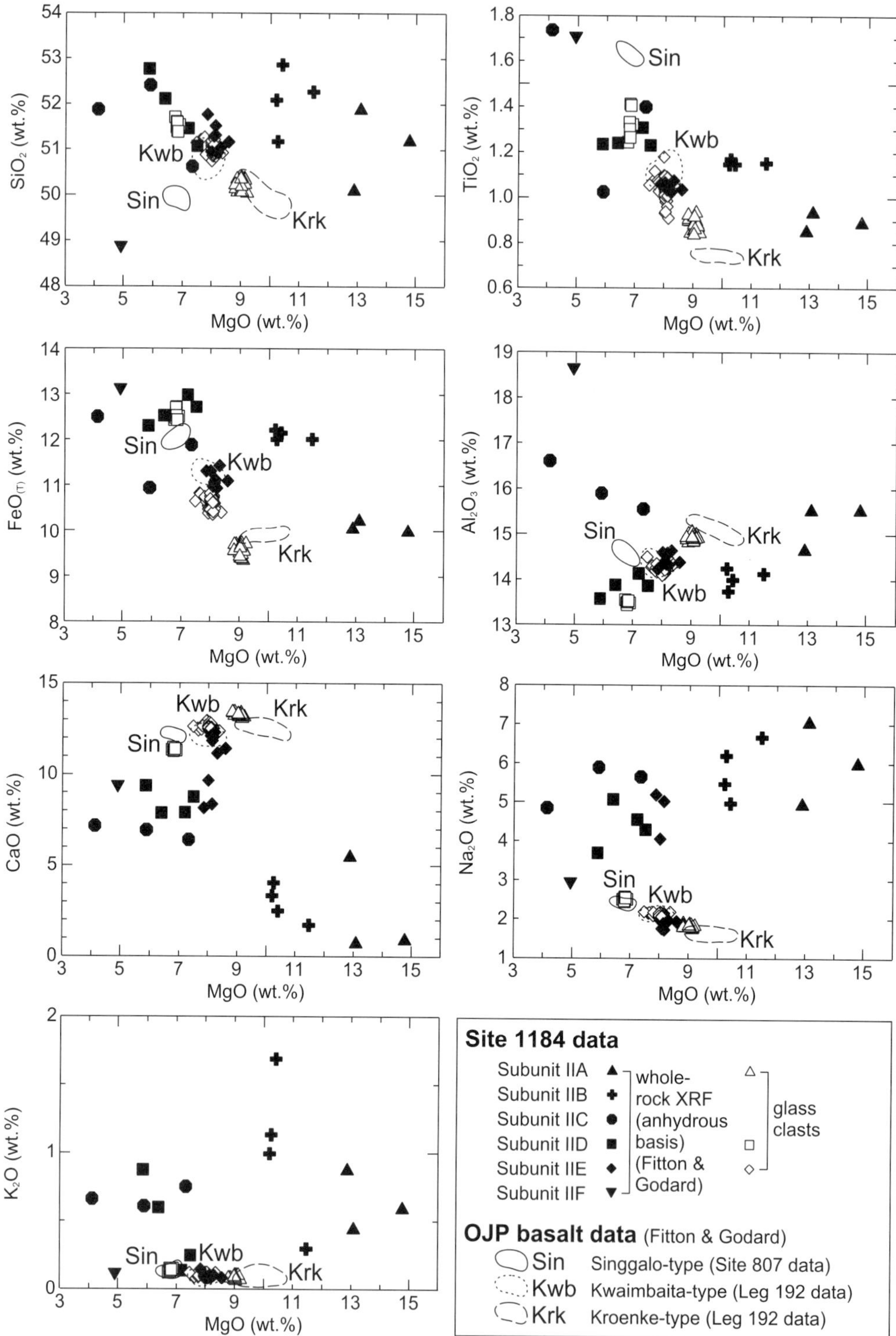

Fig. 4. Major-element plots v. MgO, comparing EMP data from glass clasts, whole-rock data for OJP basalts (reported on anhydrous basis, from analyses with <0.5 wt% LOI: Fitton & Godard 2004) and whole-rock data for Site 1184 bulk tuff samples (Fitton & Godard 2004), recalculated on an anhydrous basis with all Fe as FeO.

from 5 to 10% are included for comparison (closed symbols; Fitton & Godard 2004, recalculated on an anhydrous basis).

The glass from Subunit IIA (hereafter 'IIA glass') is broadly similar to the lower-MgO end of the Kroenke-type basalts. However, the IIA glass has marginally higher TiO_2, CaO and Na_2O, and marginally lower $FeO_{(T)}$ and Al_2O_3. In particular, the different TiO_2 contents for a given MgO content reflect subtle differences between the IIA glass and the Kroenke magma type.

Glass from Subunit IIE is very similar in composition to the widespread Kwaimbaita-type basalts of the OJP. The IIE glass has slightly lower $FeO_{(T)}$ contents, but there is good overlap with the Kwaimbaita field for all other elements analysed. Although Subunit IID glass has similar MgO, CaO, Na_2O and K_2O contents to the Singgalo-type magmas, the IID glass has significantly higher SiO_2, and significantly lower TiO_2 and Al_2O_3 contents.

In summary, the major-element compositions of glass clasts from Site 1184 support the hypothesis of a genetic relationship with the basalts of the high plateau. The predominant Kwaimbaita magma type appears to be present at Site 1184, and other Site 1184 magmas show varying degrees of similarity with other magma types recovered from the high plateau.

Comparison with Site 1184 bulk tuff compositions. Figure 4 demonstrates that the bulk tuff analyses differ substantially from analyses of unaltered glass from the centres of glass clasts. In particular, the processes of sedimentation, cementation and sea-water alteration have resulted in a marked depletion of CaO, and enrichment of Na_2O and K_2O relative to the unaltered glass (e.g. Hart *et al.* 1974; Verma 1992). In contrast, SiO_2, Al_2O_3, TiO_2 and $FeO_{(T)}$ have remained relatively unaffected. Not all subunits have been affected in the same way, e.g. MgO contents in Subunit IIA bulk tuffs are elevated by approximately 4 wt% relative to the unaltered IIA glass, whereas Subunit IID and IIE bulk tuffs are similar to those of the equivalent glass clasts. This difference in behaviour probably reflects changes in alteration style and/or degree of cementation within the volcanic pile. Although some of the difference between glass and bulk tuff compositions may be due to variability in the proportions of lithic clasts and crystal fragments within the tuffs, these components are generally volumetrically minor (Thordarson 2004), and cannot account for significant compositional differences.

Comparison between glass and bulk tuff compositions enables some qualitative estimates to

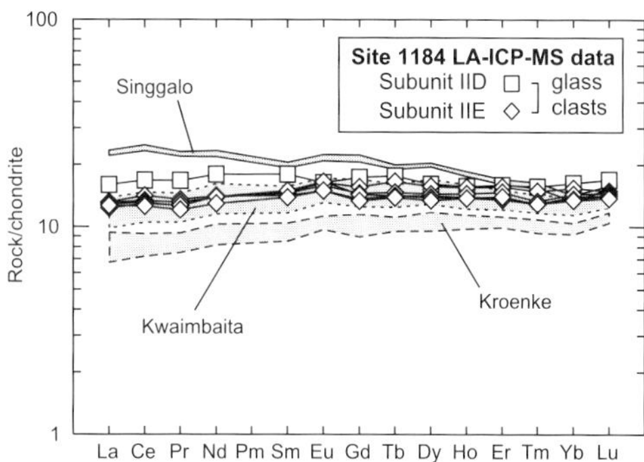

Fig. 5. Chondrite-normalized rare earth element (REE) plot (normalizing values from Sun & McDonough 1989) for Site 1184 glass clasts (LA-ICP-MS data) compared with fields for Kwaimbaita-, Kroenke- and Singgalo-type OJP basalts (data from Fitton & Godard 2004).

be made of how volcaniclastic sequences like that at Site 1184 can act as sources or sinks of particular elements in the oceanic environment. For example, the bulk tuff samples from Subunit IIA demonstrate significant elemental mobility: in particular, MgO, Na_2O and K_2O have been added to the volcanic pile and CaO has been removed.

Incompatible trace-element compositions of glass clasts

Trace-element data for glass clasts are presented in Table 2, and plots of trace-element compositions are presented in Figures 5 and 6. Because of the uneven distribution of unaltered glass in Hole 1184A, there are five samples from Subunit IIE and only one sample from Subunit IID. Figures 5 and 6 demonstrate that the homogeneity in major-element composition amongst clasts from the same subunit (Subunit IIE) extends to incompatible trace elements as well. A difference in trace-element composition between Subunits IID and IIE is just discernible – the sample from Subunit IID has marginally higher incompatible trace-element contents, particularly for the highly incompatible elements, although the overall patterns are similar.

Comparison with trace-element data from basalts of the OJP (Fitton & Godard 2004) demonstrates that the Subunit IIE glass overlaps with the Kwaimbaita magma type in trace-element compositions. Glass from Subunit IID is marginally more enriched in a range of incompatible elements than Kwaimbaita-type magmas, but it is not enriched enough to resemble the Singgalo-type magmas (Fig. 7).

Table 2. *LA-ICP-MS data (ppm) for glass clasts from Site 1184, Subunits IID and IIE*

Site	1184A	1184A	1184A	1184A	1184A	1184A	NIST612	**NIST612**
Core Section	31R-7	39R-7	41R-6	42R-5	44R-3	45R-2	measured	**recommended**
Interval (cm)	53–58	95–99	67–72	69–74	78–83	75–81	average	**value**
Depth (mbsf)	392.85	471.62	488.72	496.93	513.53	521.40		
Subunit	IID	IIE	IIE	IIE	IIE	IIE		
V	359.9 ± 21.7	339.8 ± 12.9	305.7 ± 23.1	312.8 ± 28.3	330.1 ± 30.7	306.0 ± 21.8	38.54 ± 2.74	**39.22**
Cr	329.3 ± 6.6	372.8 ± 17.4	307.9 ± 10.7	325.8 ± 15.1	358.9 ± 31.8	321.2 ± 6.33	40.42 ± 7.96	**39.88**
Co	43.7 ± 11.43	53.75 ± 5.02	53.6 ± 4.74	54.5 ± 0.57	55.9 ± 3.61	40.1 ± 3.54	34.97 ± 3.09	**35.26**
Ga	14.7 ± 2.23	17.60 ± 1.90	18.0 ± 0.06	20.4 ± 0.26	20.4 ± 2.08	15.6 ± 1.08	36.97 ± 2.56	**36.24**
Rb	3.4 ± 0.09	2.39 ± 0.35	2.3 ± 0.01	2.3 ± 0.53	2.4 ± 0.21	1.70 ± 0.09	31.00 ± 1.65	**31.63**
Sr	96.8 ± 0.44	97.65 ± 0.80	103.6 ± 5.47	102.8 ± 6.10	103.3 ± 4.50	95.3 ± 3.76	76.14 ± 3.00	**76.15**
Y	23.9 ± 0.60	19.63 ± 0.81	20.0 ± 0.28	19.7 ± 0.14	19.0 ± 1.14	18.1 ± 0.65	38.09 ± 1.36	**38.25**
Zr	55.1 ± 2.59	47.22 ± 0.50	51.9 ± 0.85	51.7 ± 1.30	50.1 ± 2.29	50.3 ± 0.92	35.84 ± 0.66	**35.99**
Nb	3.70 ± 0.12	2.86 ± 0.27	3.19 ± 0.11	3.09 ± 0.15	3.15 ± 0.36	2.72 ± 0.08	38.10 ± 1.18	**38.06**
Cs	0.49 ± 0.15	0.38 ± 0.11	0.41 ± 0.12	0.34 ± 0.15	0.30 ± 0.05	0.51 ± 0.02	41.68 ± 1.68	**41.64**
Ba	21.2 ± 0.81	16.85 ± 1.27	17.8 ± 1.06	17.7 ± 0.14	17.1 ± 1.16	16.2 ± 0.51	37.71 ± 1.18	**37.74**
La	3.79 ± 0.13	3.13 ± 0.14	3.11 ± 0.19	3.06 ± 0.26	2.93 ± 0.09	3.01 ± 0.07	35.68 ± 0.81	**35.77**
Ce	10.29 ± 0.90	8.67 ± 0.70	8.21 ± 0.24	8.05 ± 0.09	7.93 ± 0.41	7.72 ± 0.23	38.22 ± 1.14	**38.35**
Pr	1.59 ± 0.02	1.29 ± 0.17	1.25 ± 0.00	1.20 ± 0.06	1.21 ± 0.06	1.15 ± 0.04	37.20 ± 1.24	**37.16**
Nd	8.38 ± 0.50	6.59 ± 0.02	6.49 ± 0.75	6.57 ± 0.33	6.53 ± 0.34	6.06 ± 0.19	35.25 ± 1.09	**35.24**
Sm	2.75 ± 0.38	2.17 ± 0.47	2.25 ± 0.13	2.23 ± 0.45	2.29 ± 0.54	2.13 ± 0.09	36.52 ± 1.53	**36.72**
Eu	0.95 ± 0.06	0.87 ± 0.00	0.95 ± 0.18	0.92 ± 0.04	0.96 ± 0.04	0.87 ± 0.01	34.35 ± 0.57	**34.44**
Gd	3.57 ± 0.43	2.81 ± 0.30	3.21 ± 0.12	2.99 ± 0.22	2.95 ± 0.29	2.75 ± 0.06	36.78 ± 0.79	**36.95**
Tb	0.66 ± 0.12	0.53 ± 0.05	0.62 ± 0.08	0.55 ± 0.02	0.53 ± 0.05	0.52 ± 0.01	35.88 ± 0.96	**35.92**
Dy	4.09 ± 0.22	3.59 ± 0.23	4.00 ± 0.30	3.67 ± 0.28	3.53 ± 0.03	3.42 ± 0.10	36.02 ± 1.10	**35.97**
Ho	0.89 ± 0.09	0.79 ± 0.07	0.88 ± 0.02	0.82 ± 0.05	0.78 ± 0.03	0.78 ± 0.02	37.94 ± 1.05	**37.87**
Er	2.64 ± 0.00	2.29 ± 0.07	2.59 ± 0.15	2.44 ± 0.07	2.25 ± 0.01	2.32 ± 0.06	37.31 ± 0.90	**37.43**
Tm	0.40 ± 0.04	0.33 ± 0.04	0.38 ± 0.00	0.34 ± 0.01	0.33 ± 0.05	0.33 ± 0.01	37.50 ± 1.17	**37.55**
Yb	2.77 ± 0.02	2.38 ± 0.07	2.62 ± 0.52	2.45 ± 0.23	2.32 ± 0.23	2.29 ± 0.08	39.94 ± 1.44	**39.95**
Lu	0.43 ± 0.09	0.36 ± 0.02	0.38 ± 0.02	0.37 ± 0.04	0.36 ± 0.03	0.35 ± 0.01	37.81 ± 1.06	**37.71**
Hf	1.65 ± 0.06	1.40 ± 0.07	1.55 ± 0.23	1.42 ± 0.10	1.42 ± 0.16	1.43 ± 0.05	34.62 ± 1.31	**34.77**
Ta	0.23 ± 0.03	0.18 ± 0.03	0.21 ± 0.03	0.20 ± 0.02	0.19 ± 0.02	0.19 ± 0.01	39.61 ± 1.78	**39.77**
Pb	0.60 ± 0.20	1.48 ± 0.04	0.76 ± 0.02	1.55 ± 0.52	1.45 ± 0.13	0.61 ± 0.04	38.80 ± 2.65	**38.96**
Th	0.28 ± 0.04	0.26 ± 0.03	0.25 ± 0.07	0.20 ± 0.05	0.27 ± 0.04	0.15 ± 0.01	37.46 ± 1.82	**37.23**
U	0.10 ± 0.04	0.08 ± 0.04	0.08 ± 0.01	0.11 ± 0.04	0.09 ± 0.00	0.08 ± 0.01	37.31 ± 3.82	**37.15**

Errors (all reported at 2σ) are calculated on the basis of duplicate analyses and the reproducibility of the NIST612 standard glass.
The NIST612 average data are on the basis of eight analyses.

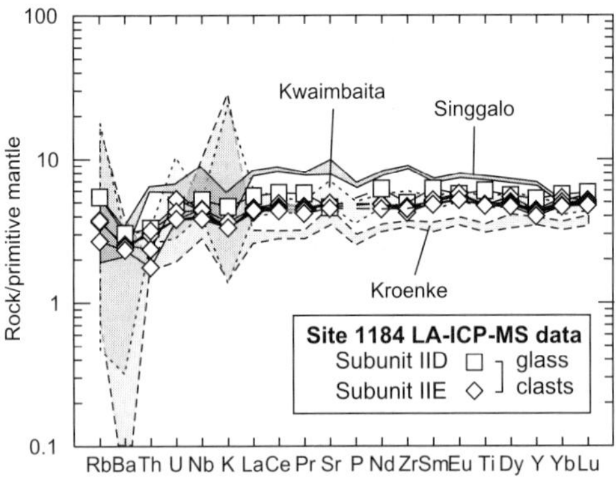

Fig. 6. Primitive-mantle-normalized multi-element diagram (normalizing values from Sun & McDonough 1989) for Site 1184 glass clasts (LA-ICP-MS data) compared with fields for Kwaimbaita-, Kroenke- and Singgalo-type OJP basalts (data from Fitton & Godard 2004).

Sr, Nd and Pb isotope compositions of bulk tuff samples

The ubiquitous presence of altered rims on the glass clasts made it impossible to obtain sufficient fresh glass for Sr-, Nd- and Pb-isotopic analysis, and, thus, four bulk tuff samples were analysed to verify the relationship between the variably altered volcaniclastic sequence at Site 1184 and Leg 192 lavas at Sites 1183, 1185, 1186 and 1187. The analyses are presented in Table 3 and Figure 8. Epsilon Nd values discussed in this section are initial values, calculated assuming an age of 120 Ma (denoted $\varepsilon_{Nd(t=120\,Ma)}$).

Sample 1184A-11R-3, 28–34 cm has the highest $^{87}Sr/^{86}Sr$ and lowest $^{143}Nd/^{144}Nd$ among the samples. It has an $\varepsilon_{Nd(t=120\,Ma)}$ that is slightly lower than those of the Leg 192 lavas ($\varepsilon_{Nd(t=120\,Ma)}$ = 5.8–6.5: Tejada *et al.* 2004) and individual basaltic lithic clasts in the volcaniclastic sequence at Site 1184 ($\varepsilon_{Nd(t=120\,Ma)}$ = 6.0–6.5:

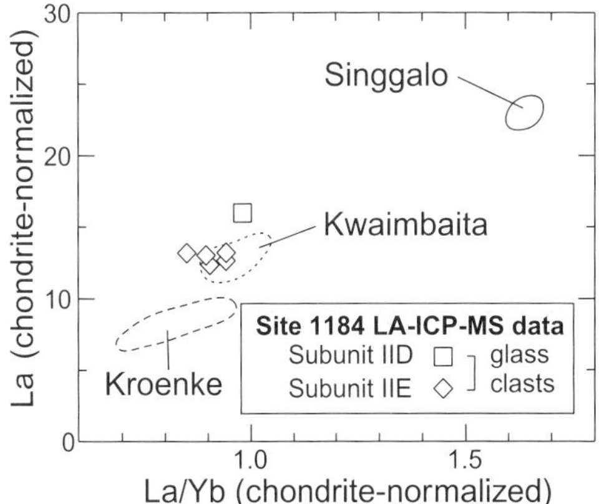

Fig. 7. Plot of La v. La/Yb (both chondrite normalized) for Site 1184 glass clasts (LA-ICP-MS data) compared with fields for Kwaimbaita-, Kroenke- and Singgalo-type OJP basalts (data from Fitton & Godard 2004).

Shafer *et al.* 2004), but falls within the range of previously analysed Kwaimbaita-type lavas (cf. Mahoney & Spencer 1991; Mahoney *et al.* 1993; Tejada *et al.* 1996, 2002; Neal *et al.* 1997). This sample belongs to Subunit IIA, which has a Kroenke-type elemental signature according to glass clasts and bulk tuff analyses. Kroenke-type lavas have a Kwaimbaita-type lava isotopic signature (Tejada *et al.* 2004). The Nd and Sr concentrations of sample 1184A-11R-3, 28–34 cm dropped dramatically after leaching (see table 4 of Fitton & Godard 2004 for unleached values), suggesting that most of its Sr and Nd contents are in the altered, leachable phases such as secondary phosphates and clay (e.g. Cheng *et al.* 1987; Mahoney & Spencer 1991; Janney & Castillo 1996, 1997). Thus, intense sea-water alteration is most probably responsible for its high $^{87}Sr/^{86}Sr$ (e.g. McCulloch *et al.* 1981; Cheng *et al.* 1987; Verma 1992). Sample 1184A-11R-3, 28–34 cm has lower Pb-isotopic ratios than the other samples analysed and all Ontong Java Plateau lavas; this is also probably due to intense sea-water alteration.

The $\varepsilon_{Nd(t\,=\,120\,Ma)}$ of sample 1184A-25R-5, 40–47 cm, from Subunit IIC, is within the ranges of Leg 192 Kwaimbaita-type lavas (Tejada *et al.* 2004) and lithic basaltic clasts from Site 1184 (Shafer *et al.* 2004). This is consistent with the similarity of most of the major- and trace-element contents of bulk tuff samples from this unit with Leg 192 Kwaimbaita-type lavas from Sites 1183, 1185 and 1186 (Fitton & Godard 2004). Age-corrected Sr- and Pb-isotopic ratios of the sample, however, are different from Leg 192 Kwaimbaita-type lavas. Although this sample is not as severely altered as sample 1184A-11R-3, 28–34 cm, its Sr- and Pb-isotope systematics may also have been modified by sea-water alteration.

Sample 1184A-42R-6, 7–18 cm is from Subunit IIE; the glass clasts from this subunit are compositionally homogeneous and overlap with the composition of Leg 192 Kwaimbaita-type lavas. The Sr-, Nd- and Pb-isotopic composition of sample 1184A-42R-6, 7–18 cm is very similar to that of Leg 192 lavas (Tejada *et al.* 2004) and overlaps with the field of the lithic basaltic clasts from Site 1184 (Shafer *et al.* 2004). It is important to note that this sample was chosen to be a geochemical reference sample for the volcaniclastic sequence at Site 1184 and, hence, was carefully chosen from the freshest section of the sequence. Sample 1184A-46R-1, 54–60 cm and the two other samples, on the other hand, were chosen not for freshness, but to cover the range of compositions.

Sample 1184A-46R-1, 54–60 cm, from Subunit IIF, has the highest Nd-isotopic ratios among the samples analysed. Consequently, it has higher $\varepsilon_{Nd(t\,=\,120\,Ma)}$ than all Leg 192 Kwaimbaita-type lavas and the lithic basaltic clasts from Site 1184. It also has high $^{207}Pb/^{204}Pb_{(pres.)}$ for a given $^{206}Pb/^{204}Pb_{(pres.)}$ ratio compared to Kwaimbaita- and Singgalo-type lavas, although not as high as the $^{207}Pb/^{204}Pb_{(pres.)}$ of the tuff layer above the basement at Site 1183 (15.563 v. 15.581: Tejada *et al.* 2004). Interestingly, the $^{206}Pb/^{204}Pb_{(pres.)}$, $^{208}Pb/^{204}Pb_{(pres.)}$ and $^{87}Sr/^{86}Sr_{(t\,=\,120\,Ma)}$values overlap with Leg 192 Kwaimbaita-type lavas. Hence the majority of the isotopic signature of sample 1184A-46R-1, 54–60 cm is similar to that of Kwaimbaita-type lavas. A possible explanation for the isotopic composition of this sample is that it came from a source that has a slightly different composition (i.e. in Nd isotopes and $^{207}Pb/^{204}Pb$) from that of the Kwaimbaita-type lavas. More likely, its source is similar to that of the rest of Leg 192 lavas and volcaniclastic rocks, but its isotopic composition is variably affected by alteration.

In summary, the Nd-isotope signature of three out of four bulk tuff samples of the variably altered volcaniclastic sequence analysed is similar to that of the Kwaimbaita-type lavas; the Nd-isotope signature of the fourth sample is also close. The Pb-isotopic systematics are similar only to a limited degree and it is hard to predict their behaviour in the variably altered volcaniclastic rocks. The Sr-isotope systematics cannot be trusted as a petrogenetic tracer for the volcaniclastic sequence. That the Nd-isotope systematics of the volcaniclastic bulk tuff samples

Table 3. *Sr-, Nd- and Pb-isotopic composition of volcaniclastic rocks from ODP Site 1184*

Sample	$^{87}Sr/^{86}Sr$	$^{143}Nd/^{144}Nd$	$^{206}Pb/^{204}Pb$	$^{207}Pb/^{204}Pb$	$^{208}Pb/^{204}Pb$	Rb	Sr	Sm	Nd	U	Th	Pb	$^{87}Sr/^{86}Sr$	$^{143}Nd/^{144}Nd$	ε_{Nd}	$^{206}Pb/^{204}Pb$	$^{207}Pb/^{204}Pb$	$^{208}Pb/^{204}Pb$
	measured	–	–	–	–								120 Ma	–	–	–	–	–
1184A-11R-3, 28–34 cm																		
Subunit IIA	0.712880	0.512925	17.503	15.481	37.176	2.27	6.95	0.29	0.92	0.21	0.21	1.39	0.71127	0.512776	5.7	17.331	15.473	37.120
duplicate		0.512918						0.29	0.92					0.512769	5.5			
1184A-25R-5, 40–47 cm																		
Subunit IIC	0.704588	0.512981	17.653	15.520	37.481	0.56	9.03	0.79	2.12	0.15	0.54	1.92	0.70428	0.512805	6.2	17.564	15.516	37.376
1184A-42R-6, 7–18 cm																		
Subunit IIE	0.703948	0.512984	18.554	15.519	38.352	0.32	76.4	1.36	3.85	0.07	0.22	0.42	0.70393	0.512818	6.5	18.357	15.509	38.151
1184A-46R-1, 54–60 cm																		
Subunit IIF	0.703269	0.513032	18.658	15.563	38.417	0.36	50.4	0.62	1.84	0.38	0.83	1.27	0.70323	0.512873	7.5	18.306	15.546	38.165
duplicate (unleached)		0.513021						0.62	1.84					0.512862	7.3			

Sr- and Nd-isotope ratios were measured on leached powders. Analytical uncertainty for $^{87}Sr/^{86}Sr$ measurements is ±0.000018 but in-run precisions were better than ± 0.000013, except for sample1184A-11R-3, 28–34 cm, which was ± 0.000025. Sr-isotope ratios were measured by dynamic multi-collection, fractionation corrected to $^{86}Sr/^{88}Sr$ = 0.1194 and normalized to $^{87}Sr/^{86}Sr$ = 0.71025 for NBS 987. Analytical uncertainty for $^{143}Nd/^{144}Nd$ measurements is ±0.000014 (0.3 ε units) but in-run precisions were better than ±0.000012, except for the first run of sample1184A-11R-3, 28–34 cm, which was ± 0.000035. Nd-isotope ratios were measured in oxide form by dynamic multi-collection, fractionation corrected to $^{146}NdO/^{144}NdO$ = 0.72225 ($^{146}Nd/^{144}Nd$ = 0.7219) and are reported relative to $^{143}Nd/^{144}Nd$ = 0.511850 for the La Jolla Standard. Pb-isotope ratios were measured by static multi-collection and are reported relative to the values of Todt *et al.* (1995) for NBS SRM 981; the long-term errors measured for this standard are ± 0.008 for $^{206}Pb/^{204}Pb$ and $^{207}Pb/^{204}Pb$, and ± 0.024 for $^{208}Pb/^{204}Pb$; the in-run precisions were better than this. Total procedural blanks are negligible: <10 picograms (pg) for Nd, <35 pg for Sr, <3 pg for Th, <5 pg for U and <60 pg for Pb. U, Th and Pb elemental concentrations used in the age-corrections for Pb-isotopes are given in the table. $\varepsilon_{Nd} = 0$ today corresponds to $^{143}Nd/^{144}Nd$ = 0.51264; for $^{147}Sm/^{144}Nd$ = 0.1967, $\varepsilon_{Nd}(t) = 0$ corresponds to $^{143}Nd/^{144}Nd$ = 0.512486 at 120 Ma.

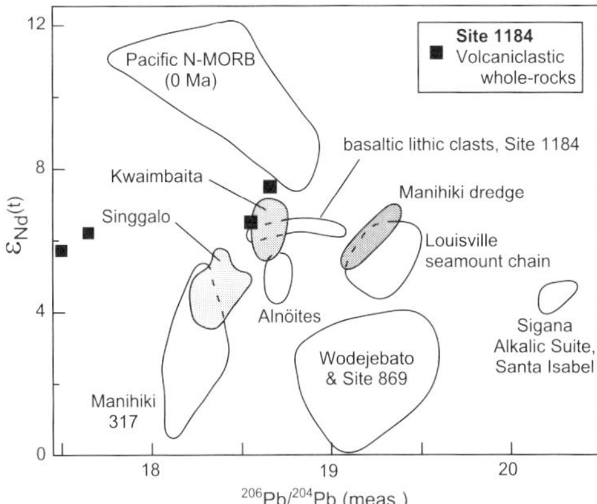

Fig. 8. Plot of ε_{Nd} v. $^{206}Pb/^{204}Pb$ for bulk tuff samples from Site 1184 compared with fields for previously analysed Kwaimbaita-, Singgalo-, Sigana-type and Alnöite lavas from the OJP (Mahoney & Spencer 1991; Tejada *et al.* 1996, 2002; Neal *et al.* 1997) and basaltic lithic clasts from Site 1184 (Shafer *et al.* 2004). Analytical error is smaller than the symbol used. Other fields shown are for the Louisville Seamount Chain, Manihiki Plateau (Mahoney 1987; Mahoney & Spencer 1991), Pacific normal-MORB (Tejada *et al.* 1996), and volcaniclastic rocks from ODP Site 869 and Wodejebato Guyot (Janney & Castillo 1999).

can be used as petrogenetic tracer for the source(s) of the volcanogenic components is also shown by results of isotopic investigations of the lithic basaltic clasts in the volcaniclastic sequence at Site 1184 (Shafer *et al.* 2004), the tuff layer above the lavas at Site 1183 (Tejada *et al.* 2004) and the volcaniclastic sequence in the Nauru Basin (Castillo 2004).

Discussion

Relationship between magma composition and stratigraphy

There is no consistent stratigraphic trend in composition between magma types that indicates evolution via magmatic differentiation processes, particularly fractional crystallization of a single magma batch. The glass clasts demonstrate that the earliest eruptions (Subunit IIE) were of a Kwaimbaita-type magma, followed by a more evolved magma (Subunit IID). The top of the succession (Subunit IIA) reverts to a more primitive magma type, similar to the Kroenke-type lavas of Sites 1185 and 1187. Furthermore, trace-element data for bulk tuff samples from all subunits of Hole 1184A show that at least five distinct magma compositions were erupted over the lifetime of the sampled portion of this volcano (Fitton & Godard 2004, fig. 7). The distribution of magma types relative to the stratigraphy indicates that the magma chamber was replenished with more primitive magma at least twice during the sampled time interval.

The dominance of Kwaimbaita- and Kroenke-type compositions in the glass clasts is consistent with observations made on the basaltic lithic clast population of the volcaniclastic rocks (Shafer *et al.* 2004). The modest heterogeneity observed at Site 1184 contrasts markedly with the majority of the OJP, in particular with the homogeneous Kwaimbaita-type magmas that dominate the sampled portions of the high plateau and the island of Malaita. The presence of complexly zoned plagioclase megacrysts (Babbs 1997) and plagioclase-rich xenoliths (Mahoney *et al.* 2001) in the Kwaimbaita-type lavas indicate that residence in magma chambers was responsible, at least in part, for this homogeneity. The juvenile components of the volcaniclastic sequence in the eastern salient of the OJP may not have had such an extensive magma storage system; alternatively, the explosivity of the eruptions may have tapped a wider range of magma compositions from a single magma reservoir, compared to effusive eruptions that may have only tapped the upper, well-mixed portion of the magma chamber (cf Kilauea: Dzurisin *et al.* 1995).

Extent of subaerial volcanism and implications for plume models

Confirming the presence and extent of volcaniclastic rocks that are part of the main plateau sequence may help to answer questions about the origin of the plateau and its subsidence history. Site 1184 has the only known example of near-source subaerially erupted volcaniclastic rocks on the OJP. Vitric tuff layers immediately above basement at DSDP Site 289 (Andrews *et al.* 1975) and ODP Site 1183 may represent shallow-water or subaerial eruptions, but these tuffs are considerably finer grained than the lapilli-dominated sequence at Site 1184, and the two layers at Site 1183 show evidence of sedimentary reworking (Mahoney *et al.* 2001). The loci of shallow-water or subaerial volcanic activity contributing to these vitric tuff layers remain unknown.

Consequently, the volume of volcaniclastic rocks making up the Ontong Java Plateau is extremely poorly constrained. Site 1184 is the only site to have sampled 'basement' on the whole of the eastern salient, a feature that constitutes approximately a quarter of the area of the OJP (Fig. 1). Two-dimensional (2D) seismic surveys do not appear to discriminate effectively between basaltic and volcaniclastic basement, and therefore only further drilling will determine the extent of subaerial volcanic activity.

The presence of non-vesicular pillow basalts rather than subaerial lavas at the current topographic crest of the high plateau has created difficulties for the plume-head model of oceanic plateau generation (e.g. Richards *et al.* 1991), which predicts dynamic uplift of 1–3 km associated with the arrival of a plume head at the base of the lithosphere (e.g. Hill 1991). This amount of dynamic uplift, coupled with the relatively thin lithosphere thought to be required to explain the high degrees of partial melting (Mahoney *et al.* 1993; Tejada *et al.* 1996; Neal *et al.* 1997; Fitton & Godard 2004), and the thick plateau crust should have been sufficient to elevate the plateau crest above sea level during volcanism. Neal *et al.* (1997), however, point out that a hidden layer of high-density cumulates would have both limited uplift and tempered post-emplacement subsidence of the OJP.

The OJP has undergone subsidence of approximately 1.5 km since eruption of its youngest (sampled) lavas (Roberge *et al.* 2004*b*). This is considerably less than would be expected from the plume-head model combined with normal thermal subsidence. In spite of this problem, however, the plume-head model remains the most widely accepted hypothesis for the generation of the OJP, mainly because it is the only model amongst several contenders that adequately explains the generation of such a large magma volume derived from a relatively undepleted (i.e. non-mid-ocean ridge basalt) and volatile-poor (Roberge *et al.* 2004*a*) source in a relatively short period of time.

If the OJP *did* form above a plume head, and the apparent lack of subaerial volcanism on the high plateau is genuine, it suggests that our models of plume-generated uplift and/or plateau-related subsidence require re-evaluation. For example, if the anomalously hot mantle material rapidly flowed laterally during the plume impingement event, as has been suggested for the North Atlantic Igneous Province (Saunders *et al.* 1997; Larsen *et al.* 1999), the hot material could form a thinner, laterally more extensive sheet than is envisaged in many mantle plume models, and the amount of uplift would be lower. Alternatively, modelling by Olson (1994) suggests that if a plume head is emplaced beneath the lithosphere as a solitary thermal diapir rather than continuing to be fed from beneath, maximum topographic uplift would precede the peak of volcanism by about 5 Ma, and by the time the final lavas were emplaced (i.e. those that are sampled by ocean drilling) the plateau could have already undergone significant subsidence due to the withdrawal of dynamic support. In this case, any subaerial lavas and/or volcaniclastic successions would be buried beneath a carapace of younger, submarine lavas. On the high plateau, drilling has only penetrated the uppermost 217 m of the 30–35 km-thick crust, so it is possible that subaerial lithologies remain hidden underneath.

However, other oceanic plateaus do contain subaerial lithologies close to the top of the volcanic stratigraphy (e.g. Caribbean Plateau: White 1999; Kerguelen: Coffin *et al.* 2000). This demonstrates that, whether or not subaerial rocks exist deeper within the volcanic pile of the OJP, its mode of formation may have been different from those of other Cretaceous Pacific plateaus such as the Caribbean Plateau. The presence of a high-density cumulate layer within the OJP, as modelled by Neal *et al.* (1997), may be one of the more significant dissimilarities between the OJP and other Pacific oceanic plateaus.

Environmental and climatic implications

The OJP represents an immense outpouring of magma, but there is no significant mass extinction temporally associated with its eruption. This contrasts with the eruption of many continental flood basalt provinces, which correlate temporally with (and may have contributed to) major mass extinctions (e.g. Wignall 2001). A likely explanation for the lack of environmental catastrophe associated with the OJP is the observation that the majority of the volcanism occurred in the submarine environment, and that the magmas are relatively volatile-poor (Michael 1999; Roberge *et al.* 2004*a*). This would have restricted the flux of climate-affecting volatile species (e.g. SO_2, Cl, F, CO_2) into the atmosphere. However, our limited sampling of the OJP means that the proportion of subaerial volcanism, and thus the flux of volatiles associated with OJP magmatism, is very poorly constrained. It should also be noted that the likelihood of a mass extinction is not a simple function of volatile output of a volcanic system, as it depends on a web of interlinked factors combined with the ability of Earth's systems to withstand climatic and environmental perturbations (e.g. White 2002). Consequently, statements that assume a direct link between extinctions and volatile fluxes should be treated with caution.

Although the OJP does not correlate with any mass extinctions, it is temporally associated (Pringle 2001) with an environmental disturbance, recorded in the sedimentary and geochemical record as the Aptian Oceanic Anoxic Event 1a ('Selli Event'). At this time, black shale deposition was associated with a negative $\delta^{13}C$ excursion, a fall in sea-water $^{87}Sr/^{86}Sr$ and an increase in metal concentrations (Larson & Erba 1999). These environmental features are consistent with their being caused by large-scale basaltic volcanism, with increased volatile input into the atmosphere combined with increased hydrothermal fluxes into the oceans.

Conclusions

- The juvenile magmatic components of the volcaniclastic rocks recovered at Site 1184 are chemically and isotopically similar to pillow basalts and sheet flows recovered elsewhere on the OJP. This corroborates the hypothesis that the subaerial, explosive volcanism on the eastern salient of the OJP is part of the main early Aptian plateau-building magmatic event. In particular, the juvenile magmas contributing to Subunit IIE of Site 1184 are indistinguishable from Kwaimbaita-type magmas in terms of major- and trace-element composition, and are also similar in terms of Nd- and, to a limited extent, Pb-isotopic composition.
- Although all the glass clasts have a low-K tholeiitic composition, each of the three subunits that contain unaltered glass records a distinct juvenile magma composition. This suggests that the eastern salient had a different type of magma plumbing system to that of the high plateau, allowing a greater degree of heterogeneity to be preserved. No intermixing of glass clasts between subunits was detected, supporting the shipboard hypothesis (arising from the discovery of wood fragments at subunit boundaries) that each subunit represents a discrete eruptive phase with intervening periods of quiescence.
- Whereas drilling at Site 1185 indicated a hiatus between eruption of Kwaimbaita-type and Kroenke-type basalts, the volcaniclastic succession at Site 1184 indicates that both magma types were present within the lifetime of a single Surtseyan volcano, albeit one that had non-eruptive intervals of sufficient length for colonization by woody vegetation.
- Bulk tuff analyses for some samples closely match juvenile glass clast compositions from the same subunit, whereas others are markedly different. This reflects variability in the degree of alteration/cementation of the tuffs, and, to a lesser extent, the proportion of non-juvenile material (e.g. lithic clasts) incorporated into the sedimentary pile.
- At least some of the Ontong Java Plateau volcanism was subaerial, and thus had the potential to inject volatile, climate-modifying, species into the Cretaceous atmosphere. The volume of volcaniclastic rocks on the eastern salient (and elsewhere on the plateau), and thus the extent of these subaerial eruptive events, remains unquantified. Although the eruption of the OJP does not appear to have had a signifcant affect on contemporaneous biota, the magmatism does correspond temporally with the anoxic Selli Event.

This research used samples provided by the Ocean Drilling Program (ODP). ODP is sponsored by the U.S. National Science Foundation (NSF) and participating countries under management of Joint Oceanographic Institutions (JOI), Inc. We thank P. Janney and A. Pietruszka for their helpful reviews, R. Wilson for assistance with electron microprobe analyses, W. Kinman for help with the LA-ICP-MS analyses, and A. Saunders for informal discussions. R. V. White is supported by a Royal Society Dorothy Hodgkin Research Fellowship, and an ODP Rapid Response Grant from the Natural Environment Research Council (NERC). C. R. Neal was supported by a grant from USSSAC.

References

ANDREWS, J.E., PACKHAM, G.H. *et al.* 1975. Wellington, New Zealand to Apra Guam, April–June 1973. *Proceedings of the Deep Sea Drilling Project, Initial Reports*, **30**. College Station, TX, 753 pp.

BABBS, T.L. 1997. *Geochemical and petrological investigations of the deeper portions of the Ontong Java Plateau: Malaita, Solomon Islands.* PhD Thesis, University of Leicester.

BIRKHOLD, A.L. 2000. *A geochemical investigation of Ontong Java Plateau basement exposed on the Island of Makira (San Cristobal), Solomon Islands, South Pacific.* PhD Thesis, University of Notre Dame.

CASTILLO, P.R. 2004. Geochemistry of Cretaceous volcaniclastic sediments in the Nauru and East Mariana basins provides insights into the mantle sources of giant oceanic plateaus. *In*: FITTON, J.G., MAHONEY, J.J., WALLACE, P.J. & SAUNDERS, A.D. (eds) *Origin and Evolution of the Ontong Java Plateau.* Geological Society, London, Special Publications, **229**, 353–368.

CHAMBERS, L.M., PRINGLE, M.S. & FITTON, J.G. 2002. Age and duration of magmatism on the Ontong Java Plateau: ^{40}Ar–^{39}Ar results from ODP Leg 192. Abstract V71B-1271. *Eos, Transactions American Geophysical Union*, **83**, F47.

CHAMBERS, L.M., PRINGLE, M.S. & FITTON, J.G. 2004. Phreatomagmatic eruptions on the Ontong Java Plateau: an Aptian $^{40}Ar/^{39}Ar$ age for volcaniclastic rocks at ODP Site 1184. *In*: FITTON, J.G.,

MAHONEY, J.J., WALLACE, P.J. & SAUNDERS, A.D. (eds) *Origin and Evolution of the Ontong Java Plateau*. Geological Society, London, Special Publications, **229**, 325–331.

CHENG, Q., PARK, K.H., MACDOUGALL, J.D., ZINDLER, A., LUGMAIR, G.W., STAUDIGEL, H., HAWKINS, J.W. & LONSDALE, P.F. 1987. Isotopic evidence for a hotspot origin of the Louisville seamount chain. *In*: KEATING, B.H., FRYER, P., BATIZA, R. & BOEHLERT, G.W. (eds) *Seamounts, Islands, and Atolls*. American Geophysical Union, Geophysical Monograph, **43**, 283–296.

COFFIN, M.F., FREY, F.A. & WALLACE, P.J. 2000. *Proceedings of the Ocean Drilling Program, Initial Reports*, **183** (CD-ROM).

DUNCAN, R.A. & CLAGUE, D.A. 1985. Pacific Plate motion recorded by linear volcanic chains. *In*: NAIRN, A.E.M., STEHLI, F.G. & UYEDA, S. (eds) *The Ocean Basins and Margins: The Pacific Ocean*, vol. **7A**. Plenum, New York, 89–121.

DZURISIN, D., LOCKWOOD, J. P, CASADEVALL, T.J. & RUBIN, M. 1995. The Uwekahuna Ash Member of the Puna Basalt; product of violent phreatomagmatic eruptions at Kilauea Volcano, Hawaii, between 2800 and 2100 ^{14}C years ago. *Journal of Volcanology and Geothermal Research*, **66**, 163–184.

FITTON, J.G. & GODARD, M. 2004. Origin and evolution of magmas on the Ontong Java Plateau. *In*: FITTON, J.G., MAHONEY, J.J., WALLACE, P.J. & SAUNDERS, A.D. (eds) *Origin and Evolution of the Ontong Java Plateau*. Geological Society, London, Special Publications, **229**, 151–178.

GLADCZENKO, T.P., COFFIN, M.F. & ELDHOLM, O. 1997. Crustal structure of the Ontong Java Plateau: Modeling of new gravity and existing seismic data. *Journal of Geophysical Research*, **102**, 22 711–22 729.

HAGEN, R.A., MAYER, L.A., MOSHER, D.C., KROENKE, L.W., SHIPLEY, T.H. & WINTERER, E.L. 1993. Basement structure of the northern Ontong Java Plateau. *In*: BERGER, W.H., KROENKE, L.W. & MAYER, L.A. (eds) *Proceedings of the Ocean Drilling Program, Scientific Results*, **130**, 23–31.

HART, S.R., ERLANK, A.J. & KABLE, E.J.D. 1974. Seafloor basalt alteration: some chemical and Sr isotopic effects. *Contributions to Mineralogy and Petrology*, **44**, 219–230.

HILL, R.I. 1991. Starting plumes and continental break-up. *Earth and Planetary Science Letters*, **104**, 398–416.

JANNEY, P.E. & CASTILLO, P.R. 1996. Basalts from the Central Pacific Basin: Evidence for the origin of Cretaceous igneous complexes in the Jurassic Western Pacific. *Journal of Geophysical Research*, **101**, 2875–2894.

JANNEY, P.E. & CASTILLO, P.R. 1997. Geochemistry of Mesozoic Pacific MORB: constraints on melt generation and the evolution of the Pacific upper mantle. *Journal of Geophysical Research*, **102**, 5207–5229.

JANNEY, P.E. & CASTILLO, P.R. 1999. Isotope geochemistry of the Darwin Rise seamounts and the nature of long-term mantle dynamics beneath the south central Pacific. *Journal of Geophysical Research*, **104**, 10 571–10 589.

KROENKE, L.W. 1972. *Geology of the Ontong Java Plateau*. Hawaii Institute of Geophysics Report HIG, 72–75.

KROENKE, L.W., RESIG, J.M. & COOPER, P.A. 1986. Tectonics of the southeastern Solomon Islands: formation of the Malaita anticlinorium. *In*: VEDDER, J.G., POUND, K.S. & BOUNDY, S.Q. (eds) *Geology and Offshore Resources of Pacific Island Arcs – Central and Western Solomon Islands*. Circum-Pacific Council for Energy and Mineral Resources, Earth Science Series **4**, 109–116.

KROENKE, L.W., BERGER, W.H., JANECEK, T.R. *et al.* 1991. *Proceedings of the Ocean Drilling Program, Initial Results*, **130**.

LARSEN, T.B., YUEN, D.A. & STOREY, M.J. 1999. Ultrafast mantle plumes and implications for flood basalt volcanism in the Northern Atlantic Region. *Tectonophysics*, **311**, 31–43.

LARSON, R.L. & ERBA, E. 1999. Onset of the Mid-Cretaceous greenhouse in the Barremian–Aptian; igneous events and the biological, sedimentary, and geochemical responses. *Paleoceanography*, **14**, 663–678.

MAHONEY, J.J. 1987. An isotopic survey of Pacific ocean plateaus: implications for their nature and origin. *In*: KEATING, B.H., FRYER, P., BATIZA, R. & BOEHLERT, G.W. (eds) *Seamounts, Islands, and Atolls*. American Geophysical Union, Geophysical Monograph, **43**, 207–220.

MAHONEY, J.J. & SPENCER, K.J. 1991. Isotopic evidence for the origin of the Manihiki and Ontong Java oceanic plateaus. *Earth and Planetary Science Letters*, **104**, 196–210.

MAHONEY, J.J., STOREY, M., DUNCAN, R.A., SPENCER, K.J. & PRINGLE, M. 1993. Geochemistry and age of the Ontong Java Plateau. *In*: PRINGLE, M.S., SAGER, W.W., SLITER, W.V. & STEIN, S. (eds) *The Mesozoic Pacific: Geology, Tectonics, and Volcanism*. American Geophysical Union, Geophysical Monograph, **77**, 233–261.

MAHONEY, J.J., FITTON, J.G., WALLACE, P.J. *et al.* 2001. *Proceedings of the Ocean Drilling Program, Initial Reports*, **192** (CD-ROM).

MCCULLOCH, M.T., GREGORY, R.T., WASSERBURG, G.J. & TAYLOR, H.P., JR. 1981. Sm–Nd, Rb–Sr, and O-18/O-16 isotopic systematics in an oceanic crustal section; evidence from the Samail ophiolite. *In*: COLEMAN, R.G. & HOPSON, C.A. (eds) *Oman Ophiolite*. *Journal of Geophysical Research*, **86**, 2721–2735.

MICHAEL, P.J. 1999. Implications for magmatic processes at Ontong Java Plateau from volatile and major element contents of Cretaceous basalt glasses. *Geochemistry Geophysics Geosystems*, **1**, 1999GC000025.

NEAL, C.R., MAHONEY, J.J., KROENKE, L.W., DUNCAN, R.A. & PETTERSON, M.G. 1997. The Ontong Java Plateau. *In*: MAHONEY, J.J. & COFFIN, M.F. (eds) *Large Igneous Provinces: Continental, Oceanic, and Planetary Flood Volcanism*. American Geophysical Union, Geophysical Monograph, **100**, 183–216.

OLSON, P. 1994. Mechanics of flood basalt magmatism. *In*: RYAN, M.P. (ed.) *Magmatic Systems*. Academic Press, New York, 1–18.

PEARCE, N.J.G., PERKINS, W.T., WESTGATE, J.A.,

GORTON, M.P., JACKSON, S.E., NEAL, C.R. & CHERNERY, S.P. 1997. New data for the National Institute of Standards and Technology 610 and 612 glass reference materials. *Geostandards Newsletter*, **21**, 115–144.

PETTERSON, M.G., BABBS, T., NEAL, C.R., MAHONEY, J.J., SAUNDERS, A.D., DUNCAN, R.A., TOLIA, D., MAGU, R., QOPOTO, C., MAHOA, H. & NATOGGA, D. 1999. Geological-tectonic framework of Solomon Islands, SW Pacific: crustal accretion and growth within an intra-oceanic setting. *Tectonophysics*, **301**, 35–60.

PRINGLE, M.S. 2001. Revised age for Chron M0r implies synchronicity of volcanism on the Ontong Java Plateau with global oceanographic events. *Eos, Transactions of the American Geophysical Union*, **82**, Fall Meeting Supplement, Abstract T21D-09.

RICHARDS, M.A., JONES, D.L., DUNCAN, R.A. & DEPAOLO, D.J. 1991. A mantle plume initiation model for the Wrangellia flood basalt and other oceanic plateaus. *Science*, **254**, 263–267.

RICHARDSON, W.P., OKAL, E.A. & VAN DER LEE, S. 2000. Rayleigh-wave tomography of the Ontong-Java Plateau. *Physics of the Earth and Planetary Interiors*, **118**, 29–51.

RIISAGER, P., HALL, S., ANDRETTER, M. & ZHAO, X. 2004. Early Cretaceous Pacific palaeomagnetic pole from Ontong Java Plateau basement rocks. *In*: FITTON, J.G., MAHONEY, J.J., WALLACE, P.J. & SAUNDERS, A.D. (eds) *Origin and Evolution of the Ontong Java Plateau*. Geological Society, London, Special Publications, **229**, 31–44.

ROBERGE, J., WHITE, R.V. & WALLACE, P.J. 2004*a*. Volatiles in submarine basaltic glasses from the Ontong Java Plateau (ODP Leg 192): implications for magmatic processes and source region compositions. *In*: FITTON, J.G., MAHONEY, J.J., WALLACE, P.J. & SAUNDERS, A.D. (eds) *Origin and Evolution of the Ontong Java Plateau*. Geological Society, London, Special Publications, **229**, 239–257.

ROBERGE, J., WALLACE, P.J., WHITE, R.V. & COFFIN, M.F. (2004*b*). Anomalous uplift and subsidence of the Ontong Java Plateau inferred from CO_2 contents of submarine basaltic glasses. Submitted to Geology.

SAUNDERS, A.D., STOREY, M., KENT, R.W. & NORRY, M.J. 1992. Consequences of plume–lithosphere interactions. *In:* STOREY, B.C., ALABASTER, T. & PANKHURST, R.J. (eds) *Magmatism and the Causes of Continental Break Up*. Geological Society, London, Special Publications, **68**, 41–60.

SAUNDERS, A.D., FITTON, J.G., KERR, A.C., NORRY, M.J. & KENT, R.W. 1997. The North Atlantic igneous province. *In*: MAHONEY, J.J. & COFFIN, M.F. (eds) *Large Igneous Provinces: Continental, Oceanic, and Planetary Flood Volcanism*. American Geophysical Union, Geophysical Monograph, **100**, 45–93.

SHAFER, J.T., NEAL, C.R. & CASTILLO, P.R. 2004. Compositional variability in lavas from the Ontong Java Plateau: results from basalt clasts within the volcaniclastic succession at Ocean Drilling Program Site 1184. *In*: FITTON, J.G., MAHONEY, J.J., WALLACE, P.J. & SAUNDERS, A.D. (eds) *Origin and Evolution of the Ontong Java Plateau*. Geological Society, London, Special Publications, **229**, 333–351.

SMITH, W.H.F. & SANDWELL, D.T. 1997. Global seafloor topography from satellite altimetry and ship depth soundings. *Science*, **277**, 1956–1962.

SUN, S.-S. & MCDONOUGH, W.F. 1989. Chemical and isotopic systematics of oceanic basalts: implications for mantle composition and processes. *In*: SAUNDERS, A.D. & NORRY, M.J. (eds) *Magmatism in the Ocean Basins*. Geological Society, London, Special Publications, **42**, 313–345.

TEJADA, M.L.G., MAHONEY, J.J., DUNCAN, R.A. & HAWKINS, M.P. 1996. Age and geochemistry of basement and alkalic rocks of Malaita and Santa Isabel, Solomon Islands, southern margin of Ontong Java Plateau. *Journal of Petrology*, **37**, 361–394.

TEJADA, M.L.G., MAHONEY, J.J., NEAL, C.R., DUNCAN, R.A. & PETTERSON, M.G. 2002. Basement geochemistry and geochronology of Central Malaita, Solomon Islands, with implications for the origin and evolution of the Ontong Java Plateau. *Journal of Petrology*, **43**, 449–484.

TEJADA, M.L.G., MAHONEY, J.J., CASTILLO, P.R., INGLE, S.P., SHETH, H.C. & WEIS, D. 2004. Pin-pricking the elephant: evidence on the origin of the Ontong Java Plateau from Pb–Sr–Hf–Nd isotopic characteristics of ODP Leg 192 basalts. *In*: FITTON, J.G., MAHONEY, J.J., WALLACE, P.J. & SAUNDERS, A.D. (eds) *Origin and Evolution of the Ontong Java Plateau*. Geological Society, London, Special Publications, **229**, 133–150.

THORDARSON, T. 2004. Accretionary-lapilli-bearing pyroclastic rocks at ODP Leg 192 Site 1184A: a record of subaerial phreatomagmatic eruptions on the Ontong Java Plateau. *In*: FITTON, J.G., MAHONEY, J.J., WALLACE, P.J. & SAUNDERS, A.D. (eds) *Origin and Evolution of the Ontong Java Plateau*. Geological Society, London, Special Publications, **229**, 275–306.

TODT, W., CLIFF, R.A., HANZER, A. & HOFMANN, A.W. 1984. $^{202}Pb+^{205}Pb$ double spike for lead isotopic analysis. *Terra Cognita*, **4**, 209.

VERMA, S.P. 1992. Seawater alteration effects on REE, K, Rb, Cs, Sr, U, Th, Pb, and Sr–Nd–Pb isotope systematics of mid-ocean ridge basalt. *Geochemical Journal*, **26**, 159–177.

WHITE, R.V. 1999. *The fundamentals of crust generation: major tonalite intrusions associated with an oceanic plateau, Aruba, Dutch Caribbean*. PhD Thesis, University of Leicester.

WHITE, R.V. 2002. Earth's biggest 'whodunnit': unravelling the clues in the case of the end-Permian mass extinction. *Philosophical Transactions of the Royal Society of London*, **360**, 2963–2985.

WIGNALL, P.B. 2001. Large igneous provinces and mass extinctions. *Earth-Science Reviews*, **53**, 1–33.

YAN, C.Y. & KROENKE, L.W. 1993. A plate tectonic reconstruction of the Southwest Pacific, 100–0 Ma. *In*: BERGER, W.H., KROENKE, L.W., MAYER, L.A. *et al.* (eds) *Proceedings of the Ocean Drilling Program, Scientific Results*, **130**, 697–709.

Phreatomagmatic eruptions on the Ontong Java Plateau: an Aptian $^{40}Ar/^{39}Ar$ age for volcaniclastic rocks at ODP Site 1184

LYNNE M. CHAMBERS[1,2], MALCOLM S. PRINGLE[3] &
J. GODFREY FITTON[1]

[1]*School of GeoSciences, University of Edinburgh Grant Institute, West Mains Road, Edinburgh EH9 3JW, UK*

[2]*Present address: NERC Isotope Geosciences Laboratory, Kingsley Dunham Centre, Keyworth, Nottingham NG12 5GG, UK*

[3]*SUERC, Rankine Avenue, Scottish Enterprise Technology Park, East Kilbride, Glasgow G75 0QF, UK*

Abstract: The discovery of a thick (337.7 m drilled) succession of volcaniclastic sediments at Ocean Drilling Program (ODP) Site 1184, on the eastern salient of the Ontong Java Plateau (OJP), shows that at least part of the plateau formed at or near sea level. All the other parts of the OJP that have been drilled or are exposed on land are composed of basaltic lava flows erupted in a deep-marine environment. The composition of the volcaniclastic rocks at Site 1184 is essentially indistinguishable from that of basaltic lava flows recovered from the other drill sites on the OJP, suggesting that the eastern salient formed during the same magmatic event (at approximately 120 Ma) that formed the main part of the plateau. A steep (–54°) magnetic inclination preserved in the volcaniclastic succession is consistent with an Early Cretaceous age, but rare nannofossils recovered from the volcaniclastic rocks suggest a much younger Eocene age. In order to resolve this paradox, we attempted to date the succession by the $^{40}Ar/^{39}Ar$ technique, even though the rocks are highly altered. Two samples of feldspathic material separated from basaltic clasts give *minimum* age estimates of *c*. 74 Ma. The basaltic clasts have compositions similar to that of their host rocks and are therefore probably cognate. A weighted average of the results of total fusion experiments on four or five small euhedral plagioclase crystals separated from the matrix of the volcaniclastic rocks at each of four levels in the lower part of the succession gave a value of 123.5 ± 1.8 Ma, and this probably represents a reasonable estimate of the age of eruption. Although not very precise, the $^{40}Ar/^{39}Ar$ results are the best that presently can be obtained from such altered rocks. They rule out an Eocene age for the volcaniclastic succession at Site 1184, and we suggest that the Eocene nannofossils were introduced through contamination, probably along fractures.

The Ontong Java Plateau consists of two geographically distinct parts: the main or high plateau, and the eastern salient. Basaltic lava flows forming the main plateau have been sampled at seven Deep Sea Drilling Project and Ocean Drilling Program (ODP) drill sites. $^{40}Ar/^{39}Ar$ analysis suggests that the main plateau formed mostly in a single magmatic episode at approximately 120 Ma (Mahoney *et al.* 1993; Tejada *et al.* 1996, 2002; Chambers *et al.* 2002). The NE lobe of the eastern salient was drilled at Site 1184 during ODP Leg 192, and penetrated 337.7 m into a succession of volcaniclastic rocks consisting of tuff and lapilli tuff. A ferro-manganese oxide crust separates the volcaniclastic sequence from the overlying Lower Miocene pelagic chalk at this site. The volcaniclastic rocks were divided into six subunits (IIA–IIF) defined by the occurrence of fossil wood and organic material, changes in grain size, and sorting and sedimentary structures (Mahoney *et al.* 2001; Thordarson 2004). A detailed volcanological study by Thordarson (2004) suggests that the volcaniclastic succession at Site 1184 was the result of large phreatomagmatic eruptions in an emergent or subaerial setting. By contrast, the basaltic flows sampled on the main plateau were all erupted under deep water (Roberge *et al.* 2004).

The volcaniclastic succession at Site 1184 is important to our understanding of the evolution of the OJP. This site is the only one to have been drilled on the eastern salient and is the only one to show evidence for extensive subaerial eruption. The NE lobe of the eastern salient forms a significant proportion (about 20%) of the total area of the OJP, so it is important that we establish whether it represents a part of the main (*c*. 120 Ma) plateau-forming magmatic event or

From: FITTON, J. G., MAHONEY, J. J., WALLACE, P. J. & SAUNDERS, A. D. (eds) 2004. *Origin and Evolution of the Ontong Java Plateau*. Geological Society, London, Special Publications, **229**, 325–331. 0305-8719/$15.00

is unrelated to it. The volcaniclastic rocks forming most of the succession at Site 1184 are chemically (Fitton & Godard 2004) and isotopically (White *et al.* 2004) indistinguishable from the basalt forming the main plateau, suggesting that their eruption was related to the main phase of OJP magmatism. Basaltic lava flows sampled on the main plateau consist mostly of low-K, moderately evolved tholeiite (Kwaimbaita-type) and high-Mg tholeiite (Kroenke-type), and both types are represented in the volcaniclastic succession at Site 1184 (Fitton & Godard 2004; White *et al.* 2004). The steep magnetic inclination (–54°) preserved in the volcaniclastic succession provides further evidence for an Early Cretaceous age (Riisager *et al.* 2004).

Despite the geochemical and palaeomagnetic evidence that the volcaniclastic rocks were erupted during the main (*c.* 120 Ma) plateau-forming event, rare nannofossils recovered from the volcaniclastic succession at Site 1184 suggest a middle Eocene (zone NP16) age (Bergen 2004). If confirmed, this would mean that at least part of the NE lobe of the eastern salient is unrelated to the main plateau, and such a result would pose the formidable problem of how to explain the generation of identical magma in two unrelated magmatic episodes 80 Ma apart. Furthermore, the eruption of relatively unfractionated, Kroenke-type magma on both the main plateau (Sites 1185 and 1187) and at Site 1184 (subunit IIA) requires that each episode would have involved large degrees (*c.* 30%; Fitton & Godard 2004) of mantle melting.

Pervasive alteration has affected the whole succession at Site 1184 (Mahoney *et al.* 2001), and the whole–rock composition of bulk samples shows high weight loss on ignition (LOI) of 3.8–9.7 wt% (Fitton & Godard 2004). Because of this, the volcaniclastic rocks would not normally be regarded as suitable for $^{40}Ar/^{39}Ar$ dating. However, in view of their importance to our understanding of the origin and evolution of the OJP, we attempted to date a suite of samples from Site 1184 in order to help resolve the paradox created by conflicting evidence for the age of eruption.

Sample selection

The volcaniclastic rocks at Site 1184 are composed mostly of ash- to lapilli-size glass shards, tachylite and crystal fragments, along with variable amounts of ash aggregates, armoured lapilli and lithic fragments, in a cement of zeolites, clay and calcite (Thordarson 2004). Coarse basaltic fragments, up to 65 mm in diameter, are present in trace to minor amounts throughout the succession. Very little volcanic material has survived unaltered in the upper 250 m of the volcaniclastic succession (above Core 38-R), but fresh glass is found below this level (Mahoney *et al.* 2001).

We consider that the most dateable materials from the volcaniclastic succession, given its highly altered state, are rare, isolated (i.e. not within clasts) plagioclase crystals and small, pebble-sized basaltic clasts. The euhedral morphology of the plagioclase crystals suggests that they are phenocrysts rather than xenocrysts, and therefore provide the best material for dating the succession. Eight or ten plagioclase crystals were separated from each of four samples from subunit IIE in the lower part of the volcaniclastic succession (Table 1).

Fourteen small pebble-sized basalt clasts, collected from throughout the succession, were analysed for major and trace elements by Shafer *et al.* (2004). Although highly altered, these clasts are similar to their volcaniclastic host rock in their incompatible trace-element contents and so are most likely to be cognate. Two of these 14 samples (both from subunit IIB) were large enough and sufficiently coarse grained to attempt $^{40}Ar/^{39}Ar$ dating. Both were composed of aphyric–plagioclase-microphyric basalt with a holocrystalline groundmass, although clays had replaced glass, augite and some plagioclase in places. The composition of the two clasts and the bulk-rock composition of their volcaniclastic host rocks is similar to that of the Kwaimbaita-type basalt found elsewhere on the plateau (Shafer *et al.* 2004).

Sample preparation

The individual plagioclase crystals were initially separated by crushing and sieving, and then concentrated using conventional magnetic separation techniques. The concentrates were then cleaned ultrasonically in warm 6 M HCl for 20–30 min and rinsed in deionized water ultrasonically for at least 15 min. Only the highest-quality plagioclase crystals (clear, no cracks or inclusions, generally 125–250 μm in diameter) were carefully picked by hand and loaded into 2 mm-diameter wells in a 22 mm-diameter pure-Al irradiation disk. Between eight and 10 plagioclase crystals were loaded into each well. Flux monitor standards (Fish Canyon Tuff sanidine) were also loaded into eight of the 20 wells spaced equally around each pan. The pans were stacked, loaded into an Al irradiation can and irradiated for 24 h in the Cd-lined RODEO facility of the 48 MW HIFAR Reactor in Petten, the Netherlands.

Table 1. *$^{40}Ar/^{39}Ar$ results for plagioclase crystals and basaltic clasts from ODP Site 1184*

Sample*	Material	Experiment type	Age spectrum analysis				Integrated or total-fusion	Isochron analysis				
			Age (Ma)	Error (1σ)	MSWD	Total ^{39}Ar (%)	age (Ma)	Age (Ma)	Error (1σ)	Intercept ± 2σ	MSWD	*N*
1184A-individual plagioclase	plagioclase	fusions	123.54	1.79	0.20	60.95	122.81					6 of 8
1184A-21R-6, 28–30 cm, clast	*feldspathic material*	*step-heating*	*73.76*	*0.77*	*216.71*	*81.87*	*75.18*	*71.85*	*1.11*	*304.9±47.3*	*162.63*	*15 of 24*
1184A-22R-1, 24–27 cm, clast	*feldspathic material*	*step-heating*	*74.06*	*0.52*	*141.69*	*97.91*	*74.19*	*71.73*	*0.79*	*371.3±56.7*	*86.00*	*13 of 19*

* Plagioclase crystals are from subunit IIE, clasts are from subunit IIB. 1184A individual plagioclase age is a weighted mean age of individual laser total fusions on single plagioclase crystals from: 1184- 40R-1, 124–127 cm; 1184- 41R-1, 65–69 cm; 1184- 42R-5, 53–58 cm; 1184- 45R-1, 51–55 cm.
Experiments shown in *italics* failed on one or more of the rigorous acceptance criteria for $^{40}Ar/^{39}Ar$ crystallization ages (Pringle 1993; Singer & Pringle 1996).
All of the ages presented here are calculated using an age of 28.02 Ma for the USGS Fish Canyon Tuff sanidine (Renne *et al.* 1998)

Samples of feldspathic material separated from the two selected clasts were prepared in a similar way to the individual plagioclase crystals. The clasts were crushed and sieved to a 125–250 μm-size fraction and the feldspathic components (mostly plagioclase) were concentrated using conventional magnetic separation techniques. The feldspathic material was then cleaned ultrasonically in warm 6 M HCl and deionized water, in accordance with our standard procedure for altered basalt, and dried. It was not possible to leach the samples more aggressively because of the very small amounts of material available. The fine- to medium-grained nature of the feldspathic concentrates and the limited amount of material precluded the rigorous hand picking used for the individual plagioclase crystals. Only the most obviously altered material was removed before the remaining sample was divided into aliquots, weighed into 5 mm-diameter Cu foil packets and then loaded into quartz vials. The quartz vials were irradiated at the Oregon State Radiation Center CLICIT reactor for 12 h. Flux monitor standards (Fish Canyon Tuff sanidine) were loaded at least every 12 mm within the vials.

$^{40}Ar/^{39}Ar$ techniques

Analysis of the samples from Site 1184 was carried out on a MAP 215 mass spectrometer using a CO_2 laser to heat the individual plagioclase crystals, and a furnace to heat the feldspathic material from the basaltic clasts. The plagioclase crystals were first degassed at 700–750°C ('first red'), with the released gas pumped away and not measured. The subsequent fusion step gas was released by fusing the individual plagioclase crystals for 30–60 s. The gas was then cleaned for 5 min using a stainless-steel cold finger cooled with an acetone–dry ice slush and two 2 SAES C50 St101 getters, one operated at 400°C and the other at room temperature.

The $^{40}Ar/^{39}Ar$ data presented here have been calculated using a value of 28.02 Ma for the Fish Canyon Tuff sanidine standard (Renne *et al.* 1998). Each experiment is only accepted as a valid crystallization age estimate if it fulfils the rigorous acceptance criteria of Pringle (1993) and Singer & Pringle (1996). The results of the experiments are given in Table 1, with those experiments that failed any of the acceptance criteria indicated by italics. Details of the complete $^{40}Ar/^{39}Ar$ analyses can be obtained from the first author.

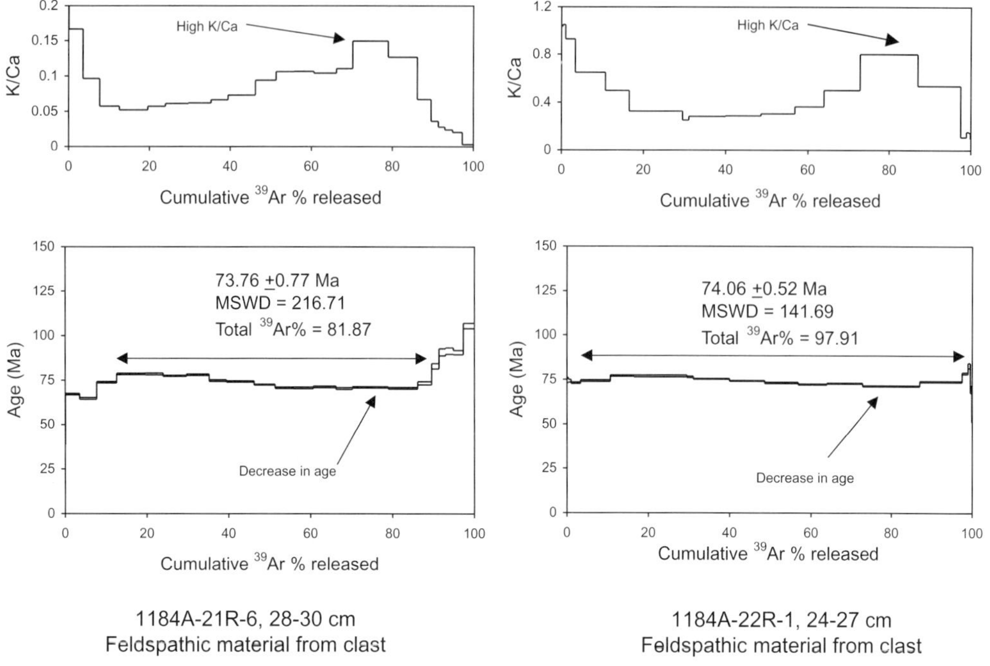

Fig. 1. K/Ca, age spectra and isochron plots for feldspathic material separated from two basalt clasts from Site 1184. The increase in K/Ca corresponds to a drop in age and results in a large MSWD for these samples. The steps included in the pseudo-plateau age estimates are marked by arrows.

Results

The age and K/Ca spectra of the two feldspathic separates from the basalt clasts found in Cores 21R-6, 28–30 cm and 22R-1, 24–27 cm are shown in Figure 1. The high total amount of K, indicated by high K/Ca, in these samples suggests addition of K through alteration. K/Ca in the least altered (lowest LOI) basaltic clasts from Site 1184 is 0.02–0.03 (Shafer *et al.* 2004), whereas the least altered basalt (LOI <0.5 wt%) from the main part of the OJP generally has K/Ca <0.01 (Fitton & Godard 2004). In addition, whole-rock LOI of the two clasts (Shafer *et al.* 2004) is high, at 3.96 (21R-6, 28–30 cm) and 6.87 wt% (22R-1, 24–27 cm). The small individual step errors shown in Figure 1 highlight the stepwise decreasing age spectrum that is typical of argon recoil. Recoil of the nucleus of ^{39}K, caused by the emission of a proton during the (n,p) reaction, results in the loss or redistribution of the ^{39}Ar during the irradiation process. Fine-grained samples, as in the case of the OJP samples or those in which the K-bearing phase is highly fractured, will lose a higher proportion of their ^{39}Ar by recoil than will coarser-grained samples.

Interestingly, the K/Ca plot (Fig. 1) shows an increase at high-temperature release that corresponds to a decrease in the age of the step. We attribute this increase in K/Ca and decrease in apparent age to the growth of low-temperature, authigenic K-feldspar with an age significantly less than the crystallization age of the basalt. The high K/Ca ratios at low temperatures corresponding to a decrease in age can be attributed to low-temperature alteration minerals. The best estimates of the *minimum* age for these samples are given by the pseudo-plateaus shown in Figure 1 of 73.8 ±0.8 and 74.1 ±0.5 Ma (1σ). The presence of recoil is reflected in the high MSWD (mean squared weighted deviations) values (>100) for these pseudo-plateaus, where MSWD is the test for concordance of the plateau points.

Plagioclase crystals separated from four intervals (40R-1, 124–127 cm; 41R-1, 65–69 cm; 42R-5, 53–58 cm; and 45R-1, 51–55 cm) in the volcaniclastic succession were analysed by total fusion with a CO_2 laser. The eight to ten crystals irradiated were divided into two and each set of four or five crystals was analysed as one total fusion. The plot shown in Figure 2 is therefore not a conventional age spectrum because each

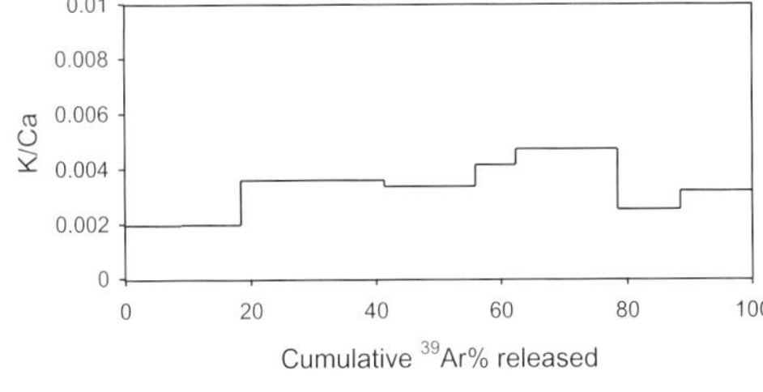

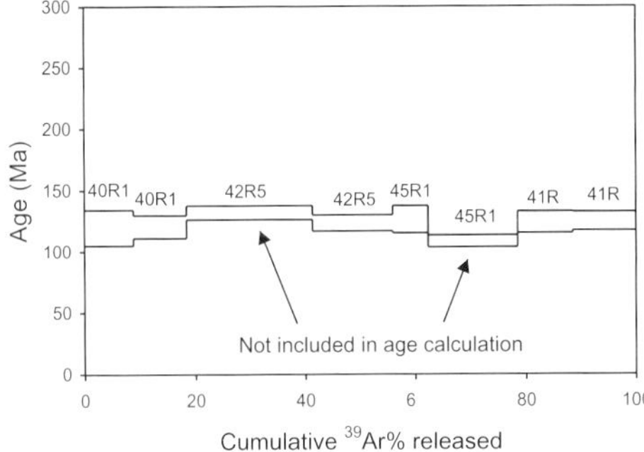

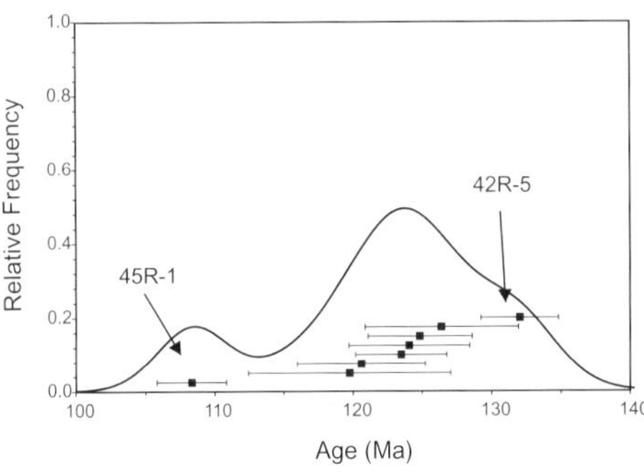

Fig. 2. A K/Ca plot, a non-conventional age 'spectrum', and an age–frequency plot for the plagioclase separates from Site 1184. Each 'step' in the 'age spectrum' represents an individual total fusion of plagioclase, and the width of each 'step' reflects the amount of ^{39}Ar released by the respective fusion, expressed as a percentage of the total released by all eight experiments. The results for the two samples not included in the age calculation are identified. The relative frequency diagram shows that the results of these two fusions are statistically different from the others.

'step' represents an individual fusion, with each labelled appropriately, and not a heating increment. The width of each 'step' reflects the amount of ^{39}Ar released by the respective fusion, expressed as a percentage of the total released by all eight experiments. The age quoted for this experiment (Table 1) of 123.5 ±1.8 Ma (1σ) is a weighted average of six out of eight total fusions and represents our preferred estimate of the age of the plagioclase in the volcaniclastic succession. Analyses of two plagioclase crystals, one from each of Cores 45R-1, 65–69 cm and 42R-5, 53–58 cm (yielding total fusion ages of approximately 132 and 108 Ma, respectively; Fig. 2), were excluded. Including these two analyses yields a weighted average age of 121.4 ± 6.8 Ma.

Discussion and conclusions

The two feldspathic separates from Cores 21R-6 and 22R-1 gave discordant ^{40}Ar/^{39}Ar plateaus with high MSWD values that can only be used as a *minimum* age estimate of 74 Ma. The ^{40}Ar/^{39}Ar results from the plagioclases gave a weighted mean total fusion age of 123.5 ±1.8 Ma, which we consider to be the most reliable estimate of the ^{40}Ar/^{39}Ar age of the volcaniclastic rocks recovered at Site 1184.

Our 123.5 Ma estimate for the Site 1184 volcaniclastic rocks is about 80 Ma older than the apparent biostratigraphic age of these rocks (Bergen 2004). Rare and poorly preserved nannofossils recovered from 45 samples (over the interval 9R-1–45R-5) from the volcaniclastic succession indicate a middle Eocene age (zone NP16 of Martini 1971; see Bergen 2004). Two Palaeocene and seven Late Cretaceous specimens were also found within the 45 samples. The sedimentary succession immediately above the volcaniclastic rocks is of lowermost Miocene age (zone NN2 of Martini 1971).

Sikora & Bergen (2004) argue that the biostratigraphic evidence is sound and points to an Eocene age for the volcaniclastic sequence. They propose that our weighted mean ^{40}Ar/^{39}Ar total fusion age estimate (123.5 ± 1.8 Ma) is based on plagioclase xenocrysts derived from older basaltic material derived from an Aptian basement during Eocene volcanism. However, all of the separates analysed, even the two not included in the weighted mean, gave rather similar Cretaceous ages. If all eight plagioclase separates contained xenocrysts, then this would imply that our separation sampled no plagioclase phenocrysts from the volcaniclastic host rock, either because it did not contain any or we missed them. Given their rather evolved composition, it seems unlikely that the host rocks contain no plagioclase phenocrysts. The euhedral morphology of the plagioclase crystals also tends to suggest that they are more likely to be phenocrysts than xenocrysts. Furthermore, the similarity between the composition of the basaltic clasts and that of their volcaniclastic hosts suggests that the clasts are cognate and not xenoliths from an older basement. The volcaniclastic rocks forming subunits B, D and E at Site 1184 are compositionally indistinguishable from Kwaimbaita-type basalt from the main plateau (Fitton & Godard 2004; White *et al.* 2004), and the clasts found in these subunits also have Kwaimbaita-type compositions (Shafer *et al.* 2004). By contrast, subunit C has a very distinctive composition, with enrichment in Nb, Ta and Th compared with other incompatible elements (Fitton & Godard 2004), and so do the clasts from this subunit (Shafer *et al.* 2004). Although the clasts yield only a minimum age estimate (74 Ma), this age is considerably older than Eocene.

We conclude that there is no $^{40}Ar/^{39}Ar$ evidence for an Eocene age. Although they are not very precise, we believe that our $^{40}Ar/^{39}Ar$ results are the best that can be achieved at present from these highly altered rocks. The result for the plagioclase crystals is consistent with the compositional and palaeomagnetic evidence that the volcanic succession was erupted during the main plateau forming phase of OJP magmatism at approximately 120 Ma. A detailed volcanological study by Thordarson (2004) shows that very little of the volcaniclastic succession has been reworked, ruling out the possibility that the large discrepancy between our $^{40}Ar/^{39}Ar$ age estimates and the biostratigraphic ages could be the result of Eocene reworking of a Cretaceous volcaniclastic succession. Also, redeposition of the succession in the Eocene would have destroyed an Early Cretaceous palaeomagnetic inclination. The discrepancy between biostratigraphic and $^{40}Ar/^{39}Ar$ ages is most probably the result of contamination of the samples with nannofossils introduced through fractures in the volcaniclastic succession. Thordarson (2004) found no trace of detrital or carbonate sediments within the succession and noted that the nannofossil-bearing intervals described by Bergen (2004) were cement-filled voids and veins.

This research used samples provided by the Ocean Drilling Program (ODP). ODP is sponsored by the US National Science Foundation (NSF) and participating countries under management of Joint Oceanographic Institutions (JOI), Inc. The work was carried out while L.M. Chambers was employed as a NERC-funded PDRA (grant NER/T/S/2000/00107), and $^{40}Ar/^{39}Ar$ analyses were approved by the NERC Isotope Geosciences Facility Steering Committee proposal IP/701/0301. J. Bergen, G. Iturrino and P. Sikora are thanked for helpful discussion, and P. Renne, A. Koppers, J. Mahoney, P. Wallace and an anonymous reviewer for their comments on the manuscript.

References

BERGEN, J. A. 2004. Calcareous nannofossils from ODP Leg 192, Ontong Java Plateau. *In*: FITTON, J.G., MAHONEY, J.J., WALLACE, P.J. & SAUNDERS, A.D. (eds) *Origin and Evolution of the Ontong Java Plateau*. Geological Society, London, Special Publications, **229**, 113–132.

CHAMBERS, L.M., PRINGLE, M.S. & FITTON, J.G. 2002. Age and duration of magmatism on the Ontong Java Plateau: $^{40}Ar–^{39}Ar$ results from ODP Leg 192. Abstract V71B-1271, *Eos, Transactions of the American Geophysical Union*, **83**, F47.

FITTON, J.G. & GODARD, M. Origin and evolution of magmas on the Ontong Java Plateau. *In*: FITTON, J.G., MAHONEY, J.J., WALLACE, P.J. & SAUNDERS, A.D. (eds) *Origin and Evolution of the Ontong Java Plateau*. Geological Society, London, Special Publications, **229**, 151–178.

MAHONEY, J.J., FITTON, J.G. WALLACE, P.J. *et al.* 2001. *Proceedings of the Ocean Drilling Program, Initial Reports*, **192**.

MAHONEY, J.J., STOREY, M., DUNCAN, R.A., SPENCER, K.J. & PRINGLE, M. 1993. Geochemistry and geochronology of the Ontong Java Plateau. *In*: PRINGLE, M., SAGER, W., SLITER, W. & STEIN, S. (eds) *The Mesozoic Pacific. Geology, Tectonics, and Volcanism*. Geophysical Monograph, American Geophysical Union, **77**, 233–261.

MARTINI, E. 1971. Standard Tertiary and Quaternary calcareous nannoplankon zonation. *In*: FARINACCI, A. (ed.) *Proceedings of the Second Planktonic Conference, Roma, 1970*, Vol. 2. Edizioni Tecnoscienza, Rome, 739–785.

PRINGLE, M.S. 1993. Age progressive volcanism in the Musicians seamounts: a test of the hot spot hypothesis for the late Cretaceous Pacific. *In*: PRINGLE, M.S., SAGER, W.W., SLITER, W.V. & STEIN, S. (eds) *The Mesozoic Pacific. Geology, Tectonics, and Volcanism*. American Geophysical Union, Geophysical Monograph, **77**, 187–215.

RENNE, P.R., SWISHER, C.C., DEINO, A.L., KARNER, D.B., OWENS, T.L. & DEPAOLO, D.J. 1998. Intercalibration of standards, absolute ages and uncertainties in $^{40}Ar–^{39}Ar$ dating. *Chemical Geology*, **145**, 117–152.

RIISAGER, P., HALL, S. ANTRETTER, M. & ZHAO, X. 2004. Early Cretaceous Pacific palaeomagnetic pole from Ontong Java Plateau basement rocks. *In*: FITTON, J.G., MAHONEY, J.J., WALLACE, P.J. & SAUNDERS, A.D. (eds) *Origin and Evolution of the Ontong Java Plateau*. Geological Society, London, Special Publications, **229**, 31–44.

ROBERGE, J., WHITE, R.V. & WALLACE, P.J. 2004. Volatiles in submarine basaltic glasses from the Ontong Java Plateau (ODP Leg 192): implications for magmatic processes and source region compositions. *In*: FITTON, J.G., MAHONEY, J.J., WALLACE, P.J. & SAUNDERS, A.D. (eds) *Origin and Evolution of the Ontong Java Plateau*. Geological Society, London, Special Publications, **229**, 239–257.

SHAFER, J.T., NEAL, C.R. & CASTILLO, P.R. 2004. Compositional variability in lavas from the Ontong Java Plateau: results from basalt clasts within the volcaniclastic succession at Ocean Drilling Program Site 1184. *In*: FITTON, J.G., MAHONEY, J.J., WALLACE, P.J. & SAUNDERS, A.D. (eds) *Origin and Evolution of the Ontong Java Plateau*. Geological Society, London, Special Publications, **229**, 333–357.

SIKORA, P.J. & BERGEN, J.A. 2004. Lower Cretaceous planktonic foraminiferal and nannofossil biostratigraphy of Ontong Java Plateau sites from DSDP Leg 30 and ODP Leg 192. *In*: FITTON, J.G., MAHONEY, J.J., WALLACE, P.J. & SAUNDERS, A.D. (eds) *Origin and Evolution of the Ontong Java Plateau*. Geological Society, London, Special Publications, **229**, 83–111.

SINGER, B.S. & PRINGLE, M.S. 1996. Age and duration of the Matuyama–Brunhes geomagnetic polarity reversal from $^{40}Ar/^{39}Ar$ incremental heating analysis of lavas. *Earth and Planetary Science Letters*, **139**, 47–61.

TEJADA, M.L.G., MAHONEY, J.J., DUNCAN, R.A. & HAWKINS, M.P. 1996. Age and geochemistry of basement and alkalic rocks of Malaita and Santa Isabel, Solomon Islands, southern margin of Ontong Java Plateau. *Journal of Petrology*, **37**, 361–394.

TEJADA, M.L.G., MAHONEY, J.J., NEAL, C.R., DUNCAN, R.A. & PETTERSON, M.G. 2002. Basement geochemistry and geochronology of Central Malaita, Solomon Islands, with implications for the origin and evolution of the Ontong Java Plateau. *Journal of Petrology*, **43**, 449–484.

THORDARSON, T. 2004. Accretionary-lapilli-bearing pyroclastic rocks at ODP Leg 192 Site 1184: a record of subaerial phreatomagmatic eruptions on the Ontong Java Plateau. *In*: FITTON, J.G., MAHONEY, J.J., WALLACE, P.J. & SAUNDERS, A.D. (eds) *Origin and Evolution of the Ontong Java Plateau*. Geological Society, London, Special Publications, **229**, 275–306.

WHITE, R.V., CASTILLO, P.R., NEAL, C.R., FITTON, J.G. & GODARD, M. 2004. Phreatomagmatic eruptions on the Ontong Java Plateau: chemical and isotopic relationship to Ontong Java Plateau basalts. *In*: FITTON, J.G., MAHONEY, J.J., WALLACE, P.J. & SAUNDERS, A.D. (eds) *Origin and Evolution of the Ontong Java Plateau*. Geological Society, London, Special Publications, **229**, 307–323.

Compositional variability in lavas from the Ontong Java Plateau: results from basalt clasts within the volcaniclastic succession at Ocean Drilling Program Site 1184

JOHN T. SHAFER[1], CLIVE R. NEAL[1] & PATERNO R. CASTILLO[2]

[1]*Department of Civil Engineering and Geological Sciences, University of Notre Dame, 156 Fitzpatrick Hall, Notre Dame, IN 46556, USA (e-mail: jshafer@nd.edu)*

[2]*Geosciences Research Division, Scripps Institution of Oceanography, University of California, San Diego, La Jolla, CA 92093-0212, USA*

Abstract: Tholeiitic basalts have been recovered from drill sites in different locations on the Ontong Java Plateau (OJP) and are remarkably homogeneous across this large igneous province. The most abundant basalt type is represented by the Kwaimbaita Formation on Malaita in the Solomon Islands, where it is capped by the isotopically distinct and slightly more incompatible-element-enriched basalt of the Singgalo Formation. Ocean Drilling Program (ODP) Leg 192 drilled five sites on the OJP, four of which penetrated basement lava successions. All basalt recovered during Leg 192 is chemically and isotopically indistinguishable from Kwaimbaita-type lavas.

Site 1184 of ODP Leg 192 is situated on the eastern salient of the OJP, and is unique because the recovered volcaniclastic succession contains the first conclusive evidence for emergence of part of the OJP above sea level. Within this succession are clasts of basaltic material. We report the major element-, trace-element and isotopic compositions of 14 moderately to highly altered basalt clasts. On the basis of incompatible-element concentrations, specifically high field strength elements (HFSE) and rare earth elements (REE), four groups of clasts are defined. Group 1 clasts are similar to basalt from the Kwaimbaita Formation. Group 2 clasts show variable composition, but the heavy rare earth element (HREE) concentrations are similar to those of basalts from the Kwaimbaita Formation. Group 3 clasts have compositions similar to the high-MgO Kroenke-type basalt recovered during ODP Leg 192. Group 4 clasts are more evolved than the Kwaimbaita or Singgalo lavas, and contain deep negative Eu and Sr anomalies on primitive-mantle (PM)-normalized diagrams, as well as high concentrations of Nb, Ta and Th. Group 4 clasts also show a large fractionation of Nb from La and have $(Nb/La)_{PM}$ ratios of approximately 2. Sr-, Nd- and Pb-isotope ratios were measured on five clasts covering all four groups. Although the Sr- and Pb-isotope ratios exhibit some variability, which we attribute to alteration, the Nd-isotope ratios are within the field defined for Kwaimbaita-type lavas.

We conclude that most of the compositional variability displayed by these clasts is a result of alteration and that Ta appears to be the most immobile incompatible trace element. All of the clasts were derived from the mantle source that produced the Kwaimbaita-type and Kroenke-type basalts. Our data emphasize the widespread nature of Kwaimbaita-type basalt and show that the source region was active under both the eastern salient and the high plateau of the OJP.

The Ontong Java Plateau (OJP) is the largest of the Earth's large igneous provinces (LIPs) and covers an area of approximately 2.0×10^6 km^2. The OJP was emplaced rapidly, primarily around 122 Ma, with a possible second plateau-building event at about 90 Ma (e.g. Mahoney *et al.* 1993; Tejada *et al.* 1996, 2002). The *c.* 122 Ma event produced two geochemically similar, but isotopically distinct, lava types that were originally recognized in Unit A and Units C–G, respectively, from the Ocean Drilling Program (ODP) Leg 130 Site 807 (Mahoney *et al.* 1993). Thicker basalt sequences with similar compositions to the Unit A and Units C–G groups were described by Tejada *et al.* (2002) from subaerial outcrops on Malaita, Solomon Islands. These sequences were called the Singgalo Formation (compositionally equivalent to Unit A) and the Kwaimbaita Formation (compositionally equivalent to Units C–G). The presence of these two groups on both the northern and southern margins of the OJP, some 1200 km apart, is a testament to the size of the magmatic event(s) that produced the OJP. Singgalo-type basalt is slightly more enriched in incompatible elements than is Kwaimbaita-type basalt, and it is also

From: FITTON, J. G., MAHONEY, J. J., WALLACE, P. J. & SAUNDERS, A. D. (eds) 2004. *Origin and Evolution of the Ontong Java Plateau*. Geological Society, London, Special Publications, **229**, 333–351. 0305-8719/$15.00

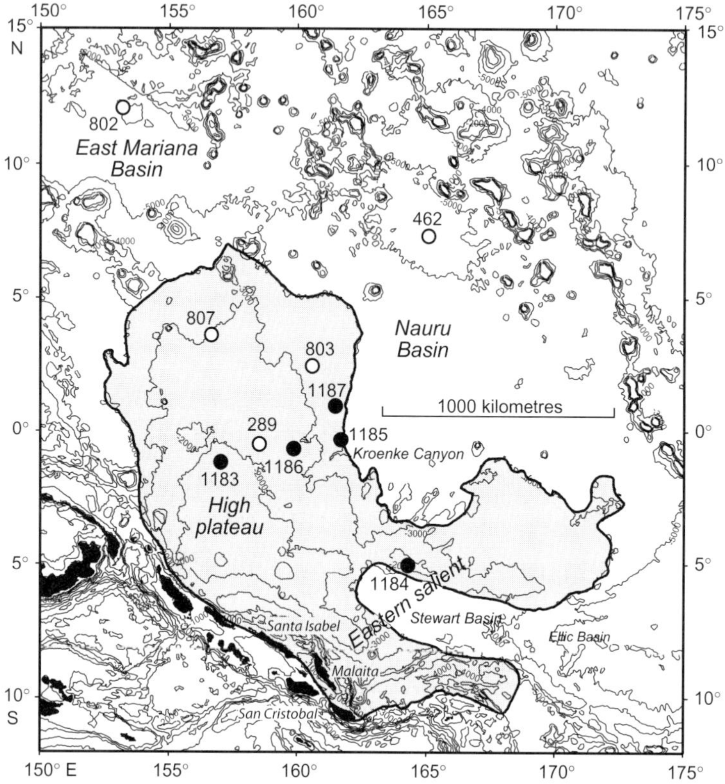

Fig. 1. Bathymetric map of the Ontong Java Plateau and surrounding area showing the locations of basement drill sites. Modified from Mahoney *et al.* (2001).

isotopically distinct. It has lower initial ε_{Nd} (relative to chondritic uniform reservoir, Chur) (+3.8–+5.4 v. +5.4–+6.5) and present-day $^{206}Pb/^{204}Pb$ (18.245–18.521 v. 18.626–18.708), and slightly higher initial $^{87}Sr/^{86}Sr$ (0.7040–0.7042 v. 0.7034–0.7039) than Kwaimbaita-type basalt (e.g. Mahoney 1987; Mahoney & Spencer 1991; Mahoney *et al.* 1993; Tejada *et al.* 1996, 2004).

Site 1184 of ODP Leg 192 is located on the unnamed northern ridge of the eastern salient of the OJP (Fig. 1). Until Leg 192, the relationship of the eastern salient to the main or high plateau was unknown (Mahoney 1987; Richards *et al.* 1989). Kroenke & Mahoney (1996) and Tejada *et al.* (1996) speculated that the eastern salient might specifically be the locus of the *c.* 90 Ma eruptions, whereas Mahoney *et al.* (2001) suggested that it could be either part of the main plateau or a product of the plume tail after main plateau emplacement. At about 90 Ma, the eastern salient may have been rifted into northern and southern lobes during more than 300 km of N–S extensional tectonism and rifting associated with the opening of the Ellice and Stewart basins (Fig. 1) (Neal *et al.* 1997). This poorly understood rifting event may also have contributed to the magmatic development of the volcaniclastic succession at Site 1184. Biostratigraphic information indicates a middle Eocene age (42–45 Ma) (Mahoney *et al.* 2001; Bergen 2004; Sikora & Bergen 2004); however, recent $^{40}Ar/^{39}Ar$ dating of Site 1184 plagioclase crystals has produced an estimated eruption age of 123.5 ± 1.8 Ma (Chambers *et al.* 2004).

One of the most significant results from ODP Leg 192 was that Site 1184 provides the first clear evidence of emergence of any part of the OJP above sea level (Mahoney *et al.* 2001; Thordarson 2004). Wood fragments were found at the boundaries between four of the six subunits within the volcaniclastic succession at Site 1184. The presence of oxidized horizons in subunit IIC probably indicates subaerial exposure (see Mahoney *et al.* 2001; Thordarson 2004).

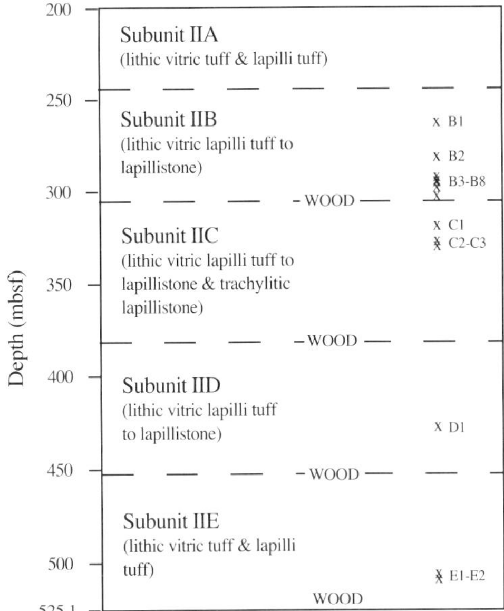

Fig. 2. Stratigraphic section of core recovered from Hole 1184A showing location of samples and of wood-bearing horizons. Modified from Mahoney *et al.* (2001) and White *et al.* (2004).

Samples and experimental methods

Samples

The drilling at Site 1184 recovered an upper interval of calcareous ooze (Unit I) and more than 330 m of volcaniclastic rocks (Unit II) (Fig. 2). Unit II is divided into six subunits (A, B, C, D, E and F) on the basis of changes in grain size, lithology and sedimentary structures (Thordarson 2004). Common lithologies include tuff, lapilli tuff and lapillistone, all of which contain occasional lithic clasts. The lithic clasts are primarily basalt, diabase or re-worked volcaniclastic material. The near absence of large lapilli suggests that the volcaniclastic material was deposited some distance from the volcanic centre (Mahoney *et al.* 2001). For the most part, the rocks recovered from Site 1184 are moderately to highly altered, although some unaltered glass is present below Core 38R. Small oxidized red lapilli are present, especially in Subunit IIC, and, as noted above, wood fragments are present at the boundaries between subunits IIB–C, IIC–D, IID–E, and IIE–F. A thinly bedded, fine-grained tuff in subunit IIC appears to be a primary ash-fall deposit (Mahoney *et al.* 2001). These features indicate emergence of the OJP in this region.

A total of 14 basaltic clasts extracted from subunits IIB, IIC, IID and IIE were selected for this study (Fig. 2). In hand sample the clasts were non-vesicular to sparsely vesicular and aphyric to sparsely plagioclase-phyric basalts. The clasts were generally small (1–4 cm) and were a greyish colour. We refer to these as Clasts B1–E2, with the letter referring to the subunit and the number referring to the position within the subunit. For example, Clast B2 is higher in the core than clast B7 (see Tables 1 and 2 for core intervals).

Analytical methods

The basalt clasts were cut from the volcaniclastic matrix using a water-cooled rock saw, then ground on a diamond grinding wheel in order to remove any excess adhering matrix and to remove any contamination from the rock saw. The biggest clasts were broken into <0.5 cm chips that were crushed further in an alumina SPEX jaw crusher. The jaw crusher was thoroughly cleaned with ultrapure water and a clean nylon scouring pad between the processing of different clasts. The chips were washed in 18 MΩ water for 30 min in an ultrasonic bath. They were then air dried and examined using a binocular microscope. Although all samples displayed varying degrees of alteration, chips exhibiting excessive alteration were removed. The chips were powdered using a SPEX Mixer/Mill in alumina shaker bombs or by hand in an agate mortar and pestle (depending upon the amount of sample available).

Major-element concentrations were determined via inductively coupled plasma-optical emission spectroscopy (ICP-OES) at the University of Notre Dame. Approximately 0.1 g of sample powder was mixed with 0.5 g of lithium metaborate and fused at 1025°C for 30 min The molten pellets were quenched in 5% HNO_3, which was transferred to polypropylene bottles and made up to 100 g with 5% HNO_3. This solution was placed in an ultrasonic bath for 1 h in order to facilitate complete dissolution of the fused glass. The geochemical reference materials analysed for major elements during this study were BHVO-2, BIR-1 and BPL-1. Drift was monitored by running a standard solution every four analyses. Calibration was performed by taking blank- and drift-corrected counts of a selected reference material and dividing by the accepted elemental concentrations. This allowed for a simple data reduction algorithm similar to the procedure used for shipboard ICP-OES reduction during Leg 192 (Mahoney *et al.* 2001). Between each sample solution, 5% nitric acid was used to remove any memory effects by flushing any remaining solution from the tubing and glassware. Blank levels were generally very low (e.g., the average raw counts per second (cps) of Ti (λ = 334.94) for blanks were 890 v. 3 768 000 cps for BHVO-2) indicating that the 5% nitric acid flush was efficient in minimizing memory effects. Standard deviations for replicate reference materials were generally less than 6%, although for K_2O and Na_2O the standard deviation could be as high as 25% due to a combination of low abundance, poor sensitivity and high background. Loss on ignition (LOI) was determined by heating between 0.1 and 0.5 g of sample powder in platinum crucibles at 1025°C for 4 h. The samples were weighed before and after removal from the oven, and total weight per cent LOI was determined.

Trace-element abundances were quantified via ICP-mass spectrometry (ICP-MS) at the University of Notre Dame (see Neal 2001 for full analytical details). If it appeared that immobile element ratios (i.e. Zr/Hf, Zr/Nb, Nb/Ta, Nb/La and Zr/Sm) deviated from typical OJP basalt ranges (Tejada *et al.* 1996, 2002; Neal *et al.* 1997), an aliquot of the sample powder was fused using sodium peroxide as a flux to ensure complete dissolution of any refractory phases (Longerich *et al.* 1990). BHVO-1 was used as the reference material for the trace-element analyses. Blank levels were below 100 cps, and standard deviations (SDs) for replicated standard reference materials were generally less than 3% (V and Cr >10%, Ni and Zn between 5 and 10%, all others below 3%).

Strontium-, Nd- and Pb-isotopic measurements were made on five basaltic clasts using a Micromass Sector 54 multi-collector thermal ionization mass spectrometer at the Scripps Institution of Oceanography (SIO). In order to minimize the effects of sea-water alteration on the Sr and Nd isotopic composition, the sample powders were subjected to a harsh multi-step leaching procedure similar to that described by Mahoney (1987) and Castillo *et al.* (1991). Unleached powders were used for Pb-isotope analysis because almost all the leached powders yielded very low concentrations of Pb, indicating that most of the Pb was in leachable phases. Both the leached and unleached powders were dissolved in clean Teflon vessels using *c.* 1 ml of a 2:1 mixture of concentrated HF and HNO_3, and heated on a hotplate at low power for 16–24 h. The resulting solutions were evaporated to dryness, redissolved in a small amount of concentrated HNO_3, evaporated to dryness and the procedure repeated. Strontium and rare earth elements (REE) were first separated in primary cation-exchange columns; Nd was separated from the rest of the REE by passing the REE aliquots through small EDTA ion-exchange columns. Lead was separated using a standard anion-exchange method (e.g. Lugmair & Galer 1992; Janney & Castillo 1996) in a $HBr–HNO_3$ medium. Concentrations of Rb, Sr, Sm, Nd, U, Th and Pb were measured on separate dissolutions using a high-resolution ICP-MS at SIO using a procedure similar to that described by Janney & Castillo (1996).

Results

Petrography

The generally small size of the clasts (sometimes <1 cm after grinding) meant that thin sections could only be made for nine of the 14 samples, but at least one thin section was made of clasts from each Site 1184 subunit. Plagioclase and clinopyroxene are the most common groundmass phases in these samples and vary from extremely fine-grained crystals (<0.05 mm) to larger crystals (up to about 0.3 mm) with a sub-ophitic–intergranular texture. In one of the most altered samples (Clast C1, LOI = 10.38 wt%), secondary zeolite is extremely abundant (identified as natrolite based on this sample having Na_2O = 5.73 wt%). A trachytic texture is still evident (Fig. 3a). Even in samples with relatively low LOI (e.g. Clast B6, LOI = 4.88 wt%), clinopyroxene is generally altered and replaced by clays (Fig. 3b). In samples with larger grain sizes (e.g. Clast E1), clinopyroxene is generally partially altered and replaced by brown clays, but the plagioclase crystals are generally unaltered (Fig. 3c). In Clasts C1, B3, B8, E1 and E2, zeolite minerals are present as vesicle fill and veins in the surrounding matrix.

Major elements

Major-element concentrations and LOI of the 14 basalt clasts analysed by ICP-OES are presented in Table 1. These data have been recalculated on a volatile-free basis. The uncorrected major element totals vary from 97.0 to 106.8 wt%. The large range of major-element totals is controlled by two samples: the consistent low totals of sample C2 (replicated multiple times on different fusions) and the high total of sample D1, for which sample powder was extremely limited and thereby precluded multiple fusions. We suggest that these data be treated with caution. The altered nature of all Site 1184 clasts samples is highlighted by the fact that the majority of samples have at least 5 wt% LOI (the total range is 1.6–11 wt%). Concentrations of CaO,

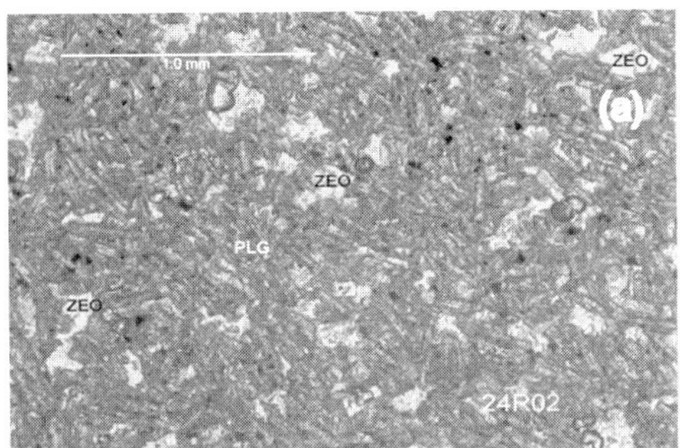

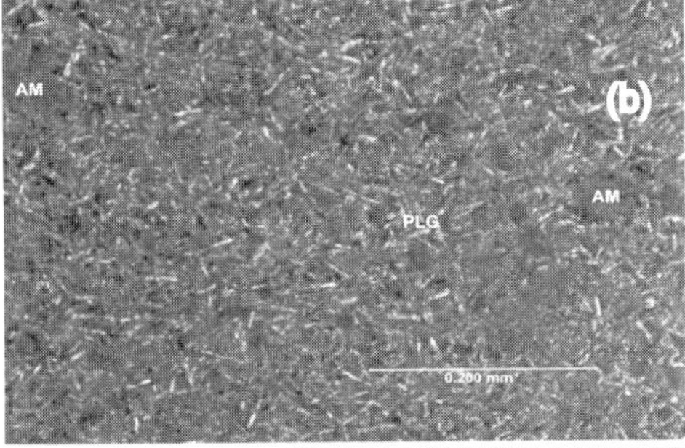

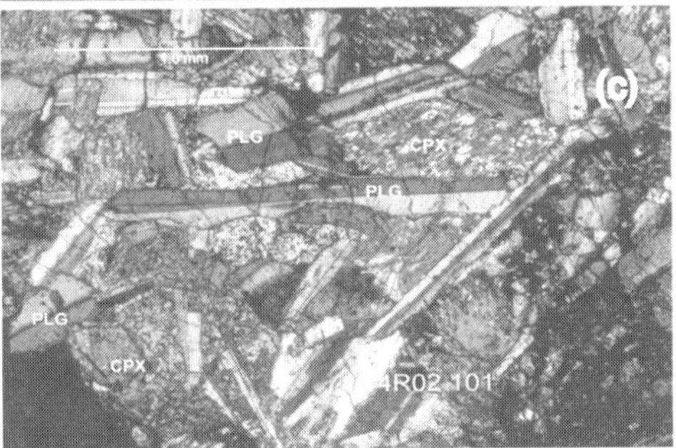

Fig. 3. Photomicrographs of basalt clasts. (**a**) Group 4 clast C1 (sample 24R-2, 68–70 cm) in plane-polarized light showing clear zeolite (probably natrolite) and small plagioclase laths. (**b**) Group 1 clast B6 (sample 21R-6, 73–77 cm) in plane-polarized light showing clinopyroxene altered to a brown mineral; small plagioclase laths define a subtrachytic texture. (**c**) Group 1 clast E1 (sample 44R-2, 101–105 cm) under crossed polars. This sample has one of the lowest LOI values in the Site 1184 suite, yet clinopyroxene still shows signs of alteration. Mineral abbreviations are as follows: ZEO, zeolite; PLG, plagioclase; CPX, clinopyroxene; and AM, alteration mineral.

SiO_2 and MgO are negatively correlated with LOI (CaO shows the best correlation as seen in Fig. 4a), and K_2O exhibits a positive correlation (Fig. 4b; Na_2O displays a poor positive correlation with LOI). In terms of the SiO_2 v. total alkalis classification diagram (Fig. 5a), these samples are transitional between alkalic and tholeiitic. However, as indicated in Figure 4b, alteration has variably increased Na_2O and K_2O levels. Fresh glass clasts analysed by White *et al.* (2004) plot as low-K tholeiites. When elements considered to be immobile under low-temperature alteration conditions are used to discriminate between alkali and tholeiitic basalts, the basalt clasts plot strictly in the tholeiitic field (Fig. 5b).

Trace elements

Considerable variability in trace-element abundances (Table 2, re-calculated to a volatile-free basis) and in element/element ratios, especially for ratios of normally immobile elements (e.g. Nb/La), is displayed by the Site 1184 basalt clasts. We have subdivided our basalt clasts into four groups on the basis of incompatible-element profiles (Fig. 6a–d). Group 1 comprises four clasts (Clasts B1, B6, E1 and E2; Fig. 6a) that have primitive-mantle-normalized profiles similar to those of the Kwaimbaita Formation lavas. These clasts also possess among the lowest LOI values of all clasts analysed in this study (1.68–4.88 wt%; Table 1). Group 2 clast compositions (Clasts B2, B3, B4, B8 and D1; Fig. 6b) show greater variability, although the heavy REE (HREE) have similar normalized abundances to those of lavas from the Kwaimbaita Formation. These clasts exhibit higher LOI values (4.01–7.85 wt%). Clast D1 exhibits a deep Zr–Hf trough and light REE (LREE) enrichment (Fig. 6b). This could reflect the incomplete dissolution of a refractory oxide or other phase, possibly zircon, during sample preparation; however, the HREE are not depleted relative to the Kwaimbaita Formation lava compositions (Fig. 6b). The extremely limited amount of basalt clast from this sample precluded multiple dissolutions (i.e. flux fusion), so the veracity of this anomaly cannot be assessed and, thus, must be interpreted with caution. However, in other aspects the sample is similar to the Kwaimbaita Formation basalts (i.e. HREE, Nb and Ta abundances). Group 3 clasts (Clasts B5 and B7; Fig. 6c) have normalized profiles similar to those of Kroenke-type basalts (Sites 1185 and 1187; Fitton & Godard 2004), although in Clast B5 there are depletions at Nb–Ta and Zr–Hf relative to the REE. These profiles have been duplicated by separate dissolution, including the sodium peroxide fusion method (Longerich *et al.* 1990). These two clasts also contain the highest Ni abundances of the 14 clasts analysed in this study (222–296 ppm; the range of Ni abundance reported from the Kroenke-type basalts is 170–230 ppm – Fitton & Godard 2004). Clast B5 also has TiO_2 abundances comparable to those

Table 1. *Major element concentrations* (wt%), loss on ignition, and core intervals for ODP Site 1184 basaltic clasts*

Core, Interval (cm)	Clast	Group	Depth (mbsf)	SiO_2	TiO_2	Al_2O_3	$Fe_2O_3^T$	MnO	MgO	CaO	Na_2O	K_2O	P_2O_5	Total†	LOI
18R-2, 102–104	B1	1	260.96	49.60	1.24	12.21	13.08	0.26	8.94	9.73	4.44	0.34	0.17	99.4	4.25
20R-4, 50–52	B2	2	282.89	51.62	1.01	14.45	7.81	0.14	7.09	13.29	3.90	0.61	0.08	99.2	6.22
21R-5, 40–47	B3	2	293.08	51.15	1.40	15.94	10.91	0.04	4.40	4.60	3.16	8.25	0.16	100.4	7.32
21R-5, 55–60	B4	2	293.23	47.19	1.26	18.59	18.69	0.10	10.62	0.85	1.75	0.81	0.14	99.2	7.85
21R-6, 28–30	B5	3	294.46	53.45	0.73	17.00	7.74	0.11	5.47	12.64	2.46	0.33	0.08	98.6	3.96
21R-6, 73–77	B6	1	294.91	49.89	1.02	14.58	11.39	0.30	8.24	9.93	4.06	0.56	0.03	100.0	4.88
22R-1, 24–27	B7	3	297.54	50.81	1.02	15.34	10.28	0.28	8.11	8.63	4.46	1.05	0.03	103.1	6.87
22R-4, 120–123	B8	2	301.77	52.41	1.01	13.83	7.16	0.16	7.75	13.36	3.37	0.75	0.21	97.3	4.01
24R-2, 68–70	C1	4	318.03	48.52	1.05	17.14	12.70	0.28	7.07	4.83	5.73	2.00	0.67	103.8	10.38
25R-5, 16–18	C2	4	330.85	49.93	2.12	13.00	18.02	0.39	6.44	5.26	3.38	1.10	0.37	97.0	7.22
25R-6, 115–117	C3	4	333.10	50.07	1.97	15.23	14.99	0.39	5.11	6.49	4.53	0.91	0.33	99.1	11.83
35R-8, 32–35	D1	2	432.30	49.49	1.18	12.83	12.99	0.18	7.60	9.88	5.22	0.56	0.07	106.8	4.87
44R-2, 101–105	E1	1	511.68	49.27	0.95	13.37	11.13	0.21	7.90	13.93	2.95	0.26	0.04	100.0	3.17
44R-2, 126–132	E2	1	512.43	48.86	0.97	13.97	11.51	0.19	7.89	13.31	2.99	0.27	0.03	100.3	1.68
‡BHVO-2 average				50.0	2.71	13.7	12.3	0.18	7.24	11.4	2.21	0.52	0.29		
Recommended values				49.9	2.73	13.5	12.3	0.13	7.23	11.4	2.22	0.52	0.27		
S.D (1σ) (n = 7)				1.5	0.07	0.5	0.2	0.01	0.09	0.3	0.12	0.07	0.04		
BIR-1 average (n = 2)				50.2	1.02	16.2	12.1	0.22	9.89	14.3	2.34	bd	0.07		
Recommended values				47.96	0.96	15.5	11.3	0.18	9.70	13.3	1.82	0.03	0.021		
BPL-1 average (n = 2)				50.8	2.49	16.8	14.1	0.24	8.43	10.6	2.85	0.30	0.72		
Recommended values				46.1	2.30	14.8	13.6	0.18	8.12	9.7	2.31	1.11	0.49		

* all major- and trace-element data presented have been normalized to 100% on a volatile-free basis.
mbsf, metres below seafloor; LOI, loss on ignition; bd, below detection limit.
Total† is the un-normalized major-element total of each clast + LOI.
‡ BHVO-2 recommended values from USGS preliminary certificate of analysis. BPL-1 recommended values from Scott Hughes, Idaho State University, pers. comm. BIR-1 values from Flanagan (1984) and Gladney & Roelandts (1988).

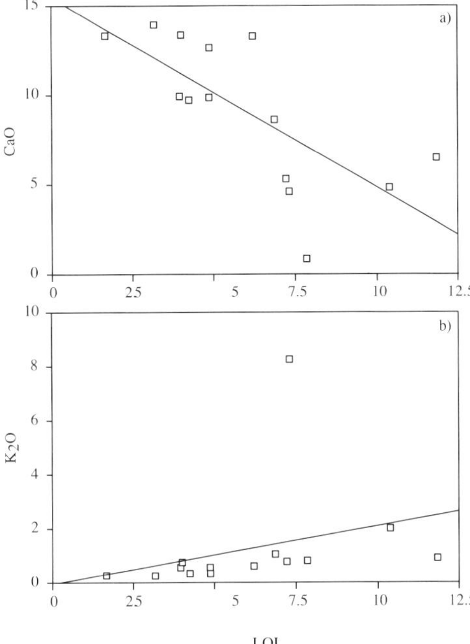

Fig. 4. (**a**) Loss on ignition (LOI) v. CaO (wt%) and (**b**) LOI v. K_2O (wt%) indicating that major elements can be mobilized and either lost (CaO) or gained (K_2O) during sea-water alteration of basalts.

of Kroenke-type basalts (TiO_2 = 0.73 wt%, v. 0.733 wt% in average Kroenke-type basalt), but has lower overall Zr and Nb contents (24.2 and 1.36 ppm, respectively, v. 40.6 and 2.18 ppm in average Kroenke-type basalt). Clast B7 contains higher Zr (56.8 ppm), Nb (2.95 ppm) and TiO_2 (1.02 wt%) abundances, and appears to be intermediate between the Kroenke-type and Kwaimbaita-type compositions. Group 4 clast compositions (Clasts C1, C2 and C3) have more enriched incompatible-element concentrations (specifically, Nb, Ta, Th and the HREE) than average Wairahito basalts (the most evolved rock type so far found on the OJP) described by Birkhold (2000) from the island of Makira (also known as San Cristobal) (Fig. 6d). Clasts C1 and C3 exhibit negative Sr and Eu anomalies that are consistent with plagioclase fractionation. In addition, the trace-element composition of the volcaniclastic matrix within which the three Group 4 clasts were contained also shows increased levels of Nb, Ta and Th over the REE (Fitton & Godard 2004). However, the Nb, Ta, Th or REE abundances of the matrix are not as enriched as in the Group 4 clasts.

Clasts B2, B5, B8, E1 and E2 all exhibit a curious positive Y anomaly of varying magnitude, which has been replicated twice using separate dissolutions, including sodium peroxide fusion. Such an anomaly is not seen in lavas typical of the Kwaimbaita-type lavas.

Fractionation of the LREE from high field strength elements (HFSE), such as Nb, is seen in Clasts B2, B3, B4, B8, C1, C2 and C3 (Fig. 6b, d). Clast B8 is Kwaimbaita-like over most of the profile and in certain incompatible-element ratios (La/Lu = 9.6 v. 10.5 for average Kwaimbaita-type basalt, La/Ce = 0.32 v. 0.34), yet has a $(Nb/La)_{PM}$ ratio of 1.7 (v. 1.09 for average Kwaimbaita-type basalt). As with the positive Y anomaly, this type of fractionation is not seen in typical Kwaimbaita Formation basalts (Tejada *et al.* 1996, 2002; Neal *et al.* 1997).

Radiogenic isotope ratios

Four samples (Clasts B4, B7, C3 and E1) were analysed for Sr-, Nd- and Pb-isotope ratios, and Clast B8 for Sr- and Nd-isotope ratios (Table 3). The five clasts analysed represent at least one sample from each of the groups defined on the basis of incompatible elements. All isotope data were age-corrected to 120 Ma (Chambers *et al.* 2004) using high-resolution ICP-MS parent–daughter data obtained from splits of the solutions analysed for isotopes, except for U and Th values of Clast B4, which were determined by high-resolution ICP-MS on leached powders. Age-corrected ε_{Nd} values ($\varepsilon_{Nd(t)}$) form a narrow range (+6.0–+6.5), whereas initial $^{87}Sr/^{86}Sr$ ($^{87}Sr/^{86}Sr_I$) and age-corrected $^{206}Pb/^{204}Pb$ ($^{206}Pb/^{204}Pb_{(t)}$) exhibit relatively more variation (0.70288–0.70480 and 18.139–18.458, respectively; Table 3). The range of $\varepsilon_{Nd(t)}$ is within that for the Kwaimbaita basalts (Fig. 7a, b). Although the wide range of Sr- and Pb-isotope compositions overlaps with the ranges for the Kwaimbaita and the Singgalo basalts (cf. Tejada *et al.* 2002) (Fig. 7a, b), we interpret the range in our data to be a result of secondary alteration effects (see later for further discussion). Therefore, we shall not use our Sr- and Pb-isotopic results in the following petrogenetic interpretation.

Discussion

The influence of alteration

The basalt clasts from Site 1184 are lithic fragments derived from coeval or possibly pre-existing volcanic rocks during explosive phreatomagmatic eruptions (Thordarson 2004). Therefore, alteration effects are a major concern

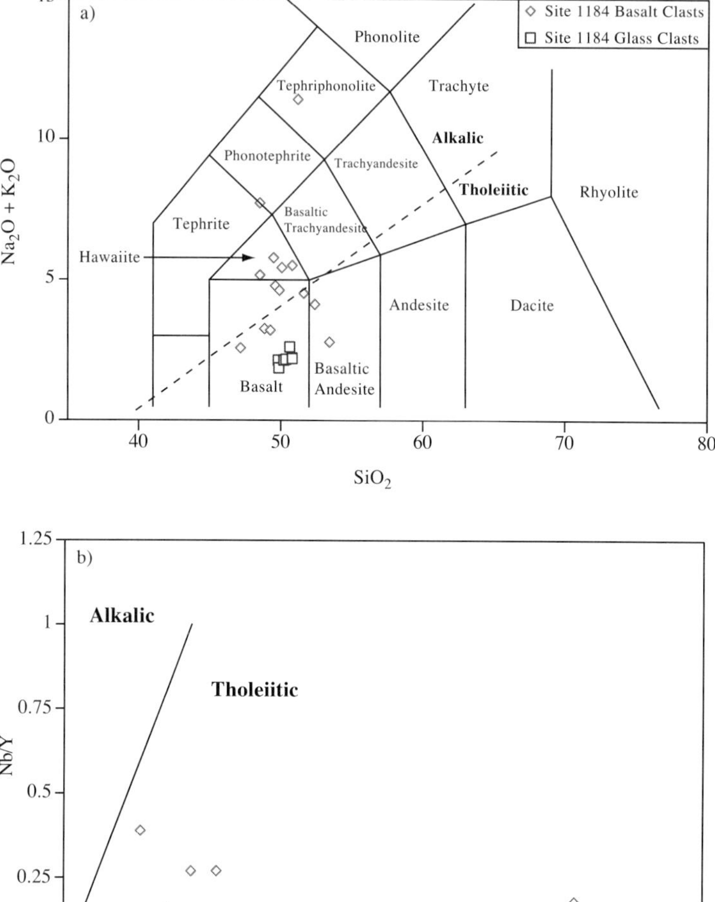

Fig. 5. Classification of the basaltic clasts. (**a**) SiO_2 v. total alkalis ($Na_2O + K_2O$), where the dashed line separates Hawaiian tholeiitic and alkali lavas (Macdonald & Katsura 1964). Data for Site 1184 glass clasts are from White *et al.* (2004). Fields of rock types are from Le Bas *et al.* (1986). (b) Nb/Y v. $Zr/(P_2O_5 \times 10^4)$ after Winchester & Floyd (1976). The Site 1184 clasts plot as transitional between alkalic and tholeiitic on the SiO_2 v. total alkali plot, yet strongly tholeiitic on the Nb/Y v. $Zr/(P_2O_5 \times 10^4)$ plot, suggesting that Na_2O and K_2O were increased during alteration of these samples.

in our study and here we evaluate these prior to any petrogenetic interpretation.

The variability of the major-element composition of the clasts relative to the OJP basalt groups is interpreted primarily to be the result of alteration. The generally high LOI values, coupled with petrographic observations, demonstrate the moderately to highly altered nature of these clasts. In Figure 4a, b it is seen that CaO is negatively correlated and K_2O is positively correlated with LOI, demonstrating that elements can be depleted or enriched (respectively) during alteration. Indeed, many studies in the last several decades (e.g. Hart 1970) have reported that during low-temperature submarine alteration, ocean-ridge tholeiites lose SiO_2, CaO, and MgO, and gain K_2O, some Fe_2O_3, MnO, Na_2O, and P_2O_5. It would appear that the Site 1184 clasts have only gained K_2O (and possibly Na_2O) and lost CaO, SiO_2 and

Table 2. *Trace-element concentrations* of Leg 192, Site 1184 basaltic clasts (all values are in ppm)*

Core	18R-2	20R-4	21R-5	21R-5	21R-6	21R-6	22R-1	22R-4	24R-2	25R-5	25R-6	35R-8	44R-2	44R-2	24R-2	25R-5	25R-6	BHVO-2‡	SD	Recommended values§
Interval (cm)	102–104	50–52	40–47	55–60	28–30	73–77	24–27	120–123	68–70	16–18	115–117	32–35	101–105	126–132	68–70	16–18	115–117	*n* = 7		
Clast	B1	B2	B3	B4	B5	B6	B7	B8	C1	C2	C3	D1	E1	E2	MC1†	MC2†	MC3†			
Group	1	2	2	2	3	1	3	2	4	4	4	2	1	1						
Li	9.09	3.69	7.06	10.5	bd	3.74	1.28	2.97	8.74	5.51	6.29	2.59	5.71	2.49	8.78	9.38	10.8	4.7	0.7	5
Be	0.16	0.15	0.08	0.11	0.23	0.26	1.16	0.12	0.20	0.23	0.28	0.31	0.20	0.25	0.49	0.52	0.54	1.05	0.1	*1.1*
Sc	66.6	39.3	41.9	56.1	74.6	52.8	221.9	48.0	17.7	37.8	42.8	48.9	46.2	45.4	49.7	51.8	53.6	33.1	3.0	32
V	568.4	192.7	522.5	771.1	1259	375.9	1785	410.3	179.8	183.2	192.6	482.1	296.6	285.5	217.0	235.8	238.5	336.3	14.9	317
Cr	646.8	298.5	177.7	243.5	337.3	378.1	946.4	210.8	13.7	15.4	33.4	273.9	325.6	362.7	177.1	194.8	192.6	284.7	48.8	280
Co	55.5	25.4	97.5	45.0	58.7	51.2	181.0	32.9	24.7	38.8	38.3	57.2	51.0	54.3	50.3	50.9	54.2	52.3	4.3	45
Ni	107.2	71.2	88.9	88.5	234.3	89.4	307.7	70.9	14.3	12.3	17.5	110.5	111.9	116.5	68.3	75.6	74.1	127.6	7.9	119
Cu	74.7	9.24	406.8	27949	232.9	169.5	804.6	51.8	49.3	61.9	69.2	49.7	153.1	162.0	138.1	144.9	146.4	139.0	2.2	127
Zn	211.3	24.0	17.9	45.6	23.3	139.5	632.2	43.3	123.4	192.0	198.7	73.2	140.8	93.6	151.3	148.1	167.0	111.0	5.4	103
Ga	18.8	16.2	15.1	26.9	23.1	20.4	81.1	16.6	32.1	24.7	18.9	19.2	17.9	18.8	35.1	24.0	34.4	23.3	1.4	*21*
Rb	4.57	2.03	16.5	12.9	1.76	1.85	5.06	2.40	5.50	12.0	7.87	0.21	0.27	0.33	10.8	9.36	11.2	9.7	1.5	9.8
Sr	82.7	105.1	49.9	31.7	369.1	96.0	92.8	123.4	148.7	89.8	222.7	88.2	102.0	96.3	145.5	117.7	189.2	399.0	22.8	389
Y	21.8	44.3	30.7	28.0	21.0	22.3	16.1	55.5	45.5	56.6	54.3	23.5	30.5	46.5	39.3	38.0	36.7	26.0	1.9	26
Zr	66.3	42.9	57.9	85.5	24.2	56.8	56.8	59.9	212.2	181.1	187.6	28.1	52.7	53.7	90.1	89.2	86.1	172.1	15.4	172
Nb	3.50	2.24	3.66	4.00	1.36	3.21	2.95	6.56	17.5	15.1	14.7	3.51	2.60	2.96	8.78	7.40	8.41	19.8	1.0	18
Ba	7.65	28.9	157.7	54.1	14.2	26.4	39.7	40.7	19.1	46.5	47.5	8.17	9.29	9.55	32.1	36.4	127.7	138.9	3.3	130
La	3.22	1.67	1.98	8.4	1.92	3.12	2.89	3.83	7.59	7.11	7.39	4.01	2.78	2.85	7.35	7.88	7.40	16.3	0.4	15
Ce	9.03	6.88	5.25	17.5	5.96	9.12	7.48	12.1	24.2	21.1	21.2	13.6	7.81	8.37	18.3	18.4	17.7	40.4	1.7	38
Pr	1.51	1.30	0.86	2.3	1.05	1.45	1.14	2.1	3.68	3.38	3.28	2.07	1.29	1.34	2.70	2.70	2.59	5.9	0.1	*5.7*
Nd	7.16	7.09	4.21	10.6	5.05	6.91	5.75	10.1	16.8	16.3	15.3	8.98	6.46	6.30	11.7	11.6	11.1	25.1	0.5	25
Sm	2.66	2.70	1.76	3.22	1.67	2.50	1.90	3.08	5.87	5.72	5.56	2.92	2.15	2.42	3.55	3.46	3.37	6.6	0.5	6.2
Eu	0.80	0.81	0.67	0.74	0.59	0.87	0.69	1.01	1.54	1.59	1.55	1.08	0.82	0.84	1.39	1.29	1.28	2.19	0.1	*2.06*
Gd	3.57	3.63	3.23	4.13	2.46	3.32	2.69	4.32	7.41	8.27	7.67	3.63	3.09	3.24	5.24	4.98	4.79	6.75	0.1	6.3
Tb	0.62	0.63	0.67	0.72	0.42	0.61	0.45	0.72	1.31	1.48	1.42	0.64	0.55	0.56	0.88	0.89	0.83	1.01	0.05	0.9
Dy	4.13	4.17	4.71	4.83	2.54	4.21	2.97	4.49	8.44	10.3	9.64	4.27	3.59	3.84	6.17	6.01	5.63	5.68	0.26	*5.2*
Ho	0.85	0.88	1.10	1.05	0.55	0.92	0.59	1.00	1.77	2.21	2.14	0.88	0.80	0.80	1.38	1.34	1.27	1.04	0.04	1.0
Er	2.45	2.58	3.45	3.31	1.65	2.66	1.69	2.92	4.93	6.81	6.44	2.50	2.28	2.36	4.22	4.11	3.87	2.77	0.12	*2.4*
Tm	0.37	0.38	0.56	0.50	0.25	0.39	0.25	0.43	0.75	0.99	1.00	0.37	0.35	0.36	0.64	0.63	0.58	0.38	0.03	*0.33*
Yb	2.57	2.55	3.44	3.38	1.40	2.62	1.62	2.65	5.29	7.53	6.74	2.25	2.25	2.45	4.06	4.04	3.80	2.15	0.1	2.0
Lu	0.33	0.34	0.53	0.51	0.22	0.39	0.21	0.40	0.75	1.06	0.97	0.31	0.32	0.33	0.60	0.59	0.56	0.29	0.02	0.28
Hf	1.79	1.54	1.80	2.45	0.75	1.65	1.56	1.36	5.31	5.01	4.65	0.99	1.55	1.51	2.41	2.39	2.25	4.48	0.12	4.1
Ta	0.21	0.25	0.25	0.23	0.08	0.21	0.16	0.20	1.13	0.89	0.91	0.22	0.16	0.18	0.46	0.42	0.42	1.30	0.1	1.4
Pb	0.38	0.14	0.06	0.04	bd	bd	bd	bd	0.64	0.32	0.31	0.06	0.08	0.05	0.76	0.90	0.71	1.54	0.07	*2.05*
Th	0.27	0.22	0.27	0.33	0.07	0.24	0.14	0.60	1.78	1.61	1.50	0.18	0.22	0.24	0.65	0.60	0.57	1.30	0.1	1.20
U	0.05	0.01	0.09	0.13	0.05	0.08	0.04	0.01	0.06	0.11	0.34	0.10	0.09	0.05	0.11	0.12	0.13	0.47	0.04	*0.42*

* All trace-element data have been normalized on a volatile-free basis.
† Concentrations of samples MC1–MC3 are of the volcaniclastic matrix surrounding clasts C1–C3, respectively. Clast names are composed of the subunit from which the clast was extracted and relative position within core (see text).
‡ BHVO-2 recommended values from USGS preliminary certificate of analysis (Wilson 1997).
§ Recommended values in italics are for BHVO-1 (Govindaraju 1994).
bd, below detection limit.

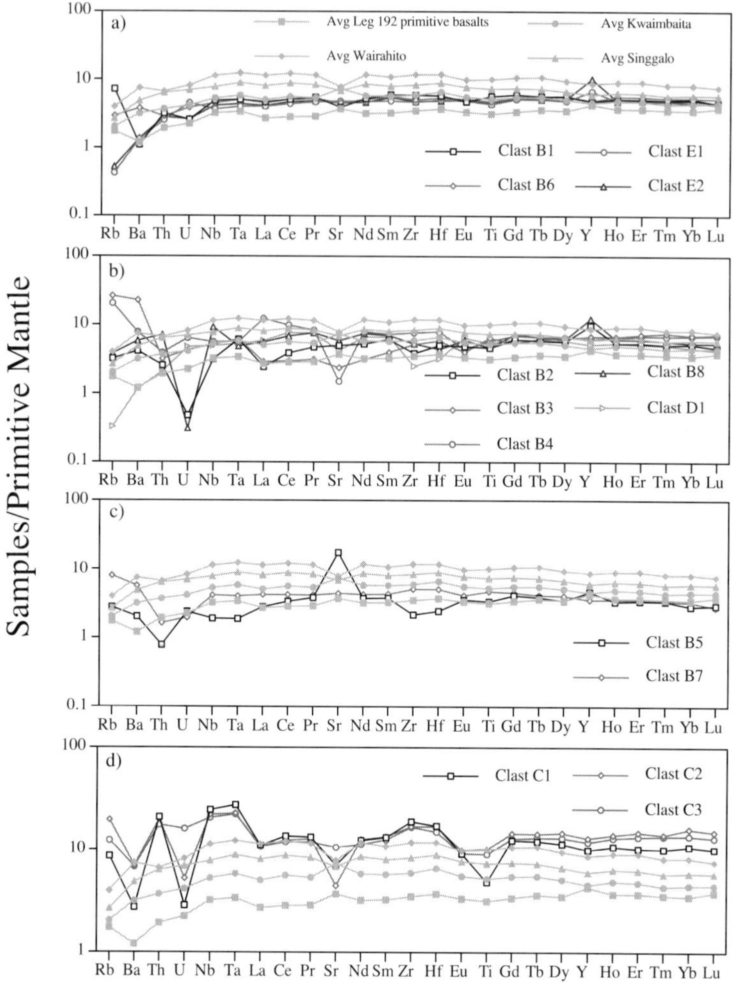

Fig. 6. Primitive-mantle-normalized incompatible-element plots of (**a**) Group 1 – Clasts B1, B6, E1 and E2; (**b**) Group 2 – Clasts B2, B3, B4, B8 and D1; (**c**) Group 3 – Clasts B5 and B7; and (**d**) Group 4 – Clasts C1, C2 and C3. The grey profiles represent Kwaimbaita-type basalt (circle), Singgalo-type basalt (triangle), Kroenke-type basalt (square) and the Wairahito Formation basalts (diamond). Group 1 clasts show similar profiles to Kwaimbaita-type basalt, Group 3 clasts show affinity to Kroenke-type basalt and Group 4 clasts are distinct from the main basalt types. Group 2 clasts exhibit more compositional variability, which is probably due to more intense alteration, although they have HREE profiles similar to Kwaimbaita-type basalt. See text for details. Data from which average OJP basalt compositions were calculated are from Neal *et al.* (1997), Tejada *et al.* (1996, 2002), Birkhold (2000) and Mahoney *et al.* (2001).

MgO to any significant degree, although the Group 4 samples with >5 wt% LOI have the highest P_2O_5 contents.

In terms of incompatible trace elements, and Sr- and Pb-isotope ratios, the basalt clasts from Site 1184 exhibit much more variation than do basalt compositions across the OJP (e.g. Fig. 7a–c). This is also probably a result of sea-water alteration, which affected the abundances of fluid-mobile trace elements such as Rb, Sr, Pb and, to a certain extent, U (e.g. Hart *et al.* 1974; Menzies & Seyfried 1979). The effect of sea-water alteration on both trace elements and isotopes is demonstrated by the negative correlation between bulk-clast Sr concentrations and isotopic compositions (measured on leached splits of sample) (Fig. 8). The negative correlation suggests that the clasts with low Sr

Table 3. *Sr-, Nd- and Pb-isotopic data for basaltic clasts from ODP Site 1184*

Core	21R-5	22R-1	22R-4	25R-6	25R-6 Dupl.	44R-2	44R-2 Dupl.
Interval (cm)	55–60	24–27	120–123	115–117	115–117	101–105	101–105
Clast	B4	B7	B8	C3	C3	E1	E1
Group	2	3	2	4	4	1	1
Rb	18.2	5.86	1.07	2.20		0.09	
Sr	43.3	113.7	110.2	18.5		64.4	
$^{87}Sr/^{86}Sr$	0.706850	0.703944	0.704852	0.703467		0.703654	
$^{87}Sr/^{86}Sr_{(t)}$	0.704779	0.703690	0.704804	0.702881		0.703647	
Sm	0.9	0.48	3.68	3.13		1.27	
Nd	3.51	1.18	11.1	6.03		2.74	
$^{143}Nd/^{144}Nd$	0.512924	0.513002	0.512976	0.513042		0.513012	
$^{143}Nd/^{144}Nd_{(t)}$	0.512803	0.512810	0.512819	0.512797		0.512793	
$\varepsilon Nd_{(t)}$	6.2	6.3	6.5	6.1		6.0	
Th	0.18	0.31		1.20		0.30	
U	0.35	0.06		0.23		0.06	
Pb	0.48	0.19		0.30		0.36	
$^{206}Pb/^{204}Pb$	19.101±8	18.831±7		19.053±3	19.037±6	18.567±2	18.566±6
$^{207}Pb/^{204}Pb$	15.541±7	15.543±6		15.577±2	15.556±5	15.519±1	15.516±6
$^{208}Pb/^{204}Pb$	38.763±17	38.614±15		39.000±6	38.941±13	38.380±4	38.366±14
$^{206}Pb/^{204}Pb_{(t)}$	18.144	18.458		18.139	18.124	18.371	18.370
$^{207}Pb/^{204}Pb_{(t)}$	15.499	15.525		15.532	15.511	15.509	15.506
$^{208}Pb/^{204}Pb_{(t)}$	38.059	38.059		37.437	37.380	38.060	38.046

Rb, Sr, Sm, Nd, Th, U and Pb concentrations (ppm) from analysis of leached powder. Age (t) is taken as 120 Ma. Duplicate analyses were made on different dissolutions of the same unleached power. Analytical uncertainty for $^{87}Sr/^{86}Sr$ measurements is ±0.000018 but in-run precisions were better than ±0.000012. Sr-isotopic ratios were measured by dynamic multi-collection, fractionation-corrected to $^{86}Sr/^{87}Sr = 0.1194$ and normalized to $^{87}Sr/^{86}Sr = 0.71025$ for NBS 987. Analytical uncertainty for $^{143}Nd/^{144}Nd$ measurements is 0.000014 (0.3 ε units) but in-run precisions were better than 0.000010. Nd-isotopic ratios were measured in oxide form by dynamic multi-collection, fractionation-corrected to $^{146}NdO/^{144}NdO = 0.72225$ ($^{146}Nd/^{144}Nd = 0.7219$) and are reported relative to $^{143}Nd/^{144}Nd = 0.511850$ for the La Jolla Standard. Pb-isotopic ratios were measured by static multi-collection and are reported relative to the values of Todt *et al.* (1996) for NBS SRM 981; the long-term errors measured for this standard are ±0.008 for $^{206}Pb/^{204}Pb$ and $^{207}Pb/^{204}Pb$, and ±0.024 for $^{208}Pb/^{204}Pb$. The within-run errors shown refer to the last significant figure. Estimated uncertainties on concentrations are *c.*1% for both Sm and Nd, and *c.*2% on Rb, Sr, U, Th and Pb. Total procedural blanks are negligible: <10 picograms (pg) for Nd, <35 pg for Sr, <3 pg for Th, <5 pg for U and <60 pg for Pb. $\varepsilon_{Nd} = 0$ today corresponds to $^{143}Nd/^{144}Nd = 0.51264$; for $^{147}Sm/^{144}Nd = 0.1967$, $\varepsilon_{Nd(t)} = 0$ corresponds to $^{143}Nd/^{144}Nd = 0.512486$ at 120 Ma.

contents and high $^{87}Sr/^{86}Sr$ ratios have exchanged significant amounts of Sr with sea-water (8 ppm Sr, present day $^{87}Sr/^{86}Sr = 0.709$, at 120 Ma $^{87}Sr/^{86}Sr = c.\ 0.707$ – Hess *et al.* 1986; DePaolo & Ingram 1985). The clasts with high $^{87}Sr/^{86}Sr$ have similar Sr-isotopic compositions to sea water at 120 Ma, indicating that most of alteration occurred during or soon after the eruptions. As noted earlier, the Pb-isotopic signature appears to have been affected by alteration. White *et al.* (2004) observed that, in the bulk volcaniclastic rocks at Site 1184, Sr- and Pb-isotope ratios have been affected by sea-water alteration. The same type of result was also seen in the Singgalo-type vitric tuff at Site 1183 (Tejada *et al.* 2004).

A number of studies have shown that the concentrations of some trace elements, such as REE and HFSE, in igneous rocks are generally not affected during mild sea-water alteration (e.g. Ludden & Thompson 1979; Bienvenu *et al.* 1990). Yet, it has also been documented that the REE are relatively mobile during more extensive alteration compared to the HFSE, and that the LREE are preferentially mobilized over the HREE (Ludden & Thompson 1979; Bach & Irber 1998; Kikawada *et al.* 2001). Overall, the Site 1184 basalt clasts exhibit large variations in element ratios such as La/Nb, La/Ta, Th/Ta, Th/Nb and Zr/Hf. Kurtz *et al.* (2000) reported enrichments in alteration-resistant, insoluble trace elements (Zr, Nb, Hf, Ta, Th) in strongly weathered Hawaiian soils over parent lava values due to extensive mass loss of more soluble major elements during soil formation. The presence of wood and oxidized horizons

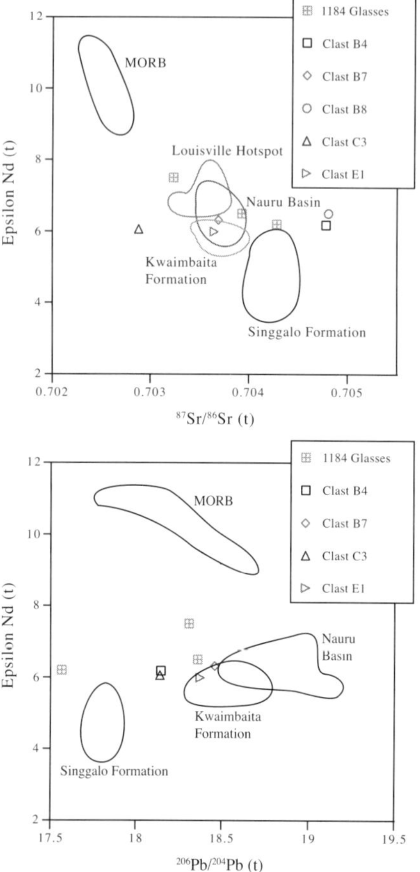

Fig. 7. (**a**) $^{87}Sr/^{86}Sr_{(t)}$ v. $\varepsilon_{Nd(t)}$; (**b**) $^{206}Pb/^{204}Pb_{(t)}$ v. $\varepsilon_{Nd(t)}$. Site 1184 samples have $\varepsilon_{Nd(t)}$ within in the range of Kwaimbaita-type basalt, yet show relatively large variations in both $^{87}Sr/^{86}Sr$ and $^{206}Pb/^{204}Pb$. We interpret these variations as being the result of secondary alteration that has changed the abundances of the fluid-mobile elements Rb, Sr, Pb and, to a certain extent, U. Site 1184 whole-rock data are from White *et al.* (2004). Data for fields taken from Mahoney (1987); Cheng *et al.* (1987); Castillo *et al.* (1991, 1994); Mahoney *et al.* (1991, 1993) and Tejada *et al.* (1996, 2002).

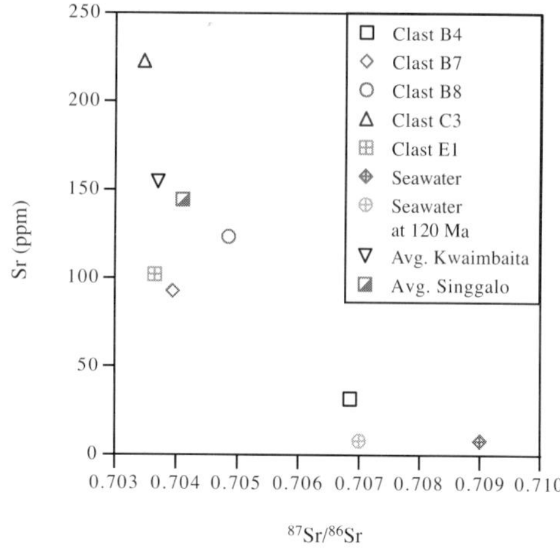

Fig. 8. Sr (unleached, ppm) v. $^{87}Sr/^{86}Sr$. The negative correlation suggests low Sr and high $^{87}Sr/^{86}Sr$ clasts have exchanged with sea water (8 ppm Sr and $^{87}Sr/^{86}Sr = 0.709$).

suggests that subaerial weathering of the volcaniclastic deposits did occur. However, Thordarson (2004) has shown that each volcaniclastic subunit probably represents a single eruptive event and, as such, soil could only have formed at the top of the subunits well away from the clasts. The Group 4 clasts were taken from a highly oxidized horizon of the core; they exhibit marked enrichments of Nb and Ta (as well as P_2O_5) and slight enrichments of Zr and Hf over the REE. Preferential mobilization of the LREE over the HREE (Ludden & Thompson 1979; Bach & Irber 1998; Kikawada *et al.* 2001) can explain the similarity between HREE abundances in the Group 2 clasts and those in average Kwaimbaita-type basalt (Fig. 6b) and the marked depletions and enrichments in the LREE and other incompatible elements. Enrichment of the LREE during alteration may also account for the Nb–Ta and Zr–Hf depletions relative to the REE in Group 2 Clast D1 (Fig. 6b) and Group 3 clast B5 (Fig. 6c).

In order to understand the nature of incompatible-element mobility during alteration, relatively immobile elements need to be identified for comparisons to be made. Kurtz *et al.* (2000) concluded that Nb and Ta were the most stable in the development of soils on volcanic terrains, but that Th, Zr and Hf were mobile during this process. The Site 1184 basalt clasts have $(Nb/Ta)_{PM}$ (primitive-mantle-normalized) ratios of 0.96 ± 0.06 (1σ) if Group 2 Clasts B2 ($(Nb/Ta)_{PM} = 0.51$) and B8 ($(Nb/Ta)_{PM} = 1.90$) are omitted. These two clasts exhibit quite dissimilar Nb abundances (2.24 and 6.56 ppm, respectively), yet similar Ta concentrations (0.25 and 0.20 ppm, respectively), and we conclude that at least in these two samples Nb was mobile. In examining the primitive-mantle-normalized patterns (Fig. 6a–d) our data set suggests that Ta was relatively immobile (note the narrow range of normalized Ta values in Fig. 6a–d, especially in the Group 2 patterns of

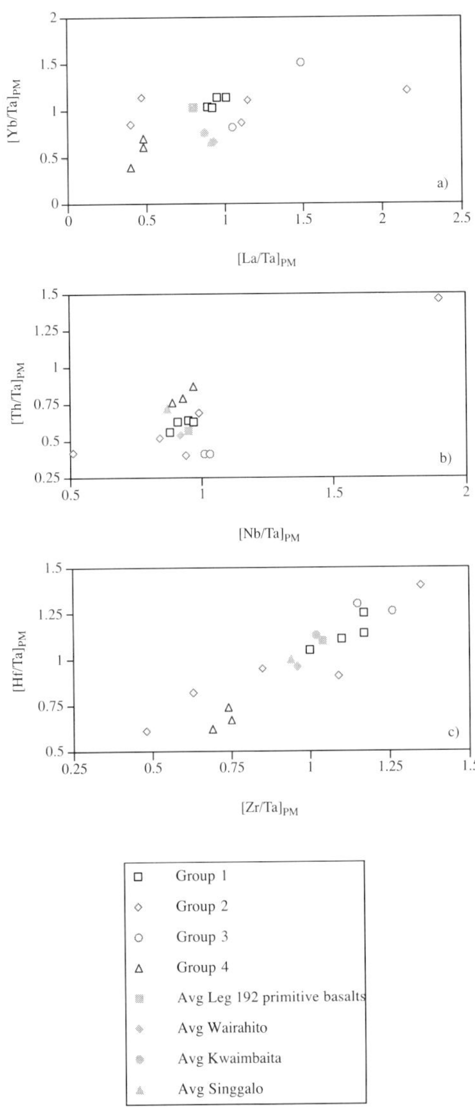

Fig. 9. (**a**) $(La/Ta)_{PM}$ v. $(Yb/Ta)_{PM}$; (**b**) $(Nb/Ta)_{PM}$ v. $(Th/Ta)_{PM}$; (**c**) $(Zr/Ta)_{PM}$ v. $(Hf/Ta)_{PM}$. The scatter in these plots away from typical OJP values is interpreted as showing the relative mobility of La, Yb, Nb, Th, Zr and Hf during alteration.

Fig. 6b). In an attempt to examine mobility, elements of interest (i.e. HFSE and REEs) have been normalized to Ta and compared with the average compositions of the Kwaimbaita, Singgalo, Wairahito and Kroenke-type basalts (Fig. 9a–c). The distribution seen in these figures is interpreted as primarily indicating mobility of La, Yb, Zr and Hf during alteration, although $(Yb/Ta)_{PM}$ values may be changed somewhat during fractional crystallization. Also, $(La/Ta)_{PM}$ ratios of most non-arc-related, mantle-derived magmas are relatively close to 1.0 (Sun & McDonough 1989; McDonough & Sun 1995). In magmas contaminated by continental crust, these values deviate from unity, but there is no evidence for such contamination in the OJP basalts (Mahoney 1987; Mahoney & Spencer 1991; Mahoney *et al.* 1993).

Assuming Ta is immobile, the relative mobility of other elements can be estimated. In terms of both light and heavy REE mobility, Group 1 clasts have similar $(La/Ta)_{PM}$ ratios to the average composition of each OJP basalt group, and slightly higher $(Yb/Ta)_{PM}$ ratios (Fig. 9a). Group 2 clasts exhibit a wide range of $(La/Ta)_{PM}$ values indicative of LREE enrichment and depletion, whereas $(Yb/Ta)_{PM}$ is more restricted. One Group 3 clast (Clast B5) is radically different from the average OJP basalts, suggesting that it may have been enriched in the REE. All Group 4 clasts have lower $(La/Ta)_{PM}$ and $(Yb/Ta)_{PM}$ values that overlap and extend to lower values than the average values for OJP basalts. Overall, these relationships are consistent with the altered nature of the Site 1184 clasts and with the observation that the LREE are more mobile than the HREE during alteration of basaltic material (e.g. Bach & Irber 1998).

The HFSE Zr, Nb, Hf and Th exhibit varying degrees of enrichment and depletion relative to Ta and the average compositions of OJP basalt types. As noted above, Nb appears immobile, except in two Group 2 clasts (Clasts B2 and B8), in which depletions and enrichments (respectively) relative to Ta are observed (Fig. 9b). All other clasts have $(Nb/Ta)_{PM}$ values *c.* 1.0, and in Figure 9b the Group 1 clasts plot directly on top of the average OJP basalt compositions, indicating that Nb/Ta ratios are unaffected by alteration. However, $(Th/Ta)_{PM}$ values for the remaining clasts are variable, with Group 2 clasts having the largest range (*c.* 0.4–1.5), which we interpret as indicating Th removal and addition relative to Ta. Whereas Th shows signs of mobility, it does not show nearly as much variation as La, Yb, Nb and Hf. Group 3 clasts have relative Th depletions, whereas those from Group 4 show slight Th enrichments (Fig. 9b). This contrasting behaviour between Groups 3 and 4 clasts is also seen in Figure 9c, where the Group 3 clasts are relatively enriched in Zr and Hf and those from Group 4 are depleted. Relative to the average OJP basalt compositions, Group 1 clasts appear to have a slight Zr enrichment (Fig. 9c). Again, Group 2 clasts exhibit a wide range of values.

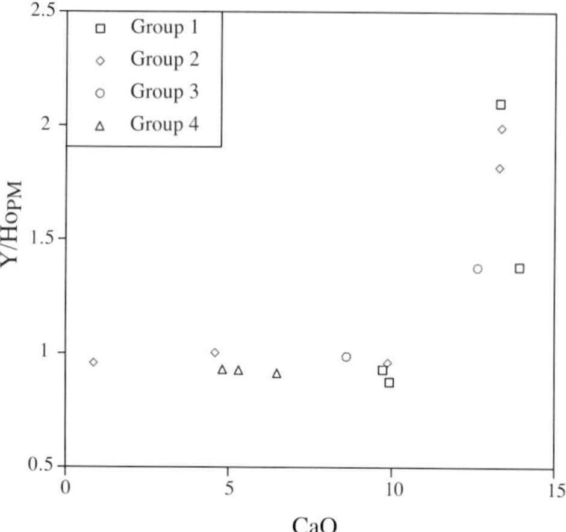

Fig. 10. CaO (wt%) v. $(Y/Ho)_{PM}$. $(Y/Ho)_{PM}$ is used to represent the positive Y anomaly present in five Site 1184 clasts. This anomaly is present in samples with CaO abundances over 10 wt%. This implies that the Y carrier is a Ca-rich phase, such as chabazite, a Ca-rich zeolite. See text for full details.

The presence of positive Y anomalies in five of the Site 1184 basalt clasts, as seen in their primitive-mantle-normalized incompatible-element plots (Fig. 6a–c), is puzzling. This could be the product of a polyatomic interference (e.g. $^{52}Cr^{37}Cl$, $^{70}Zn^{19}F$) that may have been induced through incomplete removal of HCl during our initial sample preparation or through incomplete HF removal during dissolution. However, these anomalies were replicated during different dissolutions ($HF–HNO_3$ acid and sodium peroxide fusion), so it is unlikely that an acid-induced polyatomic interference is the cause. Such positive Y anomalies are present in lavas that have undergone varying degrees of alteration (e.g. Kuschel & Smith 1992; Cotten *et al.* 1995). Past studies of element mobility during the weathering of igneous rocks have reported all of the REEs and Y as being collectively (although not uniformly) enriched or depleted in the weathered product (e.g. Duddy 1980; Marsh 1991; Price *et al.* 1991). Price *et al.* (1991) attributed these weathering effects to the presence or absence of secondary phosphates or clay minerals. Duddy (1980) concluded that the REE+Y were effectively immobilized by the development of vermiculite as a weathering product. Kuschel & Smith (1992) suggested the relative enrichment of Y was due solely to secondary phosphate and presented SEM images and mineral analyses of these phases in support of this conclusion, but generation of the positive Y anomaly in some of the weathered samples was not supported by the REE and Y contents of the secondary phosphates (i.e. no positive Y anomaly was present in these phases). However, Cotten *et al.* (1995) reported the presence of secondary rhabdophane-type REE–Y phosphates that did contain significant positive Y anomalies; the concentration of Y was of the order of 6–15 wt% Y_2O_3 (47 000–120 000 ppm Y). If such secondary phosphates were developed during the alteration of the Site 1184 basalt clasts, it is evident that only a minute amount would dramatically alter the Y abundance of the whole-rock composition. However, the REE are also enriched in these secondary phases (Cotten *et al.* 1995), but we see no concomitant REE enrichment in our clasts. In addition, there is no correlation between Y and P_2O_5 contents, and the samples with the highest P_2O_5 abundances exhibit no positive Y anomaly.

Only clasts with CaO >10 wt% exhibit a positive Y anomaly (represented as $(Y/Ho)_{PM}$) (Fig. 10). This indicates that the carrier of Y is rich in CaO. The Y carrier could, therefore, be calcite (present in thin section of some clasts) but there is no correlation between $(Y/Ho)_{PM}$ and LOI or between $(Y/Ho)_{PM}$ and Sr. Mahoney *et al.* (2001) reported the presence of zeolites in the Site 1184 volcaniclastic matrix. Our examination of thin sections of four of the samples that exhibit the positive Y anomaly (Clasts B3, B8, E1 and E2) shows that zeolite minerals are present as vesicle fill and veins both within and surrounding the clasts in the volcaniclastic matrix. If a Ca-rich zeolite (e.g. chabazite) was present in some of the basalt clasts, it could preferentially enrich the bulk composition in Y if significant interaction with sea water had occurred. For example, Wheat *et al.* (2002) noted that relative to ocean-ridge basalt, seawater exhibits a positive Y anomaly. Clast C1 contains zeolite minerals but does not show the Y anomaly. This sample has low CaO (4.83 wt%) and high Na_2O (5.73 wt%), suggesting that the zeolite present in Clast C1 is natrolite, a Na-rich zeolite. Formation of a Ca-rich zeolite during alteration by sea water in the other clasts with zeolites could produce a positive Y anomaly, as it appears that Na-rich zeolites are not Y carriers. Further work needs to be conducted in order to test this tentative conclusion.

In summary, we conclude that on the basis of primitive-mantle-normalized diagrams (Fig. 6a–d), Ta is the most immobile trace element in the suite of 14 basalt clasts studied from Site 1184. Niobium is generally immobile but two Group 2 clasts exhibit either Nb enrichment or depletion, indicating that Nb was mobilized

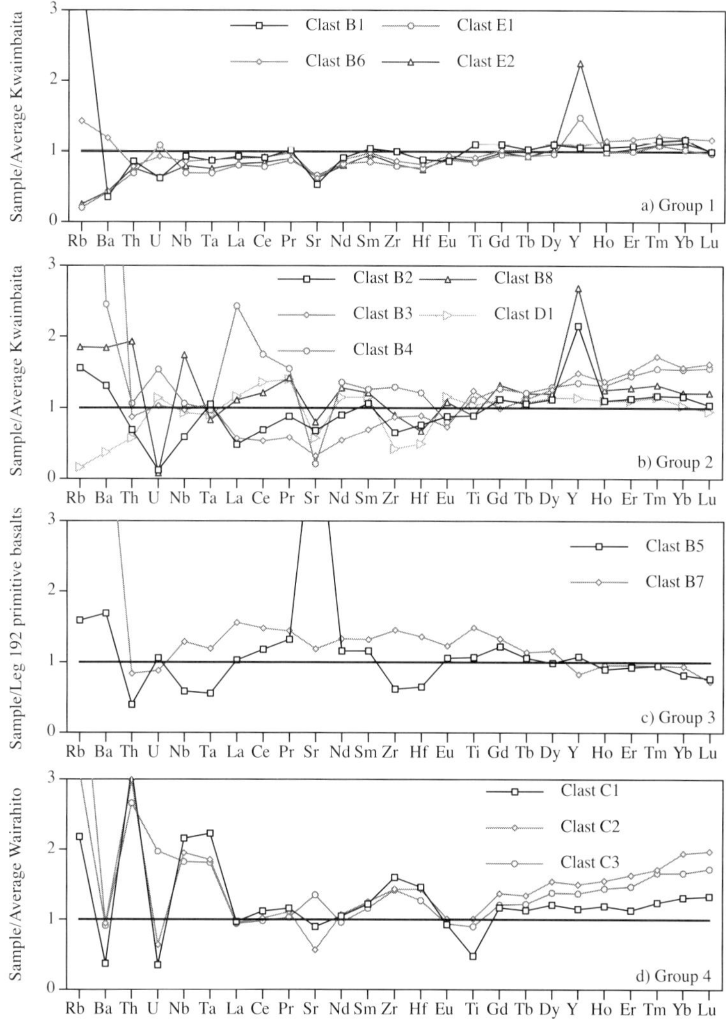

Fig. 11. (**a**) Group 1 basalt clasts normalized to average Kwaimbaita-type basalt; (**b**) Group 2 basalt clasts normalized to average Kwaimbaita-type basalt; (**c**) Group 3 basalt clasts normalized to average Kroenke-type basalt; and (**d**) Group 4 basalt clasts normalized to average Wairahito basalt. These profiles highlight the Site 1184 group affinities shown in Figure 6a–d. Data from which average basalt compositions were calculated are from Tejada *et al.* (1996, 2002), Neal *et al.* (1997), Birkhold (2000), Mahoney *et al.* (2001) and Fitton & Godard (2004).

during alteration. Zirconium and Hf also exhibit mobility and the LREE are more mobile than the HREE. Yttrium has been added to some clasts and is manifest as positive Y anomalies on Figure 6. We tentatively suggest that this is the result of secondary zeolite formation as a result of interaction of the clasts with sea water.

Petrogenetic interpretations

Although alteration has affected the major- and trace-element compositions of the Site 1184 basalt clasts, their incompatible-element profiles can help indicate similarities to other OJP basalt types (Fig. 6a–d). In order to highlight these similarities, we have also normalized data for each of the clasts from Groups 1–4 to the OJP basalt types to which they are most similar (Fig. 11a–d). Group 1 clasts show clear similarities to Kwaimbaita-type basalt (Neal *et al.* 1997; Tejada *et al.* 2002). Depletions in Rb and Ba and variable U concentrations can be attributed to alteration (Seyfried *et al.* 1998). The relatively low Sr abundances in these clasts relative to the

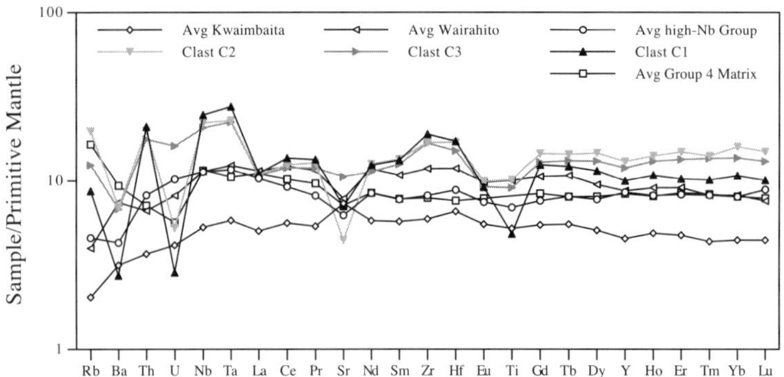

Fig. 12. Average Group 4 clast composition, average Group 4 volcaniclastic matrix composition, average high-Nb group of Fitton & Godard (2004), and average Kwaimbaita-type basalt, all normalized to primitive mantle. Note the similarities in the high-Nb group to the Group 4 volcaniclastic matrix and the significant difference of that to the Group 4 clasts. However, the incompatible-element profiles are similar (enriched Nb and Ta over the LREE), although distinct differences do occur (such as the significant depletion in Eu and Ti in the Group 4 clasts). These data are interpreted to mean that the Group 4 matrix and clasts are not from the same eruptive episode, yet they are probably related, as they both represent a distinctive high-Nb magma type not seen before on the OJP.

average Kwaimbaita-type basalt may be a result of greater amount of plagioclase fractionation in the Site 1184 magmas, but there is no corresponding Eu depletion (Figs 6a and 11a). It is much more likely to be due to preferential Sr removal during alteration. The Group 2 clasts, the most variable group, are tentatively interpreted to be akin to the Kwaimbaita-type basalt on the basis of ε_{Ndt} and HREE abundances (Figs 6b and 11b). Group 2 clasts are slightly more enriched in the HREE than are the average Kwaimbaita-type basalt, and elements more incompatible than Ti in Figure 11b are highly variable (Fig. 11b). However, Ta exhibits little variation, as does Nb (except for Clasts B2 and B8; see above), and values of these two elements are similar to the Ta and Nb values of the Kwaimbaita-type basalt (Group 2 Ta = 0.20–0.25 ppm v. 0.24 ppm for the average Kwaimbaita-type basalt). The two clasts that make up Group 3 are broadly similar to the Kroenke-type basalts (Fig. 11c), although Clast B7 has Ti, Zr and Nb concentrations similar to those of average Kwaimbaita-type basalt. Clast B5 exhibits the greatest variability with depletions in Nb, Ta, Zr and Hf, and a large enrichment in Sr. Clast B7 appears to be intermediate between Kroenke-type and Kwaimbaita-type basalt compositions. Finally, Group 4 clasts are more evolved than the Wairahito basalts (the most evolved rock type so far recorded from the OJP) of Birkhold (2000), and are significantly enriched in Th, Nb, Ta and the HREE relative to average Wairahito basalt. This group also displays negative Eu and Ti anomalies (Figs 6d and 11d), and clasts forming this group have the highest LOI values (*c.* 7.2–11.8 wt%). These samples have similarly shaped profiles to the whole-rock analyses of those of four samples of bulk tuff from Subunit IIC (high-Nb group, Fitton & Godard 2004), although they are significantly more enriched than the high-Nb group. However, analysis of the volcaniclastic matrix surrounding the clasts from Subunit IIC shows strong similarities to the high-Nb group (Fig. 12). The compositional differences between the Group 4 clasts and the high-Nb group and Group 4 matrix indicate that the clasts and matrix were not formed during the same eruptive episode, although they are probably related because they both show high Nb, Ta and Th relative to the REE, which is unique among OJP samples. The Group 4 basalt clasts and matrix and the high-Nb group of Fitton & Godard (2004) represent a magma type not seen elsewhere in the region.

Source region

Compositions of unaltered glass present in the lower sections of the Site 1184 core are similar to those of the widespread Kwaimbaita-type lavas (White *et al.* 2004). The basalt clasts studied here define a narrow range of initial $\varepsilon_{Nd(t)}$ values from +6.0 to +6.5, within the field defined by the Kwaimbaita-type basalt (Mahoney 1987; Mahoney *et al.* 1993; Tejada *et al.* 1996, 2002, 2004). Combining this observation with similarities in normalized incompatible element profiles, we suggest that Groups 1–4 were derived from the Kwaimbaita-type source, and represent basalts with affinities to both the Kwaimbaita-type and the MgO-rich Kroenke-type basalts (e.g. Mahoney *et al.* 2001; Tejada *et al.* 1996, 2002, 2004). Group 4 clasts are relatively evolved, more so than any other lava type yet recovered from the OJP, and thus are unique OJP samples. However, the similarity of Nd isotopic ratios with those of the Kwaimbaita- and Kroenke-type basalts indicates derivation from an isotopically similar or identical source (Tejada *et al.* 2002, 2004).

Summary and conclusions

Ocean Drilling Program Leg 192 recovered the first evidence that at least part of the eastern salient of the Ontong Java Plateau was erupted at or above sea level. Unlike the OJP basalt sequence cored at ODP Leg 130 Site 807, and the now subaerial outcrops on Malaita and Santa Isabel (Solomon Islands), the Singgalo type of basalt flows is absent.

Major-element, trace-element and isotope variations within the suite of basalt clasts extracted from the volcaniclastic sequence at Site 1184 show the effects of secondary alteration on basalt derived from the Kwaimbaita-type mantle source. On the basis of incompatible-element profiles (normalized to estimated primitive mantle values), four groups of clasts are defined. Most of the clasts (Groups 1 and 2) have affinities with Kwaimbaita-type basalt, but two show some similarities to the relatively Mg-rich Kroenke-type lavas (Group 3). The three Group 4 clasts are unique among samples recovered from the OJP in that their incompatible-element abundances are relatively enriched. Secondary alteration has mobilized elements that are immobile during moderate alteration, namely Zr, Hf, Th, the REE and, in two clasts, Nb. Of the HFSE, only Ta appears unaffected by this secondary alteration. Several samples also exhibit curious positive Y anomalies, which we tentatively interpret to be the result of secondary Ca-rich zeolite (chabazite?) development through interaction with sea water.

Ages estimated for the Site 1184 sequence from ^{40}Ar–^{39}Ar measurements (Chambers *et al.* 2004) strongly suggest that the eastern salient formed concurrently with the high plateau and that there is not a younging trend from the high plateau to the eastern salient (Kroenke & Mahoney 1996). Evidence for emergence of a portion of the plateau during its formation has potentially important implications for the effects of OJP emplacement on the local and, perhaps, global environment during the Aptian, the magnitude of which must be addressed by future drilling expeditions to the OJP and other parts of the western Pacific. Furthermore, the discovery that the Kwaimbaita-type source was also producing basaltic volcanism both on the main plateau and on the eastern salient has important implications for the modelling of magma dynamics and evolution during the growth of the OJP.

This research used samples provided by the Ocean Drilling Program (ODP). ODP is sponsored by the US National Science Foundation (NSF) and participating countries under management of Joint Oceanographic Institutions (JOI), Inc. Funding for this research was provided by USSSP to C. R. Neal and P. R. Castillo. We would like to thank D. Birdsell of the Center for Environmental Science and Technology at the University of Notre Dame for his help with the ICP-OES analyses. We would also like to thank reviewers P. Wallace, S. Revillon and G. Fitton, as well as one anonymous reviewer, for their helpful comments and suggestions.

References

BACH, W. & IRBER, W. 1998. Rare earth element mobility in the oceanic lower sheeted dyke complex: evidence from geochemical data and leaching experiments. *Chemical Geology*, **151**, 309–326.

BERGEN, J.A. 2004. Calcareous nanofossils from ODP Leg 192, Ontong Java Plateau. *In*: FITTON, J.G., MAHONEY, J.J., WALLACE, P.J. & SAUNDERS, A.D. (eds) *Origin and Evolution of the Ontong Java Plateau*. Geological Society, London, Special Publications, **229**, 113–132.

BIENVENU, P., BOUGAULT, H., JORON, J.L., TREUIL, M. & DMITRIEV, L. 1990. MORB alteration; rare-earth element/ non-rare-earth hygromagmaphile element fractionation. *Chemical Geology*, **82**, 1–14.

BIRKHOLD, A.L. 2000. *A geochemical investigation of Ontong Java Plateau basement exposed on the Island of Makira (San Cristobal), Solomon Islands, South Pacific.* PhD Thesis, University of Notre Dame.

CASTILLO, P.R., CARLSON, R.W. & BATIZA, R. 1991. Origin of the Nauru Basin igneous complex: Sr, Nd, and Pb isotope and REE constraints. *Earth and Planetary Science Letters*, **103**, 200–213.

CASTILLO, P.R., PRINGLE, M.S. & CARLSON, R.W. 1994. East Mariana Basin Tholeiites: Cretaceous intraplate basalts or rift basalts related to the Ontong Java plume? *Earth and Planetary Science Letters*, **123**, 139–154.

CHAMBERS, L.M., PRINGLE, M.S. & FITTON, J.G. 2004. Phreatomagmatic eruptions on the Ontong Java Plateau: an Aptian $^{40}Ar/^{39}Ar$ age for volcaniclastic rocks at ODP Site 1184. *In*: FITTON, J.G., MAHONEY, J.J., WALLACE, P.J. & SAUNDERS, A.D. (eds) *Origin and Evolution of the Ontong Java Plateau*. Geological Society, London, Special Publications, **229**, 325–331.

CHENG, Q., PARK, K.H., MACDOUGALL, J.D., ZINDLER, A., LUGMAIR, G.W., STAUDIGEL, H., HAWKINS, J.W.

& LONSDALE, P.F. 1987. Isotopic evidence for a hotspot origin of the Louisville seamount chain. *In*: KEATING, B.H., FRYER, P., BATIZA, R. & BOEHLERT, G.W. (eds) *Seamounts, Islands, and Atolls*. American Geophysical Union, Geophysical Monograph, **43**, 283–296.

COTTEN, J., LE DEZ, A., BAU, M., CAROFF, M., MAURY, R.C., DULSKI, P., FOURCADE, S., BOHN, M. & BROUSSE, R. 1995. Origin of anomalous rare-earth element and yttrium enrichments in subaerially exposed basalts: Evidence from French Polynesia. *Chemical Geology*, **119**, 115–138.

DEPAOLO, D.J. & INGRAM, B.L. 1985. High-resolution stratigraphy with Sr isotopes. *Science*, **227**, 938–941.

DUDDY, I.R. 1980. Redistribution and fractionation of rare-earth and other elements in a weathered profile. *Chemical Geology*, **30**, 363–381.

FITTON, J.G & GODARD, M. 2004. Origin and evolution of magmas on the Ontong Java Plateau. *In*: FITTON, J.G., MAHONEY, J.J., WALLACE, P.J. & SAUNDERS, A.D. (eds) *Origin and Evolution of the Ontong Java Plateau*. Geological Society, London, Special Publications, **229**, 151–178.

FLANAGAN, F.J. 1984. Three USGS mafic rock reference samples, W-2, DNC-1, and BIR-1. *US Geological Survey Bulletin*, **1623**, 54.

GLADNEY, E.S. & ROELANDTS, I. 1988. 1987 compilation of elemental concentration data for USGS BIR-1, DNC-1, and W-2. *Geostandards Newsletter*, **12**, 63–118.

GOVINDARAJU, K. 1994. 1994 compilation of working values and descriptions for 383 geostandards. *Geostandards Newsletter*, **118**, 1–158.

HART, R. 1970. Chemical exchange between sea water and deep ocean basalts. *Earth and Planetary Science Letters*, **9**, 269–279.

HART, S.R., ERLANK, A.J. & KABLE, E.J.D. 1974. Sea floor basalt alteration: Some chemical and Sr isotopic effects. *Contributions to Mineralogy and Petrology*, **44**, 219–230.

HESS, J., BENDER, M.L. & SCHILLING, J.-G. 1986. Evolution of the ratio of strontium-87 to strontium-86 in seawater from Cretaceous to present. *Science*, **231**, 979–984.

JANNEY, P.E. & CASTILLO, P.R. 1996. Basalts from the central Pacific Basin; evidence for the origin of Cretaceous igneous complexes in the Jurassic western Pacific. *Journal of Geophysical Research, B, Solid Earth and Planets*, **101**, 2875–2893.

KIKAWADA, Y., OSSAKA, T., OI, T. & HONDA, T. 2001. Experimental studies on the mobility of Lanthanides accompanying alteration by acidic hot spring water. *Chemical Geology*, **176**, 137–149.

KROENKE, L.W. & MAHONEY, J.J. 1996. Rifting of the Ontong Java Plateau's eastern salient and seafloor spreading in the Ellice Basin; relation to the 90 Ma eruptive episode on the plateau. *Eos, Transactions of the American Geophysical Union*, **77**, 713.

KURTZ, A.C., DERRY, L.A., CHADWICK, O.A. & ALFANO, M.J. 2000. Refractory element mobility in volcanic soils. *Geology*, **28**, 683–686.

KUSCHEL, E. & SMITH, I.E.M. 1992. Rare earth mobility in young arc-type volcanic rocks from northern New Zealand. *Geochimica et Cosmochimica*, **56**, 3951–3955.

LE BAS, M.J., LE MAITRE, R.W., STRECKEISEN, A. & ZANETTIN, B.A. 1986. Chemical Classification of igneous rocks based on the total alkali-silica diagram. *Journal of Petrology*, **27**, 745–750.

LONGERICH, H.P., JENNER, G.A., FRYER, B.J. & JACKSON, S.E. 1990. Inductively coupled plasma-mass spectrometric analysis of geological samples: A critical evaluation based on case studies. *Chemical Geology*, **83**, 105–118.

LUDDEN, J.N. & THOMPSON, G. 1979. An evaluation of the behavior of the rare earth elements during the weathering of sea-floor basalt. *Earth and Planetary Science Letters*, **43**, 85–92.

LUGMAIR, G.W. & GALER, S.J.G. 1992. Age and isotopic relationships among the angrites Lewis Cliff 86010 and Angra dos Reis. *Geochimica et Cosmochimica*, **56**, 1673–1694.

MACDONALD, G.A. & KATSURA, T. 1964. Chemical composition of Hawaiian lavas. *Journal of Petrology*, **5**, 82–133.

MAHONEY, J.J. 1987. An isotopic survey of Pacific oceanic plateaus: implications for their nature and origin. *In*: KEATING, B.H., FRYER, P., BATIZA, R. & BOEHLERT, G.W. (eds) *Seamounts, Islands, and Atolls*. American Geophysical Union, Geophysical Monograph, **43**, 207–220.

MAHONEY, J.J. & SPENCER, K.J. 1991. Isotopic evidence for the origin of the Manihiki and Ontong Java oceanic plateaus. *Earth and Planetary Science Letters*, **104**, 196–210.

MAHONEY, J.J., STOREY, M., DUNCAN, R.A., SPENCER, K.J. & PRINGLE, M.S. 1993. Geochemistry and age of the Ontong Java Plateau. *In*: PRINGLE, M.S., SAGER, W.W., SLITER, W.V. & STEIN, S. (eds) *The Mesozoic Pacific: Geology, Tectonics, and Volcanism*. American Geophysical Union, Geophysical Monograph, **77**, 233–262.

MAHONEY, J.J., FITTON, J.G., WALLACE, P.J. *et al.* 2001. *Proceedings of the Ocean Drilling Program, Initial Reports,* **192** (CD-ROM). Available from: Ocean Drilling Program, Texas A&M University, College Station TX 77845–9547, USA.

MARSH, J.S. 1991. REE fractionation and Ce anomalies in weathered Karoo dolerite. *Chemical Geology*, **90**, 189–194.

MCDONOUGH, W.F. & SUN, S.-S. 1995. The composition of the Earth. *Chemical Geology*, **120**, 223–253.

MENZIES, M. & SEYFRIED, W.E. 1979. Basalt–seawater interaction; trace element and strontium isotopic variations in experimentally altered glassy basalt. *Earth and Planetary Science Letters*, **44**, 463–472.

NEAL, C.R. 2001. The interior of the Moon: The presence of garnet in the primitive, deep lunar mantle. *Journal of Geophysical Research, B, Solid Earth and Planets*, **106**, 27 865–27 885.

NEAL, C.R., MAHONEY, J.J., KROENKE, L.W., DUNCAN, R.A. & PETTERSON, M.G. 1997. The Ontong Java Plateau. *In*: MAHONEY, J.J. & COFFIN, M.F. (eds) *Large Igneous Provinces: Continental, Oceanic, and Planetary Flood Volcanism*. American Geophysical Union, Geophysical Monograph, **100**, 183–216.

PRICE, R.C., GRAY, C.M., WILSON, R.E., FREY, F.A. & TAYLOR, S.R. 1991. The effects of weathering on rare-earth element, Y and Ba abundances in Tertiary basalts from southeastern Austrailia. *Chemical Geology*, **93**, 245–265.

RICHARDS, M.A., DUNCAN, R.A. & COURTILLOT, V. 1989. Flood basalts and hot-spot tracks: plume heads and tails. *Science*, **246**, 103–107.

SEYFRIED, W.E, CHEN, X. & CHAN, L.H. 1998. Trace element mobility and lithium exchange during hydrothermal alteration of seafloor weathered basalt; an experimental study at 350 degrees, 500 bars. *Geochimica et Cosmochimica Acta*, **62**, 949–960.

SIKORA, P.J. & BERGEN, J.A. 2004. Lower Cretaceous planktonic foraminiferal and nannofossil biostratigraphy of Ontong Java sites from DSDP Leg 30 and ODP Leg 192. *In*: FITTON, J.G., MAHONEY, J.J., WALLACE, P.J. & SAUNDERS, A.D. (eds) *Origin and Evolution of the Ontong Java Plateau*. Geological Society, London, Special Publications, **229**, 83–111.

SUN, S.-S. & MCDONOUGH, W.F. 1989. Chemical and isotopic systematics of oceanic basalts: implications for mantle composition and processes. *In*: SAUNDERS A.D. & NORRY, M.J. (eds) *Magmatism in the Ocean Basins*. Geological Society, London, Special Publications, **42**, 313–345.

TEJADA, M.L.G., MAHONEY, J.J., DUNCAN, R.A. & HAWKINS, M.P. 1996. Age and geochemistry of basement and alkalic rocks of Malaita and Santa Isabel, Solomon Islands, southern margin of Ontong Java Plateau. *Journal of Petrology*, **17**, 361–393.

TEJADA, M.L.G., MAHONEY, J.J., NEAL, C.R., DUNCAN, R.A. & PETTERSON, M.G. 2002. Basement geochemistry and geochronology of Central Malaita, Solomon Islands, with implications for the origin and evolution of the Ontong Java Plateau. *Journal of Petrology*, **43**, 449–484.

TEJADA, M.L.G., MAHONEY, J.J., CASTILLO, P.R., INGLE, S.P., SHETH, H.C. & WEIS, D. 2004. Pin-pricking the elephant: evidence on the origin of Ontong Java Plateau from Pb–Sr–Hf–Nd isotopic characteristics of ODP Leg 192 basalts. *In*: FITTON, J.G., MAHONEY, J.J., WALLACE, P.J. & SAUNDERS, A.D. (eds) *Origin and Evolution of the Ontong Java Plateau*. Geological Society, London, Special Publications, **229**, 133–150.

THORDARSON, T. 2004. Accretionary-lapilli-bearing pyroclastic rocks at ODP Leg 192 Site 1184: a record of subaerial phreatomagmatic eruptions on the Ontong Java Plateau. *In*: FITTON, J.G., MAHONEY, J.J., WALLACE, P.J. & SAUNDERS, A.D. (eds) *Origin and Evolution of the Ontong Java Plateau*. Geological Society, London, Special Publications, **229**, 275–306.

TODT, W., CLIFF, R.A., HANSER, A. & HOFMANN, A.W. 1996. Evaluation of a $^{202}Pb+^{205}Pb$ double spike for high-precision lead isotopic analysis. *In*: BASU, A. & HART, S. (eds) *Earth Processes: Reading the Isotopic Code*. American Geophysical Union, Geophysical Monograph, **95**, 429–437.

WHEAT, C.G., MOTTL, M.J. & RUDNICKI, M. 2002. Trace element and REE composition of a low-temperature ridge-flank hydrothermal spring. *Geochimica et Cosmochimica*, **66**, 3693–3705.

WHITE, R., CASTILLO, P.R., NEAL, C.R., FITTON, J.G. & GODARD, M. 2004. Phreatomagmatic eruptions on the Ontong Java Plateau: chemical and isotopic relationship to Ontong Java Plateau basalts. *In*: FITTON, J.G., MAHONEY, J.J., WALLACE, P.J. & SAUNDERS, A.D. (eds) *Origin and Evolution of the Ontong Java Plateau*. Geological Society, London, Special Publications, **229**, 307–323.

WILSON, S.A. 1997. *Data Compilation for USGS Reference Material BHVO-2, Hawaiian Basalt*. US Geological Survey, Open-File Report, in prep. Available online at: http//minerals.cr.usgs.gov/geo_chem_stand/basaltbhv02.html.

WINCHESTER, J.A. & FLOYD, P.A. 1976. Geochemical magma type discrimination: Application to altered and metamorphosed basic igneous rocks. *Earth and Planetary Science Letters*, **28**, 459–469.

YAN, C.Y. & KROENKE, L.W. 1993. A plate tectonic reconstruction of the Southwest Pacific, 100–0 Ma. *In*: BERGER, W.H., KROENKE, L.W., MAYER, L.A. *et al.* (eds) *Proceedings of the Ocean Drilling Program, Scientific Results*, **130**, 697–709.

Geochemistry of Cretaceous volcaniclastic sediments in the Nauru and East Mariana basins provides insights into the mantle sources of giant oceanic plateaus

PATERNO R. CASTILLO

Geosciences Research Division, Scripps Institution of Oceanography, University of California, San Diego, La Jolla, CA 92093-0212, USA (e-mail: pcastillo@ucsd.edu)

Abstract: Cretaceous volcaniclastic sediments sampled at the Nauru and East Mariana basins were chemically and isotopically analysed in order to learn more about the formation and mantle source of the Ontong Java Plateau, an oceanic large igneous province. Despite their variably altered state, the volcanogenic sedimentary components still contain petrogenetic tracers (e.g. high field strength elements, and Nd- and Pb-isotopes) that can be used to constrain the composition of their respective mantle sources. The Nauru volcaniclastics have incompatible trace-element and Nd- and Pb-isotope compositions typical of the Kwaimbaita-type tholeiitic lavas of the Ontong Java Plateau. Combined with the results of recent investigations, the presence of Kwaimbaita-type volcaniclastics in the Nauru Basin reinforces the proposal that the Kwaimbaita-type lavas comprise the bulk of the giant plateau. On the other hand, the East Mariana volcaniclastics have high incompatible trace-element concentrations and Nd- and Pb-isotope ratios typical of alkalic ocean island basalts. Their source is either the Limalok Guyot in the Ratak seamount chain of the Marshall–Gilbert Islands or the ancestral Manihiki–Hikurangi Plateau. Other geological data argue for the Manihiki–Hikurangi Plateau as the source. This implies that the mid-Cretaceous Pacific upper mantle was dominated by the sources of lavas that formed giant oceanic plateaus.

Large igneous provinces (LIPs) are unlike many of the major geological features on the surface of the Earth because their formation cannot be reconciled simply with the plate tectonics theory. The origin of LIPs, therefore, is a hotly debated topic in Earth science (e.g. Coffin & Eldholm 1991; Anderson *et al.* 1992; Smith & Lewis 1999; Hamilton 2003). Oceanic plateaus are examples of LIPs and the Ontong Java Plateau (OJP) in the western Pacific (Fig. 1) is the best example of these. It is the largest LIP, with an area of more than 1.5×10^6 km^2 and a volume of more than 5×10^7 km^3 (Coffin & Eldholm 1994). Earlier studies concluded that the OJP was formed in a single, huge magmatic event at about 122 Ma (e.g. Tarduno *et al.* 1991). Although subsequent studies recognized bimodal ages of the plateau basement lavas (*c.* 122 Ma and *c.* 90 Ma) and that several later, lesser volcanic pulses added volume to the plateau (e.g. Mahoney *et al.* 1993; Tejada *et al.* 1996), preliminary results of the recently concluded drilling of the plateau during Leg 192 of the Ocean Drilling Program (ODP) suggest that the bulk of the plateau may, indeed, have been built during the *c.* 122 Ma volcanic event (Mahoney *et al.* 2001). Another important feature of the OJP is its distinctive lava composition (e.g. Tejada *et al.* 1996, 2002, 2004; Neal *et al.* 1997). The plateau basement consists mainly of Kwaimbaita-type lavas (nomenclature is after Tejada *et al.* 2002, 2004) that have an isotopically distinct, ocean island basalt (OIB)-like composition. These characteristic LIP features, combined with the fact that the OJP was formed in an entirely oceanic setting and hence free from complications of continental lithospheric contamination, make Ontong Java an excellent site to study the formation and constrain the mantle source of an oceanic LIP.

This chapter presents geochemical data for the Cretaceous volcaniclastic sediments in the Nauru Basin and East Mariana Basin (EMB) that lie to the east and north of OJP, respectively (Fig. 1). The principal objective was to gather additional information that could help better constrain the mantle source of the giant plateau. Although Ontong Java is one of the best sampled oceanic plateaus, the igneous samples available to date barely provide a glimpse of the history and origin of the plateau (Tejada *et al.* 2004). The major- and trace-element and Nd- and Pb-isotopic compositions of the Cretaceous volcaniclastics in the two basins are used to determine whether their volcanogenic sedimentary components were produced by plateau-forming volcanism. Such

From: FITTON, J. G., MAHONEY, J. J., WALLACE, P. J. & SAUNDERS, A. D. (eds) 2004. *Origin and Evolution of the Ontong Java Plateau*. Geological Society, London, Special Publications, **229**, 353–368. 0305-8719/$15.00

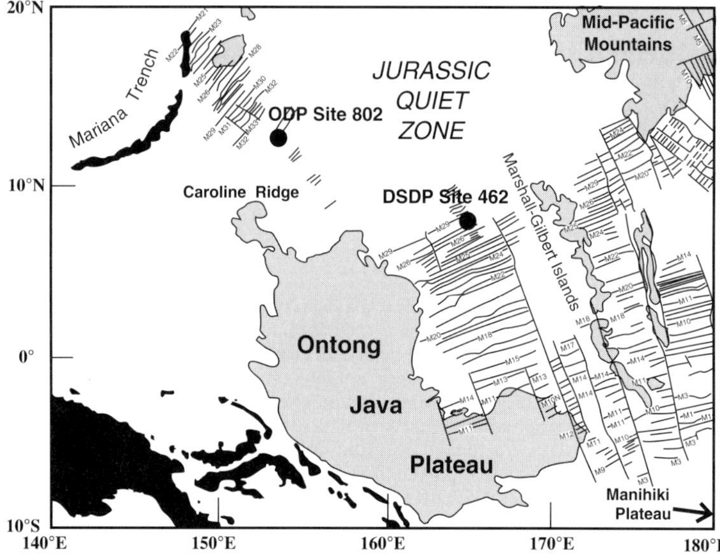

Fig. 1. Map of a section of the western Pacific Ocean showing the locations of Ontong Java Plateau, DSDP Site 462 in the Nauru Basin and ODP Site 802 in the East Mariana Basin. Original map is from Nakanishi *et al.* (1992).

information could provide additional constraints on the volume and extent of volcanism that form oceanic plateaus. These in turn would provide constraints on the overall composition and nature of the mantle source, as well as the process or processes that formed oceanic plateaus and other LIPs (e.g. Bercovici & Mahoney 1994; Larson & Kincaid 1996; Tejada *et al.* 1996, 2004; Ito & Clift 1998).

Samples studied

Nauru Basin volcaniclastics

Site 462 in the Nauru Basin was initially drilled during Deep Sea Drilling Project (DSDP) Leg 61 to sample the Jurassic Pacific crust (Fig. 1) (Larson *et al.* 1981). Drilling at the site bottomed at 1069 metres below seafloor (mbsf) in a mid-Cretaceous (111 Ma; Castillo *et al.* 1994) sequence of sills and flows, collectively called the 'mid-Cretaceous igneous complex', overlain by a thick (*c.* 450 m) sequence of Maastrichtian–Barremian volcaniclastic sediments (sedimentary Unit III and parts of Unit IV at DSDP Site 462; Larson *et al.* 1981). The igneous complex at Site 462 was revisited and deepened during DSDP Leg 89 by about another 100 m, but the Jurassic basement was still not reached (Moberly *et al.* 1986).

The mid- to Late Cretaceous volcaniclastic sediments in the Nauru Basin are generally reworked and redeposited into the basin, and consist mainly of mixed volcaniclastic sandstone and siltstone, and subordinate amounts of conglomerate and/or breccia containing cobble- and pebble-sized basaltic clasts (Larson *et al.* 1981). Most of the sand- and silt-sized epiclasts were originally hyaloclastites, with some horizons still containing clear glassy fragments (Timofeev *et al.* 1981). The other volcaniclastic components are fragments of basalt, igneous minerals, non-volcanogenic materials such as detrital carbonate, plant remains and biogenic debris, and fine-grained interstitial material ('matrix'). Previous studies have emphasized the overall similarity of the volcanogenic sedimentary components and, despite their variable alteration to clay and zeolite minerals, the tholeiitic lava affinity of their bulk chemistry (Moberly & Jenkyns 1981; Timofeev *et al.* 1981).

For this study, three samples of fairly well sorted, sand- and silt-sized hyaloclastites were selected near the bottom of the sequence to determine the bulk composition of the volcaniclastics (Table 1). Samples 462A-41R-7, 0–7 cm and 462A-41R-8, 11–117 cm were taken from a hyaloclastite section containing magnetic spherules approximately 1.5 m above the sediment–igneous complex interface, whereas 462A-80R-1, 113–118 cm was taken from a lower hyaloclastite section intercalated with layers of

Table 1. *Major and trace element compositions of representative Nauru and East Mariana Basin volcaniclastic sediments*

	462A-41R7, 143–150	462A-41R8, 48–55	462A-80R1, 113–118	802A-39R1, 52–55	802A-40R2, 120–125	802A-41R2, 50–55	802A-41R3, 26–30	802A-49R3, 18–24
(wt%)								
SiO_2	45.66	45.48	47.89	49.02	46.08	50.35	50.07	45.89
TiO_2	1.14	0.88	0.94	3.02	2.57	2.50	2.63	1.65
Al_2O_3	12.85	13.29	13.56	11.23	10.31	11.31	11.16	7.71
Fe_2O_3*	11.91	12.20	11.52	9.56	9.69	9.02	9.01	7.79
MgO	7.10	7.49	7.80	6.70	5.99	6.06	6.31	6.98
MnO	0.20	0.21	0.18	0.11	0.22	0.11	0.12	0.17
CaO	10.49	5.31	7.30	5.42	9.63	4.18	4.27	4.47
Na_2O	1.77	5.86	4.35	1.17	1.31	1.52	1.43	1.44
K_2O	0.55	0.59	0.13	0.95	1.24	1.92	1.74	2.07
P_2O_5	0.01	0.01	0.04	0.37	0.34	0.42	0.32	0.30
Total	91.66	91.32	93.72	87.54	87.37	87.39	87.06	78.47
(ppm)								
Ba	16.8	12.6	132	45.0	121	104	104	77.3
Co	49.0	34.7	34.2	66.2	62.6	53.9	46.2	43.4
Cu	183	179	137	150	98.6	134	113	164
Cr	183	282	216	290	275	267	243	450
Hf	5.6	3.4	3.5	12.8	12.1	13.0	16.0	8.6
Nb	4.1	2.0	2.1	28	29	31	27	16
Ni	120	110	88	151	177	186	164	244
Pb	1.0	1.3	0.52	2.6	2.7	3.7	3.2	2.7
Rb	3.20	2.53	1.72	13.44	20.62	34.00	22.81	28.09
Sc	52.8	37.8	43.5	29.4	24.5	25.4	22.2	23.9
Sr	89.87	12.84	404.83	201.29	338.67	381.40	284.94	266.46
Ta	0.82	0.43	0.33	2.10	2.04	2.25	2.04	1.33
Th	0.48	0.40	0.31	2.13	2.40	2.81	2.56	1.96
U	0.40	0.41	0.28	0.77	0.79	0.97	0.83	0.69
V	364.8	253.0	286.4	321.2	230.5	273.4	239.3	199.8
Y	22.46	15.01	16.11	23.18	24.47	29.40	19.42	20.17
Zn	95.4	80.7	80.5	185.6	188.8	144.4	128.3	95.9
La	3.42	2.28	2.28	22.73	23.56	27.36	19.93	20.55
Ce	8.83	5.59	6.51	46.79	51.55	55.08	42.66	41.41
Pr				6.58	7.51	8.05	6.03	5.84
Nd	6.51	4.13	4.88	26.99	30.44	32.73	24.77	22.95
Sm	2.15	1.49	1.60	5.70	6.35	6.77	5.24	4.99
Eu	0.89	0.68	0.63	1.71	1.80	1.90	1.60	1.64
Tb	0.54	0.45	0.38	0.78	0.77	0.82	0.66	0.69
Dy	3.49		2.43	4.61	4.77	5.39	3.98	3.99
Ho	0.71	0.57	0.50	0.80	0.77	0.85	0.71	0.76
Er	2.05	1.46	1.42	2.14	2.05	2.35	1.82	1.90
Tm	0.27	0.23		0.30	0.28	0.28	0.26	0.27
Yb	1.92	1.32	1.47	1.63	1.57	1.95	1.53	1.52
Lu	0.26	0.21	0.19	0.23	0.22	0.25	0.22	0.22

Replicate analyses of sediment standards show precision to be better than 2.5% for Si, Ti, Al, Fe, Mg, Mn, Ca and K; Na and P did not reproduce well. Reproducibility of standard JB-1 for most of the trace elements was satisfactorily high (to within 2.5% relative standard deviation (RSD)) except for Hf, Ho, Tm, Lu and Ta, which was as low as *c.*5% RSD. Fe_2O_3* is total Fe reported as Fe_2O_3.

fine-grained diabase or coarse-grained basalt members of the igneous flow unit (Unit IV).

East Mariana Basin volcaniclastics

Site 802 in the EMB was drilled during ODP Leg 129, also to sample the Jurassic Pacific crust (Lancelot *et al.* 1990; Larson *et al.* 1992). As in the Nauru Basin, drilling at Site 802 did not reach the Jurassic igneous basement, but bottomed at 509 mbsf in a lava-flow sequence (115 Ma; Pringle 1992; Castillo *et al.* 1994) that also belongs to the mid-Cretaceous igneous complex (Castillo *et al.* 1994; Janney & Castillo 1996). Above the lava flows are thick sequences of middle Pliocene/late Miocene–early Miocene (220 m) and late Campanian–Coniacian/Cenomanian (110 m) volcaniclastic sediments.

Both the Miocene and Cretaceous volcanogenic sedimentary components have been reworked and consist of fragments of volcanic glass, basalt clasts and igneous minerals such as olivine, spinel, pyroxene, amphibole, mica and feldspar (Lancelot *et al.* 1990; Lees *et al.* 1992). As in Nauru, both East Mariana volcaniclastic sequences contain variable amounts of non-volcanogenic components consisting of biogenic fragments, shallow-water carbonate debris and wood fragments. The East Mariana sequences are also dominated by hyaloclastites, and all the volcanogenic materials have been moderately to strongly altered to clays and zeolites. Such preponderance of hyaloclastites and presence of shallow-water carbonate and wood fragments strongly suggest that both East Mariana and Nauru volcaniclastics were derived from once emergent volcanic sources and/or from relatively shallow-water (<400 m) limbs of seamount or island (Floyd 1986; Lees *et al.* 1992). Previous analyses of several Miocene and three mid-Cretaceous bulk samples and individual basaltic clasts by Lees *et al.* (1992) indicate that the igneous components of both volcaniclastic sequences in the EMB are very similar and point toward an intra-plate alkali basaltic source. The Miocene volcaniclastic sequence can be traced thickening towards the south and indicates a possible source in the Caroline Ridge (Fig. 1) (Lancelot *et al.* 1990), but the source or sources of the mid-Cretaceous volcaniclastic sequence at Site 802 was not established (Lees *et al.* 1992). For this study, five mid-Cretaceous samples were taken from the volcaniclastic turbidites Unit V (Lancelot *et al.* 1990) for bulk-rock analysis. As above, the samples analysed are all fine-grained, containing sand- to silt-sized volcanogenic components cemented by finer-grained matrix.

Analytical methods

Analyses of major elements and Co, Cr, Ni, Sc, Sr, V and Zn were performed through inductively coupled plasma-optical emission spectrophotometry (ICP-OES; method similar to that described in Mahoney *et al.* 2001) at the Scripps Institution of Oceanography (SIO) using a Perkin-Elmer Optima 3000 DV instrument. The volcaniclastics were powdered, dried overnight at *c.* 110°C and fused in C crucibles at 1100°C for approximately 20 min inside a furnace after addition of a Li metaborate: teraborate mixed flux (1 $LiBO_2$:2 $Li_2B_4O_7$ ratio) to the powder (5 parts flux:1 part rock powder). The fused beads were dissolved, and then finally diluted by a factor of approximately 4000 in 3% HCl. Standardization for major- and trace-element analysis was conducted using international igneous standard BHVO-1 and sedimentary standards MAG-1, SGR-1 and SDO. Accuracy and precision of the analyses were monitored by repeated measurements of the sedimentary standards as unknowns; estimated errors are given in footnotes of Table 1.

Rare earth elements (REE), Ba, Hf, Nb, Pb, Rb, Ta, Th, U and Y of the volcaniclastics, as well as Sm, Nd, Rb, and Sr of the leached powders analysed for isotopes, were determined by inductively coupled plasma-mass spectrometry (ICP-MS) on a Finnigan Element 2 high-resolution ICP-MS. ^{115}In was used as an internal standard and calibration was conducted using a series of three synthetic multi-element standard solutions and one blank solution. The procedure used is similar to that described by Janney & Castillo (1996) with some modifications. Prior to analysis, rock powders (*c.* 0.025 g) were dissolved in clean Teflon vessels using approximately 1 ml of a 2:1 mixture of concentrated HF and HNO_3, and heated on a hotplate at low power for 16–24 h. The resulting solution was evaporated to dryness, resuspended in a small amount of concentrated HNO_3 and evaporated to dryness twice. The samples were finally diluted to a factor of *c.* 4000 in a 2.5% HNO_3 solution containing 1 ppb ^{115}In. Hafnium, Nb and Ta were determined immediately following dilution using the method described above. The rest of the trace elements were determined later. Accuracy and precision of the analyses were monitored by repeated measurements of the international rock standard JB-1 and an in-house standard; estimated errors are given in the footnotes of Table 1.

Sr-, Nd- and Pb-isotopic measurements were made on four samples, two from each site, using a Micromass Sector 54 multi-collector thermal ionization mass spectrometer at SIO. In order to minimize the effects of sea-water alteration on the Sr- and Nd-isotopic composition, the powders of the samples were subjected to a harsh multi-step leaching procedure similar to that of Mahoney (1987) and Castillo *et al.* (1991). Unleached powders were used for Pb- and for two duplicate Nd-isotope analyses. Both the leached and unleached powders were digested using the ICP-MS dissolution procedure described above. Strontium and REE were first separated in primary cation-exchange columns; Nd was separated from the rest of the REE by passing the REE cuts through small EDTA ion-exchange columns. Lead was separated using a

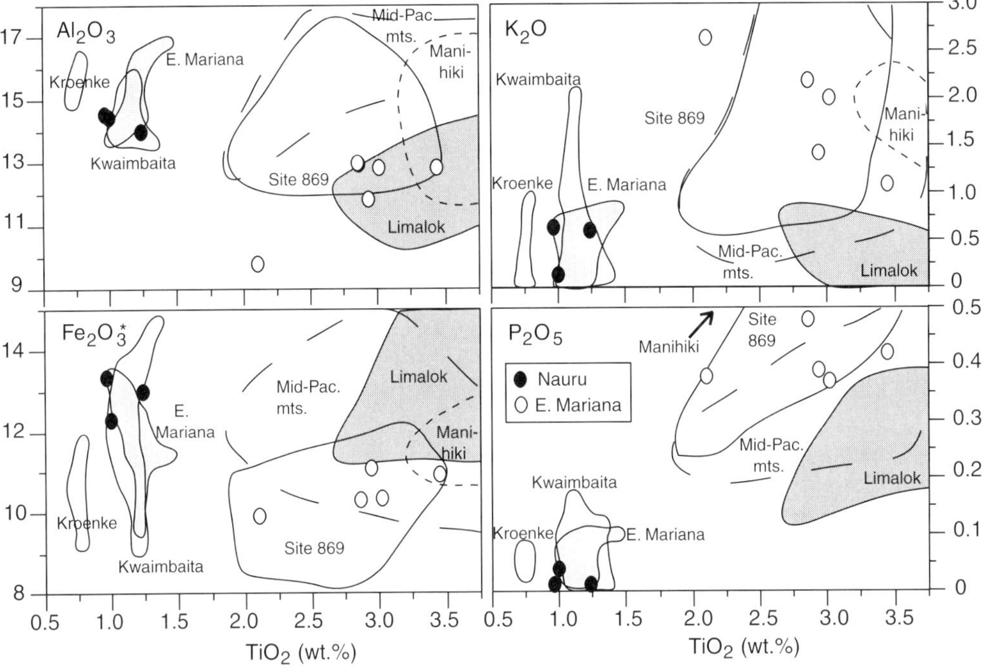

Fig. 2. TiO_2 v. Al_2O_3, Fe_2O_3* (total Fe), K_2O and P_2O_5 (all in wt%) for the Nauru (DSDP Site 462) and East Mariana (ODP Site 802) volcaniclastic sediments. Shown for reference are fields for the recently drilled Kwaimbaita- and Kroenke-type lavas from the Ontong Java Plateau (Mahoney *et al.* 2001), East Mariana Basin basement tholeiites (Janney & Castillo 1996), ODP Site 869 volcaniclastic sediments and basaltic clasts (Janney *et al.* 1995), Mid-Pacific Mountains lavas (Baker *et al.* 1995) and lavas from seamounts on the Manihiki Plateau (Beiersdorf *et al.* 1995).

standard anion-exchange method in a HBr–HNO_3 medium (e.g. Lugmair & Galer 1992; Janney & Castillo 1996).

Results

Major oxides

It can be seen that the major elements do not sum close to 100 wt%, averaging only about 92 wt% for the Nauru Basin and 86 wt% for the EMB volcaniclastic sediments (Table 1). This is due to the high volatile contents of the samples (Timofeev *et al.* 1981; Lees *et al.* 1992), mainly from hydration of the glassy fragments during devitrification, and of the basaltic clasts and igneous minerals during sea-water alteration. Additional major-element modification arises due to the diluting effects of the silica and calcium contents of the marine siliceous and calcareous components of the volcaniclastics. Thus, the major-element analyses may be unreliable for determining the petrogenetic affinity of the samples. Nevertheless, when the major elements were recalculated on a volatile-free basis, some systematic chemical characteristics can be observed. The Nauru volcaniclastics have lower contents of TiO_2, a major oxide less susceptible to sea-water alteration (e.g. Cann 1970; Pearce & Cann 1973) than East Mariana volcaniclastics (Fig. 2). For given TiO_2, the former have higher Al_2O_3, Fe_2O_3* and MgO, but lower K_2O and P_2O_5, than the latter. In terms of major oxides, the Nauru volcaniclastics plot with the Kwaimbaita-type lavas drilled during ODP Leg 192 (Mahoney *et al.* 2001) and EMB tholeiitic basalts, which are compositionally very similar to the Nauru Basin tholeiitic basalts (Castillo *et al.* 1991, 1992, 1994; Janney & Castillo 1996). Surprisingly, none of the East Mariana volcaniclastics is compositionally similar to the OJP lavas. Instead, they show a wide range of values and overlap with the Cretaceous alkalic OIB from the Allison and Resolution guyots of the Mid-Pacific Mountains (Christie *et al.* 1995) and hyaloclastites and basaltic clasts in the Cretaceous volcaniclastic sediments from ODP Site 869, which are products of the early, mildly to moderately alkalic stage of volcanism that formed the Pikini Atoll–Wodejebato Guyot of the Marshall-Gilbert Islands (Janney *et al.* 1995). They also partially overlap with the Cretaceous alkalic OIBs from the Limalok Guyot of the Marshall-Gilbert Islands (Christie *et al.* 1995) and the Manihiki Plateau (Beiersdorf *et al.* 1995).

Trace elements

Sea-water alteration is also widely known to change the concentration of many trace elements in submarine lavas (e.g. Hart *et al.* 1974; Verma 1992). However, certain trace elements such as the high field strength elements (HFSE; e.g. Nb, Y and including Ti) have been shown to be highly resistant to mobilization by sea water (e.g. Cann 1970; Pearce & Cann 1973; Hart *et al.* 1974; Bienvenu *et al.* 1990). To a certain extent, the REE (e.g. La, Sm, Yb) abundances are also not appreciably changed by interaction with sea water (Ludden & Thompson 1978; Staudigel & Hart 1983; Bienvenu *et al.* 1990). Thus, despite the alteration effects, important trace-element characteristics of the volcaniclastics can still be identified using these elements.

The volcaniclastics show wide ranges of incompatible-element contents (Table 1), but as a whole these elements correlate with one another (Fig. 3). Nauru volcaniclastics show a lesser range and total contents of incompatible elements than East Mariana volcaniclastics, not because of their small number but due to their tholeiitic compositional characteristics. Almost all East Mariana volcaniclastics have Nb/Y >1, which is typical of alkalic basalts (Winchester & Floyd 1976), whereas Nauru volcaniclastics have Nb/Y <1. Alkalic East Mariana volcaniclastics have higher Ti/Nb than do tholeiitic Nauru volcaniclastics. That the Nauru volcaniclastics have lower incompatible trace-element contents than East Mariana volcaniclastics is also clearly shown by the relatively flat REE pattern of the former and the light REE (LREE)-enriched pattern of the latter (Fig. 4). Another important geochemical feature is the similarity of the REE patterns in the volcaniclastics within each site, consistent with their distinct incompatible-element ratios.

The Nauru volcaniclastics consistently overlap with the average values for the tholeiitic basalts sampled from the basement of the OJP (e.g. Tejada *et al.* 1996, 2002; Neal *et al.* 1997) and East Mariana and Nauru basins (Saunders 1986; Castillo *et al.* 1991, 1994) in all the trace-element plots (Figs 3 and 4). Previous studies have shown that these lavas came from petrogenetically related mantle sources (Castillo *et al.* 1991,1994; Mahoney & Spencer 1991; Janney & Castillo 1996; Tejada *et al.* 1996, 2002; Neal *et al.* 1997). On the other hand, the East Mariana volcaniclastics show considerable overlap with typical Cretaceous OIB such as those from the Resolution and Allison guyots of the Mid-Pacific mountains, and the hyaloclastites and basaltic clasts from ODP Site 869 at the apron of Wodejebato Guyot.

Sr-, Nd- and Pb-isotopes

The Sr-, Nd- and Pb-isotope ratios of the volcaniclastics are presented in Table 2. Ratios of $^{143}Nd/^{144}Nd_{(meas.)}$ and $^{87}Sr/^{86}Sr_{(meas.)}$ of the Nauru volcaniclastics are similar to each other within errors; $^{143}Nd/^{144}Nd_{(meas.)}$, $^{206}Pb/^{204}Pb_{(meas.)}$ and $^{208}Pb/^{204}Pb_{(meas.)}$ values of the East Mariana volcaniclastics are also similar to each other. As a whole, the Nauru volcaniclastics have higher $^{143}Nd/^{144}Nd_{(meas.)}$ and lower $^{206}Pb/^{204}Pb_{(meas.)}$ and $^{208}Pb/^{204}Pb_{(meas.)}$ values than the East Mariana volcaniclastics. Interestingly, their $^{87}Sr/^{86}Sr_{(meas.)}$ and $^{207}Pb/^{204}Pb_{(meas.)}$ ratios overlap.

Relative to some Cretaceous samples from the western Pacific, the Pb-isotope ratios of East Mariana volcaniclastics overlap with those of the Allison, Resolution and Wodejebato guyots, volcaniclastic sediments from the apron of Wodejebato sampled at ODP Site 869 (Janney & Castillo 1999) and Manihiki Plateau (Mahoney & Spencer 1991) (Fig. 5a). The East Mariana volcaniclastics have similar $^{206}Pb/^{204}Pb_{(meas.)}$, but lower $^{207}Pb/^{204}Pb_{(meas.)}$ and $^{208}Pb/^{204}Pb_{(meas.)}$ ratios than the Limalok Guyot of the Marshall-Gilbert Islands (Koppers *et al.* 1995). Sample 462A-80R1, 113–118 cm has similar Pb-isotope ratios to both the Kwaimbaita- and Singalo-type lavas, but sample 462A-41R7, 143–150 cm has higher ratios. All the volcaniclastics have lower Pb-isotope ratios than those for the Sigana alkalic suite of the OJP.

To better compare with the Cretaceous OJP lavas, measured Sr- and Nd-isotope ratios were age-corrected to 120 Ma (Table 2; cf. Tejada *et al.* 1996, 2002, 2004). A general feature of the volcaniclastics from either basin is their relatively high Sr-isotope ratios. Even after age-correction, the volcaniclastics generally have higher $^{87}Sr/^{86}Sr_{(t)}$ than the Kwaimbaita- and Singgalo-type lavas save for sample 802-39R-1, 52–55 cm from the EMB, which overlaps with the Singgalo-type lavas (cf. Mahoney & Spencer 1991; Neal *et al.* 1997; Tejada *et al.* 1996, 2002,

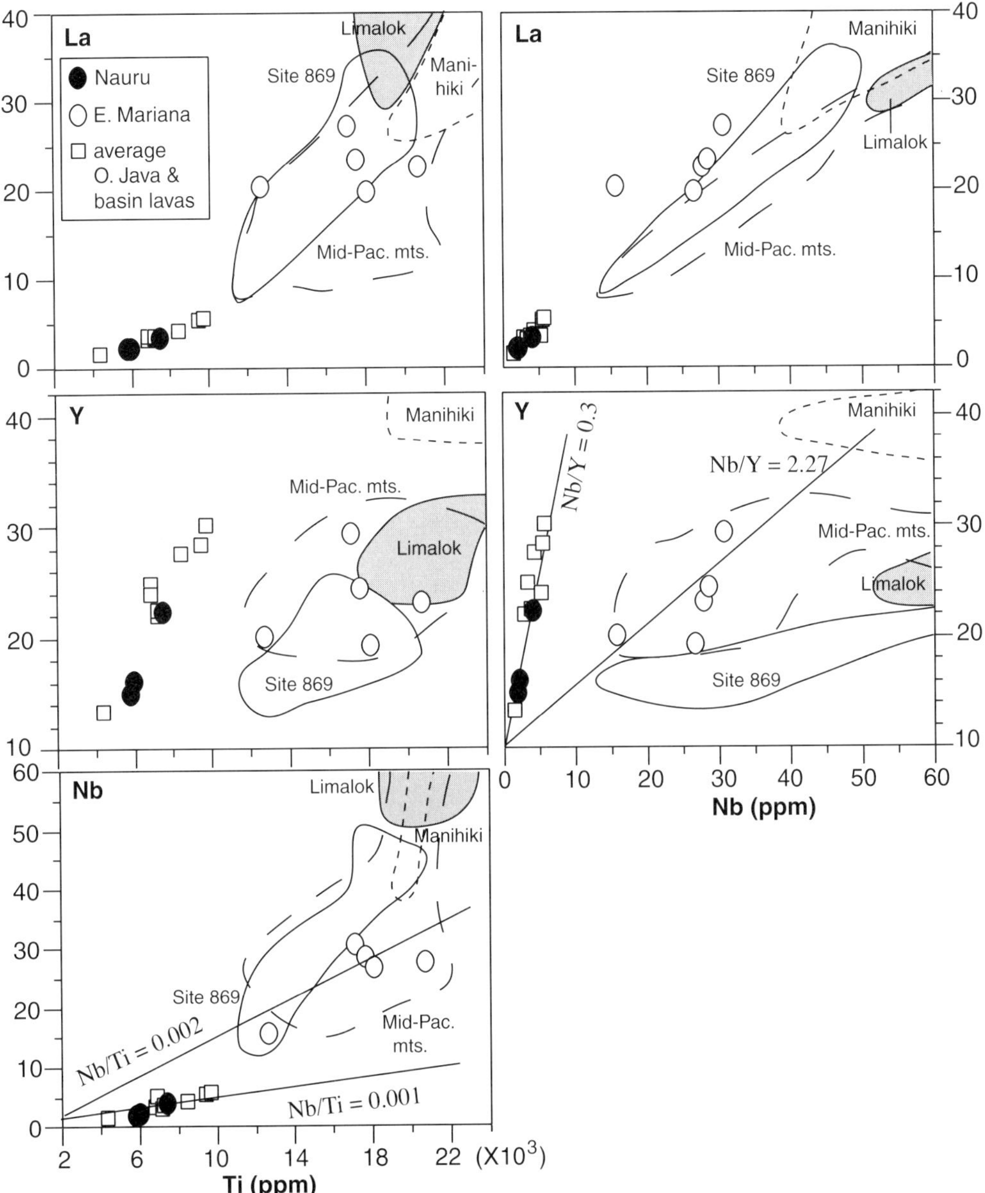

Fig. 3. Titanium v. La, Y and Nb and Nb v. La and Y (all in ppm) for the Nauru (DSDP Site 462) and East Mariana (ODP Site 802) volcaniclastic sediments. Shown for reference are average values of Cretaceous basement lavas from the Ontong Java Plateau (Neal *et al.* 1997) and Nauru and East Mariana basins (Saunders 1986; Castillo *et al.* 1991, 1994), and fields for the ODP Site 869 volcaniclastic sediments and basaltic clasts (Janney *et al.* 1995), Mid-Pacific Mountains lavas (Baker *et al.* 1995), and lavas from seamounts on the Manihiki Plateau (Beiersdorf *et al.* 1995). Lines in the Ti v. Nb and Nb v. Y plots represent average Nb/Y and Nb/Ti ratios of Nauru and East Mariana volcaniclastics.

2004). The $^{87}Sr/^{86}Sr_{(t)}$ ratios of East Mariana volcaniclastics also overlap with the silica-undersaturated and MgO- and alkali-rich alnöite lavas from the island of Malaita, which have higher Sr- for given Nd-isotope ratios than the Ontong Java tholeiitic lavas (Neal & Davidson 1989; Tejada *et al.* 1996). The volcaniclastics have higher $^{87}Sr/^{86}Sr_{(t)}$ and $^{143}Nd/^{144}Nd$ (expressed in epsilon notation $\varepsilon_{Nd(t)}$) values than the Allison and Resolution guyots, Wodejebato Guyot,

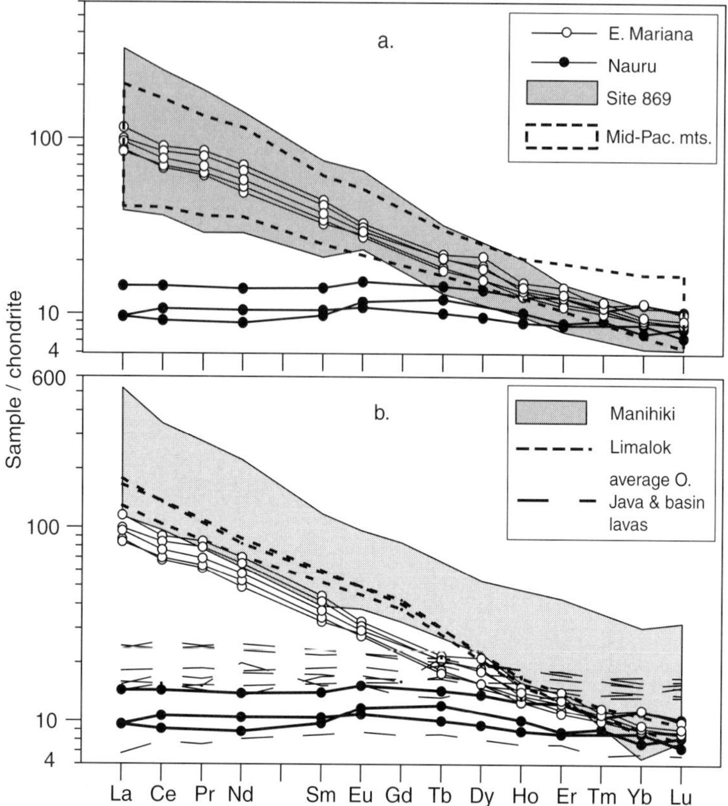

Fig. 4. Chondrite-normalized REE values for the Nauru (DSDP Site 462) and East Mariana (ODP Site 802) volcaniclastic sediments. Note the small intra-group variation in REE concentrations for each group. In (**a**), the data are plotted against the background fields for the ODP Site 869 volcaniclastic sediments and basaltic clasts, Mid-Pacific Mountains lavas (Janney *et al.* 1995), and Allison and Resolution guyots of the Mid-Pacific mountains (Baker *et al.* 1995). In (**b**), the data are plotted relative to average values of Cretaceous basement lavas from the Ontong Java Plateau (Neal *et al.* 1997) and Nauru and East Mariana basins (Castillo *et al.* 1991, 1994) (both as dashed lines), Limalok Guyot of the Marshall-Gilbert Islands (Christie *et al.* 1995) and Manihiki seamounts (Beiersdorf *et al.* 1995). Chondrite-normalizing values are from Sun & McDonough (1989).

volcaniclastic sediments from the apron of Wodejebato sampled at ODP Site 869 (Janney & Castillo 1999) and the Limalok Guyot (Koppers *et al.* 1995).

The Kwaimbaita-type lavas and average leached and unleached Nauru volcaniclastics have similar $\varepsilon_{Nd(t)}$ values (Fig. 5b). The $^{206}Pb/^{204}Pb_{(meas.)}$ ratio of sample 462A-80R-1, 113–118 cm also completely overlaps with the Kwaimbaita-type lavas. The high $^{206}Pb/^{204}Pb_{(meas.)}$ ratio of sample 462A-41R-7, 143–150 cm, however, is unlike that of any Ontong Java lavas; it partially overlaps with the Resolution Guyot. The East Mariana volcaniclastics and Singgalo-type lavas have similar $\varepsilon_{Nd(t)}$ values but the former have higher measured (and age-corrected) $^{206}Pb/^{204}Pb$ ratios than the $^{206}Pb/^{204}Pb_{(meas.)}$ ratios of the Kwaimbaita-type, Singgalo-type and Alnöite lavas. The $\varepsilon_{Nd(t)}$ and $^{206}Pb/^{204}Pb_{(meas.)}$ values of East Mariana volcaniclastics overlap with those of the dredged lavas from the Manihiki Plateau (Mahoney & Spencer 1991); they also plot very close to those of the Cretaceous Limalok Guyot (Koppers *et al.* 1995).

Discussion

Available geochemical data for lavas from widely separated drill holes and from outcrops in several islands on the OJP show some variations in the composition of the Cretaceous

Table 2. *Sr-, Nd- and Pb-isotope ratios of Nauru and East Mariana Basin volcaniclastic sediments*

Sample	$^{87}Sr/^{86}Sr$ (measured)	$^{143}Nd/^{144}Nd$ (measured)	Rb	Sr	Sm	Nd	$^{87}Sr/^{86}Sr$ (120 Ma)	$^{143}Nd/^{144}Nd$ (120 Ma)	$\varepsilon_{Nd(t)}$ (120 Ma)	$^{206}Pb/^{204}Pb$ (measured)	$^{207}Pb/^{204}Pb$ (measured)	$^{208}Pb/^{204}Pb$ (measured)	$^{206}Pb/^{204}Pb$ (120 Ma)	$^{207}Pb/^{204}Pb$ (120 Ma)	$^{208}Pb/^{204}Pb$ (120 Ma)
East Mariana Basin															
802A-39R1, 52–55 cm	0.704263	0.512836	7.5	206.7	0.94	4.24	0.704084	0.512731	4.8	19.199	15.600	39.092	18.850	15.583	38.774
802A-41R3, 26–30 cm	0.705073	0.512820	11.5	407.5	0.64	3.57	0.704933	0.512735	4.9	19.145	15.617	39.068	18.838	15.602	38.757
Nauru Basin															
462A-41R7, 143–150 cm	0.704363	0.513012	1.09	64.7	0.69	1.98	0.704280	0.512847	7.0	19.106	15.632	38.615	18.623	15.608	38.429
Duplicate		0.512958			2.15	6.51		0.512802	6.2						
462A-80R1, 113–118 cm	0.704379	0.512980	0.41	14.9	1.60	4.64	0.704243	0.512817	6.5	18.573	15.528	38.384	17.939	15.497	38.155
Duplicate		0.512950			1.60	4.88		0.512795	6.0						

Strontium- and Nd-isotope ratios were measured on leached powders. Duplicate analyses were made on unleached powders. Analytical uncertainty for $^{87}Sr/^{86}Sr$ measurements is ±0.000018 but in-run precisions were better than ±0.000012. Strontium-isotope ratios were measured by dynamic multi-collection, fractionation corrected to $^{86}Sr/^{88}Sr = 0.1194$ and normalized to $^{87}Sr/^{86}Sr = 0.71025$ for NBS 987. Analytical uncertainty for $^{143}Nd/^{144}Nd$ measurements is ±0.000014 (0.3 ε units) but in-run precisions were better than ±0.000010. Neodymium-isotope ratios were measured in oxide form by dynamic multi-collection, fractionation corrected to $^{146}NdO/^{144}NdO = 0.72225$ ($^{146}Nd/^{144}Nd = 0.7219$) and are reported relative to $^{143}Nd/^{144}Nd = 0.511850$ for the La Jolla Standard. Pb-isotope ratios were measured by static multi-collection and are reported relative to the values of Todt *et al.* (1996) for NBS SRM 981; the long-term errors measured for this standard are ± 0.008 for $^{206}Pb/^{204}Pb$ and $^{207}Pb/^{204}Pb$, and ± 0.024 for $^{208}Pb/^{204}Pb$; the in-run precisions were better than these. Total procedural blanks are negligible: <10 picograms (pg) for Nd, <35 pg for Sr, <3 for Th, <5 pg for U and <60 pg for Pb. Uranium, Th and Pb elemental concentrations used in the age-corrections for Pb-isotopes are given in Table 1. $\varepsilon_{Nd} = 0$ today corresponds to $^{143}Nd/^{144}Nd = 0.51264$; for $^{147}Sm/^{144}Nd = 0.1967$, $\varepsilon_{Nd(t)} = 0$ corresponds to $^{143}Nd/^{144}Nd = 0.512486$ at 120 Ma. Rb, Sr, Sm and Nd concentrations are in ppm.

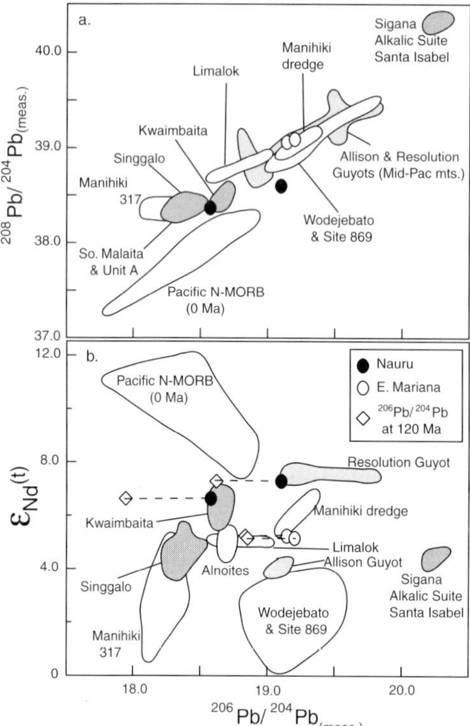

Fig. 5. Isotopic variations for the Nauru (DSDP Site 462) and East Mariana (ODP Site 802) volcaniclastic sediments. (**a**) $^{206}Pb/^{204}Pb_{(t)}$ v. $^{208}Pb/^{204}Pb_{(t)}$ and (**b**) $^{206}Pb/^{204}Pb_{(t)}$ v. $\varepsilon_{Nd(t)}$. Dashed line connects present-day (measured) and age-corrected to 120 Ma values. Fields for the Kwaimbaita- and Singgalo-type lavas, Sigana and Santa Isabel alkalic suite and alnöites of the Ontong Java Plateau (Neal *et al.* 1997; Tejada *et al.* 2002), Manihiki Plateau (Mahoney 1987; Mahoney & Spencer 1991), Resolution and Allison guyots in the Mid-Pacific Mountains (Janney & Castillo 1999), Limalok Guyot in the Marshall Islands (Koppers *et al.* 1995; Koppers 1998), and ODP Site 869 and Wodejebato Guyot in the Marshall Islands (Janney & Castillo 1999) are shown for reference. $^{206}Pb/^{204}Pb$ ratios are not corrected for age because of a lack of U, Th and Pb concentration data for Manihiki Plateau, ODP Site 869, Mid-Pacific Mountains and Marshall-Gilbert Islands lavas.

basement of the plateau. Lavas range from relatively primitive Kroenke-type tholeiitic basalts (up to 10.9 wt% MgO and 227 ppm) to fairly differentiated Kwaimbaita-type tholeiites; a few alkalic basalts were also sampled from the islands (e.g. Tejada *et al.* 1996, 2002; Neal *et al.* 1997; Mahoney *et al.* 2001; Fitton & Godard 2004). Most of the geochemical variations in the tholeiitic basalts, however, can be attributed to different degrees of partial melting of a common source followed by magmatic differentiation, mainly through fractional crystallization, of the resultant magmas (e.g. Tejada *et al.* 1996, 2002; Neal *et al.* 1997; Fitton & Godard 2004). Indeed, Sr-, Nd- and Pb-isotope ratios indicate that these lavas can be traced back to distinct OIB-like mantle sources dominated by the source of the Kwaimbaita-type lavas, and a lesser amount of the source of the Singgalo-type lavas. Thus, the voluminous plateau lavas have a remarkably narrow range of mantle source composition (see also Tejada *et al.* 2004). In terms of trace-element compositions, both lava types and their respective source are very similar, and the only significant distinction is that the Singgalo-type is slightly more enriched in highly incompatible trace elements than the Kwaimbaita-type (e.g. Tejada *et al.* 1996, 2002; Neal *et al.* 1997; Fitton & Godard 2004).

Available age data suggest that the formation of the OJP (at 122Ma) coincided with the widespread, 'mid-Cretaceous volcanic event' that formed numerous volcanic seamounts and a number of oceanic plateaus and seamounts in the Pacific (e.g. Winterer *et al.* 1973; Larson *et al.* 1981; Moberly *et al.* 1986; Larson 1991; Janney & Castillo 1996). Concurrent with this volcanic event was the shedding of volcanic debris from the slopes of the volcanoes, via mass flow and turbidity currents, onto adjacent apron and basinal areas where it was deposited as volcaniclastic sediment (e.g. Kelts & Arthur 1981; Moberly & Jenkyns 1981; Floyd 1986; Lees *et al.* 1992; Janney *et al.* 1995; Janney & Castillo 1999). For this reason, Cretaceous volcaniclastic deposits are present throughout large areas of the western Pacific and have been sampled at a number of DSDP and ODP drill sites. The volcanogenic components of many of these sedimentary deposits can be traced back to nearby, emergent volcanic edifices that are at most 150 km away (e.g. ODP Site 800 to Himu and/or Golden Dragon Seamount – Lees *et al.* 1992; DSDP Sites 199 and 585 to Ita Maitai Guyot – Moberly *et al.* 1985; ODP Site 869 to Pikini–Wodejebato Guyot – Janney *et al.* 1995; Janney & Castillo 1999).

The sources of the Cretaceous volcaniclastic sediments at DSDP Site 462 in the Nauru Basin and ODP Site 802 in the EMB are more problematic because drilling at these sites was specifically aimed at reaching the Jurassic basement, and hence both sites were deliberately located well away (at least 300 km) from volcanic seamounts or islands. The fact that there are thick volcaniclastic sedimentary deposits in these basins clearly indicates that the volcanic activity that produced the volcanogenic sedimentary components was vigorous in nature and volumetrically significant. Previous studies of the Nauru and East Mariana volcaniclastic sediments produced good descriptions of the deposits. The moderate to strongly altered states of the volcanogenic sedimentary components and the dearth of alteration-resistant petrogenetic tracers (e.g. ratios of alteration-resistant trace elements and long-lived radiogenic isotopes) in the volcaniclastics and/or the source lavas available at the time of investigations, however, made the link between the volcaniclastics and any particular source or sources tenuous at best (Timofeev *et al.* 1981; Moberly & Jenkyns 1981; Lees *et al.* 1992). For example, the gross compositional similarity of the Nauru volcaniclastics with the underlying tholeiitic basalts comprising the igneous complex led early workers to suggest that the source of the volcanogenic components is within the underlying complex itself (e.g. Moberly & Jenkyns 1981). As noted earlier, however, the presence of fragments of shallow-water carbonate and plant remains in the volcaniclastics necessitates that the source of the volcanogenic components be fairly shallow if not totally emergent (Larson *et al.* 1981). The fact that the complex is flat-lying and generally devoid of emergent structural or morphological features precludes it from being the source of the large amounts of the volcanogenic components. The source of East Mariana volcaniclastics at Site 802, on the other hand, had been identified simply as nearly emergent mid-Cretaceous seamounts (Lees *et al.* 1992).

Despite the moderate to strong alteration, some of the major- and trace-element contents of Nauru and East Mariana volcaniclastics display coherent features that are typical of comparably less altered oceanic lavas. These petrogenetic tracers, particularly the ratios of incompatible elements such as La/Nb, Nb/Ti, La/Sm and others that have been shown to be useful indicators of mantle source composition (e.g. Erlank & Kable 1976; le Roex 1986; Weaver 1991), indicate that the volcanogenic components in the Nauru Basin volcaniclastic sediments and the Ontong Java basement lavas both originated from the same mantle source. It is most likely that the volcaniclastic sediments were derived from the OJP by mass-wasting. On the other hand, the EMB volcaniclastics have an alkalic OIB affinity that resembles those of some mid-Cretaceous lavas (Figs 2–4).

There are not enough major- and trace-element analyses of the Nauru volcaniclastics to determine whether they have either Kwaimbaita- or Singgalo-type lava affinity. Moreover, the East Mariana volcaniclastics have a range in composition that suggests several possible alkalic OIB sources. Hence, the purpose of analysing the Sr-, Nd- and Pb-isotope ratios, the more conservative and also most popular petrogenetic tracers, is to determine which of the two Ontong Java mantle types is the source of Nauru volcaniclastics and which among the mid-Cretaceous lavas is geochemically closest to the East Mariana volcaniclastics.

As mentioned earlier, the Sr-isotope ratios of the volcaniclastics are relatively high despite the harsh acid-leaching procedure employed on the sample powders prior to dissolution (Table 2). This suggests that the Sr-isotope systematics of the volcaniclastics has been altered beyond mitigation by acid-leaching and cannot be used as a petrogenetic tracer of the volcaniclastic source. This is not surprising because Sr-isotopic ratios of oceanic rocks are easily increased by sea-water alteration (e.g. McCulloch *et al.* 1981; Cheng *et al.* 1987; White 1993). On the other hand, sea water is exceedingly low in Nd (4×10^{-6} ppm) and Pb (2×10^{-6} ppm) contents, and it is commonly observed that the Nd- and Pb-isotope ratios of old and variably altered oceanic lavas are generally not affected by sea-water alteration (e.g. Castillo *et al.* 1991; Mahoney *et al.* 1993, 1998; Frey & Weis 1995). Only severe sea-water alteration can affect both the Nd- and Pb-isotope ratios (e.g. Cheng *et al.* 1987). Thus, it is hoped that both the Nd- and Pb-isotope systematics of Nauru and East Mariana volcaniclastics have remained closed to alteration.

The Nd-isotope ratios of Nauru volcaniclastics may have been slightly affected by sea-water alteration because the measured and age-corrected values of the leached aliquots are slightly different and are systematically higher than the values of the unleached aliquots whose measured and age-corrected values are closer to each other. On average, the Nd-isotope ratios of unleached Nauru volcaniclastics are higher than the Singgalo-type lavas but they are similar to those of the Kwaimbaita-type lavas (Fig. 5b). Combined with the major- and trace-element data, particularly the coherent REE patterns (Fig. 4), it is thus safe to conclude that the Nd-isotopic systematics of Nauru volcaniclastics has not been severely affected by sea-water alteration, and that the Nauru volcaniclastics most probably were originally Kwaimbaita-type lavas from the OJP. Sample 462A-80R1, 113–118 cm also has a Kwaimbaita-type Pb-isotope signature. On the other hand, sample 462A-41R7, 48–55 cm, which has the slightly higher leached Nd-isotope ratios, also has higher Pb-isotope ratios than the Kwaimbaita-type lavas. The data suggest that the Pb-isotope systematics of Nauru volcaniclastics may have been disturbed, more so than the Nd-isotopes, by sea-water alteration. This is most probably not so surprising because of the variably altered state of most of these volcaniclastics. Although the Pb content of oceanic rocks may indeed be sea-water-alteration-resistant, U and, to a limited extent, Th contents of oceanic rocks can be mobilized by sea water (e.g. Mahoney *et al.* 1998). If such mobilization of the radiogenic parent elements occurred early in the history of Nauru volcaniclastics (e.g. during the erosion of the Kwaimbaita-type lavas from the OJP and their subsequent deposition into the basin), then both the measured and age-corrected Pb-isotope ratios may not be representative of the original isotope ratios of the eroded lavas. That the Nd-isotope systematics is conservative, whereas Sr- and Pb-isotope systematics are open to sea-water alteration, is also observed in both the Singgalo-type vitric ash at ODP Site 1183 (Tejada *et al.* 2004) and in the Kwaimbaita-type lava clasts in the volcaniclastic sediments at ODP Site 1184 (Shafer *et al.* 2004; White *et al.* 2004) on the OJP.

As with the measured and age-corrected values of the unleached Nauru volcaniclastics, the Nd-isotope ratios of the leached East Mariana volcaniclastics are very similar, indicating that the Nd-isotope systematics of East Mariana volcaniclastics is unaffected by sea-water alteration. Unlike the former, the Pb-isotope systematics of the latter also appears to be unaffected by sea-water alteration because the age-corrected analyses are also very similar. The less altered nature of the East Mariana Nd- and Pb-isotope systematics is most probably due to the higher contents of Nd, Sm, Pb, Th and U in the alkalic East Mariana volcaniclastics compared with those of the tholeiitic Nauru volcaniclastics (by a factor of *c.* 2–8; Tables 1 and 2). Thus, the East Mariana volcaniclastics were less vulnerable to the effects of sea-water alteration than the Nauru volcaniclastics. Assuming that the Nd- and Pb-isotope ratios are both unaffected by alteration, the source of East Mariana volcanogenic sedimentary components can be narrowed down to either the Limalok Guyot of the Marshall-Gilbert Islands or the Manihiki Plateau.

Occupying a basin that has an area of approximately 10^6 km^2, the Nauru Basin volcaniclastics represent approximately an additional 0.5×10^6 km^3 of Kwaimbaita-type magma. This lends some support to the conclusion of most recent studies that the source of the bulk of the OJP basement lavas is a homogeneous, Kwaimbaita-type mantle (e.g. Mahoney *et al.* 2001; Tejada *et al.* 2004). Note that this mantle source may not only be volumetrically significant, but temporally persistent as well. The same mantle pocket was the source of the Kwaimbaita-type lavas erupted at 122 and 90 Ma that formed the plateau and the igneous complex in the EMB (115 Ma) and Nauru Basin (111 Ma) (Castillo *et al.* 1991, 1994; Mahoney *et al.* 1993; Janney & Castillo 1996; Tejada *et al.* 1996, 2002; Neal *et al.* 1997), and possibly in the NW Central Pacific Basin (Late Albian) (Janney & Castillo 1996).

Although the Limalok Guyot has similar alkalic rock affinity and major- and trace-element and isotopic composition to East Mariana volcaniclastics (Figs 2–4), available age data suggest that the crystallization age of Limalok lavas is only about 68 Ma (Koppers 1998), younger than most of the East Mariana volcanogenic components. Moreover, because of its position in the middle of the Ratak (eastern) seamount chain of the Marshall-Gilbert Islands, the Ralik (western) seamount chain of the Marshall-Gilbert Islands, some volcanoes of which are older than Limalok, was a formidable barrier between Limalok and EMB. The relatively small size of Limalok Guyot indicates relatively weak volcanism that could not have deposited its volcanic products beyond the Ralik seamount chain. The isotopic signature of Limalok Guyot is also unusual compared to the samples from the rest of the Ratak seamount chain (cf. Davis *et al.* 1989; Staudigel *et al.* 1991), which suggests that the Limalok magma source is volumetrically a minor constituent of the Marshall-Gilbert mantle source (Koppers 1998).

The Aptian age of the Manihiki Plateau (Lanphere & Dalrymple 1976; Sliter 1992; Mahoney *et al.* 1993) makes it a more realistic source of EMB volcaniclastics. Although the published isotope data for the Manihiki Plateau are for tholeiitic basalts (Fig. 5a, b), there are >69–110 Ma alkalic seamounts on top of the plateau that have a geochemical signature that indicates that they came from the same mantle source as the tholeiites (Figs 2–4) (Beiersdorf *et al.* 1995). More importantly, these alkalic seamounts have produced an extensive layer of volcaniclastics that covers a large portion of the Manihiki Plateau and surrounding areas (Schlanger *et al.* 1976; Coulbourn & Hill 1991; Beiersdorf *et al.* 1995). The major complication with the Manihiki Plateau being the source of EMB volcaniclastics is the great distance between the plateau and the basin, and the presence of huge barriers such as the Marshall-Gilbert Islands between them (Fig. 1).

It is important to note that the Manihiki

Plateau and Hikurangi Plateau in the SW Pacific may have once existed as a single, huge oceanic plateau during the mid-Cretaceous. An E–W-trending fossil spreading centre in the Pacific plate roughly between the Manihiki and Hikurangi plateaus, the Osbourn Trough on the seaward side of the Tonga–Kermadec Trench, has recently been discovered and this spreading centre may have separated the plateaus (Billen & Stock 2000). Thus, early volume estimates of the Manihiki Plateau at about 2×10^7 km^3 (e.g. Mahoney & Spencer 1991) is an absolute minimum. It follows that the volcanic activity or activities that formed the ancestral Manihiki–Hikurangi Plateau, as with those that formed Ontong Java and other LIPs, must have been intense, and erupted huge volumes of lavas in relatively short periods of time (e.g. Coffin & Eldholm 1991, 1994). It is thus possible that the volcanogenic sediments in the EMB were derived from nearby, nearly emergent mid-Cretaceous seamounts (Lees *et al.* 1992) that are distally associated with the intense Manihiki plateau-building magmatism. However, these seamounts are now overlain by the Miocene Caroline Ridge, or have been subducted in the Marianas Trench (Fig. 1). An alternative possibility is that the volcaniclastics shed from the plateau were deposited directly into the basin because the EMB and Manihiki–Hikurangi Plateau were relatively closer to each other during the mid-Cretaceous than they are today. This notion is based on the proposal that the mid-Cretaceous igneous complex forming the basement of the East Mariana, Nauru and Central Pacific basins represents oceanic crust formed along roughly E–W-trending spreading centres (Janney & Castillo 1996), which in turn suggests that the plate tectonic configuration then was different from today. In either possibility, the East Mariana volcaniclastics originated from the mantle source of another giant, mid-Cretaceous oceanic plateau. This implies that the proposed origins and histories of the Ontong Java and Manihiki plateaus are closely intertwined because of the closeness of their sizes, gross morphological features, geochemical signatures and formational ages (see also Mahoney 1987; Mahoney & Spencer 1991; Janney & Castillo 1996, 1999). Finally, this hypothesis also implies that the mantle sources of lavas that formed giant oceanic plateaus must have filled a large volume of the upper mantle during the mid-Cretaceous, creating the anomalous composition of the mid-Cretaceous, relative to Jurassic and modern, Pacific oceanic crust (Janney & Castillo 1997).

Conclusions

LIPs are large enigmatic features on the surface of the Earth. Constraining the origin and history of the Ontong Java Plateau, the largest and best example of a LIP, is crucial to our understanding of LIP formation. Previous and on-going investigations of the OJP, however, are barely scratching the surface of the giant plateau – more time, effort and data are needed to fully understand the geology of the plateau. This study presents the geochemistry of volcaniclastic sediments in the Nauru Basin and EMB in the hope of better understanding the mantle source of the OJP. The volcanogenic sedimentary components were deposited in the Nauru Basin and EMB penecontemporaneously with the formation of the OJP. These volcaniclastic sediments are different from similar deposits in other western Pacific basins in that they are located relatively far from emergent volcanic edifices, so their provenance must be large and emergent volcanic edifices.

Although the volcaniclastic sediments are variably altered, contents of select major and trace elements and isotope ratios of Nd and to a certain extent, Pb, indicate that the source of the volcaniclastic sediments in the Nauru Basin is the Kwaimbaita-type lavas from the OJP. This volcaniclastic deposit adds to the volume of Kwaimbaita-type lavas comprising the bulk of the plateau basement and the basements of the Nauru Basin and EMB. Hence, the mid-Cretaceous mantle responsible for the formation of the OJP was dominated by the source of this type of lava. The mantle source of the Singgalo-type lavas from Ontong Java is probably volumetrically minor.

Data indicate two possible sources of the volcaniclastic sediments in the EMB – the Limalok Guyot in the Ratak seamount chain of the Marshall-Gilbert Islands and the ancestral Manihiki–Hikurangi Plateau. Other geological information argues for the Manihiki–Hikurangi Plateau as the most likely source. However, the large distance between Manihiki Plateau and EMB is a major problem. All that can be said for now is that either the intense volcanic activities that formed the ancestral Manihiki–Hikurangi Plateau may have generated local sources for the volcaniclastics in the EMB or plate tectonic configuration of the western Pacific during the mid-Cretaceous may have made it possible for the volcanogenic components from the plateau to be deposited directly into the basin. Either hypothesis posits that the East Mariana volcaniclastics were also produced from a mantle source responsible for the formation of another giant oceanic plateau in the Pacific during the mid-Cretaceous.

The assistance of C. Mahn, C. MacIsaac and K. Walda with the ICP-OES, ICP-MS and TIMS analyses, and helpful reviews and comments by P. Floyd, R. Keller, G. Fitton and J. Mahoney are greatly appreciated. Samples used in this study were provided by ODP; the analyses were funded through JOI-USSSP grant 418928-EA260. ODP is sponsored by the National Science Foundation and participating countries under management of Joint Oceanographic Institutions, Inc.

References

ANDERSON, D.L., ZHANG, Y.-S. & TANIMOTO, T. 1992. Plume heads, continental lithosphere, flood basalts and tomography. *In*: STOREY, B.C., ALABASTER, T. & PANKHURST, R.J. (eds) *Magmatism and the Causes of Continental Break-up*. Geological Society, London, Special Publications, **68**, 99–124.

BAKER, P.E., CASTILLO, P.R. & CONDLIFFE, E. 1995. Petrology and geochemistry of igneous rocks from Allison and Resolution Guyots, ODP Sites 865 and 866. *Proceedings of the Ocean Drilling Program, Scientific Results*, **143**, 245–262.

BEIERSDORF, H., BACH, W., DUNCAN, R., ERZINGER, J. & WEISS, W. 1995. New evidence for the production of EM-type ocean island basalts and large volumes of volcaniclastites during the early history of the Manihiki Plateau. *Marine Geology*, **122**, 181–205.

BERCOVICI, D. & MAHONEY, J. 1994. Double flood basalts and plume head separation at the 660-kilometer discontinuity. *Science*, **266**, 1367–1369.

BIENVENU, P., BOUGAULT, H., JORON, J.L., TREUIL, M. & DMITRIEV, L. 1990. MORB alteration: Rare-earth element/non-rare-earth hygromagmaphile element fractionation. *Chemical Geology*, **82**, 1–14.

BILLEN, M.I. & STOCK, J. 2000. Morphology and origin of the Osbourn Trough. *Journal of Geophysical Research*, **105**, 13 481–13 489.

CANN, J.R. 1970. Rb, Sr, Y, Zr and Nb in some ocean floor basaltic rocks. *Earth and Planetary Science Letters*, **10**, 7–11.

CASTILLO, P.R., CARLSON, R.W. & BATIZA, R. 1991. Origin of Nauru Basin igneous complex: Sr, Nd and Pb isotope and REE constraints. *Earth and Planetary Science Letters*, **103**, 200–213.

CASTILLO, P.R., FLOYD, P.A. & FRANCE-LANORD, C. 1992. Isotope geochemistry of Leg 129 basalts: implications for the origin of the widespread Cretaceous volcanic event in the Pacific. *Proceedings of the Ocean Drilling Program, Scientific Results*, **129**, 405–513.

CASTILLO, P.R., PRINGLE, M.S. & CARLSON, R.W. 1994. East Mariana Basin tholeiites: Jurassic ocean crust or Cretaceous rift basalts related to the Ontong Java plume? *Earth and Planetary Science Letters*, **123**, 139–154.

CHENG, Q.C., PARK, K.-H., MACDOUGALL, J.D., ZINDLER, A., LUGMAIR, G.W., HAWKINS, J.W., LONSDALE, P. & STAUDIGEL, H. 1987. Isotopic evidence for a hot spot origin of the Louisville Seamount chain. *In*: KEATING, B., FRYER, P., BATIZA, R. & BOEHLERT, G. (eds) *Seamounts, Islands, and Atolls*. American Geophysical Union, Geophysical Monograph, **43**, 283–296.

CHRISTIE, D.M., DIEU, J.J. & GEE, J.S. 1995. Petrologic studies of basement lavas from northwest Pacific guyots. *Proceedings of the Ocean Drilling Program, Scientific Results*, **144**, 495–512.

COFFIN, M.F. & ELDHOLM, O. 1991. *Large Igneous Provinces*. JOI/USSAC Workshop Report. University of Texas at Austin Institute of Geophysics Technical Report, **114**.

COFFIN, M.F. & ELDHOLM, O. 1994. Large igneous provinces: crustal structure, dimensions and external consequences. *Reviews of Geophysics*, **32**, 1–36.

COULBOURN, W.T. & HILL, P.J. 1991. A field of volcanoes on the Manihiki Plateau: mud or lava? *Marine Geology*, **98**, 367–388.

DAVIS, A.S., PRINGLE, M.S., PICKTHORN, L.G., CLAGUE, D.A. & SCHWAB, W.C. 1989. Petrology and age of alkalic lava from the Ratak Chain of the Marshall Islands. *Journal of Geophysical Research*, **94**, 5757–5774.

ERLANK, A.J. & KABLE, E.J.D. 1976. The significance of incompatible elements in Mid-Atlantic Ridge basalts from 45°N with particular reference to Zr/Nb. *Contributions to Mineralogy and Petrology*, **54**, 281–291.

FITTON, J.G. & GODARD, M. 2004. Origin and evolution of magmas on the Ontong Java Plateau. *In*: FITTON, J.G., MAHONEY, J.J., WALLACE, P.J. & SAUNDERS, A.D. (eds) *Origin and Evolution of the Ontong Java Plateau*. Geological Society, London, Special Publications, **229**, 151–178.

FLOYD, P.A. 1986. Glassy and basaltic fragments within graded volcaniclastic sediments, East Mariana Basin, Deep Sea Drilling Project Leg 89. *Initial Reports of the Deep Sea Drilling Project*, **89**, 433–447.

FREY, F.A. & WEIS, D. 1995. Geochemical constraints on the origin and evolution of the Ninetyeast Ridge: a 5000 km hotspot trace in the eastern Indian Ocean. *Contributions to Mineralogy and Petrology*, **121**, 18–28.

HAMILTON, W.B. 2003. The closed upper-mantle circulation of plate tectonics. *In*: STEIN, S. & FREYMUELLER, J.T. (eds) *Plate Boundary Zones*. American Geophysical Union Geodynamics Monograph, **30**, 359–410.

HART, S.R., ERLANK, A.J. & KABLE, E.J.D. 1974. Sea floor basalt alteration: Some chemical and Sr isotopic effects. *Contributions to Mineralogy and Petrology*, **44**, 219–230.

ITO, G.T. & CLIFT, P.D. 1998. Subsidence and growth of Pacific Cretaceous plateaus. *Earth and Planetary Science Letters*, **161**, 85–100.

JANNEY, P.E., BAKER, P.E. & CASTILLO, P.R. 1995. Petrology and geochemistry of basaltic clasts and hyaloclastites from volcaniclastic sediments at ODP Site 869. *Proceedings of the Ocean Drilling Program, Scientific Results*, **143**, 263–276.

JANNEY, P.E. & CASTILLO, P.R. 1996. Basalts from the Central Pacific Basin: Evidence for the origin of Cretaceous igneous complexes in the Jurassic Western Pacific. *Journal of Geophysical Research*, **101**, 2875–2894.

JANNEY, P.E. & CASTILLO, P.R. 1997. Geochemistry of Mesozoic Pacific MORB: constraints on melt generation and the evolution of the Pacific upper mantle. *Journal of Geophysical Research*, **102**, 5207–5229.

JANNEY, P.E. & CASTILLO, P.R. 1999. Isotope geochemistry of the Darwin Rise seamounts and the nature of long-term mantle dynamics beneath the south central Pacific. *Journal of Geophysical Research*, **104**, 10 571–10 589.

KELTS, K. & ARTHUR, M.A. 1981. Turbidites after ten years of Deep-Sea Drilling-wringing out the mop? *In*: WARME, J.E., DOUGLAS, R.G. & WINTERER, E.L. (eds) *The Deep-Sea Drilling Project: A Decade of Progress*. SEPM, Special Publication, **32**.

KOPPERS, A.A.P. 1998. *$^{40}Ar/^{39}Ar$ geochronology and isotope geochemistry of the west Pacific seamount province*. PhD Thesis, Vrije Universiteit.

KOPPERS, A.A. P., STAUDIGEL, H., CHRISTIE, D.M., DIEU, J.J. & PRINGLE, M.S. 1995. Sr–Nd–Pb isotope geochemistry of Leg 144 west Pacific guyots: implications for the geochemical evolution of the 'SOPITA' mantle anomaly. *Proceedings of the Ocean Drilling Program, Scientific Results*, **144**, 535–545.

LANCELOT, Y., LARSON, R.L. *et al.* 1990. *Proceedings of the Ocean Drilling Program, Initial Reports*, **129**.

LANPHERE, M.A. & DALRYMPLE, G.B. 1976. K–Ar ages of basalts from Leg 33: Sites 315 (Line Islands) and 317 (Manihiki Plateau). *Initial Reports of the Deep Sea Drilling Project*, **33**, 649–653.

LARSON, R.L. 1991. Latest pulse of Earth: evidence for a mid-Cretaceous superplume. *Geology*, **19**, 547–550.

LARSON, R.L. & KINCAID, C. 1996. Onset of mid-Cretaceous volcanism by elevation of the 670 km thermal boundary layer. *Geology*, **24**, 551–554.

LARSON, R.L., SCHLANGER, S.O. *et al.* 1981. *Initial Reports of the Deep Sea Drilling Project*, **61**.

LARSON, R.L., LANCELOT, Y. *et al.* 1992. *Proceedings of the Ocean Drilling Program, Scientific Results*, **129**.

LE ROEX, A.P. 1986. Geochemical correlations between southern African kimberlites and South Atlantic hotspots. *Nature*, **324**, 123–456.

LEES, G.J., ROWBOTHAM, G. & FLOYD, P.A. 1992. Petrography and geochemistry of graded volcanoclastic sediments and their clasts. *Proceedings of the Ocean Drilling Program, Scientific Results*, **129**, 137–150.

LUDDEN, J.N. & THOMPSON, G. 1978. Behavior of rare earth elements during submarine of tholeiitic basalts. *Nature*, **274**, 147–149.

LUGMAIR, G.W. & GALER, S.J.G. 1992. Age and isotopic relationships among the angerites Lewis Cliff 86010 and Angra dos Reis. *Geochimica et Cosmochimica Acta*, **56**, 1673–1694.

MAHONEY, J.J. 1987. An isotopic survey of Pacific oceanic plateaus: implications for their nature and origin. *In*: KEATING, B., FRYER, P., BATIZA, R. & BOEHLERT, G. (eds) *Seamounts, Islands, and Atolls*. American Geophysical Union Geophysical Monograph, **43**, 283–296.

MAHONEY, J.J. & SPENCER, K.J. 1991. Isotopic evidence for the origin of the Manihiki and Ontong Java oceanic plateaus. *Earth and Planetary Science Letters*, **104**, 196–210.

MAHONEY, J.J., STOREY, M., DUNCAN, R.A., SPENCER, K.J. & PRINGLE, M. 1993. Geochemistry and age of the Ontong Java Plateau. *In*: PRINGLE, M.S., SAGER, W.W., SLITER, W.V. & STEIN, S. (eds) *The Mesozoic Pacific: Geology, Tectonics and Volcanism*. American Geophysical Union Geophysical Monograph, **77**, 233–261.

MAHONEY, J.J., FREI, R., TEJADA, M.L. G., MO, X.X., LEAT, P.T. & NÄGLER, T.F. 1998. Tracing the Indian Ocean mantle domain through time: isotopic results from old West Indian, East Tethyan, and South Pacific seafloor. *Journal of Petrology*, **39**, 1285–1306.

MAHONEY, J.J., FITTON, J.G., WALLACE, P.J. *et al.* 2001. *Proceedings of the Ocean Drilling Program, Initial Reports*, **192**. Available from World Wide Web: http://www-odp.tamu.edu/publications/192_IR/192ir.htm

MCCULLOCH, M.T., GREGORY, R.T., WASSERBURG, G.J. & TAYLOR, H.T. 1981. Sm–Nd, Rb–Sr, and $^{18}O/^{16}O$ isotopic systematics in an oceanic crustal section: evidence from the Samail ophiolites. *Journal of Geophysical Research*, **86**, 2721–2735.

MOBERLY, R. & JENKYNS, H.C. 1981. Cretaceous volcanogenic sediments of the Nauru Basin, Deep Sea Drilling Project Leg 61. *In*: LARSON, R. *et al.* (eds) *Initial Reports of the Deep Sea Drilling Project*, **61**, 533–548.

MOBERLY, R., SCHLANGER, S.O. *et al.* 1986. *Initial Reports of the Deep Sea Drilling Project*, **89**.

NAKANISHI, M., TAMAKI, K. & KOBAYASHI, K. 1992. Magnetic anomaly lineations from Late Jurassic to Early Cretaceous in the west-central Pacific Ocean. *Geophysical Journal International*, **109**, 701–719.

NEAL, C.R. & DAVIDSON, J.P. 1989. An unmetasomatized source for the Malaitan alnoite (Solomon Islands): Petrogenesis involving zone refining, megacryst fractionation, and assimilation of oceanic lithosphere. *Geochemica et Cosmochimica Acta*, **53**, 1975–1990.

NEAL, C.R., MAHONEY, J.J., KROENKE, L.W., DUNCAN, R.A. & PETTERSON, M.G. 1997. The Ontong Java Plateau. *In*: MAHONEY, J.J. & COFFIN, M.F. (eds) *Large Igneous Provinces: Continental, Oceanic, and Planetary Flood Volcanism*. American Geophysical Union, Geophysical Monograph, **100**, 183–216.

PEARCE, J.A. & CANN, J.R. 1973. Tectonic setting of basic volcanic rocks determined using trace element analyses. *Earth and Planetary Science Letters*, **19**, 290–300.

PRINGLE, M.S. 1992. Radiometric ages of basaltic basement recovered at Sites 800, 801 and 802, western Pacific Ocean. *Proceedings of Ocean Drilling Project, Scientific Results*, **129**, 361–388.

SAUNDERS, A.D. 1986. Geochemistry of basalts from

the Nauru Basin, Deep Sea Drilling Projects 61 and 89. *Initial Reports of the Deep Sea Drilling Project*, **89**, 499–518.

SCHLANGER, S.O., JACKSON, E.D. *et al.* 1976. *Initial Reports of the Deep Sea Drilling Project*, **33**.

SHAFER, J.T., NEAL, C.R. & CASTILLO, P.R. 2004. Compositional variability in lavas from the Ontong Java Plateau: results from basaltic clasts within the volcaniclastic succession at Ocean Drilling Program Site 1184. *In*: FITTON, J.G., MAHONEY, J.J., WALLACE, P.J. & SAUNDERS, A.D. (eds) *Origin and Evolution of the Ontong Java Plateau*. Geological Society, London, Special Publications, **229**, 333–351.

SLITER, W.V. 1992. Cretaceous planktonic foraminiferal biostratigraphy and paleoceanographic events in the Pacific Ocean with emphasis on indurated sediments. *In*: ISHIZAKI, K. & SAITO, K. (eds) *Centenary of Japanese Micropaleontology*. Terra Science, Tokyo, 261–299.

SMITH, A.D. & LEWIS, C. 1999. The planet beyond the plume hypothesis. *Earth Science Reviews*, **48**, 135–182.

STAUDIGEL, H. & HART, S.R. 1983. Alteration of basaltic glass: Mechanics and significance for the oceanic crust-seawater budget. *Geochimica et Cosmochimica Acta*, **47**, 337–350.

STAUDIGEL, H., PARK, K. H., PRINGLE, M., RUBENSTONE, J.L. SMITH, W.H.F. & ZINDLER, A. 1991. The longevity of the South Pacific isotopic and thermal anomaly. *Earth and Planetary Science Letters*, **102**, 24–44.

SUN, S.-S. & MCDONOUGH, W.F. 1989. Chemical and isotopic systematics of oceanic basalts: implications for mantle composition and processes. *In*: SAUNDERS, A.D. & NORRY, M.J. (eds) *Magmatism in the Ocean Basins*. Geological Society, London, Special Publications, **42**, 313–345.

TARDUNO, J.A., SLITER, W.V., KROENKE, L., LECKIE, M., MAYER, H., MAHONEY, J.J., MUSGRAVE, R., STOREY, M. & WINTERER, E.L. 1991. Rapid formation of Ontong Java Plateau by Aptian mantle plume volcanism. *Science*, **254**, 399–403.

TEJADA, M.L.G., MAHONEY, J.J., DUNCAN, R.A. & HAWKINS, M.P. 1996. Age and geochemistry of basement and alkalic rocks of Malaita and Santa Isabel, Solomon Islands, southern margin of Ontong Java Plateau. *Journal of Petrology*, **37**, 361–394.

TEJADA, M.L.G., MAHONEY, J.J., NEAL, C.R., DUNCAN, R.A. & PETTERSON, M.G. 2002. Basement geochemistry and geochronology of Central Malaita, Solomon Islands, with implications for the origin and evolution of the Ontong Java Plateau. *Journal of Petrology*, **43**, 449–484.

TEJADA, M.L.G., MAHONEY, J.J., CASTILLO, P.R., INGLE, S.P., SHETH, H.C. & WEIS, D. 2004. Pin-pricking the elephant: evidence on the origin of the Ontong Java Plateau from Pb–Sr–Hf–Nd isotopic characteristics of ODP Leg 192 basalts. *In*: FITTON, J.G., MAHONEY, J.J., WALLACE, P.J. & SAUNDERS, A.D. (eds) *Origin and Evolution of the Ontong Java Plateau*. Geological Society, London, Special Publications, **229**, 133–150.

TIMOFEEV, P.P., KOPORULIN, V.I., VARENTSOV, I.M., EREMEEV, V.V. & CHAPOROV, D.YA. 1981. Composition and conditions of genesis of sedimentary rocks in the lower part of Site 462, Deep Sea Drilling Project Leg 61. *Initial Reports of the Deep Sea Drilling Project*, **61**, 567–581.

TODT, W., CLIFF, R.A., HANSER, A. & HOFFMAN, A.W. 1996. Evaluation of a ^{202}Pb–^{205}Pb double spike for high-precision lead isotope analysis. *In*: BASU, A. & HART, S. (eds) *Earth Processes: Reading the Isotopic Code*. American Geophysical Union, Geophysical Monograph, **95**, 429–437.

VERMA, S.P. 1992. Seawater alteration effects on REE, K, Rb, Cs, Sr, U, Th, Pb and Sr–Nd–Pb isotope systematics of mid-ocean ridge basalt. *Geochemical Journal*, **26**, 159–177.

WEAVER, B.L. 1991. The origin of ocean island basalt end-member compositions: trace element and isotopic constraints. *Earth and Planetary Science Letters*, **104**, 381–397.

WHITE, R.V., CASTILLO, P.R., NEAL, C.R., FITTON, J.G. & GODARD, M. 2004. Phreatomagmatic eruptions on the Ontong Java Plateau: chemical and isotopic relationship to Ontong Java Plateau basalts. *In*: FITTON, J.G., MAHONEY, J.J., WALLACE, P.J. & SAUNDERS, A.D. (eds) *Origin and Evolution of the Ontong Java Plateau*. Geological Society, London, Special Publications, **229**, 307–323.

WHITE, W.M. 1993. ^{238}U/^{204}Pb in MORB and open system evolution of the depleted mantle. *Earth and Planetary Science Letters*, **115**, 211–226.

WINCHESTER, J.A. & FLOYD, P.A. 1976. Geochemical magma type discrimination: application to altered and metamorphosed basic igneous rocks. *Earth and Planetary Science Letters*, **28**, 459–469.

WINTERER, E.L., EWING, J. *et al.* 1973. *Initial Reports of the Deep Sea Drilling Project*, **17**.

Index

Page numbers in italic, e.g. *40*, refer to figures. Page numbers in bold, e.g. **12**, signify entries in tables.

accretionary-lapilli tuff 291–293, *294–295*
Alite Limestone Formation (ALF), Malaita 74–75
apparent wander polar path (APWP), Pacific 40–41, *40*, *41*
  interpretation as true polar wander or motion between hot spots 41–42
Austral hot spot, location **12**

Banaba Island *10*, *16*
basalts, submarine 321
  environmental and climatic implications 320–321
  extent of subaerial volcanism and implications for plume models 319–320
  isotope studies 133–136, 147, 310
    analytical methods 136
    basement thickness and magma types *135*
    drill sites *134*
    Hf isotopes **137**, 141, *141*, **143**, *144*
    incompatible trace-element compositions 315, *315*, *316*, **316**, *317*
    major-element composition of glass clasts 311–315, *311*, **312–313**, *314*
    Sr, Nd and Pb compositions of bulk tuff 316–319, **318**, *319*
    Sr, Nd and Pb compositions of site 1184 samples 136–141, **137**, **138**, *139*, *140*, *141*, **143**, *144*
  low-temperature alteration 259, 270, 272
    alteration characteristics 260–269, *261*, *262*, *263*, *264*, *266*, *269*
    black and dusky green halos 270–271
    brown halos 271
    geological setting and samples 259–260
    pervasively altered dark grey basalt 270
    stable isotopes 269–270, *270*
    study methods 260
    temperatures of alteration 271–272
    veins and pore space fillings 271
  relationship between magma composition and stratigraphy 319
  volatiles in basaltic glasses 239–241, 254
    analytical methods 241–243
    depths of crystallization 248
    geological setting and sample characteristics 241
    high chlorine content, causes of 248–250, *249*
    major element compositions *243*, **244–245**
    study results 243–247, *246*, *247*, *248*
    volatile release during OJP formation 253–254
    water in mantle source region 250–253, *251*, *252*
basement lavas obtained from OPD Leg-192, experimental petrology 185, 214–215
  analytical methods 187
  depth and temperature of magma chamber 208–210, *209*
  geological setting and samples 185–187, *186*
  magma differentiation model 213–214
  melting experiments 195–208
  mineralogy of phenocrysts 192–195, *193*, *194*
  origin of high-An parts of plagioclase 210–211, *210*, *211*
  origin of reverse-zoned olivine 211–213, *212*, *213*
  whole rock and basaltic glass compositions 188–192, **188**, *189*, **190–191**
Beer's law 242
Bougainville *64*

Caroline Basin *10*
Caroline hot spot, location **12**
Caroline Islands *10*
Central Pacific Fault Zone *14*
Choiseul *64*
Cobb chain *11*
Cobb hot spot, location **12**
Coleman volcano *64*
Colville–Lau Ridge *10*
Coral Sea Basin *10*, *64*

Deep Sea Drilling Project (DSDP) Leg-30 83
  drill sites 2, *2*
  site-288 biostratigraphy 105–107, *107*, *108*
  site-289 biostratigraphy 103–105, *104*
D'entrecasteaux Basin *10*

East Mariana Basin *2*, *10*, *14*
  Cretaceous volcaniclastic sediments 353–354, 360–366
    analytical methods 356–357
    major and trace element compositions **355**, *357*
    major oxides 357–358
    Sr, Nd and Pb isotopes 358–360, *359*, **361**, *362*
    study samples 355–356
    trace elements 358, *360*
  free-air gravity field *15*
  predicted bathymetry *16*, *32*
Eastern Salient *2*, *14*
  predicted bathymetry *16*, *32*
Eauripik Ridge *10*
Ellice Basin *10*, *14*
  free-air gravity field *15*
  predicted bathymetry *16*
Ellice Islands *10*
Emperor chain *11*, *14*

Fiji *10*
Florida *16*, *64*
foraminiferal biostratigraphy 83, 107–110
  biozonation 84–85, *84*, **86**, **87**
  carbon isotope stratigraphy 85–89
  DSDP Leg-30
    site-288 105–107, *107*, *108*
    site-289 103–105, *104*
  methods 83–84
  ODP Leg-192
    site-1183 89–95, **89**, *93*, *97*
    site-1184 102–103

site-1185 98–101, *99*
site-1186 **90**, **91**, **92**, 95–98, *96*, *97*
site-1187 101–102, *101*
Foundation hot spot, location **12**
France, calcareous microfossil biostratigraphy *117*

Ghizo Ridge *64*
Gilbert Islands *10*
Guadalcanal *64*

Haruta Limestone Formation (HLF), Malaita 75–77
Hawaii hot spot *11*
location **12**
Hawaiian chain *11*, *14*
Hawaiian Islands *10*
Hawaiian Lineation set *14*
Hawaiian–Emperor seamount chain (HEB) 9–10
Hess Rise *14*
High plateau *2*
hyaloclastite *267*

isotope studies
implications for origin of OJP
mantle reservoir hypothesis 141–145
meteorite impact rather than plume impact hypothesis 145–147
plate separation and ridge-centred hot-spot hypotheses 145
ODP Leg-192 basalts 133–136, 147
analytical methods 136
basement thickness and magma types *135*
drill sites *134*
Hf isotopes **137**, 141, *141*, **143**, *144*
Pb, Nd and Sr isotopes 136–141, **137**, **138**, *139*, *140*, *141*, **143**, *144*

Japanese Lineation set *14*

Kana Keoki volcano *64*
Kashima Fault Zone *14*
Kavachi volcano *64*, *66*
Kerguelen Plateau 2
Kodiak hot spot, location **12**
Kroenke Canyon *2*, *32*
Kuril Trench *14*
Kwaraae Mudstone Formation (KMF), Malaita 73–74

large igneous provinces (LIPs) 1
magma petrogenesis study 219–220, 234
analytical techniques 221–222
geological background 220–221, *221*
low-temperature alteration and effect of sulphide immiscibility 226–231
model results 233–234, *233*
partitioning and fractionation 232–233, **232**
results 222–226, *223*, **224–225**, *226*, *227*, *228*
source modelling 231–232, *231*
Lau Basin *10*
Line Islands *14*
Louisade Plateau *64*
Louisville chain *11*
Louisville hot spot *17*, 21–22, 27–28
effect of motion on palaeolatitudes 27, *27*
location **12**
modelling motion 22–24, 26–27
modelled drift *24*
radial viscosity models *23*
results 24–25, **25**
Lyra Basin *2*, *14*
free-air gravity field *15*
predicted bathymetry *16*

Magellan Lineation set *14*
magmas of the OJP, origin and evolution 151–153, 174
geographical distribution of magma types 165–167, *167*
magma types 154
magmatic evolution 167–168
mantle temperature 171–173, *172*
nature of mantle source and degree of melting 168–171, *170*
possible mantle plume origin for magmatism 173–174
stratigraphic sections *153*
study analytical methods 154–155, **155**
study results 155–165, **156–158**, *159*, **160–161**, **162–163**, *164*, *165*, *166*
Makira *64*
Malaita *2*, *16*, 63, *64*
Alnoite intrusions 78
fractured Solomon terrain collage 65
generalized stratigraphy 68–69, *68*
geology 66–67, *70*
Late Tertiary–present-day arc volcanism in Solomon Islands 65–6
local stratigraphic formations above Middle Pliocene unconformity 77–78
previous stratigraphic work 67–68, **69**
regional geo-tectonics 63–65, *64*
value as a study site for OJP 78–79
Malaita Anticlinorium *64*
Malaita Volcanic Group (MVG) 67, 68, *68*
cover sequence geology
Alite Limestone Formation (ALF) 74–75
Haruta Limestone Formation (HLF) 75–77
Kwaraae Mudstone Formation (KMF) 73–74
Maramasike Volcanic Formation (MVF) 75
Suafa Limestone Formation (SLF) 77
eruptive environment 73
geological map *70*
lithological characteristics
basalts 71, *72*
dolerite and gabbro/microgabbro 71–73, *72*
orbicular/spheruloidal basalt and gabbro *72*, 73
location 69
petrography 73
thickness 69–71
Mamua Fault Zone *14*
Manihiki Plateau *14*
Manus Basin *64*
Manus Trough *10*
Maramasike *64*
Maramasike Volcanic Formation (MVF), Malaita 75
Maramuni–Trobriand Arc *10*
Mariana Trench *14*
Marquesas hot spot, location **12**

Mellish Plateau *64*
melting experiments on basement lavas
experimental and analytical methods 199–201, *201*, **202**
experimental results 201
attainment of equilibrium 202–203, *203*
mineral appearance and melt compositions 203–208, **204–206**, *207*
preliminary considerations 201–202, **202**
materials 195–199, **199**
miarolitic cavities in basalts 264–265, *265*, 267–268, 269
mid-Pacific Mountain Lineation set *14*
mid-Pacific Mountains *14*
motion of OJP hot spot 9–10, 18, 27–28
effect on palaeolatitudes 27
flow line 12, *17*
modelling 22–24, 26–27
radial viscosity models *23*
sequence of events 12–18, *13*, *17*
trail 10–12, *11*, *17*
Musicians hot spot, location **12**
Mussau Trough *14*

nannofossil studies 83, 107–110
basement ages 121
Aptian 121–124
Eocene **122–123**, *124*
biostratigraphy 113–114
biozonation 84–85, *84*, **86**, **88**
carbon isotope stratigraphy 85–89
Cenozoic
Eocene–Miocene 120–121
Palaeocene–Eocene **118–119**, 120–121, *120*
Cretaceous
lower 114–116, *114*
summary **115–116**, *119*
upper 116–120
DSDP Leg-30
site-288 105–107, *107*, *108*
site-289 103–105, *104*
methods 83–84, 113
ODP Leg-192 113
site-1183 89–95, **89**, *93*, *97*
site-1184 102–103
site-1185 98–101, *99*
site-1186 **90**, **91**, **92**, 95–98, *96*, *97*
site-1187 101–102, *101*
systematic palaeontology 125–126
Cenozoic species 126–129
Cretaceous species 129–131
Nauru Basin *2*, *10*, *13*, *14*
Cretaceous volcaniclastic sediments 353–354, 360–366
analytical methods 356–357
major and trace element compositions **355**, *357*
major oxides 357–358
Sr, Nd and Pb isotopes 358–360, *359*, **361**, *362*
study samples 354–355
trace elements 358, *360*
free-air gravity field *15*
predicted bathymetry *16*, *32*
Nauru Island *10*, *16*
New Britain *10*, *64*
New Britain Trench *64*
New Georgia group *64*
New Hebrides Trench *10*
New Ireland *10*, *64*
North Fiji Basin *10*
Nosappu Fault Zone *14*
Nova–Canton Trough *14*

Ocean Drilling Program (ODP) Leg-192 42, 83, 113
accretionary-lapilli-bearing pyroclastic rocks 275–276
depositional environment and age 303–304
geological setting 276–277, *276*
lithofacies analysis 284–299, **291**, *292*
lithologies 282–283
lithostratigraphic associations 299–301
lithostratigraphic members 283–284, *284–285*, **286–290**, 299–301
methods and terminology 277–278, *277*, **277**
number and size of eruptions 303
primary v. secondary eruption processes 302–303
significance 302
volcaniclastic components 278–282, *278*, *279*, **280–281**
age of site 1184 rocks 325–326, 329–330
$^{40}Ar/^{39}Ar$ analysis 327, **327**
samples 326–327
study results 328–329, *328*, *329*
compositional variability of site 1184 rocks 333–334
analytical methods 335–336
influence of alteration 339–346, *345*, *346*
major-element composition 336–337, **338**, *339*, *340*
petrogenic interpretations 346–348, *347*, *348*
petrography 336, *337*
radiogenic isotope ratios 339, **343**, *344*
samples 334–335, *335*
source region 348
trace-element composition 337–339, **341**, *342*
drill locations *92*
drill sites 2, *2*, *32*, *39*, *46*, **47**
origin of site 1184 rocks 301, 308–310, *309*, 321
environmental and climatic implications 320–321
extent of subaerial volcanism and implications for plume models 319–320
incompatible trace-element compositions 315, *315*, *316*, **316**, *317*
major-element composition of glass clasts 311–315, *311*, **312–313**, *314*
relationship between magma composition and stratigraphy 319
Sr, Nd and Pb compositions of bulk tuff 316–319, **318**, *319*
rock magnetic results 45–47, 55–58
Curie temperature determination 49–51, **50–51**, *52*
hysteresis loop parameters 51–52, *52*, **53–54**, *55*, *56*
laboratory methods 48–49
low-temperature properties **50–51**, 52–55, *57*
site setting and basement lithology 47–48
basalt alteration characteristics
site-1183 262–265, *263*, *264*
site-1185 *263*, *264*, 265–267, *266*

site-1186 *263*, *264*, 267–268
site-1187 *263*, *264*, 268–269, *269*
biostratigraphy
site-1183 89–95, **89**, *93*, *97*, *99*
site-1184 102–103
site-1185 98–101, *99*
site-1186 **90**, **91**, **92**, 95–98, *96*, *97*
site-1187 101–102, *101*
Oceanic anoxic event (OAE) 85
Ontong Java Plateau (OJP) 1–2, *2*, *10*, 21–22, 307–308
age and biostratigraphy, overview 3
apparent Pacific polar wander path (APWP) 40–41, *40*, *41*
interpretation as true polar wander or motion between hot spots 41–42
biostratigraphy 83, 107–110, 113
basement ages 121–125, **122–123**, *124*
biozonation 84–85, *84*, **86**, **87**, **88**
carbon isotope stratigraphy 85–89
Cenozoic **118–119**, 120–121, *120*
Cretaceous 114–120, *114*, *119*
DSDP Leg-30 103–107, *104*, *107*, *108*
methods 83–84, 113
nannofossils 113–114
ODP Leg-192 89–103, **89**, **90**, **91**, **92**, *93*, *96*, *97*, *101*
early Cretaceous Pacific palaeomagnetic pole 31
emplacement 45 47, 55 58
basement lithology 47–48
Curie temperature determination 49–51, **50–51**, *52*
hysteresis loop parameters 51–52, *52*, **53–54**, *55*, *56*
laboratory methods 48–49
low-temperature properties **50–51**, 52–55, *57*
free-air gravity field *15*
geological evolution and palaeomagnetism, overview 3
geological setting 31–32
isotopic characteristics of ODP Leg-192 basalts 133–136, 147
analytical methods 136
basement thickness and magma types *135*
drill sites *134*
Hf isotopes **137**, 141, *141*, **143**, *144*
implication for origin of OJP 141–147
Pb, Nd and Sr isotopes 136–141, **137**, **138**, *139*, *140*, *141*, **143**, *144*
lava compositional variability 333–334
analytical methods 335–336
influence of alteration 339–346, *345*, *346*
major-element composition 336–337, **338**, *339*, *340*
petrogenic interpretations 346–348, *347*, *348*
petrography 336, *337*
radiogenic isotope ratios 339, **343**, *344*
samples 334–335, *335*
source region 348
trace-element composition 337–339, **341**, *342*
location at 125Ma after plate reconstruction *22*
low-temperature alteration of submarine basalts 259, 270, 272
alteration characteristics 260–269, *261*, *262*, *263*, *264*, *266*, *269*
black and dusky green halos 270–271
brown halos 271
geological setting and samples 259–260
pervasively altered dark grey basalt 270
stable isotopes 269–270, *270*
study methods 260
temperatures of alteration 271–272
veins and pore space fillings 271
magma differentiation processes 185, 214–215
depth and temperature of magma chamber 208–210, *209*
geological setting and samples 185–187, *186*
magma differentiation model 213–214
origin of high-An parts of plagioclase 210–211, *210*, *211*
origin of reverse-zoned olivine 211–213, *212*, *213*, *214*
magma petrogenesis study 219–220, 234
analytical techniques 221–222
geological background 220–221, *221*
low-temperature alteration and effect of sulphide immiscibility 226–231
model results 233–234, *233*
partitioning and fractionation 232–233, **232**
results 222–226, *223*, **224–225**, *226*, *227*, *228*
source modelling 231–232, *231*
magmatic processes and source region compositions 239 241, 254
analytical methods 241–243
depths of crystallization 248
geological setting and sample characteristics 241
high chlorine content, causes of 248–250, *249*
study results 243–247, *243*, **244–245**, *246*, *247*, *248*
volatile release during OJP formation 253–254
water in mantle source region 250–253, *251*, *252*
modelling hot-spot motion 22–24, 26–27, 27–28
radial viscosity models *23*
origin and evolution of magmas 151–153, 174
geographical distribution of magma types 165–167, *167*
magma types 154
magmatic evolution 167–168
mantle temperature 171–173, *172*
nature of mantle source and degree of melting 168–171, *170*
possible mantle plume origin for magmatism 173–174
stratigraphic sections *153*
study analytical methods 154–155, **155**
study results 155–165, **156–158**, *159*, **160–161**, **162–163**, *164*, *165*, *166*
palaeolatitudes
effect of hot-spot motion 27
effect of true polar wander (TPW) 27, *28*
partial melting below surface 179, 183
accumulated fractional melting 181–183, **182**
equilibrium melting 183
modelling method 179–181, *180*, *181*, *182*
petrology and geochemistry, overview 3–5
possible mantle plume origin, overview 6–7
predicted bathymetry *16*, *32*, *152*
site 1184 of the ODP
early Cretaceous Pacific palaeomagnetic pole 38–40

early Cretaceous palaeomagnetic
palaeocolatitudes for the Pacific 33–34, **38**, *39*
geological setting 32, *32*
rock and palaeomagnetic properties of
volcaniclastics 33, *33*, *34*, **35–37**, *38*
systematic palaeontology 125–126
Cenozoic species 126–129
Cretaceous species 129–131
value of Malaita as a study site 78–79
volcaniclastic rocks, overview 5–6
Ortelius Fault Zone *14*
Osborn Trough *10*

Pacific absolute plate motion (APM) 9–10, 18
finite rotations parameters **12**
hot-spot locations **12**
hot-spot trails *11*
magnetic anomaly lineations and fracture zones *14*
sequence of events 12–18, *13*, *17*
Pacific palaeomagnetic pole, early Cretaceous 31,
38–40, 42
apparent polar wander path (APWP) 40–41, *40*, *41*
interpretation as true polar wander or motion
between hot spots 41–42
geological setting of OJP 31–32
predicted bathymetry of OJP area *32*
site 1184 of the ODP on OJP
geological setting 32, *32*
palaeocolatitudes 33–34, **38**, *39*
rock and palaeomagnetic properties of
volcaniclastics 33, *33*, *34*, **35–37**, *38*
Pacific tectonics, interpreted from ODP samples
45–47, 55–58
Curie temperature determination 49–51, **50–51**, *52*
hysteresis loop parameters 51–52, *52*, **53–54**, *55*, *56*
laboratory methods 48–49
low-temperature properties **50–51**, 52–55, *57*
study site setting and basement lithology 47–48
Parece Vela Basin *10*
partial melting model 179, 183
computational method 179–181
peridotite composition projection *180*, *181*, *182*
modelling results
accumulated fractional melting 181–183, **182**
equilibrium melting 183
phenocrysts, mineralogy
clinopyroxene 194–195, *194*, **199**, *200*
olivine 192, *193*, *194*, *195*, **196**
origin of reverse zoning 211–213, *212*, *213*, *214*
plagioclase 192–194, *193*, **197**, *198*
origin of high-An parts 210–211, *210*, *211*
Phoenix Fault Zone *14*
Phoenix Lineation set *14*
phreatomagmatic eruptions 275–276, 307–308
age of ODP Leg 192 site 1184 rocks 325–326,
329–330
$^{40}Ar/^{39}Ar$ analysis 327, **327**
samples 326–327
study results 328–329, *328*, *329*
depositional environment and age 303–304
geological setting 276–277, *276*
lithofacies analysis 284–291, **291**, *292*
lithologies 282–283
lithostratigraphic associations 299
lithostratigraphic members 283–284, *284–285*,
286–290, 299
accretionary-lapilli tuff 291–293, *294–295*
fines-poor massive tuff and normally graded tuff
293–295
fines-rich massive tuff and normally graded tuff
295–296
massive lapilli tuffs 296–297, *296*
member IIA 299–300
member IIB 300
member IIC 300
member IID 300–301
member IIE 301
member IIF 301
normally graded lapilli tuffs 297
normally graded, symmetrically graded and
massive lapillistone 297–299, *298*
thinly bedded tuff 296
methods and terminology 277–278, *277*, **277**
number and size of eruptions 303
origin of ODP Leg 192 site 1184 rocks 301,
308–310, *309*, 321
environmental and climatic implications 320–321
extent of subaerial volcanism and implications
for plume models 319–320
incompatible trace-element compositions 315,
315, *316*, **316**, *317*
major-element composition of glass clasts
311–315, *311*, **312–313**, *314*
relationship between magma composition and
stratigraphy 319
Sr, Nd and Pb compositions of bulk tuff 316–319,
318, *319*
primary v. secondary eruption processes 302–303
significance of accretionary lapilli 302
volcaniclastic components 278–282, *278*, *279*,
280–281
Pitcairn hot spot, location **12**
platinum-group elements (PGEs) in OJP basalts 220,
234
analytical techniques 221–222
low-temperature alteration and effect of sulphide
immiscibility 226–231
model results 233–234, *233*
partitioning and fractionation 232–233, **232**
results 222–226, *223*, **224–225**, *226*, *227*, *228*
source modelling 231–232, *231*
plume conduits 23–24
plume conduits 26
Pocklington Rise *64*

Queensland Plateau *64*

Ranonga *64*
Ratak Islands *10*
Ratak–Gilbert–Ellice chain *13*
Russells *64*

Samoa hot spot *11*, *17*
location **12**
San Cristobal *2*, *16*
San Cristobal Trench *64*
Santa Isabel *2*, *16*, *64*
Savo volcano *64*

Shatsky (N) hot spot, location **12**
Shatsky (S) hot spot, location **12**
Shatsky Rise *14*
Simbo volcano *64*
Society hot spot, location **12**
Solomon Islands
  fractured terrain collage 65
  Late Tertiary–present-day arc volcanism 65–6
  regional geo-tectonics 63–65, *64*
Solomon Sea Basin *64*
  predicted bathymetry *16*
Solomon Trench *10*
South Fiji Basin *10*
South Shatsky Fault Zone *14*
South Solomon Trench *10*
Stewart Basin *2*, *10*, *14*
  predicted bathymetry *16*, *32*
Suafa Limestone Formation (SLF), Malaita 77
systematic palaeontology 125–126
  Cenozoic species 126–129
  Cretaceous species 129–131

Taba–Feni Arc *64*
Tasman Basin *10*
Tokelau (N) hot spot, location **12**
Tokelau (S) hot spot, location **12**
Tokelau chain *13*
Tokelau Islands *10*
Tonga Trench *10*
true polar wander (TPW) 21–22
  effect on palaeolatitudes 27, *28*
Tuamotu hot spot, location **12**
Tuvalu hot spot, location **12**

Vanuatu Arc *64*
Viti Levu *10*
Vitiaz Trench *10*, *64*

Waghenaer Fault Zone *14*
Waghenaer South Fault Zone *14*
Wake (N) hot spot, location **12**
Wake (S) hot spot, location **12**
Wallis hot spot *13*
Wallis Islands *10*
West Torres Plateau *64*
Woodlark Basin *64*
  free-air gravity field *15*
  predicted bathymetry *16*